3rd International Conference on Nanotechnology Applications in Africa: Opportunities and Constraints Enabling Quality Research at the Smallest Scale 2019

IOP Conference Series: Materials Science and Engineering
Volume 805

Ogbomoso, Nigeria
22 - 24 October 2019

Editors:

Agbaje Lateef
Musibau A. Azeez
Yekinni K. Sanusi
Modiu O. Durowoju
Taofeek A. Yekeen
Evariste B. Gueguim-Kana
Tesleem B. Asafa
O. Adedokun

ISBN: 978-1-7138-2800-6
ISSN: 1757-8981

Printed from e-media with permission by:

Curran Associates, Inc.
57 Morehouse Lane
Red Hook, NY 12571

Some format issues inherent in the e-media version may also appear in this print version.

Additional copies of this publication are available from:

Curran Associates, Inc.
57 Morehouse Lane
Red Hook, NY 12571 USA
Phone: 845-758-0400
Fax: 845-758-2633
Email: curran@proceedings.com
Web: www.proceedings.com

TABLE OF CONTENTS

IOP Conf. Series: Materials Science and Engineering **805** (2020) 011001 doi:10.1088/1757-899X/805/1/011001

Preface

Conference proceedings on papers presented at the conference on 3rd International Conference on 'Nanotechnology Applications in Africa: Opportunities and Constraints' held on 22-24 October, 2019 at the Ladoke Akintola University of Technology, Ogbomoso, Nigeria.

It gives me a great pleasure to write the preface for this conference proceeding. It is more gratifying to have the conference proceeding published in partnership with the *Institute of Physics Conference Series: Materials Science and Engineering*; a reputable and prestigious platform renowned for the publication of conference proceedings worldwide.

The organizers of the conference, the LAUTECH Nanotechnology Research Group (*NANO*$^+$) (www.lautechnanotech.com) is a multidisciplinary research group with the focus of conducting cutting-edge research in nanotechnology and dissemination of information about the discipline in Nigeria and beyond. The group was formed on September 4, 2014 with the track record of publications and organization of workshop and conferences. In 2017, the group organized a workshop on '*Synthesis, characterization and applications of nanoparticles*', while a workshop

IOP Conf. Series: Materials Science and Engineering **805** (2020) 011001 doi:10.1088/1757-899X/805/1/011001

cum conference tagged '*Nanotechnology for sustainable development: opportunities for Africa*' was organized in 2018. Among the papers presented at the conference, fourteen of them were published in Volume 23 issue 2 of *Science Focus,* the flagship journal of the Faculty of Pure and Applied Sciences, Ladoke Akintola University of Technology, Ogbomoso, Nigeria (https://sciencefocuslautech.org/).

Leveraging on the successes of the two earlier activities, the group organized her 3rd workshop cum conference on October 22-24, 2019 with the theme '*Nanotechnology applications in Africa: opportunities and constraints*' at LAUTECH, Ogbomoso, Nigeria. The conference was attended by over one hundred and fifty participants from about forty-five institutions including scholars, policy makers, students, representatives from the ministry of science and technology led by the Chairman of the national steering committee on nanotechnology policy and development in Nigeria, Mr. Akinyemi O. Oyefeso, and the honorable minister of science and technology, Dr. Ogbonnaya Onu who was represented by Dr. Kelechi Ebisike. Over ninety people attended the workshop on green synthesis and applications of nanoparticles, while ninety-two abstracts were submitted for the plenary session of the conference. The opening ceremony held on 23 October, 2019 drew large audience from the academia and members of the public that included about 200 hundred students from primary and secondary schools. It was chaired by the Vice-Chancellor of LAUTECH, Ogbomoso, Prof. Michael O. Ologunde. It was unarguably the largest gathering on nanotechnology discourse in Nigeria. The plenary speakers were Prof. A.S. AbdulKareem (Federal University of Technology, Minna, Nigeria) and Dr. E.E. Elemike (Federal University of Petroleum Resources, Effurun, Nigeria).

Among other things, sub-Saharan Africa region especially Nigeria was requested to key into nanotechnology initiatives as a strategy towards improving her socio-economic development and the transformation of important sectors. The conference underscored the need for the prioritization of funding of nanotechnology research by Tertiary Education Trust Fund (TETFund) and other funding agencies in Nigeria and Africa to bridge the gap in human capital development and acquisition of sophisticated analytical equipment for research activities in nanotechnology. The conference also called for the development of national policy on nanotechnology research and development to engender rapid developments in the field of

IOP Conf. Series: Materials Science and Engineering **805** (2020) 011001 doi:10.1088/1757-899X/805/1/011001

nanotechnology. The conference recognized LAUTECH, Ogbomoso as a key player in nanotechnology research in Nigeria through her nanotechnology research group and thus sought for the establishment of a national centre of excellence in nanotechnology at the University.

Amongst eighty abstracts that were presented at the conference in diverse areas of agriculture, science, engineering and medical fields that can engender development in Africa through the multidisciplinary field of nanotechnology, forty-four full-length research and review articles were submitted by authors for publication in conference proceeding of *IOP Conference Series: Materials Science and Engineering.* These articles were peer reviewed by seasoned researchers in the field of nanotechnology in compliance with the policy of the publisher. The articles covered wide range of fields that include applications in solar cell technology, pharmaceuticals, biomedical applications (antimicrobial, larvicidal, antioxidant, anti-inflammatory, anticoagulant, thrombolytic and anti-diabetic), science diplomacy for nanotechnology development, nanotechnology in the built environment, mathematical modeling of nanofluids, thin film development, composite and materials science, nanosensors and nanoelectronics technology, nanofertilizer, nanosanitizer, nanonutrients, nanoremediation, nanotoxicity and wastewater treatment, corrosion inhibition, biosynthesis and optimization of production of metallic nanoparticles, and production of nanocelluloses and nanosilica from biomasses.

These articles depicted the cosmopolitan applications of nanomaterials, which can be exploited for the advancement of economy of a nation; thus, justified the theme of the conference. Undoubtedly, the collection of articles in this conference proceeding is a rich compendium of research activities on nanotechnology in Nigeria, which would be of interest to broad-based readers. I congratulate the authors, whose works are herein published, having chosen the prestigious platform to project their research activities to the world. The open access nature of the platform would facilitate wider readership of the articles and their dissemination which would be of immense benefits to the authors and the world academic community.

I thank the members of editorial board for the painstaking efforts to review all the articles and ensuring that they meet acceptable standards in international publishing. Members of *NANO*$^+$ are specially commended for their efforts at promoting nanotechnology research in Nigeria. The

IOP Conf. Series: Materials Science and Engineering **805** (2020) 011001 doi:10.1088/1757-899X/805/1/011001

supports offered by the management of Ladoke Akintola University of Technology, Ogbomoso and the Federal Ministry of Science and Technology, Abuja, Nigeria at organizing the conference are highly appreciated.

I hereby welcome readers to the exciting articles published in this proceeding.

Professor Agbaje LATEEF
Professor of Microbiology,
Head, Nanotechnology Research Group (*NANO*$^+$),
Ladoke Akintola University of Technology, Ogbomoso, Nigeria.
www.lautechnanotech.com
alateef@lautech.edu.ng

IOP Conf. Series: Materials Science and Engineering **805** (2020) 011002 doi:10.1088/1757-899X/805/1/011002

3rd International Conference on 'Nanotechnology Applications in Africa: Opportunities and Constraints' held on 22-24 October, 2019 at the Ladoke Akintola University of Technology, Ogbomoso, Nigeria.

Organizing committee

Prof. A. Lateef
Chairman of Nanotechnology Research Group

Prof. M.A. Azeez
Chairman of Organizing Committee

Members of the organizing committee

Dr. I.C. Oladipo

Prof. T.A. Yekeen

Prof. M.O. Durowoju

Dr. E.A. Adebayo

IOP Conf. Series: Materials Science and Engineering **805** (2020) 011002 doi:10.1088/1757-899X/805/1/011002

Dr. T.B. Asafa

Prof. Y.K. Sanusi

Dr. M.K. Awodele

Dr. O. Adedokun

Dr. J.A. Badmus

Plenary Speakers

Prof. A.S. AbdulKareem
Federal University of Technology, Minna

Dr. E.E. Elemike
Federal University of Petroleum Resources, Effurun

Mr.A.O. Oyefeso
Chairman, National Steering Committee on Nanotechnology, Federal Ministry of Science and Technology, Abuja

Guests at the Conference

Prof. A.T. Oladipo
Dean, Faculty of Pure & Applied Sciences,
LAUTECH, Ogbomoso
Host

Prof. O. Olorode (Rtd)
Formerly of Obafemi Awolowo University,
Ile-Ife

Prof. M.O. Ologunde
Vice-Chancellor, LAUTECH, Ogbomoso
Chairman of the conference

Dr. Kelechi Ebisike
Representative of Honourable Minister of
Science and Technology, Abuja

IOP Conf. Series: Materials Science and Engineering **805** (2020) 011002 doi:10.1088/1757-899X/805/1/011002

Participants at the Conference

Dignitaries at the Conference

Editors of Proceeding

Prof. Agbaje Lateef (Chair)	Department of Pure and Applied Biology, Ladoke Akintola University of Technology, Ogbomoso, Nigeria; alateef@lautech.edu.ng
Prof. Musibau A. Azeez (Editor)	Department of Pure and Applied Biology, Ladoke Akintola University of Technology, Ogbomoso, Nigeria; maazeez@lautech.edu.ng
Prof. Yekinni K. Sanusi	Department of Pure and Applied Physics, Ladoke Akintola University of Technology, Ogbomoso, Nigeria; yksanusi@lautech.edu.ng
Prof. Modiu O. Durowoju	Department of Mechanical Engineering, Ladoke Akintola University of Technology, Ogbomoso, Nigeria; modurowoju@lautech.edu.ng
Prof. Taofeek A. Yekeen	Department of Pure and Applied Biology, Ladoke Akintola University of Technology, Ogbomoso, Nigeria; tayekeen@lautech.edu.ng
Prof. Evariste B. Gueguim-Kana	Department of Microbiology, University of KwaZulu-Natal, Pietermaritzburg, South Africa; kanag@ukzn.ac.za
Dr. Tesleem B. Asafa	Department of Mechanical Engineering, Ladoke Akintola University of Technology, Ogbomoso, Nigeria; tbasafa@lautech.edu.ng
Dr. O. Adedokun	Department of Pure and Applied Physics, Ladoke Akintola University of Technology, Ogbomoso, Nigeria; oadedokun@lautech.edu.ng

IOP Conf. Series: Materials Science and Engineering **805** (2020) 011003 doi:10.1088/1757-899X/805/1/011003

Peer review statement

All papers published in this volume of *IOP Conference Series: Materials Science and Engineering* have been peer reviewed through processes administered by the proceedings Editors. Reviews were conducted by expert referees to the professional and scientific standards expected of a proceedings journal published by IOP Publishing.

IOP Conf. Series: Materials Science and Engineering **805** (2020) 012001 doi:10.1088/1757-899X/805/1/012001

Synthesis and characterization of CZTS thin films from compound target deposited by RF sputtering method

S Abdullahi[1*], M Momoh[1], AU Moreh[1], AM Bayawa[2] and A Saidu[1]
[1]Department of Physics, Usmanu Danfodiyo University, Sokoto, Nigeria.
[2]Department of Pure and Applied Chemistry, Usmanu Danfodiyo University, Sokoto, Nigeria.
*E-mail: abdullahi.sanusi@udusok.edu.ng

Abstract. A potential way to improve the quality of Copper-Zinc-Tin-Sulfide (Cu_2ZnSnS_4(CZTS) absorber thin film by one step process of Radio Frequency (rf) sputtering using a single compound target is proposed for thin film solar cells. Structural and Optical characterizations were carried out using X-ray Diffractometer, Raman spectrophotometer and UV-VIS spectrophotometer. The X-ray diffraction analysis showed *hkl* planes of (002), (110), (112) and (212). Raman results confirmed the existence of secondary phases of SnS, $Cu_{2-x}S$ and CTS. The optical absorption coefficient of the thin film was found to be greater than $10^4 cm^{-1}$ indicating a direct band gap nature of the samples. The optical band energy was calculated and found to be in the range of 1.51eV to 2.4 eV. The band gap energy obtained is more appropriate for photovoltaic applications.

Keywords: Film thickness, annealing temperature, rf sputtering, compound target, XRD, absorption coefficient

1. Introduction

Copper-Zinc-Tin-Sulfide (CZTS) made entirely from abundant materials, has attracted a great interest due to its potential applications in sustainable thin-film solar cell devices. It is a p-type semiconductor with a favorable band gap (~ 1.5 eV) and a high absorption coefficient ($>10^4 cm^{-1}$) [1] which minimizes the quantity of material used in devices. In other words, a thin layer of this material can absorb a greater number of photons from solar radiation and can therefore be used as an absorber layer for solar cell. A variety of deposition techniques were

IOP Conf. Series: Materials Science and Engineering **805** (2020) 012001 doi:10.1088/1757-899X/805/1/012001

employed for the deposition of CZTS. These techniques include but not limited to thermal and electron-beam evaporation [2], co-evaporation [3], direct current (DC) and radio frequency (RF) magnetron sputtering [4-7], hybrid sputtering [8], sol-gel spin-coated deposition, and electro deposition (co-electroplating) [9-10], chemical bath deposition [11], One temperature method [12] and SILAR [13]. Sputtering is one of the most viable methods for depositing thin films. It has the advantage over other vacuum deposition techniques of being simple as all the deposition parameters can be controlled [14-17]. So far, most of the reported CZTS thin film depositions are in the order of 1000 nm or more in terms of the film thickness. Annealing of thin films on the other hand is known to improve the properties of thin films [18-20]. Reports of depositions below 1000 nm thick and annealing at 450 °C or below for CZTS thin films are scarce. For this purpose, CZTS thin films of 550 nm thickness were deposited on corning glass substrates by radio frequency (rf) sputtering technique. The influence of annealing temperature on the thin films was investigated.

2. Experimental Procedures

2.1 Preparation and annealing of the CZTS thin films

CZTS thin films were grown by rf sputtering technique on bare corning glass 7059. The substrates were pre-cleaned with acetone and ethanol followed by rinsing with deionized water and drying in flowing nitrogen (N_2) of 99.99% purity. The sputtering chamber was evacuated to$10^{-4}\ Pa$. A pre-sputtering was carried out for 1 minute. The Zn-rich target is a single disk composed of CZTS of 4N purity at ratios of 1.1:1.8:0.9:1.0 (AJA International) having a diameter of about 2 cm and a thickness of about 0.25 cm. The target/substrate distance was maintained at 7 cm while the rf power was kept constant at 75 W. To ensure uniform deposition, the substrate was rotated at 35 rpm throughout the deposition period. After the deposition of the 550 nm thick thin films, the samples were rinsed with deionized water, dried in flowing nitrogen and then heated under nitrogen atmosphere for 30 min at 250, 350 and 450 °C in a tubular furnace, while the dwelling period was maintained at 60 min. The samples were then allowed to naturally cool to room temperature. Out of the four samples deposited, three were annealed and the fourth sample was reserved as the reference or as-deposited sample.

2.2 Characterization

The crystalline properties of the films were analyzed by high- resolution X-ray diffraction (XRD – SIEMENS Diffractometer D5000) operated at 40 kV and 30 mA. Raman scattering spectroscopy (Reinshawinvia Raman Microscope) with an Olympus microscope equipped with a 100X magnification lens and in the backscattering configuration. The excitation source was a green Argon ion laser operating at 532 nm and 220 mW output powers. The morphology of the films was inspected by Field Emission Scanning Electron Microscope (FE-SEM, Model: JBM-7000F, Japan). The optical properties were analyzed by spectrophotometer (AVANTES AVASPEC-2048 UV-VIS-NIR) operated in the wavelength range of 180-1100 nm at room temperature.

3. Results and Discussion

3.1 X-ray diffraction

The stacked XRD diffractograms of all the samples are displayed in Figure 1. The spectra of the samples appeared polycrystalline which may be attributed to the film thickness and the nature of the substrate. Some of the samples exhibited kesterite structure with diffraction peaks at various angles. The as-deposited sample shows diffraction peaks at $2\theta = 15.94^o$, 26.82^o and 34.46^o corresponding to (002), (110) and (212) according to ICSD 01-075-412. Diffraction peaks observed at $2\theta = 30.66^o$ and 43.86^o for the sample annealed at 250 ^oC indicates the existence of secondary phases of SnS_2 and Cu_2SnS_2. The presence of a single peak at $2\theta = 43.78^o$ for the sample annealed at 350 ^oC also indicates the presence of Cu_2SnS_2.

After annealing at 450 ^oC, diffracting planes of (110), (112) and (212) at $2\theta = 26.32$, 28.5 and 33.82^o were observed.

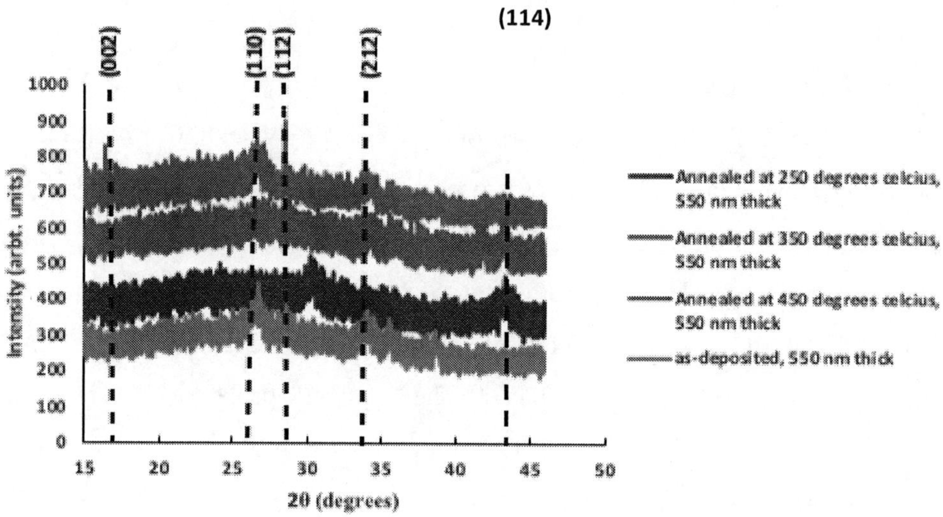

Figure 1. X-ray diffraction patterns of the as-deposited sample and the samples annealed at 250, 350 and 450 °C.

Table 1 gives the structural parameters for the thin films of CZTS. The parameters in the table are the diffraction peak (*hkl*), lattice parameters (*a* and *c*), full width at half maximum (FWHM), grain size (D), dislocation density (δ) and micro strain (ε_m). The *hkl* plane of (110) observed for the as-deposited sample is also reported elsewhere [15, 21]. The *hkl* plane shifted to higher angle at the annealing temperature of 250 °C. The annealing temperature of 350°C and 450 °C resulted in the emergence of the *hkl* planes of (112) and (212) of the kesterite CZTS phase. The standard value for *a* as given by the Joint Committee on Powder Diffraction Standards (JCPDS) is 0.5421 nm along the (002) plane. The value obtained for the as-deposited samples is very close to that obtained in the literature [21-22]. Annealing the samples at 250 °C did not produce much change in the lattice parameter '*a*'. The lattice parameter '*a*' obtained for the sample is very close to those obtained by researchers [23-25]. Lattice parameter '*a*' for the samples annealed at 350 and 450 °C are within the reported range by other researchers [26-28].

The standard lattice parameter *c* for CZTS is 1.0848 nm as referenced in the JCPDS. A low value of 0.88 has been determined for the as-deposited sample. Annealing the sample at 250

IOP Conf. Series: Materials Science and Engineering **805** (2020) 012001 doi:10.1088/1757-899X/805/1/012001

°C led to the decrease in the value of *c* compared to the as-deposited sample. Increasing the annealing temperature to 350 °C further changed the value of *c*.

At the annealing temperature of 450 °C, the sample deviated from the standard value of *c* which might be associated with high concentration of ions which lowers ionic radii as reported [29]. Variations of *c* observed in all the samples might also be connected to nano-grained materials whose crystallites were under strain. Such strain creates local deviation of lattice constants from its bulk value which is size dependent [30]. Changes in the values of *c* more especially indicates that the crystallites are experiencing compression in the *c* direction implying that defects can cause changes in the lattice constant. Variations in the lattice parameters *a* and *c* are also observed elsewhere [29-30]. The values reported here are close to those obtained in previous studies [22, 31].

The FWHM (obtained from x-ray software calculations) for the as-deposited sample is 0.6927. The value of FWHM decreased to 0.1732 when the sample is annealed at 250 °C. As the annealing temperature is increased to 350 °C, the FWHM stood at 0.6927. At the annealing temperature of 450 °C, the FWHM is 0.2598. A decrease in the FWHM has been observed for the sample annealed at 450 °C compared to the sample annealed at 350 °C. The FWHM obtained is very close to that reported elsewhere [30, 32-33].

The grain size was estimated from the full width at half maximum (FWHM) of the (112) peak using Scherrer's formula [21, 34]

$$D = \frac{0.9\lambda}{\beta cos\theta} \qquad (1)$$

where λ is the wavelength of the X-ray used, β is the FWHM and θ is the Bragg's angle

The crystallite size (Table 1) observed for the as-deposited sample is 13.45 and 59.81nm for the sample annealed at 250 °C. The crystallite size for the sample annealed at 350 °C is lower than that of 250 °C which however increased for higher temperature which is similar to those previously reported [34-35]. As the annealing temperature is increased to 450 °C, the crystallite size increased if compared with the samples annealed at 350 °C. Increase or

IOP Conf. Series: Materials Science and Engineering **805** (2020) 012001 doi:10.1088/1757-899X/805/1/012001

decrease in the grain size reflects either the increase or decrease in the concentration of lattice imperfections due to the change in the internal micro-strain within the films.

The as-deposited sample has the highest dislocation density (δ) of $5.53\times 10^{-15} lines/m^2$. Annealing temperature of 250 °C resulted in the decrease of the dislocation density. An increase in the dislocation density has also been observed for the sample annealed at 350 °C if compared with the sample annealed at 250 °C. At the annealing temperature of 450 °C, the dislocation density decreased if compared with the sample annealed at 350 °C. The dislocation density reported here is very close to earlier reports [33, 35].

The micro strain (ε_m) of the as-deposited sample is determined to be 0.16. At the annealing temperature of 250 °C, the micro strain decreases if compared with the as-deposited sample. At the annealing temperature of 350 °C, the micro strain increases if compared with the sample annealed at 250 °C. Micro strain from sample annealed at is 450 °C is 0.06. Micro strain of the sample annealed at 450 °C reduced when compared with the sample annealed at 350 °C. The change in the value of the micro-strain because of the increase in annealing temperature may be due to the movement of interstitial sulfur atoms from inside the crystallites to the grain boundary, which leads to a reduction in the concentration of lattice imperfections.

Table 1. Observed structural parameters of CZTS thin films

Annealing temp. (°C)	**Thickness (nm)**	***hkl***	**a (nm)**	**C (nm)**	**FWHM (*β*)**	***D* (nm)**	***δ* ($\times 10^{-15}$lines/m²)**	ε_m
As-deposited	550	110	0.550	0.88	0.6927	13.45	5.53	0.16
250	550	114	0.530	1.06	0.1732	59.81	0.28	0.04
350	550	212	0.550	1.00	0.6927	27.53	1.32	0.08
450	550	112	0.530	1.06	0.2598	36.70	0.74	0.06

3.2 Raman characterization

The phase purity and composition of the samples was analyzed by Raman spectroscopy. Figure 2 shows the measured Raman spectra for the as-deposited sample and the samples annealed at 250, 350 and 450 °C under nitrogen atmosphere. The spectra show several peaks characteristics of vibrational modes of CZTS and some of its secondary phases. The as-deposited sample showed a strong peak at 297 cm^{-1} which is a phase of CTS [36]. Annealing at 250 °C resulted in the disappearance of the peak at 297 cm^{-1} and appearance of CZTS peak at 288 cm^{-1}.There are no visible CZTS peaks observed for the sample annealed at 350 °C. A low CZTS peak at 337 cm^{-1} has been observed for the sample annealed at 450 °C. This peak corresponds to the A_1 mode associated with S-cation (322 to 377 cm^{-1}). There are some arm shoulder peaks for all the samples. These shoulder peaks are related to defects [37].

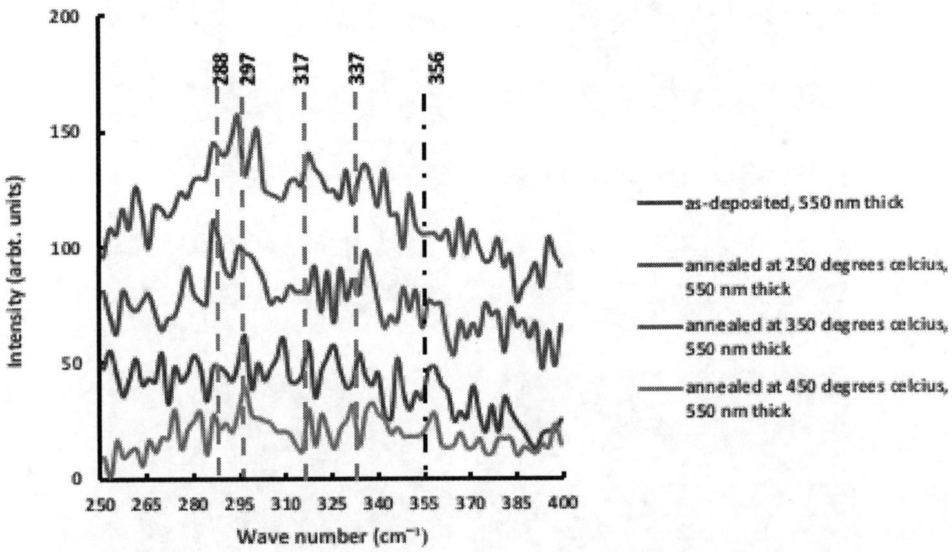

Figure 2. Raman spectra of the as-deposited sample and the samples annealed at 250, 350 and 450 °C.

3.3 SEM characterization

Figure 3 (a, b, c and d) exhibits the micrographs of the CZTS thin films magnified at x30,000. Figure 3a is for the as-deposited sample which shows grains sizes with spherical shaped nanoparticles with uniform distribution. The direct contact of Cu and Zn may be responsible for this [38]. Fig. 3b (sample annealed at 250 °C) shows a cloudy surface in which some of the cloudy structures are brighter than others. The annealing temperature of 250 °C has transformed the spherical shaped nanoparticles of the as-deposited sample to a cloudy surface seen in Figure 3b.

Figure 3c (sample annealed at 350 °C) shows granular particles covering the entire surface of the substrate. The needle-like structures seen in Figure 3c may be attributed to the presence of CuO in the film. These CuO needles are formed after annealing. The increase in the annealing temperature has transformed the cloudy surface of the sample annealed at 250 °C to the needle-like structures. The surface of Figure 3d (sample annealed at 450 °C) is marked by void less compacted structure. There are many transformations observed when the annealing temperature has been increased from 350 to 450 °C. The surface of the sample annealed at 350 °C changed from void less structure to a surface with granular particles seen in Fig. 3d.

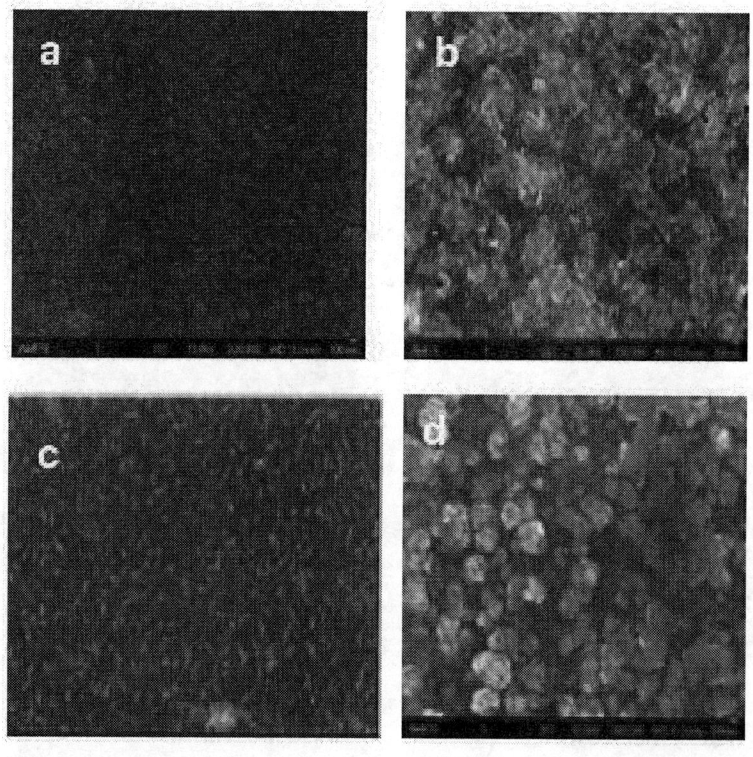

Figure 3. SEM micrographs of the as-deposited sample (a) and the samples annealed at 250 (b), 350 (c) and 450 °C (d)

IOP Conf. Series: Materials Science and Engineering **805** (2020) 012001 doi:10.1088/1757-899X/805/1/012001

3.4 Optical characterization

(a) Transmittance

Figure 4 presents the transmittance spectra of the as-deposited and annealed CZTS thin films taken at room temperature. With average transmittance of 0.0092%, the as-deposited sample is practically opaque. It is observed that this film exhibits low transmittance, indicating that the film grew with a high density of native defects (vacancies, interstitial and antisites). The sample annealed at 250 °C is having an average transmittance of 2.25%. The annealing temperature may be responsible for the increase in transmittance if compared with the as-deposited sample. Low transmittance indicates high absorption [39]. The scattering of the incident light on the surface of the films affects the film transmittance by reflecting the incoming light and thus reducing the transmittance. This explains the difference in transmittance of the CZTS films [21, 25, 40]. The average transmittance of the sample annealed at 350 °C stands at 13.40%. The increase in the annealing temperature to 350 °C has resulted in the increase in transmittance for this sample. Transmittance spectra of thin film of CZTS thin film annealed at 450 °C is also shown in Fig 4. The spectra show that the sample is practically opaque having an average transmittance of 1.02 %. There is a decrease in the transmittance if this sample is compared with sample annealed at 350 °C of the same thickness.

This decrease is attributed to the increase in the annealing temperature. The absorption edge observed for the samples annealed 250 °C and 350 °C ranges from 380 nm to 480 nm.

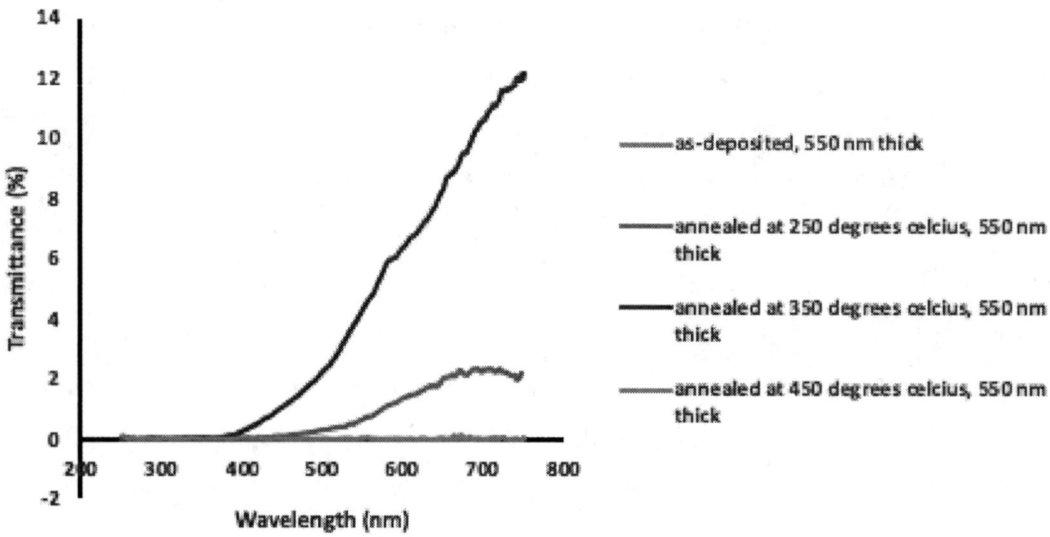

Figure 4. Transmittance spectra of the as-deposited sample and the samples annealed at 250, 350 and 450 °C.

(b) Energy band gap

The optical energy band gap can be calculated as follows [15]:

$$(\alpha h\upsilon)^n = A(h\upsilon - Eg) \quad (2)$$

where h is Planck's constant; υ is the frequency of the incidentphoton; Eg is the optical band gap; A is a constant; and n is an index characterizing the type of optical transition. For a direct allowed transition such as that of CZTS, n is equal to 2. For direct band gap materials, the optical energy gap is determined by extrapolating the straight-line segment of the $(\alpha h\upsilon)^2$ versus $(h\upsilon)$ plot to the $h\upsilon$ -axis.

Figure 5a, b, c and d show the optical band gap of the as-deposited sample and the samples annealed at 250, 350 and 450 °C respectively. The band gap of the as-deposited (Figure 5a) sample is determined to be 1.4eV which is very close to the standard. This band gap has been reported [21]. The band gap of sample annealed at 250 °C has also been exhibited in Figure 5b. This sample shows a deviation (2.5 eV) from the standard (1.5 eV) in the value of band gap compared with the bulk CZTS.

The increase in the band gap is attributed to the presence of CTS and ZnS in the samples as

IOP Conf. Series: Materials Science and Engineering **805** (2020) 012001 doi:10.1088/1757-899X/805/1/012001

confirmed by XRD and Raman. The increase may also be attributed to the presence of a quantum confinement in band structure due to size or strain effect [41]. Furthermore, a decrease in sulfur or variation of composition may also be responsible for the deviation [42]. The shift in the value of the band gap because of annealing has also been reported [43].

The sample annealed at 350 °C is having an average band gap of 1.4 eV. The increase in the annealing temperature from 250 °C to 350 °C improved the band gap of the sample. In general, the distortion of the band gap from the standard is related to structural defects such as lattice strain [24]. The band gap of the sample annealed at 450 °C (1.5eV) is close to the value reported in the literature [23, 44]. The annealing temperature of 450 °C led to the improvement of the band gap of the sample annealed at 350 °C.

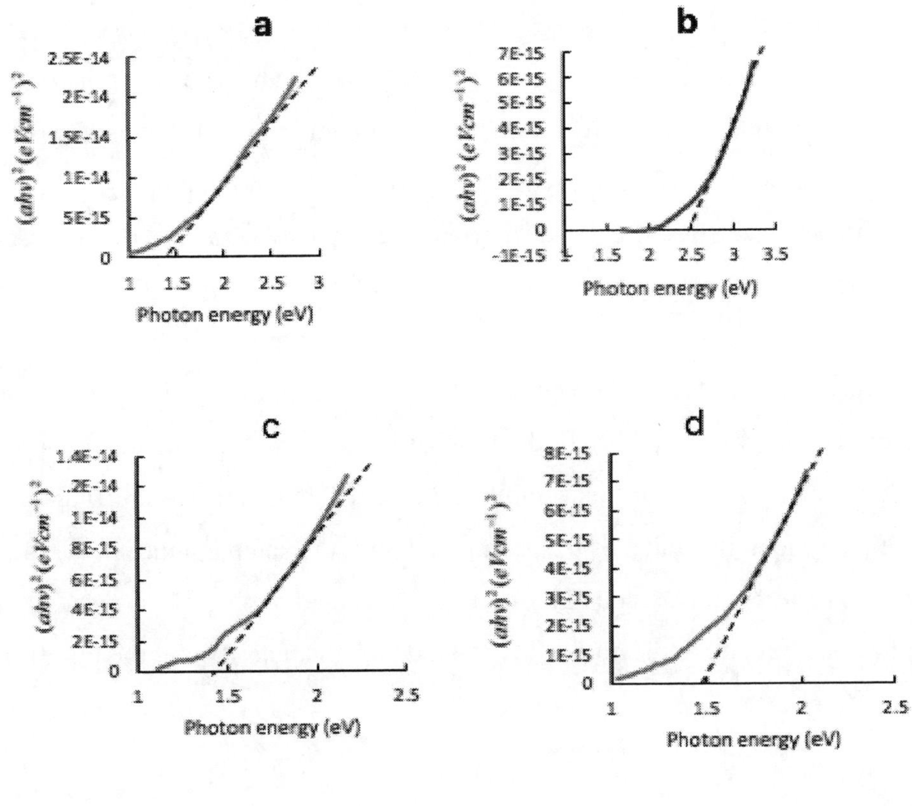

Figure 5. $(\alpha h\nu)^2$ versus Photon energy (eV) for the as-deposited sample and the samples annealed at 250, 350 and 450 °C.

IOP Conf. Series: Materials Science and Engineering **805** (2020) 012001 doi:10.1088/1757-899X/805/1/012001

(c) Absorption coefficient (α)

Based on the transmittance (T) and reflectance (R) data, the absorption coefficient can be determined near the band edge [15, 23]:

$$\alpha = \frac{1}{t}\ln\left[\frac{(1-R)^2}{T}\right] \qquad (3)$$

where α is the absorption coefficient; R is the reflectance; T is thetransmittance; and t is the thickness of the film.

The absorption coefficient determines how strongly the specific wavelength of light is absorbed in the material. Figure 6 shows the absorption coefficient (α) versus wavelength for as-deposited CZTS thin films. All the samples showed absorption coefficient greater than 10^4 cm^{-1} in the visible and near infrared spectral range showing that even the as-deposited sample have the property of absorbing photons. The absorption coefficient determines how strongly the specific wavelength of light is absorbed in the material [45]. The average absorption coefficient for the as-deposited sample is $3.50\times 10^4 cm^{-1}$. The value obtained is in good agreement with that obtained [33, 46]. The absorption coefficient of the sample annealed at 250 ^oC is also of the order of 10^4 cm^{-1}, which supports direct band gap nature of the material. The absorption coefficient has slightly decreased if compared with that of the as-deposited sample. The absorption coefficient for this sample is $3.21\times 10^4 cm^{-1}$. From Figure 6, it is evident that the film annealed at 350 ^oC is having high absorption coefficient. According to the Hopfield quasi-cubic model for copper-containing compounds, this complex near-edge spectrum is determined by the p–d hybridization of the valence band of the compound [12]. The absorption coefficient (α) for this sample is $3.55\times 10^4 cm^{-1}$. The annealing temperature has led to the increase in the value of α if compared with the sample annealed at 250 ^oC. The optical absorption coefficient of the CZTS thin film annealed at 450 ^oC was also greater than 10^4 cm^{-1} in the visible region indicating a direct band gap characteristic of the thin film.

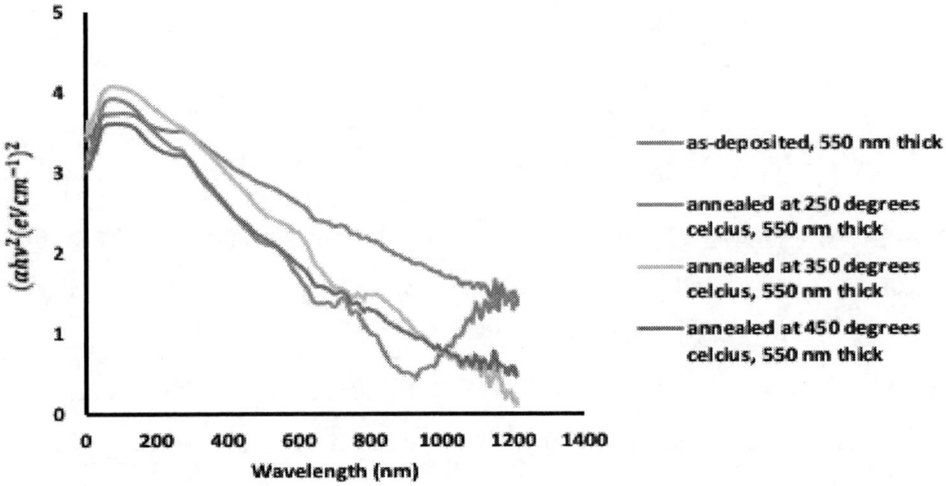

Figure 6. Absorption coefficient versus wavelength for as-deposited sample and the samples annealed at 250, 350 and 450 °C.

Conclusion

CZTS thin films have been deposited by rf sputtering technique from compound target. The annealing temperature was varied from 250 to 450 °C. The structural and optical properties of the samples were investigated. From XRD and RAMAN analysis, some of the samples exhibited the expected secondary phases of ZnS and CTS, while some showed the kesterite structure in preferred orientation of (002), (110), (112) and (212) respectively. The XRD spectra of all the samples appeared amorphous but that did not affect the magnitude of the absorption coefficient if compared with the literature.This shows that even though the film thickness is less than 700 nm and the annealing temperature is less than 500 °C, the absorption coefficient is similar to what is obtained in the literature. The band gap of some of the samples is within the range suggested for absorber layer of a solar cell. The absorption coefficient was greater than $1 \times 10^4 cm^{-1}$. Further investigations are necessary to control the emergence of secondary phases in CZTS thin films.

Acknowledgement

The authors wish to thank TETfund through the management of Usmanu Danfodiyo University Sokoto, Nigeria for sponsoring this research. We also wish to thank the Director and the entire staff of Advanced Materials and Engineering Research Institute (AMERI) of the Florida International University Miami, Florida USA for technical assistance.

References

[1] Kurokawa M, Tanaka K, Moriya K and Uchiki H 2012 Fabrication of three-dimensional-structure solar cell with Cu_2ZnSnS_4. *Jpn J. Phys.* **51** 10NC33. https://doi.org/10.1143/JJAP.51.10NC33.

[2] Zhou Y, Zhou W, Du Y, Li M and Wu S 2011 Sphere-like kesterite Cu_2ZnSnS_4 nanoparticles synthesized by a facile solvothermal method. *Mater. Lett.* **65** (11) pp 1535-1537.

[3] Tanaka T, Yoshida A, Saiki D, Saito K, Guo Q, Nishio M and Yamaguchi T 2011 Influence of composition ratio on properties of Cu_2ZnSnS_4 thin films fabricated by co-evaporation. *Thin Solid Films* **518** (21)pp S29-S33.

[4] Nukala H, Johnson J L, Bhatia A E A, Lund W M, Hlaing O_o, M. M. Nowell M M, Rieth L W and Scarpulla M A 2010 Synthesis of optimized CZTS thin films for photovoltaic absorber layers by sputtering from sulfide targets and sulfurization. *Symp. EE – Defects Inorg. Photovoltaic Mater.* **1268** pp EE03-04. https://doi.org/10.1557/PROC-1268-EE03-04.

[5] Hong C W, Shin S W, Gurav K V, Vanalakar S A, Yeo S. J, Yang H S and Kim J H 2015 Comparative study on the annealing types on the properties of Cu_2ZnSnS_4thin films and their application to solar cells. *Appl. Surf. Sci.* **334** pp 180-184.

[6] Fernandes P. A, Salomé P M P and Da Cunha A F 2011 Study of polycrystalline Cu_2ZnSnS_4 films by Raman scattering. *J. Alloys Compds.* **509** (28) pp 7600-7606.

[7] Liu F, Li Y, Zhang K, Wang B, Yan C, Lai Y and Liu Y 2010 *In situ* growth of Cu_2ZnSnS_4 thin films by reactive magnetron co-sputtering. *Sol. Energ. Mater. Sol. Cells* **94** pp 2431-2434.

[8] Tanaka T, Takeshi N, Daisuke K, Mitsuhiro N, Qixin G, Akihiro W, Akira Y, and Hiroshi O 2005 Preparation of Cu_2ZnSnS_4 thin films by hybrid sputtering. *J. Phys. Chem. Solids* **66** (11) pp 1978-1981.

[9] Chan C P, Lam H, Wong K Y and Surya C 2009 Electrodeposition of Cu_2ZnSnS_4 thin films using Ionic Liquids. *Symp. Photovoltaic Mater. Manuf. Issues* **1123** pp 05-06.

[10] Riha S C, Fredrick S J, Sambur J B, Liu Y, Prieto A L and Parkinson B A 2011 Photoelectrochemical Characterization of nanocrystalline thin-Film Cu_2ZnSnS_4 photocathodes. *ACS Appl. Mater. Interf.* **3** (1) pp 58-66.

[11] Shyju T, Anandhi S, Suriakarthick R, Gopalakrishnan R and Kuppusami P 2015 Mechanosynthesis, deposition and characterization of CZTS and CZTSe materials for solar cell applications. *J. Solid State Chem.* **227** pp 165-177.

[12] Bodnar I V, Telesh E V, Gurieva G and Schou J 2015 Transmittance spectra of Cu_2ZnSnS_4 thin films. *J. Electron. Mater.* **44** (10) pp 3283-3287.

[13] Sun R, Zhao M, Zhuang D, Gong Q, Xie M, Ouyang L and Zhang L 2015 Effects of selenization on phase transition and S/(S+Se) ratios of as-deposited Cu_2ZnSnS_4 absorbers sputtered by a quaternary target. *Mater. Lett.* **164** pp 140-143.

[14] Kubart T, Ericson T, Scragg J, Edoff M and Platzer-Björkman M C 2014 Reactive sputtering of Cu_2ZnSnS_4thin films-Target effects on the deposition process stability. *Surf. Coat. Technol.* **240** pp 281-285.

[15] Feng J, Huang X, Chen W, Wu J, Lin H, Cheng Q, and Zhang F 2016 Fabrication and characterization of Cu_2ZnSnS_4 thin films for photovoltaic application by low-cost single target sputtering process. *Vacuum* **126** pp 84-90.

[16] He J, Sun L, Zhang K, Wang W, Jiang J, Chen Y, Yang P and Chu J 2013 Effect of post-sulfurization on the composition, structure and optical properties of Cu_2ZnSnS_4 thin films deposited by sputtering from a single quaternary target. *Appl. Surf. Sci.* **264** pp 133-138.

[17] Hwa J Y, Woo J J, Chandra M B, ByulKangand H and Soo CY 2015 Improved photovoltaic and grain boundary characteristics of single elementary target-sputtered $Cu_2ZnSnSe_4$ thin films by post sulfurization/selenization process. *J. Phys. D: Appl. Phys.* **48** (24) 245103.

[18] Seboui Z, Gassoumi A, Cuminal Y and KamounTurki N 2013 Physical properties of Cu_2ZnSnS_4 thin films deposited by spray pyrolysis technique. *J. Renew. Sustain. Energ.* **5** 023113. https://doi.org/10.1063/1.4795399.

[19] Sripan C, Madhavan V E, Kasi Viswanath A and Ganesan R 2017 Sulfurization and annealing effects on thermally Evaporated CZTS films. *Mater. Lett.* **189** pp 110-113.

[20] Kumar Swami S, Chaturvedi N, Kumar A and Dutta V 2016 Effect of sulphurization on the activation energy of spray deposited kesterite (Cu_2ZnSnS_4) films. *AIP Phys. Conf. Proc.* **1731** pp 120001-3.

[21] Rabeh M B, Touatti R and Kanzari M 2013 Substrate temperature effects on structural optical and electrical properties of vacuum evaporated Cu_2ZnSnS_4 thin films. *Int. J. Engr. Pract. Res* **2** (2) pp 71-76.

[22] Sanchez T G, Mathew X and Mathews N R 2016 Obtaining phase-pure CZTS thin films by annealing vacuum evaporated CuS/ZnS stack *J. Cryst. Growth* **445** pp 15-23.

[23] Isah KU, Yabagi JA, Ahmadu U, Kimpa MI, Kana MG and Oberafo AA 2013 Effect of different copper precursor layer thickness on properties of Cu_2ZnSnS_4 (CZTS) thin films prepared by sulfurization of thermally deposited stacked metallic layers. *IOSR J. Appl. Phys* **2** (6) pp 14-19.

[24] Tablaoui M, Derbal M and Lebbou K 2016 Discussion on the purity and crystallization of CZTS compound. *J. Cryst. Growth* **452** pp 131-134.

[25] Touatti R, Rabeh M B and Kanzari M 2014 Advanced inorganic materials and structures for structural and optical properties of the new absorber Cu_2ZnSnS_4, thin films grown by vacuum evaporation method. *Energ. Procedia* **44** pp 44-51.

[26] Giaccherini A, Baldassarre A, Donini L, Lepore GO, Caneschi A, De Luca A, Innocenti M, Montegrossi G, Giuseppe C, Oberhauser W and Pardi L 2018 Sustainable synthesis of quaternary sulphides: The problem of the uptake of zinc in CZTS. *J. Alloys Compd.* **775** pp 1221-1229.

[27] Hergert F and Hock R 2007 Predicted formation reactions for the solid-state syntheses of the semiconductor materials $Cu_2SnX_3$and Cu_2ZnSnX_4(X = S, Se) starting from binary chalcogenides. *Thin Solid Films* **515** pp 5953-5956.

[28] Patel K K, Shah D V and Kherai V 2013 Effects of annealing on structural properties of copper zinc tin sulphide (CZTS) material. *J. Nano-Electron. Phys.* **5** (2) pp 02031.

[29] Sun Y, Zong K, Zheng H, Wang H, Liu J, Yan H and M. Zhu M 2013 Ethylene glycol-based dip coating route for the synthesis of Cu_2ZnSnS_4 thin film. *Mater. Lett.* **92** pp 195-197.

[30] Moreh A U, Momoh M, Hamza B, Abdullahi S, Yahya H N, Namadi S and Umar S 2014 Influence of substrate temperature on electrical resistivity and surface morphology of $CuAlS_2$ thin films prepared by thermal evaporation method. *Int. J. Nano Mater. Sci.* **3** pp 47- 55.

[31] Kattan N, Hou B, Fermín D J and Cherns D 2015 Crystal structure and defects visualization of Cu_2ZnSnS_4 nanoparticles employing transmission electron microscopy and electron diffraction. *Appl. Mater. Today* **1** (1) pp 52-59.

[32] Singh O P, Muhunthan N, Singh V N and Singh B P 2015 Effect of annealing time on the composition , microstructure and band gap of copper zinc tin sulfide thin films. *Adv. Mater. Lett.* **6** (1) pp 2-7.

[33] Jheng B T, Huang K M, Chen S F and Wu M C 2014 Effects of substrate temperature on the Cu_2ZnSnS_4 films deposited by radio-frequency sputtering with single target. *Thin Solid Films* **564** pp 345-350.

[34] Mkawi E M, Ibrahim K, Ali M K M and Mohamed A S 2013 Dependence of copper concentration on the properties of Cu_2ZnSnS_4 thin films prepared by electrochemical method. *Int. J. Electrochem. Sci.* **8** pp 359-368.

[35] Khalate S A, Kate R S, Kim J H, Pawar S M and Deokate R J 2017 Effect of deposition temperature on the properties of Cu_2ZnSnS_4 (CZTS) thin films. *Superlattices Microstruct.* **103** pp 335-342.

[36] Berg D. M, Alexandre C, Jérôme G, Victor I, Alejandro P, Shafaat A, Hariklia D, Susanne S and Phillip J D 2014 Simplified formation process for Cu_2ZnSnS_4 based solar cells. *Thin Solid Films* **573** pp 148-158.

[37] Zargar R A, Bhat M A, Parrey I R, Arora M, Kumar J, and Hafiz A K 2017 Optical properties of ZnO/SnO_2 composite coated film. *Optik - Int. J. Light Electron Optics* **127** (17) pp 6997-7001.

[38] Wang H 2011 Progress in thin film solar cells based on Cu_2ZnSnS_4. *Int. J. Photoenerg.* Article ID 801292. http://doi.org/10.1155/2011/801292.

[39] ElAnzeery H, El Daif O, Buffiere M, Oueslati S, Ben Messaoud K, Agten D, Brammertz G, Guindi R, Kniknie B, Meuris M and Poortmans J 2015 Refractive index extraction and thickness optimization of $Cu_2ZnSnSe_4$ thin film solar cells. *Phys. Status Solidi a* **212** (9) pp 1984-1990.

IOP Conf. Series: Materials Science and Engineering **805** (2020) 012001 doi:10.1088/1757-899X/805/1/012001

[40] Amal M I and Kim K H 2013 Crystallization of kesterite Cu_2ZnSnS_4 prepared by the sulfurization of sputtered Cu-Zn-Sn precursors. *Thin Solid Films* **534** pp 144-148.

[41] Yang H, Jauregui L A, Zhang G, Chen Y P and Wu Y 2012 Non-Toxic and abundant copper zinc tin sulfide nanocrystals for potential high temperature thermoelectric energy harvesting. *Nano Lett.* 1**2** (2) pp 540-545.

[42] Kodigala S R 2014 Thin film solar cells from earth abundant materials (Elsevier, London NW1 7BY, UK pp 67-140).

[43] Yeh MY, Lei PH, Lin SH and Yang CD. 2016 Copper-Zinc-Tin-Sulfur thin film using spin-coating technology. *Mater.* **9** (7) 526. https://doi.org/10.3390/ma9070526.

[44] Zakaria Z, Chelvanathan P, Rashid M J, Akhtaruzzaman M, Alam M M, Al-Othman A, Alamoud A, Sopian K and Amin N 2015 Effects of sulfurization temperature on Cu_2ZnSnS_4thin film deposited by single source thermal evaporation method. *Jpn J. Phys.* **54** 08KC18. https://doi.org/10.7567/JJAP.54.08KC18.

[45] Adeyinka A D, Olopade M A and Chendo M A 2017 Enhancement of the conversion efficiency of Cu_2ZnSnS_4 thin film solar cell through the optimization of some device parameters. *Optik – Int. J. Light Electron Optics* **133** pp 122-131.

[46] Maheshawari B U and Kumar S V 2015 Phase transformation of solution-based p-type Cu_2ZnSnS_4 thin film: applicable for solar cell. *Int. J. Energ. Res.* **39** (6) pp 771-777.

Evaluation of impact and hardness properties of Al 6063-AgNPs composites produced by stir cast technique

L Mudashiru[1,2], A Adeodu*[1,2], A Awodoyin[2], A Bello[2], O Oloyede[2], and P Aiyedun[2]
[1]Department of Mechanical Engineering, Ladoke Akintola University of Technology, Ogbomoso, Nigeria
[2]Department of Mechanical and Mechatronics Engineering, Afe Babalola University, Ado-Ekiti, Nigeria
*Email Address: (adeoduac@abuad.edu.ng)

Abstract
The effect of silver nanoparticles (AgNPs) on the mechanical and microstructural properties of aluminium 6063 alloy was investigated. Aluminium 6063 matrix was homogenously mixed with AgNPs impregnated with calcium carbonate at 2, 4 and 6% weight fractions. The mixture was used to produce refined samples through stir casting method. Measurement of impact strength and hardness property of the produced samples at ambient temperature was done by using Charpy impact and Brinell hardness testing machine in accordance with ASTM E23 and ASTM E384 respectively. The impact energy values (234-247 J) of all the samples with AgNPs/$CaCO_3$ reinforcements were higher than the as-cast Al alloy (228 J) while the sample with 6% AgNp has the highest energy impact (247 J). Similarly, the hardness values (56.49-183.70 BHN) of all the samples with AgNPs/$CaCO_3$ reinforcements were higher than the as-cast Al alloy (38.05 BHN) while the sample with 6% AgNPs has the highest (183.70 BHN). The magnitude of impact and hardness increased evidently with increase in percentage weight fraction of the AgNPs. The fracture images of the impact test samples revealed that the surface of AgNPs ($CaCO_3$)-Al6063 composites were with dimples and there was a micro-void formation in $CaCO_3$-Al6063 composite. The use of stir-casting technique influenced the homogeneity and microstructure of the AgNPs composites positively. The Al 6063-Ag nanocomposites possess better qualities in terms of impact energy and hardness, likewise can replace conventional aluminium alloy based on its performance.

Keywords: Al6063 alloy; Brinell hardness; calcium carbonate; impact strength; silver nanoparticles

1. Introduction

Aluminium metal, alloys and composites are becoming more relevant in engineering applications as a result of their excellent physical and mechanical properties [1]. The usage of aluminium alloy has gained serious attention in the area of automobile, aerospace and marine due to it higher strength, better resistance to corrosion, better ductility and light weight [2]. Its application in aerospace and marine industries is as a result of minimal density, increased strength, stiffness, resistance to fatigue and stable dimension at high temperature [3]. The adoption in metal matrix composites is also an emerging trend owing to its excellent mechanical properties. It has ability to solidify at a large varying temperature and as well adapted perfectly for treatment in the semi-solid state [4, 5]. Aluminium matrix composites are particularly influential due to best combination of strength, ductility, toughness and most especially can be process via conventional method [6].

Reinforcement type plays a key role in the property reliability of the metal matrix composites, especially aluminium matrix composite [6]. The reinforcement must not be a reactive type and should be steady in the specified temperature. The frequently used reinforcements are silicon carbide (SiC) and aluminium oxide (Al_2O_3) [6], resulting in increase in tensile strength, hardness and wear resistance when silicon carbide is used and likewise Al_2O_3 offers good compressive strength and wear resistance [6]. In the situation where the reinforcements are particles with nanometric sizes, the composite or alloy is known to be nanocomposites [7] which are more stable at medium temperature [8]. Alloying with silver nanoparticles (AgNPs) is gaining attention in the field of engineering owing to its uniqueness in optical, electrical and thermal properties as its being delivered into products ranging from photovoltaic to biological and chemical sensors [9, 10]. Emerging application areas of AgNPs are in antimicrobial coatings, textiles and biomedical devices [11].

Stir casting technique has been adopted for its commercial production due to its simplicity, flexibility in the control of matrix and most importantly, cost effectiveness [3]. It offers greater matrix reinforcement bonding than the other conventional methods due to the stirring action, couple with appropriate processing parameters [12]. Optimal mechanical property of the composite materials is a function of the level of homogeneity, which must be reasonably high [3]. The porosity level must also be minimal [13, 14]. Different studies by authors on the wear rate of the materials were as a result of the distance between the loading and sliding [15, 16]. The sliding wear resistance of the composite was examined at different sliding speeds over different loading and distance levels. There is increase in wear behaviour as the loads and distances are increasing while the material exhibits less wear with increase in speed of sliding. Another study on the microhardness of composite of aluminium 7075 alloy with AgNPs by Flores-Campos *et al.* [17] shows that microhardness value increases with increasing AgNPs in relation with the aluminium alloy, both in the powder and extruded conditions. In this study, the effects of AgNPs impregnated with calcium carbonate on impact strength and hardness properties of aluminium 6063 alloy were examined in order to ascertain the responsiveness of the matrix to the varying weight fractions of reinforcement.

2. Materials and Methods

The materials considered in the study are AgNPs impregnated with calcium carbonate, incorporated into aluminium 6063 alloy as matrix. This type of alloy is good for production of lightweight metal castings [3]. Al 6063 was obtained as scrap aluminium profile sourced from some construction sites within Afe Babalola University, Ado-Ekiti, Nigeria. The percentage mass fraction of the compositions of Al6063 alloy is shown in Table 1. Silver nanoparticles of about 80 nm in size used for this study was synthesized using cobweb [18] and obtained from the Laboratory of Industrial Microbiology and Nanobiotechnology, Department of Pure and Applied Biology, Ladoke Akintola University of Technology, Ogbomoso, Nigeria. Also, calcium carbonate was sourced locally from a supplier at Ado-Ekiti, Nigeria. Table 2 shows the property values of aluminium matrix and AgNPs. Figure 1 shows the average size distribution of silver nanoparticles measured with particle analyzer.

Table 1. Chemical composition of Al 6063 (Mass Fraction %)

Si	Fe	Cu	Mn	Mg	Cr	Zn	Ti	Al
0.86	0.19	≤0.01	0.04	0.51	0.01	0.01	0.008	Bal.

Source: [3]

IOP Conf. Series: Materials Science and Engineering **805** (2020) 012002 doi:10.1088/1757-899X/805/1/012002

Table 2. Properties of Al6063 and AgNPs

Material	Elastic Modulus (Gpa)	Density Kg/m3	Poisson Ratio	Hardness (HB 500)	Tensile Strength (mPa)	Melting point (°C)	Boiling Point (°C)
Al 6063	70-80	2.7	0.33	95	310	652	2519
AgNPs	72-76	10.49	0.37	206	360	962	2212

Source: [3]

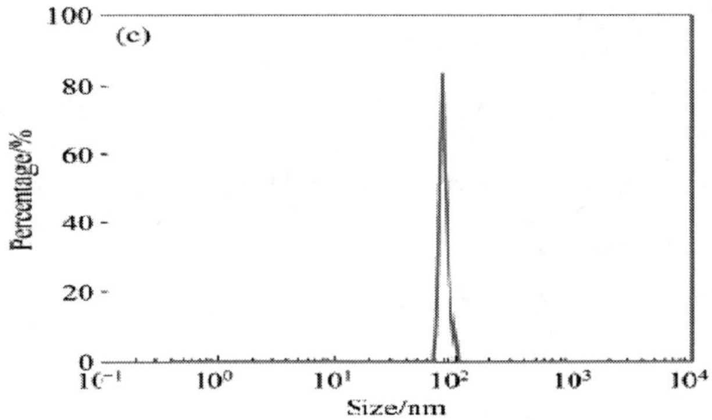

Figure 1. Average size of the AgNPs

2.1 Sample Preparation

The sample preparation was divided into three stages, namely: impregnation AgNPs with $CaCO_3$, recycling of scrap aluminium profile and preparation of Al-AgNPs composite from Al6063 alloy.

2.1.1 Impregnation of AgNPs with $CaCO_3$

Exactly 95 ml of AgNPs was measured, mixed with 158 g of $CaCO_3$ thoroughly using Heidolph rzr 20121 automatic stirrer in a beaker at 1783 rpm into slurry form. The slurry was then sundried for 5 h for solidification, after which it was ground into powder using crucible, sieve through a 425 μm mesh to obtained even particle size distribution.

IOP Conf. Series: Materials Science and Engineering **805** (2020) 012002 doi:10.1088/1757-899X/805/1/012002

Figure 2. (a) Measuring of $CaCO_3$ (b) Mixing of AgNPs with $CaCO_3$

2.1.2 Recycling of Aluminium 6063 Scrap

The recycling of Al 6063 scraps followed the sequence described in Figure 3. The first step in the preparation of the scrap Al 6063 profiles include washing and drying in open air to remove stains and sand contaminants. Aluminium scrap is then weighed, packed into diesel fired crucible furnace, heated to 900 ^{o}C to remelt. Salt flux was added to absorb the contaminants and protect the aluminium metal against oxidation. Refining of the melted aluminium was undertaken to separate the slag from the pure metal. Thereafter, aluminium billets were produced by reheating the pure metal to 900 ^{o}C, and subsequently pouring into a preheated mould at 600 ^{o}C prepared in a sand core.

Figure 3. (a) Remelting of Al6063 scraps (b) Melted Al6063 (c) Al6063 billets

2.1.3 Preparation of Aluminium Composites

The modified silver AgNPs was used to prepare 850 g metal matrix composite of aluminium-silver nanoparticles (Al-AgNPs) by melt stirred technique. The aluminium billets were divided into seven portions, measured and labelled as samples 1-6 and C (control sample). Each billet portion was first preheated to 450 °C before finally melted at 750 °C in a crucible furnace. The modified AgNPs of specified percentage weight fraction of 2, 4 and 6% were weighed and preheated at 100 °C and incorporated into the melted aluminium billets, thoroughly mixed at 200 rpm for 10 min and poured into 30 × 200 mm cylindrical mould heated to 600 °C. These procedures were repeated for Al-$CaCO_3$ only at the same specified percentage weight fraction. Masses of the base metal matrix and percentage weight fractions of the reinforcements are described in Table 3.

Table 3. Mixing ratio of base metal and reinforcements

Sample	Percentage fraction of Reinforcement (%)	Mass fraction of Al 6063 (g)	Mass fraction of AgNPs/$CaCO_3$ (g)	Mass fraction of $CaCO_3$ (g)	Mass of Composite (g)
1	2.00	833.00	17.00	-	850
2	4.00	816.00	34.00	-	850
3	6.00	799.00	51.00	-	850
4	2.00	833.00	-	17.00	850
5	4.00	816.00	-	34.00	850
6	6.00	799.00	-	51.00	850
7	C	850	-	-	-

One sample each was prepared from the casting samples to analyze the microstructure. The samples were etched using abrasive paper of grade 800 and finally polished to obtain a reflective surface finish. The microstructure of each sample was observed by optical microscope. Figure 4(a) reveals micrograph for the grain structure of Al6063 Matrix. Figure 4(b) shows the micrograph of Al-AgNPs. The silver nanoparticles were spread within the interior of the grain also along the grain boundaries of the Al6063 alloy. Figure 4(c) shows the micrograph of Al-$CaC0_3$. In Figures 4(b) and (c), there was even distribution of AgNPs and $CaCO_3$ particles within the Al6063 alloy.

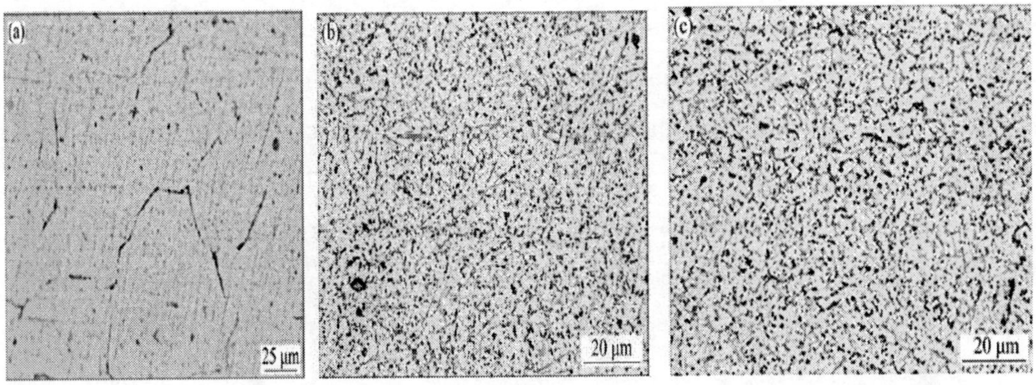

Figure 4. (a) Matrix grain structure of Al6063 (b) Micrograph of 2 % Al-AgNPs (c) 2 % Al-$CaCO_3$

2.2 Mechanical Properties of the Samples

The samples were machined after solidified, to sizes accordingly to the required mechanical and microstructural analysis to be carried out in accordance with ASTM E23 and E384 Standards.

2.2.1 Impact Analysis

The impact analysis was carried out using Charpy Impact Tester to determine the behaviour of the samples under sudden load which gives the impact strength or energy require to cause deformation or break of the sample. Seven samples from all the combinations of the percentage weight fractions of the composites and control were machined to 55 × 10 × 5 mm charpy specimens with centre V-notch according to ASTM E23. The specimen was placed on the impact tester's vice such that the notch faces away from the hammer. The scale pointer was set to zero position and safety considerations were considered. The pendulum was released by the release mechanism and the impact energy (E) was recorded directly from the scale. The exercise was repeated twice for each sample and impact strength (IS) was determine by finding the average of the impact energies.

IOP Conf. Series: Materials Science and Engineering **805** (2020) 012002 doi:10.1088/1757-899X/805/1/012002

2.2.2 Hardness Analysis

The hardness analysis was carried out using brinell hardness testing machine with 1. 5 mm diameter ball and 100 kg load. This is to measure the extent of a localized penetration by a standardized round or pointed indenter in accordance with ASTM E384 standard. The specimens were placed one after the other on the hardness testing table and elevated gradually for the specimen to make contact with the indenter. When secured contact was established, the load lever was moved from point A to B to load the machine, left for 10 seconds. The diameter of the indentation was measured using microscope and micrometer. The expression for the Brinell hardness number is given in equation 1

$$BHN = \frac{2P}{\pi D(D\sqrt{D^2 - d^2})} \qquad (1)$$

Where D is the diameter of the indenter, d is the indent diameter and P is the load applied

3. Results and Discussion

The results of the influence of AgNPs impregnated with calcium carbonate on mechanical properties of AA6063 alloy are discussed in this section. These were compared with influence of only calcium carbonate on AA6063.

3.1 Impact Test

Figure 5 shows the variation of energy absorbed by Al6063 alloy with increase in the percentage weight fractions (2, 4 and 6 %) of AgNPs-($CaCO_3$) and $CaCO_3$ respectively. It was observed that there is increase in impact energy absorbed as the percentage weight fractions of reinforcements increase. Significant increase was observed in the impact strength of AgNPs-($CaCO_3$)-Al6063 composite compared to $CaCO_3$-Al6063 composite. The addition of reinforcements increases the ductility and deformation capabilities of Al6063 composites [19, 20]. The reinforcement promotes the dislocating motion which is required for plastic deformation of the Al6063 matrix [20]. Impact properties of the composites are also explained as a function of force of cohesion between the metal matrix and the reinforcements [21]. This is obvious in AgNPs-($CaCO_3$)-Al6063 compared to $CaCO_3$-Al6063 due to the sizes of the AgNPs to $CaCO_3$. Figures 6(a-c) show the SEM images of the fracture surfaces to explain impact energy behaviour of the as-cast Al6063, AgNPs-($CaCO_3$)-Al6063 and $CaCO_3$-Al6063 composites respectively. The as-cast Al6063 matrix has very less capacity to reduce the dislocating motion required for plastic deformation, while there are ductile modes fractures in AgNPs-($CaCO_3$) and $CaCO_3$-Al6063 fractograph counterparts [3].

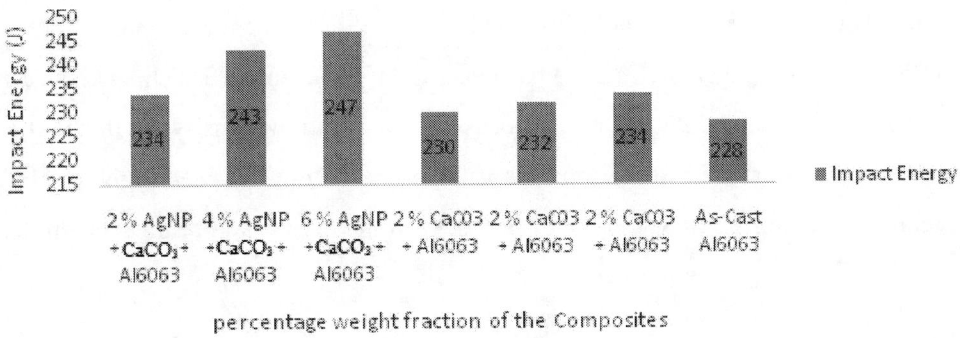

Figure 5. The response of composites to impact energy

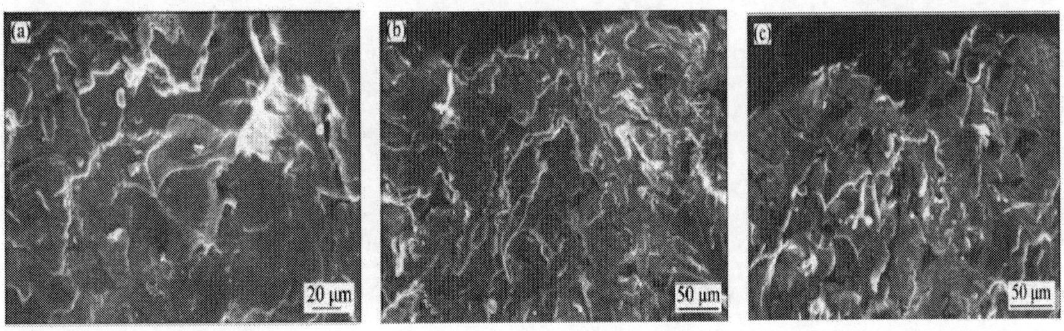

Figure 6. (a) SEM fractograph of As-cast Al6063 (b) 2%AgNPs-($CaCO_3$)-Al6063 (c) 2% $CaCO_3$-Al6063

3.2 Hardness Test

Hardness testing using the normal Brinell hardness measurement was conducted to determine the level of brittleness of AgNPs-($CaCO_3$) and $CaCO_3$-Al6063 composites as compared to as-cast Al6063 alloy. Figure 7 shows the variation of Brinell hardness numbers the composites with increase in percentage weight fractions of the reinforcements. From the Figure 7, there is continuous increase in the hardness number as the percentage weight fractions of the reinforcements increased. The increase is due to substantial homogeneous dispersion of the reinforcements in the aluminium matrix [17] and therefore acts as problems to the motion of dislocation [22]. Hardness of the composites is a function of the reinforcement and matrix [20].

The presence of silicon in excess as part of the composition of aluminium matrix increases the brittleness and hardness of Al6063 composites [20]. Also, as observed in Figure 7, there is conspicuous margin of increase in hardness between AgNPs-($CaCO_3$)-Al6063 and $CaCO_3$-Al6063 composites. This increase is linked to increase in the percentage weight fraction of hard and brittle phases of the AgNPs-($CaCO_3$) particles in Al6063 alloy [23]. The presence of $CaCO_3$ mixed with AgNPs in the Al6063 alloy raises the dislocation density at the particle-matrix interface due to differences in the thermal expansion coefficient between AgNPs reinforcement and Al metal matrix causing plastic incohesion between the matrix and reinforcement [24-27].

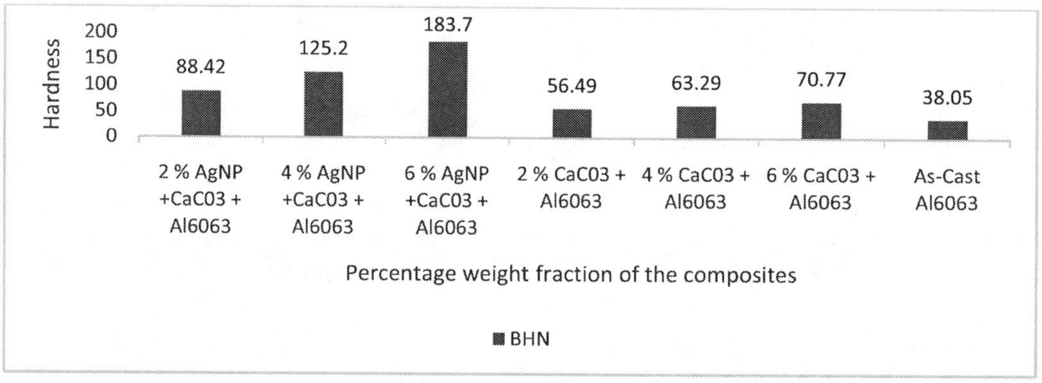

Figure 7. The effect of percentage weight fraction of the composites on hardness

4. Conclusion

Al6063 composites were successfully fabricated via stir casting process using silver nanoparticles synthesized from cob web and calcium carbonate. The effects of silver nanoparticles impregnated with calcium carbonate on mechanical properties of Al6063 alloy were examined. Impact energies and hardness of the composites increase with increase in percentage weight fraction of the reinforcement. Evidence of dimples surfaces was observed in AgNPs-($CaCO_3$)-Al6063 and micro-void formation was observed in $CaCO_3$-Al6063 composite. Application of the AgNPs-($CaCO_3$)-Al6063 composite in window frame will be very good in terms of impact and hardness properties especially when strength to weight ratio is a deciding factor.

References

[1] Mondolfo L F 1976 Aluminum alloys: Structure and properties [M]. Boston, London: Butterworths.

[2] Kalaiselvan K, Murugan N and Siva P 2011 Production and characterization of AA6061−B4C stir cast composite. *Mater. Design* **32** (7) pp 4004-4009.

[3] Pitchayyapillai G, Seenikannan P, Balasundar P and Narayanasamy P 2017 Effect of Nano-Silver on Microstructure, Mechanical and Tribological Properties of Cast 6061 Aluminium Alloy. *Trans. Nonferrous Met. Soc. China* **27** (10) pp 2137-2145.

[4] Chen K H, Fang H C, Zhang Z, Chen X and Liu G 2008 Effect of Yb, Cr and Zr additions on recrystallization and corrosion resistance of Al-Zn-Mg-Cu alloys. *Mater. Sci. Eng.: A* **497** (1-2): 426-431.

[5] Yi M, Zhi-hao Z, Jian-zhong C 2013 Effect of minor Zr and Sc on microstructures and mechanical properties of Al-Mg-Si-Cu-Cr-V alloys. *Trans. Nonferrous Met. Soc. China* **23** (7) pp 1882-1889.

[6] Igbai A A and Nuruzzama D M 2016 Effect of reinforcement on the mechanical properties of aluminium matrix composites: review. *Int. J. Appl. Eng. Res.* **11** (21) pp 10408-10413

[7] Esawi AM, Morsi K, Sayed A, Gawad AA and Borah P 2009 Fabrication and properties of dispersed carbon nanotube–aluminum composites. *Mater. Sci. Eng.: A.* **508** (1-2) pp 167-173.

[8] Cardoso KR and Rodrigues CA 2004 Processing of aluminium alloys containing titanium addition by mechanical alloying. *Mater. Sci. Eng.: A.* **375** pp 1201-1205.

[9] Auld J H, 1986 Structure of metastable precipitate in some Al-Cu-Mg-Ag alloys. *Mater. Sci. Technol..* **2** (8) pp 784-787.

[10] Song M, Chen KH and Huang LP 2006 Effects of Ag addition on mechanical properties and microstructures of Al-8Cu-0.5 Mg alloy. *Trans. Nonferrous Met. Soc. China* **16** (4) pp 766-771.

[11] Li W R, Xie X B, Shi Q S, Zeng H Y, Ouyang Y S and Chen Y B 2010 Antibacterial activity and mechanism of silver nanoparticles on *Escherichia coli*. *Appl. Microbiol. Biotechnol.* **85** pp 1115-1122.

[12] Shabani M O and Mazahery A 2013 Application of GA to optimize the process conditions of Al matrix nano-composites . *Compos. Part B: Eng.* **45** (1) pp 185-191.

[13] Shabani M O and Mazahery A 2012 The performance of various artificial neurons interconnections in the modelling and experimental manufacturing of the composites. *Mater. Tehnol.* **46** (2) pp 109-113.

[14] Mazahery A and Shabani M O 2013 Microstructural and abrasive wear properties of SiC reinforced aluminum-based composite produced by compocasting. *Trans. Nonferrous Met. Soc. China* **23** pp 1905-1914.

[15] Selvakumar N and Narayanasamy P 2016 Optimization and effect of weight fraction of MoS_2 on the tribological behavior of Mg-TiC-MoS_2 hybrid composites. *Tribol. Trans.* **59** (4) pp 733-747.

[16] Ravindran P, Manisekar K, Narayanasamy P, Selvakumar N and Narayanasamy R 2012 Application of factorial techniques to study the wear of Al hybrid composites with graphite addition. *Mater. Design* **39** pp 42-54.

[17] Flores-Campos R, Mendoza-Ruiz DC, Amézaga-Madrid P, Estrada-Guel I, Miki-Yoshida M, Herrera-Ramírez JM and Martínez-Sánchez R 2010 Microstructural and mechanical characterization in 7075 aluminum alloy reinforced by silver nanoparticles dispersion. *J. Alloys Compds.* **495** (2) pp 394-398.

[18] Lateef A, Ojo SA, Azeez MA, Asafa TB, Yekeen TA, Akinboro A, Oladipo IC, Gueguim-Kana EB and Beukes LS 2016 Cobweb as novel biomaterial for the green and eco-friendly synthesis of silver nanoparticles. *Appl. Nanosci.* **6** (6) pp 863-874.

[19] Nishida Y 2013 Introduction to Metal Matrix Composites: Fabrication and Recycling. *Springer Science & Business Media*: Berlin, Germany.

[20] Razzaq A M, Majid D L, Ishak M R and Basheer U M 2017 Effect of fly ash addition on the physical and mechanical properties of AA6063 alloy reinforcement. *Metals* **7** (11) 477. https://doi.org/10.3390/met7110477.

[21] Srivastava A K and Das K 2009 Microstructural and mechanical characterization of in situ TiC and (Ti, W) C-reinforced high manganese austenitic steel matrix composites. *Mater. Sci. Eng. A*: **516** (1-2) pp 1-6.

[22] Basak A, Pramanik A, Islam M N and Anandakrishnan V 2015 Challenges and recent developments on nanoparticle-reinforced metal matrix composites. *In Fillers and Reinforcements for Advanced Nanocomposites; Elsevier Science & Technology*: Amsterdam, The Netherlands; p. 349.

[23] Hassan S B and Aigbodion V S 2015 Effects of egg shell on the microstructures and properties of Al-Cu-Mg/egg shell particulate composites. *J. King Saud Univ.-Eng. Sci.* **27** (1), pp 49-56.

[24] Aigbodion V S and Hassan S B 2017 Effects of silicon carbide reinforcement on Microstructure and properties of cast Al-Si-Fe/SiC particulate composites. *J. Mater. Sci. Eng.: A* **447** (1-2) pp 355–360.

[25] Aigbodion VS 2010 Potential of using bagasse ash particle in metal matrix composite. PhD Thesis, Department of Metallurgical and Materials Engineering, Ahmadu Bello University, Samaru, Zaria, Nigeria. pp. 34-50.

[26] Rajan TP, Pillai RM, Pai BC, Satyanarayana KG and Rohatgi PK 2007 Fabrication and characterisation of Al–7Si–0.35 Mg/fly ash metal matrix composites processed by different stir casting routes. *Compos. Sci. Technol.* **67** (15-16) pp 3369-3377.

[27] Siva Prasad D, Rama Krishna A 2011 Production and Mechanical Properties of A356.2/RHA Composites. *Int. J. Adv. Sci. Technol.* **33** pp 51-58.

IOP Conf. Series: Materials Science and Engineering **805** (2020) 012003 doi:10.1088/1757-899X/805/1/012003

Green Synthesis, characterization of Zinc oxide nanoparticles using *Chromolaena odorata* extract and evaluation of its properties for photoanode of solar cells

SW Balogun*[1], YK Sanusi[2], and BA Agboluaje[2]

[1]Department of Materials Science and Engineering, Kwara State University, Malete, Ilorin, Nigeria

[2]Department of Pure and Applied Physics, Ladoke Akintola University of Technology, Ogbomoso, Nigeria.

*Corresponding author: sbalogun94@gmail.com

Abstract

Synthesis of Zinc oxide nanoparticles (ZnONPs) using plant extract was carried out in this work because of its advantages over other method of synthesis namely; simplicity, environmental friendliness and elimination of toxic chemicals. Fresh Siam leaf extract was characterized over a period of 10 days to study the effect of time on the optical photon energy absorption properties. ZnONPs was synthesized and deposited on substrate by spin-coating method and characterized by UV-vis spectroscopy, Fourier transform infra-red spectroscopy, Scanning Electron microscopy, and Energy Dispersive X-ray. ZnONPs thin film device was then fabricated to study the electrical properties. From UV-vis spectroscopic result, as the time (day) increased, there was an increase in the value of the transmittance and corresponding reduction in photon energy absorption. FTIR result gives the functional groups and absorption number at 3448.04cm^{-1} O-H single bond, 2524.85cm^{-1} C-H bond, 1437.15 cm^{-1}, 880.59 cm^{-1}, 729.06 cm^{-1}, 433.44 cm^{-1} Zn-OH and Zn-O variety of single bond. The surface morphology shows large grain size. Energy band gap of ZnONPs was approximately 3.7 eV. The fabricated ZnONPs thin film device under illumination has efficiency of 2.01$^0/_0$. From the morphology, optical, and electrical properties of the ZnONPs thin film and device, it could be a suitable material for crystalline solar cells as photoanode.

Keywords: Zinc oxide nanoparticles, *Chromolaena odorata,* photoanode, solar cell, optical properties

1. Introduction

Challenges in provision of eco-friendly and environment-friendly sources of energy have led to intensive research in nanomaterials. Nanotechnology is an arm of technology that studies different materials at a nanometric scale and its application is found in materials science, engineering, and electronics. Nanoparticles have different magnetic and optoelectronics properties that are regulated by their size and shape distribution. Synthesis of nanoparticles using plant extract is gaining importance because this method of synthesis eliminated the need of toxic chemicals method of synthesis and also, because of its simplicity and environmental friendliness. The use of zinc oxide (ZnO) is because of its unique properties and its exciton binding energy [1-5]. Siam weed leaf (*Chromolaena odorata*) extract was used in this study because of its abundant availability. The use of different plant extracts to synthesize ZnO nanoparticles has been reported [6-8]. Zhang *et al.* [9] reported the use of ZnONPs in solar cells fabrication. Little or nothing has been reported on green synthesis of zinc oxide nanoparticles using Siam weed leaf extract as a material for solar cells application. The objectives of this research are to synthesis and characterize zinc oxide nanoparticles using Siam weed leaf extract as well studies on the application of zinc nanoparticles thin film device in solar cells fabrication.

2. Materials and Methods

2.1 Materials

Zinc acetate was purchased from Sigma-Aldrich, while other materials used for the research were sourced from local vendors of laboratory wares and reagents.

2.2 Methods

2.2.1 Preparation of the plant extract

Glass slides of dimension 18 mm by 18 mm were used as substrates. The glass substrates were washed with detergent solution for 10 to 15 min in ultrasonic sonicator and rinsed in distilled water for 15 min at 27 ^oC. The substrate was cleaned with isopropanol alcohol in ultrasonic bath for 15 min at 27 ^oC and dried in a stream of nitrogen gas. Siam plant was plucked from the surroundings, and the leaves were detached from the stem (Figure 1a). These leaves were thoroughly washed with clean water and then rinsed with distilled water. The fresh leaves were cut into pieces and pounding using the laboratory mortar and pestle. After the pounding process, the liquid content was extracted with the use of Whatman filter paper placed on a

IOP Conf. Series: Materials Science and Engineering **805** (2020) 012003 doi:10.1088/1757-899X/805/1/012003

clean conical flask over a period of 8 h. Brownish aqueous solution of the extract precipitated at the bottom of the conical flask (Figure 1b). The extracted aqueous leaf content was sealed and stored for further studies.

Figure 1 . (a) Siam weed leaf after plucking **(b)** Siam weed leaf extract

2.2.2 Green Synthesis of ZnONPs and deposition on substrate

Synthesis of ZnONPs was done by mixing 25 ml of Zinc acetate dihydrate solution with 5 ml of Siam weed extract solution and stirred with magnetic stirrer for 12 h for homogenous mixture. After the reaction, the solution was kept overnight for the nanoparticles to precipitate. Precipitate that settled at the bottom of the sealed beaker was obtained by removal of excess unwanted floating solution. The precipitate was washed with deionized water. The obtained extracted precipitate was heated at the temperature of 400 ^{o}C for 30 min to evaporate the solvent under high purity argon gas in a carbolite tubular furnace (model Srw 21-501042 Type-CT17). The pellet-like substance which is white brown was ground to powder particles and stored for further deposition and characterization purposes.

To deposit ZnONPs onto substrate, 0.5 g of synthesized ZnONPs was mixed with 5 ml of ethanol and stirred with the use of magnetic stirrer for 3 h for homogenous mixture, after which it was deposited on substrate with the use of spin-coater at spin coat speed of 3000 rpm for 30secs. Spin-coating techniques enhance uniform deposition of solution on substrate at set speed. Characterization was carried out on Siam weed leaf extract and ZnONPs nanoparticles after deposition. Characterization tools employed were Scanning Electron Microscopy (SEM) to study the morphology, Electron Dispersive X-ray (EDX) to determine the elemental composition, Fourier Transform Infra-red (FTIR) spectroscopy to reveal the chemical bond present in the sample.

2.2.3 ZnONPs thin film device fabrication and I-V Characterization

To fabricate ZnONPs thin films device, ZnONPs solution was deposited by spin-coating method on Indium tin oxide (ITO) coated glass substrate at 3000 rpm for 30 secs. Silver paste was deposited on ZnONPs as the electrode and ITO coated glass substrate as the counter electrode. ZnO films were heated at 100 °C in an oven for 30 min to evaporate solvent. Then current-voltage characterization of thin films device was carried out under illumination of solar Simulator with Kiethley 2400 SMU. Short-circuit current (I_{SC}) and open-circuit voltage (V_{OC}) were noted. Fill factor (FF) and efficiency (η) of the thin film device were computed using equations (1) and (2) respectively. The input photon power (P_{in}) was obtained by multiplying the measured light intensity with the active cell area.

$$FF = \frac{Pmax}{1sc\ Voc} = \frac{1mp\ Vmp}{1sc\ VoC} \quad (1)$$

$$\eta = \frac{I_{SC}.V_{OC}.FF}{P_{in}} \quad (2)$$

3. Results and Discussion

3.1 Optical properties of Siam weed extract and biosynthesized ZnONPs

Figure 2 shows the effect of time on the optical properties of Siam weed extract. At day 1, peak optical transmittance values of about 55% from 400-500 nm in visible spectrum of wavelength band and 55% from 350-400 nm in the UV spectrum band were obtained. The extract was then characterized after 5 days and 10 days respectively to observe the effect of time on the optical properties of the plant extract. It was observed that the value of transmittance was about 85% both in UV and Visible spectrum of wavelength band. It can be deduced that as the time (day) increases, there is an increase in the value of the transmittance. It also shows that as transmittance increases, there is a decrease in photon energy absorption by the material. Figure 3 shows the graph of absorption of photon energy versus wavelength of Siam weed extract at different days of characterization. The result shows that optical properties characterization carried out on day 1 has the highest optical absorption of photon energy. Figure 4 shows graph of absorbance versus wavelength of ZnONPs. Efficient conversion of solar energy requires materials to absorb strongly in the visible region of the spectrum. Figure 5 shows Tauc Plot of $(\alpha h\nu)^2$ versus ($h\nu$) to obtain the value of the energy band gap of ZnONPs as a function of photon energy absorption. The energy band gap obtained from straight line plot of $(\alpha h\upsilon)^2$ vs. $h\upsilon$ by extrapolation of the line to base line is 3.7 eV.

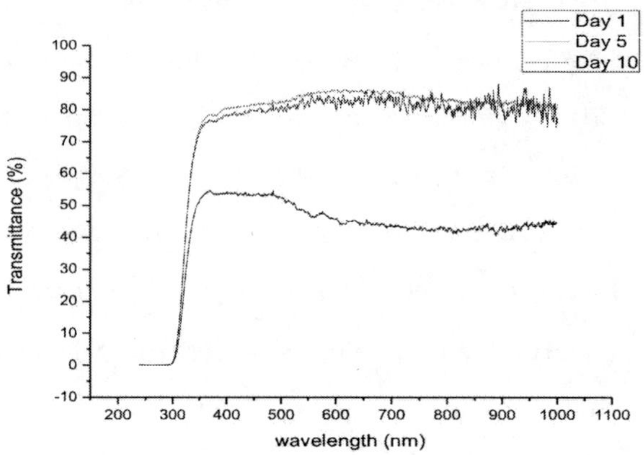

Figure 2. The transmittances of Siam weed leaf extract at different days

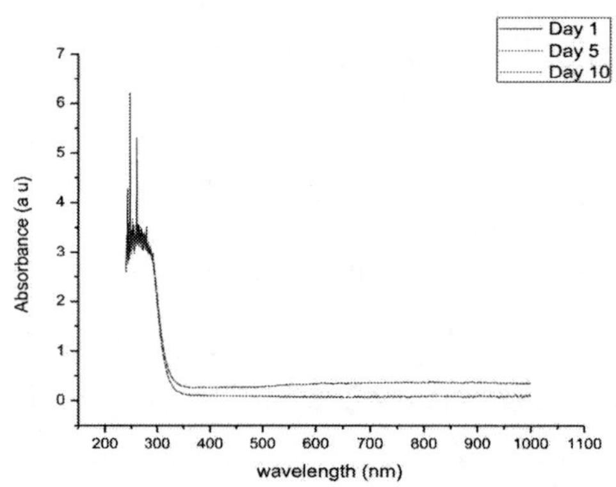

Figure 3. The absorbance readings of Siam leaf extract at different days

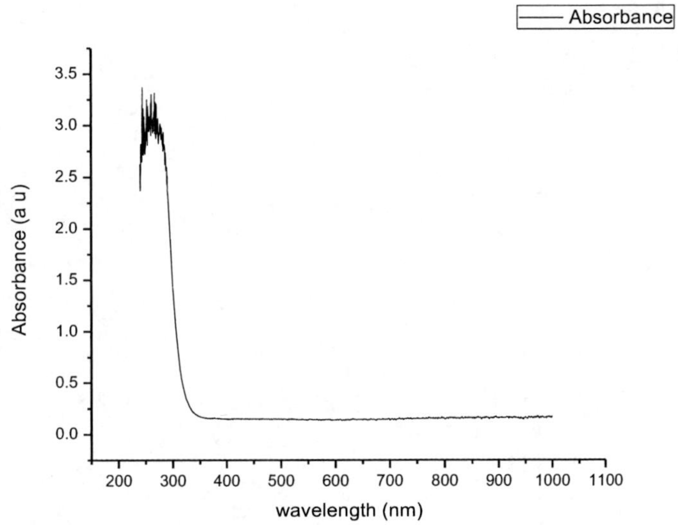

Figure 4. The absorption spectrum of ZnONPs

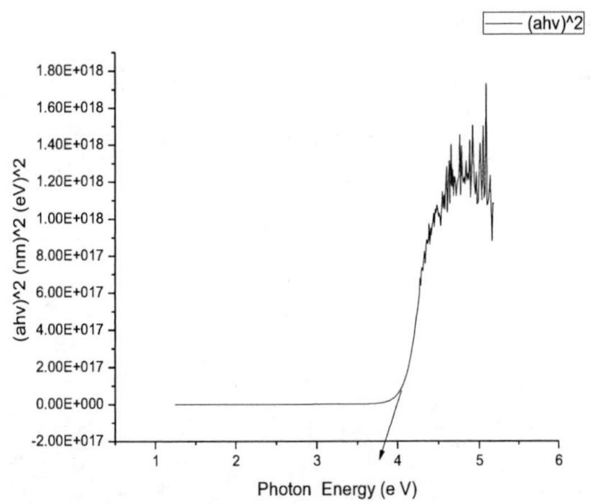

Figure 5. TaucPlot of $(\alpha h\nu)^2$ versus $(h\nu)$ energy band gap

3.2 FTIR analysis of ZnONPs

Figure 6 shows the IR spectrum of the sample of ZnONPs with absorption number occuring at 3448.04, 2524.85, 1437.15, 880.59, 729.06 and 433.44 cm^{-1}. The synthesized ZnONPs shows Zn-O, Zn-OH, and variety of single bonds at low wavenumbers as shown in the figure, while O=O, C-H occurs at 1437.15 cm^{-1} in the fourth region. The first region peaks correspond to absorption caused by C-H, O-H and N-H single bonds, and the results are in consistence with similar results reported elsewhere [10-12].

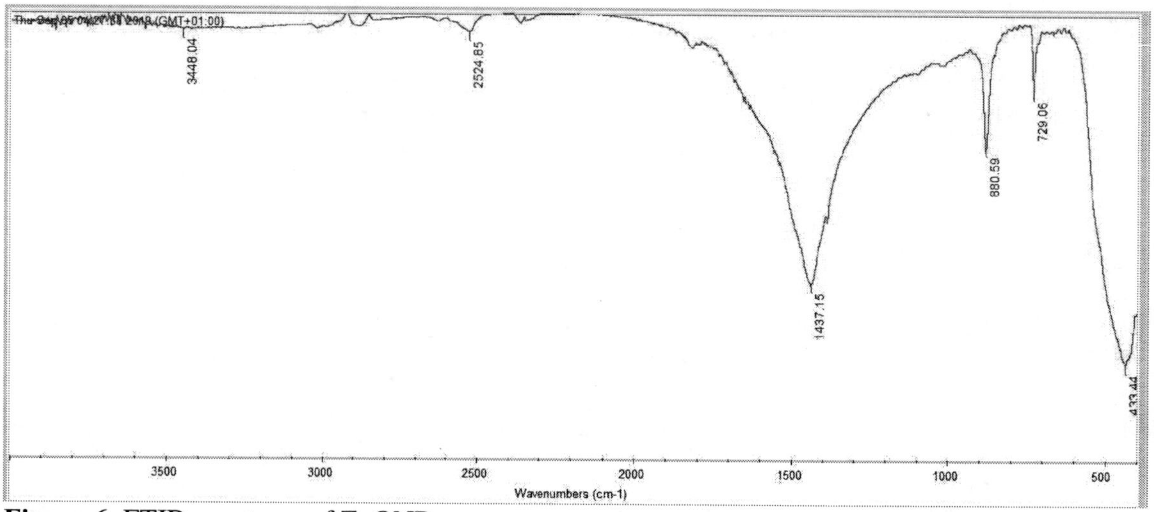

Figure 6. FTIR spectrum of ZnONPs

3.3 Morphological analysis of ZnONPs

Figures 7 and 8 show the surface morphology of synthesized ZnONPs using Scanning Electron Microscope (SEM). In Figure 7, the ZnONPs nanoparticles have large grain size which would be good for the absorption of photon energy.

IOP Conf. Series: Materials Science and Engineering **805** (2020) 012003 doi:10.1088/1757-899X/805/1/012003

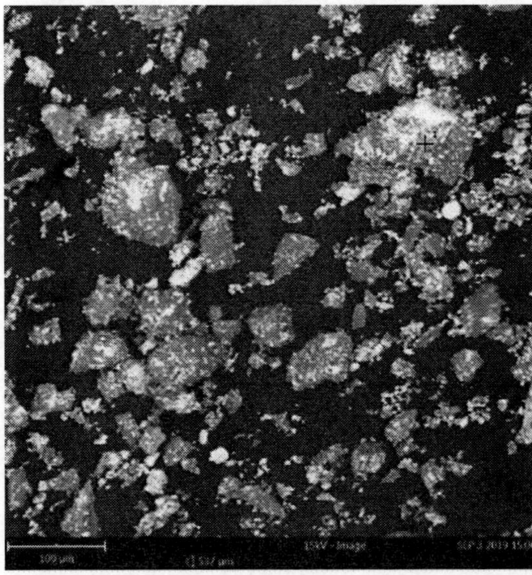

Table 1. The elemental composition of ZnONPs

Element Number	Element Symbol	Element Name	Atomic Conc.	Weight Conc.
6	C	Carbon	68.35	42.47
30	Zn	Zinc	7.92	26.79
20	Ca	Calcium	4.14	8.59
11	Na	Sodium	6.78	8.07
8	O	Oxygen	6.77	5.60
12	Mg	Magnesium	3.47	4.37
14	Si	Silicon	0.68	0.98
47	Ag	Silver	0.13	0.75
13	Al	Aluminum	0.41	0.57
17	Cl	Chlorine	0.29	0.53
7	N	Nitrogen	0.55	0.40
19	K	Potassium	0.17	0.35
15	P	Phosphorus	0.19	0.31
16	S	Sulfur	0.13	0.22

Figure 7. Surface morphology of ZnONPs at 100 μm and elemental composition

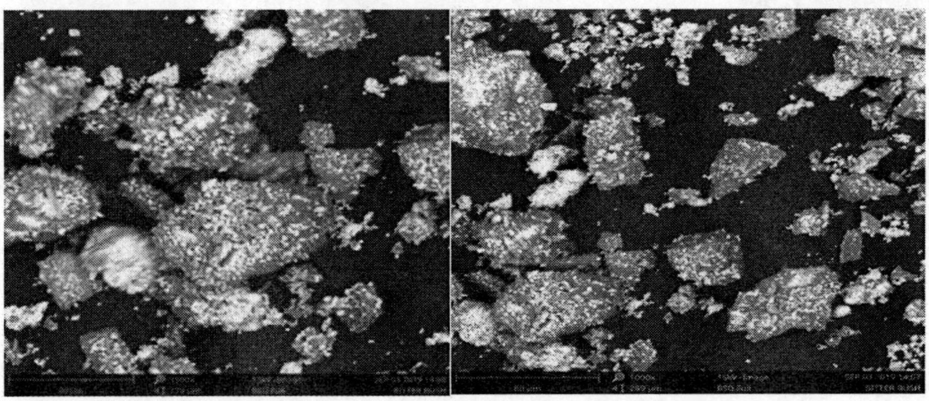

Figure 8. Surface morphology of ZnONPs at 50 μm **(b)** at 80 μm

3.4 Energy-dispersive X-ray spectroscopy

The EDX spectra revealed the elemental composition of the ZnONPs as shown in Figure 9. Atomic concentration and weight concentration of the elements as tabulated in Table 1. The EDX result shows the presence of some impurities. Carbon has a high concentration, Zn is 26.79% and O is 5.60%. These values might be attributed to high composite ratio of Siam solution to Zinc acetate dehydrate solution or it might be due to high baking temperature of ZnONPs at 400 °C for 30 min to evaporate the solvent. Further studies will be carried out using

IOP Conf. Series: Materials Science and Engineering **805** (2020) 012003 doi:10.1088/1757-899X/805/1/012003

higher Zinc acetate dehydrate solution-composite ratio and low temperature will be employed to evaporate the solvent.

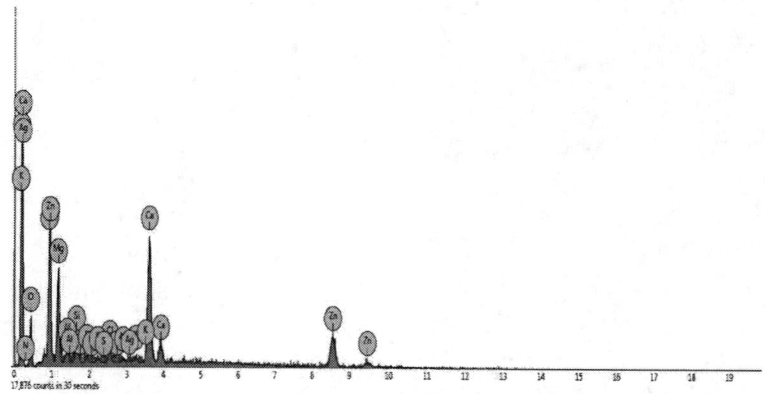

Figure 9. Energy-dispersive X-ray signals of ZnONPs showing elemental composition

3.5 Electrical properties analysis

Figure 10 shows the I-V characteristics from where the Fill-factor (FF) and efficiency (η) were computed. The thin film device Fill Factor (FF) is 0.6623, while ZnONPs thin film device efficiency under illumination is 2.01%.

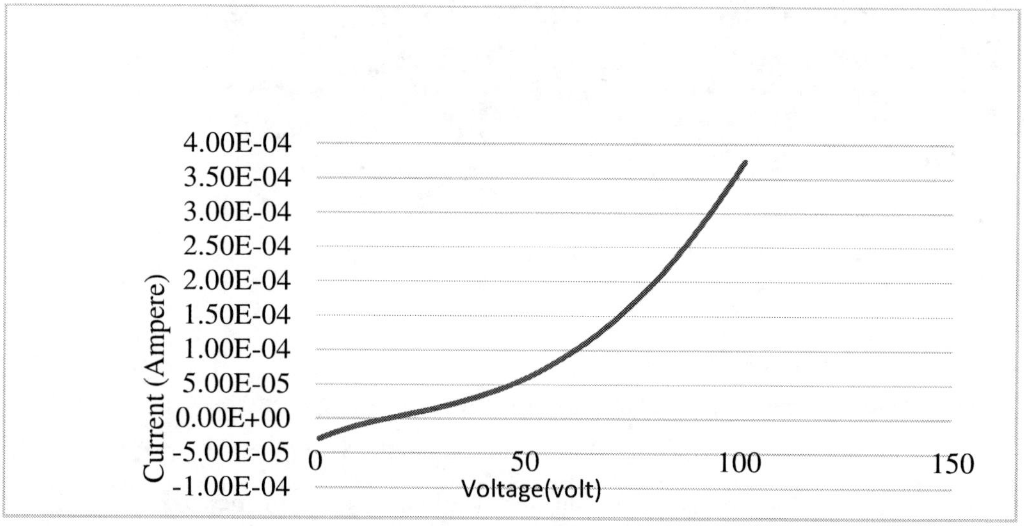

Figure 10. Current- Voltage characteristics for ZnONPs thin film device

Conclusion

Synthesis of ZnONPs using plant extract eliminated the need of toxic chemicals method of synthesis of nanoparticles. ZnONPs was deposited by spin- coating method and characterized by UV-vis spectroscopy, Fourier transform infra-red spectroscopy, Scanning Electron microscopy, and Energy Dispersive X-ray. Energy band gap of ZnONPs was 3.7 eV. FTIR result revealed the functional groups and absorption number of ZnONPs which shows that single bonds were found at low wavenumber. The surface morphology shows large grain size which means more adsorption of photon energy by ZnONPs. Efficiency of ZnONPs thin film device fabricated under illumination is $2.01^{0}/_{0.}$ Results obtained can lead to performance enhancement when used as photoanode of crystalline solar cells.

Conflict of Interest

The authors declare that there is no conflict of interest regarding the publication of this article.

Acknowledgment

The authors' appreciation is extended to Materials Science and Engineering Laboratory, Kwara State University, Malete-Ilorin, Nigeria for assistance on the analyses.

References

[1] Coleman VA and Jagadish C 2006 Basic properties and applications of ZnO. In Zinc Oxide Bulk, Thin films and Nanostructures (Eds.: C. Jagadish and S. Pearton). Elsevier

[2] Rouhi J, Mahmud S, Naderi N, Ooi CR and Mahmood MR 2013 Physical properties of fish gelatin-based nanocomposites films incorporated with ZnO nanorods. *Nanoscale Res. Lett.* **8**, 364.

[3] Balogun S W, Yekini K S and Adebayo O A 2017 Impact of post-deposition heat treatment on the morphology and optical properties of Zinc Oxide (ZnO) thin film prepared by spin-coating technique. *J. Photonic Mater. Tech.* **3** (3) pp 20-28.

[4] Balogun S W, Yekini K S and Adebayo O A 2018 Structural and optical properties of titanium dioxide thin film deposited by spin-coating technique. *Inter. J. Dev. Res* **08** (01) pp 18486-18490.

[5] Ramesh P, Rajendran A and Subramanian A 2014 Synthesis of zinc oxide nanoparticles from

fruit of *Citrus aurantifolia* by chemical and green method. *Asian J. Phytomed. Clin. Res.* **2**(4) pp 189-195.

[6] Geetha M S, Nagabhbana H and Shivananjaiah H N 2016 Green mediated synthesis and characterization of ZnO nanoparticles using *Euphorbia Jatropa* latex as reducing agent. *J. Sci.: Adv. Mater. Devices* **1** (3) pp 301-310.

[7] Suliman A E, Tang Y and Xu L 2007 Preparation of ZnO nanoparticles and nanosheets and their application to dye- sensitized solar cells. *Sol. Energ. Mater. Sol. Cells* **91** 1658-1662.

[8] Zhang Q, Chou T P, Russo B and Jenekhe S A, Cao G 2008 Polydisperse aggregates of ZnO nanocrystallites: A method for energy-conversion-efficiency enhancement in dye-sensitized solar cells. *Adv. Funct. Mater.* **18** (11) pp 1654-1660.

[9] Taufiq A, Ulya H N, Utomo J, Sunaryono N, Hidayat, Susanto H, Mufti N and Soontaranon M S 2018 Structural, optical, and antifungal characters of zinc oxide nanoparticles prepared by Sol-gel method. *J. Phys.: Conf. Ser.* **1093** (01) 012001. https://doi.org/10.1088/1742-6596/1093/1/012001.

[10] Dobrucka R and Dugaszewka J 2016 Biosynthesis and antibacterial activity of ZnO nanoparticles using *Trifolium pratense* flower extract. *Saudi J. Biol. Sci.* **23** pp 517-523.

[11] Becheri A, Dürr M, Nostro P L and Baglioni P 2008 Synthesis and characterization of zinc oxide nanoparticles: Application to textiles as UV-absorbers. *J. Nanopart. Res.* **10** (4) pp 679-689.

[12] Xiong G, Pal U, Serrano J G, Ucer K B and Williams R T 2006 Photoluminesence and FTIR study of ZnO nanoparticles: the impurity and defect perspective *Phys. Status Solidi C* **3** pp 3577-3581.

IOP Conf. Series: Materials Science and Engineering **805** (2020) 012004 doi:10.1088/1757-899X/805/1/012004

Antioxidant properties of silver nanoparticles biosynthesized from methanolic leaf extract of *Blighia sapida*

AO Akintola[1*], BD Kehinde[2], PB Ayoola[1], AG Adewoyin[1], OT Adedosu[2], JF Ajayi[3], and SB Ogunsona[1]

[1]Department of Science Laboratory Technology, Ladoke Akintola University of Technology, Faculty of Pure and Applied Sciences, Ogbomoso, Nigeria.

[2]Department of Biochemistry, Faculty of Basic Medical Sciences, Ladoke Akintola University of Technology, Ogbomoso, Nigeria.

[3]Department of Physiology, Faculty of Basic Medical Sciences, Ladoke Akintola University of Technology, Ogbomoso, Nigeria.

*Corresponding Author: Email: aoakintola@lautech,edu.ng; Telephone: +2348023807230

ABSTRACT

Synthesis of nanoparticles by biological methods using microorganisms, enzymes or plant extracts has been suggested as possible ecofriendly alternative to chemical and physical methods which involve the use of harmful reducing agents. In the current study, silver nanoparticles (AgNPs) were synthesized by green approach from methanolic leaf extract of *Blighia sapida*. The synthesized AgNPs were characterized by UV-visible (UV-vis) spectroscopy, Fourier Transform Infrared (FTIR) spectroscopy and Scanning Electron Microscopy (SEM). The antioxidant activity was evaluated using DPPH radical scavenging assay, determination of total reductive potential, total phenolics content (TPC) and total flavonoids content (TFC) of the synthesized AgNPs. SEM analysis revealed that the size of the synthesized silver nanoparticles ranged from 50-70 nm with maximum UV-vis absorbance at 413 nm. DPPH radical scavenging activity, reducing power, total phenolic and total flavonoid contents of the synthesized AgNPs increased in a dose dependent manner as compared to ascorbic acid the standard reference used. This result confirmed that *Blighia sapida* is a potential biomaterial for synthesizing AgNPs which can be exploited for its antioxidant activity.

Keywords: *Blighia sapida*, antioxidant activity, silver nanoparticles, DPPH radical scavenging activity, free radicals

1. Introduction

Free radicals play a key role in oxygen dependent (aerobic) living systems. They are part of cell respiration and other important cellular processes but are also involved in aging and disease development [1, 2]. Free radicals are unstable molecules with free unpaired electrons.

IOP Conf. Series: Materials Science and Engineering **805** (2020) 012004 doi:10.1088/1757-899X/805/1/012004

They are highly reactive because the free electrons always attempt to bond with other electrons and form covalent pairs. In this process, the free radicals remove the electrons from other molecules. Beyond affecting regulation of cells, it can also damage molecules such as carbohydrates, fats, proteins and nucleic acids [3, 4]. There are different sources of free radicals within the cells and its environment. In aerobic organisms, free radicals are produced during normal metabolic processes [5]. The major sources include electron transfer in the plasma membrane and cell respiration in the mitochondria membrane. Mitochondria are the main source of the oxidative damage because free radicals such as superoxide can escape from the electron transport chain [6, 7]. In order to prevent intracellular damage by free radicals, cells have built up intracellular antioxidant system. This process converts free electrons into a non-reactive form by proteins.

Antioxidants control oxidative reactions by inhibiting, delaying or hindering the oxidation of the biomolecules [8, 9]. The key antioxidant enzymes possess certain elements that shield and protect proteins [10, 11]. Non enzymatic antioxidants can also neutralize radicals for example water soluble substances such as Vitamin C, glutathione or fat-soluble substances such as Vitamin E, β-carotene [8, 12, 13]. Synthetic antioxidants such as butylated hydroxyl toluene (BHT) and butylated hydroxyl anisole (BHA) have newly been reported to be harmful for human health [14, 15]. Thus, the search for effective, non-toxic, natural compounds with antioxidative activity has been increased in recent years. Recently nanomaterials have started playing a fundamental role in human life and health owing to their substantial benefits of biomedical applications [16] like medical imaging, drug delivery, diseases diagnosis [17], cancer treatment [18, 19], treatment of infectious diseases [20, 21], treatment of neurodegenerative disorder including Parkinson disease [22] and so on.

Moreover, the strong antioxidant property exhibited by some nanomaterials is opening exciting potential to develop new regimens with enhanced and targeted actions. For example, gold, silver and selenium nanoparticles have been shown to possess ability to reduce oxidative stress due to their efficient redox-active radical-scavenging properties [23, 24, 25, 26]. *Blighia sapida* is a soap berry perennial herbaceous plant that is prominently found in Western Tropical Africa (Figure 1). It belongs to the family Sapindaceae. In Nigeria, it is commonly known as Ackee (English), Gwanja Kusa (Hausa), Isin (Yoruba) and Okpu (Igbo) [27]. It is an evergreen tree which grows to a height of between 7 and 25m. It occurs naturally from Senegal, Cameroon and Equatorial Guinea. The tree is often cultivated for its edible fruit in many areas of tropics and subtropics. When ripe, Ackee fruit splits to expose a fleshy cream colour pulp aril attached to a shiny black oblong seed [28]. The fleshy fruit can also be

IOP Conf. Series: Materials Science and Engineering **805** (2020) 012004 doi:10.1088/1757-899X/805/1/012004

curried, used in soups and stews as a meat substitute [29]. The fruit must be allowed to open fully before it is detached from the tree because unripe or overripe arils as well as the seed contain hypoglycin and its derivatives which are strongly toxic [30]. *Bligha sapida* is useful in Africa traditional medicine. In Cote d' Ivore, it is commonly used for the treatment of yellow fever, epilepsy, oedema and as a laxative and diuretic [31]. Sap from the terminal bud is instilled in the eyes to treat opthalmia and conjunctivitis [32]. The pulp of grand up leafy twig is rubbed on the forehead to treat migraine/headache [31, 33]. The root, bark, leaves, capsules and seeds were identified for the cure of 22 diseases in Benin [34]. The plant was reported to have antioxidant activity [35, 36, 37]. Therefore the aim of this study was to synthesize silver nanoparticles from the leaves of *Blighia sapida* and to investigate its possible antioxidant potentials.

a b c

Figure 1. *Blighia sapida* plant (a), *Blighia sapida* fruit (b), *Blighia sapida* ripe fruit showing a fleshy cream colour pulp aril and shiny black oblong seed (c)

2 Materials and Methods

2.1 Reagents

All reagents were of analytical grade and obtained from Sigma Aldrich Chemical, Germany.

2.2 Plant materials

Fresh leaves of *Blighia sapida* were collected from the Teaching and Research Farm of Ladoke Akintola University of Technology, Ogbomoso. The plant was authenticated at the herbarium unit of the Department of Pure and Applied Biology, Ladoke Akintola University of Technology Ogbomoso (Voucher number : LHO - 538). Healthy leaves with no sign of damage were air dried under shade at room temperature for four weeks and pulverized into powdered form.

2.3 Preparation of plant extract

Methanolic leaf extract of *Blighia sapida* was prepared by soaking 2 kg of the pulverized plant material in 6 l of methanol for 4 days. The liquid portion was then separated from the leaf shaft with the aid of Whatman (No. 1) filter paper and the filtrate was concentrated using a rotary evaporator and dried on a water bath set at 40 °C to obtain a dry residue which is the crude methanolic extract (yield weight 150 g) which was stored in a desiccator until further use.

2.4 Phytochemical screening

The methanolic extract was screened for the presence or absence of alkaloids, flavonoids tannins, phenols and steroids using standard methods [38, 39].

2.5 Synthesis of silver nanoparticles

The extract was centrifuged at 4,000 rpm for 20 min. The supernatant obtained was used to synthesize silver nanoparticles as described by Lateef *et al.* [40]. Exactly 1 ml of the supernatant was added to the reaction vessel containing 40 ml of 1 mM silver nitrate ($AgNO_3$) solution for the reduction of silver ion. The reaction was carried out at room temperature (30±2 °C) for 2 h. The formation of silver nanoparticles was monitored through visual observation of the change in colour.

2.6 Characterization of silver nanoparticles

The formation of silver nanoparticles from *Blighia sapida* methanolic leaf extract was also determined by measurement of the absorbance spectrum of the reaction mixture using UV-visible spectrophotometer (UV-2450 Shimadzu Model, USA). The spectrum data recorded was then plotted.

The silver nanoparticles solution was centrifuged at 10,000 rpm for 20 min. The solid residue was then dried at room temperature and the powder obtained was used for FTIR measurement using KBr pellet with an LF-45 fluorescence spectrophotometer FTIR instrument (Shimadzu Model IR, USA) in the wavelength 4000-500 cm^{-1} [40, 41].

Scanning electron microscopy was done using a Hitachi scanning electron microscope (Model, S-2600N-Tokyo Japan) operating in the high vacuum anode with an acceleration voltage of 20 KV. Thin films of the sample were prepared on a carbon coated copper grid by just dropping a very small amount of the sample on the grid, extra solution was removed

IOP Conf. Series: Materials Science and Engineering **805** (2020) 012004 doi:10.1088/1757-899X/805/1/012004

using a blotting paper and the film on the SEM grid were allowed to dry by putting it under a mercury lamp for 5 min.

2.7 Antioxidant assay

2.7.1 DPPH radical scavenging assay

The 1,1, dipheny 1-2 picrylhydrazyl (DPPH) scavenging potential of the silver nanoparticles from *Blighia sapida* methanolic leaf extract was determined using the method of Mensor *et al.* [42]. Different concentrations (50, 75, 100, 125 and 150 µg/mL) of the silver nanoparticles and the standard (ascorbic acid) were taken in different test tubes. Thereafter, 1 ml of freshly prepared DPPH (0.3 mM) was added to each test tube and vortexed thoroughly. Finally, the solution was incubated in dark place for 30 min. For control, 2 ml of methanol was added instead of silver nanoparticles and run simultaneously with the test. The percentage DPPH radical scavenging activity of the silver nanoparticles was calculated using the formula.

$$\% \text{ DPPH Radical Scavenging Activity} = \frac{Ac - As}{Ac} \times 100$$

Ac is the control absorbance of DPPH radical + methanol; **As** is the sample absorbance of DPPH radical + sample AgNPs / Ascorbic acid.

2.7.2 Reducing power assay

The reducing power of the silver nanoparticles was determined by the method of Linn *et al.* [43] with slight modification. Different concentrations of AgNPs (50, 75, 100, 125 and 150 µg/ml) were prepared and separately mixed with 2.5 ml phosphate buffer (0.2 M, pH 6.6) and 2.5 ml of potassium ferricyanide (1%). The mixture was incubated at 50 $^{\circ}$C for 20 min. Further, 2.5 ml of tricholoroacetic acid (10%) was added to the mixture, which was then centrifuged at 3,000 rpm for 10 min. Finally, the supernatant solution (2.5 ml) was mixed with 2.5 ml of distilled water and 0.5 ml $FeCl_3$ (0.01%) and absorbance were measured at 700 nm in UV-visible spectrophotometer.

2.7.3 Determination of total phenolics content

The total phenolics content (TPC) of the silver nanoparticles synthesized from methanolic leaf extract of *Blighia sapida* was determined using the Folin Ciolcateau assay [44]. About 1 ml of different concentrations (50, 75, 100, 125 and 150 µg/mL) of the AgNPs was taken into a 25 ml volumetric flask and 10 ml of distilled water and 1.5 ml of Folin Ciolcateau reagent were added to it. The mixture was kept for 5 min and then 4 ml of 20% sodium carbonate was added and made up to 25 ml with distilled water and allowed to incubate for 30 min at room

IOP Conf. Series: Materials Science and Engineering **805** (2020) 012004 doi:10.1088/1757-899X/805/1/012004

temperature. The total phenolic content was determined using spectrophotometry at 765 nm and was expressed as gallic acid (GA) equivalent (µg/g).

2.7.4 Determination of total flavonoids content

Total flavonoids content (TFC) of the silver nanoparticles from methanolic leaf extract of *Blighia sapida* was estimated using the aluminum chloride colorimetric method of Chang *et al.* [45]. About 0.1 ml of different concentrations of the silver nanoparticles (50, 75, 100, 125 and 150 µg/ml) were separately mixed with 0.1 ml of 10% $AlCl_3.6H_2O$, 0.1 ml of 1M sodium acetate and 2.8 ml of distilled water and incubated at room temperature for 30 min. Total flavonoid content was determined using spectrophotometry at 415 nm and was expressed as µg/g quercetin equivalent.

3. Results and Discussion

Qualitative phytochemical screening revealed the presence of phenolics, alkaloids, flavonoids, tannins, saponins and steroids. Preliminary identification of nanoparticles formation was determined by observing the colour change. The AgNPs were characterized with colour formation produced from the reduction of silver ion by the plant extract. The intensity of the colour increases as the bio-reduction of silver ion progresses and become stable when the reduction is completed. The colour of the solution changed from light yellow to dark brown (Figure 2). Varying colours including yellow, brown and dark brown has been reported by several authors [24, 40, 46].

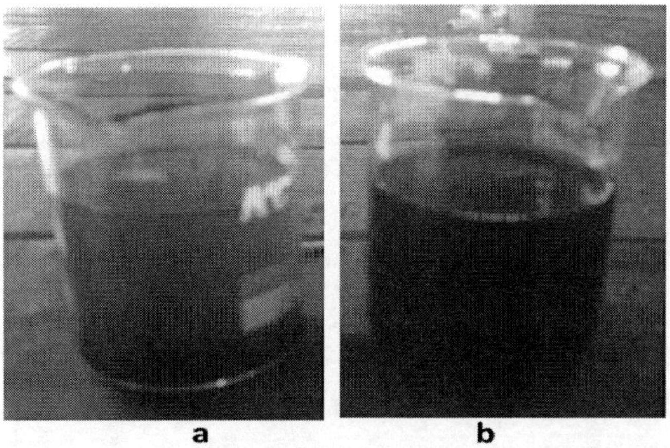

Figure 2. Colour change in the reaction mixture (silver nitrate + *Blighia sapida* extract) (a) Colour of the mixture after mixing the extract with silver nitrate solution (b) Colour change of the mixture after 2 h

The reduction of Ag to AgNPs was further confirmed using UV-Vis spectroscopy. UV-vis spectroscopy is a valuable technique used to detect the characteristics surface plasmon resonance (SPR) pattern of the metal nanoparticles. Metal nanoparticles exhibited SPR phenomena when metal electrons in the conduction band collectively oscillate in resonance with certain wavelength of incident light. Figure 3 depicts the UV-visible spectrum of the biosynthesized AgNPs from methanolic leaf extract of *Blighia sapida* with absorption spectrum at 413 nm which is a characteristic band for silver. No other peak was observed in the spectrum indicating the formation of AgNPs. These values are within the range reported for AgNPs [40, 47].

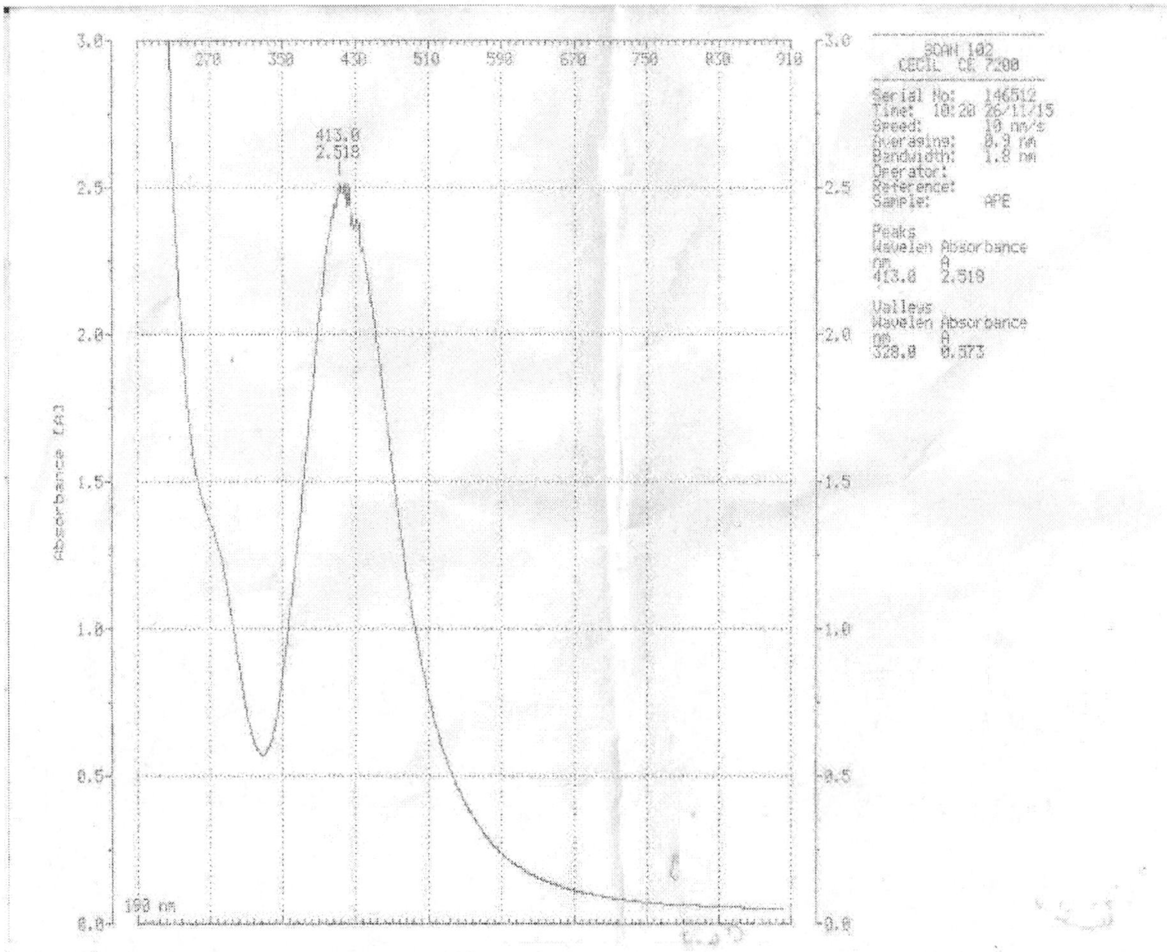

Figure 3. UV-vis spectrum of silver nanoparticles synthesized from methanolic leaf extract of *Blighia sapida*

FTIR spectroscopy was used to identify the possible biomolecules present in *Blighia sapida* leaf extract which are responsible for reducing the Ag^+ to Ag^0, capping and stabilizing the

IOP Conf. Series: Materials Science and Engineering **805** (2020) 012004 doi:10.1088/1757-899X/805/1/012004

silver nanoparticles. FTIR spectra recorded for the biosynthesized AgNPs is presented in Figure 4. The FTIR spectrum of the nanoparticles produced showed peaks at 3325, 2902, 2382, 2117, 1635, 1288, 1145, 505, 457 and 422 cm^{-1}. The vibrational peak around 3325 cm^{-1} is assigned to O-H that could possibly emanate from carbohydrate or phenolics. The peak at 2902 cm^{-1} is ascribed to C-H of the alkyl group, the peak around 2117 cm^{-1} is due to C=C from alkene or aromatic amine, 2382 cm^{-1} corresponds with C=C from alkyne, the peak at 1635 cm^{-1} suggest C=O from carboxyl group or esters and N-H binding vibration of proteins, the peak at 1145 cm^{-1} is likely due to C-O from phenols while the corresponding peaks at 505, 457 and 422 cm^{-1} can be ascribed to O-H from phenols or conjugated chromophore. The appearance of these peaks suggested the presence of protein, carbohydrates, hydrocarbons and phytochemicals such as phenolics in the AgNPs which may be responsible for stabilizing and capping effects on the nanoparticles. The biomolecules may have interacted with the silver ions through their oxygen donor atoms and are adsorbed on the surface of the metal ions. The tentative assignment of the peak is supported by other literature reports [48].

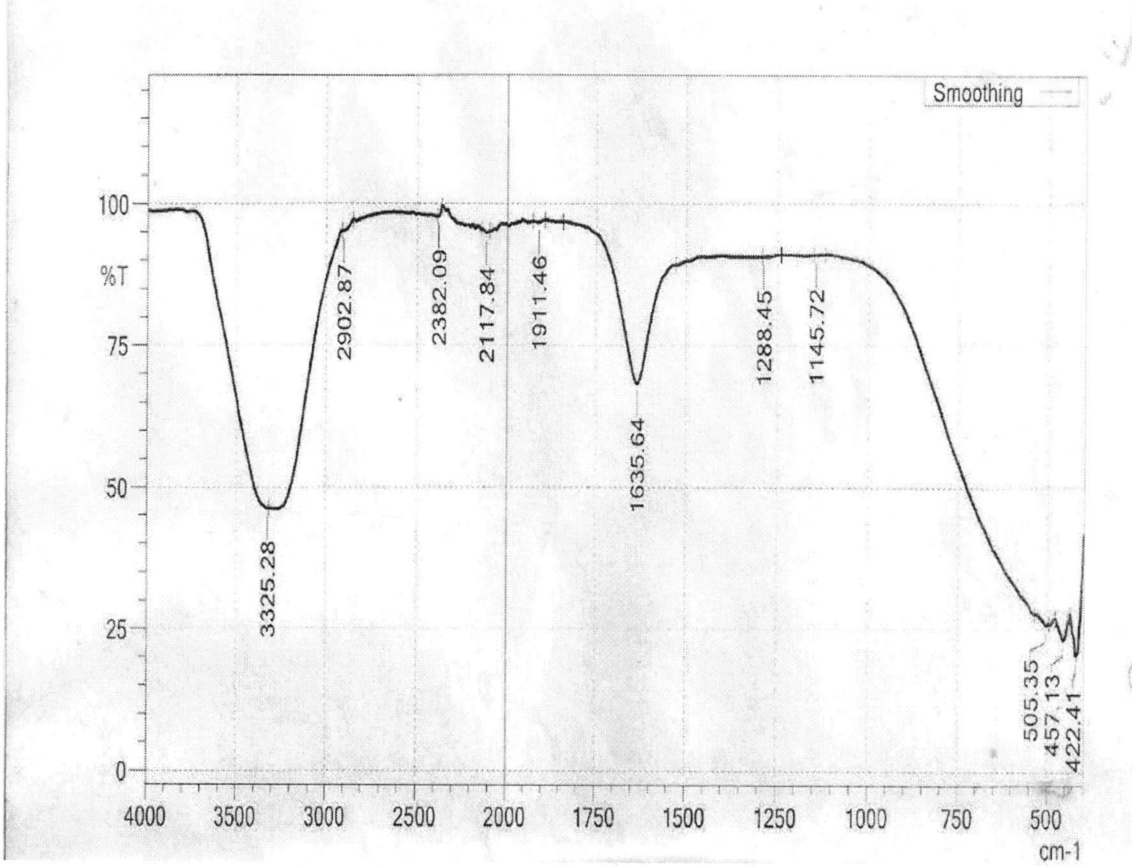

Figure 4. FTIR spectrum of silver nanoparticles synthesized from methanolic leaf extract of *Blighia sapida*

The size, shape and morphology of the synthesized AgNPs were determined by scanning electron microscopy. The synthesized AgNPs were spherical in shape with diameter range of 50-70 nm (Figure 5).

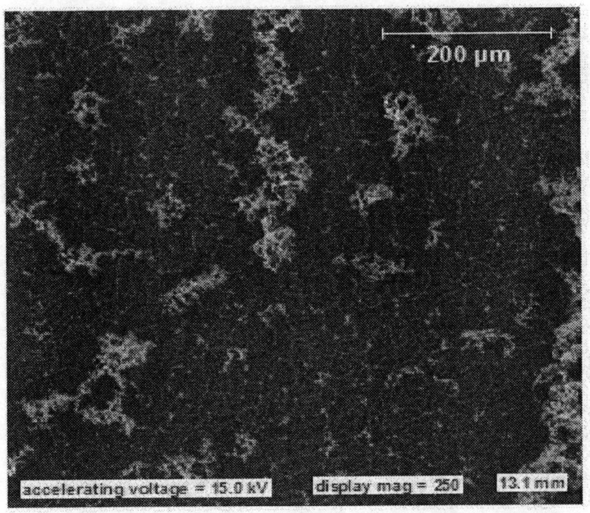

Figure 5. SEM image of silver nanoparticles synthesized from methanolic leaf extract of *Blighia sapida*

In vitro antioxidant activity of the synthesized AgNPs was studied by analyzing four antioxidant capacities which are indicative of the antioxidant potential of the synthesized AgNPs. The studied parameters were DPPH radial scavenging activity, reductive potential, total phenolics and total flavonoids contents of the synthesized AgNPs at different concentrations from 50-150 µg/ml. DPPH is a stable organic free radical that has been used for investigating the free radical activities and thus antioxidant activity of various natural products [49, 50, 51]. Figure 6 shows the dose response for the DPPH scavenging activity of the synthesized AgNPs. The AgNPs synthesized from the methanolic extract of *Blighia sapida* are potential free radicals scavengers with effective inhibition activity in a dose dependent manner. The varying concentration of the AgNPs (50, 75, 100, 125 and 150 µg/ml) significantly scavenged DPPH by 58.10, 59.26, 62.33, 71.24 and 75.42% respectively. However, these activities are less than that of ascorbic acid, the standard reference used. The reducing power of a compound is related to its electron transfer ability and therefore may serve as a significant indicator of its potential antioxidant activity [52]. The reductive capabilities of the biosynthesized AgNPs is shown in Figure 7. The reducing power of the AgNPs increased with increasing amount of sample. The biosynthesized AgNPs exhibited a maximum reducing capability of 53.52% at 150 µg/ml which is lowered than that of ascorbic acid (70.19%). Figures 8 and 9 depict the total phenolics and total flavonoids contents of the biosynthesized AgNPs. The AgNPs contained phenols and flavonoids in a concentration dependent manner. At a concentration of 150 µg/ml, the AgNPs has total phenolic content of 36.52 µg/g and total flavonoid content of 10.14 µg/g respectively. However, these values are less than that of phenolic and flavonoid contents of ascorbic acid, the standard antioxidant.

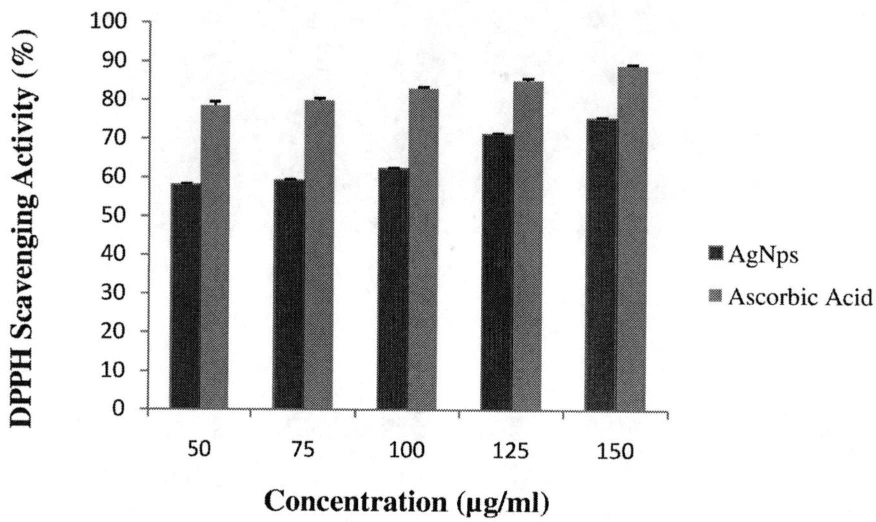

Figure 6. DPPH radical scavenging activity of AgNPs and ascorbic acid

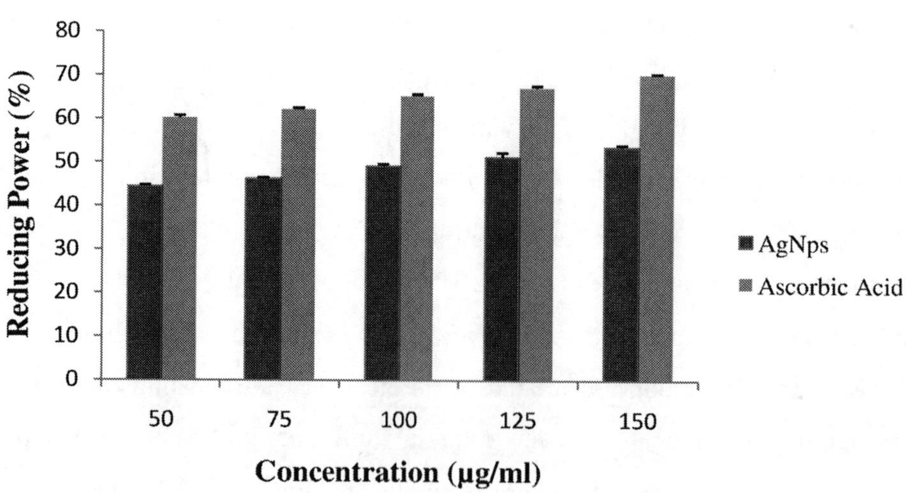

Figure 7. The reducing power assay of the AgNPs and ascorbic acid

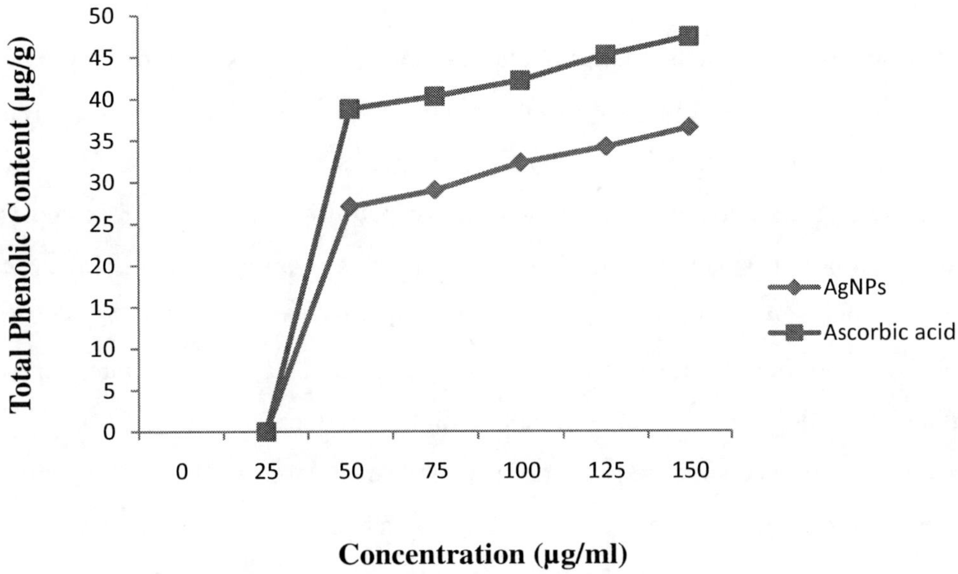

Figure 8. Total phenolics content of the AgNPs and ascorbic acid

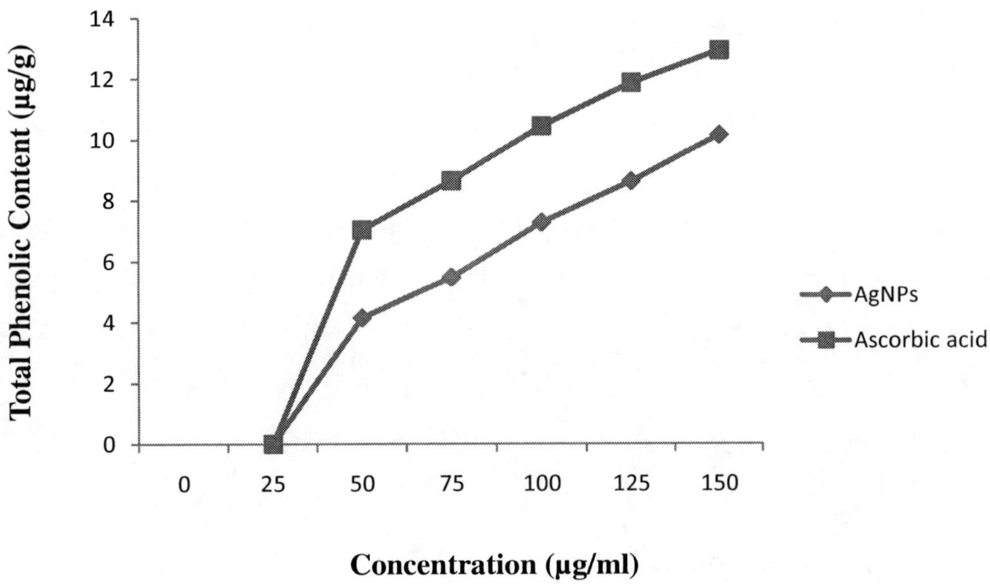

Figure 9. Total flavonoids content of the AgNPs and ascorbic acid

Phytochemicals like flavonoids, phenolics, anthocyanins have been of great interest as the sources of natural antioxidants. Plant materials such as leafy vegetables, fruits, seeds, cereals

IOP Conf. Series: Materials Science and Engineering **805** (2020) 012004 doi:10.1088/1757-899X/805/1/012004

have been reported to contain natural antioxidants [53, 54]. Flavonoids are a class of secondary plant phenols with powerful antioxidant properties [55]. Phenols are regarded as the most important oxidative components of plants hence, correlation between the concentration of total phenolics and antioxidant capabilities have been reported [56]. The antioxidant activity of phenol is mainly due to their redox properties which play an important role in scavenging and neutralizing free radicals. In the current study, the antioxidant activity of the AgNPs may be attributed to the high phenolics and flavonoids content in the plant. These plant phenolics are strong antioxidants with high reducing capacity [57]. The ability of *Blighia sapida* to reduce silver ions or to form silver nanoparticles by reducing silver ions is likely due to the presence of phenolic compounds which are electron donors.

4 Conclusion

This study reported the green synthesis of silver nanoparticles by simple and ecofriendly manner using the methanolic leaf extract of *Blighia sapida*. The leaf extract contains protein, carbohydrates and phenolic compounds which might be responsible for the bio-reduction of silver ion and capping/stabilization of the silver nanoparticles as was confirmed by FTIR studies. The biosynthesized silver nanoparticles from leaf extract of *Blighia sapida* showed antioxidant activity through its DPPH radical scavenging activity, its reductive potential, high phenolics and flavonoids contents. This study provides an insight into the usage of *Blighia sapida* leaf as a good source of naturally occurring antioxidant and could have great importance as therapeutic agent in preventing or ameliorating oxidative stress related diseases.

References

1. Harrman D 1992 Role of free radicals in aging and diseases. *Ann. N. Y. Acad. Sci.* **673** (1) pp 126-141.
2. Bagchi K and Puris S (1994) Free radical and antioxidants in health and disease. *East. Mediterr. Health J.* **4** pp 350-360.
3. Young I S and Woodside J V 2001 Antioxidants in health and disease. *J. Clin. Pathol.* **54** (3) pp 176-186.

4. Valko M, Lebfritz D, Monal J, Cronin MTD, Mazur M and Tesler J 2007 Free radicals and antioxidants in normal physiological functions and human disease. *Int. J. Biochem. Cell Biol.* **39** (1) pp 44-84.
5. Mohammed AA and Ibrahim AA 2004 Pathological roles of reactive oxygen species and their deference mechanism. *Saudi Pharmaceut. J.* **12** (6) pp 1-18.
6. Shigenaga M K, Hager T M and Ames B N 1994 Oxidative damage and mithochondria decay in aging. *Proc. Natl. Acad. Sci. USA* **91** (23) pp 10771-10778.
7. Liu T, Stern A and Roberts L J 1999 The isoprostanes: Novel prostanglandin-like products of the free radical catalyzed peroxidation of arachidonic acid. *J. Biomed. Sci.* **6** (4) pp 226-235.
8. Sies H 1997 Oxidative stress oxidants and antioxidants. *Exp. Physiol.* **82** (2): 291-295.
9. Kumar S 2011 Free radicals and antioxidants: Human and food system. Adv. Appl. Sci. Res. **2** (1) pp 129-135.
10. Harris ED 1992 Regulation of antioxidant enzymes: *FASEB J.* **6** (9) pp 2675-2683.
11. Harris ED 1992 Copper as a cofactor and regulation of copper, zinc superoxide dismutase. *J. Nutr.* **122** (suppl_3) pp 636-640.
12. Trombino S, Serini S, Di-Nicuolo F, Celleno L, Ando S, Picci N, Calviello G and Palloza P 2004 Antioxidant effect of ferulic acid in isolated membranes and intact cells: synergistic interactions with alpha-tocopherol, beta-carotene and ascorbic acid. *J. Agric. Food Chem.***52** (8) pp 2411-2420.
13. Hudson BJF and Mahgoub SEO 2006 Naturally occurring antioxidants in leaf lipids. *J. Sci. Food Agric.* **31** (7) pp 646-650.
14. Abramovič H. and Abraham N 2006 Effect of added rosemary extract on oxidative stability of *Camelina sativa* oil. *Acta Agric. Slov.* **87** (2) pp 255-261.
15. Kowalski R 2007 GC analysis of changes in the fatty acid composition of sunflower and olive oils heated with quercetin, caffeic acid, protocatechuic acid and butylated hydroxyaniosle. *Acta Chromatogr.* **18** pp 15-23.
16. Zolnik BS and Sadrich N 2009 Regulatory perspective on the importance of ADME assessment of nanoscale material containing drugs. *Adv. Drug Deliv. Rev.* **61** (6) pp 422-427.
17. Miyazaki K and Islam N 2007 Nanotechnology systems of innovation. An analysis of industry and academia research activities. *Technovation* **27** (11) pp 661-675.

18. Garde D 2012 Chemo bomb nanotechnology effective in halting tumors. http://www.fiercedrugdelivery.com/story/chemo-bomb-nanotechnology-effective-halting-tumors/. Accessed 17 January 2020.
19. Peiris PM, Bauer L, Toy R, Tran E, Pansky J, Doolittle E, Schmidt E, Hayden E, Mayer A, Keri RA and Griswold MA 2012 Enhanced delivery of chemotherapy to tumor using a multicomponent nanochain with radiofrequency tunable drug release. *ACS Nano* **6** (5) pp4157-4168.
20. Banoee M, Seif S, Nazari ZE, Jafari-Fesharaki P, Shahverdi HR, Moballegh A, Moghaddam KM, Shahverdi AR 2010 ZnO nanoparticles enhanced antibacterial activity of ciprofloxacin against *Staphylococcus aureus* and *Escherichia coli*. *J. Biomed. Mater. Res B: Appl. Biomater.* **92** (2) pp 557-561.
21. Radovic-Moreno AF, Lu TK, Puscasu VA, Yoon CJ, Langer R, Farokhzad OC 2012 Surface charge-switching polymeric nanoparticles for bacterial cell wall-targeted delivery of antibiotics. *ACS Nano* **6** (5) pp 4279-4287.
22. Wong HL, Wu XY and Bendayan R 2012 Nanotechnology advances for the delivery of central nervous system therapeutics. *Adv. Drug Deliv. Rev.* **64** (7) pp 686-700.
23. Saad AM, Abdel-Aleem AA, Ghareeb MA, Hamed MM, Abdel-Aziz MS and Hadad AH 2017 *In vitro* antioxidant, antimicrobial and cytotoxic activities and green biosynthesis of silver and gold nanoparticles using *Callistemon citrus* leaf extract. *J. Appl. Pharmaceut. Sci.* **7** (6) pp 141-149.
24. Sood R and Chopra DS 2017 Improved yield of green synthesized crystalline silver nanoparticles with potential antioxidant activity. *Int. Res. J. Pharm.* **8** (4) pp 4-17.
25. Bhakya S, Muthukrishnan S, Sukumaran M and Muthukumar M 2016 Biogenic synthesis of silver nanoparticles and their antioxidant and antibacterial activity. *Appl. Nanosci.* **6** (5) pp 755-766.
26. Thilagavathi T, Kathirav G and Srinivasan K 2016 Antioxidant activity and synthesis of silver nanoparticles using the leaf extract of *Limonia acidissima*. *Int. J. Pharm. Biosci.* **7** (4): 201-205.
27. Aderinola OA, Farinu GO, Akinlade JA, Olayeni TB, Ojebiyi OO and Ogunniyi PO 2007 Nutritional potential of *Blighia sapida* K Konig (Ackee ackee) leaves as a dry season feed resource for West African dwarf goats in the derived savanna zone of Nigeria. *Livestock Res. Rural Devel.* **19** (6) paper 78.

28. Blake OA, Jackson JC, Jackson MA and Gordon CA 2002 Assessment of dietary exposure to the natural toxin hypoglycin in ackee (*Blighia sapida*) by Jamaican consumers. *Food Res. Int.* **37** (8) pp 833-838.
29. Arbonnier M 2002 Arbes, arbutus et liaries des zones seches d' Afrique de l'Quest. Edition 2, CIRAD-MNH pp 173-179.
30. Joskow R, Belson M, Vesper H, Backer L and Rubin C 2006 Ackee fruit poisoning: an outbreak investigation in Haiti 2000-2001 and review of the literature. *Clin. Toxicol.* **44** (3) pp 267-273.
31. Abbiw DK 1990 The useful plants of Ghana. Intermediate Technology Publications and the Royal Gardens, Kew, UK.
32. Irvin FR 1965 Botany and Medicine West African. University Press Ibadan, Nigeria. pp 102-130.
33. Natalini B, Capodiferro V, De Luca C, Espinal R. Isolation of pure (2S, 1′ S, 2′ S)-2-(2′-carboxycyclopropyl) glycine from *Blighia sapida* (Akee) 2000 *J. Chromatogr.* **873** (2) pp 283-286.
34. Ekué MR, Sinsin B, Eyog-Matig O and Finkeldey R 2010 Uses, traditional management, perception of variation and preference in Ackee (*Blighia sapida* K. Koenig) fruit traits in Benin. Implications for domestication and conservation. *J. Ethnobiol. Ethnomed.* **6** Article Number 12. https://doi.org/10.1186/1746-4269-6-12.
35. Olawale AE, Adeduntan SA, Oyerinde OV and Adeniyi MM 2016 Antioxidant properties and proximate composition of *Blighia sapida* D. Koenig (Aril) in rainforest and savannah zone of South West Nigeria. *Int. J. Eng. Sci.* **5** (3) pp 5-15.
36. Amirah PO and Oloyede HO 2017 Antioxidant properties and proximate composition of *Blighia sapida* stem bark in alloxan induced diabetic rats. *Global J. Med. Res. B: Pharm. Drug Discov. Toxicol. Med.* **17** (2) pp 1-8.
37. Ojo OA, Ajiboye BO, Imiere OD, Adeyonu O, Olayide I and Fadaka A 2018 Antioxidative properties of *Blighia sapida* K.D Koenig stem bark extract and inhibitory effects on carbohydrate hydrolyzing enzymes associated with non-insulin dependent diabetes. *Pharmacogn. J.* **10** (2) pp 376-383.
38. Banso A and Ngbede JE 2006 Phytochemical screening and *in vitro* antifungal properties of *Fagara zanthoxyloides. J. Food Agric. Environ.* **4** (3/4) pp 8-9.
39. Ngbede JE, Yakubu RA and Nyam DA 2008 Phytochemical screening for active compound in *Camarium schlveinfurthil* (Atile) leaves from Jos North, Plateau State, Nigeria. *Res. J. Biol. Sci.* **3** (9) pp 1076-1078.

40. Lateef A, Azeez MA, Asafa TB, Yekeen TA, Akinboro A, Oladipo IC, Ajetomobi FE, Gueguim-Kana EB and Beukes LS 2015 *Cola nitida*-mediated biogenic synthesis of silver nanoparticles using seed and seed shell extracts and evaluation of antibacterial activities. *BioNanosci.* 2015; 5: 196-205.
41. Bhat R, Deshpannde R, Ganachari SV, Huh DO and Venkataraman A 2011 Photo-irradiated biosynthesis of silver nanoparticles using edible mushroom *Pleurotus florida* and their antibacterial activity studies. *Bioinorg. Chem. Applicat.* Article ID 650979. https://doi.org/10.1155/2011/650979.
42. Mensor LL, Menez FS, Leitão GG, Reis AS, Dos-Santos TC, Coube CS and Leitao SG 2001 Screening of Brazilian plant extracts for antioxidant activity by the use of DPPH free radical method. *Phytoter. Res.* **15** pp 127-130.
43. Linn ES, Yang CT, Chou HJ and Chang T 2009 Screening of antioxidant activities by the edible Basidomycete *Antrodia cinnamomea* strain in submerged culture *J. Food Biochem.* **34** (6) pp 1141-1156.
44. Harbone JB 2000 Physiochemical methods. A Guide to Modern Techniques of plant Analysis 3rd ed. Chapman and Hall, pp. 42-43.
45. Chang C, Yang M, Wen H and Chern J 2002 Estimation of total flavonoids content in proposils by two complementary colorimeter methods. *J. Food Drug Anal.* **10** (3) pp 178-182.
46. Rajakannu S, Shankar S, Perumal S, Subramanian S and Dhakshinamoorthy GP 2015 Biosynthesis of silver nanoparticles using *Garcinia mangostana* fruit extract and their antibacterial, antioxidant activity. *Int. J. Curr. Microbiol. Appl. Sci.* **4** pp 944-52.
47. Mahmoud W, Elazzazy AM and Danial EN 2017 *In vitro* evaluation of antioxidant, biochemical and antimicrobial properties of biosynthesized silver nanoparticles against multidrug-resistant bacterial pathogens. *Biotechnol. Biotechnol. Equip.* **31** (2) pp 373-379.
48. Verma A and Mehata MS 2016 Controllable synthesis of silver nanoparticles using Neem leaves and their antimicrobial activity. *J. Rad. Res. Appl. Sci.* **9** (1) pp 109-115.
49. Kumar SR, Ashish J and Satish N 2011 *Momordica charantia* Linn. A mini review. Int. J. Biomed. Res. **11** (2) pp 579-87.
50. Rushender R, Eerika M, Madhusudhanan N and Konda MV 2012 *In vitro* antioxidant and scavenging activity of *Nymphaea pubescens*. *J. Pharmaceut. Res.* **5** (7) pp 3804-3806.

51. Olamide EA, Adebiyi FO, Tan NH and Zeng GZ 2017 *In vitro* antioxidant activity, total phenolics and total flavonoids content of ethanol extract of stem and leaf of *Grenia carpinifolia. Beni-Suef Univ. J. Basic Appl. Sci.* **6** (1) pp 10-14.
52. Gülçin I, Büyükokuroğlu ME, Oktay M and Küfrevioğlu Öİ 2003 Antioxidant and analgesic activities of turpentine of *Pinus nigra* Arn. subsp. *pallsiana* (Lamb.) Holmboe. *J. Ethnopharmacol.* **86** (1) pp 51-58.
53. Akintola AO, Ayoola PB and Ibikunle GJ 2012 Antioxidant activity of two Nigerian green leafy vegetables. *J. Pharmaceut. Biomed. Sci.* **14** (15) pp 1-5.
54. Aksoy L, Kolay E, Ağılönü Y, Aslan Z and Kargıoğlu M 2013 Free radical scavenging activity total phenolic content, total antioxidant status of endemic *Thermopsis turcica Saudi J. Biol. Sci.* **20** (3) pp 235-239.
55. Tarusco TG, Barney DL and Exon TJ 2004 Content and profile of flavonoid and phenolic acid compounds in conjunction with antioxidant capacity for a variety of North West *Vaccinium* berriers. *J. Agric. Food Chem.* **52** (10) pp 3169-3176.
56. Moskotivz J, Yim KA, and Choke PB 2002 Free radicals and disease. *Arch. Biochem. Biophys*. **397** (2) pp 354-359.
57. Pieta PG 2000 Flavonoids as antioxidants. *J. Nat. Prod.* 2000 **63** (7) pp 1035-1042.

Improved power conversion efficiency in Perovskite solar cell using silver nanoparticles modified photoanode

D Eli [1, 4*], MY Onimisi[1], S Garba[2], PM Gyuk[3], T Jamila[3], and HP Boduku[3]

[1]Department of Physics, Nigerian Defence Academy, Kaduna, Nigeria
[2]Department of Chemistry, Nigerian Defence Academy, Kaduna, Nigeria
[3]Department of Physics, Kaduna State University, Kaduna, Nigeria
[4]Department of Physical Sciences, Greenfield University, Kaduna, Nigeria

*Corresponding author: E-mail: danladielibako@gmail.com

Abstract

Metal nanoparticles have demonstrated outstanding properties in photovoltaic devices through introducing localized surface plasmon effects. The performance of Perovskite solar cell (PSC) by incorporating Ag@P_4VP NPs was investigated systematically. The plasmonic enhancement effects are explored based on the combination of UV-visible absorption spectroscopy and Scanning Electron Microscope (SEM). The performance, especially the short circuit current density (*Jsc*), and open circuit voltage (*Voc*) of the PSC containing Ag@P_4VP NPs was significantly affected. The power conversion efficiency (*PCE*), *Jsc* and *Voc* of the reference device shows a value of 3.80 %, 11.04 $mAcm^{-2}$ and 0.85 V. Upon introducing AgNPs@P_4VP, a *PCE* of 5.69 %, *Jsc* of 12.61 $mAcm^{-2}$ and *Voc* of 0.88 V were recorded, which improved the *PCE* ~ 39.4 % over that of the standard device. The improvement is attributed to an increase in photocurrent density due to enhanced light harvesting by silver nanoparticles.

Keywords: Metal nanoparticles, surface plasmon, AgNPs@P_4VP

1 Introduction

PSCs have demonstrated remarkable breakthrough almost a decade now since after its invention by Kojima and group [1], due to its advantages of low cost, high efficiency and simple fabrication process. Its efficiency has grown from 3.9 % in 2009 to over 23 % in late 2018 [2, 3].

IOP Conf. Series: Materials Science and Engineering **805** (2020) 012005 doi:10.1088/1757-899X/805/1/012005

Despite its remarkable attainment, further improvement of PSCs performance is still necessary for accelerating its commercialization [4]. The use of metal nanoparticles to trap and confine light energy in designated places in PSC can lead to enhanced and superior optoelectronic properties and has been demonstrated by several researchers [4-7].

One of the important issues in plasmonic PSCs is the stability of the plasmonic NPs. When introduced metal nanoparticles (NPs) make direct contact with the perovskite crystal, problems associated to recombination, and corrosion of metal NPs are witnessed. Reports have demonstrated that TiO_2, and SiO_2 were applied as encapsulants [4, 6-8] to prevent the effects above, which however, do not allow easy transfer of carriers generated by perovskite nanocrystal surrounding porous semiconductor. The use of polymer as an encapsulating agent has seldom been reported. In this work, we developed a new and efficient plasmonic composite structure to effectively enhance the PCE of PSCs by incorporating AgNPs with poly(4-vinyl pyridine) as encapsulating material. The polymer has the ability of adhesion to small perovskite molecules, increasing exciton lifetime, increasing diffusion and delocalization lengths which will make considerable contribution to the enhancement of the device performance.

2 Materials and methods

2.1 Synthesis of Ag nanoparticles

To fabricate the silver NPs, a modified two-step reduction synthesis procedure was implemented, which was developed based on the conventional reduction method [9-11]. We first heat the mixture containing sodium borohydride ($NaBH_4$) and tri-sodium citrate (TSC) at the ratio of 2:7 (1×10^{-3} mol dm^{-3}: 3.5×10^{-3} mol dm^{-3}) to 60 °C at 300 rpm for 30 min under vigorous stirring to ensure a formation of homogenous solution. 45 min. Later, 4 ml of an aqueous solution of $AgNO_3$ (4×10^{-3} mol dm^{-3}) was added drop-wise to the mixture, and the temperature was further raised to 100 °C to make the solution boil quickly. The reaction proceeded for another 45 min. Finally, the solution was cooled to room temperature with stirring, and the NPs were collected by centrifugation at 4000 rpm and redispersed in ethanol via sonication for 15 min.

2.2 Preparation of Ag@P_4VP

An aqueous solution of P_4VP (purchased from Sigma-Ardrich) was added to the silver formed above, followed by Acetone treatment to modify the silver NPs surface. To facilitate P_4VP

IOP Conf. Series: Materials Science and Engineering **805** (2020) 012005 doi:10.1088/1757-899X/805/1/012005

coating, the solution was stirred for 24 h at room temperature using a magnetic stirrer plate 78-1 (PEC medicals USA). The P_4VP-coating silver NPs are collected by centrifugation and redispersed in deionized water by sonication.

2.3 Preparation of Photoanodes Precursors

A pure perovskite-precursor solution (1.20 mol/l) was prepared following a method earlier used by Luo et al. [7]

2.4 Preparation of Photoanodes

The FTO were cleaned with cotton wool and ethanol. The blocking layer and the mesoporous layer for the control electrode was deposited and sintered following a method earlier demonstrated by Eli et al. [12]. The perovskite was spin coated on the formed mesoporous TiO_2, it was annealed at 100°C for 10 min. For the plasmon modified cell (S1), we put into consideration, the melting point of the P_4VP polymer (about 170 °C) which is ~2.94 time less than the annealing temperature for TiO_2. To disperse the Ag@P_4VP NPs into the TiO_2 layer homogeneously, spin coating technique was utilized to introduce Ag@P_4VP NPs (in ethanol solutions) onto TiO_2 layer at 600 rpm for 20 seconds. Thereafter, the ethanol was removed by drying naturally. The formed precursor was deposited onto the blocking layer using spin coating technique followed by the deposition of the perovskite absorber via a successive two-step spin-coating process, at 3000 rpm for 30 seconds.

2.5 Preparation of the Counter Electrode

The counter electrode was prepared by employing screen printing procedure. It involves the use of 120 mesh screen printer to screen print a conductive carbon paste (Elcocarb B/SP, Solaronix) to obtain the carbon back contact with area of 1 cm^2. It was followed by drying at 160 °C for 5 min, under the monitoring of an IR thermometer and thereafter, sintered at 400 °C for 30 min.

2.6 Solar Cell Assembly

Ethylene-Vinyl Acetate (EVA) was used to thermally laminate the counter electrode to the appropriate photoanode. The EVA consists of 11% vinyl acetate thereby resulting in a non-toxic thermoplastic polymer useful as hot melting adhesives for thin films in electrical application.

2.7 Characterization and Measurement

The surface morphologies were observed by Scanning Electron Microscopy (Phenom Pro X model, Eindhoven de Netherlands) operated at an acceleration voltage of 10 KV. At that point, ISolution Image Analysis was used to measure the size of Ag nanoparticles that was deposited on the TiO_2 to be 32 nm. Optical spectra were examined and characterized using ultraviolet–visible light (UV-vis) spectrometer (Axiom Medicals UV752 UV-vis-NIR). The current voltage performance of PSCs are estimated under AM 1.5G irradiation using a setup involving a Xenon lamp, an AM 1.5 light channel, and an Electrochemical Analyzer (Keithley 2400 source meter).

3 Results and Discussion

3.1. Scanning Electron Micrographs

The image of the unmodified TiO_2 film in Figure 1(a) demonstrates a dense surface, and there are no shinning particles seen. Figure 1(b) demonstrates the SEM picture of Ag@P_4VP NPs nanoparticles incorporated into the porous TiO_2. From the picture, AgNPs@P_4VP were distributed both on the surface of the TiO_2 and inside the TiO_2. After careful inspection, Ag signals were clearly detected with dense surface and shining particles distributed within. The shining surface is indicative that AgNPs can scatter incident light to increase light absorption surface area.

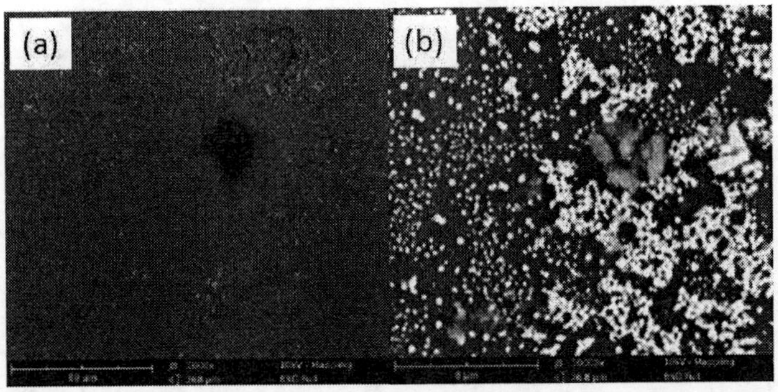

Figure 1. SEM image of: (a) bare TiO_2 (b) TiO_2/AgNPs@P_4VP

3.2. UV-visible spectroscopy

Figure 2(a) shows the absorption of the methylammonium lead halide perovskite absorber within the wavelength range of 300-1200 nm. From the Figure, the material absorbed from 350 to 800 nm with absorption peaks observed at ~385 nm, with absorbance height of ~1.81, ~443 nm, with absorbance height of ~1.73, and ~511 nm, with absorbance height of ~1.67. The absorption at the visible region is an indication that this material meets the requirements for its use as light harvester in this research.

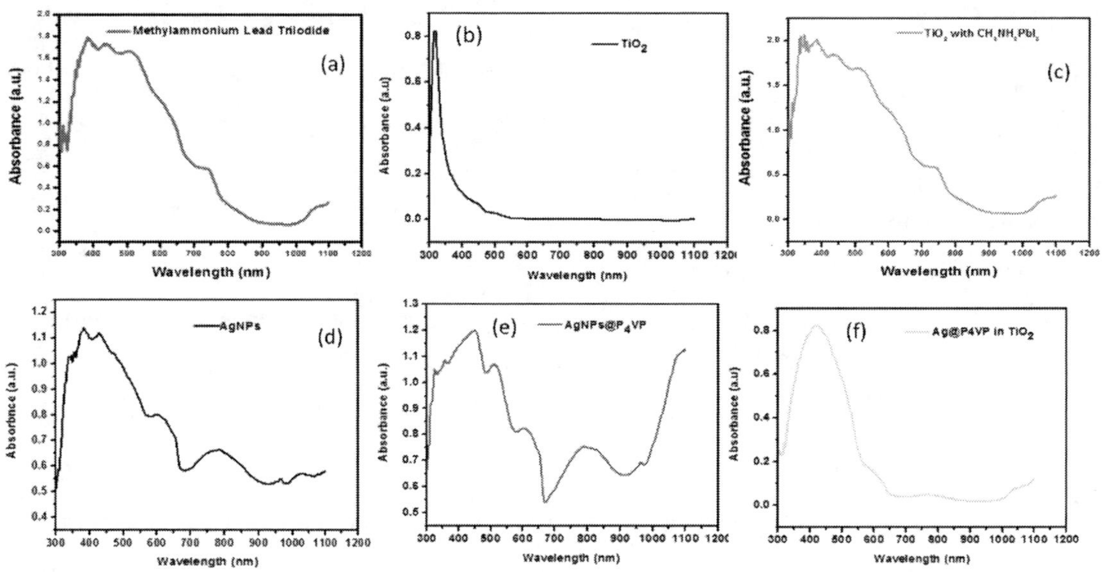

Figure 2. The UV-vis spectroscopy of (a) the pure Methylammonium lead triiodide (b) the bare m-TiO_2, (c) m-TiO_2 with Methylammonium lead triiodide, (d) AgNPs (e) AgNPs@P_4VP, and (f) AgNPs@P_4VP blended within the p-TiO_2 layer

Figure 2(b) shows the absorbance-wavelength plot for TiO_2 in the absence of methyl ammonium lead halide. As seen, no obvious absorption peak was observed within the Vis-NIR region with TiO_2, but a sharp peak was noticed around 316 nm at absorbance height of ~0.82 within the UV region. This peak is due to electronic transitions between molecules having an intermediate ionic degree, which conform to the synthesized molecular materials. This peak suggests that the modification of TiO_2 is necessary to make it active under visible light. Figure 2(c) shows the UV-vis absorption of TiO_2 after sensitization with the methylammonium lead halide. The optical absorption enhancement was noticed in the $CH_3NH_3PbI_3$ modified TiO_2 film with peaks at 385 nm with absorbance height of ~2.01, 438 nm with absorbance height of ~1.86, and 516 nm with absorbance height of ~1.72 absorbing at the range of 330-750 nm within the visible region.

The properties exhibited above shows that, TiO_2 in the presence of $CH_3NH_3PbI_3$ absorber can absorb more photons, possibly because of the large deformations in the network of lead and iodine atoms caused by light in allowing electric charges to migrate through defects and protecting them from being trapped in the materials. A shift of the extinction spectra was observed when the P_4VP shell was coated on AgNPs. Absorption peaks were observed at 453, 496, 679, 855 and 1031 nm with absorbance heights at ~1.14, ~1.11, ~0.80, ~ 0.66 and ~ 0.54 for AgNPs (Figure 2(d)). Also resonances were exhibited at 395, 425, 518, 581, 679, 865 and 1031 nm after encapsulating the AgNPs with P_4VP shell (Figure 2(e)) which shows a slight blue shift at the resonance. As shown in Figure 2(f), the slightly red shifted extinction spectrum of Ag@P_4VP modified samples as compared to that of TiO_2 is expected from the change in the LSPR which slightly shifts to longer wavelength upon addition of the AgNPs and P_4VP. Also, the absorption intensity increase as AgNPs@P_4VP is modified on the reference sample. This behavior was usually ascribed to Ag^0nanoparticles inducing visible light absorption [13, 14]. Moreover, absorption bands at 400-500 nm may be also attributed to Ag clusters.

3.3 Photovoltaic performance

Since the five PSCs function in a way similar to a junction solar cells, therefore, their performance parameters was resolved from the *J-V* and *P-V* curves following equations (1) and (2) respectively.

$$FF = \frac{J_{\max} \times V_{\max}}{J_{SC} \times V_{OC}} \quad (1)$$

$$\eta = \frac{FF \times J_{SC} \times V_{OC}}{P_{IRRADIANCE}} .100\% \quad (2)$$

FF is Fill Factor, η is solar cell efficiency, V_{max} is maximum voltage, J_{max} is maximum current density, *Jsc* is short circuit current density, *Voc* is open circuit voltage and $P_{IRRADIANCE}$ is light intensity.

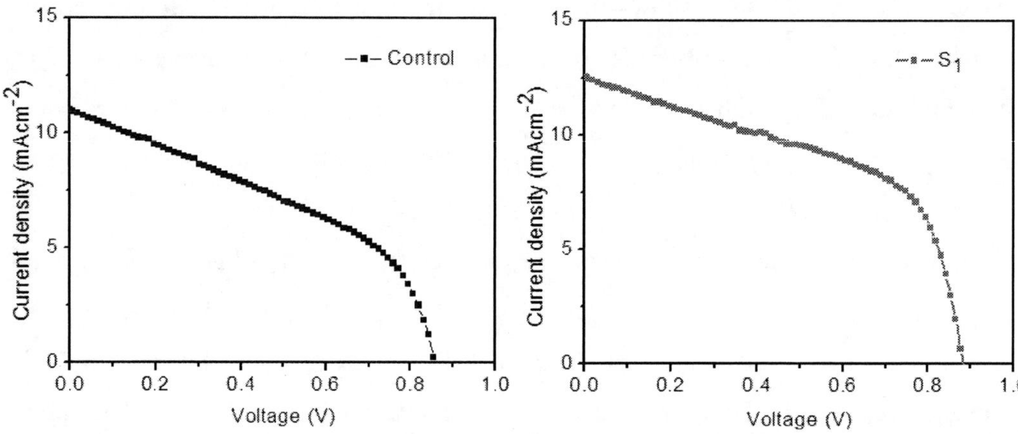

Figure 3. J–V characteristic curves of Ag@P_4VP modified and unmodified PSCs

Figure 3 shows the current density–voltage (*J–V*) curves of the performing PSC devices with different modification. The photovoltaic parameters which include J_{SC}, V_{OC}, PCE and FF are summarized in Table 1.

Table 1. Photovoltaic performance of Ag@P_4VPNPsmodified and unmodified PSCs under 100 mWcm^{-2}

Device	*Jsc* (mAcm^{-2})	*Voc* (V)	*FF*	*PCE* (%)
Control Device	11.04	0.85	0.41	3.80
S1	12.61	0.88	0.51	5.69

From Table 1, the control device which is the reference device exhibits a V_{OC} of 0.85 V, J_{SC} of 11.04 mAcm^{-2} and *FF* of 0.41, which result in a total PCE of 3.80 %. When Ag@P_4VPNPs is introduced into the m-TiO_2 layer, a greatly enhanced performance was exhibited. The device S1 exhibits a PCE of 5.69 % with *Jsc*=12.61 mAcm^{-2}, *Voc*=0.88 V, and *FF*=0.51 which shows an enhancement of ~ 1.50 times in PCE from the control device. The obvious increase demonstrated in *Jsc* is connected to the enhanced light absorption and broadened light absorption range of the perovskite nanocrystal resulting from the SPR of AgNPs, which stimulates the absorber to generate more charge carriers. Furthermore, the Schottky barrier at the TiO_2(Ag@P_4VP)/absorber interface will form electron-hole separation centers which would be beneficial in improving the movement of photogenerated electrons shuttling through the porous TiO_2 network, reducing recombination of electrons and holes, thus increasing *Jsc*. As well known, silver has long been regarded as being catalytically active [15,16]. However, AgNPs dispersed on metal-oxide supports are active catalysts for a variety of chemical reactions [12, 16-18]. In our work, when AgNPs was added, the catalytic effect of AgNPs sited on TiO_2 support could increase the reaction rate between the TiO_2 and $CH_3NH_3PbI_3$, thereby increasing electron mobility. As a consequence, the quenching losses probability between electrons and holes may reduce, thus increasing *Jsc* and *Voc*. The photocurrent enhancement in the perovskite solar cells suggested that localized surface plasmon resonance (LSPR) and electrical effects of Ag NPs enhance the photovoltaic response [12, 15, 19, 20].

4 Conclusion

The effect of plasmonic nanoparticles (NPs) with poly (4-vinylpyridine) as shell nanostructure was investigated systematically. The presence ofAg@P_4VP NPs significantly improved the performance of the PSCs by increasing Jsc, Voc, FF and η. The PSC gave a *Jsc* of 12.61 $mAcm^{-2}$, *Voc* of 0.88 V and FF of 0.51, yielding a PCE of 5.69 %. The increase of *Jsc* is connected to the enhanced light absorption and spectral enhancement of the composite photoanode due to the Surface Plasmon Resonance (SPR) of the AgNPs, while the improved *Voc* may be reasonably related to the shift in quasi-Fermi energy towards the negative potential of the composite system as a result of the enhanced electron capture and storage capability resulting from AgNPs addition.

References

[1] Kojima A, Teshima K, Shirai Y, and Miyasaka T 2009 Organometal halide Perovskites as visible-light sensitizers for photovoltaic cells. *J. Am. Chem. Soc.* **131** (17) pp 6050- 6051.

[2] Li S, Zhu X, Wang B, Qiao Y, Liu W, Yang H, Liu N, Chen M, Lu H and Yang Y 2018 Influence of Ag nanoparticles with different sizes and concentrations embedded in a TiO_2 compact layer on the conversion efficiency of Perovskite solar cells. *Nanoscale Res. Lett.* **13** (1) 210. https://doi.org/10.1186/s11671-018-2626-y.

[3] Green MA, Emery K, Hishikawa Y, Warta W, Dunlop E. D, Levi D H and Ho-Baillie AWY 2017 Solar cell efficiency tables (version 49). *Prog. Photovoltaics* **25** (1) pp 3-13.

[4] Wu R, Yang B, Zhang C, Huang Y, Cui Y, Liu P, Zhou C, Hao Y, Gao Y and Yang J 2016 Prominent efficiency enhancement in Perovskite solar cells employing silica-coated gold nanorods. *J. Phys. Chem C* **120** (13) pp 6996-7004.

[5] Cheng Y, Chen C, Chen X, Jin J, Li H, Song H and Dai Q. 2016. Considerably enhanced Perovskite solar cells via the introduction of metallic nanostructures. *J. Mater. Chem. A* **5** (14) pp 6515-6521.

[6] Saliba M, Zhang W, Burlakov VM, Stranks SD, Sun Y, Ball JM, Johnston MB, Goriely A, Wiesner U and Snaith HJ 2015 Plasmonic-induced photon recycling in metal halide Perovskite solar cells. *Adv. Funct. Mater.* **25** (31) pp 5038-5046.

[7] Luo Q, Zhang C, Deng X, Zhu H, Li Z, Wang Z, Chen X and Huang S 2017 Plasmonic effects of metallic nanoparticles on enhancing performance of Perovskite solar cells. *ACS Appl. Mater. Interf.* **9**(40) pp 34821-34832.

[8] Zhang W, Saliba M, Stranks SD, Sun Y, Shi X, Wiesner U and Snaith HJ 2013 Enhancement of Perovskite-based solar cells employing core-shell metal nanoparticles. *Nano Lett.* **13** (9) pp 4505-4510.

[9] Eli D and Gyuk PM 2019 High efficiency dye sensitized solar cells by excitation of localized surface plasmon resonance of AgNPs. *Sci. World J* **14** (2) pp 125-130.

[10] Nourolahi H, Bolorizadeh MA and Behjat A 2016. Light absorption with branched gold cauliflower-like nanostructure arrays. *Vacuum* **123** pp 29-34.

[11] Agnihotri S, Mukherji S and Mukherji S 2014 Size-controlled silver nanoparticles synthesized over the range 5-100 nm using the same protocol and their antibacterial efficacy. *RSC Adv.* **4** (8) pp 3974-3983.

[12] Eli D, Ibeh G J, Ige O O, Owolabi J A, Ugbe R U, Sherifdeen B O, Onimisi M Y and Ali H 2019 Silver nanoparticles as nano antenna for TiO_2 activation and its application in DSSC for enhanced performance. *J. Theor. Appl. Phys.* **1** (3) pp 88-98.

[13] Tian Y and Tatsuma T 2004 Plasmon-induced photoelectro-chemistry at metal nanoparticles supported on nanoporous TiO_2. *Chem. Comm.* **10** (16) pp 1810- 1811.

[14] Castro CA, Jurado A, Sissa D and Giraldo SA 2012. Performance ofAg-TiO_2 photocatalysts towards the photocatalytic disinfection of water under interior-lighting and solar-simulated light irradiations. *Int. J. Photoenerg.* ID 261045. https://doi.org/10.1155/2012/261045.

[15] Eli D, Onimisi M Y, Abdu S G, Gyuk P M and Ezeoke J 2016 Enhanced performance of a dye sensitized solar cell using silver nanoparticles modified photoanode. *J. Scientif. Res. Rep.* **10** (4) pp 1-8.

[16] Guo K, Li M, Fang X, Liu X, Sebo B, Zhu Y, Hu Z and Zhao X 2013 Preparation and enhanced properties of dye-sensitized solar cells by surface plasmon resonance of Ag nanoparticles in nanocomposite photoanode. *J. Power Sources* **203** pp 155-160.

[17] Eli D, Gyuk PM, Oluwaseyi BS 2018 Size controlled silver nanoparticles for enhanced performance of dye sensitized solar cells. *Nig. J. Phys.* **27** S pp 265-272.

[18] Onimisi M Y, Eli D, Abdu S G, Aboh H O and Ezeoke J 2016 Size effects of silver nanoparticles on the photovoltaic performance of dye sensitized solar cells. *Am. Chem. Sci. J.* **13** (3) pp 1-8.

[19] Wang D H, Kim D Y, Choi K. W, Seo J H, Im S H, Park J H, Park O O and Heeger A J 2011 Enhancement of donor-acceptor polymer bulk heterojunction solar cell power conversion efficiencies by addition of Au nanoparticles. *Angew. Chem.* **50** (24) pp 5519-5523.

[20] Catchpole K R and Polman A 2008 Plasmonic solar cells. *Opt. Express* **16** (26) pp 21793-21800.

Preliminary investigation of the surface activation of diposable screen printed carbon electrodes using synthesized gold nanoparticles

SA Fowotade*[1,2], NA Yusof[1, 3], J Abdullaha[1] and Y Sulaiman[1]
[1]Department of Chemistry, Faculty of Science, Universiti Putra Malaysia, 43400 UPM Serdang, Selangor, Malaysia; [2]Department of Science Laboratory Technology, Hussaini Adamu Federal Polytechnic, Kazaure, Jigawa, Nigeria; [3]Institute of Advanced Technology, Universiti Putra Malaysia, 43400 UPM Serdang, Selangor, Malaysia
*E-mail: fowotades14@gmail.com; sulayman75@hafedpoly.edu.ng +2348036570430I

Abstract
Gold nanoparticles, GNPs, are known to be the most stable metal nanoparticles. They exhibit many unique and interesting physical and optical properties with profound potentials for diverse applications. However, they are highly delimited in terms of size distribution. Different sizes of gold nanoparticles, synthesized through simple laboratory method, were characterized using zetasizer nano series particle size analyzer and ultraviolet-visible spectrophotometer. The synthesized gold nanoparticles were used to modify screen printed carbon electrodes and electrochemical characterization was conducted. The synthesized GNPs sizes range from 42.54-1730.00 nm. The 49.27 nm sized GNPs displayed the highest oxidative peak current in the cyclic voltammogram obtained. All the GNPs modified electrodes gave higher peak over the bare electrode. This implied that the GNP/SPCEs allowed the flow of current or offered least resistance to the flow of current as compared to bare/SPCE. The high peaks indicate successful oxidation process at the anode electrode of the electrochemical process. The present work showcased a simple approach of coating the surface of the disposable SPCEs with GNPs unlike other previously employed pathways. The work also showed that GNPs can successfully be assembled onto the surface of SPCE through simple dipping of the SPCE in the GNPs solution. In addition this approach is cheap, simple and environmentally friendly. The fabricated electrode, GNP/SPCE could be a promising electrode for future detections and determinations.

Keywords: Gold nanoparticles, peak current, Zetasizer, screen printed carbon electrode, spectrophotometry

IOP Conf. Series: Materials Science and Engineering **805** (2020) 012006 doi:10.1088/1757-899X/805/1/012006

1 Introduction

A number of studies have been conducted on the preparation, property and developmental utilization of gold nanoparticles (GNPs) which is among the most researched nanomaterials. GNPs have stable physical and optical qualities [1] making them suitable for applications in drug delivery, electronics [2], sensor (biosensor and/or electrochemical sensor) [3] among others. A number of capping and stabilizing agents have been utilized in the synthesis of GNPs. Each approach is determined by the specific area of usage. The production of these interesting nanoparticles is dependent on the amount of sodium citrate and gold salt, pH and temperature [2]. The size and numbers of the particles give rise to the varying architectural structures like rod-, polygon-, starlike and branched GNPs that can be produced from a solution-governed synthetic pathway. Usually, nanomaterials of multiples sizes and shapes are bound to evolve by toiling with the aforementioned factors [4]. The inherent physicochemical properties of GNPs which make them to be utilized in numerous applications are attributed to their non-toxicity, nobility, easy synthetic protocols and stress-free functionalization. Gold cores of 1-150 nm in diameters had been successfully synthesized [5]. GNPs alone and/or GNPs with other nanoparticles of interest are often supported on disposable screen printed electrodes.

Screen printed carbon electrodes (SPCEs) are the products of microfabrication screen printing technology involving gradual and controlled deposition of thick film of ink comprising graphite, binder and solvent on unreactive supports like ceramics or plastics. In comparison to the screen printed gold or platinum electrodes, the SPCEs have profited tremendously from the technological breakthrough [6]. They can be electrochemically or chemically activated to enhance their charge transfer traits [7–10]. However, some measures may be employed to modify the surface of the electrodes to obtain better performance. These may include doping with catalysts [11] or surface coating with carbon nanotubes [12–14].

GNPs can be used to decorate the surface of activated MWCNTs in a number of ways. They can be connected to the oxidized nanotube through the use of bifunctional thiols acting as covalent linkages [3]. The connection via hydrophobic and/or hydrogen bonds had also been reported. Acetone can hydrophobically inter play between MWCNTs and GNPs protected with octanethiols. While the alkyl chain on the alkanethiol interact with MWCNTs surface in a non-aqueous manner, the hydrogen bonds interaction occurred between the carboxylic groups on the MWCNTs and those on the GNPs surface resulting in a rigid nanocomposite [5]. Another approach is pie-stacking using pyrene derivatives and lastly is the electrostatic interaction which is often utilized in layer-by-layer formulations. In this regard ionic polyelectrolytes often links charged GNPs onto oxidized MWCNTs surface. The often used conductive ionic polyelectrolytes include poly(diallyldimethylammonium chloride), PDDA with positive charge and poly(sodium 4-styrenesulfonate), PSS with negative charge. However, charged GNPs can interact solely with activated MWCNTs surface, producing a rather low density binding compared to that involving the ionic polyelectrolytes [15]. The present work employed simple dipping method to observe the binding electrostatic force of attraction between the carbon of SPCE and the GNPs. The present study synthesized GNPs using a modified technique reported elsewhere [16]. The resulting GNPs were subjected to electrochemical and chemical characterization and particle size determination. These methods constitute the preliminary investigation of the activation of SPCEs using various sizes of synthesized GNPs.

2 Materials and Methods

2.1 Reagents

Gold (III) chloride trihydrate (≥99.9%) and sodium citrate dehydrate (≥99%) were purchased from Sigma-Aldrich (St. Louis, USA). Sulphuric acid (95-97%) was procured from Friendemann Schmidt (Parkwood, Australia). Potassium ferric cyanide, $K_3Fe(CN)_6$ was purchased from Bendosen (Kajan, Selangor, Malaysia) and potassium chloride, KCl (99.5%) was bought from R & M Chemicals (Selangor, Malaysia). Other chemicals are of the qualitative analytical grade. Deionized water from a Millipore Milli-Q purifier was used throughout the experiments. All glass wares, plastic wares and ceramics were washed thoroughly with liquid detergent and properly rinsed with deionized water and oven dried prior to the initiation of the experiments.

2.2 Equipment

Cyclic voltammetry (CV) was performed via a DropSens potentiostat, μStat 8000 electrochemical workstation (Asturias, Spain). A three-electrode system was used for all electrochemical experiments. Screen-printed carbon electrodes, comprising carbon-based working electrode (4 mm diameter) made of carbon, platinum-based counter electrode and silver-based reference electrode were provided by MIMOS (Bukit Jalil, Selangor, Malaysia). The particle size was measured with Zetasizer Nano 5 system (Malvern Instruments Ltd., Malvern, UK). UV-Visible spectrophotometer (Perkin Elmer, Lambda 35, Massachusetts, USA) was utilized for the routinely quantitative determination of absorbance in the synthesized sample solutions. A digitalized pH meter purchased from Thermo Fisher Scientific (Chicago, IL, USA), was employed in the adjustment of pH values of prepared solutions in the voltammetric experiments. A digital balance used to accurately measure the weight of solids was supplied by A&D Company (Tokyo Japan). Hot plate with stirrer used for heating was obtained from Hwashin Technology Company (Seoul, Korea)

2.3 Preparation of colloidal gold nanoparticles solution

Exactly 0.01% $HAuCl_4.3H_2O$ was prepared from the solid stock and 1% sodium citrate was equally prepared from the stock. These solutions were used for the production of gold nanoparticles (GNPs) colloidal solution. Synthesis of GNPs was carried out based on the description of Manso *et al.* [16] with slight modification. Then, 30 ml of $HAuCl_4$.3H2O solution and 0.2 ml of sodium citrate solution were mixed together in a beaker, thoroughly stirred and heated to boiling. The heat source was removed after 10 min, but stirring continued for another 10 min to produce GNPs colloidal solution. The experiment was performed with varying volumes of sodium citrate solution (0.3, 0.4, 0.5 and 0.6 ml) producing different sizes of GNPs solutions. In this study, the prepared GNPs are tagged 30/0.2, 30/0.3, 30/0.4, 30/0.5 and 30/0.6. These ratios implied the constant volume (30 ml) of gold salt solution in combination with the variable volumes of citrate solution.

2.4 Characterization of prepared GNPs colloidal solutions

The prepared GNPs were characterized for size and absorption using Malvern zetasizer nano 5 particles size analyzer and UV-visible spectrophotometry, respectively. The

IOP Conf. Series: Materials Science and Engineering **805** (2020) 012006 doi:10.1088/1757-899X/805/1/012006

electrochemical characterization was achieved by coating the surface of the SPCE with various sizes of GNPs. Each bare/SPCE was immersed into the particular GNPs solution for a period of 2 min afterwards the GNPs coated SPCE tagged nGNPs/SPCE was air dried and stored in the desiccator for cyclic volumetric determination. The working electrode of the SPCE was modified with the GNPs.

2.5 Electrochemical characterization of modified SPCEs

The electrolyte employed for determination of effective electroactive surface area (EESA) of modified electrodes, comprised of 1mM $K_3Fe(CN)_6$ and 0.1M KCl at variable scan rates. The voltammetric analyzer used for this study comprises of DropSens potentiostat, μStat 8000 electrochemical workstation (Asturias, Spain) attached to a computer system. The instrument was operated at the following conditions for a cyclic voltammetry (CV) method; activating potential -0.6V to +0.6V, scan rate $0.05Vs^{-1}$, in the presence of 1 mM $K_3Fe(CN)_6$/0.1MKCl electrolytic medium at room temperature. The gold nanoparticle SPC electrodes were prepared by dipping the bare SPCE in respective GNPs colloidal suspension for a time interval of 2 min and then on removal each GNPs coated electrode was air dried. The prepared electrodes for this study are; 0.2GNPs/SPCE, 0.3GNPs/SPCE, 0.4GNPs/SPCE, 0.5GNPs/SPCE and 0.6GNPs/SPCE.

3 Results and Discussion

The size distribution of GNPs is shown in Figure 1. The DLS plot shows a bimodal distribution for 30/0.2 GNP (Figure 1A) with peaks at 10 and 90 nm respectively, a unimodal distribution for 30/0.3 GNP (Figure 1B) with peak around 1000 nm, a quadrimodal distribution for 30/0.4 (Figure 1C) with peaks estimated around 5, 60, 700 and 7000 nm respectively, a trimodal distribution for 30/0.5 GNP (Figure 1D) with peaks approximated around 2, 8 and 90 nm respectively and a bimodal distribution for 30/0.6 GNP (Figure 1E) with peaks at 50 and 100 nm respectively. The genesis of these peaks is difficult to explain and may be attributed to the intricacy of the design of the DLS method [19].

IOP Conf. Series: Materials Science and Engineering **805** (2020) 012006 doi:10.1088/1757-899X/805/1/012006

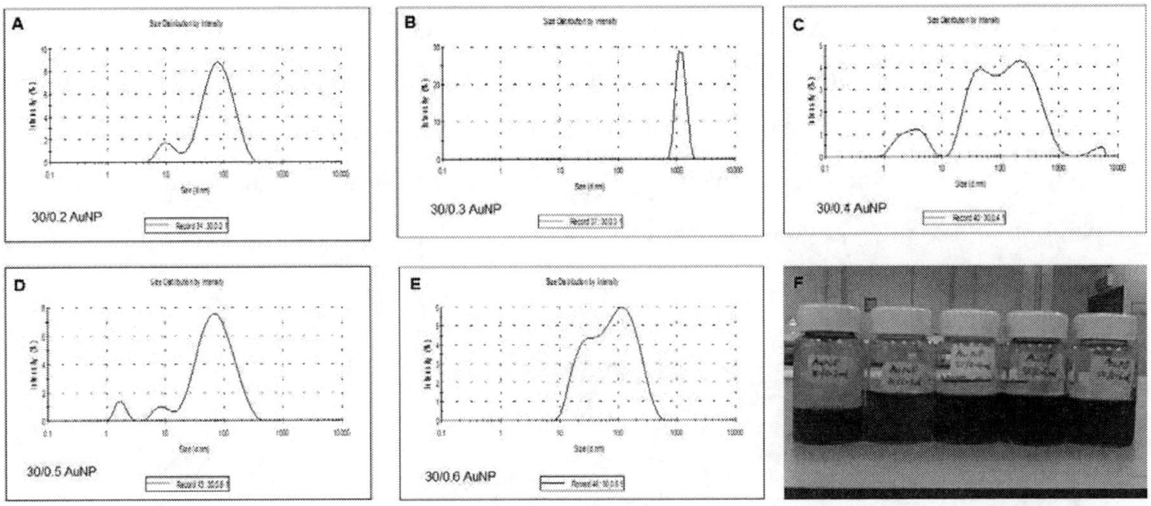

Figure 1. Particle size distribution (PSD) of GNPs with different sizes (A), (B), (C), (D), (E) from Zetasizer nano series using DLS technique and picture of the 5 synthesized GNPs dispersions (F)

Also, the multiple peaks may arise from other species present with GNPs, since single peak and a somewhat narrow distribution is expected for a pure gold solution. These outcomes show similarity to those reported elsewhere [19] where gold sizes were measured in the presence and absence of plasma. Table 1 is the extrapolation of the intensity versus particle size diameter of Figure 1. The modal size of 30/0.3GNP is the highest (1281 nm), while the least modal value (1.294 nm) is jointly shared between 30/0.4 and 30/0.5 GNPs. The mean volume range of the all the GNPs approximate to 0.1 - 22.0 with the exception of 30/0.3GNP. However, among all the modal sizes, size 30/0.2 GNP (7.531 nm) corresponded to the reported value in the literature [18]. DLS is an approach that makes use of particulate assembly in the determination of particle size distribution [19]. The z-average of the synthesized GNPs over the diameter range of particle sizes are displayed in Table 1. The 30/0.2 GNPs also satisfied both conditions of average size (49 nm) being within the nanoscale region and acceptable polydispersity index (46%). This choice is also in agreement with previous findings [20] which posited that polydispersity index should not be above 50% for a monodisperse medium.

Figure 2 shows the GNPs particle size distribution with the mean volume obtained from the zetasizer nano series analyzer. Four out of the five GNPs are within the diameter less than or equal to 20 nm (Figure 2A). The odd of the synthesized GNPs is 30/0.3 which is apparently displayed as curve e (Figure 2B). However, the 30/0.3 GNPs is represented as curve a (Figure 2C) with size ranging from 500- 2500 nm. This value is well outside the nanoscale range of 1-100 nm. The reason for this anomaly may be attributed to particle agglomeration during the preparation of the colloidal gold nanoparticle solution.

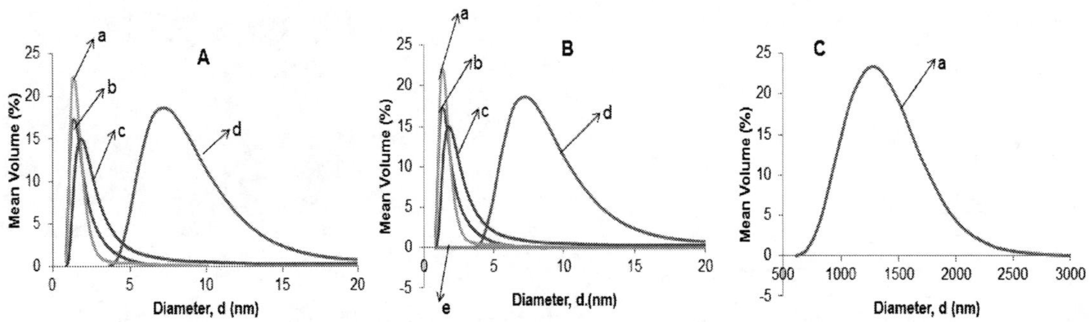

Figure 2. A plot of Mean volume (%) against diameter (nm) for PSD of synthesized GNPs, curve a, 30/0.5; curve b, 30/0.4; curve c, 30/0.6, curve d, 30/0.2(A) and curve e, 30/0.3(B), curve a, 30/0.3 (C)

Table 1. Particle size distribution of synthesized GNPs from Zetasizer Nano series using DLS technique

Parameters	30/0.2GNPs	30/0.3 GNPs	30/0.4 GNPs	30/0.5 GNPs	30/0.6 GNPs
Z-average (nm) ($\overline{x}$ & α)	49.27± 0.245	1730± 1.013	43.35± 0.585	42.54 ±0.170	46.95± 0.050
Polydispersity index ($\overline{x}$ & α)	0.457 ±0.003	0.419± 0.083	1.00± 0.000	0.560± 0.003	0.538± 0.005
Diameter range (nm)	4.187-164.2	712.4-6439.0	0.8332-7.531	0.8332-8.721	7.531-396.1
Mean volume range (%)	0.100-18.50	0.300-23.40	0.100-17.20	0.100-22.10	0.100-14.90
Modal size (nm)	7.531	1281	1.294	1.294	13.54

NB: $\overline{x}$ = mean, α = standard deviation

The UV-visible absorption spectra of synthesized GNPs are shown in Figure 3. The absorbance peaks of GNPs appear at wavelength range approximating 520-530nm, depending on

IOP Conf. Series: Materials Science and Engineering **805** (2020) 012006 doi:10.1088/1757-899X/805/1/012006

the size of GNPs. The wavelength is a measure of the size of nanoparticles. As displayed in Figure 3, the peaks of the absorption spectra shifted towards the short wavelength side and the full breadth at mid peak of the spectra decreased with soaring volume of sodium citrate from 0.2 to 0.6 ml. This is a realization that citrate solution is a determinant of the size of GNPs. The faster the capping of the GNPs by citrate molecules, the smaller the ensuing nanoparticles. Similar results have been reported in the literature [21].

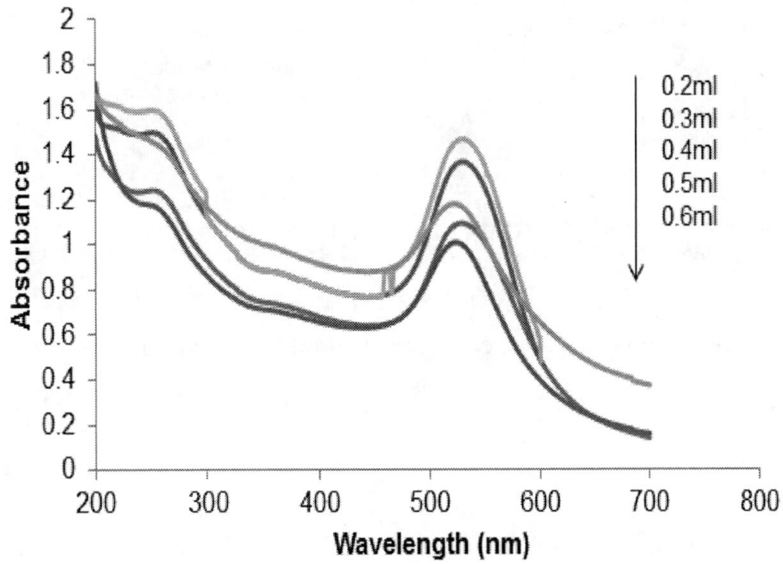

Figure 3. UV-visible absorption spectra of synthesized spherical GNPs

The cyclic voltammograms of GNPs modified screen-printed carbon electrodes and bare SPC electrodes are shown in Figure 4. All the voltammograms exhibited excellent reversible redox process, GNPs modified and bare electrode alike. This justifies the establishment of equilibrium at the electrode surface involving the redox couple $Fe^{3+}//Fe^{2+}$ from the ferric cyanide electrolyte. However, the anodic oxidative peak current (Ipa) for each modified electrode slightly differs, approximately, with 0.2GNPs/SPCE providing the highest peak current of 10.2 μA and 0.4GNPs/SPCE giving the least amount of 9.1 μA (Figure 4).

IOP Conf. Series: Materials Science and Engineering **805** (2020) 012006 doi:10.1088/1757-899X/805/1/012006

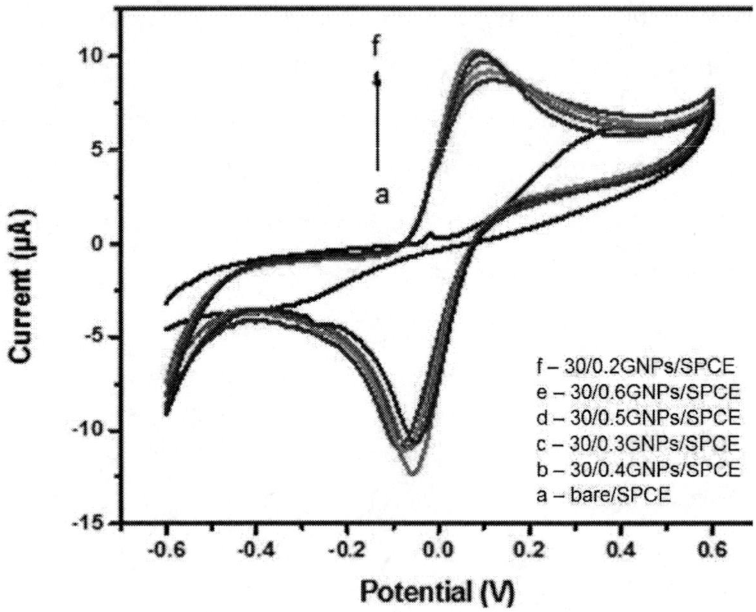

Figure 4. Cyclic voltammograms of overlay forms of all the voltammograms for all the different sizes of GNPs/SPCE in comparison with the bare/SPCE in 1mM $K_3Fe(CN)_6$/0.1MKCl at a scan rate of 0.05V/s

The oxidative peak current of each of the GNPs/SPCE is higher than that of the bare/SPCE (Figure 4). The high value of Ipa recorded from the modified electrodes may be due to the catalytic property of GNPs which enhances rapid electron transfer at the surface of SPC electrodes. The 30/0.2GNPs/SPCE gives the highest oxidative peak current, while 30/0.4GNPs/SPCE gives the least oxidative peak current. The electrochemical study outcome also supported previous findings regarding the 30/0.2GNPs. Thus 30/0.2 sized gold nanoparticles may be utilized in the future development of the new electrode.

The effective electro-active surface area (EESA) for both bare and modified electrodes is also evaluated with the aid of Randles-Sevick equation [22, 23].

$$i_p = (2.69\ X\ 10^5)n^{3/2}A_{eff}D^{1/2}C\upsilon^{1/2} \quad (1)$$

Where n is the number of electron(s) participating in the redox process, A is the effective electro-active surface area (cm^2), D is the diffusion coefficient of the molecules in the solution ($cm^2\ s^{-1}$), C is the molar concentration of $[Fe(CN)_6]^{3+}$ and v is the scan rate (Vs^{-1}).

The effective electro-active surface area of each of the electrode can be obtained by substituting the value of the respective slope in the plot of I_p vs $\upsilon^{1/2}$ (Figure 5) and other constants for the ferrocyanide electrolyte, Fe^{3+}/Fe^{2+} redox couple such as n= 1, D = 7.6 x 10^{-6}, C= 1x 10^{-3} into the equation

$$A_{eff} = \frac{slope}{2.69\,X10^5\,n^{3/2}\,D^{1/2}\,C} \tag{2}$$

The calculated effective electro-active surface area (A_{eff}) for the bare/SPCE is 0.01163327 cm^2. There was an increase in A_{eff} to 0.055007407 cm^2 when bare/SPCE was modified with colloidal solution of GNPs films. The A_{eff} of GNPs/SPCE is approximately 4.7 folds greater than bare/SPCE. These results suggest that the surface modification of bare/SPCE using GNPs provided much more effective surface areas than the bare/SPCE. The increase in the EESA of the GNPs modified SPCE may lead to corresponding increase in oxidative peak current as earlier discussed, suggesting rapid rate of electron transfer at the surface of modified electrode over the bare electrode.

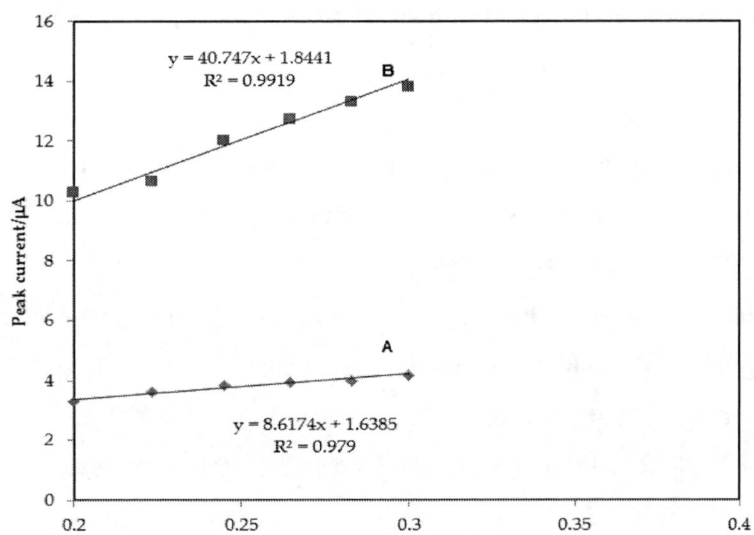

Figure 5. A plot of peak current (A) versus square root of scan rate (vs^{-1}) for bare/SPCE (A) and GNPs/SPCE (B)

4 Conclusion

Various sizes of gold nanoparticles were successfully prepared through simple laboratory techniques. Physicochemical and electrochemical characterizations of the prepared GNPs were carried out via basic instrumental means. Enlargement of effective surface area of modified electrode over bare electrode was also confirmed. All the modified electrodes produced higher oxidative peak current over bare electrodes confirming the relevance of surface active agents on SPCE. The present findings may be the much needed harbinger for the future development of sensitive, selective, cheap and environmentally friendly electrodes for all kinds of pertinent detections and determinations.

Acknowledgment

I wish to sincerely appreciate the Ministry of Higher Education (MOHE), Malaysia for funding this research work through the Nanomite grant (No. 9443101). We are also indebted to the management of Universiti Putra Malaysia for the laboratories used in carrying out this project. I also acknowledge a colleague, Late Benjamin for his technical contribution to this work.

Conflicts of Interest: The authors declare no conflict of interest

References

[1] Ghosh D and Chattopadhyay N 2013 Gold nanoparticles: acceptors for efficient energy transfer from the photoexcited fluorophores. *Opt. Photon. J.* **3** (01) 18. http://doi.org/10.4236/opj.2013.31004.

[2] Ahmad T and Khan W 2013 Size variation of gold nanoparticles synthesized using tannic acid in response to higher chloroauric acid concentrations. *World J. NanoSci. Eng.* **3** (3) 62. http://doi.org/10.4236/wjnse.2013.33009.

[3] Zeng S, Yong KT, Roy I, Dinh XQ, Yu X and Luan F 2011. A review on functionalized gold nanoparticles for biosensing applications. *Plasmon.* **6** (3) 491. http://doi.org/10.1007/s11468-011-9228-1.

[4] Sau T K and Murphy C J 2004 Room temperature, high-yield synthesis of multiple shapes of gold nanoparticles in aqueous solution. *J. Am. Chem. Soc.* **126** (28) pp 8648-8649.

[5] Ghosh P, Han G, De M, Kim CK and Rotello VM 2008. Gold nanoparticles in delivery applications. *Adv. Drug Deliv. Rev.* **60** (11) pp 1307-1315.

[6] Fanjul-Bolado P, Hernández-Santos D, Lamas-Ardisana PJ, Martín-Pernía A and Costa-García A 2008 Electrochemical characterization of screen-printed and conventional carbon paste electrodes. *Electrochim. Acta* **53** (10) 3635-3642.

[7] Ghamouss F, Tessier PY, Djouadi A, Besland MP and Boujtita M 2007 Screen-printed carbon electrode modified on its surface with amorphous carbon nitride thin film: Electrochemical and morphological study. *Electrochim. Acta* **52** (15) pp 5053-5061.

[8] Ghamouss F, Tessier PY, Djouadi MA, Besland MP and Boujtita M 2007. Examination of the electrochemical reactivity of screen printed carbon electrode treated by radio-frequency argon plasma. *Electrochem. Comm.* **9** (7) pp 1798-1804.

[9] Pchelintsev NA, and Millner PA 2007 Development of surface activated screen-printed carbon transducers for biosensors application. *Analy. Lett.* **40** (7) pp 1317-1332.

[10] Wei H, Sun JJ, Xie Y, Lin CG, Wang YM, Yin WH and Chen GN 2007 Enhanced electrochemical performance at screen-printed carbon electrodes by a new pretreating procedure. *Anal. Chim. Acta* **588** (2) pp 297-303

[11] Chikae M, Idegami K, Kerman K, Nagatani N, Ishikawa M, Takamura Y and Tamiya E 2006 Direct fabrication of catalytic metal nanoparticles onto the surface of a screen-printed carbon electrode. *Electrochem. Comm.* **8** (8) pp 1375-1380.

[12] Guan WJ, Li Y, Chen YQ, Zhang XB and Hu GQ 2005 Glucose biosensorbased on multi-wall carbon nanotubes and screen printed carbon electrodes. *Biosens. Bioelectron.* **21** (3) pp 508-512.

[13] Lin Y, Lu F and Wang J 2004 Disposable carbon nanotube modified screen-printed biosensor for amperometric detection of organophosphorus pesticides and nerve agents. *Electroanal.* **16** (1-2) pp 145-149.

[14] Wang, J., & Musameh, M. (2004). Carbon nanotube screen-printed electrochemical sensors. *Analyst* **129** (1) pp 1-2.

[15] Georgakilas V, Gournis D, Tzitzios V, Pasquato L, Guldi DM and Prato M 2007 Decorating carbon nanotubes with metal or semiconductor nanoparticles. *J. Mater. Chem.* **17** (26) pp 2679-2694.

[16] Manso J, Agüí L, Yáñez-Sedeño P and Pingarrón JM 2004 Development and characterization of colloidal gold-cysteamine-carbon paste electrodes. *Analy.Lett.* **37** (5) pp 37-41.

[17] Horiba Scientific 2013 A guidebook to particle size analysis. *Horiba Instruments, Inc.* http://www.horiba.com/fileadmin/uploads/Scientific/Documents/PSA/PSA_Guidebook.pdf. Accessed 2 June, 2017.

[18] Polte J, Ahner TT, Delissen F, Sokolov S, Emmerling F, Thünemann AF and Kraehnert R 2010. Mechanism of gold nanoparticle formation in the classical citrate synthesis method derived from coupled *in situ* XANES and SAXS evaluation. *J. Am. Chem. Soc.* **132** (4) pp1296-301.

[19] Montes-Burgos I, Walczyk D, Hole P, Smith J, Lynch I and Dawson K 2010. Characterisation of nanoparticle size and state prior to nanotoxicological studies. *J. Nanopart. Res.* **12** (1) pp 47-53.

[20] Kaszuba M, McKnight D, Connah MT, McNeil-Watson FK and Nobbmann U 2008 Measuring sub nanometre sizes using dynamic light scattering. *J. Nanopart. Res.* **10** (5) pp 823-829.

[21] Long NN, Kiem CD, Doanh SC, Nguyet CT, Hang PT, Thien ND and Quynh L 2009 Synthesis and optical properties of colloidal gold nanoparticles. *J. Phys.: Conf. Series* **187** (1) p. 012026. https://doi.org/10.1088/1742-6596/187/1/012026.

[22] Muhammad A, Yusof N, Hajian R and Abdullah J 2016 Construction of an electrochemical sensor based on carbon nanotubes/gold nanoparticles for trace determination of amoxicillin in bovine milk. *Sens.* **16** (1) 56. https://doi.org/10.3390/s16010056.

[23] Muhammad N, Abdullah J, Sulaiman Y and Ngee LH 2017 Voltammetric determination of nitrophenol using PEDOT decorated graphene oxide as composite film. *Int. J. Electrochem. Sci.* **12** (10) pp 9432-9444.

IOP Conf. Series: Materials Science and Engineering **805** (2020) 012007 doi:10.1088/1757-899X/805/1/012007

Spectroscopic, microscopic and electrical characterization of nanoscopic polyindole DNA-templated nanomaterials

M I Yahaya
Department of Chemistry, Sokoto State University, Nigeria

Email: ymansur64@gmail.com

Abstract
There has been an increasing demand for in expensive, accurate, movable and reliable nanomaterials for nanoelectronics devices and other applications. Deoxyribonucleic acid has robust nature, therefore it self-fabrication with indole can produce highly organised functional nanostructures that are of great interest for chemical industry applications. This research work is concerned with the synthesis, spectroscopic, microscopic and electrical characterisation of hybrid polyindole (Pln) DNA templated nanowires. Polyindole (PIn) has been templated on λ-DNA via oxidative polymerisation of indole using $FeCl_3$ to produce conductive PIn/DNA nanowires. The formation of PIn/DNA nanowires were verified by FTIR, UV-vis and XPS spectroscopy techniques. AFM, SEM and TEM techniques were used to characterise the nanowires dimensions. AFM studies revealed an average height of 1.60 nm for free DNA and the Pln/DNA nanowires have diameters in the range 2–15 nm with the dominance of 3-4 nm mean diameter range. The electrical properties of Pln/DNA nanowires as drop-cast films were investigated by two-terminal current voltage (I-V) measurements on a probe station. The nanowires were drop-cast (5 μL of as-prepared dispersion) onto platinum microband electrodes. The conductance of these films at 20 °C was of the order of 10-100 μS. In addition, the conductance of PIn/DNA nanowires exhibits Arrhenius behaviour (Ea = 0.80 $\pm$ 0.06 eV) as a function of temperature. The above results have revealed the potentials of the Pln/DNA nanowire in nanoelectronics applications.

Keywords: Polyindole, Nanomaterials, Template, Characterisation, DNA

1 Introduction
In recent times, there has been an increasing demand for in expensive, accurate, movable and reliable nanomaterials for nanoelectronics devices and other applications. Conductive polymers (CPs) such as indole are being explored as promising materials for such applications, due to their ability to form good basis for premise chemical sensors either as a sensing element or as a matrices to deactivate specific analytes [1].Their charge transport properties are based on chemical redox reactions and provide various chemo-electrical signal transduction mechanisms. Combining them with other functional materials has provided opportunities to tailor their major morphological and physicochemical properties; improve sensitivity and selectivity at low temperature that often result in enhanced sensing performance. These correspond to one of the greatest advantages of using CPs as the sensing material [2].

It was reported that while electro polymerisation of indole which have substituent on the benzene ring produced polymers and indole while derivatives with substitution at the pyrrole ring do not electropolymerised [3]. Therefore, it is considered that polymerisation does not involve coupling via the benzene ring. But work by Wadatkar and co-workers, [4] showed that coupling of 5-substituted indole trimers take place at the ring nitrogen.

IOP Conf. Series: Materials Science and Engineering **805** (2020) 012007 doi:10.1088/1757-899X/805/1/012007

Deoxyribonucleic acid has robust nature, therefore it self-fabrication with indole can produce highly organised functional nanostructures that are of great interest for chemical industry applications. Its utilization as template to direct the self-fabrication or supramolecular polymerization of artificial molecules is a bioinspired concept in chemistry approach, as control over it length and sequence permits the achievement of ideally monodisperse and sequence-controlled polymers [5-6].

In this research work we report the production of conductive polymers/DNA nanomaterials based on polyindole DNA templated nanowires that were fabricated by rapid and cost-effective wet chemistry method. The chemical properties of the nanowires were investigated using different spectroscopic techniques such as Ultra-Violet Visible spectroscopy, X-ray Photoelectron Spectroscopy and Fourier Transform Infrared spectroscopy, Electron microscopy (SEM and TEM) and Atomic force were employed to characterise the nanowire dimensions. These techniques demonstrated the formation of a supramolecular hybrid polymer comprising conductive polymers and DNA

2 Experimental

2.1 Chemicals and Materials

All chemicals were purchased from Sigma-Aldrich Company Ltd. Lambda DNA (λ- DNA, Cat no. N3011S) was purchased from New England Biolabs (UK) Ltd. n-Si (100) wafers were purchased from Compart Technology Ltd. Platinum microband electrodes were supplied by Windsor Scientific Ltd (Pt-MB4000) and Copper grids (HC300Cu) for TEM were purchased from EM Resolutions Ltd. UK. Deionised water (18 M Ω cm resistivity) was obtained from a NANOpure Diamond Life Science ultrapure water system equipped with a DIamond RO Reverse Osmosis System.

2.2 Substrates Cleaning

n-Si (100) wafers was cut into ~1 x 1 cm^2 pieces using a diamond scribe and immersed in acetone for 50 minutes. The chips were then rinsed with copious deionised water and dried in a gentle stream of N2 gas before further drying in an oven for 5 - 10 minutes. Final cleaning with FEMTO low pressure plasma system, using oxygen plasma oxidation (90W, 15 sccm, 15 min) at 40% power was carried out.

2.3 Polymerisation of Indole in DNA-Containing Solutions

Polyindole DNA templated (PIn/DNA) nanowires were chemically synthesised using $FeCl_3$ as an oxidant; by adding 5 μL of freshly prepared indole (C_8H_7N) solution (6.24 mM) to 20 μL of λ-DNA (500 ng μL^{-1}) in the presence of 5 μL of $MgCl_2$ (0.5 mM), then 5 μL of $FeCl_3$ (1 mM) drop-wise into the solution. The solution was thoroughly mixed and allowed to react for at least 1h at room temperature prior to analysis [7].

2.4 Samples preparation for Fourier Transform Infrared Spectroscopy

For all FTIR measurements in this work about 5μL of the prepared Pln/DNA nanowires films were drop-cast on clean Si (100) substrate and left to dry for 1h prior to measurements. FTIR spectra were acquired using a Shimadzu FTIR (IRAffinity-1S). Spectra were recorded in the range 400–4000 cm^{-1}, with 128 scans at 4 cm^{-1} resolution and a clean Si (100) substrate was used as a background.

IOP Conf. Series: Materials Science and Engineering **805** (2020) 012007 doi:10.1088/1757-899X/805/1/012007

2.5 Samples preparation for Ultraviolet–Visible Spectroscopy

For all UV-vis measurements, samples solution was made using calf thymus DNA (CT-DNA). Typically, 0.125 mL freshly prepared samples solution was added to 0.5 mL of CT-DNA solution (162.5 μg mL^{-1}, 10mM Tris-HCl pH 8 + 1 mM ethylene diaminetetraacetic acid (EDTA)) in the presence of 0.125 ml of (0.5mM) $MgCl_2$. Then, 0.125 mL of $FeCl_3$ (1 mM) was added drop-wise to the solution. The mixture was stirred and allowed to react at room temperature for at least 1 h. UV-vis absorption spectra were recorded on a CARY 100 BIO spectrophotometer at room temperature. Diluted sample volume of 1.5 μL was used with water background. The absorbance values reported have been scaled to a standard path length of 1 cm by the instrument software.

2.6 Samples preparation for X-ray Photoelectron Spectroscopy

Samples were prepared for XPS by drop-casting ~20 μL of solution onto a clean Si(100) substrate and then left to dry in air at room temperature in a laminar flow hood. A Theta Probe photoelectron spectrometer was used to collect photoemission spectra. The binding energies obtained in the XPS analysis were calibrated using Carbon (284.8 eV) as a reference. Photoemission spectra were fitted with a combination of mixed singlet components using the CasaXPS software version 2.3.16 from Casa Software Ltd. (Teignmouth, UK) Backgrounds were subtracted by the Shirley method [8].

2.7 Samples Preparation for Transmission Electron Microscopy

Samples (5 μL) were cast dropped on copper grids and placed on a laminar flow hood to dry for 5 h. Transmission electron microscopy experiments in this research were carried out with Philips CM100 TEM (FE1) using Holey carbon films in a mesh of Copper grids as substrate and camera images were acquired with AMT CCD (Deben).

2.8 Samples Preparation for Scanning Electron Microscopy

Five micro litters (5 μL) of samples were cast dropped on treated n-Si (100) wafers. The deposited samples were dried in a C-Flow vertical unidirectional (Laminar) Airflow cabinet over night before SEM analysis. The SEM analysis was carried out using JEOL JSM-5610LV Scanning Electron Microscope and some samples were coated with few nanometers of gold film using sputter coater (Bio-rad SC500, Sputter coater) machine.

2.9 Electrical Measurements

Two-terminal conductivity measurements were conducted using Pt microband electrodes deposited on clean silanized Si/SiO_2 chips. A 2 μL drop of an aqueous solution of Pln/DNA nanowires was placed on these electrodes and aligned across the gap between the Au fingers by molecular combing. A Cascade Microtech 110008-M probe station with Agilent B1500A Semiconductor Device Analyzer controlled by B1500A's EasyEXPERT software was used for the measurements. Prior to electrical testing, the two-terminal device undergoes a heating/cooling cycles from 223 to 423 K by using a heating/cooling chuck, under nitrogen gas supply to drive up any water bound to the Pln/DNA nanowire. All the electrical measurements were carried out under dry nitrogen without illumination in the shielded sample compartment of the probe station. Current-voltage measurement from 223 to 423 K was performed on the probe station using a thermal chuck system. For each of the electrical tests, the current was measured for applied voltages from -2 to 2 V in steps of 0.2 V.

IOP Conf. Series: Materials Science and Engineering **805** (2020) 012007 doi:10.1088/1757-899X/805/1/012007

The slopes of the Arrhenius plots in Figure 11 were analysed using equation (1) to determine the activation energies associated with the hopping of charge in the Pln/DNA nanowire samples.

$$lnG = lnG_o - \frac{E_a}{K_B T} \quad (1)$$

The slopes are $-E_a/k_B$ for E_a in J and $-eE_a/k_B$ where e is the charge on a proton, KB is the average kinetic energy and E_a is in eV.

3 Results and Discussion

3.1 FTIR Characterisation

FTIR spectra in the spectral range 3600–600 cm^{-1} were used to identify the polymerisation of polyindole and its interaction with DNA. Comparing the IR transmittance spectra of indole and polyindole (Figure 1) a weak peak can be observed at 1362 cm^{-1} in the indole spectrum which is characteristic of the C-H rocking mode; the peak shifted to 1366 cm^{-1} in the PIn spectrum [9]. Similarly, the peak observed at 1694 cm^{-1} in the indole spectrum is a result of –C = C – ring modes which shifts to 1674 cm^{-1} in indole polymer. Other bands observed include one at 1516 cm^{-1} in indole due to C - C stretch (in-ring)/N-H bend that shifts to 1500 cm^{-1} in Pln spectra and another broad peak at 964 cm^{-1} in indole that is attributed to N –H wag/C-H bend, the peak shift to 972 cm^{-1} in Pln. Two sharp peaks at 675 cm^{-1} seen in both indole and Pln, are due to C - H bending vibrations [10 – 11].

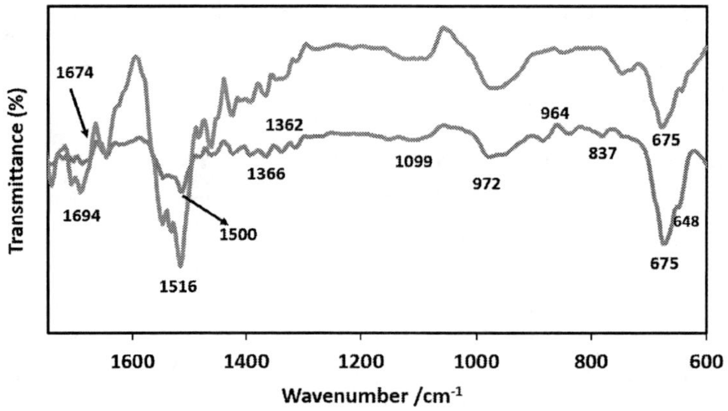

Figure 1. FTIR Transmission spectra of indole (orange i.e. top graph line) and polyindole (blue i.e. second graph line) 128 scans co-added and averaged, 4 cm^{-1} resolution.

From the same Figure 1, peak at 1099 cm^{-1} in polyindole is usually assigned to C-N stretch of aliphatic amine, spectra band 837 $cm^{-1}$is due to vibration of =CH2 out-of-plane wagging vibrations in the polymeric matrix and the band located at 648 cm^{-1} is attributed to the cis CH=CH bending vibration modes; all of which were absent in indole spectra [12]. These changes in peaks positions in indole and polyindole have indicated the polymerisation of the monomer to polyindole.

According to literature [13] there are three main possibilities for the structure of poly(indole) – coupling via the 1, 2, or 3-positions. It is thought that polymerisation does not involve coupling via the benzene ring. But in another work by Mount and co-workers [14], they showed that coupling of 5-substituted indoletrimers takes place at the ring nitrogen. The structure of polyindole is more complex than that of other conductive polymers, such as polypyrrole, because of the different coupling schemes that are

IOP Conf. Series: Materials Science and Engineering **805** (2020) 012007 doi:10.1088/1757-899X/805/1/012007

possible. However, the presence of the N-H stretching band in the PIn spectrum suggests that the 1-position (N) is not involved in coupling [15].

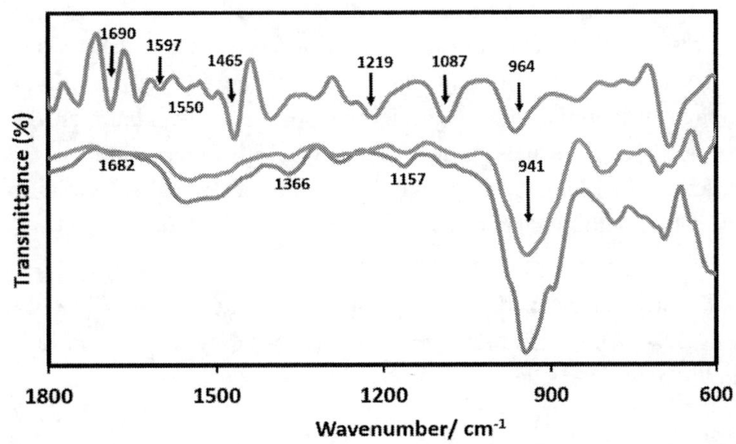

Figure 2. FTIR Transmission spectra of Pln/DNA (orange) vs. controls spectra of DNA (green) and Pln (blue). The spectra are off set for clarity.

Figure 2 compares FTIR transmission spectra of DNA, Pln and Pln/DNA in region between 600 and 1800 cm^{-1}. Several changes observed in the DNA nucleobase region (1200-1700 cm^{-1}) of the spectra upon DNA/Pln formation indicate that the polymer also interacts with the nucleobases. For instance, peaks at 1219, 1465, 1550, 1597 and 1690 cm^{-1} present in free DNA were absent in Pln/DNA spectra.

Previous works [16–17] have shown that templating conductive polymer on DNA relies on the non-covalent interaction of the nascent polymer chains with the template. Watson *et al.* [18] states that the characteristic bands from the DNA structure are still apparent in the FTIR transmittance spectrum of DNA/iron oxide after sequential treatments with Fe^{3+}/Fe^{2+} ions and NaOH, though several notable shifts in their peak positions and intensities are observed as a consequence of the interactions between the DNA and iron oxide material which take place.

Careful examination of the finger print region reveals that some of the DNA related bands are slightly shifted relative to the pure DNA spectrum as can be seen in Table 1.

Table 1: Selected bands in DNA and PIn/DNA FTIR transmission spectra and their assignments

Wave number in DNA (cm^{-1})	**Wave number in PIn/DNA (cm^{-1})**	**Peak shift (cm^{-1})**	**Assignment** [19-21]
964	941	-23	C-C deoxyribose stretch
1087	1157	+70	symmetric PO_2^- stretch
1219	Not observed	-	Asymmetric PO_2^- stretch
1319	1366	+47	C-N stretch of cytosine/guanine
1465	Not observed	-	In-p lane vibration of cytosine
1550	Not observed	-	stretching vibration of purine ring (N7)
1690	1682	-8	Thymine (C2 = O stretching)

IOP Conf. Series: Materials Science and Engineering **805** (2020) 012007 doi:10.1088/1757-899X/805/1/012007

For example the C – C deoxyribose stretch, at 964 cm^{-1}; the PO_2^- symmetric stretches of the phosphate backbone at 1087 cm^{-1}; the asymmetric PO_2^- stretch at 1219 cm^{-1} and C-N stretch of cytosine/guanine at 1319 cm^{-1} shows a major increase in their intensities and shifted by -23, +70 and +47 cm^{-1} to lower and higher frequencies respectively. The peak, centred at 1091 cm^{-1} comprises several unresolved components. The rich chemical functionality presented by the DNA offers several types of sites that allow it to bind a range of materials. The anionic phosphodiester backbone and the nucleobases of the nucleosides for instance provide routes to seed the DNA with other chemical species. The negatively charged backbone can bind cationic species via electrostatic interactions, whilst the nucleobase groups are capable of coordinating with metal cations.

It can be concluded that changes in the peak position and intensity of FTIR transmittance spectrum indicated that the PIn/DNA sample is not a simple mixture of DNA and PIn but rather an intimate mixing of DNA with PIn in the hybrid polymer.

3.2 UV-vis Characterisation

The UV-vis absorption spectra of PIn, free DNA, PlnDNA and PIn/DNA - DNA recorded in aqueous solution are shown in Figure 3. The UV absorption spectrum of a freshly prepared solution of Pln shows the characteristic sharp and broad n–π* andπ–π* absorption peaks at 216 and 274 nm respectively, typical of indole and both correspond to the polymer chain transition [22].

Free DNA has a characteristic broad absorption band at 260 nm and compared to the optical spectra of PIn, this absorption peak appears to be broader and slightly shifted to higher wavelength at 280 nm in PIn/DNA solution, a combination of these peaks due to the presence of some unreacted monomer is observed. In addition, the PIn and PIn/DNA spectra have a peak at about 273 nm and substantial tails extending into the visible region; these aspects of the spectra are very similar to previous reports for polyindole. These features are not present in the monomer but are characteristic of this type of polymer and originate from the extended conjugation and the midgap states in the doped polymer formed during oxidative polymerisation [23].

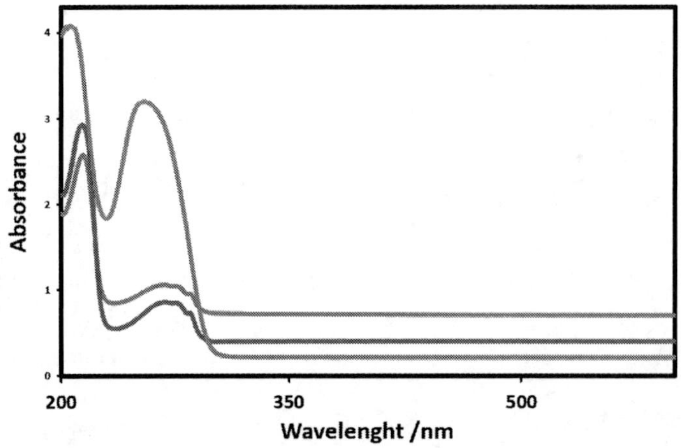

Figure 3. UV-vis absorption spectra at different stages of the synthesis process: absorption spectra of aqueous 6.24 mM Pln (red); free DNA (green) and PlnDNA (blue)

IOP Conf. Series: Materials Science and Engineering **805** (2020) 012007 doi:10.1088/1757-899X/805/1/012007

Previous studies have shown that, π-π^* bands of polymers in different solvents also show a hypsochromic shift with increase in dielectric constant of the solvent and the exciton band due to the inter/intra chain charge transfer is because of absorption from the highest occupied molecular orbital (HOMO) band centered on the benzenoid units to the lowest unoccupied molecular orbital (LUMO) band centered on the quinoid units [24-25]. In conclusion, it can be said that Pln/DNA is supramolecular structure containing DNA and Pln.

3.3 X-ray Photoelectron Spectroscopy studies

XPS survey spectra of Pln/DNA samples (Figure 4) revealed the presence of the elements C, N, O, Cl (which originated from the oxidant used or from $MgCl_2$ that was used in the preparation), and (weakly) P. Presence of DNA in the sample material was obtained from the P2p signal at 133.2 eV, arising from the phosphorus in the phosphodiester backbone of the DNA. Iron was not observed in the survey spectra which confirm that the $FeCl_3$ was used only to drive the polymerisation without any oxidative damage to DNA.

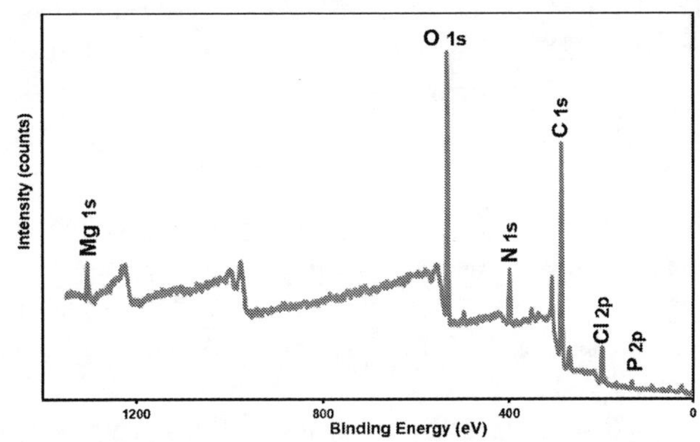

Figure 4. XPS Survey scan chart of Pln/DNA nanowire at pass energy of 20 eV and the step size of 0.3 eV. Some higher resolution spectra were recorded with pass energy of 40 eV and a step size of 0.1 eV. The N1s and C1s spectra of PIn/DNA samples examined in this study are presented in Figure 5a and b respectively.

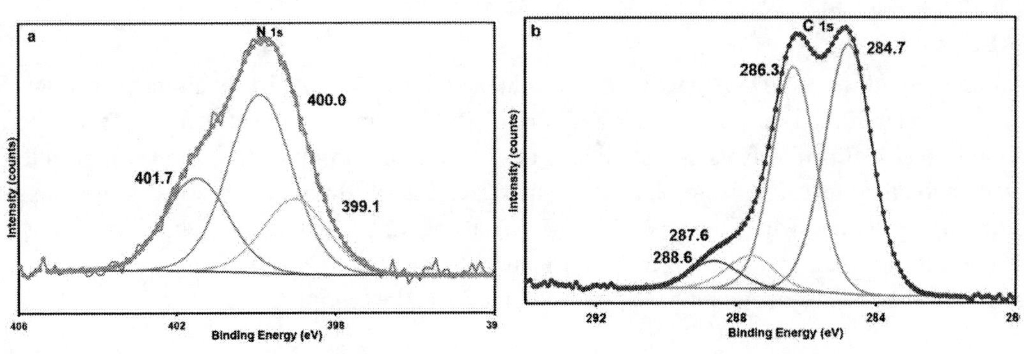

Figure 5. a) N_{1S} XPS Spectrum and b) C_{1S} XPS Spectrum of PlnDNA nanowires.

Three main peaks are observed in the N1s spectrum of PIn/DNA (401.7, 400.0 and 399.1 Ev). Previous workers have observed two components in the N1s spectra of DNA alone: the lower binding energy component at 398.8 eV and another at 400.6 eV. The lower energy peak is attributed to neutral nitrogen atoms. The component at 400.0 is observed for peptidic N atoms and is also typical of the N atoms in the pyrimidines of DNA [26].

While it is likely that the two lowest energy peaks in the N1s spectra of PIn/DNA contain similar contributions from the DNA, there will also be a contribution to these features from the N atoms of polyindole according to the extent of protonation and doping level. This is in agreement with previous reports on polyindole [27] in which it was proposed that the higher binding energy components are due to nitrogen atoms that bear a positive charge and are in chemically or structurally in equivalent environments.

The N1s peaks at 399.1 and 400.0 eV are typical of both polyindole and DNA. However, the third feature at 401.7 eV in the spectrum is definitely not present in the N1s spectrum of DNA and can be assigned to polyindole alone. N1s spectra of polyindole have similarly been reported to contain a major peak at 400.6 eV and a higher binding energy signal at 402.5 eV; these were attributed to ($-NH^+$ (polaron)) and ($=NH^+$(bipolaron)), respectively [28]. Therefore, the third component in PIn/DNA samples with a binding energy of 401.7 eV is assigned to positively charged nitrogen atoms associated with the charge carriers in PIn.

The C1s spectrum can be resolved into four different components centred at 284.7, 286.3, 287.6, and 288.6 eV. Four components have also been observed in C1s spectra of pure DNA by previous workers, although the peak at 284.9 eV is dominant in pure DNA. Two of the lowest binding energy components are assigned to C–H, C–C, and C–N species from PIn and DNA. The third peak at 287.6 eV is attributed to carbons in functional groups of the type, C=N or $C-N^+$, mainly from PIn because this feature is more intense relative to the 286.3 and 284.7 eV components in PIn/DNA samples than in pure DNA. Since the fourth peak at 288.6 eV is comparable to that at 284.7 eV in these samples, but is very weak in pure DNA, it is assigned to $C=N^+$ carbons in PIn. In general, the XPS spectra of PIn/DNA show large contributions from species assigned to Pln.

3.4 Morphological characterisation of Pln/DNA nanowires

3.4.1 AFM studies

The morphology of the PIn/DNA nanowires was characterized using AFM by aligning them on Si/SiO_2 surface. The water diluted images revealed network of DNA bundles with some rope like structures. The formation of well-defined stable DNA structures on the surface, toroid's, rods, bended molecules and so on is well studied by researchers, for example, Japaridze *et al.* [29] always observed shortening of DNA molecules upon drying, when they deposited DNA using divalent cations, which was consistent in magnitude with a partial B- to A-form conformational transition [29]. But in our work, both $MgCl_2$ addition and substrate pre-treatment have effects on the structures formed.

The mechanism of the network formation in DNA is still not yet understood, although it has been noticed that the formation depends on variety of factors, for instance, type and concentration of DNA and counterions in the solution, surface properties and network treatment after formation. Proposed mechanisms for network formation from cyclic DNA include overlapping of the circles where divalent charged positive ions are serving as salt bridges not only between it and the surface, but also between its chains. A similar mechanism for network formation by overcrossing of linear DNA chains is proposed to involve double strand splitting and triple strand organisation, while junction formation has been observed in short DNA networks of short strands [30-31].

The effective deposition and alignment of DNA - based nanowires onto surfaces is highly dependent upon the hydrophobicity of the substrate surface; hydrophilic surfaces such as freshly cleaved mica and silicon dioxide have been proven to cause nanowires to coil up or align in random directions. When the Pln/DNA nanowires were imaged on treated Si/SiO_2 (Figures 6a-d) nanostructures were clearly seen which show a regular and smooth structure with complete and continuous coverage of the duplex DNA by the polymer material. The nanostructures have shown relatively similar morphologies to other polymer nanowires prepared using DNA as a template [32-33] .

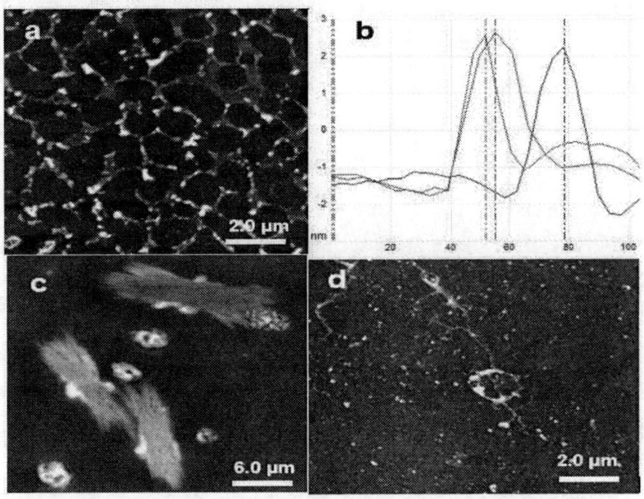

Figure 6 Pln/DNA AFM height images: (a) hexagonal network of Pln/DNA films (b) AFM height profile of the aligned nanowires(c) dense Pln/DNA film (d) thread like Pln/DNA with branches and some spherical shape.

Although complete, regular and smooth appearance of the Pln/DNA nanowires was achieved at shorter reaction time in comparison with the earlier mentioned examples of polymer nanowires, a large number of agglomerated materials were also observed on the substrate surface. Santos *et al.* [34] explained the mechanism that leads to height formation in ambient AFM and employed them to describe how the apparent height of DNA molecules is affected by their local hydrophilicity/hydrophobicity in relation to the supporting surface.

Further statistical analysis of the 150 Plm/DNA nanowires (figure 7) size have indicated the dominance of 3-4 nm diameter range of wires. Larger structures with diameter greater than 13 nm were also recorded indicating the continuation of the polymerisation reaction after the nanowires formation.

IOP Conf. Series: Materials Science and Engineering **805** (2020) 012007 doi:10.1088/1757-899X/805/1/012007

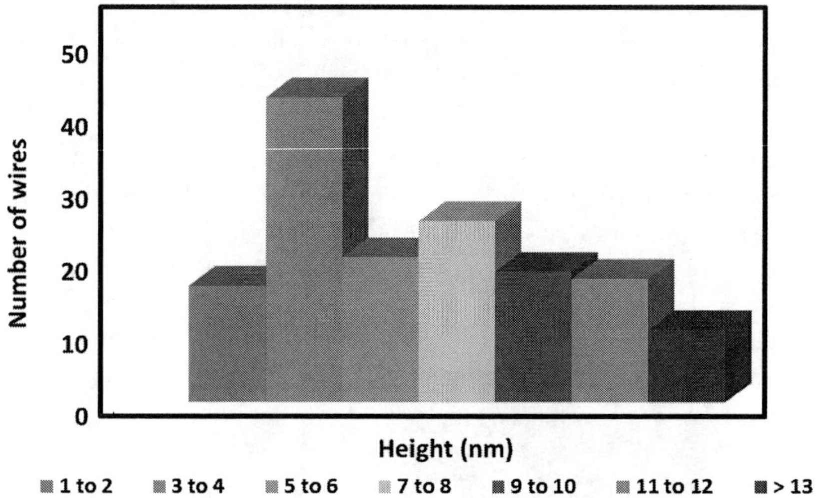

Figure 7. A histogram displaying the heights of 150 Pln/DNA nanowires

3.4.2 TEM Studies

In Figure 8 Pln/DNA composite TEM images were displayed showing the mesh/network like morphology (a-b) and floccule-like structures with semi-spherical white un-templated polymers (c-d), all the images showed the DNA-templated Pln/DNA nanowires to be one dimensional structure with different thickness as seen in their AFM images. As can be seen the surface of the nanowires structures were not all smooth, which indicted characteristic structures that can provide large surface that may enhance the performance of the conductive network formation for charge transport when the particles are compacted and can be good for gas sensing test.

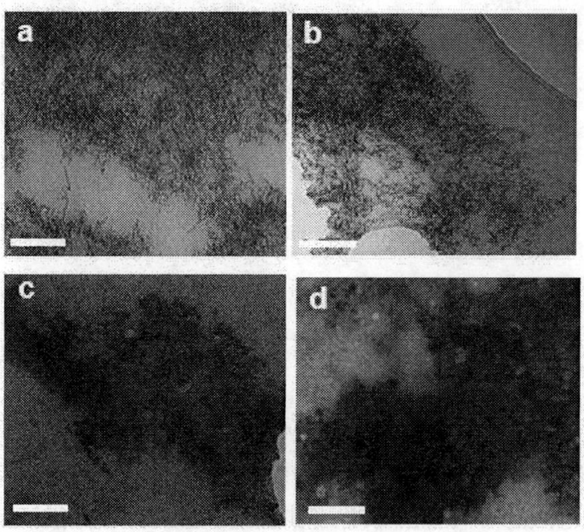

Figure 8. Pln/DNA TEM images: (a-b) Pln/DNA nanowires network (c-d) images showing floccule-like structure with semi-spherical white un-templated polymers. Scale bar = 100 nm

3.4.3 SEM Studies

From Previous studies, bare DNA SEM images reveals a thick ordered film of the material while in the diluted form the images have a network like shape with white spot of some undiluted DNA materials, which may be due to the DNA being more free in the diluted form that network of the free DNA can be seen clearly in contrast to the tick film in the undiluted form [34].

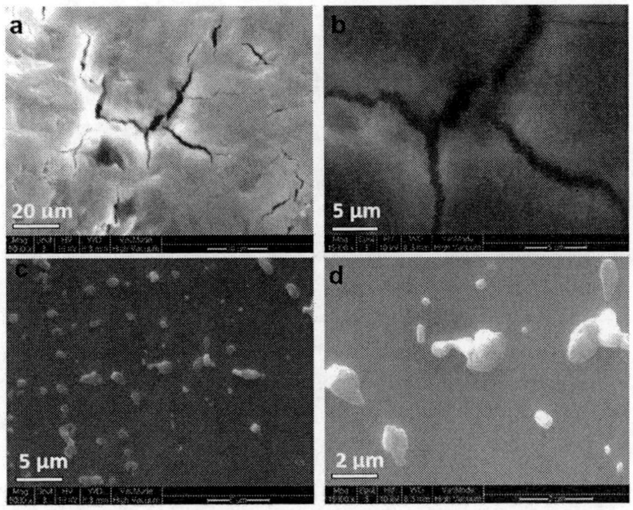

Figure 9. Free and water diluted Pln/DNA sputtered with gold, SEM images: (a-b) thick films (c-d) individual thick structures of diluted samples. Scale bar = 5 nm.

The diluted Pln/DNA SEM images (Figures 9c-d) show individual thick material structures that are likely because of aggregation its particles.

Mudila *et al.* [35] reported that, when a Lewis acid, adds to the N atom in the indole, it is on C_3 atom of the ring, which in turn affects both the conductivity and polymerisation reaction of the indole compound. In Chhattise *et al.* [36] work, the synthesized polyindole reveals the formation of irregularly shaped particles with size ranging from 0.2 to 6 micron. The high resolution image depicts the presence of spherical particles with smooth surface which they believed are fused together to form chunk like morphology of the compound. The surface morphology of PIn powder sample with 0.6 M of $FeCl_3$ was analyzed by FE-SEM the results shows macro-granular structure believe to be due to the aggregation of small globular structures. The irregular structure of particles has reflects its definite amorphous morphology. This is different from our Pln/DNA images that reveal thick films of the materials [37].

3.5 Electrical Characterization of Pln/DNA nanowires

To test if the device is conductive between electrodes and surface (short-circuit) the probe needles were positioned at points 1-1,1-2 and 1-3 on the micro electrode surface and I-V sweepings in the range -5 to +5 V at a constant temperature of 20 ^{o}C was measured.

IOP Conf. Series: Materials Science and Engineering **805** (2020) 012007 doi:10.1088/1757-899X/805/1/012007

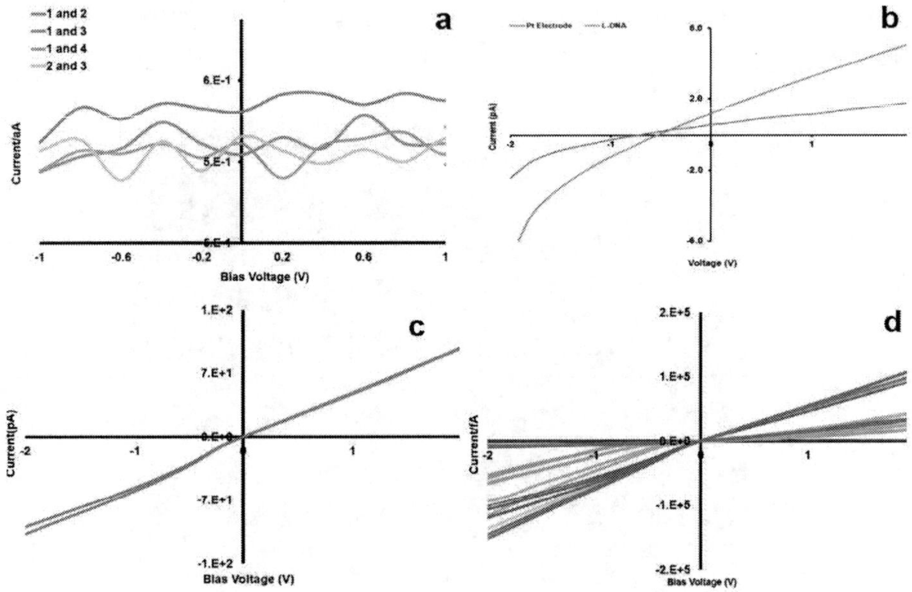

Figure 10. Current–Voltage plots for Pln/DNA: (a) plots acquired by placing the probe needles at different pairs on the microelectrode chip (b) Pt and free lambda-DNA plots as control (c) plots of the sample at room temperature (d) curves of a two-point contact Pln/DNA at temperature range from 223 K to 423 K.

Figure 10a shows the I-V curve of the device from the sweeps at different probe needles point, the plots are noisy and show negligible current (<1pA) around zero bias indicating high resistivity. To check the instrument for comparison and proved that the current measured is from the Pln/DNA, measurements were made on a single platinum microelectrode point and free lambda DNA aligned across the Pt electrodes. The result is displayed in Figure 10b; the curves for both the Pt electrode and the lambda DNA displayed Ohmic behaviour in the pA regime at close to zero bias, but the currents are very low and may include contribution from ion transport.

Additionally to test the claim that the Pln/DNA- templated nanowire's bridges are responsible for the electrical conduction, a 2 µL drop of an aqueous solution of Pln DNA-templated nanowires were placed on these electrodes and aligned across the gap between the Pt fingers by molecular combing. Current-Voltage sweepings, in the range of -5 to +5 V at a constant temperature of 20 °C were measured.

As shown in Figure 10c, the Pln/DNA bulk nanomaterials show a significant current when compared to Pt and DNA. The control measurements support the assertion that the Pn/DNA bridges are responsible for the electrical conduction, therefore they have electrical conductivity.

The conductivity of any material changes with temperature and based on the direction of this change the nature of this material can be identified. The conductivity of an inherent semiconductor increases with increasing temperature, because more valence electrons are exited into the conduction band, whereas it decreases in the case of metals (because of electron-phonon scattering) but in polymers the situation is different.

Current-voltage studies over a range of temperatures were performed to provide useful qualitative and quantitative information regarding I-V behaviour and to elucidate details of conduction mechanism. This method relies on the alignment of the nanowires by molecular combing across two micro fabricated Pt electrodes on a thermally oxidized Si chip. Variable-temperature I-V studies of the two-terminal device were performed over a temperature range of 223 to 423 K in order to elucidate details of the conduction mechanism.

Current – voltage curves of Pln/DNA mesh of nanowires device were recorded under nitrogen at sequence of temperatures in the range of 223 to 423 K. In Figure 10d the resulting curves are presented. As can be seen the current-voltage curves show linearity at low bias voltages and exhibit a reproducible, linear response at a range of temperature. In addition, displayed an increase in current output with increase temperature, which is same as results of similar measurements on PIn/DNA and PPy/DNA nanowires in other studies [38].

Many researchers have suggested that another way to view the metal/polymer or nanowire/metal system is to consider the charge transport process in the same framework as that proposed for thin films of redox polymers. They behave very differently to inorganic semiconductors because of high doping level therefore; the ion transport cannot be explained by Nernst Plank equation [39].

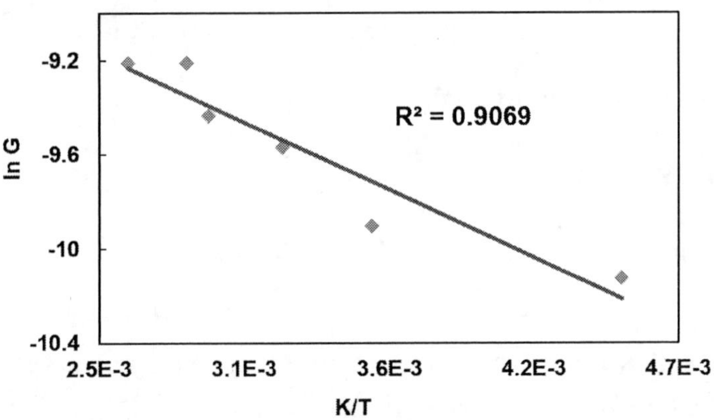

Figure 11. Arrhenius plot for the conductance G of Pln/DNA nanowires.

Figure 11 shows the corresponding Arrhenius plot of the conductance of the Pl/DNA nanowire. The plot explained the temperature dependence of the rate constant and suggests simple electron hopping as the dominant mechanism for electron transport, shown through the exponential behaviour of the conductance when the temperature is increased.

Using equation 1, the activation energy (Ea) for the Pln/DNA nanowires was 1.08 x 10^{-20}J which is equivalent to 6.76 x 10^{-02} + 0.52eV. The band gap of most conductive polymers is of order 3 eV that would predict Ea equal to 1.5 eV if the process were limited by thermal excitation across the gap. Instead, in our studies Ea is <1 eV and lower than similar studies (0.35 ± 0.002 eV) which may be because the process is thermally assisted tunneling between localized sites. Many researchers believed that the conduction pathway and therefore level of conductance in conductive polymers is dependent upon their planarity and anisotropy ratio; the doping percentage; the alignment of the polymer chains; conjugation length and the purity of the sample [40].

4 Conclusion

Polyindole (PIn) has been templated on λ-DNA via oxidative polymerisation of indole using FeCl3 to produce conductive PIn/DNA nanowires. The formation of PIn/DNA nanowires were verified by FTIR, UV–vis and XPS spectroscopy techniques. AFM, SEM and TEM techniques were used to characterize the nanowires dimensions. AFM studies revealed an average height of 1.60 nm for free DNA and the Pln/DNA nanowires have diameters in the range 2–15 nm with the dominance of 3-4 nm mean diameter range. The electrical properties of Pln/DNA nanowires as drop-cast films were investigated by two-terminal current voltage (I-V) measurements on a probe station. The nanowires were drop-cast (5 μL of as-prepared nanowire dispersion) onto platinum microband electrodes (10 μm gap, 10 μm width). The conductance of these films at 20 °C was of the order of 10-100 μS. In addition, the conductance of PIn/DNA nanowires exhibits Arrhenius behaviour (Ea = 0.80 ± 0.06 eV) as a function of temperature. The above results have revealed the potentials of the Pln/DNA nanowire in nanoelectronics applications.

References

[1] Tang N, Jiang Y, Qu H and Duan X 2017 Conductive polymer nanowire gas sensor fabricated by nanoscale soft lithography. *Nanotechnol.* **28** (48) 485301. https://doi.org/10.1088/1361-6528/aa905b.

[2] Le TH, Kim Y and Yoon H 2017 Electrical and electrochemical properties of conducting polymers. *Polym.* **9** (4) pp 150. https://doi.org/10.3390/polym9040150.

[3] Park CS, Lee C and Kwon OS 2016 Conducting polymer based nanobiosensors. *Polym.* **8** (7) pp 249.

[4] Wadatkar NS, and Waghuley SA 2015 Complex optical studies on conducting Polyindole as Synthesized through chemical route. *Egyptian J. Basic Appl .Sci.* **2** (1) pp 19-24.

[5] Wang ZG and Ding B 2013 DNA-based self-assembly for functional nanomaterials. *Adv. Mater.* **25** (28) pp 3905-3914.

[6] Slatko BE, Heinrich P, Nixon BT, and Eckert RL 1993 Preparation of templates for DNA sequencing. *Curr. Protocols Mol. Biol.* **21**(1) pp 7.3.1-7.3.10.

[7] Pruneanu S, Al-Said SA, Dong L, Hollis TA, Galindo MA, Wright NG, Houlton A, and Horrocks BR 2008 self-assembly of DNA-templated polypyrrole nanowires: spontaneous formation of conductive nanoropes. *Adv. Funct. Mater.* **18** (16) pp 2444-2454.

[8] Shirley DA 1972 High-resolution X-ray photoemission spectrum of the valence bands of gold. *Phys. Rev. B* **5** (12) pp 4709.

[9] Chhattise P, Handore K, Horne A, Mohite K, Chaskar A, Dallavalle S, Chabukswar V 2016 Synthesis and characterization of polyindole and its catalytic performance study as a heterogeneous catalyst. *J. Chem. Sci.* **128** (3) pp 467-475.

[10] Watson SM, Hedley JH, Galindo MA, Al-Said SA, Wright NG, Connolly BA, Horrocks BR, and Houlton A 2012 synthesis, characterisation and electrical properties of supramolecular DNA-templated polymer nanowires of 2, 5-(Bis-2-thienyl)-pyrrole. *Chem: Euro J.* **18** (38) pp 12008-12019.

[11] Wasnik HR and Kelkar DS 2015 Study of DC electrical conductivity, magnetic susceptibility and thermal behaviour of pyrrole aniline copolymer. *Mater. Sci. Eng.* **73** (1) pp 012026.

[12] Tyagi G, Charak S and Mehrotra R 2012 Binding of an indole alkaloid, vinblastine to double stranded DNA: A spectroscopic insight into nature and strength of interaction. *J. Photochem. Photobiol. B: Biol.* **108** (1) pp 48-52.

[13] Talbi H, Ghanbaja J, Billaud D and Humbert B 1997 Vibrational properties and structural studies of doped and dedoped polyindole by FTIR, Raman and EEL spectroscopies. *Polym.* **38** (9) pp 2099-2106.

[14] Chang G, Yang L, Liu S, Luo X, Lin R, and Zhang L 2014 Synthesis of indole-based functional polymers with well-defined structures via a catalyst-free C-N coupling reaction. *RSC Adv.* **4** (58) pp 30630-30637.

[15] Roy S, and Chakraborty C 2019 Nanostructured metallo-supramolecular polymer-based gel-type electrochromic devices with ultrafast switching time and high colouration efficiency *J. Mater. Chem.* C **7** (10) pp 2871-2879.

[16] Cheng CC, Wang JH, Chuang WT, Liao ZS, Huang JJ, Huang SY, Fan WL, and Lee DJ 2017 Dynamic supramolecular self-assembly: Hydrogen bonding-induced contraction and extension of functional polymers. *Polym. Chem.* **8** (21) pp 3294-3299.

[17] Dovbeshko GI, Gridina NY, Kruglova EB, and Pashchuk OP 2000 FTIR spectroscopy studies of nucleic acid damage. *Talanta* **53** (1) pp 233-246.

[18] Watson SM, Mohamed HD, Horrocks BR, and Houlton A 2013 Electrically conductive magnetic nanowires using an electrochemical DNA-templating route. *Nanoscale* **5** (12) pp 5349-5359.

[19] Whelan DR, Bambery KR, Heraud P, Tobin MJ, Diem M, McNaughton D, and Wood BR 2011 Monitoring the reversible B to A-like transition of DNA in eukaryotic cells using Fourier transform infrared spectroscopy. *Nucleic Acids Res.* **39** (13) pp 5439-5448.

[20] Marty R, N'soukpoé-Kossi CN, Charbonneau D, Weinert CM, Kreplak L, and Tajmir-Riahi HA 2009 Structural analysis of DNA complexation with cationic lipids. *Nucleic Acids Res.* **37** (3) pp 849-857.

[21] Taillandier IE and Liquier J 1992 Infrared spectroscopy of DNA. *Methods Enzymol.* **211** pp 307-335.

[22] Bhagat DJ and Dhokane GR 2015 Electro-optical properties of one pot synthesized polyindole in the presence of poly(vinyl acetate). *Electr. Mater. Lett.* **11** (3) pp 346-351.

[23] Trivedi MK, Branton A Dahryn T, Nayak G Saikia G and Jana S 2015 Physical and structural characterisation of biofield treated imidazole derivatives. *Nat. Prod. Chem. Res.* **3** pp 5.

[24] Huang J, Virji S, Weiller BH, and Kaner RB 2003. Polyaniline nanofibers: Facile synthesis and chemical sensors. *J. Am. Chem. Soc.* **125** (2) pp 314-315.

[25] Wetzel C, Vogt A, Rudnick A, Mena-Osteritz E, Köhler A, and Bäuerle P 2017 Thiophene–Pyrrole containing *S*, *N*-Heteroheptacenes: Synthesis, and optical and electrochemical characterisation. *Org. Chem. Front.* **4** (8) 1629-1635.

[26] Gong P, Lee CY, Gamble LJ, Castner DG and Grainger DW 2006 Hybridization behavior of mixed DNA/alkylthiol monolayers on gold: characterization by surface plasmon resonance and ^{32}P radiometric assay. *Anal. Chem.* **78** (10) pp 3326-3334.

[27] Talbi H, Maarouf EB, Humbert B, Alnot M, Ehrhardt JJ, Ghanbaja J, and Billaud D 1996 Spectroscopic studies of electrochemically doped polyindole. *J. Phys. Chem. Solids* **57** (6-8) pp 1145-1151.

[28] Wang X, Wang M, Ju X, Zhang Y, Zhao J, and Zhang J 2016 Poly(indole-6-carboxylicacid)/Carbon composite as the matrix for the immobilization of the phenanthroline metal complexes, and the use of the composites as electrocatalyst for oxygen reduction reaction. *Int. J. Electrochem. Sci.* **11** (3) pp 1960 -1974.

IOP Conf. Series: Materials Science and Engineering **805** (2020) 012007 doi:10.1088/1757-899X/805/1/012007

[29] Japaridze A, Vobornik D, Lipiec E, Cerreta A, Szczerbinski J, Zenobi R, and Dietler G 2016 Toward an effective control of DNA's submolecular conformation on a surface. *Macromol.* **49** (2) pp 643-652.

[30] Ding Y, Chai J, Centrella PA, Gondo C, DeLorey JL, and Clark MA 2016 Development and synthesis of DNA-encoded benzimidazole library. *ACS Comb. Sci.* **20** (5) pp 251-255.

[31] Doi M and Edwards SF 1988 *The Theory of Polymer Dynamics*, Vol. 73. Clarendon Press, Oxford,

[32] LeBlanc S, Wilkins H, Li Z, Kaur P, Wang H, and Erie DA 2017 Using atomic force microscopy to characterize the conformational properties of proteins and protein–DNA complexes that carry out DNA repair. *Methods Enzymol.* **592** pp 187-212.

[33] Buechner CN and Tessmer I 2013 DNA substrate preparation for atomic force microscopy studies of protein–DNA interactions. *J. Mol. Recog.* **26** (12) pp 605-704.

[34] Santos S, Verdaguer A and Chiesa M 2012 The effects of adsorbed water layers on the apparent height of nanostructures in ambient amplitude modulation atomic force microscopy. *J. Chem. Phys.* **137** (4) pp 044201.

[35] Mudila H, Prasher P, Kumar M, Kumar A, Zaidi MG and Kumar A 2019 Critical analysis of polyindole and its composites in super capacitor application. *Mater. Renew. Sustain. Energ.* **8** (2) pp 9.

[36] Chhattise P, Handore K, Horne A, Mohite K, Chaskar A, Dallavalle S and Chabukswar V 2016 synthesis and characterization of polyindole and its catalytic performance study as a heterogeneous catalyst. *J. Chem. Sci.* **128** (3) 467-475.

[37] Phasuksom K and Sirivat A 2016 Synthesis of nano-sized polyindole via emulsion polymerization and doping. *Synthetic Metals* **219** pp 142-153.

[38] Dong L, Hollis T, Connolly BA, Wright NG, Horrocks BR, Houlton A 2007 DNA-templated semiconductor nanoparticle chains and wires. *Adv. Mater.* **19** (13) pp 1748-1751.

[39] Heeger J 2001 Nobel Lecture: Semiconducting and metallic polymers: The fourth generation of polymeric materials. *Rev. Modern Phys.* **73** (3) pp 681-700.

[40] Gojny FH, Wichmann MH, Fiedler B, Kinloch IA, Bauhofer W, Windle AH and Schulte K 2006 Evaluation and identification of electrical and thermal conduction mechanisms in carbon nanotube/epoxy composites. *Polym.* **47** (6) pp 2036-2045.

Synthesis and characterization of nanostructured DNA-templated polyimidazole nanowire

M I Yahaya
Department of Chemistry, Sokoto State University, Nigeria

Email: ymansur64@gmail.com

Abstract

In this research work, nanostructured DNA-templated polyimidazole nanowires synthesized using simple and low-cost fabrication method that employed DNA as a template on which to carry out the polymerisation was reported. Their chemical properties were probed using different spectroscopic techniques (Ultra-Violet Visible (UV-vis) spectroscopy, X-ray Photoelectron Spectroscopy (XPS) and Fourier Transform Infrared (FTIR) spectroscopy). Atomic force (AFM) and electron microscopy (SEM and TEM) were used to characterize the nanowire dimensions. The techniques proved the formation of a supramolecular hybrid polymer containing DNA and imidazole polymers. The electrical property was measured using probe station. Morphological results exhibit different morphologies with agglomerate nanostructures for the diluted film, while dense rope like network of nanostructures were observed for undiluted templated polymers with diameter of 2-13 nm. Activation energy (E_a) including the uncertainty on it for the Plm/DNA bulk nanowires calculated was $10.6 \pm 0.5 \times 10^{-3}$ J mol^{-1} equivalents to 0.110 ± 0.005 eV which is not surprising because the process is thermally assisted tunneling between localized sites.

Keywords: Nanowire, polymer, template, DNA, Characterization

1. Introduction

Conductive polymers and their derivatives have continuously been used as active layers in nanoelectronics devices. Composites of conjugated polymers are getting increasingly utilized in organic electronics devices, thanks to the electronic and optical properties of the polymers, which are like those of semiconductors. However, the mechanical properties and processability of polymers makes them more attractive than crystalline inorganic semiconductors for future applications: their light weight, flexibility and inexpensive preparation procedures are drawing the attention of many researchers [1-2].

IOP Conf. Series: Materials Science and Engineering **805** (2020) 012008 doi:10.1088/1757-899X/805/1/012008

They have lots of advantages over their metal oxides counterparts that operate at higher temperatures [3]. They are easy to synthesize by different electrochemical and chemical methods and copolymerisation/structural derivations can refine their molecular chain structure. The formation of charge carriers upon oxidizing or reducing their conjugated backbone is responsible for their inherent conductivity [4].

One dimension (1D) nanomaterials have high surface to volume ratios resulting in higher sensitivity, straight conduction pathways, fast electron transport and adjustability of their surface modification [5-6].

Electrochemical and electronic devices are easier to manufacture and miniaturize than optical ones because they can use microelectronic fabrication techniques [7-9]. This motivated the present research that synthesized polyimidazole DNA-templated (Plm/DNA) nanowires through low cost chemical method and their chemical, morphological and electrical characterizations.

2. Experimental section

2.1 Materials

All chemicals were purchased from Sigma-Aldrich Company Ltd. Lambda DNA (λ- DNA, Cat no. N3011S) was purchased from New England Biolabs (UK) Ltd. n-Si (100) wafers were purchased from Compart Technology Ltd. Platinum microband electrodes were supplied by Windsor Scientific Ltd (Pt-MB4000) and Copper grids (HC300Cu) for TEM were purchased from EM Resolutions Ltd. UK. Deionised water was obtained from a NANOpure Diamond Life Science ultrapure water system equipped with a DIamond RO Reverse Osmosis System.

2.2 Substrates Cleaning

The n-Si (100) wafers was cut into ~1 x 1 cm^2 pieces using a diamond scribe and immersed in acetone for 50 min. The chips were then rinsed with copious deionised water and dried in a gentle stream of N2 gas before further drying in an oven for 5 - 10 minutes. Final cleaning with FEMTO low pressure plasma system, using oxygen plasma oxidation (90W, 15 sccm, 15 min) at 40% power was carried out.

2.3 Chemical Synthesis of PIm/DNA Nanowires

DNA-templated Polyimidazole nanowires were chemically synthesized using $FeCl_3$ as an oxidant. Typically, 5 μL of freshly prepared imidazole ($C_3H_4N_2$) (solution (3 mM) was added to 20 μL of λ-DNA (500 ng μL^{-1}) in the presence of 5 μL of $MgCl_2$ (0.5 mM), then 5 μL of $FeCl_3$ (1 mM) was added drop-wise into this solution. The solution was thoroughly mixed and allowed to react for at least 2 h at room temperature prior to analysis [10].

IOP Conf. Series: Materials Science and Engineering **805** (2020) 012008 doi:10.1088/1757-899X/805/1/012008

2.4 Samples Preparation for Fourier Transform Infrared Spectroscopy (FTIR)

For all FTIR measurements in this work about 5μL of the prepared Plm/DNA nanowires films were drop-cast on clean Si (100) substrate and left to dry for 1h prior to measurements. FTIR spectra were acquired using a Shimadzu FTIR (IRAffinity-1S). Spectra were recorded in the range 400–4000 cm^{-1}, with 128 scans at 4 cm^{-1} resolution and a clean Si (100) substrate was used as a background.

2.5 Samples Preparation for Ultraviolet–Visible Spectroscopy (UV-vis)

For UV-vis measurements, 2 mL of aqueous CT-DNA (162.5 ng μL^{-1}) solution was mixed with 0.5 mL freshly prepared imidazole solution in the presence of 0.5ml $MgCl_2$ (0.5 mM). Then, 0.5 mL of $FeCl_3$ (1 mM) was added drop-wise to the solution. The mixture was stirred and allowed to react at room temperature for 1 h. UV-vis absorption spectra were recorded on a CARY 100 BIO spectrophotometer at room temperature. Water was used as the background. The absorbance values reported have been scaled to a standard path length of 1 cm by the instrument software.

2.6 Samples Preparation for X-ray Photoelectron Spectroscopy (XPS)

Samples were prepared for XPS by drop-casting ~20 μL of solution onto a clean Si(100) substrate and then left to dry in air at room temperature in a laminar flow hood. A Theta Probe photoelectron spectrometer was used to collect photoemission spectra. The binding energies obtained in the XPS analysis were calibrated using Carbon (284.8 eV) as a reference. Photoemission spectra were fitted with a combination of mixed singlet components using the CasaXPS software version 2.3.16 from Casa Software Ltd. Backgrounds were subtracted by the Shirley method [11].

2.7 Samples Preparation for Transmission Electron Microscopy (TEM)

Samples (5 μL) were cast dropped on copper grids and placed on a laminar flow hood to dry for 5 h. Transmission electron microscopy experiments in this research were carried out with Philips CM100 TEM (FE1) using Holey carbon films in a mesh of copper grids as substrate and camera images were acquired with AMT CCD (Deben).

2.8 Samples Preparation for Scanning Electron Microscopy (SEM)

Five microlitres (5 μl) of samples were cast dropped on treated n-Si (100) wafers. The deposited samples were dried in a C-Flow vertical unidirectional over night before SEM analysis. The SEM analysis was carried out using JEOL JSM-5610LV Scanning Electron Microscope and some samples were coated with few nanometers of gold film using sputter coater machine.

IOP Conf. Series: Materials Science and Engineering **805** (2020) 012008 doi:10.1088/1757-899X/805/1/012008

2.9 Electrical Measurements

Two-terminal conductivity measurements were conducted using Pt microband electrodes deposited on clean silanized Si/SiO_2 chips. A 2 μl drop of an aqueous solution of Pln/DNA nanowires was placed on these electrodes and aligned across the gap between the Au fingers by molecular combing. A Cascade Microtech 110008-M probe station with Agilent B1500A Semiconductor Device Analyzer controlled by B1500A's EasyEXPERT software was used for the measurements. Prior to electrical testing, the two-terminal device undergoes a heating/cooling cycles from 223 to 423 K by using a heating/cooling chuck, under nitrogen gas supply to drive up any water bound to the Pln/DNA nanowire. All the electrical measurements were carried out under dry nitrogen without illumination in the shielded sample compartment of the probe station.

Current-voltage measurement from 223 to 423 K was performed on the probe station using a thermal chuck system. For each of the electrical tests, the current was measured for applied voltages from -2 to 2 V in steps of 0.2 V.

3. Results and Discussion

3.1 FTIR characterization of PIm/DNA nanowires

Figure 1 display the joined IR spectra of both Plm and imidazole, in PIm nanostructures spectra, the band located at 1640 cm^{-1} is attributed to N-H bend/ - C = C - and the one at 1527 cm^{-1} is due to C-H stretching in ring. The peak at 1327 cm^{-1} in both Plm and imidazole, are due to C-N stretching in aromatic compounds. In same region, bands at 1527 cm^{-1} in Plm spectra and the one at 1535 cm^{-1} in imidazole are both ascribed to C-C stretching in ring just like the bands at 1640 and 1647 cm^{-1} respectively, are assigned to N-H bending and –C=C- stretching. In contrast, peaks at 1065 and 1258 cm^{-1} were observed in imidazole but disappeared in polyimidazole, which may be due to the polymerization of the imidazole monomer [12-14].

IOP Conf. Series: Materials Science and Engineering **805** (2020) 012008 doi:10.1088/1757-899X/805/1/012008

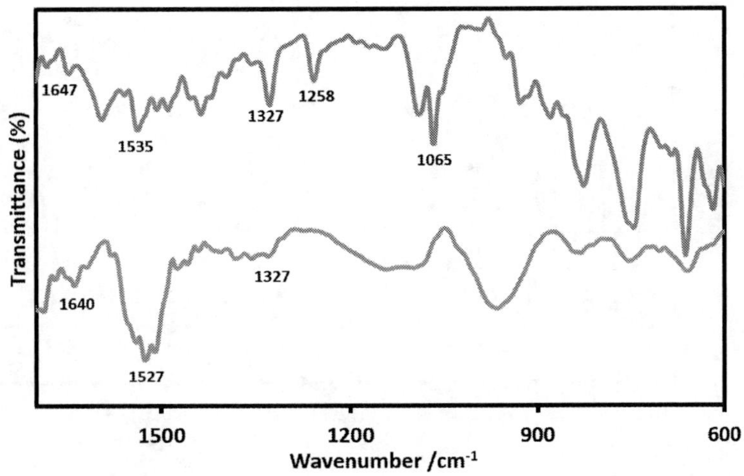

Figure 1. Transmission FTIR spectra of imidazole (blue) and polyimidazole (orange

According to previous studies [15-17] imidazole molecular structure shows the presence of C-H stretching vibrations in the regions greater than 3000 cm^{-1} which is characteristic region for the ready identification of SP^2 C-H stretching vibrations and at this region, the bands are not affected considerably by the nature of the substituents. But an N–H stretching band is observed in the 3500-3200 cm^{-1} region; this position depends upon the degree of hydrogen bonding. They exhibit characteristic band patterns in the 1600-1300 cm^{-1} ring stretching region which pattern is affected by the nature of the substituent. At 800-600 cm^{-1} region the pattern of out-of-plane C–H bending bands are the characteristic of these compounds.

In order to provide evidence that intimate interaction of polyimidazole and DNA chains in the self-assembled nanowires is indicated by FTIR, comparative transmission spectra of Plm/DNA and controls (free DNA and Plm) is shown in Figure 2, the strong sharp peak at 748 cm^{-1} spotted in the imidazole spectrum is characteristic of the N-H wag mode; it becomes broad and shifts to 732 cm^{-1} in Plm/DNA nanowire. The C-C stretch peaks due to imidazole (1527 cm^{-1}) have remained in the Plm/DNA material (1527 cm^{-1}) spectrum, indicating that the hybrid polymer contains both the polymeric materials and DNA [15].

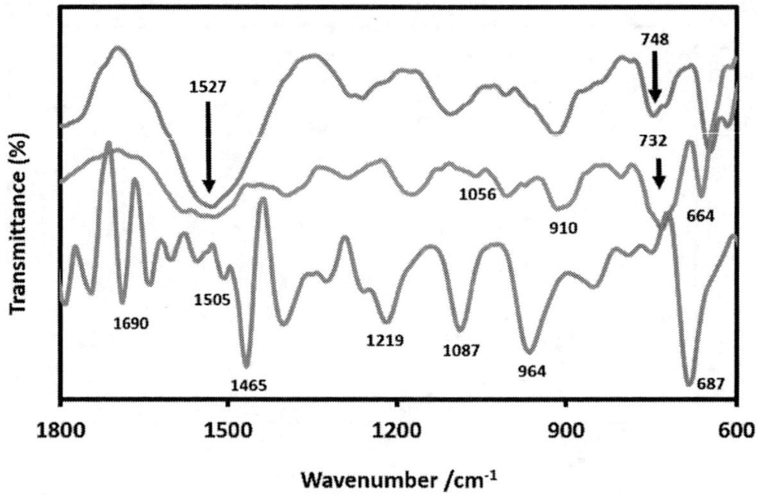

Figure 2. FTIR Transmission spectra of Plm/DNA (orange) vs. spectra of free λ-DNA (green) and Plm (blue). 128 scans co-added and averaged 4 cm^{-1} resolutions. The spectra are offset for clarity.

Finger print region FTIR transmission spectra comparison of DNA and PlmDNA nanowires from same figure above reveals, a secondary amine N-H wag frequency between 700-800 cm^{-1}; C-N stretching between 1100-1200 cm^{-1} and the C=O stretches of cytosine, guanine and thymine between 1600-1800 cm^{-1}. These data demonstrated that there were slight changes in the hydrogen bonding environment between PlmDNA nanowires and DNA samples. The transmittance peaks that were observed from 900-1000 cm^{-1} for both DNA and PlmDNA samples can be assigned to bending vibration of water molecules due to moisture absorption by samples [18].

Table 1. Selected finger print bands in DNA and PIm/DNA spectra and their assignments

Wave number in DNA (cm^{-1})	ve number in PIm/I (cm^{-1})	Peak Shift (cm^{-1})	Assignment [19-20]
687	664	-23	C-C bending vibrations
964	910	-54	C-C deoxyribose stretch
1087	1056	-31	Symmetric PO_2^- stretch
1505	NO	-	In-plane vibration of cytosine and Guanine
1550	1527	-23	Purine stretching (N7)
1690	NO	-	Thymine (C2 = O stretching)
1219	NO	-	Asymmetric PO_2^- stretch
1465	NO	-	In-plane vibration of cytosine

NO, not observed

A more detailed tabulated analysis of the finger print region of DNA and PlmDNA (Table 1) reveals that C-C bending vibrations were assigned to the peaks at 687 and 664 cm^{-1} respectively, while C-C deoxyribose stretch for both DNA and PlmDNA were observed at 964 and 910 cm^{-1} individually. The symmetric PO^{2-} stretch; asymmetric PO^{2-} stretch; in-plane vibration of cytosine and Guanine; Purine stretching (N7); in-plane vibration of cytosine and Thymine C2 = O stretching (1087,1505, 1550, 1464, 1219 and 1690 cm^{-1}) were found to shifts to lower frequencies or absent in the hybrid material (910,1056 and 1527 cm^{-1} and absent).

It can be concluded that from the IR transmission spectra results that some changes in the DNA-related bands to different values in Plm/DNA spectra after the synthesis of the nanowires, proves the interaction of the Plm and DNA during the templating and that the prepared nanowires are not mere mixtures of Plm and DNA molecules which is similar to other studies [21].

3.2UV-vis characterization of Plm/DNA nanowires

Figures 3 show the UV-vis absorption spectra of free DNA, polyimidazole and Plm/DNA. Imidazole derivatives absorb UV light because of the presence of conjugated pi (π) bonding systems (π-π* transition) and nonbonding electron system (n-π* transition). There are certain energy gaps between π-π* and n-π* orbitals. From the figure, it was observed that structural changes occurred in synthesized imidazole nanowires as compared to controls (Plm and free DNA), which were not sufficiently large to alter UV-visible spectral properties in solution state. Polyimidazole DNA-templated nanowires compared to free DNA exhibited both hypsochromic and bathochromic effect in absorbance maxima (λ_{max}) 267 and 290 nm respectively. Whereas, it exhibited only bathochromic shift at the same absorbance maxima (λ_{max}) compared to 277 nm for Plm. These π-π* and n-π* electron transitions are as a results of the presence of conjugated chromophores in the compound [22].

The continuous variation of wavelength and intensity of UV-vis bands may result from the copolymerization effect of polyimidazole with DNA. Put differently, the polymer formed by oxidative polymerization of imidazole with DNA is a complex mixture of two substances rather than the mixture of two homopolymers as shown in our UV-vis results [23]. This is similar to FTIR results.

IOP Conf. Series: Materials Science and Engineering **805** (2020) 012008 doi:10.1088/1757-899X/805/1/012008

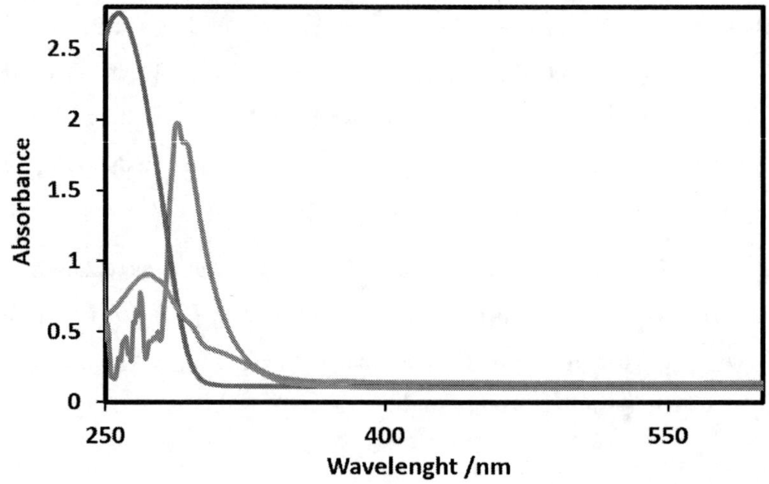

Figure 3: UV- Vis absorption spectra at different stages of the synthesis process: absorption spectra of aqueous 3 mM Plm (green); free DNA (red) and PlmDNA (blue)

3.3XPS characterization of Plm/DNA nanowires

XPS survey spectra of Plm/DNA samples (Figure 4) exhibit the presence of the elements C, N, O, Cl, and (weakly) P. Clear evidence for the presence of DNA in the sample material was obtained from the P2p signal at 133.2 eV, arising from the phosphorus in the phosphodiester backbone of the DNA. No signature of the presence of iron was observed in the survey spectra which confirm that the $FeCl_3$ was used only to drive the polymerization without any oxidative damage to DNA (the Cl observed is likely from $MgCl_2$)

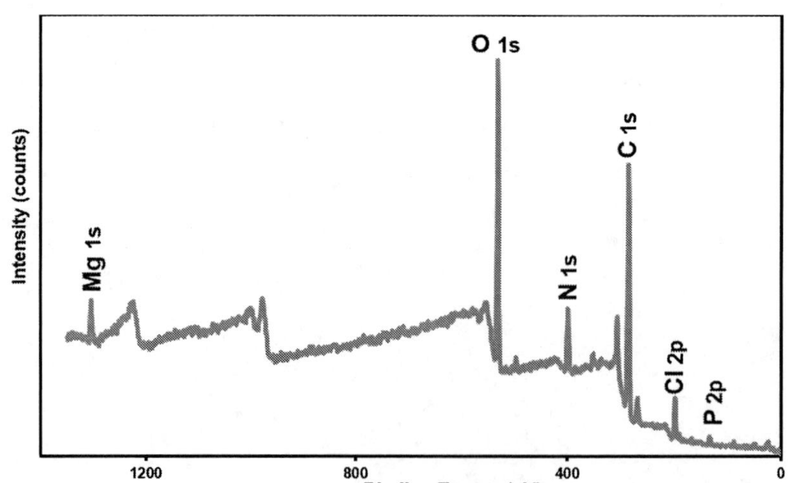

q xza

Figure 4: XPS Survey scan chart of Plm/DNA nanowire at pass energy of 20 eV and the step size of 0.3 eV. Some higher resolution spectra were recorded with pass energy of 40 eV and a step size of 0.1 eV.

The N1s and C1s spectra of PIm/DNA samples examined in this study are presented in Figures 5a and b respectively. Three main peaks were observed in the N1s spectrum of PIm/DNA (399.3, 400.3 and 401.7 eV). Previous researchers have observed two components in the N1s spectra of DNA alone- the lower binding energy component at 398.8 eV and another at 400.6 eV. The lower energy peak is ascribed to neutral N atoms. The component at 400.6 eV is observed for peptidic N atoms and is also typical of the N atoms in the pyrimidines of DNA [24-25].

While it is likely that the two lowest energy peaks in the N1s spectra of PIm/DNA contain similar contributions from the DNA, there will also be a contribution to these features from the N atoms of polyimidazole according to the extent of protonation and doping level. This is consistent with previous reports on polyindole [26] and polypyrrole [27] in which it was proposed that the higher binding energy components are due to nitrogen atoms that bear a positive charge and are in chemically or structurally inequivalent environments. The third component at 401.7 eV in our spectra is not present in N1s spectra of DNA and can be assigned to polyimidazole alone. N1s spectra of polypyrrole have similarly been reported to contain a major peak at 400.6 eV and a higher binding energy signal at 402.5 eV; these were attributed to (- NH^+) and (= NH^+), respectively [28]. Therefore the third component in our PIm/DNA samples, with a binding energy of 401.7 eV, can be assign to positively charged nitrogen atoms associated with the charge carriers in PIm.

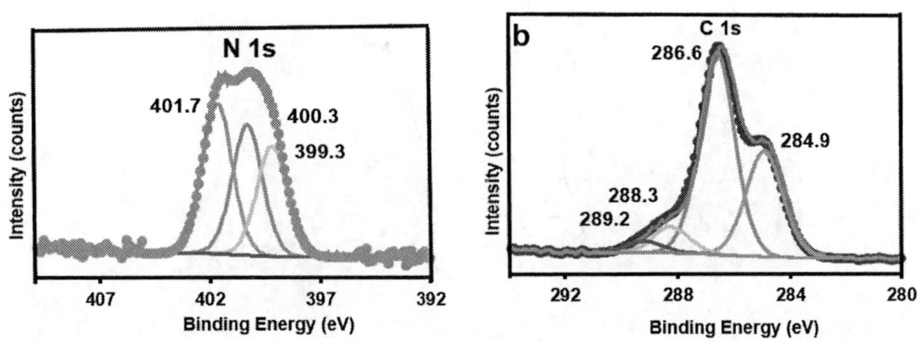

Figure 5. (a) N_{1S} XPS Spectrum and (b) C_{1S} XPS Spectrum of PlmDNA nanowires.

The C1s spectra (Figure 5b) can be resolved into four different components centered at 284.9, 286.6, 288.3, and 289.2 eV. Four components have also been observed in C1s spectra of pure DNA by previous workers, although the peak at 284.4 eV is dominant in pure DNA [29]. The two components at the lowest binding energy were assigned to C-H, C-C, and C-N species from PIm and DNA. The third peak at 287.3 eV is ascribed to carbons in functional groups of the type, C=N or C-N^+, mainly from PIm because this feature is more intense relative to the 284.9 and 286.6 eV components in our samples than in pure DNA.

Considering that the fourth peak at 289.0 eV is comparable to that at 284.9 eV in our samples, but is very weak in pure DNA, it was assigned to $C=N^+$ carbons in PIm.

In general, the XPS spectra of PIm/DNA show large contributions from species assigned to Plm, which indicate that the PIm/DNA nanowires can be visualized as a thick coat of PIm on the DNA template instead of small number of PIm chains wrapping around the DNA while chloride ions make up the required counter charge.

3.4 AFM characterization of Plm/DNA nanowires

Figures 6a-d delineate AFM height images of Plm/DNA nanostructures placed on plasma treated Si/SiO_2 wafers surfaces in this study. The images exhibit different morphologies with agglomerate nanostructures for the diluted film, while dense rope like network of nanostructures were observed for undiluted templated polymers.

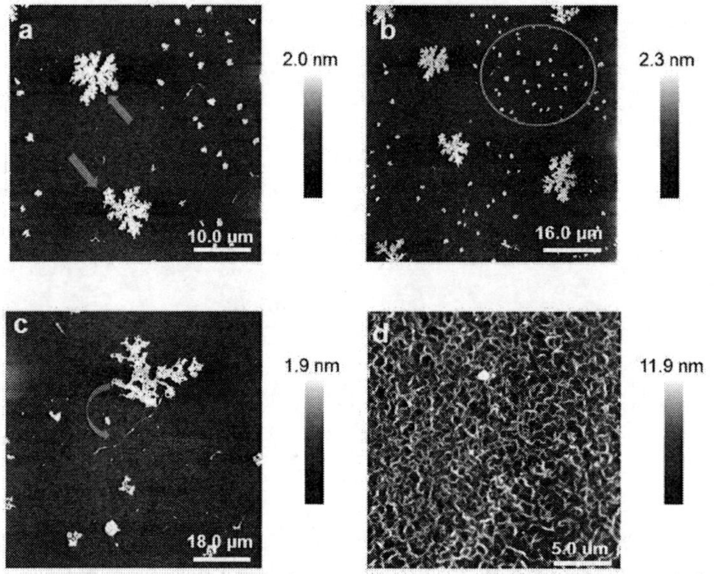

Figure 6. (a) Tapping mode™ AFM height images of Plm/DNA nanostructures aligned on Si/SiO_2 substrate after 24 hours incubation (a) Plm/DNA agglomerate structures (b) Individual spherical shape disperse nanostructures (c) individual wires unfolding from the dense film (d) dense rope like network of undiluted Plm/DNA ropes like nanostructures.

This agglomeration is likely due to either free polymer or the presence of residuals of buffer salts or both. Additional possibility is condensation of DNA strands due to shielding of charges on the phosphate groups of the DNA by the action of metal ions [30-31].

IOP Conf. Series: Materials Science and Engineering **805** (2020) 012008 doi:10.1088/1757-899X/805/1/012008

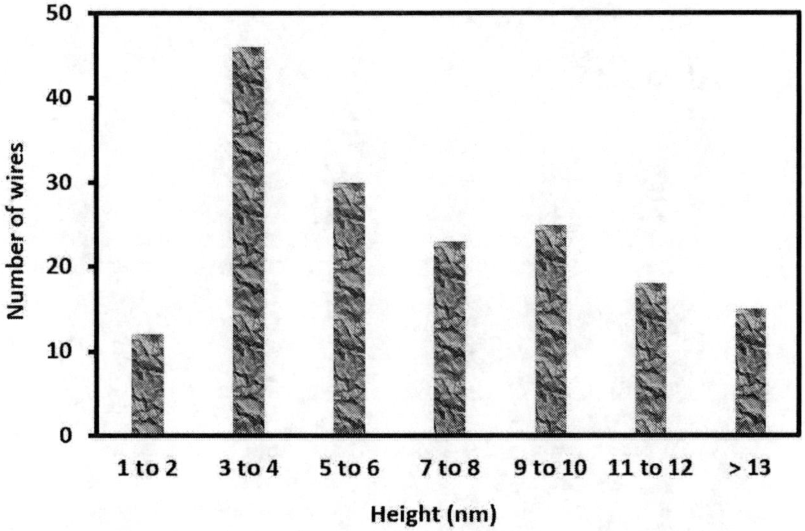

Figure 7. A histogram displaying the heights of 150 Plm/DNA nanowires

Statistical study carried out on 150 Plm/DNA nanowires shows the distribution of the diameters, with a modal diameter of 3-4 nm (Figure 7). Small number of larger structures can be seen in the chart with diameters greater than 13 nm. Previous studies attributed these larger structures to the wrapping/bundling of the Plm/DNA structures into larger rope-like assemblies formed in the solution and concluded that the formation of polymer on DNA takes place initially through the low density of binding spherical-shaped conducting polymer particles on the duplex DNA templates [32].

3.5 TEM characterization of Plm/DNA nanowires

Figure 8 shows the TEM images of Plm/DNA nanowires absorbed on carbon coated copper grid with different morphologies comprising of (a) dense networks, (b-c) align rope/thread like Plm/DNA nanomaterials and (d) DNA-templated nanomaterials bridging the holes in the substrate. These images further revealed the nanowires to be continuous and uniform in the bounding of polymers to the DNA. The images shows similar morphologies as shown in the AFM images, even though they are not as explicit as AFM since DNA is not easily observed by TEM due to a lack of contrast and usually requires a stain to image the nanostructure or using other innovative techniques.

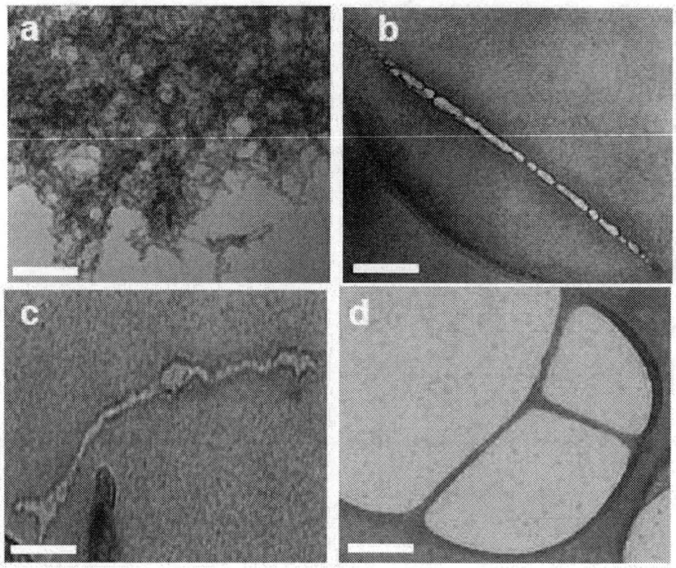

Figure 8: TEM images of Plm/DNA absorbed on copper grid: (a) nanostructures films/network (b) aligned Plm/DNA film showing the joining/bridging of the nanomaterials (c) thread/rope like image (d) TEM image bridging across hole in a copper grid. Scale bar = 100 nm

3.6 SEM Characterization of Plm/DNA Nanowires

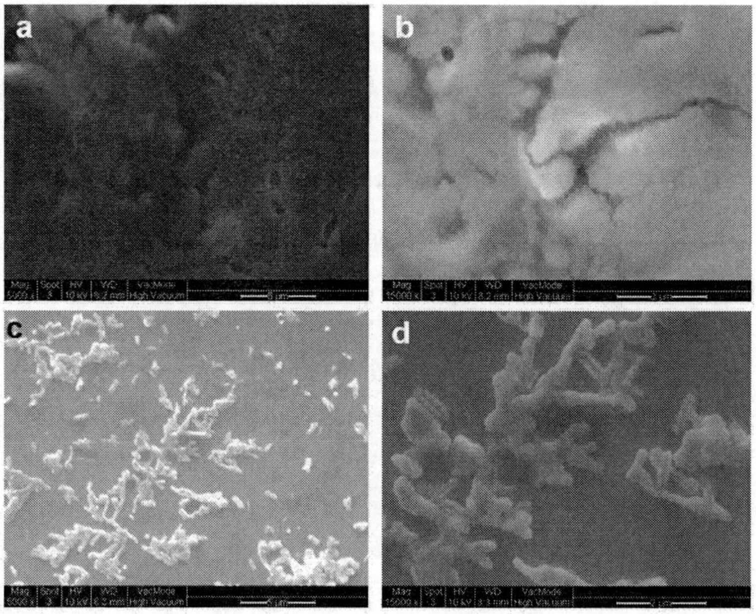

Figure 9: Plm/DNA and water diluted Plm/DNA sputtered with gold, SEM images on Si/SiO_2 surface: (a-b) thick films of the Plm/DNA materials (c-d) fern like network structures of diluted Plm/DNA. Scale bar = 5 nm.

IOP Conf. Series: Materials Science and Engineering **805** (2020) 012008 doi:10.1088/1757-899X/805/1/012008

Like other polymers, the SEM micrographs for the undiluted Plm/DNA in Figures 9(a-b) revealed a dense film of the DNA template polymer composite which may results from the agglomeration of the material. But the diluted Plm/DNA materials are loosely bond, that the SEM images in Figures 9(c-d) indicate a fern like morphologies for the Plm/DNA [33].

3.7 Electrical characterization of Plm/DNA nanowires

In other to investigate the conduction mechanism and electrical behaviour of the Plm/DNA nanomaterials current voltages measurements were carried out using a four terminal probe. Sweeps at different probe needles point, gives noisy and show negligible current (<1pA) around zero bias indicating high resistivity.

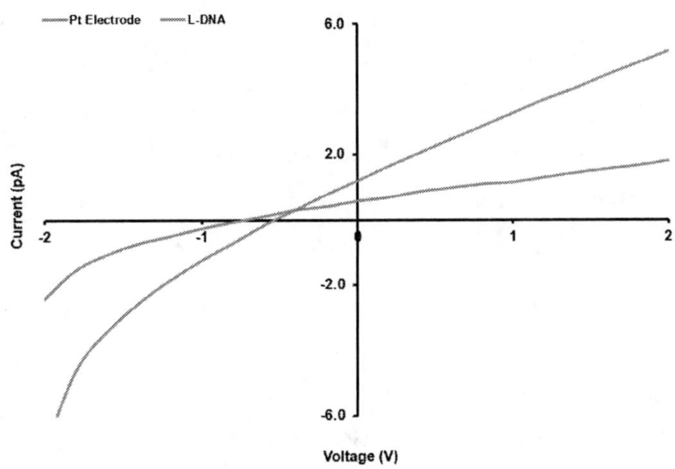

Figure 10. Current -Voltage plots for Pt and free lambda-DNA aligned across Pt electrodes as control

To check the instrument for comparison and proved that the current measured is from the Plm/DNA, measurements were made on a single platinum microelectrode point and free lambda DNA aligned across the Pt electrodes. The result is displayed in Figure 10; the curves for both the Pt electrode and the lambda DNA which displayed Ohmic behaviour in the pA regime at close to zero bias, but the currents are very low and may include contribution from ion transport.

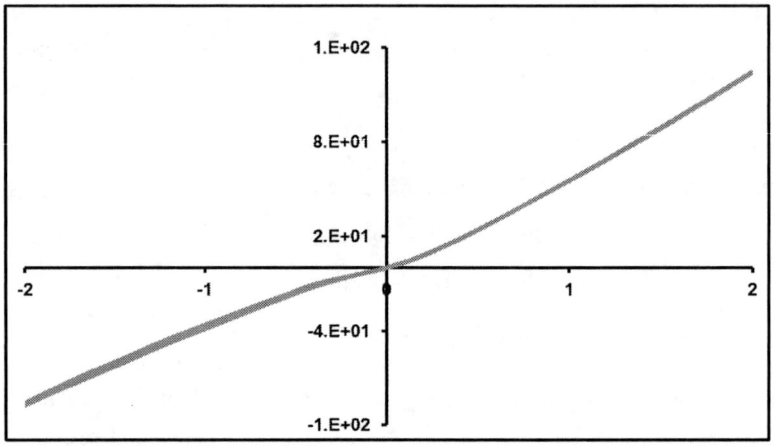

Figure 11. Current-Voltage plots for Plm/DNA-templated nanowires

Additionally to test the claim that the polyimidazole DNA-templated nanowire's bridges were responsible for the electrical conduction. Current-voltage sweepings in the range of -2 to +2 V at a constant temperature of 20 °C were measured. As shown in Figure 11, the polyimidazole DNA-templated bulk nanomaterials shows a significant current value when compared to Pt and DNA. The current voltage control measurements support the assertion that the Plm/DNA bridges are responsible for the electrical conduction, therefore they have electrical conductivity.

Current-voltage measurements over a range of temperatures were performed to provide useful qualitative and quantitative information regarding I-V behaviour and to elucidate details of conduction mechanism. This method relies on the alignment of the nanowires by spreading molecular combing method across two micro fabricated Pt electrodes on a thermally oxidized Si chip. Variable-temperature I-V studies of the two-terminal device were performed over a temperature range of 223 to 423 K to elucidate details of the conduction mechanism.

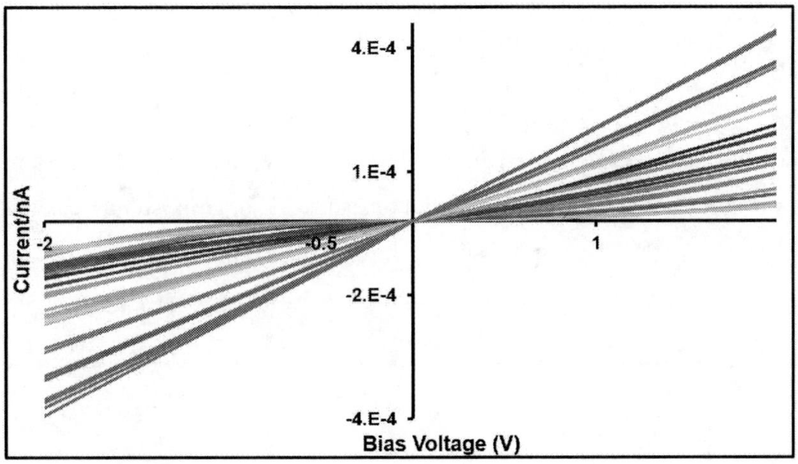

Figure 12. Current-voltage curves of a two-point contact for Plm/DNA mesh of nanowires at a temperature range from 223 to 423 K

In Figure 12, the resulting I-V curves are presented, which shows linearity at low voltages and exhibit a reproducible, linear response at a range of temperatures at low bias and displayed an increase in current output with increase temperature which is consistent with results of similar measurements on DNA-templated conductive polymers nanowires in other studies [34-37].

Many researchers have suggested that another way to view the metal/polymer or nanowire/metal system is to consider the charge transport process in the same framework as that proposed for thin films of redox polymers like ours. They behave very differently to inorganic semiconductors because of high doping level therefore; the ion transport cannot be explained by Nernst Plank equation, which described the flux of ions under the influence of both an ionic concentration gradient and an electric field [38-40].

Analysis of the temperature dependence of conductance in conjugated polymers gives a more in-depth of the conduction mechanism and is typically carried out through fitting conductance (G) values to the following equation:

$$lnG = lnGo - (\frac{To}{T})^{\beta} \quad (1)$$

This expression is used when describing the conductance behaviour of a conjugated polymer system through a variable range hopping (VRH) model. This in turn describes the low temperature conductance behaviour in strongly disordered systems where electronic states are localized. In this approach, the outcome is dependent upon the dimensionality of the system which is described by the parameter $\beta=1/(1+D)$, where 'D' is the dimensionality of the system.

Figure 13 shows the corresponding Arrhenius plot of the conductance of the Plm/DNA. The plot displayed the temperature dependence of the rate constant and suggests simple electron hopping is the dominant mechanism for electron transport, shown through the exponential behaviour of the conductance upon increasing the temperature.

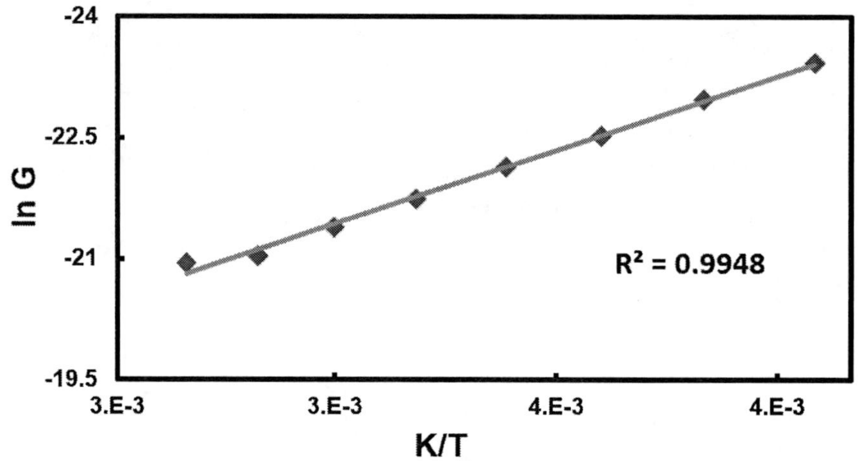

Figure 13. Arrhenius plot for the conductance G of Plm/DNA nanowires

The slopes of the Arrhenius plots in Figure 13 was analyzed to determine the activation energies associated with the hopping of charge in these polymer samples using the following equation:

$$lnG = lnG_0 - \frac{E_a}{K_B T} \qquad (2)$$

The slopes are $-E_a/k_B$ for E_a in J mol^{-1} and $-eE_a/k_B$ where e is the charge on a proton, KB is the average kinetic energy and E_a is in eV.

Activation energy (E_a) including the uncertainty on it for the Plm/DNA bulk nanowires calculated was $10.6 \pm 0.5 \times 10^{-3}$ J mol^{-1} equivalents to 0.110 ± 0.005 eV which is lower than 0.43 ± 0.02 eV reported in similar studies. The band gap of most conductive polymers is of order 3 eV that would predict Ea equal to 1.5 eV if the process was limited by thermal excitation across the gap. Instead, in our studies Ea is <1 eV. This is because the process is thermally-assisted tunneling between localized sites [41].

IOP Conf. Series: Materials Science and Engineering **805** (2020) 012008 doi:10.1088/1757-899X/805/1/012008

4. Conclusion

In this research work, nanostructured DNA-templated polyimidazole nanowires were synthesized by simple and low-cost fabrication method that applied DNA as a template for the polymerization. From IR results, changes in the DNA-related bands to different values in Plm/DNA spectra after the synthesis of the nanowires, proves the interaction of the Plm and DNA during the templating and that the prepared nanowires are not mere mixtures of Plm and DNA molecules. Similarly, UV-vis results have shown that, the polymer formed by oxidative polymerization of imidazole with DNA is a complex mixture of two substances rather than the mixture of two homopolymers. The XPS spectra of PIm/DNA show large contributions from species assigned to Plm, which indicate that the PIm/DNA nanowires can be visualized as a thick coat of PIm on the DNA template instead of small number of PIm chains wrapping around the DNA while chloride ions make up the required counter charge. Morphological (AFM, SEM and TEM) studies exhibited different morphologies with agglomerate nanostructures for the diluted film, while dense rope like network of nanostructures were observed for undiluted templated polymers with diameter of 2-13 nm.

Activation energy (E_a) including the uncertainty on it for the Plm/DNA bulk nanowires calculated was $10.6 \pm 0.5 \times 10^{-3}$ J mol^{-1} equivalents to 0.110 ± 0.005 eV which is not surprising because the process is thermally-assisted tunneling between localized sites.

References

1. Yoon H 2013 Current trends in sensors based on conducting polymer nanomaterials. *Nanomater.* **3** (3) pp 524-529.
2. Šetka M, Drbohlavová J and Hubálek J 2017 Nanostructured polypyrrole-based ammonia and volatile organic compound sensors. *Sens.* **17** (3) pp 562.
3. Nguyen DN and Yoon H 2016 Recent advances in nanostructured conducting polymers: from synthesis to practical applications. *Polym.* **8** pp 118-157.
4. Xie H, Xiong NN, Wang YH, Zhao YZ and Li JZ 2016 Sintering behaviour and effect of silver nanowires on the electrical conductivity of electrically conductive adhesives. *J. Nanosci Nanotechnol.* **16** (1) pp 1125-1137.
5. Feng C, Li W, Li C L, Zhu H, Zhang Y, Ruan S, Chen W and Yu L 2012 Highly efficient rapid ethanol sensing based on $In_{2-x}Ni_xO_3$ nanofibers. *Sens. Actuators B Chem.* **166** pp 83-88.
6. Li Y, Ban H and Yang M 2016 Highly sensitive NH_3gas sensors based on novel polypyrrole-coated SnO_2 nanosheet nanocomposites. *Sens. Actuators B Chem.* **224** pp 449-457.
7. Hosseini ZS, Irajizad A and Mortez A 2015 Room temperature H_2S gas sensor based on rather aligned ZnO nanorods with flower-like structures. *Sens. Actuators B Chem.* **207** pp 865-871.
8. Jones A P 1999 Indoor air quality and health. *Atmos. Environ.* **33** (28) pp 4535-4564.

IOP Conf. Series: Materials Science and Engineering **805** (2020) 012008 doi:10.1088/1757-899X/805/1/012008

9. Burge PS 2004 Sick building syndrome. *Occup. Environ. Med.* **61** pp 185-190.
10. Mariam NA, Reda H, Nicholas GW, Alton BH, Andrew H and Benjamin RH 2013 Networks of DNA-templated palladium nanowires: structural and electrical characterisation and their use as hydrogen gas sensors. *Faraday Discuss.* **164** pp 71-91.
11. Hadjarab F and Erskine JL 1985 Image properties of the hemispherical analyzer applied to multichannel energy detection. *J. Electron Spectr. Relat. Phenomena* **36** (3) pp 227-243.
12. Whelan DR, Bambery KR, Heraud P, Tobin MJ, Diem MD, McNaughton M and Wood BR 2011 Monitoring the reversible B to A-like transition of DNA in eukaryotic cells using Fourier transform infrared spectroscopy. *Nucleic Acids Res.* **39** (13) pp 5439-5448.
13. Marty R, N'soukpoé-Kossi CN, Charbonneau D, Weinert CM, Kreplak L and Tajmir-Riahi HA 2009 Structural analysis of DNA complexation with cationic lipids. *Nucleic Acids Res.* **37** (3) pp 849-857.
14. Da Silva EN, da Silva PAB, Graminha AE, de Oliveira PF, Damasceno JL, Tavares DC, Batista AA and Poelhsitz GV 2017 Synthesis, characterization, cytotoxic activity, and interactions with CT-DNA and BSA of cationic Ruthenium(II) complexes containing Dppm and quinoline carboxylates. *Bioinorg. Chem. Appl.* Article ID 2562780. https://doi.org/10.1155/2017/2562780.
15. Ramasamy R 2015 Vibrational spectroscopic studies of imidazole. *Armenian J. Phys.* **8** (1) pp 51-55.
16. Ramasamy R 2013 Analysis of vibrational spectra of pyridoxazinone based on density functional theory calculations. *J. Appl. Spectr.* **80** (4) pp 492-498.
17. Stuart, B., 2000. Infrared spectroscopy. *Kirk-Othmer Encyclopedia of Chemical Technology*, pp.1-18.
18. Trivedi MK, Branton A, Dahryn T, Nayak G, Saikia G and Jana S 2015 Evaluation of physical, thermal and spectroscopic properties of biofield treated p-hydroxyacetophenone. *Nat. Prod. Chem. Res.* **3** (5) 190. https://doi.org/10.4172/2329- 6836.1000190.
19. Dovbeshko GI, Gridina NY, Kruglova EB and Pashchuk OP 2000 FTIR Spectroscopy studies of nucleic acid damage. *Talanta* **53** (1) pp 233-246.
20. Ouameur AA and Tajmir-Riahi HA 2004 Structural analysis of DNA interactions with biogenic polyamines and cobalt(III)hexamine studied by Fourier transform infrared and capillary electrophoresis. *J. Biol. Chem.* **279** (40) pp 42041-42054.
21. Christopher FM and Adam TW 2003 DNA-templated construction of copper nanowires. *Nano Lett.* **3** (3) pp 359-363.
22. Maxut A, Nozière B, Fenet B and Mechakra H 2015 Formation mechanisms and yields of small imidazoles from reactions of glyoxal with NH_4^+ in water at neutral pH. *Phys. Chem. Chem. Phys.* **17** (31) pp 20416-20424.
23. An S, Abdiryim T, Ding Y and Nurulla I 2008 A comparative study of the microemulsion and interfacial polymerization for polyindole. *Mater. Lett.* **62** (6-7) pp 935-938.

24. Givehchi M, Polton A, Tardi M, Moreau M, Sigwalt P and Vairon JP 2000 The indanyl cation: a real time observation. *Macromol. Symp.* **157** (1) pp 77-92.

25. Gong P, Lee CY, Gamble LJ, Castner DG and Grainger DW 2006 Hybridization behavior of mixed DNA/alkylthiol monolayers on gold: characterization by surface plasmon resonance and ^{32}P radiometric assay. *Anal. Chem.* **78** (10) pp 3326-3334.

26. Cancado LG, Pimenta MA, Neves BRA, Dantas MSS and Jorio A 2004 Influence of the atomic structure on the Raman spectra of graphite edges. *Phys. Rev. Lett.* **93** (24) pp 247401.

27. Talbi H, Maarouf EB, Humbert B, Alnot M, Ehrhardt JJ, Ghanbaja J and Billaud D 1996 Spectroscopic studies of electrochemically doped polyindole. *J. Phys. Chem. Solids* **57** (6-8) pp 1145-1151.

28. Carquigny S, Sanchez JB, Berger F, Lakard B and Lallemand F 2009 Ammonia gas sensor based on electrosynthesized polypyrrole films. *Talanta* **78**(1) pp 199-206.

29. Deng Z, Stone DC and Thompson M 1996 Effect of redox state on the response of poly-N-(2-cyanoethyl) pyrrole coated thickness-shear mode acoustic wave sensors to organic vapours. *Analyst* **121** (9) pp 1341-1348.

30. Finšgar M 2015 The first X-ray photoelectron spectroscopy surface analysis of 4-methyl-2-phenyl-imidazole adsorbed on copper. *Anal. Methods* **7** (16) pp 6496-6503.

31. Pang D, Thierry AR and Dritschilo A 2015 DNA studies using AFM: Capabilities for measurement of short DNA fragments. *Front. Mol. Biosci.* **2**:1. https://doi.org/10.3389/fmolb.2015.00001.

32. Watson SMD, Mohamed HDA, Horrocks BR and Houlton A 2013 Electrically conductive magnetic nanowires using an electrochemical DNA-templating route. *Nanoscale* **5** (12) pp 5349-5359.

33. Marini M, Falqui A, Moretti M, Giugni A, Gentile F, Candeloro P and Fabrizio ED 2015 The structure of DNA by direct imaging. *Sci. Adv.***1** (7) e1500734. https://doi.org/10.1126/sciadv.1500734.

34. Hassanien R, Al-Hinai M, Farha Al-Said SA, Little R, Šiller L, Wright NG, Houlton A and Horrocks BR 2000 Preparation and characterization of conductive and photoluminescent DNA-templated polyindole nanowires. *ACS Nano* **4** (4) pp 2149-2159.

35. Pruneanu S, Al-Said SA, Dong L, Hollis TA, Galindo MA, Wright NG, Houlton A and Horrocks BR 2008 Self-assembly of DNA-templated polypyrrole nanowires: Spontaneous formation of conductive nanoropes. *Adv. Funct. Mater.* **18** (16) pp 2444-2454.

36. Dong L, Hollis T, Connolly BA, Wright NG, Horrocks BR, and Houlton A 2007 DNA-templated semiconductor nanoparticle chains and wires. *Adv. Mater.* **19** (13) pp 1748-1751.

37. Xing Y 2004 Synthesis and electrochemical characterization of uniformly dispersed high loading Pt nanoparticles on sonochemically-treated carbon nanotubes *J. Phys. Chem. B*, **108** (50) pp 19255-19259.

38. Xie SJ, Mei LM and Lin DL 1994 Transition between bipolaron and polaron states in doped heterocycle polymers. *Phys. Rev. B* **50** (18) pp 13364-13370.

39. Kang SD and Snyder GJ 2017 Charge-transport model for conducting polymers. *Nature Mater.* **16** (2) pp 252-257.

IOP Conf. Series: Materials Science and Engineering **805** (2020) 012008 doi:10.1088/1757-899X/805/1/012008

40. Heeger J 2001 Nobel lecture: Semiconducting and metallic polymers: The fourth generation of polymeric materials. *Rev. Modern Phys.* **73** (3) pp 681-700.
41. Al-Hinai M, Hassanien R, Watson SMD, Wright NG, Houlton A and Horrocks BR 2016 Metal-conductive polymer hybrid nanostructures: Preparation and electrical properties of palladium-polyimidazole nanowires. *Nanotechnol.* **27** (19) 095704. https://doi.org/10.1088/0957-4484/27/9/095704.

IOP Conf. Series: Materials Science and Engineering **805** (2020) 012009 doi:10.1088/1757-899X/805/1/012009

Assessment of micro-morphology of the leaves of *Solanum nigrum* (L.), *Amaranthus hybridus* (L.) and *Celosia argentea* (L.) cultivated using kola pod extract-mediated AgNPs as biofertilizer

*AF Ogundola, MO Liasu, ATJ Ogunkunle, OS Ajao and MO Jimoh
Department of Pure and Applied Biology, Ladoke Akintola University of Technology, Ogbomoso, Oyo State, Nigeria

*Corresponding author: afpogundola@gmail.com

In this study, assessment was made on the growth response and micro-morphology of the leaves of *Amaranthus hybridus*, *Celosia argentea* and *Solanum nigrum* cultivated using super grow (Golden Neo-Life Diamite-(GNLD) bio-fertilizer, kola pod extract-mediated AgNPs with super grow. AgNPs was synthesized biologically making use of pod extract of *Cola nitida.* A pot experiment was conducted to cultivate the three vegetables on soil from botanical gardens, LAUTECH in 5 kg soil capacity pots and super grow (biofertilizer) and mixture of AgNPs and super grow were applied at the second week after planting. Data were collected from the plant growth parameters. Random ccollection of the leaves was done at the onset of flowering and prepared for viewing with a binocular light microscope (Olympus) at 100X and 400X magnifications. Photographs were obtained digitally with Android phone using Microsoft image programme for windows. Statistical analysis of variance was done used MINITAB release 17, and means were separated with Fisher's Least significant different at $p<0.05$. Leaves of plant treated with AgNPs have their micro organs (stomata and subsidiary/epidermal cells) bigger in sizes and increased in densities. This situation was observed in the leaves of the three plant species used in this study. Conclusively, Kola pod extract-mediated AgNPs strengthened the super grow and influenced the growth response as there was significant difference among the values recorded from the treatments. Also, micro-morphology of the AgNPs treated plant samples recorded significant values.

Keywords: Micro-morphology, vegetables, Kola pod extract-AgNPs, light microscopy

1 Introduction

Plants being the most dependable group of living organisms in space; they have vital roles they play in the existence of other living organisms that make them indispensable. Vegetable among plants is recognized as a cheap source of micronutrients needed by man [1]. Among the leafy vegetables that are common on consumers' table are *Amaranthus hybridus* and *Celosia argentea,* while *Solanum nigrum* rarely appear in the markets. However, their spread nearly round the world bring about various forms and structures as a result of living conditions, genetic structure and phylogenetic characteristics [2]. Micro-morphological parameters of different plants have been used as aids in the taxonomical recognition of plant species [3]. Therefore, the possibility of alteration in micro-morphology of the plant (especially the leaf organs) when cultivated on different media is necessary for investigation. Cultivation of edible leafy vegetables like *Solanum nigrum* (L.), *Amaranthus hybridus* (L.), and *Celosia argentea* (L.), concentrates on leaves as the essential part for consumption which is believed to alleviate malnutrition problems, simply because of their nutrient composition, immune-boosting activities and so on [4-6]. Highest priority is therefore given to leaf production of the vegetables in this study; this is due to the fact that leaf reflects the consequence of nutrient deficiency in vegetables [7].

Generally the use of organic fertilizer is believed to be safe, cost effective and thus available within the reach of every one [8]. Application of green synthesized nanoparticles in agriculture with suitable results has been to the knowledge of many researchers. Nanotechnology is believed to be one of the key events in biotechnologies of the 21st century. The promise of nanotechnology to improve current agricultural practices through the enhancement of management and conservation of inputs in crops has permitted its expansion to different areas of specialization [9]. The use of nanofertlizers has shown many advantages over others as mentioned: (i) increased bioavailability (Easy dissolution; improved penetration/permeation through membranes); (ii) lower doses; (iii) lower dose-dependent toxicity; (iv) controlled release; (v) targeted biodistribution; (vi) reduction of the influence of environment on bioavailability variability [9-12]. For these reasons, acceptability is rapid even though the hazards assessment can never be overemphasised. The application of nanofertilizers as reported by DeRosa [13] has been found to have reduced nitrogen loss due to leaching, emission and long-term assimilation by soil microorganisms [13]. Other credibility to the use of nanofertilizers has been its accessibility to seed and root tissue and modulatory effects of phytochemicals [14-15]. This has established the possibility of developing new nutrient delivery systems that exploit the nanoscale porous domains on plant surfaces and show sustained release of nutrients on demand, while preventing them from premature conversion into chemical/gaseous forms that cannot be absorbed by plants [16]. The new trend brought by the application of green synthesized nanoparticles in agriculture has permitted the use of biosynthesized AgNPs mixed with super grow (biofertilizer) to cultivate three vegetables in this study. Therefore, this study aim at assessing the growth response and micromorphology (stomata) of the three vegetables in this study to foliar application of super grow (GNLD organic fertiliser) when mixed with Kola pod extract-mediated AgNPs and when applied alone.

2 Materials and Methods

2.1 Study area and plant preparation

The study was carried out in the Research farm of Botanical gardens, LAUTECH, Ogbomoso on Lat. 8.15977° N Long. 4.27426° E. Soil was collected from the same place to cultivate the three (3) vegetables: *Amaranthus hybridus, Celosia argentea* and *Solanum nigrum.* Super grow (Organic liquid fertilizer made from poultry droppings and sea bird guano) was bought from Golden Neo-Life Diamite (GNLD) International and the stock concentration was diluted at 1ml/litre of water. Kola pod extract-mediated AgNPs was prepared and characterized as explained by Lateef *et al.* [17]. Exactly 12.5 µl/ml was added to the diluted super grow. Foliar application of the treatments started from the second week after planting till the onset of the flower. Application was done twice in a week as biofertilizer for the 3 vegetables. The control plant samples of each of the three species were treated with water alone and each of the treatments was replicated 5 times.

2.2 Data collection and microscopic study

Data were taken from growth parameters: plant height, stem girth, number of leaves, and leaf area. At flowering stage of each of the vegetables, 3 freshly harvested leaves from 3 pots of each plant species were randomly plucked; cut into 10 mm-sizes and prepared for observation under light microscope. A strip of lower and upper epidermis from the middle portion of the leaves was carefully peeled off to show the transparent part [18]. Images showing the number, arrangement of stomata with guard cells and subsidiary cells were captured with the Android phone (Infinix Hot 6) fitted in the light microscope as described by Otang *et al.* [19]. The images were captured digitally using Microsoft image programme for windows. Leaf stomatal density was determined and expressed as the number of stomata per unit area [11, 19]. This was achieved by counting the number of full or half shown stomata within 50 µm unit area at 400X magnification on the lower and upper (adaxial and abaxial) surfaces of the specimens. Density of stomata per mm^2 was calculated for each plant species and recorded for each treatment. With the assistance of the concept of Silva *et al.* [20], assessment was made based on the number, type of arrangement and position of accessory cells surrounding the guard cells of the stomata.

2.3 Statistical analysis

All data on growth parameters and stomata density were subjected to statistical analysis using MINITAB Release 17. A one-way analysis of variance (ANOVA) was used to compare the mean values among the plant treatments. Means were segregated using Fisher's Least Significant Difference (LSD) paired wise comparison. The means were treated as significantly different at $p < 0.05$.

3 Results and Discussion

From the growth parameters measured from the 3 types of vegetables in this study, exponential increases was recorded in the growth values (number of leaves, length of median mid-rib and leaf area) of the treated plant samples cultivated with mixture of AgNPs and super grow, and super grow alone. The increase was observed immediately after the application of the growth promoters in the second week of plant growth as presented on Tables 1a-3d. Plant samples cultivated with mixture of AgNPs and super grow recorded significant increase at $P< 0.05$ compared to samples treated with super grow alone and the control samples (Figures 1a–3d). These results corroborate those of Ghahremani *et al.* [21] on the exponential increase in their results from foliar application of nano-calcium chelated fertilizers on *Ocimum basilicum*. However, the control plant samples recorded significantly low values in growth characteristics of the 3 plant types.

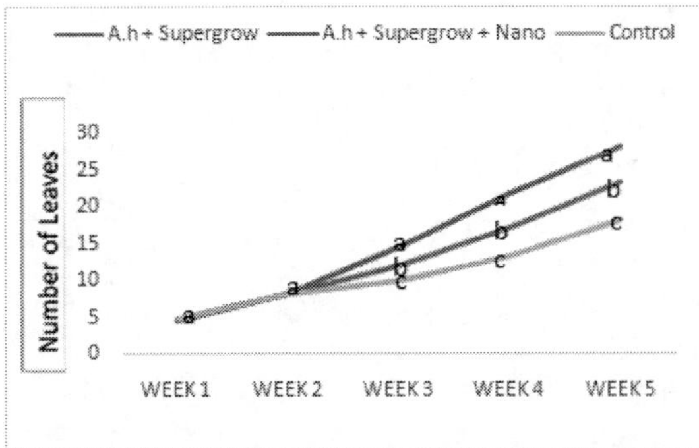

Fig. 1a: Effect of growth promoters on number of leaves of *A. hybridus*

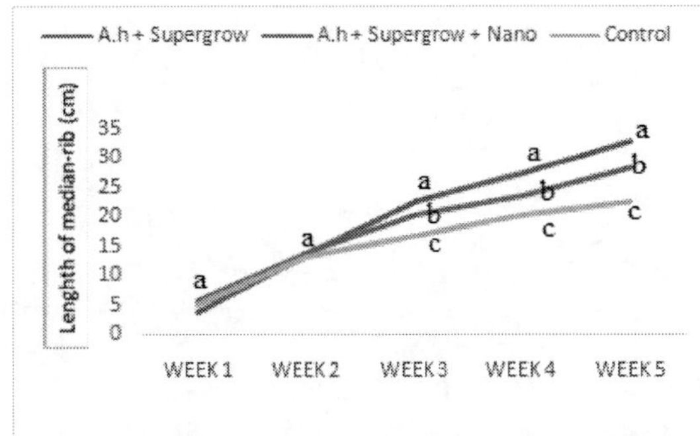

Fig. 1b: Effect of growth promoters on length of median-rib of *A. hybridus*

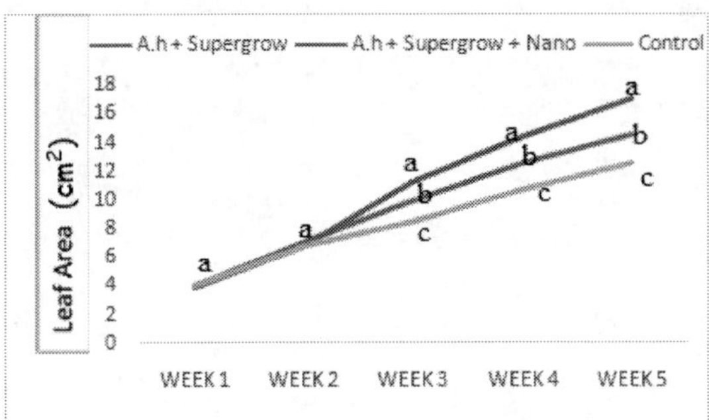

Fig. 1c: Effect of growth promoters on leaf area of *A. hybridus*

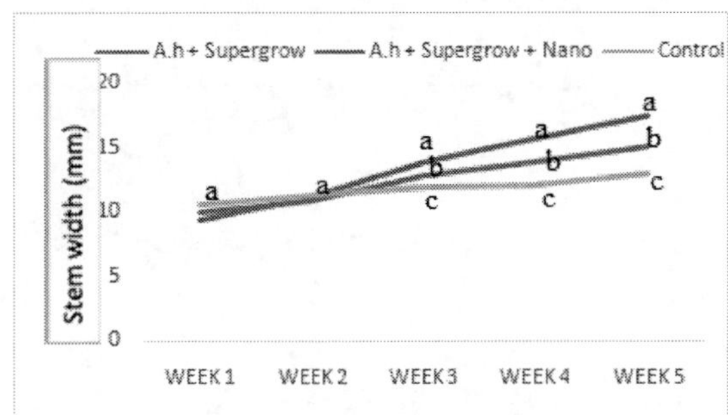

Fig. 1d: Effectof growth promoters on stem width of *A. hybridus*

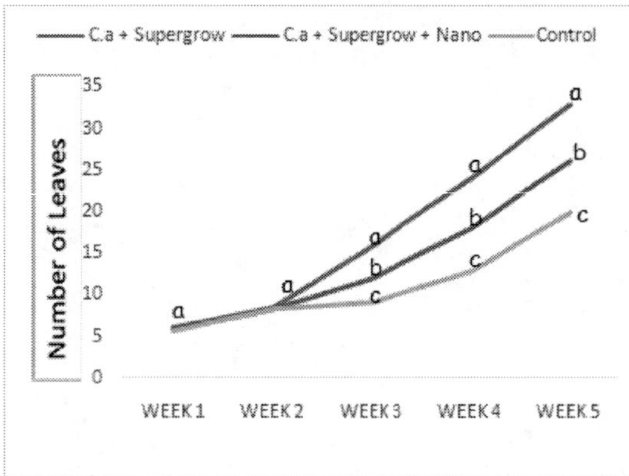

Fig. 2a: Effectof growth promoters on number of leaves of *C. argentea*

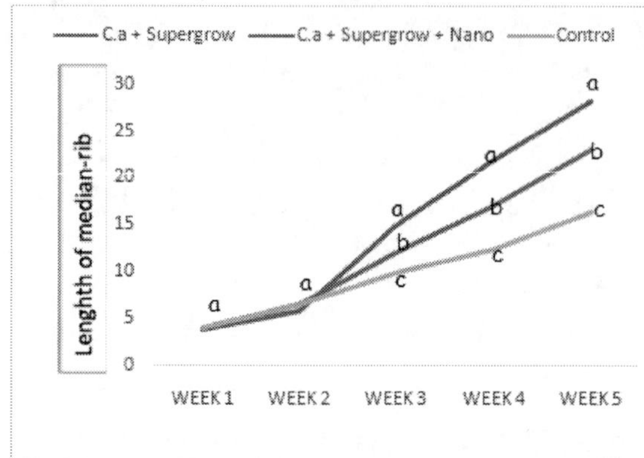

Fig. 2b: Effect of growth promoters on number of leaves of *C. argentea*

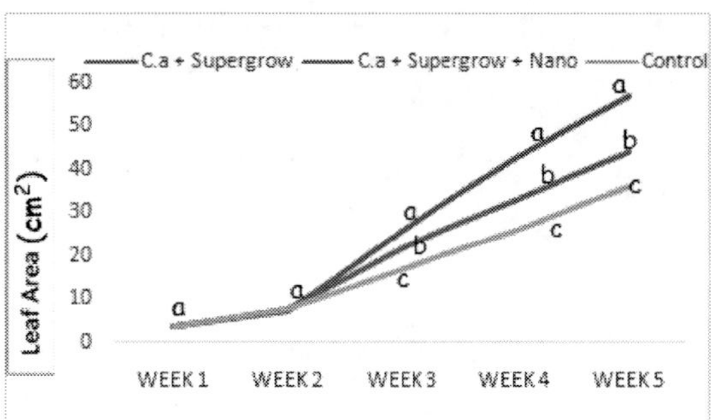

Fig. 2c: Effect of growth promoters on leaf area of *C. argentea*

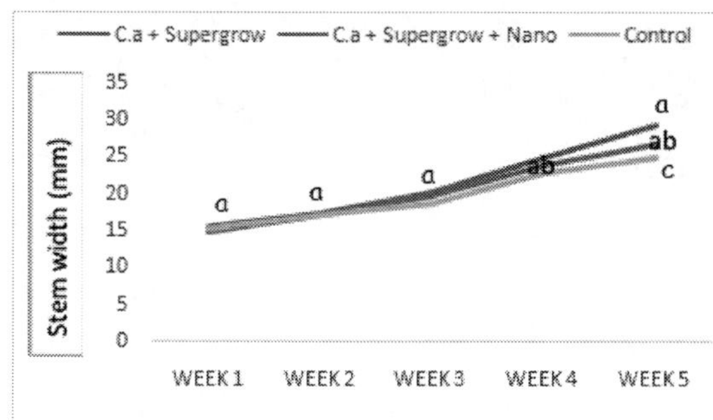

Fig. 2d: Effect of growth promoters on stem width of *C. argentea*

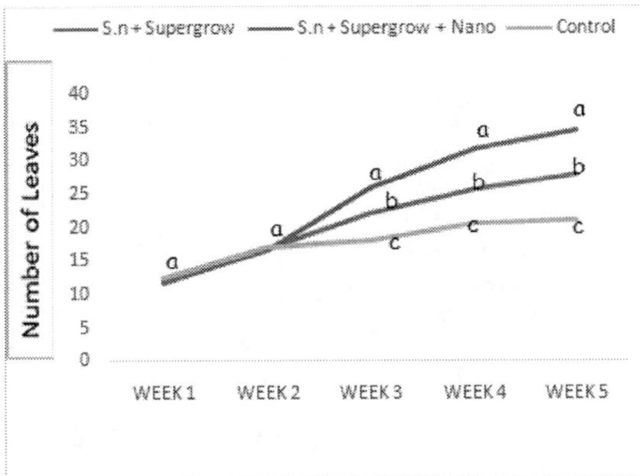

Fig. 3a: Effect of growth promoters on leaf area of *S. nigrum*

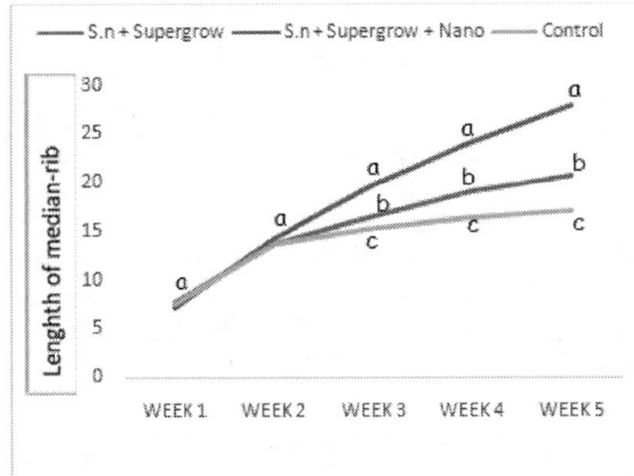

Fig. 3b: Effect of growth promoters on length of median-rib of *S. nigrum*

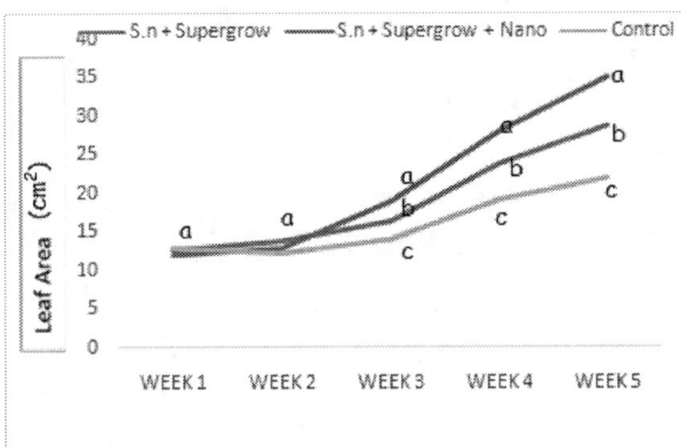

Fig. 3c: Effect of growth promoters on leaf area of *S. nigrum*

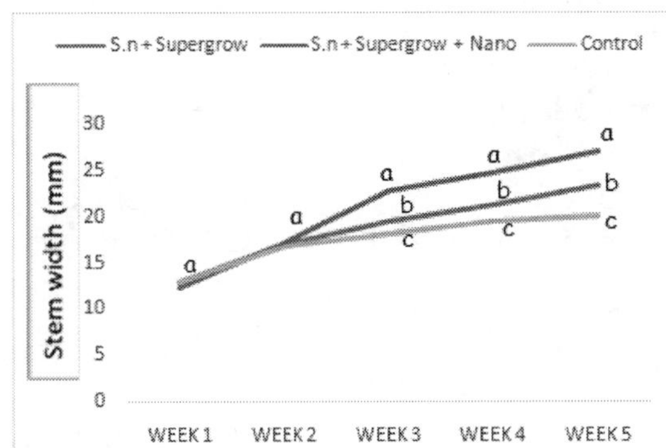

Fig. 3d: Effect of growth promoters on Stem width of *S. nigrum*

The distribution of the stomata in all the 3 vegetables is hypo-amphistomatic as stomata were present in both adaxial and abaxial surfaces of the leaves but with higher densities in the adaxial surface (Figures 4-6). This result is in line with earlier studies [22-23]. It was observed in the 3 vegetables that stomata densities on both adaxial and abaxial surfaces of the leaf samples treated with mixture of AgNPs and super grow were significantly higher at $p < 0.05$ compared to those of super grow alone and the control samples (Table 1). Paracytic stomatal arrangement which is a simple type of arrangement is of highest frequency in plant samples with mixture of AgNPs and super grow. Control plant leaves recorded high frequency of anisocytic and tetracytic stomatal arrangement (Table 1). The paracytic stomatal arrangement in those leaves treated with AgNPs may be the potential mechanism of the AgNPs to increase the stomatal densities on the leaf surfaces. It accommodates more stomatal complex which might have given the chances of the AgNPs treated leaves to increase the photosynthetic activities and produce more carbohydrates. Furthermore, the possibilities of

IOP Conf. Series: Materials Science and Engineering **805** (2020) 012009 doi:10.1088/1757-899X/805/1/012009

the plants' growth increase at faster rate, as there are evidences of improvement of plant growth under the influence of application of AgNPs [9, 15, 24].

Table 1. Comparative assessment of stomatal arrangement in three studied vegetables

Vegetables	*Amaranthus spinosus*	*Celosia argentea*	*Solanum nigrum*
Control abaxial stomata density	16^b	16^b	15^b
AgNPs + Super grow abaxial stomata density	24^a	26^a	22^a
Control adaxial stomata density	28^b	18^b	18^b
AgNPs + Super grow adaxial stomata density	37^a	32^a	29^a
Types of stomata arrangement abaxial	Control: 16 anisocytic NANO: 12 tetracytic; 5 paracytic; 7 anisocytic	Control: 16 paracytic NANO: 26 paracytic	Control: 13 anisocytic; 2 paracytic NANO: 22 paracytic
Types of stomata arrangement adaxial	Control: 18 paracytic; 6 tetracytic; NANO: 24 paracytic; 13 anisocytic	Control: 18 paracytic NANO: 32 paracytic	Control: 12 paracytic; 6 anisocytic NANO: 29 paracytic

IOP Conf. Series: Materials Science and Engineering **805** (2020) 012009 doi:10.1088/1757-899X/805/1/012009

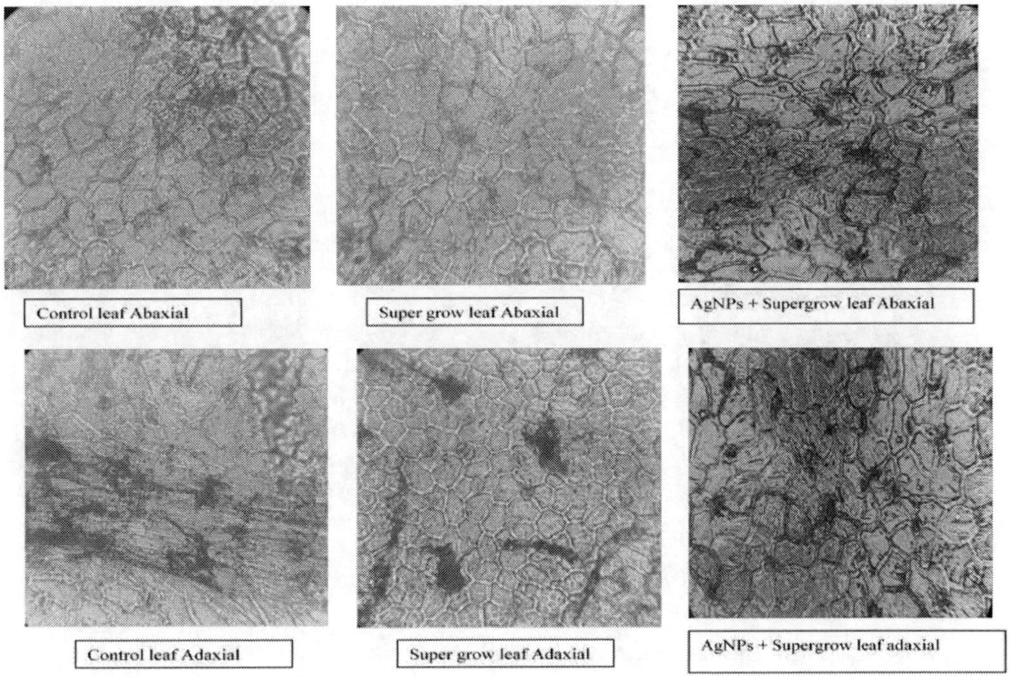

Figure 4. Stomata arrangement in *Amaranthus hybridus*

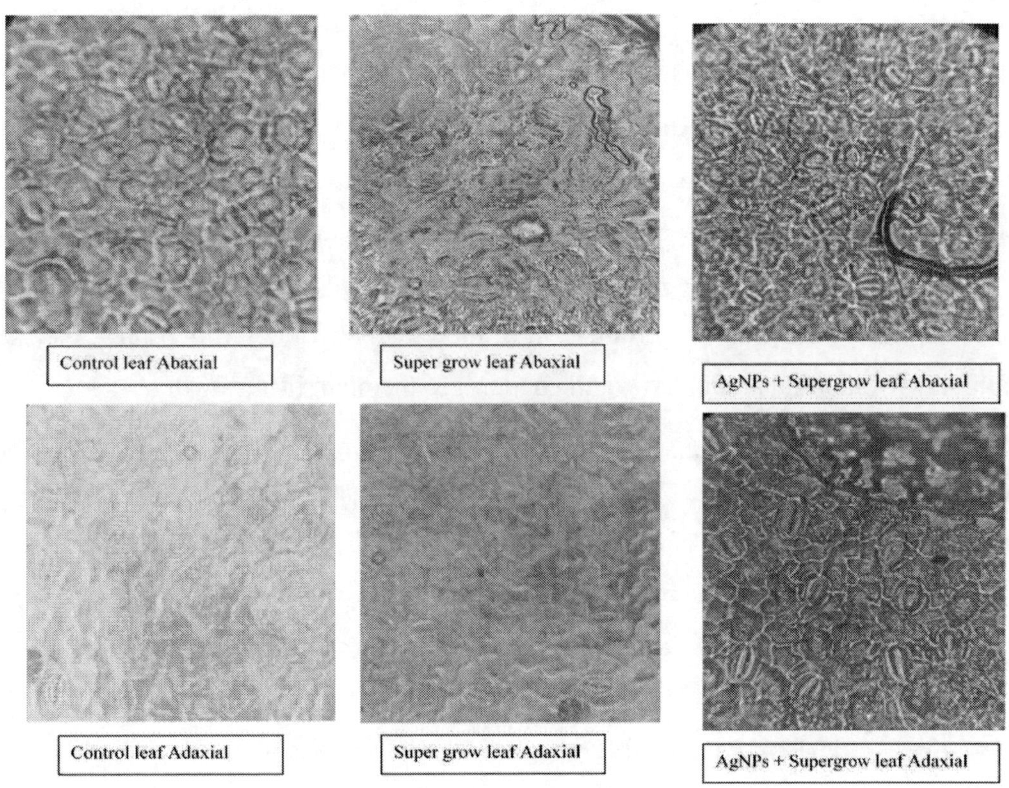

Figure 5. Stomata arrangement in *Celosia argentea*

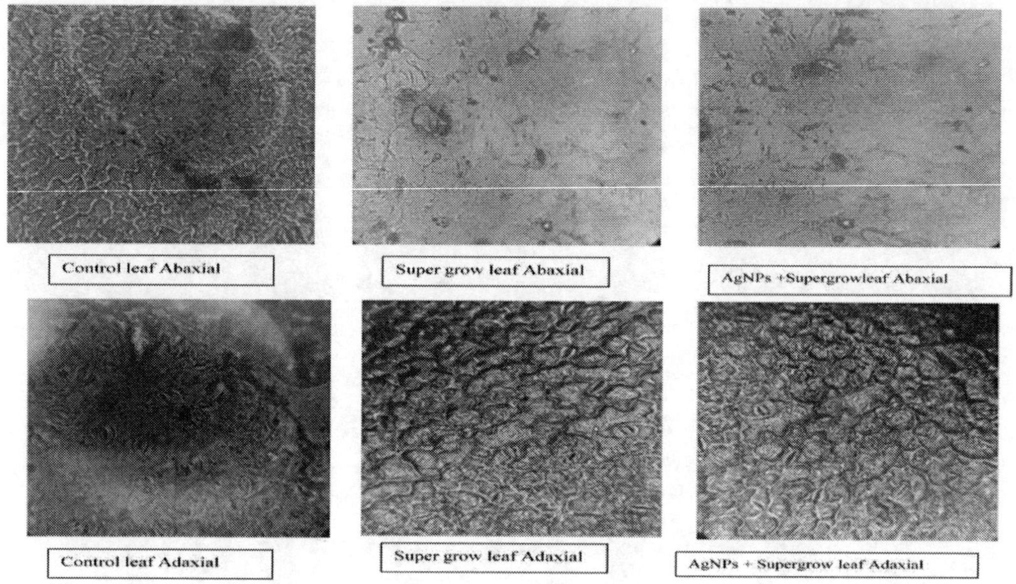

Figure 6. Stomata arrangement in *Solanum nigrum*

4 Conclusion

In all the 3 vegetables in this study, significant improvements of production characteristics was recorded with Kola Pod extract-mediated AgNPs and super grow mixture. The mixture also increased the stomata densities, and influenced the stomata arrangement in the leaves. The paracytic type of stomata arrangement is to the maximum. This type of arrangement accommodates more stomata on the leaf surface which increases photosynthetic potentials of the stomata thereby ensuring the production of more carbohydrates which may likely lead to accumulation of more plant biomass, and encourages the plant fast growth.

Acknowledgement

The authors thank the Head of Nanotechnology Research Group (*NANO*$^+$), LAUTECH, Ogbomoso for the supply of the biosynthesized AgNPs.

References

[1] Akkemik Ü 2011 Bitki Materyali (Gymnospermae) Ders Notları, İ. Ü, Orman Fak.

[2] Prabhu K, Karar PK, Hemalatha S and Ponnudurai K 2011 Comparative micromorphological and phytochemical studies on the roots of three *Viburnum* (Caprifoliaceae) species. *Turk. J. Bot.* **35** (6) pp 663-670.

[3] Odhav B, Beekrum S, Akula US and Baijnath H 2007 Preliminary assessment of nutritional value of traditional leafy vegetables in KwaZulu-Natal, South Africa. *J. Food Compos. Anal.* **20** (5) pp 430-435.

[4] Harikrishnan R, Balasundaram C, Jawahar S and Heo MS 2011 *Solanum nigrum* enhancement of the immune response and disease resistance of tiger shrimp, *Penaeus monodon* against *Vibrio harveyi*. *Aquacult.* **318** (1-2) pp 67-73.

[5] Emebu PK and Anyika JU 2011 Proximate and mineral composition of kale (*Brassica oleracea*) grown in Delta State, Nigeria. *Pak. J. Nutri.* **10** (2) pp 190-194.

[6] Mohammed M I and N Sharif. 2011. Mineral composition of some leafy vegetables consumed in Kano, Nigeria. *Nig. J. Basic Appl. Sci.* **19** (2) pp 208-211.

[7] Aluko OA, Olanipekun TO, Olasoji JO, Abiola IO, Adeniyan ON, Olanipekun SO, Omenna EC, Kareem KO and Douglas AI 2014 Effect of organic and inorganic fertilizer on the yield and nutrient composition of jute mallow. *Global J. Agric. Res.* **2** (3) pp 1-9.

[8] Nalwa HS 2004 Encyclopedia of Nanoscience and Nanotechnology. Volume 10. Valencia, CA, USA: American Scientific Publisher.

[9] Azeez L, Adejumo AL, Lateef A, Adebisi SA, Adetoro RO, Adewuyi S, Tijani KO and Olaoye S 2019 Zero-valent silver nanoparticles attenuate Cd and Pb toxicities on *Moringa oleifera* via immobilization and induction of phytochemicals. *Plant Physiol. Biochem.* **139** pp 283-292.

[10] Rai V, Acharya S and Dey N 2012 Implications of nanobiosensors in agriculture. *J. Biomater Nanobiotechnol.* **3** (2A) Article ID: 18992. https://doi.org/10.4236/jbnb.2012.322039.

[11] Bhushan B 2004 Handbook of Nanotechnology. Berlin-Heidelberg, Germany: Springer-Verlag;.

[12] DeRosa MC, Monreal C, Schnitzer M, Walsh R and Sultan Y 2010 Nanotechnology in fertilizers. *Nature Nanotechnol.* **5** 91. https://doi.org/10.1038/nnano.2010.2.

[13] Khodakovskaya M, Dervishi E, Mahmood M, Xu Y, Li Z, Watanabe F and Biris AS 2009 Carbon nanotubes are able to penetrate plant seed coat and dramatically affect seed germination and plant growth. *ACS Nano.* **3** (10) pp 3221-3227.

[14] Lin D and Xing B 2008 Root uptake and phytotoxicity of ZnO nanoparticles. *Environ. Sci. Technol.* **42** (15) pp 5580-5585.

[15] Azeez L, Lateef A, Wahab AA, Rufai MA, Salau AK, Ajayi IO, Ajayi EM, Maryam AK and Adebisi B 2019 Phytomodulatory effects of silver nanoparticles on *Corchorus*

olitorius: its antiphytopathogenic and hepatoprotective potentials. *Plant Physiol. Biochem.* **136** pp 109-117.

[16] Rao CN, Müller A and Cheetham AK (eds) 2006 The chemistry of nanomaterials: synthesis, properties and applications. Weinheim, Germany: Wiley-VCH.

[17] Lateef A, Azeez MA, Asafa TB, Yekeen TA, Akinboro A, Oladipo IC, Azeez L, Ajibade SE, Ojo SA, Gueguim-Kana EB and Beukes LS 2016 Biogenic synthesis of silver nanoparticles using a pod extract of *Cola nitida*: Antibacterial, antioxidant activities and application as a paint additive. *J. Taibah Univ. Sci.* **10** (4) pp 551-562.

[18] Radoglou KM and Jarvis PG 1990 Effects of CO_2 enrichment on four poplar clones. II. Leaf surface properties. *Ann. Bot.* **65** (6) pp 627-632.

[19] Coopoosamy RM and Naidoo KK 2011 Anatomical features of the medicinal importance of *Aloe excelsa* (Berger). *Afr. J. Biotechnol.* **10** (39) pp 7622-7632.

[20] Silva O, Serrano R and Gomes ET 2008 Botanical characterization of *Guiera senegalensis* leaves. *Microsc. Microanal.* **14** (5) pp 398-404.

[21] Ghahremani A, Akbari K, Yousefpour M and Ardalani H 2014 Effects of nano-potassium and nano-calcium chelated fertilizers on qualitative and quantitative characteristics of *Ocimum basilicum. Int J Pharm Res Schol.* **3** 00167.

[22] Mbagwu FN, Nwachukwu CU and Okoro OO 2008 Comparative leaf epidermal studies on *Solanum macrocarpon* and *Solanum nigrum. Res. J. Bot.* **3** (1) pp 45-48.

[23] Adedeji O, Ajuwon OY and Babawale OO 2007 Foliar epidermal studies, organographic distribution and taxonomic importance of trichomes in the family Solanaceae. *Int. J. Bot.* **3** pp 276-282.

[24] Azeez L, Lateef A, and Adebisi SA 2017. Silver nanoparticles (AgNPs) biosynthesized using pod extract of *Cola nitida* enhances antioxidant activity and phytochemical composition of *Amaranthus caudatus* Linn. *Appl. Nanosci.* **7** (1-2) pp 59-66.

IOP Conf. Series: Materials Science and Engineering **805** (2020) 012010 doi:10.1088/1757-899X/805/1/012010

Antimicrobial and antioxidant activities of algal-mediated silver and gold nanoparticles

I O Omomowo[1], V O Adenigba[2*], S B Ogunsona[1], G C Adeyinka[3], O O Oluyide[4], A A Adedayo[1] and B A Fatukasi[2]

[1]Department of Pure and Applied Biology, Ladoke Akintola University of Technology, Ogbomoso, Nigeria

[2]Department of Science Laboratory Technology, LadokeAkintola University of Technology, Ogbomoso, Nigeria.

[3]Department of Chemical Engineering, Faculty of Engineering, Mangosuthu University of Technology, Umlazi, KwaZulu-Natal, South Africa.

[4]Department of Biomedical Engineering, College of Health Sciences and Technology, Ijero-Ekiti, Nigeria

E-mail address: moadesiyan@lautech.edu.ng

Abstract

This study focused on the synthesis and application of nanoparticles using *Neodesmus pupukensis* (MG257914). Cell free extracts of the microalga was employed to synthesize both silver (AgNPs) and gold nanoparticles (AuNPs). The nanoparticles were characterized using UV-visible spectroscopy, Fourier transform infrared spectroscopy (FTIR), and Scanning electron microscopy (SEM). The nanoparticles were screened for their antimicrobial potential and free radical scavenging activity against stable free radical (2,2-diphenyl-1-picrylhydrazyl). The UV absorption spectra showed absorbance peaks of 430 nm and 530 nm for AgNPs and AuNPs respectively. The FTIR spectra at 3283, 2102.87, 1636.02 for AgNPs and 3264.86, 2104.49, 1636.62, 1232.39, 1028.97 cm^{-1} for AuNPs confirms the participation of proteins in the capping and stabilization of the nanoparticles. The zone of inhibition indicating the activity of the NP-AgNPs were: *Pseudomonas* sp (43 mm); *Escherichia coli* (24.5 mm); *Klebsiella pneumoniae* (27 mm), *Serratia marcescens* (39 mm) while AuNPs showed activity to only *Pseudomonas* sp (27.5 mm) and *Serratia marcescens* (28.5 mm). The antifungal potency of NP-AgNPs was confirmed with mycelial inhibition of 80.6, 57.1, 79.4, 65.4 and 69.8% against *Aspergillus niger, A. fumigatus, A. flavus, Fusarium solani* and *Candida albicans* respectively, while NP-AuNPs had 79.4, 44.3, 75.4, 54.9 and 66.4% against *A. niger, A. fumigatus, A. flavus, F. solani* and *C. albicans* respectively. Appreciable free radical scavenging properties was obtained with NP-AuNPs (68.9%) and NP-AgNPs (41.21%). The nanoparticles of *Neodesmus pupukensis* showed appreciable potential as antimicrobial and antioxidant agents and could be explored for various applications in biotechnology.

Keywords: *Neodesmus pupukensis*, Nanoparticles, Antimicrobial activity, Antioxidant activity

1. Introduction

Multidrug resistance of pathogenic organisms and the occurrence of different human diseases emerging with time have called for researches in search of compounds from inorganic sources which can serve as potential bioactive agents [1]. Various pathogenic organisms have developed resistance to some of the available drugs due to drug abuse. The alarming occurrence of diseases caused by these multidrug resistance microorganisms has called for the need to obtain biologically active compounds from biological materials. These compounds must possess the potential that inhibit the growth of pathogenic organisms and not harmful to human. Similarly, relative oxygen species (superoxide anion, hydroxyl radicals and hydrogen peroxide) which are produced when metabolism takes place starts up a process where membrane lipids and other biomolecules becomes peroxidized; this results in several diseased conditions in the body [2]. There is a need to introduce antioxidants that have the ability to stop and scavenge free radicals in the body thereby resulting in the protection of the body against diseases. The body is being protected against diseases by antioxidants; they are able to achieve this by scavenging ROS thereby preventing oxidant-induced damaging of cell structures and molecules [3]. Antioxidants are important in building the body against diseases in that they enhance the physiology of the human body [4]. Although the human system contains some effective antioxidants which are able to handle oxidative stress against the body, there is need for additional antioxidants that are non-indigenous [4].

Microalgae are organisms that contain rich metabolites reservoir of various chemical components and biologically active substances [5]. They contain materials which are biochemically active such as polyunsaturated fatty acids, polyphenols, carotenoids, vitamins, and phycobiliproteins which have been identified as antioxidants responsible for getting rid of the action of ROS that are produced when photosynthesis takes place [6-9]. Microalgae possess mechanisms with which they defend themselves against the resulting photo-induced oxidative damage which is caused by the antioxidative mechanics to detoxify and remove highly reactive oxygen species [10].

Exogenous antioxidants are present in photosynthetic microalgae and these include: polyphenols, carotenoids, sterols or vitamins [11]. Microalgae and macroalgae contain carotenoids such as: fucoxanthin, diatoxanthin, siphonaxanthin, and diadinoxanthin [8,12]. Apart from carotenoids with high bioactivity, phycobiliprotein which are also present in microalgae possess both antioxidant and antimicrobial activity [13]. Aside from the known applications of microalgae biomass as a potent antimicrobial and antioxidant agent, green synthesis of nanoparticles from microalgae is also a promising research area which has been underexplored [14-15]. Microalgae are able to produce nanoparticles using various metal ions as precursors; these metals include silver, gold, platinum, and cadmium [16-19].

Phyco-nanotechnology which involves nanoparticles synthesis from algae as bioagent is gaining attention because of the potential applications of these algal mediated nanoparticles [20-24]. Synthesis from algae has shown some advantages which are reduction in duration of synthesis [25-26], its ease of application in optimization of silver nanoparticles and absence of culture and maintaining process [27]. Algae are abundantly rich in proteins, lipids, carbohydrates; these molecules play vital roles in the reduction of metals [28]. Stability and capping of these metallic nanoparticles in solution is the major responsibility of proteins [29]. In the current study, an attempt was made to synthesize both silver nanoparticles (AgNPs) and gold nanoparticles (AuNPs) from the cell-free extracts of *Neodesmus pupukensis* (MG257914). The synthesized AuNPs and AgNPs of each microalgae were screened for both antimicrobial and antioxidant potentials.

2 Materials and Methods

2.1Maintenance and Harvesting of Microalgal cultures

Neodesmus pupukensis (MG257914) which was previously isolated from fish pond water in our laboratory was cultured in BG 11 medium. The algal biomass was harvested on day 21 using a centrifuge at 5000 rpm for 20 min. Sterile distilled water was used five times to achieve the removal of BG II medium from the microalgal biomass [22]. Five pathogenic bacteria and five fungi were collected from the Department of Science Laboratory Technology (SLT), LAUTECH, Ogbomoso, Oyo State, Nigeria. These test organisms include: bacteria (*E. coli, Klebsiella pneumoniae*, *Pseudomonas aeruginosa, S. aureus* and *S. marcescens*) and fungi (*A. niger, A. flavus, A. fumigatus, F. solani* and *C. albicans*).

2.2 Synthesis of Silver and Gold Nanoparticles from Microalgae

Algae powder was obtained from the microalga by drying the filtrate obtained above in a glass Petri dish at 80 ^oC. One gram of algal powder was dissolved in 100 ml of distilled water, boiled at 100 ^oC for 20 min [22]. The cell-free extract of the microalga was separated from the mixture in a centrifuge at 5000 rpm for 10 min. Synthesis of silver nanoparticles

IOP Conf. Series: Materials Science and Engineering **805** (2020) 012010 doi:10.1088/1757-899X/805/1/012010

from the extract was done by adding 90 ml of 1 mM $AgNO^3$ to 10 ml of the extract while the same quantity of 1 mM gold chloride was used to synthesize gold nanoparticles. The mixtures were incubated and the colour change was monitored until it stabilized [30]. The nanoparticles were designated with identification codes according to the precursor used for synthesis as follow: *Neodesmus pupukensis*-based silver nanoparticles (NP-AgNPs) and *Neodesmus pupukensis*-based gold nanoparticles (NP-AuNPs)

2.3 Characterization of nanoparticles

Bioreduction of silver nanoparticles was first observed with the change in colour and the spectra of the reaction mixtures were measured using Double beam spectrophotometer with a quartz cuvette of 1 cm optical path length with spectra range from 300 and 1000 nm.
The functional groups present in each nanoparticle were evaluated with the use of FT-IR spectrophotometer. The sample solution was centrifuged at 7000 rpm for 10 min. The resulting pellets were washed in double distilled water twice. The algal pellets were freeze dried and ground to powder with KBr. The FT -IR analysis was done on avatar 330 FT-IR instrument in the region of 4000 to500 cm^{-1}at a resolution of 4 cm^{-1}.
The silver nanoparticles of the algal were prepared for SEM analysis in order to determine the size and morphology of the nanoparticles.The microphotograph was recorded using Concise FEGSEM 6100 Zeiss Ultra Plus (Germany) at an accelerated voltage of 20.0 kV with secondary electrons in low vacuum mode. The gold nanoparticles were also prepared and a drop of each sample was deposited on copper grids already coated in carbon, the solution was dried and the micrograph of each sample was taken with JEOL JEM 2100 HRTEM at accelerating voltage of 200 kV. The elemental compositions of the silver and gold nanoparticles were measured with the same equipment.

2.4 Antimicrobial activity

The Antimicrobial activity of gold and silver nanoparticles of microalgae was determined by agar well diffusion method. One hundred microliter of each bacterium was spread uniformly on Mueller Hinton Agar (MHA) plate, and five different wells were made on the plate using a cork borer. Four different volumes (25, 50, 75, and 100 μl) of the nanoparticles were prepared. Each well represented each concentration while the fifth hole served as control where no nanoparticles were added. Standard antibiotic disc were used as control.

IOP Conf. Series: Materials Science and Engineering **805** (2020) 012010 doi:10.1088/1757-899X/805/1/012010

2.5 Antifungal assay

The antifungal potency of AgNPs and AuNPs were evaluated by adding graded concentration of each AgNPs and AuNPs into potato dextrose agar plates which were subsequently inoculated with 6 mm agar plug of 48-h old cultures of each fungus [29]. In the control experiments, fungal plugs were inoculated on PDA plates without the incorporation of nanoparticles. All the plates were incubated at 28 °C for 72 h. The diameter of fungal growth in all the plates were measured and used to determine the percentage growth inhibitions as follows:

$$\text{Mycelia inhibition (\%)} = \frac{\text{Dcontrol} - \text{Dtest}}{\text{Dcontrol}} \times 100$$

D is the diameter of fungal growths on the potato dextrose agar plates.

2.6 Antioxidant assay

The antioxidant property of the NPs was done using DPPH (2,2-diphenyl-2-picrylhydrazyl hydrate) assay. DPPH (1.97 g) was measured and dissolved in 5 ml of methanol. One millilitre of this solution was added to 3 ml of hydrogel solution (1 g of prepared hydrogel in 10 ml of distilled water). The mixture was shaken and left to stand for 30 min at room temperature. The absorbance at 517 nm was measured by using a UV-visible spectrophotometer. Antioxidant activity was estimated by calculating the % inhibition by the following formula:

$$\text{DPPH Scavenging effect(\%)} = \frac{\text{Control}_{\text{absorbance}} - \text{Sample}_{\text{absorbance}}}{\text{Control}_{\text{absorbance}}} \times 100$$

3 Results and Discussion

3.1 Synthesis of nanoparticles and UV- visible spectroscopy

The colour change indicates the synthesis of nanoparticles. It was monitored immediately after the addition of silver nitrate and gold chloride to the cell free extract of *N. pupukensis*. There was colour change from pale green to brown after 25 min of incubation for NP-AgNPs with the UV-visible absorption spectrum at 436 nm (Figure 1a). The colour change in NP-AgNPs is in agreement with previous reports on the synthesis of nanoparticles from algae and also the surface plasmon resonance is in agreement with earlier reports (22, 31). Similarly, for gold nanoparticles, the colour change from pale green to purple was observed within 2 h of incubation and the surface plasmon resonance peak was obtained at 554 nm (Figure 1b). The absorbance peak obtained for NP-AuNPs is similar to previous

IOP Conf. Series: Materials Science and Engineering **805** (2020) 012010 doi:10.1088/1757-899X/805/1/012010

reports [18-19, 23]. The absorption spectrum of UV-visible is a strong confirmation that metallic nanoparticles have been formed in the mixture.

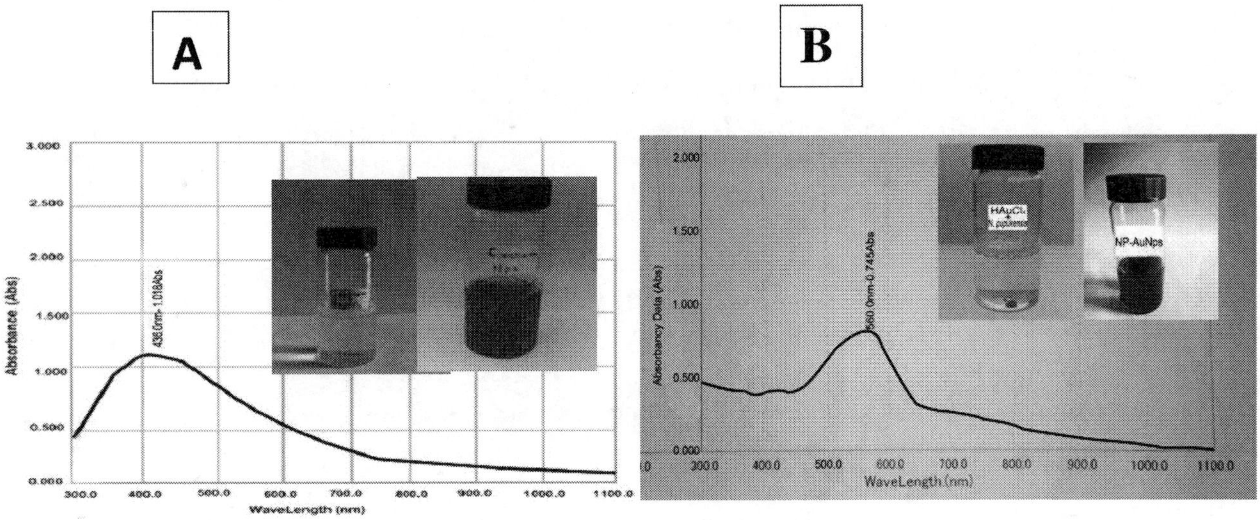

Figure 1. UV-visible absorption spectra of NP-AgNPs (A) and NP-AuNPs (B)

3.2 Electron microscopy and elemental composition of nanoparticles

The scanning electron microscopic image of NP-AgNPs showed spherical shape with particle size ranging from 52-179 nm (Figure 2a), while the transmission electron microscopy of NP-AuNPs showed circular shaped particles with size ranging from 5-34 nm (Figure 2b). The micrographs obtained confirm the bioreduction of silver ions. The nanoparticles of both silver and gold have broad absorption peaks and this accounted for the polydispersed nature of the nanoparticles.

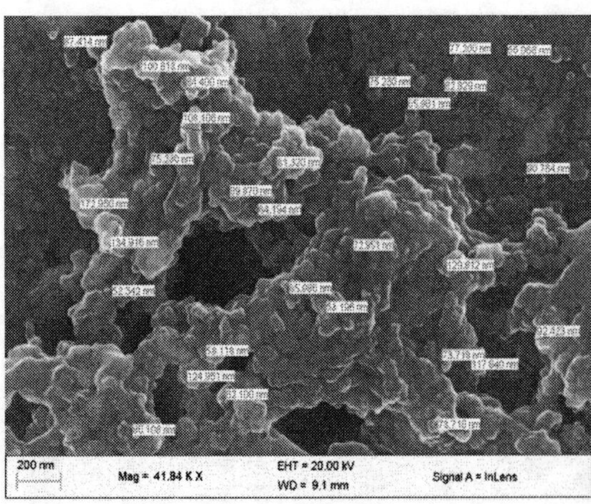

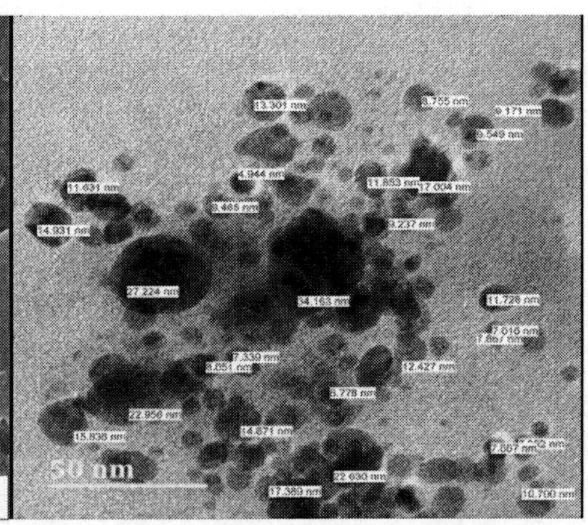

Figure 2a. SEM image of NP-AgNPs **Figure 2b**. TEM image of NP-AuNPs

The EDX spectra of NP-AgNPs revealed that silver was the most abundant with weight composition of 33.33% with a strong energy signal strength between 3-3.2keV (Figure 3a) while a strong signal strength was obtained for gold in NP-AuNPs with a weight composition of 96.62% (Figure 3b). The absorption peak displayed at energy signal strength between 3 and 4 keV is similar to previous report for metallic silver nanoparticles [14].

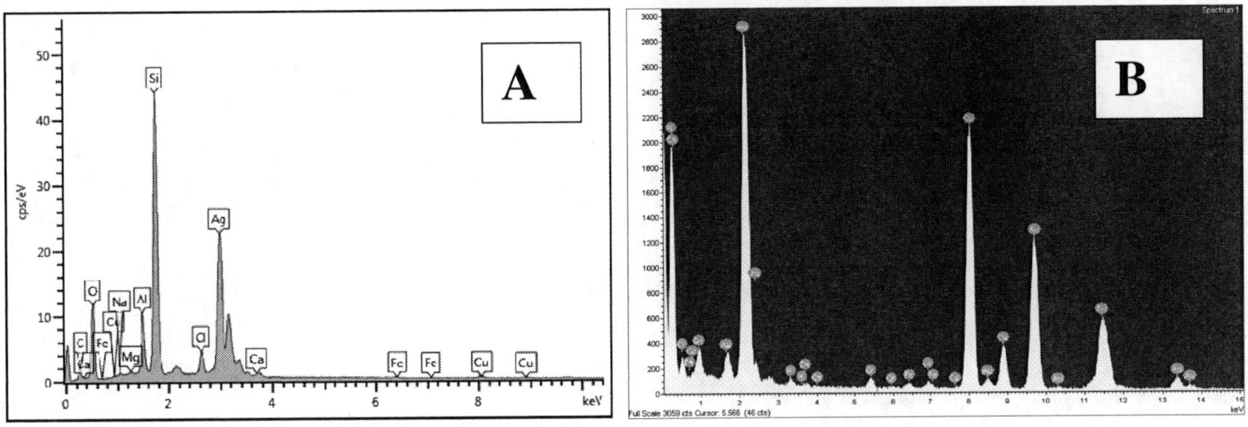

Figure 3. EDX of NP-AgNPs (A) and NP-AuNPs

3.3 Fourier Transform Infrared Spectroscopy of Algal Nanoparticles

The FTIR spectroscopy revealed prominent peaks at 3283.0, 2102.87, and 1636.02 cm^{-1} for NP-AuNPs (Figure 4a), while NP-AuNPs had prominent peaks at 3264.86, 2104.49, 1636.62, 1232.39 and 1028.97 cm^{-1} (Figure 4b). The absorbance bands present in the nanoparticles at 3283, 2102, 1636, 3264.86, 2104.49 and 1636.62 cm^{-1} indicated that alcohol and unsaturated hydrocarbon (O-H stretching vibrations in polyphenols and stretching vibrations of the (NH)=O group) were present, while absorbance band at 1232 cm^{-1} showed Aryl and Vinyl ethers (C-O stretching vibrations), and 1028 cm^{-1} shows the presence of Diakyl ether (C-O stretching vibrations). The different peaks which represent various functional groups were responsible for the reduction and capping of resulting nanoparticles.

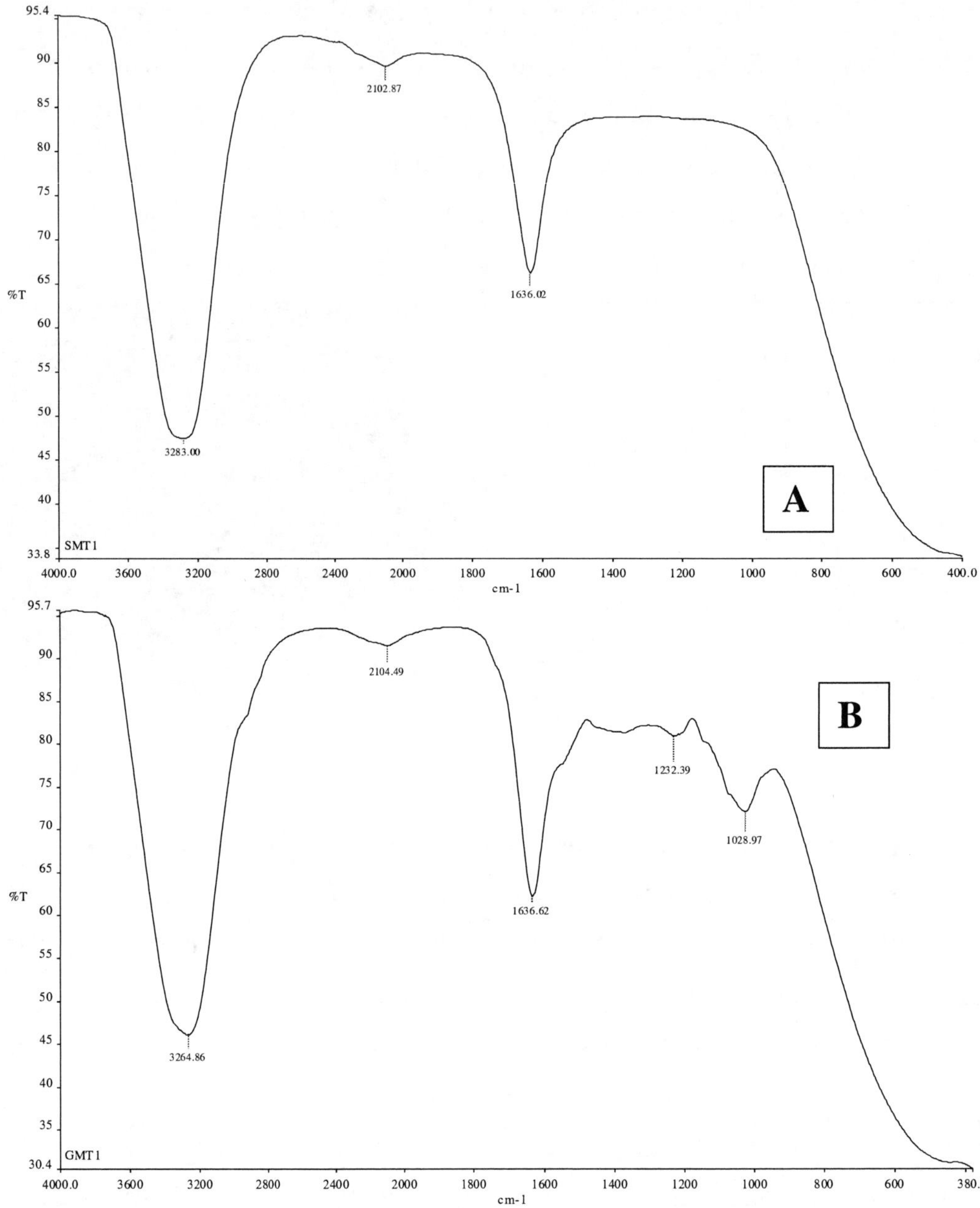

Figure 4. FTIR spectra of NP-AgNPs (A) and NP-AuNPs (B)

3.4 Antimicrobial activities of nanoparticles

The antimicrobial potential of both NP-AgNPs and NP-AuNPs was investigated against five pathogenic bacteria and toxigenic fungi. NP-AgNPs showed a high antibacterial potency of 43 mm, 24.5 mm, 27 mm and 39 mm against *Pseudomonas aeruginosa*, *E. coli*, *K. pneumoniae* and *S. marcescens* respectively, while NP-AuNPs was mildly active against both *Pseudomonas* sp and *S. marcescens* with zone of inhibition of 27.5 mm and 28.5 mm respectively (Figure 5a). The activity of NP-AgNPs and NP-AuNPs against *Staphylococcus aureus* was 24.5 and 17.5 mm respectively (Figure 5b). The comparison of standard antibiotic disc with nanoparticles of *N. pupukensis* shows high activity of silver nanoparticles against the five test pathogens. The highest zone of inhibition was obtained against *Pseudomonas aeruginosa* (43 mm), while the lowest was obtained with *E. coli* (24.5 mm). The silver nanoparticles had higher antibacterial activity compared to their gold counterparts which was fairly active against *E. coli* and *S. marcescens*. Gold nanoparticles of *Glacilaria corticata* were reported to be inactive against some bacteria but a high activity was reported when combined with antibiotics [19]. Although the AgNPs have been reported to be dose dependent in their antibacterial activity [22], the report in this work shows that silver nanoparticles of *N. pupukensis* were active at all doses compared against the test pathogens (Figure 6). However, the cell-free extract *N. pupukensis* did not inhibit any of the test isolates.

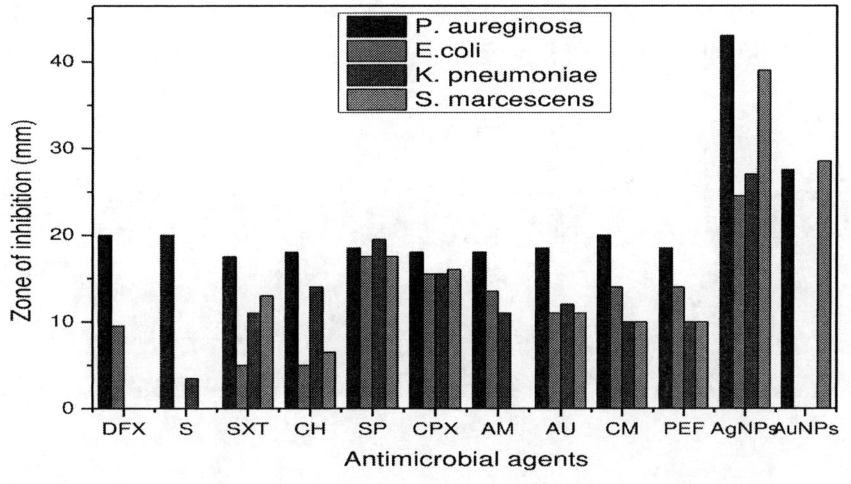

Figure 5a. Antibacterial activity of some antibiotics and nanoparticles against Gram negative pathogenic bacteria {OFX (10 μg Tarivid), S (Streptomycin 30 μg), SXT (Septrin 30 μg), CH (Chloramphenicol 30 μg), SP (Sparfloxacin 10 μg), CPX (Ciprofloxacin 10 μg), AM (Amoxacillin 30 μg), AU (Augmentin 30 μg), CM (Gentamycin 10 μg) PEF, (Pefloxacin 10 μg), AgNPs (Silver nanoparticles) and AuNPs (Gold nanoparticles)}.

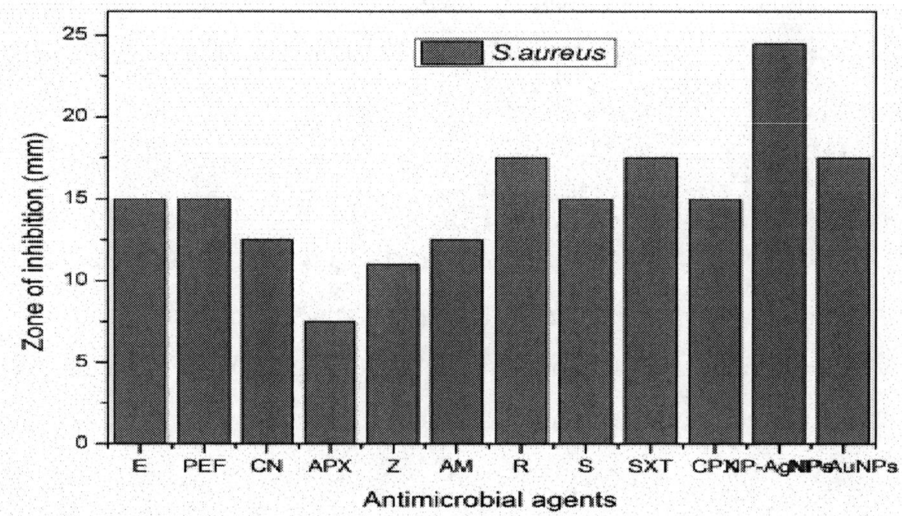

Figure 5b. Antibacterial activity of some antibiotics and nanoparticles against *Staphylococcus aureus* {E (10 µg Erythromycin), PEF, (Pefloxacin 10 µg), CN (Gentamycin 10 µg), APX (Ampiclox 30 µg),Z (Zinnacef 20 µg), AM (Amoxacillin 30 µg), R(Rocephin 25 µg), S (Streptomycin), SXT (Septrin 30 µg), CPX (Ciprofloxacin 10 µg), AgNPs (Silver nanoparticles) and AuNPs (Gold nanoparticles)}.

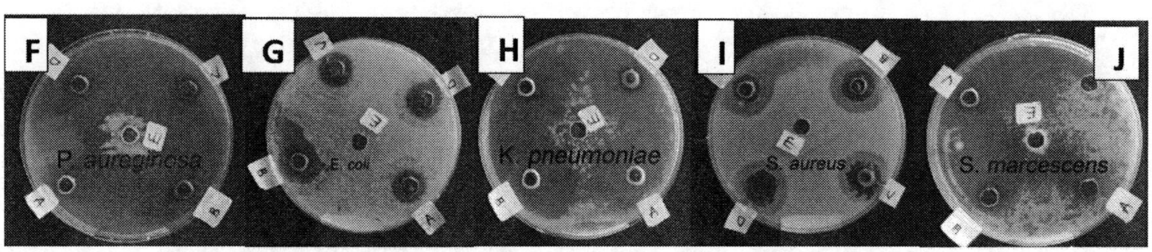

Figure 6. Antibacterial activity of silver nanoparticles synthesized from *N. pupukensis* against (F) *P. aeruginosa*; (G) *E. coli*; (H) *K. pneumoniae*; (I) *S. aureus*, and (J) *S. marcescens* at different concentrations of A (25 µl), B (50 µl), C (75 µl), D (100 µl) and E (cell free extract of *N. pupukensis*).

The antifungal potency of NP-AgNPs was confirmed with mycelial inhibition of 80.6, 57.1, 79.4, 65.4 and 69.8% against *A. niger, A. fumigatus, A. flavus, F. solani* and *C. albicans* respectively, while NP-AuNPs had 79.4, 44.3, 75.4, 54.9 and 66.4% against *A. niger, A. fumigatus, A. flavus, F. solani* and *C. albicans* respectively (Figure 7). The highest mycelial inhibition was obtained against *A. niger* by NP-AgNPs (80.6%) and the least was against *F.*

solani by NP-AuNPs (44.3 %). Both silver and gold nanoparticles of *N. pupukensis* were highly potent against the toxigenic fungi investigated and some of the plates showing mycelial inhibition are shown in Figure 8.

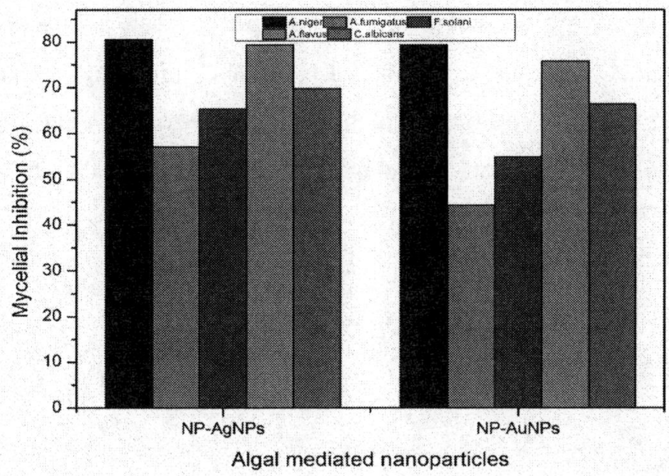

Figure 7. Antifungal activities of silver and gold nanoparticles of *Neodesmus pupukensis*

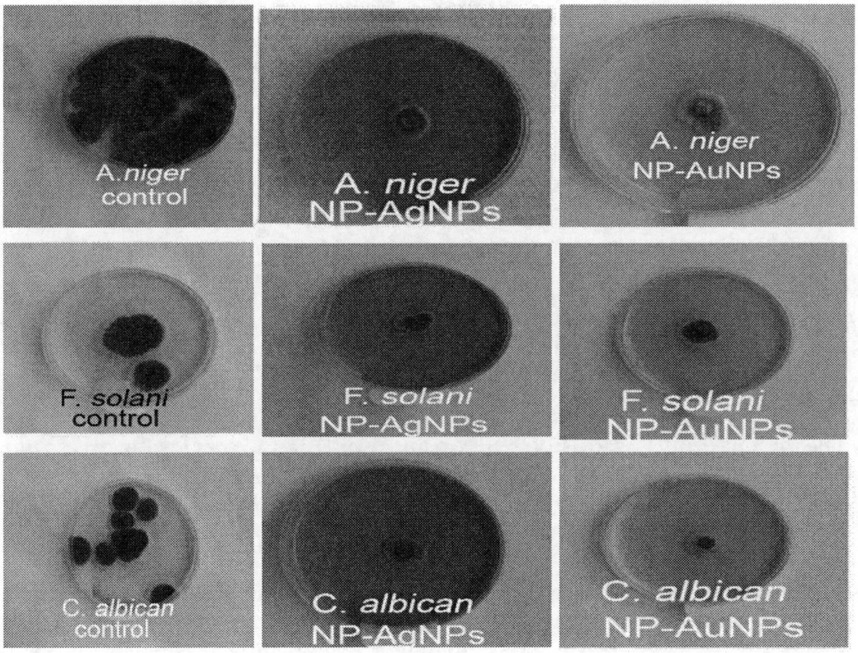

Figure 8. Potato dextrose agar plates showing mycelial inhibition of silver and gold nanoparticles of *Neodesmus pupukensis.*

3.5 Antioxidant activities of gold and silver nanoparticles

The ability of NP-AgNPs and NP-AuNPs to scavenge free radical against 2,2-diphenyl-1-picrylhydrazyl was high for NP-AuNPs at 68 .96%, followed by NP-AgNPs (41.21 %), while the least of 37.56% was obtained from the cell free extract of *N. pupukensis* (Figure 9). The ease and rapidity that comes with the use of DPPH has presented it as a tool in the determination of antioxidant potential of a compound. An increased antioxidant activity was obtained with the nanoparticles than the cell free extract of *N. pupukensis*; this agrees with a report on the antioxidant activity of silver nanoparticles of *Isochrysis* sp and the cell extract [31].

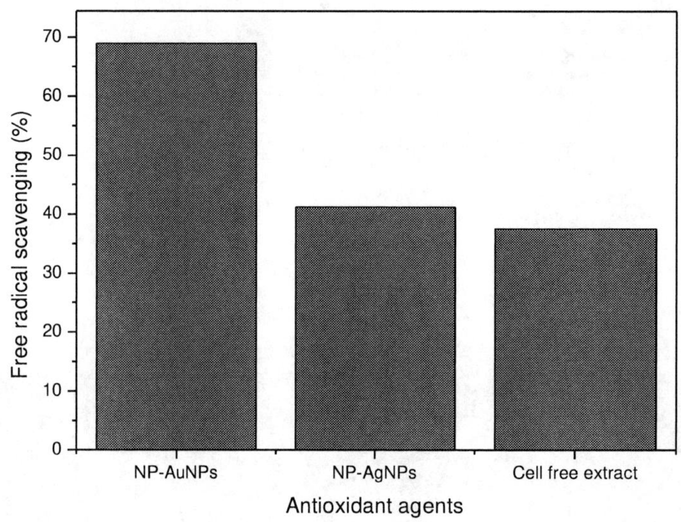

Figure 9. Free radical scavenging activity of silver nanoparticles, gold nanoparticles and cell-free extract of *N. pupukensis*

4.0 Conclusion

The green synthesis of both silver and gold nanoparticles from *N. pupukensis* was carried out. It was demonstrated that the silver nanoparticles were able to actively inhibit the growth of both Gram positive and negative bacteria, while the gold nanoparticle was fairly active against two of the five pathogens. The antifungal and antioxidant activities of the nanoparticles were also outstanding and this presents nanoparticles of *N. pupukensis* as a prospective candidate for several biomedical and biotechnological applications.

References

[1] Lauritano C, Andersen J H, Hansen E, Albrigtsen M, Escalera L, Esposito F, Helland K, Hanssen K, Romano G and Ianora A 2016 Bioactivity screening of microalgae for antioxidant, anti-inflammatory, anticancer, anti-diabetes, and antibacterial activities. *Front. Mar. Sci.* **3** 68. https://doi.org/10.3389/fmars.2016.00068.

[2] Uma R, Sivasubramanian V and Devaraj S N 2011 Evaluation of *in vitro* antioxidant activities and antiproliferative activity of green microalgae, *Desmococcus olivaceous* and *Chlorococcum humicola. J. Algal Biomass Utln.* **2** (3) pp 82 -93.

[3] Halliwell B 2007 Biochemistry of oxidative stress. *Biochem. Soc. Trans.* **35** (5) pp 1147-1150.

[4] Sansone C and Brunet C 2019 Promises and challenges of microalgal antioxidant production. *Antioxidants* **8** (7) pp 199-207.

[5] Skulberg O M M 2006 Bioactive chemicals in microalgae. In: Richmond A. (ed.) Handbook of microalgal culture, biotechnology and applied phycology. Oxford: Blackwell pp 485-512.

[6] Pugh N S A, Ross H N, ElSohly M A and ElSohly D S 2001 Isolation of three high molecular weight polysaccharide preparations with potent immunostimulatory activity from *Spirulina platensis, Aphanizomenon flos-aquae* and *Chlorella pyrenoidosa*. *Planta Med.* **67** (08) pp 737-742.

[7] Scalbert A, Manach C, Morden C, Rémésy C, Jiménez L 2005 Dietary polyphenols and prevention of diseases. *Crit. Rev. Food Sci. Nutri.* **45** (4) pp 287-306.

[8] Takaichi S 2011 Carotenoids in algae: distributions, biosyntheses and functions. *Mar. Drug* **9** (6) pp1101-1118.

[9] Galasso C, Corinaldesi C and Sansone C 2017 Carotenoids from marine organisms: Biological functions and industrial applications. *Antioxidants* **6** (4) 96. https://doi.org/10.3390/antiox6040096.

[10] Ali HE, Shanab SM, Shalaby EA, Eldmerdash U, Abdullah MA Screening of microalgae for antioxidant activities, carotenoids and phenolic contents. *Appl. Mech. Mater.* **625** pp 156-159.

[11] Xu D P, Li Y, Meng X, Zhou T, Zhou Y, Zheng J and Li H B 2017 Natural antioxidants in foods and medicinal plants: extraction, assessment and resources. *Int. J. Mol. Sci.* **18** (1) 96. https://doi.org/10.3390/ijms18010096.

[12] Guiry M D 2005 How many species of algae are there? *J. Phycol.* **48** (5) pp 1057-1063.

IOP Conf. Series: Materials Science and Engineering **805** (2020) 012010 doi:10.1088/1757-899X/805/1/012010

[13] Manirafasha E, Ndikubwimana T, Zeng X, Lu Y and Jing K 2016 Phycobiliprotein: Potential microalgae derived pharmaceutical and biological reagent. *Biochem. Eng. J.* **109** pp 282-296.

[14] Kathiraven T, Sundaramanickam A, Shanmugam N and Balasubramanian T 2015 Green synthesis of silver nanoparticles using marine algae *Caulerpa racemosa* and their antibacterial activity against some human pathogens. *Appl. Nanosc.* **5** (4) pp 499-504.

[15] Vincy W, Jasmine M, Sukumaran S and Jeeval S 2017 Algae as a source for synthesis of nanoparticles- A Review. *Int. J. Latest Trends Eng. Technol.* Pp 005-009.

[16] Brayner R, Barberousse H, Hemadi M, Djedjat C, Yéprémian C, Coradin T, Livage J, Fiévet F and Couté A 2007 Cyanobacteria as bioreactors for the synthesis of Au, Ag, Pd, and P nanoparticles via an enzyme-mediated route. *J. Nanosci. Nanotech.* **7** (8) pp 2696-2708.

[17] Parial D, Patra H K, Dasgupta A K and Pal R 2012 Screening of different algae for green synthesis of gold nanoparticles. *Eur. J. Phycol.* **47** (1) pp 22-29.

[18] Castro L Blázquez M L Muñoz J A González F and Ballester A 2013 Biological synthesis of metallic nanoparticles using algae. *IET Nanobiotechnol.* **7** (3) pp 109-116.

[19] Naveena B E and Prakash S 2013 Biological synthesis of gold nanoparticles using marine algae *Gracilaria corticata* and its application as a potent antimicrobial and antioxidant agent *Asian J. Pharm. Clin. Res.* **6** (2) pp 179-182.

[20] Thakkar K N, Mhatre S S and Parikh R Y 2010 Biological synthesis of metallic nanoparticles. *Nanomed: Nanotechnol. Biol. Med.* **6** (2) pp 257-262.

[21] Sathishkumar R S, Sundaramanickam A, Srinath R, Ramesh T, Saranya K, Meena M and Surya P 2019 Green synthesis of silver nanoparticles by bloom forming marine microalgae *Trichodesmium erythraeum* and its applications in antioxidant, drug-resistant bacteria, and cytotoxicity activity. *J. Saudi Chem. Soc.* **23** (8) pp.1180-1191.

[22] Aziz N, Fatma T, Ajit V and Ram P 2014 Biogenic Synthesis of silver nanoparticles using *Scenedesmus abundans* and evaluation of their antibacterial activity. *J. Nanopart.* Article ID 689419. doi.org/10.1155/2014/689419.

[23] Geetha S, Sathakkathulzariya J, Arthi R and Blessie H 2014 Green synthesis of gold nanoparticle using marine cyanobacteria *Gloeocapsa* sp and the antitumour potential. *J. Chem. Pharm. Sci.* **4** pp 172-174.

[24] Rauwel P Küünal S Ferdov S and Rauwe E 2015A review on the green synthesis of silver nanoparticles and their morphologies studied via TEM. *Adv. Mater. Sci. Eng.* Article ID 682749. https://doi.org/10.1155/2015/682749.

[25] Singh V P, Srivastava P K and Prasad S M 2012 Differential effect of UV-B radiation on growth, oxidative stress and ascorbate-glutathione cycle in two cyanobacteria under copper toxicity. *Plant Physiol. Biochem.* **61** pp 61-70.

[26] Ely E W, Shintaniu A, Truman B, Speroff T, Gordon S M, Harrell F E Jr, Inouye S K, Bernard G R and Dittus R S 2004 Delirium as a predictor of mortality in mechanically ventilated patients in the intensive care unit. *J. Am. Med. Assoc.* **291** (14) pp1753-1762.

[27] Singaravelu G, Arockiyamari J, Ganesh K V and Govindaraju K 2007 A novel extracellular synthesis of monodisperse gold nanoparticles using marine alga, *Sargassum wightii* Greville. *Colloids Surf. B: Biointerf.* **57** (1) pp 97-101.

[28] Shahverdi AR, Fakhimi A, Shahverdi HR and Minaian S 2007 Synthesis and effect of silver nanoparticles on the antibacterial activity of different antibiotics against *Staphylococcus aureus* and *Escherichia coli*. *Nanomed: Nanotechnol. Biol. Med.* **3** (2) pp 168-71.

[29] Khatami M, Pourseyedi S, Khatami M, Hamidi H, Zaeifi M and Soltani L Synthesis of silver nanoparticles using seed exudates of *Sinapis arvensis* as a novel bioresource, and evaluation of their antifungal activity. *Bioresour. Bioprocess.* **2** 19. https://doi.org/10.1186/s40643-015-0043-y.

[30] Vivek M, Kumar P S, Steffi S and Sudha S 2011 Biogenic silver nanoparticles by *Gelidiella acerosa* extract and their antifungal effects. *Avicenna J. Med. Biotechnol.* **3** (3) pp 143-148.

[31] Tevan R, Jayakumara S, Sahimia N H A, Iqbala N F A, Zapria I, Govindana N, Rahima M H A, Ichwanb S J A and Maniama G P 2017 Biosynthesis of silver nanoparticles using marine microalgae *Isochrysis* sp. *J. Chem. Eng. Ind. Biotech.* **2** (1) pp 1-12.

IOP Conf. Series: Materials Science and Engineering **805** (2020) 012011 doi:10.1088/1757-899X/805/1/012011

Antioxidant and antimicrobial activities of nanosilver-mycomeat composite produced through solid state fermentation of tigernut waste and cassava pulp by *Pleurotus pulmonarius*

C O Bamigboye[#], J A Amao, I A Fadiora*, J D Adegboye, O E Akinola, A A Alarape, O R Oyeleke and E A Adebayo

Microbiology Unit, Department of Pure and Applied Biology, Ladoke Akintola University of Technology, P. M. B. 4000, Ogbomoso, Nigeria.

*E-mail: iafadiora@student.lautech.edu.ng; [#]coadegunlola@lautech.edu.ng

Abstract

Bacterial deterioration is one of the most critical subjects in the production, processing, transport, and storing of food. The search for novel materials for the synthesis of nanoparticles is unending and this is likely to continue in the upcoming years. In this study, mycomeat samples were produced using cassava pulp (C) and tiger nut aggregate waste (T). The mycomeat extracts were used to produce nanosilver-mycomeat by catalytic conversion of Ag^{+} to Ag^{0}. The antioxidant and antimicrobial activities of the nanosilver-mycomeat composites were determined using established methods. The nanosilver-mycomeat composites of both C and T showed increased antioxidant and antimicrobial activities compared to the ordinary mycomeat extract. This can find applications in the packaging of food, thereby preventing food samples from deterioration.

Keywords: Nanoparticles, nanosilver-mycomeat, antioxidant, antibacterial, medicinal mushroom, tiger nut

1 Introduction

Nanotechnology deals with materials ranging in size from 1 to 100 nm, mainly nanoparticles and nanostructures [1, 2]. It has found applications in multiple areas, including medical, pharmaceutical, textile, chemical sectors, paper conservation [3], removal of wastes such as dyes [4] and photocatalytic degradation [5]. The importance of nanomaterials was realized when researchers found that size can influence the physiochemical properties of a substance, including its optical properties. Nanoparticles can be synthesized at lowcost by using the extracts from plants and products from microbes [6]. The application of

IOP Conf. Series: Materials Science and Engineering **805** (2020) 012011 doi:10.1088/1757-899X/805/1/012011

nanotechnology in food is fairly recent [7], and has experienced a continuous upsurge since 2003 [2, 8].

Tested nanomaterials in food industry include natural product nanoparticles (NPs) and metal and metal oxide nanoparticles. Among all, silver nanoparticles have enjoyed increasing commercialization as antimicrobial compounds. On the other hand, gold nanoparticles are commonly used as sensors, while Titanium dioxide NPs are used as disinfecting agents, food additives and flavour enhancer [7]. Natural product nanoparticles are used as a delivery system, as a supplement in the food industry and in food contact materials. Nanoparticles are reported to be used in food and feed packaging industries as biosensors, antimicrobials and for extending the shelf-life of packaged foods [1, 9]. However, any material intended to be in contact with food and active and intelligent materials (AIM) must comply with the provision of the framework Regulation (EC) that defines the three requirements to ensure safe and quality food [8]. These requirements are that food contact materials must not transfer their components into food in quantities that could endanger human health, must not change the composition of the food in an unacceptable way and shall not deteriorate the taste, odour or texture of the food.

Tigernut (*Cyperus esculentus*) is commonly found in all continents except Antarctica [10]. They are considered as a weed in 15 countries [10], while they are highly nutritious, medicinal, but underutilized crops in other parts of the world [11]. *Cyperus esculentus* are a rich source of essential amino acids (oleic acid, leucine, methionine, glutamic acid and aspartic acid) and minerals (iron, calcium among others) [11, 12]. Medicinally, they are useful for treating indigestion, flatulence, dysentery and diarrhoea [11]. The tubers also function as a tonic, aphrodisiac, diuretic and as a stimulant [11, 13]. In addition, consumption of the tubers help in the treatment of poliomyelitis, boils, common cold, urinary tract infections and can prevent heart diseases [12, 14]. Cassava pulp is a rich source of energy and approximately 1.5 million tonnes of cassava pulp are produced annually [15]. Cassava pulp is a waste product from the '*fufu*' (a cassava meal) processing industry and it constitutes about 30% of the original weight of cassava roots.

Food quality and safety are of great concern and as a result, preservation of food against spoilage by microbial agents and undesirable physical changes is of paramount importance. Therefore, this study is centred on the production of nanosilver-mycomeat as an antibacterial and antioxidant for food packaging materials.

2 Materials and methods

2.1 Sample Collection

The Spawn of *Pleurotus pulmonarius* also known as Oyster mushroom was purchased from Osogbo, Osun State, Nigeria. Cassava pulp was obtained from cassava processor in Ogbomoso, while tigernut was purchased from Sabo, Ogbomoso, Oyo State, Nigeria.

2.2 Preparation of the mycomeat extract

The tigernut aggregate waste and cassava pulp were each used to prepare mycomeat following the methods of Bamigboye *et al.* [16]. Briefly, the tigernut was manually sorted and cleaned to remove particles and unwanted materials. Dates together with coconut were added to the tigernut in the ratio 1:1:10 and then soaked in clean water for 5 h to soften the aggregate. The tigernut aggregate was milled into slurry in the ratio 1:3 of tigernut to water and then sieved using cheese cloth to extract waste from the aggregate. The cassava pulp was pounded using mortar and pestle. Precisely 100 g of each of the aggregate waste and cassava pulp was filled into four transparent polythene bags each. The bags were autoclaved at 121 °C for 15 min, allowed to cool and a set of two were inoculated with the spawn of *Pleurotus pulmonarius* and incubated till full ramification. The remaining two sets were left unfermented. After full ramification, it was dried at 80 °C for 8 h and milled to powder. Precisely 0.1 g of the milled powder for the fermented cassava pulp (FC), fermented tigernut aggregate waste(FK), non-fermented cassava pulp (C) and non-fermented tiger nut aggregate waste (K), were each suspended in 10 ml of distilled water in a transparent 50 ml bottle, labelled and heated in a digital dry bath (Labnet, USA) at 60 °C for 1 h. The extract was filtered using Whatman No.1 filter paper and then centrifuged at 4000 rpm for 20 min. The filtrate was collected and stored in the refrigerator at 4 °C for further use.

2.3 Biogenic synthesis of AgNPs and effect of varying the reaction mixture

For the preparation of nanosilver-mycomeat, 1 ml of each mycomeat extract (C, FC, K and FK) was added to 5, 10 and 20 ml of 1 mM $AgNO_3$. The reaction mixture was placed under sunlight exposure for the photo activation of the silver ions and identification of the concentration with the highest colour intensity. A change in the colour of the reaction mixture were visually observed, followed by the measurement of its absorbance spectrum using UV-visible spectrophotometer operating at the range of 200-800 nm wavelength range.

2.4 DPPH radical-scavenging activities of the nanosilver-mycomeat

The free radical-scavenging activity of the AgNPs was assayed according to Guntur *et al.* [17]. Precisely1 ml of each nanosilver-mycomeat solution at different concentrations (1, 2, 5, 10, 20, 40 μg/ml) were mixed with 4 ml of methanol solution of DPPH in slant bottles.

Controls were set up comprising 4 ml of DPPH solution and 1 ml of the supernatant extract (for both the fermented and non-fermented) or $AgNO_3$ in transparent bottles, incubated in the dark at room temperature for 30 min. The setups were measured at 517 nm using a spectrophotometer. Antioxidant activity was estimated by calculating the % inhibition by following the formula [18]:

$$\% \text{ DPPH radical scavenging activity} = \frac{(A0 - A1)}{A0} \times 100$$

Where, A0 is the absorbance of the control and A1 is the absorbance of the extract.

2.5 Antimicrobial activities of the synthesized AgNPs

Four clinical isolates were obtained from LAUTECH Teaching Hospital, Ogbomoso, Oyo state, Nigeria. The synthesized nanosilver-mycomeat were evaluated as antimicrobial agent using point inoculation method and agar well diffusion technique [19]. Selected isolates included *Staphylococcus aureus*, *Staphylococcus epidermidis*, *Escherichia coli* and *Proteus* species.

3 Results and Discussion

3.1 Green synthesis and characterization of silver nanoparticles

The formation of AgNPs by reduction of the aqueous silver during exposure to the mycomeat supernatant showed dark brown colour, which suggested the formation of AgNPs in solution. The colour changed from colourless to dark brown and the reaction stabilized after 1hour of exposure to sunlight (Figure 1). The colour changed due to excitation of surface Plasmon vibrations of silver nanoparticles, which gave rise to the surface Plasmon band [20]. This means that the mycomeat extract can be used in green synthesis of silver nanoparticles. The formation of nanosilver-mycomeat was confirmed by the UV-vis spectroscopy (Figure 2). The UV-vis spectra showed an absorption peak at a wavelength of approximately 410 and 420 nm for FK and FC respectively (Figure 2). Alsalhi *et al.* [21] prepared AgNPs which exhibited UV-vis absorption within the range 410-430 nm. In another study by Sriramulu and Sumathi [22], absorbance peaks were observed at 434 and 414 nm for AgNPs synthesized using extracts of *Ganoderma lucidum* and *Agaricus bisporus* respectively. Abdel-Aziz *et al.* [23] reported a broad absorption peak at 400 nm for a nanocomposite that was synthesized.

IOP Conf. Series: Materials Science and Engineering **805** (2020) 012011 doi:10.1088/1757-899X/805/1/012011

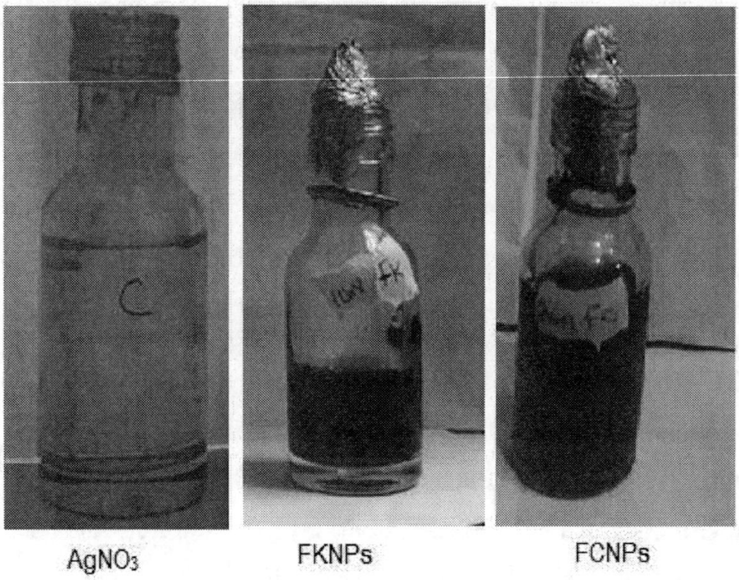

Figure 1. Visual observation of the nanosilver-mycomeat compostes synthesized using fermented tigernut aggregate waste (FKNPs) and fermented cassava pulp (FCNPs) after 1 h of reaction.

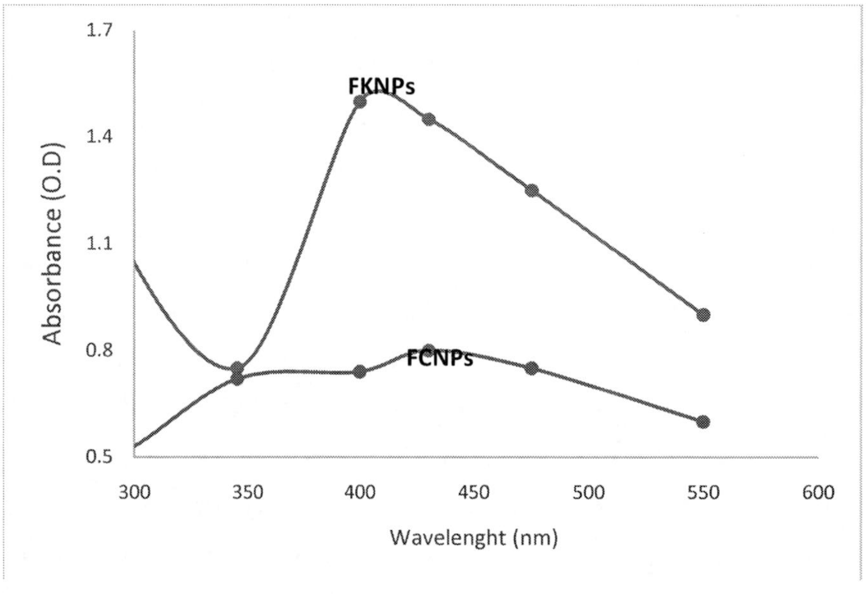

Figure 2. UV-vis absorption spectra of nanosilver-mycomeat composites synthesized using tigernut aggregate waste (FKNPs) and cassava pulp (FCNPs)

3.2 Antioxidant Activities

The 1, 1-diphenyl-2-picrylhdrazyl (DPPH) has been widely used to evaluate the free radical scavenging capacity of antioxidants. DPPH free radical is reduced to the corresponding hydrazine when it reacts with hydrogen donors. DPPH can make stable free radicals in aqueous or methanol solution. Resulting DPPH produces purple colour in methanol solution and fades to shades of yellow colour in the presence of antioxidants. The scavenging efficacy of the nanosilver-mycomeat composites is presented in Figure 3. The present study showed that nanosilver-mycomeat composites can effectively scavenge DPPH radicals, with 1 µg/ml of FCNPs, and FKNPs scavenging up to 66 and 70% of DPPH respectively. This is higher compared to the activities obtained for the ordinary mycomeat extracts. Only 41.5 and 56.4% of DPPH were scavenged by FC and FK respectively when nanoparticles were not synthesized with the extracts. The study showed that the nanosilver-mycomeat composites exhibited better DPPH scavenging activities compared to the plain extracts. In an earlier study by Boonsong *et al.* [24], 500 µg/ml of ethanolic extracts of *Lentinus edodes, Pleurotus eous* and *Auricularia auricula* scavenged only 50, 58 and 22% of DPPH radicals respectively. Synthesized silver nanoparticles of *Agaricus bisporus* extract had antioxidant activity of 75% [22], while 70% was reported for aqueous extracts of *Cymbopogon citratus* [25]. Antioxidant activity can be linked with polyphenolic content, as well as flavonoids [26].

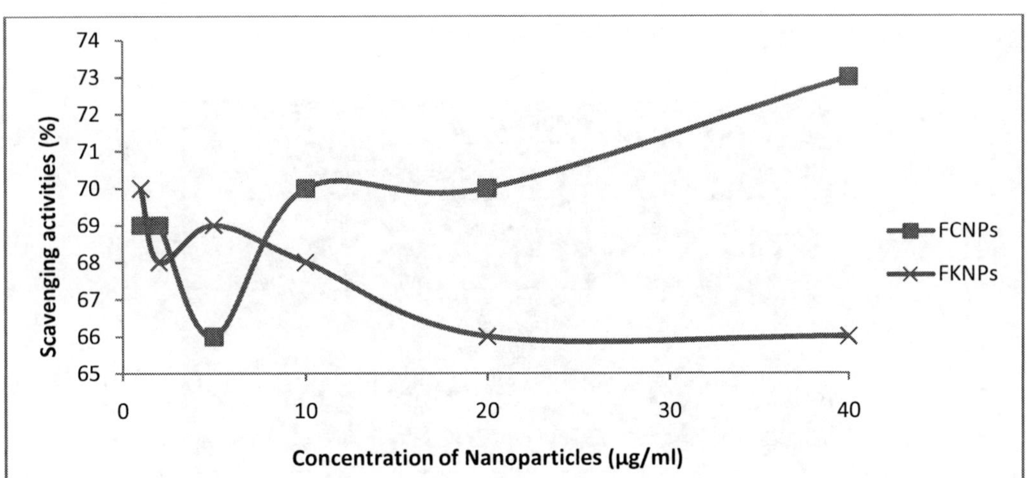

Figure 3. DPPH radical scavenging activities of nanosilver-mycomeat synthesized using tigernut aggregate waste (FKNPs) and cassava pulp (FCNPs)

IOP Conf. Series: Materials Science and Engineering **805** (2020) 012011 doi:10.1088/1757-899X/805/1/012011

3.3 Antimicrobial activities of the synthesized silver nanoparticles

The biosynthesized nanosilver-mycomeat composites displayed mild inhibitory activity against selected clinical isolates of bacteria. At 10 µg/ml, FK and FC mildly inhibited strains of *Escherichia coli, Proteus* spp and *Staphylococcus epidermidis* with the zone of inhibition being 14, 12 and 11 mm respectively. The point inoculation method strictly inhibited the growth of the organism to the point of inoculation (Figure 4). Interestingly, both silver nitrate and ordinary mycomeat had no inhibitory effect on the clinical isolates. The inhibitory activities displayed by AgNPs is attributed to its increased surface area at nanometer scale, thus the amount of silver ions released is directly linked to antimicrobial efficacy [27]. The antibacterial activities of AgNPs obtained in this study revealed the efficacy of the nanosilver-mycomeat composite to inhibit multi-drug resistant microbes in the environment. Other authors have demonstrated the antimicrobial effects of biosynthesized silver nanoparticles [28].

Main mechanism enabling the nanoparticles to act on bacteria includes the interaction of nanoparticles with membrane proteins and subsequent accumulation in the cell membrane, thereby affecting its permeability. Silver nanoparticles notably interact with the cell surface of some bacteria. This leads to structural changes and damage to cell membrane, thereby increasing the permeability of the bacteria [23]. It was also postulated that the uptake of free silver ions lead to a disturbance in ATP production and DNA replication [28, 29]. Biosynthesized AgNPs were reported to show 2 and 1.6 mm zones of inhibition against *E. coli* and *S. aureus* respectively [30]. This is similar to what was obtained in this report.

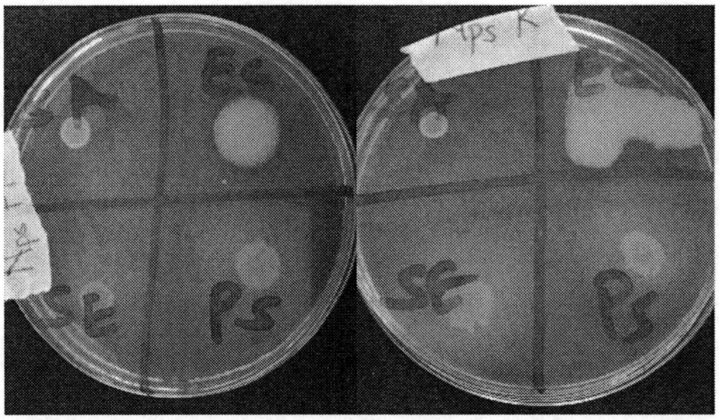

Figure 4. The antibacterial activities of nanosilver-mycomeat synthesized from the tigernut waste against the selected clinical bacteria isolates (*Escherichia coli, Staphylococcus aureus, S. epidermidis,* and *Proteus* sp) using pin point inoculation method.

4 Conclusion

AgNPs was successfully synthesized from aqueous extract of fermented cassava pulp and tigernut aggregate waste. This study has showed that the extract obtained from cassava pulp and tigernut aggregate which are important agro waste, can be used for the biogenic, green and eco-friendly synthesis of AgNPs. The particles showed mild activities against multi-drug resistant clinical bacteria strains. Fermented nanosilver-mycomeat extract (FCNPs) synthesized from cassava pulp exhibited potent higher antioxidant activities compared to that prepared from tigernut aggregate waste (FKNPs).

References

[1] Pathakoti K, Manubolu M and Hwang H-M 2017 Nanostructures: Current uses and future applications in food science. *J. Food Drug Anal.* **25 (2)** pp 245-253.

[2]He X and Hwang H-M 2016 Nanotechnology in food science: Functionality, applicability, and safety assessment. *J. Food Drug Anal.* **24** (2) pp 671-681.

[3] Fouda A, Abdel-Maksoud G, Abdel-Rahman, M A, Salem S S, Hassan SE-D and El-Sadany MA-H 2019 Eco-friendly approach utilizing green synthesized nanoparticles for paper conservation against microbes involved in biodeterioration of archaeological manuscript. *Int. J. Biodeterior. Biodegrad.* **142** pp 160-169.

[4] Naghizadeh A and Ghafouri M 2017 Synthesis and performance evaluation of chitosan prepared from Persian gulf shrimp shell in removal of reactive blue 29 dye from aqueous solution (isotherm, thermodynamic and kinetic study). *Iran J. Chem. Eng.* **36** (3) pp 25-36.

[5] Kamranifar M, Masoudi F, Naghizadeh A and Asri M 2019 Fabrication and characterization of magnetic cobalt ferrite nanoparticles for efficient removal of humic acid from aqueous solutions. *Desalin. Water Treat.* **144** pp 233-242.

[6] Aswathy A S and Philip D 2012 Green synthesis of gold nanoparticles using *Trigonella foenum-graecum* and its size-dependent catalytic activity. *Spectrochim. Acta Part A: Mol. Biomol. Spectr.* **97** pp 1-5.

[7] Hamad A F, Han J-H, Kim B-C and Rather I A 2018 The intertwine of nanotechnology with the food industry. *Saudi J. Biol. Sci.* **25** (1) pp 27-30.

[8] US DOA, 2003 Nanoscale science and engineering for agriculture and food systems: a report submitted to cooperative state research, education and extension service.

Washington, DC: Department of Agriculture; The United States Department of Agriculture: National Planning Workshop

[9] McClements D J and Xiao H 2012 Potential biological fate of ingested nanoemulsions: influence of particle characteristics. *Food Funct.* **3** (3) pp 202-220.

[10] Riemens MM, van der Weide RY and Runia WT 2008 Biology and control of *Cyperus rotundus* and *Cyperus esculentus*, review of a literature survey. Plant Research International BV Wageningen.

[11] Adejuyitan JA 2011 Tigernut processing: its food uses and health benefits. *Am. J. Food Technol.* **6** (3) pp 197-201.

[12] Maduka N and Ire FS 2018 Tigernut plant and useful application of tigernut tubers (*Cyperus esculentus*)-A review. *Curr. J. Appl. Sci.Technol.* **30** pp 1-23.

[13] Belewu MA and Abodunrin OA 2008 Preparation of kunnu from unexploited rich food source: tiger nut (*Cyperus esculentus*). *Pak. J. Nutr.* **7** (1) pp 109-111.

[14] Okorie SU and Nwanekezi EC 2014 Evaluation of proximate composition and antinutritional factors of *Cyperus esculentus* (tigernut) as influenced by boiling. *IOSR J. Environ. Sci. Toxi. Food Tech.* **7** (2) pp 70-73.

[15] Norrapoke T, Wanapat M, Cherdthong A, Kang S, Phesatcha K and Pongjongmit T (2018) Improvement of nutritive value of cassava pulp and *in vitro* fermentation and microbial population by urea and molasses supplementation. *J. Appl. Anim. Res.* **46** (1): 242-247.

[16] Bamigboye CO, Oloke JK, Adebayo EA, Majolagbe ON, Ayandiran TA and Adegunlola EO 2013 Mycomeat production through the solid state fermentation of soymilk waste by *Lentinus subnudus*. *J. Nat. Sci. Res.* **3** (11) pp 88-93.

[17] Guntur SR, Kumar NS, Hegde MM and Dirisala, VR 2018 *In-vitro* studies of the antimicrobial and free-radical scavenging potentials of silver nanoparticles biosynthesized from the extract of *Desmostachya bipinnata. Anal. Chem. Insights* **13**. https://doi.org/10.1177/1177390118782877.

[18] Biswas M, Haldar PK, and Ghosh AK 2010 Antioxidant and free-radical-scavenging effects of fruit of *Dregea volubilis. J. Nat. Sci. Biol. Med.* **1** (1) pp 29-34.

[19] Jayammal D and Sivakumar T 2013 Antibacterial activity of protein extract of marine *Pseudomonas aeruginosa* against bacterial pathogens. *Int. J. Curr. Microbiol. App. Sci.* **2** (7) pp 207-216.

[20] Salem W, Leitner DR, Zingl FG, Schratter G, Prassl R, Goessler W, Reidl J and Schild S 2015 Antibacterial activity of silver and zinc nanoparticles against *Vibrio cholerae* and enterotoxic *Escherichia coli. Int. J. Med. Microbiol.* **305** (1) pp 85-95.

[21] AlSalhi MS, Elangovan K, Ranjitsingh AJ, Murali P, Devanesan S. Synthesis of silver nanoparticles using plant derived 4-N-methyl benzoic acid and evaluation of antimicrobial, antioxidant and antitumor activity. *Saudi J. Biol. Sci.* **26** (5) pp 970-978.

[22] Sriramulu M and Sumathi S 2017 Photocatalytic, antioxidant, antibacterial and anti-inflammatory activity of silver nanoparticles synthesised using forest and edible mushroom. *Adv. Nat. Sci.: Nanosci. Nanotechnol.* **8** (4) 045012.

[23] Abdel-Aziz MS, Shaheen MS, El-Nekeety AA and Abdel-Wahhab MA 2014 Antioxidant and antibacterial activity of silver nanoparticles biosynthesized using *Chenopodium murale* leaf extract. *J. Saudi Chem. Soc.* **18** (4) pp 356-363.

[24] Boonsong S, Klaypradit W and Wilaipun P 2016 Antioxidant activities of extracts from five edible mushrooms using different extractants. *Agric. Nat. Res.* **50** (2) pp 89-97.

[25] Ajayi E and Afolayan A 2017 Green synthesis, characterization and biological activities of silver nanoparticles from alkalinized *Cymbopogon citratus* Stapf. *Adv. Nat. Sci.: Nanosci. Nanotechnol.* 2017 **8** (1) 015017.

[26] Vitta Y, Figueroa M, Calderon M and Ciangherotti C 2020 Synthesis of iron nanoparticles from aqueous extract of *Eucalyptus robusta* Sm and evaluation of antioxidant and antimicrobial activity. *Mater. Sci. Energ. Technol.* **3** pp 97-103.

[27] Schneider, G 2017 Antimicrobial silver nanoparticles-regulatory situation in the European Union. *Mater. Today: Proc.* **4** pp S200-S207.

[28] Jafari A, Pourakbar L, Farhadi K, Mohamadgolizad L and Goosta Y 2015 Biological synthesis of silver nanoparticles and evaluation of antibacterial and antifungal properties of silver and copper nanoparticles. *Turk. J. Biol.* **39** (4) pp 556-561.

[29] Mirzajani F, Ghassempour A, Aliahmadi A and Esmaeli MA 2011 Antibacterial effect of silver nanoparticles on *Staphylococcus aureus. Res. Microbiol.* **162** (5) pp 542-549.

[30] Husseiny MI, El-Aziz MA, Badr Y and Mahmoud MA 2007 Biosynthesis of gold nanoparticles using *Pseudomonas aeruginosa. Spectrochim. Acta Part A: Mol. Biomol. Spectr.* **67** (3-4) pp 1003-1006.

IOP Conf. Series: Materials Science and Engineering **805** (2020) 012012 doi:10.1088/1757-899X/805/1/012012

Effect of particle size of rice husk ash on aluminium/graphene composites

AA Yekinni*[1,2], MO Durowoju[2], JO Agunsoye[3] and LO Mudashiru[2]

[1,2]Department of Mechanical Engineering, Lagos State Polytechnic, Ikorodu, Nigeria

[2]Department of Mechanical Engineering, Ladoke Akintola University of Technology, Ogbomoso, Nigeria

[3]Department of Metallurgical and Materials Engineering, University of Lagos, Akoka, Nigeria

*E-mail: kunleyk2003@yahoo.com

Abstract

Aluminium/graphene (Al/G) composite has gained wider engineering applications because of its unique properties such as lightweight, excellent corrosion resistant, improved strength and enhanced thermal conductivity. However, difficulty in dispersing graphene in molten aluminium remains one of the major optimization challenges. This study developed, by stir casting technique, composites of aluminium alloy from recycled aluminium can, 0.4 wt. % graphene (G) and 1.6 wt.% rice husk ash (RHA) with particle sizes of 150, 300 and 600 μm. The tensile, hardness, impact and fatigue properties were analyzed using Instron Extensometer, Vickers hardness tester, Charpy Impact machine and rotating fatigue machine respectively. Furthermore, the cast samples were characterized using scanning electron microscope (SEM) and X-ray diffractometer (XRD). The morphology of the microstructure of all composites showed the retention and uniform distribution of G and RHA particles in the Al matrix devoid of presence of harmful aluminium carbide with improved mechanical properties. The study established a new approach of dispersing graphene in molten aluminium through a stir cast method with 150 μm particle size of RHA given the best mechanical and morphological properties.

1. Introduction

Aluminium is the most commonly used matrix for the metal matrix composites (MMCs) as a result of several beneficial physical and mechanical properties [1-4]. The alloys are quite appealing due to their low density, their capability to be strengthened by precipitation, high thermal and electrical conductivity, their good corrosion resistance and their high damping

IOP Conf. Series: Materials Science and Engineering **805** (2020) 012012 doi:10.1088/1757-899X/805/1/012012

capacity. However, defects such as low temperature capability, high thermal expansion coefficient and inadequate mechanical and tribological characteristics limit the use of monolithic aluminium alloy in automobile and aerospace industries.

Reinforcement increases strength, stiffness, temperature resistance capability, but generally lowers the density of metal matrix composites. The prime role of reinforcement is to carry the load that of the matrix to transfer the load to the fibres with maximum efficiency. Sometimes, the reinforcement surface can be coated to prevent a chemical reaction with the matrix. Common reinforcements are silicon carbide (SiC), alumina (Al_2O_3), titanium carbide (TiC), barium carbide (B_4C), barium (B), graphite, graphene, rice husk ash among others. Reinforcement are generally tailored to have superior properties such as high specific strength and stiffness, increased wear resistance, corrosion resistance, improved thermal shock resistance, increase in Young's modulus and enhanced high temperature performance, better mechanical fatigue and creep resistance than those of monolithic alloys [5].
It has been affirmed that graphene is the strongest material ever measured to have the highest stiffness, the crystal most elastic and most thermally conductive material [6, 7]. Production of Al composite from graphene despite its influence on mechanical and thermal properties is confronted with many challenges especially where increase in tensile property of the composite is required.

A serious challenge is that, carbon nanomaterials like graphene are initially entangled or agglomerated via Van der Waal bonds, and hence they are not easily dispersed in the melt of the Al matrix. The use of solid-state techniques therefore has been considered as more promising routes, thus enabling mechanical dispersion of such nano-materials into metals via severe plastic deformation of the metal matrix [8-14]. Choi *et al.* [14] emphasized that without solving the critical problem in dispersion, the Giga-Pascal-level strength of the graphene would not be of importance value for automotive, aerospace and other industrial/structural composite applications.

From the foregoing, this work is aimed at solving dispersion problems in the production of aluminium/graphene composites using rice husk ash as carrier and produce Al/RHA/G composites with improved mechanical properties, devoid of harmful intermetallic compounds.

IOP Conf. Series: Materials Science and Engineering **805** (2020) 012012 doi:10.1088/1757-899X/805/1/012012

2 Materials and Methods

2.1 Preparation of casting

About 18 kg of waste aluminium cans were melted in a diesel-fired crucible furnace using 30 kg crucible pot following the procedure adopted by earlier workers [15, 16]. The rice husk ash was produced by calcinating the rice husk in a muffle furnace at about 700 °C for about 2 h. The husk was well lagged with cotton wool to prevent excessive oxidation. The ash was sieved and graded into three particle sizes (150, 300 and 600 μm) (Figure 1). Graphene powder (NanoCarbon platelets) was purchased from Graphitene Ltd., Manchester United Kingdom. The powder has thickness less than 5 μm and lateral size ranging between 0.5-5 μm [15].

Two blends of composites namely: Al/RHA and Al/RHA/G were produced according to literature [15,16] from recycled aluminium alloy, rice husk ash particulate between 150-600 μm at 1.6 wt.% and 0.4 wt.% graphene of less than 5 μm particle size. Each blend of composites was mixed in accordance with the design of experiment in a fuel fired crucible furnace melted to 750 ± 2 °C. The particulates were preheated to 600 °C for three hours before dispersing into the molten bath of Al and stirred for homogeneity using a developed and improvised motorized mobile stirrer at 140 rpm for 2 min. The molten bath was poured into a pre-heated mould coated permanent die cavity steel mould and allowed to cool to room temperature before knockout. The cast samples produced were characterized using scanning electron microscope (SEM) and X-ray diffractometer (XRD). Furthermore, the cast were machined in accordance with ASTM standards into appropriate coupon for density measurement and analyses of tensile, impact, fatigue and hardness properties.

2.2 Preparation of test specimens

The microstructures were studied using scanning electron microscope equipped with an Oxford INCA™ energy dispersive spectroscopy system using a polished sample firmly held on the sample holder of a double-sided carbon tape before putting the inside the sample chamber. The SEM operation was carried out at an accelerating voltage of 5 to 20 kV. The operation was carried out at Spectrum Analytical Facility, Department of Physics, University of Johannesburg, South Africa.

IOP Conf. Series: Materials Science and Engineering **805** (2020) 012012 doi:10.1088/1757-899X/805/1/012012

XRD analyses with Cu-Kα radiation monochromatic multilayered mirrors were used for the analysis, using a PANalyticalX'Pert PRO. The machine was operated at a voltage of 45 kV and 40 mA current. A Rietveld refinement software, TOPAS™ was used to quantitatively analyze the results.

Specimens of dimensions 50 × 10 × 10 mm were prepared to determine the density. Digital Pioneer weighing balance by Ohaus Corporation, USA was used to determine the mass while the volume was obtained by Archimedes principle using graduated measuring cylinder. The rise in volume of water was recorded and used together with the mass to calculate the densities using equation 1.

$$\text{Density} = \frac{Mass}{Volume} \qquad (1)$$

The Vickers diamond test was carried out using LECO AT700 micro hardness tester at University of Lagos. A fine-grained emery polishing papers was used to prepare the samples for accurate measurement of microhardness. The specimen was further prepared using phenolic powder, grinded and then polished to produce a hardness specimen with a smooth surface finish. Then, applied load of about 50 Kg was used to indent the specimen with a dwelling time of about 10 seconds at three different positions on the specimen noting the depth of indentation to obtain the microhardness of each specimen.

Tensile test was conducted using Instron Extensometer, (model: Instron 3369), system ID: 3369S3457 at Engineering Materials Development Institute (EMDI) of Akure, Ondo State, Nigeria. The tensile test specimen was prepared in accordance with the American Society for Testing and Materials E8.

The impact energies of the control and produced composites were determined using aids of Avery Denison Universal Impact Testing Machine at Mechanical Engineering Department, University of Lagos. Specimen was prepared in accordance to ASTM A370.

Fatigue test was measured using SM 1090 Rotating Fatigue Machine at Department of Mechanical Engineering, Lagos State Polytechnic, Ikorodu. Specimen was prepared in line with ASTM E466 standard for fatigue measurement.

IOP Conf. Series: Materials Science and Engineering **805** (2020) 012012 doi:10.1088/1757-899X/805/1/012012

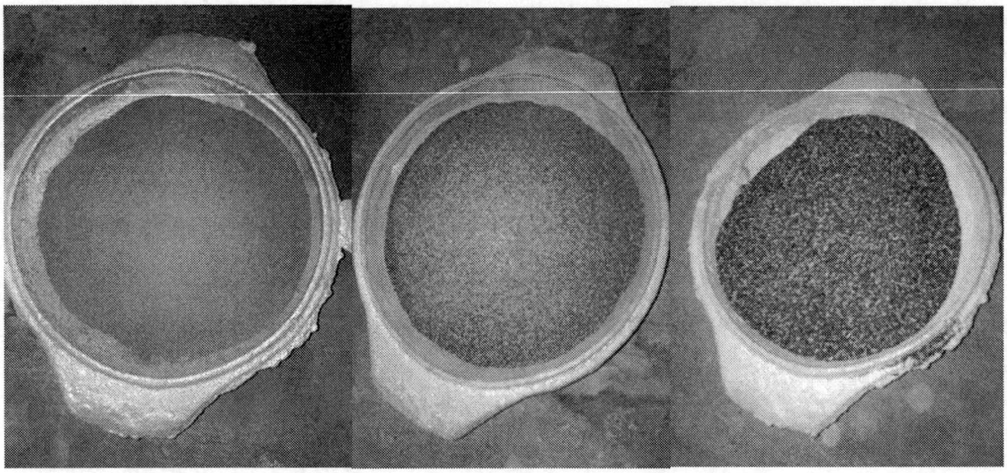

(a) (b) (c)

Figure 1. Graded rice husk ash into (a) 150 (b) 300 and (c) 600 μm

3. Results and Discussion

3.1 Compositional Analysis

The elemental composition of the unreinforced aluminium alloy sample as obtained from the spectrometric analysis shown in Table 1 revealed that the waste aluminium can is very rich in aluminium with reasonable amount of Zn, Cu, Ti and Si. These elements are expected to greatly influence the phase reactions at various particle sizes, including physical and mechanical properties of the developed composites. Table 2 shows the results of X-ray fluorescence (XRF) analyses of the rice husk ash. Oxides like SiO_2, Al_2O_3 and Fe_2O_3 present in the ash could make the matrix harder and more durable and therefore result in improved physical-mechanical properties of the composites. Al_2O_3 and SiO_2 as previously stated [17-18] are well-known reinforcing additives for improving wear resistance and strength for Al composites. CaO could react with alumina and silica to form aluminates and calcium silicates, which have good adhesive properties and improve the load bearing capability of the composites. The presence of MgO which is a refractory material, can improve the high temperature capability with low thermal conductivity [19]. Formation of several intermetallic compound is also expected in the XRD analysis due to chemical reactions between these elements/compounds and Al alloy during casting.

Table 1. The elemental composition (wt.%) of the unreinforced cast aluminium alloy ingot.

Al	**Si**	**Fe**	**Cu**	**Mn**	**Mg**	**Zn**	**P**	**Ni**	**Ti**
94.0218	0.825	0.332	1.065	0.770	0.0046	2.014	0.021	0.053	0.912
S	**Cr**	**Sn**	**Mo**						
0.016	0.005	0.0023	0.006						

Table 2. Chemical compound (wt. %) composition of rice husk ash burnt at 700 °C

SiO_2	Al_2O_3	Fe_2O_3	CaO	MgO	SO_2	K_2O	Na_2O	Others	LOI
97.095	1.135	0.316	0.073	0.825	0.146	0.181	0.092	Balance	0.965

3.2 Scanning Electron Microscopy Analysis

It can be seen from Figures 2a-g that the control sample and all composites at different particle sizes showed reasonably uniform distribution and retention of RHA and G particles due to proper stirring adopted in the use of improvised mobile stirrer. The presence of reinforcements in grain refinement and reduction in porosity can be observed for all composites when compared to the control sample. The effect of particle sizes of RHA is clearly seen in all composite with micrographs of 150 μm having the fineness grain structure and fewer grain boundaries both for Al/RHA and Al/RHA/G composites. However, the effect of graphene in further refining the grain structure of the composites and the greyish colour of the matrix of all Al/RHA/G is also obvious as a result of changes in the intermetallic compounds when compared to the composites of Al/RHA.

IOP Conf. Series: Materials Science and Engineering **805** (2020) 012012 doi:10.1088/1757-899X/805/1/012012

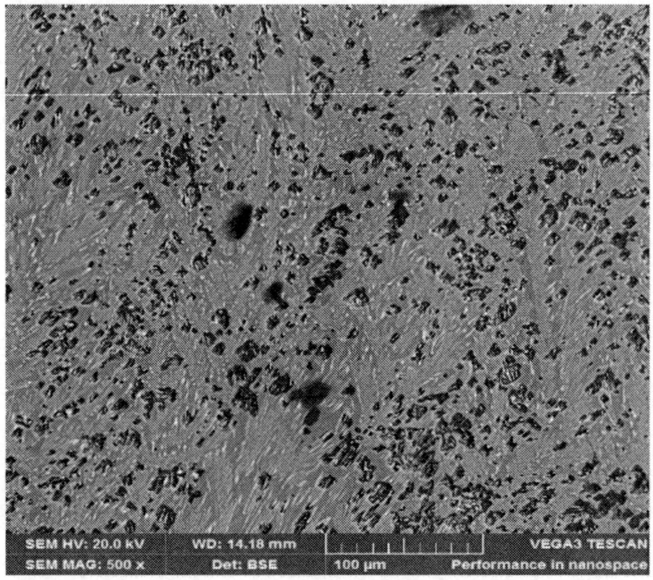

(a)

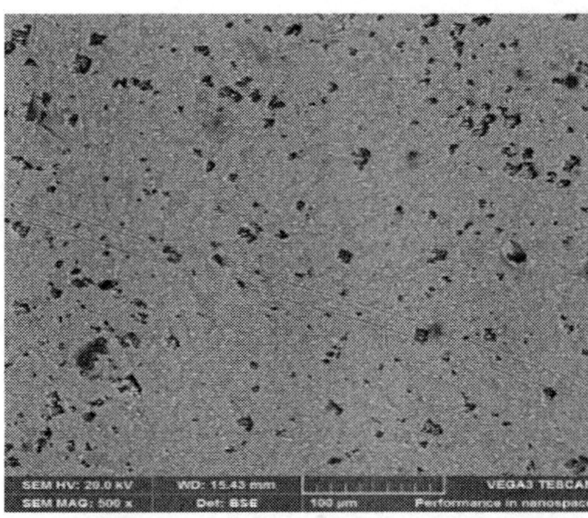

(b)

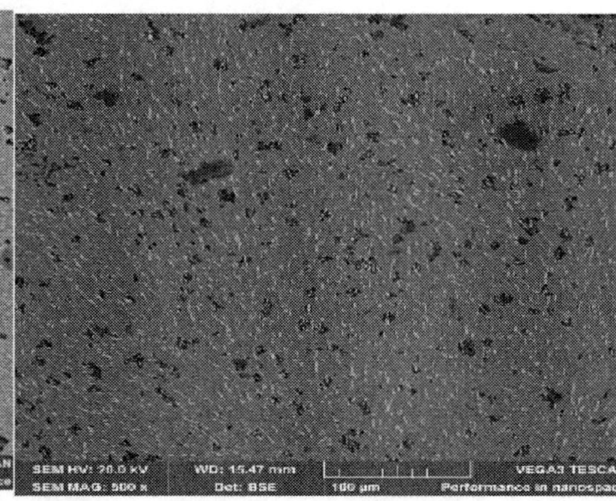

(c)

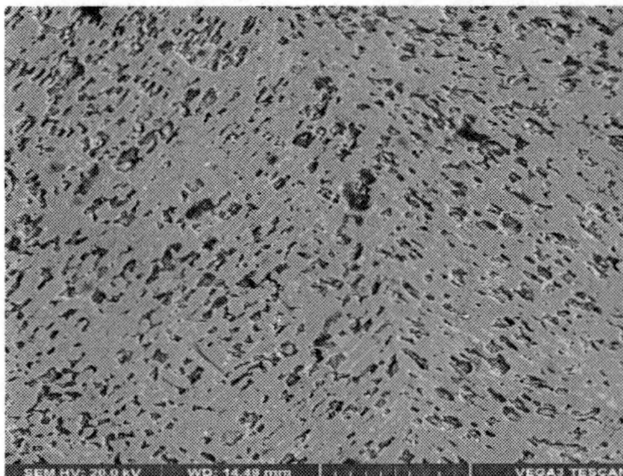

(d)

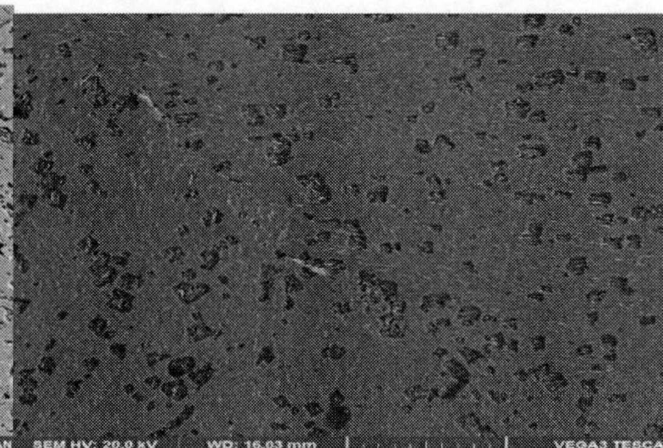

(e)

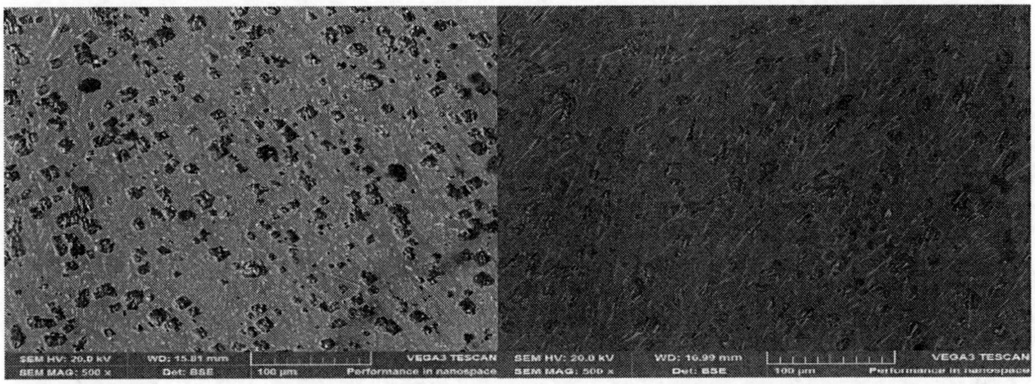

(f) (g)

Figure 2. SEM micrographs at 500X of the (a) control sample (b) Al/150 μm RHA (c) Al/150 μm RHA/0.4 % G (d) Al/300 μm RHA (e) Al/300 μm RHA/0.4 % G (f) Al/600 μm RHA (g) Al/600 μm RHA/0.4 % G

3.3 Phase Analysis of the control sample, rice husk ash and composites

It can be seen from the XRD diffractograms (Figure 3a) of the unreinforced Al that the major diffraction peaks occurred at diffracting angles (2θ) of 38.85° and 45.07° with inter-planar distance of 2.32 Å and 2.01 Å and relative X-Ray diffracting intensities of 27756.48 and 11197.12. As a result of chemical reactions formed between aluminium and other major elements in the recycled Al cans during melting and solidification, intermetallic bond formed around Al. X-ray diffractometric profiles of RHA particulate is shown in Figure 3b such that major diffraction peaks occurred at diffracting angles (2θ) of 29.60°, 39.65°, 43.40° and 48.73° with inter-planar distance of 3.02 Å, 2.27 Å, 2.09 Å and 1.91 Å and relative X-Ray diffracting intensities of 1256.10, 355.38, 341.56 and 515.81. The phases present are in agreement with the XRF analysis of RHA.

In all composites, reactions between Al, Ti, Si, Cu, Zn and SiO_2 to form compounds are clearly seen from various intermetallic bond formed within the composites. These intermetallic compounds enhanced the mechanical properties of the composites as revealed in the mechanical test results. However, for all samples, there is a shift in the diffracting angles due to reactions of reinforcements with Al alloy. As a result, major and common diffraction peaks occurred at diffracting angles (2θ) of about 39°, 42°, 43° 45°, 65°, 78° and 82° with inter-planar distance of about 2.05 Å, 2.15 Å, 2.09 Å, 2.03 Å, 1.43 Å, 1.22 Å and 1.17 Å

IOP Conf. Series: Materials Science and Engineering **805** (2020) 012012 doi:10.1088/1757-899X/805/1/012012

respectively. In the combined XRD of the control sample, Al/RHA and Al/RHA/G at different particle sizes, formation of SiC from RHA was revealed at 1.6 wt. % RHA at all particle sizes with metal-metal, metal-SiC and metal-oxide inter atomic bond for Al/RHA composites. However, reaction of graphene with RHA also revealed breakdown and transformation of graphene flakes to smaller molecules called fullerene of C_{60} type in all particle sizes of RHA. Al/RHA/G composites are characterized with metal-oxide, C_{60}-Si, metal-metal and metal-SiC atomic bonds. Phase reactions of this type go a long way to confirm that Al/RHA/G composites possess good mechanical properties when compared to monolithic Al alloy and Al/RHA composites (Figure 4).

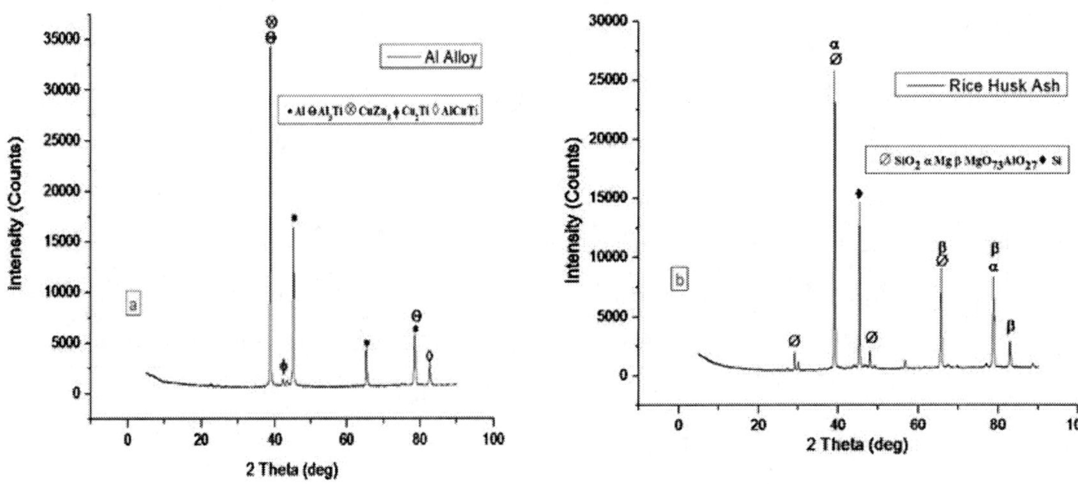

Figure 3. XRD of (a) Control sample (b) Rice Husk Ash

IOP Conf. Series: Materials Science and Engineering **805** (2020) 012012 doi:10.1088/1757-899X/805/1/012012

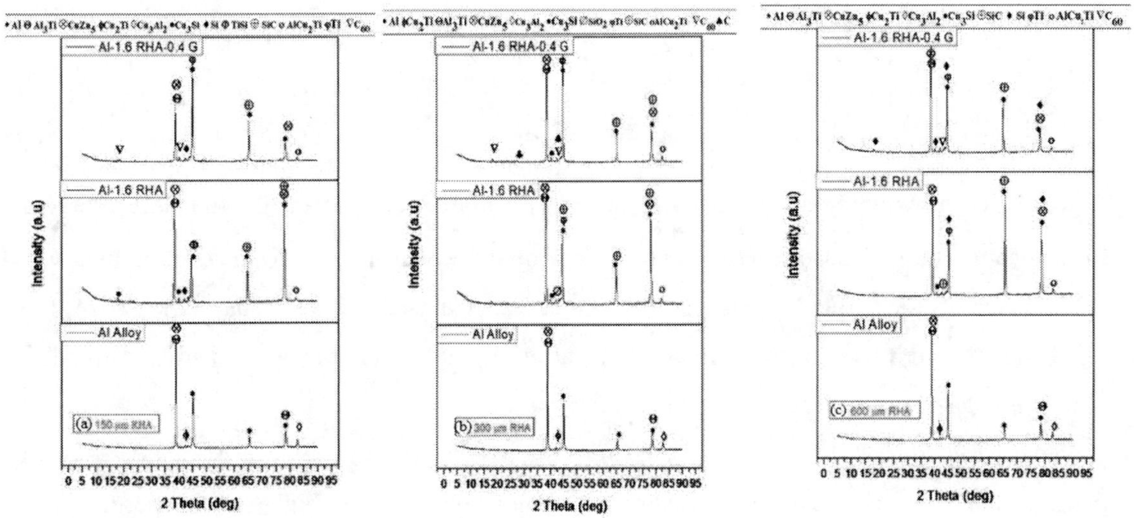

Figure 4. Combined XRD of control sample, Al/RHA and Al/RHA/G composites at (a) 150 µm (b) 300 µm (c) 600 µm particle sizes of RHA

3.4 Density and Mechanical Properties

Figure 5 shows the variation of density and mechanical properties at different particle sizes. The density of the composites of Al/RHA and Al/RHA/G increased gradually with increase in particle size when compared to the control sample (Figure 5a). However, all composites of Al/RHA/G composites have lower density than composites of Al/RHA. This occurs as a result of replacement of soft high density Al matrix with hard and low density RHA and graphene.

From Figure 5b, it is observed that the hardness increases as the particle size decreases for both Al/RHA and Al/RHA/G composites. The highest hardness obtained at 600 µm, 300 µm and 150 µm mesh sizes are 97.00 HV, 94.20 HV and 115.60 HV respectively for Al/RHA/G composites (Figure 5b).

The ultimate tensile strength according to Figure 5c decreases with increase in particle size of RHA for 150 µm, 300 µm and 600 µm mesh sizes accordingly. Al/RHA/G also showed the same trend but with higher ultimate tensile strength. These results are in agreement with the

IOP Conf. Series: Materials Science and Engineering **805** (2020) 012012 doi:10.1088/1757-899X/805/1/012012

Hall-Petch relationship (equation 2), that the grain size is inversely proportional to tensile strength.

$$\sigma_y = \sigma_o + KD^{-1/2} \qquad (2)$$

Where σ_y is the yield stress of the material; σ_o is a materials constant for the starting stress fordislocation movement of the material; K is a measure of resistance to dislocation and D is thediameter of the particulate size. As D decreases, that is decrease in particulate size, the repulsion stress felt by a grain boundary dislocation decreased and the applied stress needed to propagate dislocations through the material increased.

However, for both Al/RHA and Al/RHA/G composites, the impact energy increases with particle size. (Figure 5d). This is because the composites with the highest filler particle size showed the highest resistance to impact but with a corresponding low hardness, fatigue, yield and tensile strength values. Hence, it is recommended that for applications where impact energy is of utmost importance, increasing the particle size of the reinforcement within the composite matrix is favourable.

It can also be seen from Figure 5e that the fatigue increases as the particle size decreases. However, with addition of graphene nanoplatelets of high stiffness, the fatigue strength of the Al/RHA.G composites has further improved by about 1.2 %.

IOP Conf. Series: Materials Science and Engineering **805** (2020) 012012 doi:10.1088/1757-899X/805/1/012012

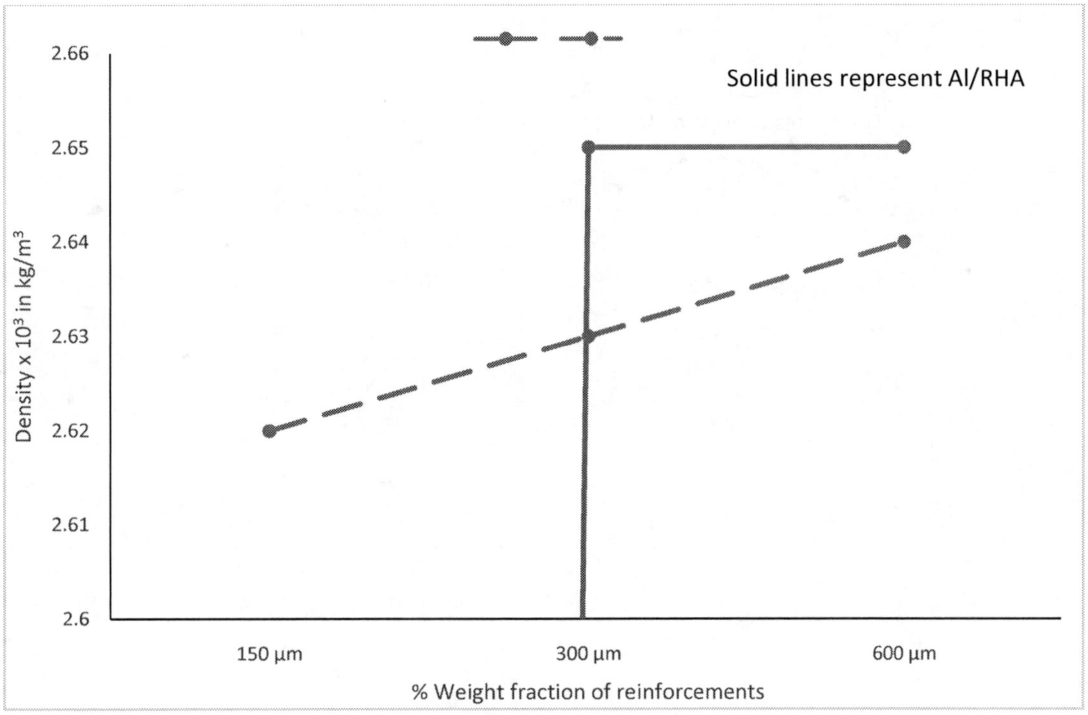

(a)

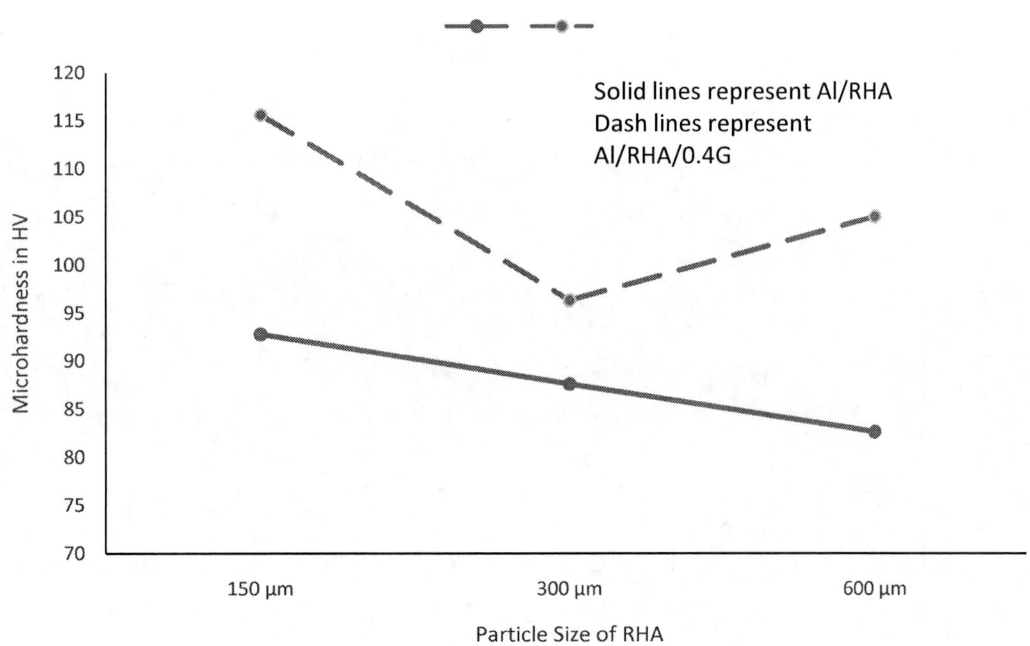

(b)

IOP Conf. Series: Materials Science and Engineering **805** (2020) 012012 doi:10.1088/1757-899X/805/1/012012

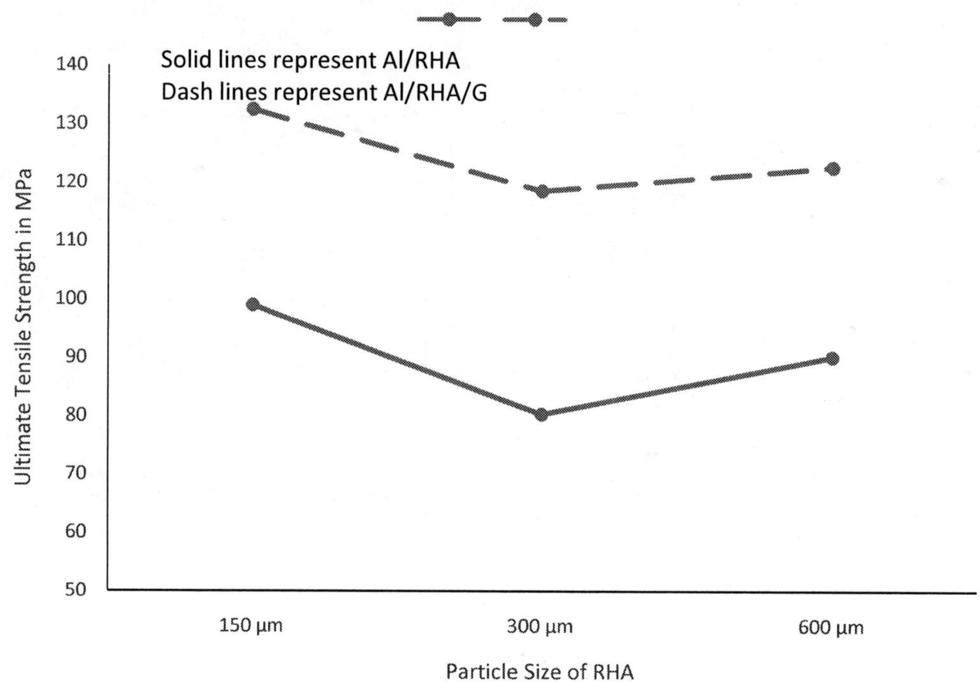

(c)

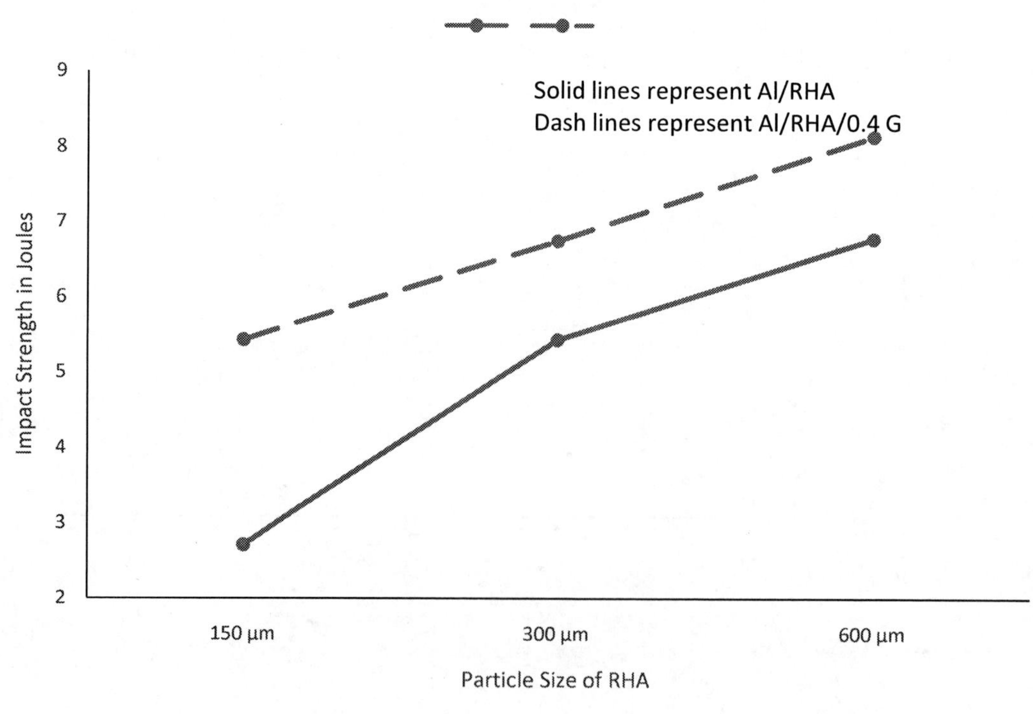

IOP Conf. Series: Materials Science and Engineering **805** (2020) 012012 doi:10.1088/1757-899X/805/1/012012

(d)

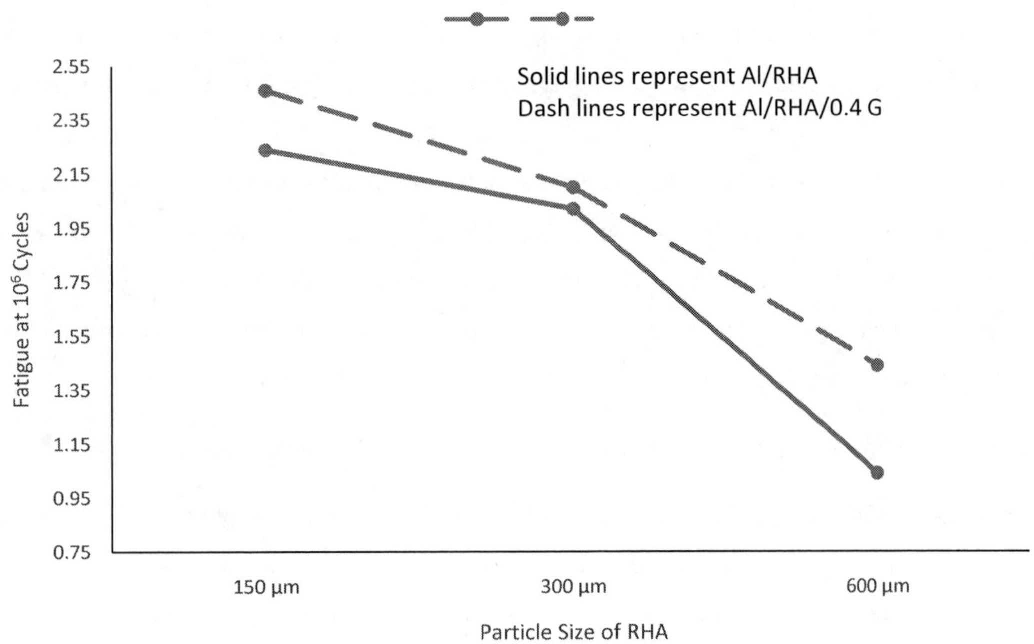

(e)

Figure 5. Variation of density and mechanical properties with particle size of RHA

3.5 Mechanism of formation of Fullerene from RHA and Graphene

Reactions between graphene, RHA and aluminium alloy after solidification did not reveal retention of graphene in its original form from the XRD results, rather, formation of fullerene of C_{60} was revealed together with other intermetallic compounds. Earlier workers [20, 21] have explained the mechanism behind the formation of fullerene from graphene without addition of RHA to graphene. In this present study, formation of fullerene could be justified from modification of earlier submission [21]. That is, the reactions between amorphous carbon from RHA, graphene and aluminium alloy could have been powered by high melting temperature of about 750 °C and shear stress of mixing/stirring of the mixture in the crucible pot, coupled with high liquid phase-solid phase transformation pressure in the die cavity steel mould during casting. These reactions could have led to the breakdown of graphene to fullerene. There is also possibility of the graphene flakes used for reinforcement to have vacancy at the middle. Such graphene flakes accordingly are energetically unstable [20], and

IOP Conf. Series: Materials Science and Engineering **805** (2020) 012012 doi:10.1088/1757-899X/805/1/012012

the probability of return motion back to the middle is negligible. Therefore, this effect of middle vacancy of graphene flakes together with high temperature and pressure of synthesizing RHA and graphene could have instigated the etching of graphene for subsequent transformation to proceed to formation of fullerene in line with the mechanism earlier suggested [21]. It was also asserted that oxidation of porous structure of RHA may occur during synthesis process due to high temperature which makes the mixture to be highly reactive [22]. Thus, confirming the ability of rice husk ash to react very actively with graphene when heat and temperature act as catalyst. The mechanisms for such transformation are described as shown in Figure 6.

That is, rice husk ash (amorphous carbon) + graphene (carbon nanoplatelets) → fullerene

Choi *et al.* [14] affirmed that fullerenes with a spherical structure may possibly be considered a more fascinating reinforcing agent when compared to carbon nanotubes or graphene, as a result of their zero-dimensional geometric characteristics. This is because they could easily disperse and difficult to destroy during severe mechanical dispersion processes in the metal matrix. This was further corroborated by earlier workers [23] that improvement of yield strength of aluminium from 150 to 250 MPa by reinforcing fullerenes via severe plastic deformation is possible. Khalid *et al.* [24] also developed aluminium-based composites by liquid metal infiltration technique, where the micrographs of the composites revealed good interfacial bonding between aluminium and fullerenes but with mechanical properties of the composites still below the theoretical expectation when compared to the actual stiffness/strength of fullerenes.

IOP Conf. Series: Materials Science and Engineering **805** (2020) 012012 doi:10.1088/1757-899X/805/1/012012

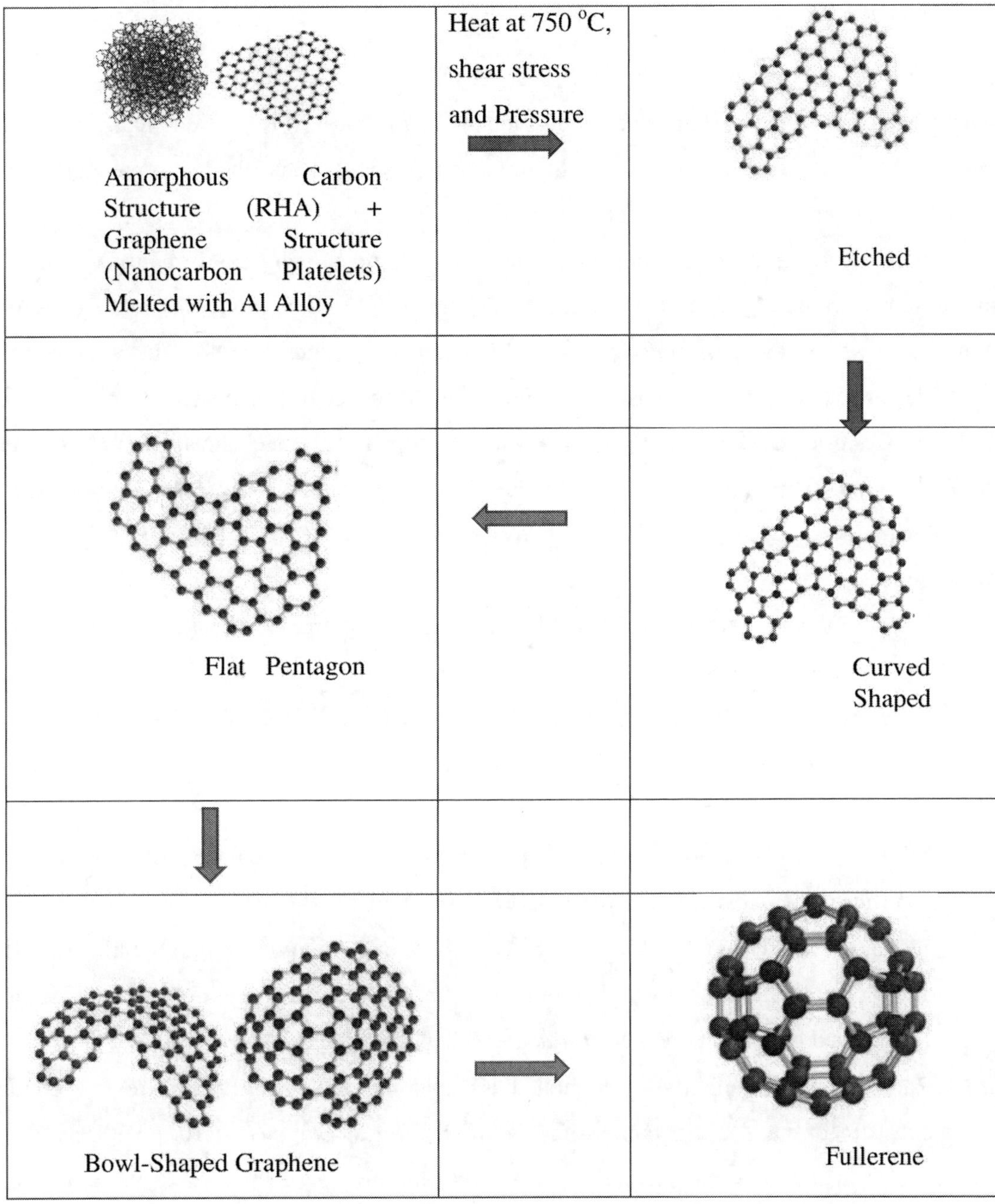

Figure 6. Modified mechanisms of formation of fullerene from rice husk ash and graphene [21].

4. Conclusion

Aluminium alloy was successfully cast from recycled waste aluminium cans and reinforced with RHA of different particle sizes and graphene to produce Al/RHA/G composites by stir casting technique. The morphology of the composites showed that the reinforcements were uniformly dispersed to give composite castings with isotropic grain structure. The phase analysis revealed composite castings with inter atomic bond showing no formation of harmful aluminium carbide. The best morphological and mechanical properties of Al/RHA/G composites were obtained from RHA having 150 μm particle sizes, therefore the most suitable to disperse graphene freely in molten aluminium. Modified mechanisms of formation of fullerene from rice husk ash and graphene was developed. The mechanical properties of the Al/RHA/G greatly improved when compared to the control sample and Al/RHA composites.

References

[1] Prasad SV and Asthana R 2004 Aluminum metal–matrix composites for automotive applications: Tribological considerations. *Tribology Lett.* **17** (3) pp 445-453.

[2] Stojanovic B and Babic M 2014 Optimization of A356/10SiC/3Gr hybrid composite wear using Taguchi method. *8th International Conference on Tribology* Sinaia, Romania, pp 708-715.

[3] Miracle DB 2005 Metal matrix composites – From science to technological significance. *Composites Sci. Technol.* **65** (15-16) pp 2526-2540.

[4] Stojanović B, Babić M, Mitrović S, Vencl A, Miloradović N and Pantić M 2013 Tribological characteristics of aluminum hybrid composites reinforced with silicon carbide and graphite. A review. *J. Balkan Tribol. Assoc.* **19** (1) pp 83-96.

[5] Ramesh CS, Hirianiah A, Harishonad KS and Noronha NP 2012 A review on hot extrusion of metal matrix composites (MMC'S). *Int. J. Eng. Sci.* **1** (10) pp 30-35.

[6] Lee C, Wei XD, Kysar JW and Hone J 2008 Measurement of the elastic properties and intrinsic strength of monolayer graphene. *Sci.* **321** (5887) pp 385-388.

[7] Prashantha KHG and Xavior MA 2014 Graphene reinforced metal matrix composite (GRMMC): A review. *Procedia Eng.* **97** (321) pp 1033-1040.

[8] Lim DK, Shibayanagi T and Gerlich AP 2009 Synthesis of multi-walled CNT reinforced aluminium alloy composite via friction stir processing. *Mater. Sci. Eng.* **507** (1-2) pp 194-199.

[9] Esawi AMK and Borady MA 2008 Carbon nanotube-reinforced aluminium strips. *Composites Sci. Technol.* **68** (2) pp 486-492.

[10] Choi HJ, Kwon G, Lee G and Bae D 2008 Reinforcement with carbon nanotubes in aluminum matrix composites. *Scripta Materialia* **59** (3) pp 360-363.

[11] Esawi AMK, Morsi K, Sayed A, Gawad AA and Borah P, 2009 Fabrication and properties of dispersed carbon nanotube–aluminum composites. *Mater. Sci. Eng.* **508** (1-2) pp 167-173.

[12] Choi HJ, Shin JH and Bae DH 2012 The effect of milling conditions on microstructures and mechanical properties of Al/MWCNT composites. *Composites Part A: Appl. Sci. Manufactur.* **43** (7) pp 1061-1072.

[13] Pérez-Bustamante R, Gómez-Esparza CD, Estrada-Guel I, Miki-Yoshida M, Licea-Jiménez L and Pérez-García SA 2009 Microstructural and mechanical characterization of Al–MWCNT composites produced by mechanical milling. *Mater. Sci Eng.* **502** (1-2) pp 159-163.

[14] Choi K, Seo J, Bae, D and Choi H. 2014. Mechanical properties of aluminum-based nanocomposite reinforced with fullerenes. *Trans.. Nonferrous Metals Soc. China* **24** pp s47-52.

[15] Durowoju MO, Agunsoye JO, Mudashiru LO, Yekinni AA, Bello SK and Rabiu TO 2017 Optimization of stir casting process parameters to improve the hardness property of Al/RHA matrix composites. *Eur. J. Eng. Res. Sci.* **2** (11) pp 5-12.

[16] Yekinni AA, Durowoju MO, Agunsoye JO and Mudashiru L.O 2018 Effects of rice husk ash and graphene on morphological and mechanical properties of recycled aluminium cans hybrid composites. *Sci. Focus* **23** (2) pp 140-155.

[17] Kenny M and Oates T 2007 Lime and Limestone, in *Ullmann's Encyclopedia of Industrial Chemistry*, Weinheim, Germany: Wiley-VCH Verlag GmbH & Co. KGaA,

[18] Raju RSS and Rao GS 2017 Assessment of tribological performance of coconut shell ash particle reinforced Al-Si-Fe composites using Grey-Fuzzy application. *Tribol. Ind.* **39** (3) pp 364-377.

[19] Baghchesara MA, Abdizadeh H and Baharvandi HR 2012 Effects of MgO nanoparticles on microstructural and mechanical properties via powder metallurgy route. *Int. J. Modern Phys.: Conf. Ser.* **5** pp 607- 614.

IOP Conf. Series: Materials Science and Engineering **805** (2020) 012012 doi:10.1088/1757-899X/805/1/012012

[20] Santana A, Popov AM and Bichoutskaia E 2013 Stability and dynamics of vacancy in graphene flakers: Edge effects. *Chem. Phys. Lett.* **557** pp 80-87.

[21] Chuvilin A, Kaiser U, Bichoutskaia E, Besley NA and Khlobystov AN 2010 Direct transformation of graphene to fullerene. *Nature Chem.* **2** (6) pp 450-453.

[22] Deshmukh P, Bhatt J, Peshwe D and Pathak S 2012 Determination of silica activity index and XRD, SEM and EDS studies of amorphous SiO_2 extracted from husk ash. *Trans. Indian Inst. Metals* **65** (1) pp 63-70.

[23] Tokunaga T, Kaneko K, Sato K and Horita Z 2008 Microstructure and mechanical properties of aluminum–fullerene composite fabricated by high pressure torsion. *Scripta Mater.* **58** (9) pp 735-738.

[24] Khalid FA, Beffort O, Klotz UE, Keller BA, Gasser P and Vaucher S 2003 Study of microstructure and interfaces in an aluminium–C_{60} composite material. *Acta Mater.* **51** (15) pp 4575-4582.

IOP Conf. Series: Materials Science and Engineering **805** (2020) 012013 doi:10.1088/1757-899X/805/1/012013

2^k Optimization approach for the removal of Pb^{2+} ions from water contaminated from oil-well drilling point on to MWCNTs

[1*]AT Gbajabiamila, [2]I Kariim and [1]BO Ighobesuo

[1]Department of Chemistry, Federal University Otuoke, Yenagoa, Nigeria

[2]Department of Chemical Engineering, Federal University of Technology, Minna, Nigeria

*Email: afeezgt@fuotuoke.edu.ng

Abstract

The release of Pb^{2+} into the surrounding ecosystem has adverse effects on man and animal, hence, removing such harmful heavy metal from water system calls for urgent attention. The potential application of Multi -walled carbon nanotubes, MWCNTs, towards the lead (II) ions removal from oil polluted water body was studied and optimized via 2^k factorial experimental approach. The MWCNTs produced was characterized via High Resolution Transmission Electron Microscope (HR-TEM), High Resolution Scanning Electron microscope (HR-SEM), Thermogravimetric Analysis (TGA), Fourier Transform Infrared (FTIR) and Brenuer-Emmette-Teller (BET) to determine the internal morphology, surface morphology, thermal stability, functional group and surface area respectively. The results of the HR-TEM depict the nano-structural diameter of the produced MWCNTs with continuous strands of tubes as depicted by SEM micrograph. The MWCNTs surface area produced was obtained to be 540 m^2/g. The sorption of Pb^{2+} ions was optimized using 2k design approach where speed (rpm), absorbent dosage (g), temperature (^oC) and time (min) were studied in a batch condition. The optimum conditions that give maximum sorption of Pb^{2+} ions on to the surfaces and the pores of the developed MWCNTs adsorbent were obtained at 60 min contact time, 59.89 ^oC, 0.20 g of adsorbent dosage and 599.98 rpm of agitation. Thus, the highest percentage sorption of Pb^{2+} ions by the MWCNTs from the oil-polluted water body was 99.5%. Hence, this study has demonstrated the potentials of MWCNTs as an effective and efficient sorbent for Pb^{2+} sorption from oil polluted wastewater.

1. Introduction

Several pollutants from non-point and point sources have contributed significantly to contamination of drinking and groundwater [1] and amongst commonly observed contaminants; inorganic lead pollutants stand second on profile of most hazardous metals with significant health risks for the living beings. These beings and by extension water bodies are possibly exposed to lead in various forms to include pipe corrosion, faucets, household plumbing systems, old paints, mining, smelting, battery manufacturing, and other industrial and urban wastes [2-4]. Wastewater discharges from mining, metallurgical, tannery, battery, chemical and plastics manufacturing industries have contributed immensely to the quantity of lead (II) ions in water bodies whose concentration has raised a huge concern [5]. The types of emerging pollutants are also increasing while the challenges posed by traditional pollutants are not efficiently solved yet.

Generally, toxic metal-polluted waters and particularly accumulation of lead, cause several disease problems such as damage to the central nervous induced damages, renal kidney disease, mental retardation, cancer and anemia including cardiovascular disorders [6-7]. Removal of toxic metals has been identified as one of the major challenges in ensuring safe water for all in addition to protecting the environment [8-9]. It is important to note that there exist different conventional processes used as purification techniques but most offer some disadvantages [3]. Among the conventional and most commonly employed means of water purification is the application of carbon nanotubes; MWCNTs as novel sorbents in an adsorption process.

MWCNTs are identified to possess excellent properties; physical, chemical, mechanical and electrical, which lead to great potential application in various fields to include adsorption in water purification. The large specific surface area (SSA) and the appreciable chemical and thermal stability of MWCNTs associated to their easy large-scale synthesis make them good candidates for adsorption kinetics study [10]. MWCNTs possess high sorption capacity compared to the activated carbon (AC) during a sorption process. The observed high sorption capacity exhibited by MWCNTs is as a result of strong interaction between MWCNTs and sorbates which exist on the surfaces of the adsorbent during the sorption process [11], dioxin for example. MWCNTs have an abundant potential application of removing pollutants like lead (Pb) ions, copper (Cu) ions from water bodies [12-13].

IOP Conf. Series: Materials Science and Engineering **805** (2020) 012013 doi:10.1088/1757-899X/805/1/012013

Sorption process is categorized as surface phenomenon where sorbate molecules adhere to the sorbent surface. The size of sorbent molecules that interact ranges from microns to nanometer. The nanomaterials (<100 nm) possess larger surface for sorption and thereby provides better sorption efficiency compared to bulk sorbents. The process of separation using the nanosized adsorbents is termed as nano-adsorption. Nanoparticles are applied individually or in the form of composites and more obviously, the application of nanocomposites give high sorption efficiencies owing to the presence of simultaneously acting adsorption mechanism; physical and chemical sorption mechanisms [14]. Specifically for precise occasion of sorption of lead, the principal mechanisms govern the sorption process are ranging from ion exchange, complexation, chemisorption, electrostatic interaction and mono-ion layer physical adsorption [15-17].

Several adsorbents such as activated carbon, alumina, silica, zeolites, titania have been extensively applied for the sorption of pollutants from wastewater. The application of MWCNTs as sorbent has attracted appreciable interest due to its ability to be functionalized as a result of defects walls which aid to achieving any intended objectives. The controlling mechanism is characteristically dependent on the nature of the nanosorbent used and the conditions of operation [14]. It has been established that various factors which include pH, pH*zpc*, adsorbent dose, size and shape were found responsible for causing changes in adsorption efficiency.

This study focuses on the production of MWCNTs in a CVD reactor and optimization of adsorptive process parameter for Pb^{2+} ions removal of pollutants from a contaminated water body with Fe-Co/kaolin MWCNTs as adsorbents using 2^k factorial design of experimental.

2. Materials and Methods

All the chemicals used in the work were of analytical grade with very high percentage purity of 95-99.99%, and they were used without any prior purification. The characteristics properties of catalyst used in this research work have been presented elsewhere [18].

2.1 MWCNTs Synthesis

The MWCNTs used was produced in a catalytic vapour deposition (CVD) reactor. Exactly 1.0 g of the sourced catalyst was measured in a crucible and placed in the center of the CVD reactor. The CVD was purged using N_2 at a flow rate of 50 ml/min at constant heating rate. Immediately the temperature reached 700 ^oC, the reactant's flow, acetylene was released to the reactor at 100 ml/min while the nitrogen flow was increased to 100 ml/min to serve as carrier gas for a reaction time of 45 min. The flow of nitrogen gas was then reduced to 50 ml/min at the completion of the reaction time and the acetylene flow was then stopped. The equipment was allowed to cool after which the produced nanomaterial was removed and kept for further analysis.

2.2 Wastewater collection and analysis

The contaminated water sample was collected from 5 different points into separate polythene bottles from a water body situated behind an oil-well drilling point in Delta State. Exactly 5.0 cm^3 of concentrated nitric acid was added to the collected wastewater sample. Further, 10 cm^3 of the water sample was digested with 5 cm^3 of HNO_3 concentrated and 2 cm^3 of HCl (5:2). Then, 15 cm^3 of deionized water was added and filtered using Whatman paper No. 45. The final filtrate was made up to 50 cm^3 of the volumetric flask with deionized water. The filtrate was analyzed using atomic absorption spectrophotometer for initial concentration of Pb^{2+}.

2.3 Adsorption Process

The adsorption process was carried out via batch adsorption method designed using 2^k factorial whereby 16 different runs were obtained for the study. Wastewater (50 ml) was transferred into 100 ml beakers for each experimental batch following 2^k optimization design approach (Table 1) and placed in a water batch shaker at specified temperature and time. After the completion of the adsorption process, the aliquot obtained was centrifuged at 300 and 600 rpm for 60 min and then filtered using a filter unit into a sample bottle for spectrophotometric analysis. Adsorbent dosage, temperature, speed, and residence time affecting the process were studied using 2^k optimization design approach. The percentage sorption of Pb^{2+} was calculated using the equation as presented in equation 1:

$$\% R = \left(\frac{C_u - C_v}{C_u}\right) \times 100 \qquad (1)$$

Where Cu and Cv are the Pb^{2+} concentrations in grams per litre initially and at time, t respectively while %R denotes percentage sorption of Pb^{2+} ions from solution.

IOP Conf. Series: Materials Science and Engineering **805** (2020) 012013 doi:10.1088/1757-899X/805/1/012013

Table 1. Experimental factors and levels used in the factorial design

Independent variables	Coded symbol	Range and	Level
		-1	+1
Contact time (min)	A	0.1	0.2
Temperature (°C)	B	30	60
Adsorbent dosage (g/l)	C	300	600
Speed (rpm)	D	30	60

3 Results and Discussion

3.1 Characterization of the as-produced MWCNTs

The internal morphology of the produced MWCNTs adsorbent as presented in Figure 1, which shows the nano-structure of the MWCNTs formed over the applied catalyst.

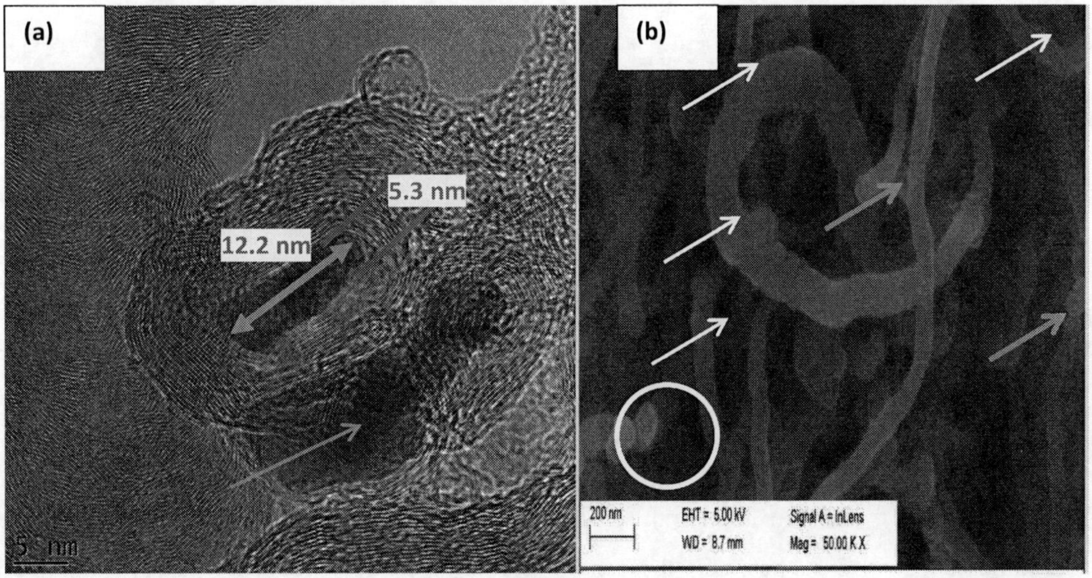

Figure 1. (a) TEM micrograph and (b) SEM micrograph for the developed MWCNTs

The internal morphology of the developed MWCNTs with unique diameter having several concentric tubes' walls was as depicted in Figure 1 (a). The produced MWCNTs have a tube dimension of 5.3 × 12.2 nm architecture confirming the nanoscale nature of the produced material (Figure 1a). The surface morphologies of the MWCNTs produced comprise of the continuous strands of nanotubes with varied diameter as presented in Figure 1 (b) with yellow arrows. The reason for the observed MWCNTs of varied diameter could be as a result of the inherent varied particle sizes of the catalyst used during the MWCNTs production. It was also observed that the produced MWCNTs have numerous branch tubes which could be as a result of the synthesis condition employed.

The thermal analysis and the surface functional group of the developed MWCNTs were determined using TGA and FTIR respectively and the results are as shown in Figure 2(a-b).

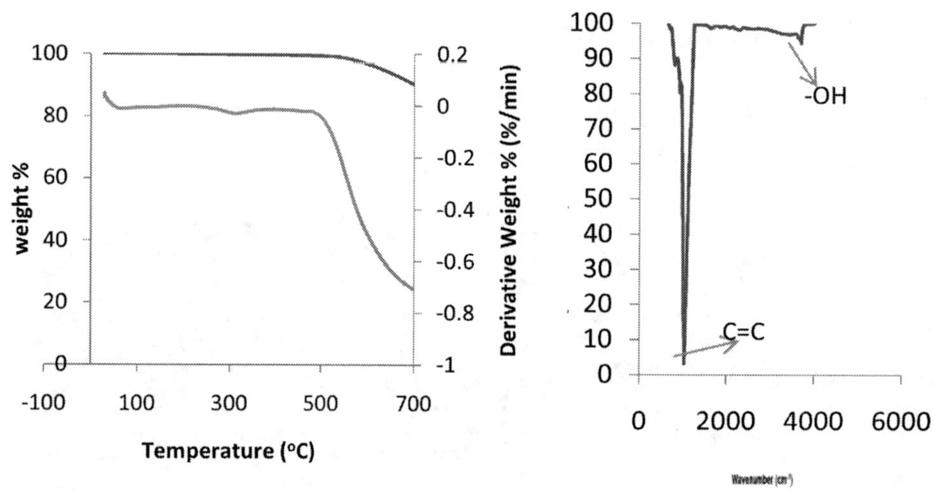

Figure 2. (a) TGA thermograph and (b) FTIR spectra of the developed MWCNTs

The result of the thermal stability as depicted in Figure 2a shows that the produced MWCNTs possess an improved thermal stability with onset degradation at 500.21 °C amounts to about 0.992 % weight loss from room temperature. The percentage degradation of the MWCNTs was also obtained to be 6.32 % between the temperature ranges of 29.67-700 °C. The obtained percentage weight loss in the MWCNTs weight could be linked to the presence of active parts of the catalyst used as well and the possible presence of amorphous formation during the MWCNTs nucleation process. Figure 2b shows the occurrence of –OH and C=C bound formation at the wavelength of 3700 and 1056 cm^{-1} respectively.

IOP Conf. Series: Materials Science and Engineering **805** (2020) 012013 doi:10.1088/1757-899X/805/1/012013

The surface area of the produced MWCNTs was also determined via BET technique and it was obtained to be 540 m^2/g. The obtained surface area is highly appreciable for effective adsorption process.

3.2 Optimization Studies for sorption of $Pb2^+$

The characterized MWCNTs were applied as sorbents for the sorption of Pb^{2+} from oil polluted water in a batch adsorption process via optimization approach. The effect of temperature, contact time, sorbent dosage and agitation were optimized via 2^k optimization approach as depicted in Table 2.

Table 2. Basic experimental runs obtained for the optimization study and parameters

Run No.	Contact time (min)	Temperature (°C)	Dosage (g)	Speed (rpm)	% Pb Removal
1.	30	60	0.10	600	82.15
2.	30	30	0.10	600	78.32
3.	60	30	0.10	300	28.43
4.	30	60	0.10	300	39.16
5.	60	30	0.20	600	84.54
6.	60	60	0.10	600	44.58
7.	30	30	0.20	600	84.98
8.	60	30	0.20	300	74.94
9.	60	60	0.10	300	66.04
10.	60	30	0.10	600	60.29
11.	30	60	0.20	300	89.09
12.	30	60	0.20	600	96.72
13.	30	30	0.10	300	18.12
14.	30	30	0.20	300	69.49
15.	60	60	0.20	300	47.20
16.	60	60	0.20	600	99.50

IOP Conf. Series: Materials Science and Engineering **805** (2020) 012013 doi:10.1088/1757-899X/805/1/012013

Analysis of variance (ANOVA) was used for the estimation of the interacting factors and the main effects for sorption of Pb^{2+} (Table 3). The significance of the process parameters were quantified via the sum of squares, SS. It was observed that as the value of SS increases, the significance of the corresponding factor in the undergoing process also increased. The relating effects of each independent process factors were statistically significant at $P < 0.05$ [19]. The Pareto chart shows the number of factors that has significant effect on the sorption process. Hence, any bar that cuts the line indicates that such effects are significant. Therefore, only C and D are significant for the sorption of Pb^{2+} on to MWCNTs sorbent as presented in Figure 3.

Table 3. ANOVA for selected factorial model

Sources	Sum of squares	df	Mean square	F Value	P-Value Prob>F	Remark
Model	7067.96	4	1766.99	5.60	0.0104	significant
A-Contact Time	66.06	1	66.06	0.21	0.6563	
B-Temperature	128.43	1	128.43	0.41	0.5367	
C-Dosage	3886.59	1	3886.59	12.31	0.0049	
D-Speed	2986.90	1	2986.90	9.46	0.0106	
Residual	3472.99	11	315.73			
Cor Total	10540.96	15				

After discarding insignificant terms from the model generated, the resultant models can be expressed as:

Final Equation in Terms of Coded Factors:

$$\%\,removal = 65.22 - 2.03A + 2.83B + 15.59C + 13.66D$$

Final Equation in Terms of significant Coded Factors:

$$\%\,removal = 65.22 + 15.59C + 13.66D$$

Final Equation in Terms of Actual Factors:

IOP Conf. Series: Materials Science and Engineering **805** (2020) 012013 doi:10.1088/1757-899X/805/1/012013

$$\% \, removal = -24.92813 - 0.13546 \times contact\ time + 0.18887 \times temperature + 311.71250 \times dosage + 0.091088 \times speed$$

Final Equation in Terms of significant Actual Factors:

$$\% \, removal = -24.92813 + 311.71250 \times dosage + 0.091088 \times speed$$

The optimum points of the sorption process was determined via numerical analysis using desirability test. The optimum conditions give maximum sorption of Pb^{2+} on to the surfaces and the pores of the developed MWCNTs adsorbent were obtained at 60 min contact time, 59.89 °C temperature, 0.20 g of adsorbent dosage and 599.98 rpm of agitation speed. The above conditions led to the maximum desirability of 0.9899 as presented in Figure 4(a). The 3-D plots showing effects of parameters on adsorption of Pb^{2+} on to MWCNTs adsorbent are as shown in Figure 5.

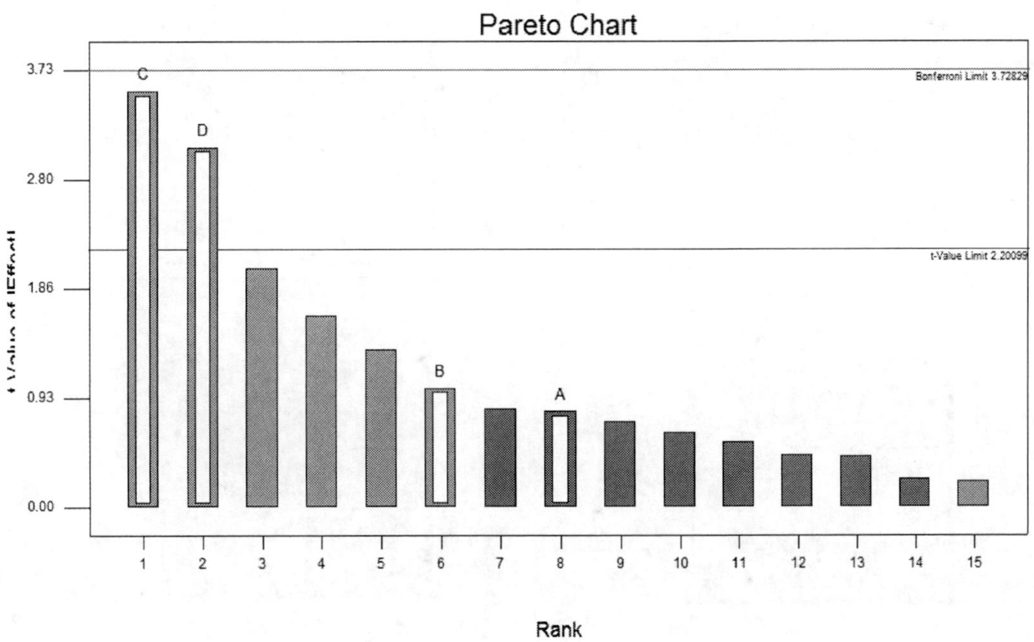

Figure 3. A Pareto chart showing the significant effect having effects on the sorption process

IOP Conf. Series: Materials Science and Engineering **805** (2020) 012013 doi:10.1088/1757-899X/805/1/012013

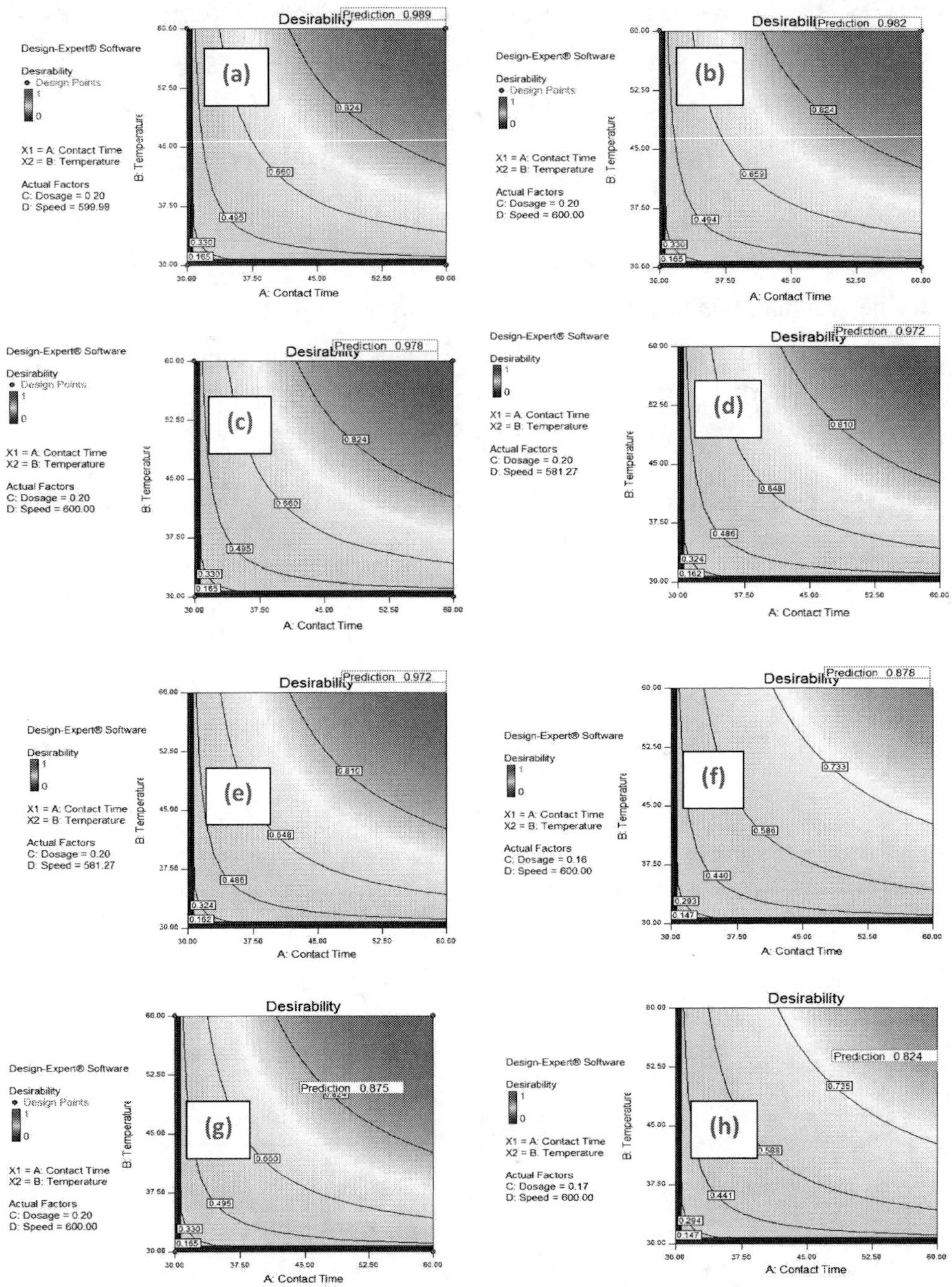

Figure 4. Desirabilty graphs for optimum condition determination

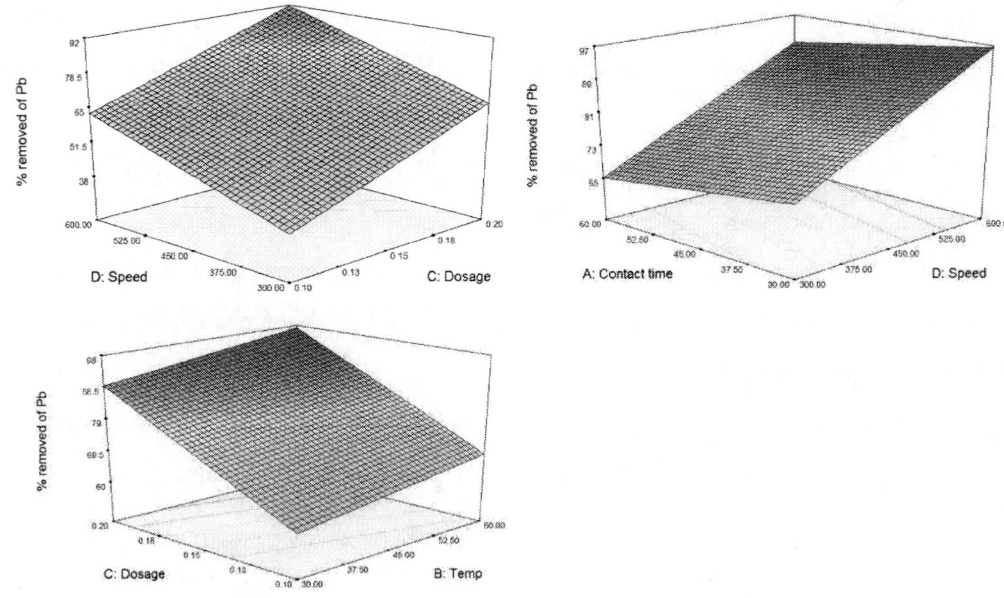

Figure 5. 3-D Plots showing effects of parameters on the lead adsorption on to MWCNTs adsorbent

4. Conclusion

Adsorption process as widely regarded as one of the effective and efficient heavy metal ions removal technology. The developed MWCNTs adsorbent possesses a diameter in the matrix of 5.3 × 12.2 nm with continuous strands of tubes as depicted by the TEM and SEM respectively. The MWCNTs adsorbent possesses a relatively large surface area of 540 m^2/g making it had the adsorptive potential towards the removal of impurities. The sorption of Pb^{2+} ions on the surface of the MWCNTs was optimized via 2^k factorial design. The optimum process that give maximum sorption of Pb^{2+} ions on to the surfaces and the pores of the developed MWCNTs adsorbent were obtained at 60 min contact time, 59.89 °C temperature, 0.20 g of adsorbent dosage and 599.98 rpm of agitation speed. Thus, this work on sorption of lead ions on Fe-Co/ Kaolin MWCNTs was successfully carried out for process optimization. The utilized MWCNTs were effective for Pb^{2+} ions sorption from oil-polluted water body.

References

[1] Ali I 2014 Water treatment by adsorption columns: evaluation at ground level. *Sep. Purif. Rev.* **43** (3) pp 175-205.

IOP Conf. Series: Materials Science and Engineering **805** (2020) 012013 doi:10.1088/1757-899X/805/1/012013

[2] Kumari M, Pittman CU and Mohan D 2015 Heavy metals [chromium (VI) and lead (II)] removal from water using mesoporous magnetite (Fe_3O_4) nanospheres. *J. Colloid Interf. Sci.* **442** pp 120-132.

[3] Mubarak N, Sahu J, Abdullah E and Jayakumar N 2014 Removal of heavy metals from wastewater using carbon nanotubes. *Sep. Purif. Rev.* **43** (4) pp 311-338.

[4] Salam MA 2013 Coating carbon nanotubes with crystalline manganese dioxide nanoparticles and their application for lead ions removal from model and real water. *Colloids Surf. A: Physichochem. Eng. Aspect* **419** pp 69-79.

[5] Kocabaş-Ataklı ZÖ and Yürüm Y 2013 Synthesis and characterization of anatase nanoadsorbent and application in removal of lead, copper and arsenic from water. *Chem. Eng. J.* **225** pp 625-635.

[6] Calderon J, Navarro ME, Jimenez-Capdeville ME, Santos-Diaz MA, Golden A, Rodriguez-Leyva I, Borja-Aburto V and Dıaz-Barriga F 2001 Exposure to arsenic and lead and neuropsychological development in Mexican children. *Environ. Res.* **85** (2) pp 69-76.

[7] Mortada W, Sobh M, El-Defrawy M and Farahat S (2001) Study of lead exposure from automobile exhaust as a risk for nephrotoxicity among traffic policemen. *Am. J. Nephrol.* **21** (4) pp 274-279.

[8] Bansal RC and Goyal M 2005 *Activated Carbon Adsorption. Taylor and Francis Group*, London, pp+ 351–353.

[9] Muyibi SA, Ambali AR and Eissa GS 2008 Development-induced water pollution in Malaysia: policy implication from an econometric analysis. *Water Policy* **10** (2) pp 193-206.

[10] Wang Y, Wei F, Luo G, Yu H and Gu G 2002 The large-scale production of carbon nanotubes in a nano-agglomerate fluidized-bed reactor. *Chem. Phys. Lett.* **364** (5-6) pp 568-572.

[11] Long QR and Yang RT 2001 Carbon nanotubes as superior sorbent for dioxin removal. *J. Am. Chem. Soc.* **123** (9) pp 2058-2059.

[12] Peng X, Luan Z, Di Z, Zhang Z and Zhu C 2005 Carbon nanotubes–iron oxides magnetic composites as adsorbent for removal of Pb(II) and Cu(II) from water. *Carbon* **43** pp 880-883.

[13] Li YH, Zhu Y, Zhao Y, Wu D and Luan Z 2006 Different morphologies of carbon nanotubes effect on the lead removal from aqueous solution. *Diam. Relat. Mater.* **15** (1) pp 9094.

[14] Bhatia M, Babu RS, Sonawane SH, Gogate PR, Girdhar A, Reddy ER and Pola M 2016 Application of nanoadsorbents for removal of lead from water. *Int. J. Environ. Sci. Technol.* **14** (5) pp 1135-1154.

[15] Zhan X-M, and Zhao X 2003 Mechanism of lead adsorption from aqueous solutions using an adsorbent synthesized from natural condensed tannin. *Water Res.* **37** (16) pp 3905-3912.

[16] Choi SB, and Yun Y-S 2006 Biosorption of cadmium by various types of dried sludge: an equilibrium study and investigation of mechanisms. *J. Hazard Mater.* **138** (2) pp 378-383.

[17] Yu B, Zhang Y, Shukla A, Shukla SS and Dorris KL 2001 The removal of heavy metals from aqueous solutions by sawdust adsorption—removal of lead and comparison of its adsorption with copper. *J. Hazard Mater.* **84** (1) pp 83-94.

[18] Kariim I, Abdulkareem AS, Abubakre OK, Mohammed IA, Bankole MT and Jimoh TO 2015 Studies on the suitability of alumina as bimetallic catalyst support for MWCNTs growth in a CVD reactor. *Int. Eng. Conf.* pp 296-305.

[19] Abdel-Ghani NT, Hegazy AK, El-Chaghaby GA and Lima EC 2009 Factorial experimental design for biosorption of iron and zinc using *Typha domingensis* phytomass. *Desalination* **249** (1) pp 343-347.

IOP Conf. Series: Materials Science and Engineering **805** (2020) 012014 doi:10.1088/1757-899X/805/1/012014

Assessment of growth performance, nutrient utilization and haematological profile of *Clarias gariepinus* fed with nanoselenium formulated diets

OD Owolabi and MK Babarinsa*
Department of Zoology, University of Ilorin, Ilorin, Nigeria
*Correspondence email: kayodebabarinsa@gmail.com

Abstract

A 63-day experiment was conducted to determine the effects of nanoselenium formulated diets on growth performance, nutrient utilization and haematological profile of *Clarias gariepinus*. Fish were acclimatized for 14 days, 10 fish each were randomly introduced into different troughs. Feed supplemented with nanoselenium at 2 mg/kg, 4 mg/kg and the control without nanoselenium were fed to group of fish each in triplicate. The feed intake, weight gain, food conversion ratio (FCR), specific growth rate (SGR), protein efficiency ratio (PER) and survival rate (SR) were determined, also the haematological profile of *Clarias gariepinus* fed were analyzed. There were significant increase ($p<0.05$) in the weight gain of fish fed 2 mg/kg of nanoselenium as compared with those fed 4 mg/kg of nanoselenium and the control. Similarly, PER and SR were significantly higher ($p<0.05$) in 2 mg/kg when compared with other formulated diets. The FCR of fish fed 2 mg/kg was significantly lower compared with 4 mg/kg nanoselenium formulated diets. WBC, RBC, haemoglobin, and haematocrit significantly increased in fish fed 2 mg/kg of nanoselenium as compared with those fed 4 mg/kg nanoselenium and the control group while there was variation in the values of other haematological parameters such as platelets, MCV, MCH and MPV. The research indicated that dietary inclusion of nanoselenium at 2 mg/kg may enhance better growth performance and nutrient utilization of *Clarias gariepinus*.

Keywords: *Clarias gariepinus*, nanoselenium, growth performance, nutrient utilization, haematology

1. Introduction

Minerals are essential micronutrients needed for the normal functioning of the body system. For the enzymatic system of fish to undergo catalytic reaction, adequate amount of micronutrients is needed. Fish cannot synthesis nutrients on their own; therefore, the need for supplementary micronutrients such as selenium in the fish diets is essential. Selenium is an important dietary micronutrient for animals required for the normal body functions and metabolisms [1, 2]. It has also been known that selenium plays important roles in antioxidant defense systems, regulation of thyroid hormone metabolism and cell growth [3]. Csilla and Miklos (2013) reported that selenium plays vital roles in antioxidant defence systems, prevents cell damage and is necessary for growth, fertility, and immune system in farm animals. Nanoparticles are known to be playing additive and functional roles in fish when incorporated in feed and fed directly or included in the culturing media of fish. Nanoparticles have an inherent tendency to grow into larger particles [4], which is an evidence that nanoparticles such as nanoselenium when use as feed additives are more active and potent than selenium at the same concentration.

Clarias gariepinus is of the family Clariidae, genus *Clarias* [5]. They are omnivorous in nature, efficient in feed utilization and have ability to withstand adverse environmental condition. It is popularly refer to as African catfish because of its hardness and ability to withstand stress to a tolerable range [6]. It is the most cultured fish in Nigeria and is widely distributed throughout the country. It is an important aquaculture candidate with faster growth rate and global market demands. In spite of the adaptability of the catfishes to cultivations and commercial importance attached to its culture in developing countries like Nigeria, there are still some gaps in meeting the demand of the consumers. This is associated with the increasing population and anthropogenic effects on water bodies causing population decline of this important species of fish in most of their natural habitat. To overcome these challenges, there is need to technologically improve the aquaculture practice of *Clarias gariepinus.*

Most of the earlier research revealed the role that nanoselenium played on growth performance in animal and various fish species [4, 7, 8], while insufficient literature is available on the role that dietary nanoselenium supplementation played on nutrients utilization and growth in *Clarias gariepinus* which led to this study. The study, therefore, aimed to assess growth performance, nutrient utilization and haematological profile of *Clarias gariepinus* fed different dietary inclusions of nanoselenium formulated diets.

2. Materials and methods

2.1 Fish diets and experimental design

Healthy and uniformly sized 150 post-fingerlings (average weight of 12.30±0.01 g and average length of 6.8±1.0) were procured from a commercial fish farm and were transferred to the laboratory. The fish were acclimatized for 14 days and fed with a commercial feeds. After acclimatization, 10 fish were randomly selected and stocked per trough. Already engineered nanoselenium of size 90 nm was procured from Hebei Tianyin Biotech Co. Ltd. (Hebei, China). Then it was mixed with other feed additives at the required inclusion level, before mixing it with the variable ingredients. Three diets, consisting of control diet (without nanoselenium) and two treatment diets at 2 mg/kg nanoselenium inclusion level and 4 mg/kg nanoselenium inclusion level were compounded (Table 1). The fish were fed twice daily (08:00 h and 17:00 h) at 3% body weight for 63 days [9, 10].

2.2 Water quality parameters

The water qualities were monitored and measured weekly. The water temperature was measured using mercury-in-glass thermometer (100 ^{o}C), pH was measured using Jenway pH meter (model E 512) and the dissolved oxygen (DO) was measured using Milwaukee DO_2 meter (model MW600). The water in the rearing media was changed weekly.

2.3 Growth parameter and nutrient utilization

The weekly weight gain (WG) was measured using Camry sensitive digital scale (model SF 400) , feed intake (FI), feed conversion ratio (FCR), specific growth rate (SGR), protein intake (PI), protein efficiency ratio (PER) and survival rate (SR) were also determined [11, 12].

2.4 Samples collection

After 63 days of the experiment, fish blood was collected. Blood samples of fish from each treatment were collected through tail ablation method [8, 11]. One ml of blood was collected in sampling EDTA bottle and shaken together to allow the blood mix with the EDTA anticoagulant and labeled for identification for further analysis.

2.5 Haematological parameters

The blood samples of fish fed with nanoselenium formulated diets were analyzed for haematological parameters. Blood parameters such as red blood cell (RBC), white blood cell (WBC), plateletcrit, mean platelet volume (MPV), hemoglobin (Hb), mean corpuscular volume (MCV), lymphocytes, and haematocrit (HCT) were determined using Diatek vet haematological analyser (model-D-3125plus).

2.6 Statistical analysis

The data obtained from the study were analyzed using analysis of variance (ANOVA) followed by Duncan's Multiple Range Test (DMRT) and expressed as mean ± SE [13]. Statistical significance for all the tests was set at $p<0.05$. All statistical analyses were done using Microsoft Excel 2007 and IBM SPSS statistics 20.0.

IOP Conf. Series: Materials Science and Engineering **805** (2020) 012014 doi:10.1088/1757-899X/805/1/012014

Table 1. Percentage composition (%) of the formulated diets

	Inclusion level of dietary nanoselenium (mg/kg)		
Ingredient*	0 (Control)	2	4
Fish meal	220	220	220
Toasted soya bean	220	220	220
Groundnut cake (GNC)	220	220	220
Maize	200	200	200
Wheat offal	90	89.98	89.96
Salt	5	5	5
Fish premix	10	10	10
Vitamin	10	10	10
Lysine	10	10	10
Methionine	10	10	10
Bone meal	5	5	5
Nanoselenium	0.00	0.02	0.04
Total	1000	1000	1000

*All the ingredients are measured in gram

3. Results and Discussion

3.1 Water quality parameters

The water temperature in the rearing media ranged from 26.4 to 27.5 °C, the pH ranged from 7.2 to 7.9, while the dissolved oxygen ranged between 6.01 and 6.37 mg/l. The normal physiological profile of fish may be subjected to changes due to the alterations in the physicochemical setup of the rearing media. Fish is known to experience different environmental diseases, stress and haematological changes as a result of physicochemical changes of certain water parameters. As a result, physicochemical water parameters were studied and well monitored in order to nullify their effects on survival, growth as well as their effect on the blood composition of the experimental fish. Casillas and Smith [14] and Tomasso *et al.* [15] reported that haematological variables are stress indicators which are caused by environmental changes. From the findings and effective management of the water quality revealed that the temperature, pH and DO are all at the tolerable range throughout the experimental period.

3.2 Growth performance and nutrient utilization parameters

The growth performance and nutrient utilization parameters of *Clarias gariepinus* fed different inclusions level of dietary nanoselenium formulated diets is presented in Table 2. Supplementation of dietary nanoselenium in the fish diets showed significant effects on growth and nutrient utilization of *Clarias gariepinus.* Higher weight gain (WG), better feed conversion ratio (FCR), lower feed intake (FI), high survival rate (SR) and protein efficiency ratio (PER) were observed in fish fed 2 mg/kg nanoselenium formulated diets as compared with those fed 4 mg/kg nanoselenium formulated diets and the control diets. Selenium and nanoselenium are of great importance to fish [8, 16, 7], although inclusion concentration of both selenium and nanoselenium varies with fish species. The better nutrient utilization in fish fed 2 mg/kg nanoselenium which positively increased the growth and feeding profile of *Clarias gariepinus* may be that at this concentration their system make up responded well to the feed as compared with 4 mg/kg formulated diets as revealed in the feed intake and weight gain.

Dietary selenium (Se) is very important. It has been demonstrated that the growth performance of channel catfish (*lctalurus punctatus*) was affected by Se level [17]. Very similar results were observed in this present study. It clearly indicated that the supplementation of nanoselenium (Nano-Se) in the fish diets could improve the final weight, weight gain and the survival rate of African catfish, *Clarias gariepinus*. Selenium deficiency generally resulted in growth depression and in some cases reproduction associated problems when not managed well [18, 19, 20]. The significance increase in the weight gain of *Clarias gariepinus* fed 2 mg/kg nanoselenium formulated diets align with the statement of [8, 21, 22] and that adequate amount of dietary selenium is essential for the proper body growth of fish and other animals [12].

3.3 Haematological parameters

Haematological parameters of fish fed different inclusions levels of nanoselenium formulated diets are presented in Table 3. The RBC, WBC, haemoglobin and haematocrit of fish fed 2 mg/kg nanoselenium formulated diets significantly increased as compared to fish fed 4 mg/kg nanoselenium formulated diets and fish fed the control diets. The nutritive status of fish can be linked to the health status of animal and potential way they handle stress resulting from their surrounding environment [23, 24]. In this study, there is reasonable interrelationship between the increased weight gain and increased number of RBC. The fish fed 2 mg/kg nanoselenium formulated diet exhibited the highest RBC and weight gain as compared with fish fed 4 mg/kg nanoselenium formulated diet. This possible correlation between RBC and weight gain may be due to the fact that the supplementation of dietary nanoselenium at 2 mg/kg concentration enhance better performance and growth. This is in agreement with previous findings [9] which reported that supplementation of selenium in tilapia feed at the rate of 2 mg/kg of feed does not alter its haematological profile of the fish as well as normal

IOP Conf. Series: Materials Science and Engineering **805** (2020) 012014 doi:10.1088/1757-899X/805/1/012014

physiological activities rather promotes better performance and productivity. This is also backed up by [25] which stated that RBC counts are mostly affected by dietary treatments fed to fish because the composition of the blood can change during malnutrition and/or stress condition [26].

Table 2. Growth response, nutrient utilization and survival parameters of *Clarias gariepinus* fed different inclusions level of dietary nanoselenium formulated diets

	Inclusion level of dietary nanoselenium (mg/kg)		
Parameter	0 (Control)	2	4
Mean Initial weight(g)	124.50± 0.01^{b}	124.00± 0.01^{b}	121.50± 0.01^{a}
Mean final weight (g)	177.00± 0.05^{a}	180.50± 0.01^{b}	175.50± 0.02^{a}
Mean weight gained (g)	52.50± 0.03^{a}	56.50± 0.01^{b}	54.00± 0.02^{a}
Feed Intake (g)	45.21± 0.58^{a}	44.41± 0.58^{a}	45.03± 0.58^{a}
Food Conversion Ratio (FCR)	0.86± 0.05^{a}	0.79± 0.02^{a}	0.83± 0.03^{a}
Fraction of the crude protein fed	0.34± 0.01^{a}	0.37± 0.01^{a}	0.40± 0.01^{a}
Protein Intake	15.22± 0.04^{a}	16.36± 0.04^{a}	17.85± 0.02^{a}
Protein Efficiency Ratio (PER)	3.45± 0.01^{a}	3.45± 0.01^{a}	3.02± 0.01^{a}
Specific Growth Rate (SGR)	0.24± 0.02^{a}	0.26± 0.02^{a}	0.26± 0.02^{a}
Survival Rate (%)	95.00± 0.10^{a}	95.00± 0.01^{a}	90.00± 0.01^{a}
Mean final standard length (cm)	13.25± 0.50^{a}	13.78± 0.01^{a}	13.05± 0.01^{a}

Means within the row with different superscripts are significantly different ($P<0.05$)

Table 3. Haematological profile of *Clarias gariepinus* fed different inclusion levels of nanoselenium formulated diets for 9 weeks (63 days)

	Inclusion level of dietary nanoselenium (mg/kg)		
Parameter	0 (Control)	2	4
RBC	2.62±0.40[a]	3.89±0.33[a]	3.30±0.99[a]
HGB	14.65±0.35[a]	16.25±0.49[a]	11.15±0.21[a]
HCT	40.40±5.09[a]	58.60±0.85[b]	48.40±14.42[a]
MCV	154.85±4.03[a]	151.55±15.20[a]	146.55±0.07[a]
MCH	44.45±0.07[a]	41.95±2.33[a]	42.90±0.57[a]
MCHC	287.00±7.07[a]	277.50±12.02[a]	293.00±4.24[a]
WBC	107.95±5.44[a]	115.85±10.25[a]	93.20±9.19[b]
LYM	79.25±1.06[a]	67.60±6.51[b]	65.25±1.63[c]
PLT	18.00±4.24[a]	43.00±7.07[b]	49.00±12.73[c]
MPV	7.85±0.35[a]	7.30±0.14[a]	7.80±0.28[a]
PCT	0.02±0.01[a]	0.04±0.01[a]	0.04±0.01[a]

RBC, Red blood cell; PLT, Platelets; HG, Hemoglobin; MPV, mean platelet volume; HCT, Haematocrit; WBC, White blood cell; MCV, Mean corpuscular volume; MCH, Mean corpuscular haemoglobin; PCT, Plateletcrit; LYMP, Lymphocytes; MCHC, Mean corpuscular hemoglobin concentration; Means within the rows with different superscripts are significantly different ($p<0.05$).

The main function of white blood cell (WBC) is to defend the body against foreign pathogenic organisms such as disease and stress. The fish fed 4 mg/kg nanoselenium formulated diets have low immune system and vulnerable to pathogenic activity and which possibly led to death of some of the fish. The amount of WBC, RBC and haemoglobin (HGB) concentration were greater in fish fed 2 mg/kg nanoselenium formulated diet which affirm that this could be the possible reason for the enhanced growth. WBC result provide vital information that dietary 2 mg/kg nanoselenium formulated diet possibly provided the fish fed with defensive mechanism making them as empowered body guards against the impending possible

IOP Conf. Series: Materials Science and Engineering **805** (2020) 012014 doi:10.1088/1757-899X/805/1/012014

pathogenic attacks or any possible ailment that might arise, which were evident in the survival rate of the fish. As a result of this, for better defensive mechanism, dietary inclusion of nanoselenium at 2 mg/kg nanoselenium in the fish diets of African catfish, *Clarias gariepinus* will be ideal. The significant reduction in WBC in fish fed 4 mg/kg nanoselenium formulated diet may not possibly be able to withstand stress experienced by the fish which reflected in the survival rate. The significant reduction of the lymphocytes of fish fed 2 mg/kg nanoselenium formulated diets may be one of those factors responsible for an enhanced growth which possibly made them secured from the impending ailment that might rise.

4. Conclusion

In conclusion, the study indicated that dietary inclusion at 4 mg/kg nanoselenium formulated diets is good as supplementary concentration but 2 mg/kg nanoselenium formulated diets enhanced better fish growth performance and nutrient utilization of *Clarias gariepinus*. Further research work need to be carried out on the two dietary concentrations (2 mg/kg and 4 mg/kg nanoselenium formulated diets) on *Clarias gariepinus* for a prolong period to reveal their impacts more on growth performance, nutrient utilization and organ function.

References

[1] Eisler R 2000 Selenium Handbook of Chemical Risk Assessment: Health Hazards to Humans, Plants, and Animals, vol.3. Lewis Publishers, CRC Press, Boca Raton, FL, pp.1649-1705.

[2] Hamilton SJ 2004 Review of selenium toxicity in the aquatic food chain. *Sci. Total Environ.* **326** (1-3) pp 1-31.

[3] Patching SG and Gardiner PHE 1999 Recent developments in Se metabolism and chemical speciation: a review. *J. Trace Elem. Med. Biol.* **13** (14) pp 193-214.

[4] Csilla P and Miklos M 2013 Myths and facts about the effects of nanoselenium in farm animals-mini review. *Eur. Chem. Bull.* **2** (12) pp 1049-1052.

[5] Olaosebikan BD and Raji A 2013 Field guide to Nigerian freshwater fishes. Revised edition. Federal College of Freshwater Fisheries Technology, New-Bussa, Nigeria. 143pp

[6] Ololade IA and Oginni O 2010 Toxic stress and haematological effects of nickel on African catfish; *Clarias gariepinus*, fingerlings. *J. Environ. Chem. Ecotoxicol.* **2** (2) pp 014-019.

[7] Zhou X, Wang Y, Gu Q and Li W 2009 Effects of different dietary selenium sources (selenium nanoparticle and selenomethionine) on growth performance, muscle composition and

glutathione peroxidase enzyme activity of crucian carp (*Carassius auratus gibelio*). *Aquacult.* **291** (1-2) pp 78-81.

[8] Khan KU, Zuberi A, Nazir S, Fernandes JBK, Jamil Z and Sarwar H 2016 Effects of dietary selenium nanoparticles on physiological and biochemical aspects of juvenile *Tor putitora*. *Turk. J. Zool.* **40** (5) pp 704-712.

[9] Falayi BA 2009 Tropical feedstuffs composition tables and some biological catalogues in fish and livestock production. (A crude, in nutrition technology). Matt Printing Press, Benin, Edo State Nigeria, 77p.

[10] Khan KU, Zuberia A, Nazirb S, Ullaha I, Jamila Z and Sarwar H 2017 Synergistic effects of dietary nanoselenium and vitamin C on growth, feeding, and physiological parameters of mahseer fish (*Tor putitora*). *Aquacult. Rep.* **5** pp 70-75.

[11] Khan K.U, Zuberi A, Ullah I and Sajjad 2015 Effects of graded level of dietary L-ascorbyl-2-polyphosphate on growth performance and some hematological indices of juvenile mahseer (*Tor putitora*). *Int. J. Agric. Biol.* **17** (4) pp 821-827.

[12] Igba S, Atique U, Mughal M.S, Khan N, Haider M.S, Iqbal KJ and Akmal M 2017 Effect of Selenium incorporated in feed on the Hematological profile of Tilapia (*Oreochromis niloticus*). *J. Aquacult. Res. Dev.* **8** 513. https://doi.org/10.4172/2155-9546.1000513.

[13] Duncan DB 1955 Multiple ranges and multiple F-tests. *Biometr.* **11** pp 1-42.

[14] Casillas E and Smith LS 1977. Effect of stress on blood coagulation and haematology in rainbow trout (*Salmo gairdneri*). *J. Fish Biol.* **10** (5) pp 481-491.

[15] Tomasso J.R, Simco BA and Davis KB 1983 Circulating corticosteroid and leucocyte dynamic in channel catfish *(Ictalurus puntatus)* during net confinement. *Tex. J. Sci.* **35** (1) pp 83-88.

[16] Jamil Z 2013 Effects of inorganic and nanoform of selenium on growth performance and biochemical indices of Mahseer (*Tor putitora*). MPhil. Thesis. Department of Animal Sciences, Faculty of Biological Sciences, Islamabad, Pakistan.

[17] Gatlin III DM and Wilson RP 1984 Dietary selenium requirement of fingerling channel catfish. *J. Nutr.* **114** (3) pp 627-633.

[18] World Health Organization 1996 Trace elements in human nutrition and health. *World Health Organization*, Geneva.

[19] Oldfield JE 1999 Selenium World Atlas, Selenium-Tellurium Development Association, Grimbergen, Belgium.

IOP Conf. Series: Materials Science and Engineering **805** (2020) 012014 doi:10.1088/1757-899X/805/1/012014

[20] Shreenath AP and Dooley J 2019 Selenium deficiency. Statpearls Publishing LLC. https://www.ncbi.nlm.nih.gov/books/NBK482260/. Accessed 19 January, 2020.

[21] Hoffmann PR and Berry MJ 2008 The influence of selenium on immune responses. *Mol. Nutr. Food Res.* **52** (11) pp 1273-1280.

[22] Raza A 2012 Effects of graded levels of dietary selenium supplementation on the growth of juvenile Mahseer (*Tor Putitora*). MPhil. Thesis. Department of Animal Sciences, Faculty of Biological Sciences, Islamabad, Pakistan

[23] Hemre GI, Hjeltnes B, Aksnes A and Waagab R 1996 Effect of gelatinized wheat and maize in diet for large Atlantic salmon (*Salmo sar* L.) on glycogen retention, plasma glucose and fish health. *Aquacult. Nutr.* **2** (1) pp 33-39.

[24] Mehboob A, Khan N, Atiq U, Iqbal KJ and Tayyab R 2017 Effect of fenugreek as a feed additive on the growth, body composition and apparent nutrients digestibility of striped catfish *Pangasius hypophthalmus* fry. *Pak. J. Zool.* **49** (6) pp 2037-2042.

. [25] Noori A, Nasir, Amar Y and Jassim A 2013 Effect of different dietary proteins and fats on some biochemical blood parameters in common carp fingerlings (*Cyprinus carpio* L.) reared in floating cages. *Asian J. Exp. Biol. Sci.* **4** pp 293-298.

[26] Feist SW and Longshow M 2000 Myxosporiodiosis of fish and bryozoan link with proliferate kidney diseases (PKD) of salmonids. *J. Fish Dis.* **1** pp 91-108.

Enhancing the thermal performance of solar collectors using nanofluids

M I Omisanya[1*], AK Hamzat[2], SA Adedayo[3], IA Adediran[4] and TB Asafa[5]

[1]School of Mechanical Engineering, Power and Energy Department, Tianjin University, 300350 Tianjin, China.

[2] Department of Mechanical Engineering, King Fahd University of Petroleum and Minerals, Dhahran, Saudi Arabia.

[3]Biomedical Engineering Department, University of Ibadan, Ibadan, Nigeria.

[4]Department of Mechanical Engineering, Ankara Yildirim Beyazit University, Ankara, Turkey.

[5] Department of Mechanical Engineering, LAUTECH, Ogbomoso, Nigeria

*E-mail: omisanyaibrahim02@gmail.com

Abstract

Solar energy remains the most ubiquitous and inexhaustible source of energy. This energy can be utilized by several approaches including the use of solar collectors. Several studies have illustrated that the efficiency of solar collectors can be significantly improved by the introduction of nanofluids which have shown improved thermal conductivity up to 160% with a subsequent reduction in greenhouse gases such as CO_2. To produce nanofluids, nanoparticles such as Al_2O_3, hybrid ZnO + Al_2O_3, and metals (Al, Cu) particles are dispersed into the based fluids such as water, glycerol, and bio-fluids. The added nanoparticles enhance the viscosity, absorption rate, convective heat transfer coefficient and heat losses of the fluid. The performance of details the applications and effectiveness of different nanofluids in four types of solar collectors - parabolic trough, flat plate, direct absorption, and evacuated tube. In addition, the work sheds light on the future trend and challenges of nanofluids (including toxicity) in solar collectors. Regardless of its toxicity, researchers have shown more interest in nanofluids use in solar collectors because of its strong sustainability to a safe environment and the exploration of hybrid nanofluids to better enhance solar collectors. Solar collectors can also be modulated by using different nanofluids at varying concentrations.

Keywords: Nanofluids, Solar collectors, Parabolic trough solar collector, Flat plate solar collector (FPSC), Direct absorption solar Collector (DASC), Evacuated tube solar collector (ETSC), Thermal Efficiency

1. Introduction

Energy researchers and policy makers are shifting focus from non-renewable energy, such as oil and coal, to renewable sources such as solar, wind and geothermal. This is strongly connected to the fact that the combustion of oil/coal produces environmental pollutants or greenhouse gasses and subsequent leads to depletion of the ozone layer. Various solutions have been proposed to ease the transition between these energy sources with solar energy utilization occupying the forefront [1]. Although solar energy has been utilized from the time immemorial, achieving maximum energy conversion with minimum waste remains a challenge [2-3]. Different solar technologies have been developed to convert sun radiation into thermal energy (e.g. thermal collectors) or electrical energy (e.g. solar panels). However, the cost of development and operation coupled with low conversion efficiency remains a huge challenge [4-7]. Solar collectors, which convert solar irradiance into thermal via heat exchanging fluid, have low efficiency with relatively high initial cost. To enhance the collectors' efficiency, studies have shown that the addition of nanoparticles such as ceramics oxide (Al_2O_3, CuO), non-metals (Graphite and Carbon nanotubes), metals (Ag, Fe) with associated high thermal conductivity can significantly increase thermal conversion efficiency [8-9]. This review examines the progress made thus far with the application of nanofluids in various solar collectors.

1.1 Solar Collectors

A solar collector is a device that actively gathers solar radiation, convert it to heat and transform such heat to a receiving medium such as air, water or solar fluid which is then used for domestic purposes. Solar collectors are divided into two main categories. The concentrating solar collectors are designed to track the sunlight to fully absorb radiation from the sun. This collector can be one axis such as parabolic trough, cylindrical trough and linear Fresnel collector, etc or two axes sun tracking collector such as parabolic tower, parabolic dish, circular Fresnel solar collector etc. The non-concentrating solar collectors use liquid or air as the absorbing medium. Flat-plate solar, vacuum tube, compound parabolic, and bowl collectors make use of heated liquids while ventilated cartridge and uncoated transparent solar collectors utilize heated air.

IOP Conf. Series: Materials Science and Engineering **805** (2020) 012015 doi:10.1088/1757-899X/805/1/012015

1.1.1 Parabolic Trough Solar Collector

Parabolic Trough Solar Collector (PTSC) is desirable for Medium-Temperature applications usually in the range of 100 °C to 400 °C [10]. Of the available concentrating solar power technologies, parabolic trough collectors arefavored to be the most commercially proven technology owing to its numerous advantages such as high storage capacity, hybrid concept, modular system, low operating cost, and land use factor [11]. PTSCs are "U" shaped concentrators constructed as a linear parabolic mirror to concentrate sunlight on to the absorber tube that is situated along its focal line (Figure 1). The heat transfer fluid absorbs solar radiation energy and recirculates it through the pipe located along withthe focal point. The receiver surface is usually covered with a selective coating that has low irradiance and high solar radiation absorption for thermal radiation. To reduce the convective heat loss in the vacuum region between the absorber and the glass cover, the absorber tube is wrapped with a transparent material [12]. By employing a tracking mechanism, the movement of the sun from east to west is successfully tracked as PTSC is aligned to the north-south axis. The tracking mechanism employed in parabolic trough could either be single-axis or dual-axis.
Parabolic Trough Solar Collect has a concentration ratio of 30-100 and receiver temperature of about 400 °C. It allows for efficient solar power generation and is currently used in many solar thermal generation plants. PTSC accounts for about 90% of the total available energy from the concentrator solar panel system [13]. Hence, the desire to improve its performance by enhancing the functionality of the components such as reflector, receiver tube, support structure, and tracking system.

The critical section of a parabolic trough solar collector is the heat transfer fluid flowing within the tube, which is typically a mixture of water or hot oil and other additives such as nanoparticles that increase the heat transfer rate. The temperature at which the fluid is injected through the tube and absorbs heat from the sun exceeds 200 °C. The heat exchanger is then used to transfer the heated fluid for domestic use. Compared to the traditional parabolic trough concentrators, nanofluidic concentrated parabolic solar collectors (NCPSC) have better energy absorption and conduction heat transfer performance [14]. The application of nanofluids as working fluids in parabolic trough solar collectors (PTSC) is receiving increasing attention.

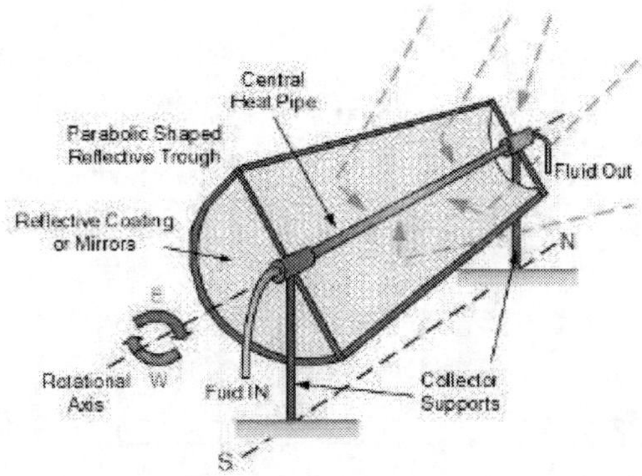

Figure 1. Diagrammatic representation of a Parabolic Trough Solar Collector [15]

1.1.2 Flat Plate Solar Collector

Flat plate solar collector was developed in the 1950s, and several techniques and methods have been deployed to enhance its thermal performance and efficiency [16]. A flat plate solar collector consists of various components such as aluminium rails, insulation, absorber, glass, back sheet, riser and header pipes. The surface is covered with glass for optimal heat transmission to the absorber. For high efficiency, absorbers are made from aluminium coated with a high absorbing material [17-18].

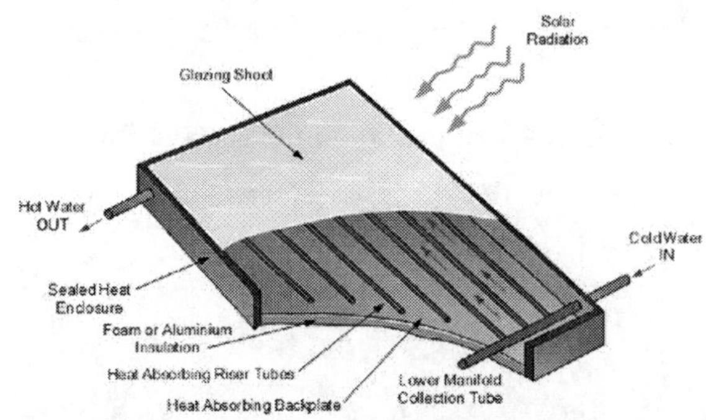

Figure 2. Typical representation of a flat plate solar collector

1.1.3 Direct Absorption Solar Collector

Direct Absorbed Solar Collector is a simplified approach to solar energy collection. This technique was originally proposed in the 1970s as a way of absorbing solar energy directly within the fluid media. Recently, direct absorption solar collectors (DASC) based on nanofluids have been proposed [19-29] and it is considered as the newer class of collectors with enhanced energy utilization capability [30].

DASC has shown potentials for efficient solar-to-thermal energy conversion with excellent stability and high solar energy absorption capabilities [31]. DASC uses nanofluid for both

IOP Conf. Series: Materials Science and Engineering **805** (2020) 012015 doi:10.1088/1757-899X/805/1/012015

solar absorption medium and heat-transfer fluid [32] and has no absorber plate [33]. Unlike solid absorber,this technique eliminates intermediate heat transfer medium [31] and can utilize parabolic trough and dish, linear Fresnel collector and heliostat field collector (solar tower) among others [34].

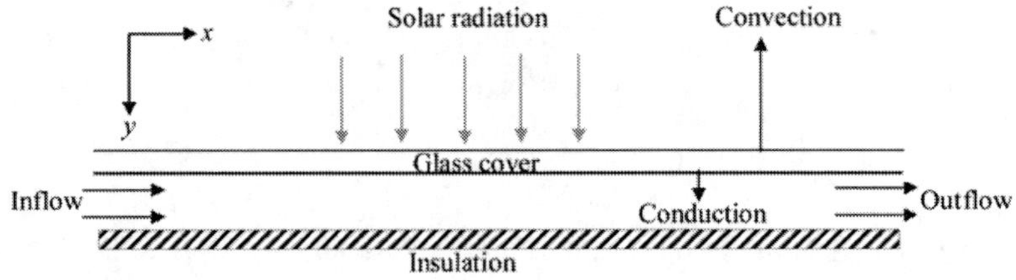

Figure 3. Schematic representation of direct absorption solar collector

1.1.4 Evacuated tube solar collector

This solar collector is made up of evacuated glass pipes with outer and inner tubes. The transparent form of the outer tube allows light rays to penetrate reaching the inner tube where it is absorbed. For maximal absorption, a copper pipe is welded and coated to an absorber plate before being inserted (Figure 3). Due to the minimal reflection of both tubes, sun rays penetrate through the outer tube while the heat isreserved inside the inner tube. Solar radiation trapped between two tubes by creating a vacuum thatis then fused on top of each other with no trapped air. The radiated heat from the sun stays inside the inner tube where it is effectively utilized. Therefore, ETSC is considered as a great device for an efficient collection of solar energy [35].

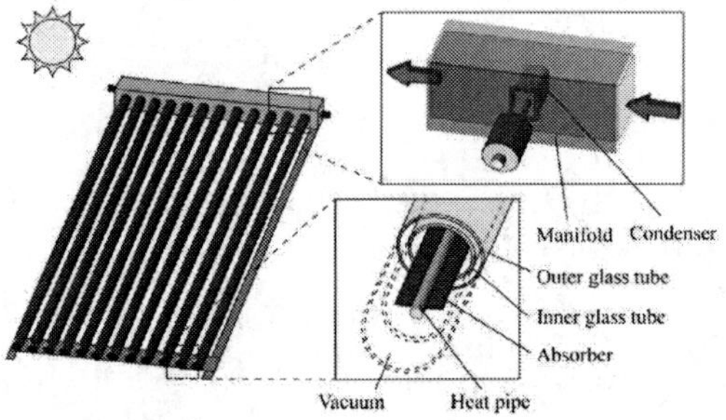

Figure 4. Evacuated tube Collector with the Inner and Outer tube [35]

IOP Conf. Series: Materials Science and Engineering **805** (2020) 012015 doi:10.1088/1757-899X/805/1/012015

2. Nanofluids

Nanofluids (NFs) are fluidscontaining nanoparticles (particles of size less 100 nm). They are produced by suspending nanoparticles in a basefluid usually water. Its use has been found to produce desired heat transfer when compared with conventional base fluid [36]. Most nanoparticles used in producing NFs are metals, non-metals, carbon nanotubes among others. Nanofluids are very useful in heat transfer and have found applications in fuel cells, boiler flue gas temperature, heat exchanger reduction, and vehicle thermal management. NFs have shown good thermophysical properties like viscosity, optical absorption, thermal diffusivity which has helped to improve the performance of solar collectors.

2.1 Nanofluids preparation

2.1.1 One step method

Generally, the preparation of nanofluids requires a setup comprising a cylindrical drum with an upper section, a lower section, and an inner surface. The base liquid is poured into the lower section forming the base liquid pool. A material that is capable of evaporating is then provided to serve as the nanoparticle. A stirrer is then used to create a speed equal to or exceeding the speed required to drag part of the liquid in the liquid pool creating a rotating liquid film along the inner surface of the drum. The evaporating material is positioned in proximity to the rotating liquid film with heating that is sufficient to evaporate the evaporantto form nanoparticles, the formation of the nanoparticles inside the rotating liquid film forms the nanofluid (Figure 5). A one-step method is simply the process by which particles in nanocrystalline forms are produced by deposition onto a low vapor-liquid. Nanoparticles produced via a one-step method avoids at a lot of processes such as dispersing, drying storage which is viewed with massive potential but its shortcoming is that nanoparticles less than 10 nm cannot be processed to form nanofluid.

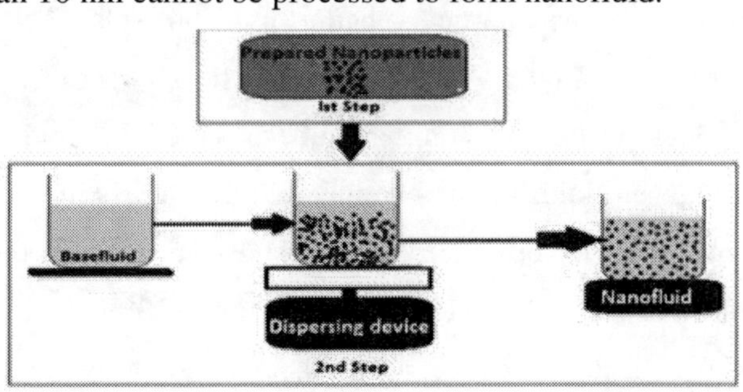

Figure 5. Schematic representation of the one-step method for the production of nanofluid

IOP Conf. Series: Materials Science and Engineering **805** (2020) 012015 doi:10.1088/1757-899X/805/1/012015

2.1.2 Two-step method

This is the widely accepted method used in the production of nanofluids. The steps involve the production of nanoparticles, either metallic oxides or from pure metals which are then dispersed into the base fluids with some degree of agitation (Figure 6). The problem associated with this method is the continuous agglomeration of the nanoparticles after mixing. The clustering can be minimized by adding surfactant as seen from various studies. However, surfactants can break down at very high temperatures. A combination of two-step and one-step method has been reported to form stable nanofluids.

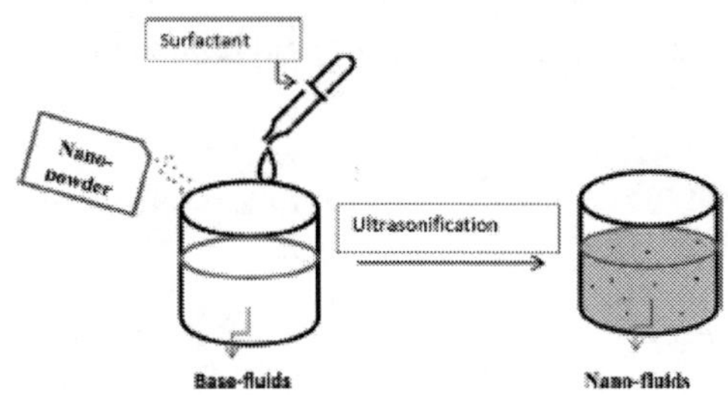

Figure 6. Schematic representation of the two-step method for the production of nanofluids [33]

2.2 Properties of Nanofluids

The addition of nanoparticles enhances convective heat transfer coefficient about tenfold especially when they are properly dispersed. Furthermore, nanofluids with great volume fraction have a high thermal conductivity which increases with temperature. Also, both optical and absorption rate can be improved by using nanofluids. This is in addition to higher thermal diffusivity and viscosity.

3. Nanofluid based parabolic trough solar collector

Bellos *et al.* [37] classify nanoparticles being used in solar collectors into metallic nanoparticles such as Fe, Cu, Zn, Al, Au, and non-metallic nanoparticles such as Al_2O_3, Fe_2O_3, SiO_2, TiO_2, ZnO, single-walled carbon nanotubes (SWCNT), and multi-walled carbon nanotubes (MWCNT). Nanofluids thermal conductivityis majorly influenced by the diameter and dispersion of constituent nanoparticles. About 41% of nanofluids based research utilized Al_2O_3, which is followed by CuO and then TiO_2. About 30% of the published works on nanofluid based solar collectors are experimental. In these studies, the impact of different nanofluidsand the percentage of nanoparticles have been the major focus. For instance, Sharma *et al.* [38] compared the influence of Al_2O_3/water and CuO/water at 0.01 and 0.05

IOP Conf. Series: Materials Science and Engineering **805** (2020) 012015 doi:10.1088/1757-899X/805/1/012015

vol. % on the thermal performance of PTSC. The results show that 0.05 vol.% CuO has the maximum overall thermal efficiency of 15.25% compared to 12.39% and 10.58% for Al_2O_3/water and water, respectively. This translates to a 44% increase in efficiency when 0.05 vol.%CuO/water nanofluid is used compared to water as the base fluid.

Compared to Sharma *et al.* [38], moderate increase in concentration of Al_2O_3 to 0.06 vol.% improved thermal efficiency by 28.95% [40]. Similarly, the heat transfer coefficient, thermal efficiency and Nusselt number were enhanced by 32%, 7%, and 32%, respectively for 1 vol% Al_2O_3 [42]. In addition, Subramani *et al.* [45] reported an 8.54% increase in the thermal behaviour of the PTSC at 0.5 vol % of Al_2O_3. Kandwal *et al.* [41] studied two nanofluids, CuO/water, and CuO/ethylene glycol at different concentrations (0.01, 0.05, and 0.1 vol. %) and flow rates. The results show that the highest thermal efficiencies of 11.01%, 8.03%, 7.46%, and 6.04% were obtained for CuO/water, CuO/ ethylene glycol, water, and ethylene glycol, respectively at 160 L/h. This implies 82% increased thermal efficiency forCuO/water nanofluid in comparison with ethylene glycol. Menbari *et al.* [44] conducted both experiment and numerical research to study the effect of CuO/water nanofluid on thermal efficiency and temperature distribution in a direct absorption parabolic trough collector for a different range of volume fraction. The results show that thermal efficiency improves by 18-52% for 0.002–0.008 vol.% CuO nanoparticles.

Coccia *et al.* [43] reported no significant change in thermal properties when Fe_2O_3, SiO_2, TiO_2, and ZnO nanoparticles were used. However, 0.01vol.% Au/water nanofluids raised convection heat transfer by approximately 50%. Another experimental study carried out on the effect of MWCNT/mineral oil (with 0.2 and 0.3 wt %) on the performance of PTSC indicates a 5-7% increase in thermal efficiency [39]. In another study, however, 0.3 vol.% MWCNT/ethylene glycol improved thermal efficiency by 30.4% compared to 14% for 0.3 vol.% SiO_2/ EG [46]. The summary of previous works is shown in Table 1.

IOP Conf. Series: Materials Science and Engineering **805** (2020) 012015 doi:10.1088/1757-899X/805/1/012015

Table 1. Summary of experimental works conducted on the nanofluid usage in PTSC.

Study	Nanofluid	Nanoparticle			Max. Increase%			Finding(s)
		Type	ϕmax	Size (nm)	ηth	h	ΔP	
Sharma *et al.* [38]	Water	Al_2O_3, CuO	0.05 vol. %	20–40	44	-	-	CuO had the highest thermal efficiency.
Kasaeian *et al.* [39]	Mineral oil	MWCNT	0.3 wt %	10	7	-	-	The first study uses carbon nanotube as nanoparticle in PTSC
Siva Reddy *et al.* [40]	Water	Al_2O_3	0.06 vol. %	<50	28.98	-	-	Minimization of losses and prevention of nanoparticles settlementcan augment the thermal efficiency of PTSC
Kandwal *et al.* [41]	EG, Water	CuO	0.1 vol. %	25–55	82	-	-	CuO/water nanofluid had better thermal efficiency in comparison with CuO/EG nanofluid
Chaudhari *et al.* [42]	Water	Al_2O_3	0.1 vol. %	40	7	32	-	Nusselt number increases by application of nanofluid Although, an asymptotic trend was observed after nanopartclevolumetric concentration of 0.2%, the maximum efficiency achieved at 0.5% volume fraction.
Subramani *et al.* [45]	Water	Al_2O_3	0.5 vol. %	40–50	8.54	-	-	The optimal nanoparticle concentration was proposed to be 0.2% by parametric study (Semi) empirical equations were obtained for Nusselt number, friction factor and collector efficiency using experimental data.
Kasaeian *et al.* [46]	EG	MWCNT, SiO_2	0.3 vol. %	4,6	30.4	-	-	MWCNT/EG was more efficient nanofluid in term of higher thermal efficiency thanSiO_2/EG
Sekhar *et al.* [47]	Water	CeO_2, Al_2O_3, TiO_2	3 vol. %	-	27	-	-	(CeO_2)/water nanofluid had the highest thermal efficiency enhancement followed by alumina and titania nanofluids.
Coccia *et al.* [43]	Water	Fe_2O_3,	20 wt %	-	~0	~50	-	Application of nanofluid didn't cause thermal

		SiO_2,TiO_2, ZnO, Al_2O_3, Au						efficiency enhancement.
Rehan *et al.* [48]	Water	Al_2O_3,Fe_2O_3	0.30 wt %	20-40	13	–	–	Al_2O_3/water had a higher thermal efficiency enhancement compared to Fe_2O_3/water nanofluid.
Subramani *et al.* [49]	Water	TiO_2	0.5 vol. %	20	9	23	-	Nanofluid has higher absorbed energy factor compared with the base fluid.
Marefati *et al.* [50]	Water	Al_2O_3,CuO,SiC	5 vol. %	-	-	~32	-	CuO/water nanofluid is the best option for thermal efficiency enhancement followed by alumina and SiC nanofluid.
Okonkwo *et al.* [51]	Therminol VP-1	Al_2O_3, CuO, Fe_3O_4	3 vol. %	-	0.22	-	-	Al_2O_3/Therminol VP-1 nanofluid is the best practical fluid to achieve the highest efficiency followed by CuO/ Therminol VP-1, and Fe_3O_4/Therminol.
Alsaady *et al.* [52]	-	Magnetic nanofluids	0.05 vol. %		25			The application of ferrofluid in PTSC seems to be a promising field of study.
De Los Rios *et al.* [53]	Water	Al_2O_3	3 vol. %	10	28	-	-	An experimental study was conducted on the thermal efficiency of parabolic trough linear concentrator collector The incident angle of the collector had an important role in enhancing the thermal efficiency of the collector (a small increase in incident angle can cause thermalefficiency enhancement)
Okonkwo *et al.* [54]	Pressurized water	OLE-TiO_2, BH-SiO_2	Up to 0.05 vol.%	70, 100	0.08	138	14.25	This study was the first effort in the application of green-synthesized nanofluid in PTSC. Application of water/OLE-TiO_2 nanofluids led to higher thermal efficiency increase, compared with the base fluid.
Menbari *et al.* [44]	Water	CuO	Up to 0.008 vol.%	<100 nm	52%	-	-	A general trend of increasing thermal efficiency (ranging from 18–52%) having nanoparticle volume fraction increase (from 0.002% to 0.008%) when water/CuO nanofluid was used in direct absorption parabolic trough collector bycomparison with the base fluid

IOP Conf. Series: Materials Science and Engineering **805** (2020) 012015 doi:10.1088/1757-899X/805/1/012015

Another approach being employed to study the influence of nanofluids on the thermal performance of solar collectors is numerical analysis. Numerical studies account for about 60-70% of published articles in this area and is largely a cost saving approach. Numerical analysis is mostly done using commercial software such as ANSYS Fluent, COMSOL, SolTrace and EES among others. The first numerical study, conducted by Kasaeian and Sokhansefat [55], utilized CFD to study the role of Al_2O_3/synthetic oil nanofluid in PTSC at 1-5 vol. % of Al_2O_3. It was found that 5 vol.% Al_2O_3 enhanced the heat transfer coefficient by ~14% at 300K. At 500K however, nanoparticles effect was insignificant. Similarly, Sokhansefat *et al.* [56] studied the effect of Al_2O_3/synthetic oil by simulating flow through an absorber tube at 1–5 vol.%. At 5 vol.%, the convection heat transfer in PTSC was increased by 14–15%. The influence of higher temperature is similar to that reported elsewhere [55]. As an improvement over the previous studies, Zadeh *et al.* [57] investigated the effect of using Al_2O_3/synthetic oil (with 0-6.5 vol. %) using a hybrid optimized model coupled with CFD analysis. The proposed model improved the maximum convection heat transfer coefficient by 36% at 300K. In another study [36], 4 vol.%, 6 vol.% and 8 vol.% Al_2O_3/synthetic oil nanofluids enhanced convective heat transfer coefficient by 35, 54 and 76%, respectively. About 40% increase in pressure drop at 4 vol.% was also reported implying more pumping power.

For 6 vol.% CuO/syltherm 800 nanofluid, the maximum increase in the heat transfer coefficient and thermal efficiency are 38 and 15%, respectively [58]. Pressure drop was also reportedby Bellos and Tzivanidis [37], where they simulated Al_2O_3/water nanofluid by minimizing entropy generation and maximizing different volume fractions of Reynolds numberfor Al_2O_3 located at different sections of a circular tube. The results show that for a given Reynolds number, the heat transfer performance (Nusselt number) and the value for friction factor are almost the same regardless of the volume fraction. Also, it was verified that there is an optimum cross-section of the tube that causes a minimum generation of entropy. An 18% increase in convective heat transfer was reported by Basbous *et al.* [61] in their numerical investigationof Al_2O_3/syltherm 800 nanofluid in the range of 1-5%. For the same system, Bellos *et al.* [62] reported a 4.25% increase at 2 vol.% Al_2O_3. They also reported a 6.34% increase in thermal efficiency when pressurized water was used asa working fluid. However, the application of pressurized water requires more effort which makes its utilization less acceptable.

The effects of different volume fractions (0.5-5%) of Al_2O_3/hot oil nanofluid on the performance of PTSC were studied by CFD simulations [63]. It was reported the efficiency enhancement around 1.2% in their model (at 5% volume fraction). They also found that increasing the nanofluid volume fraction can lead to a slight decrease in the longitudinal displacement of the absorber. Similarly, the horizontal displacement of the absorber was reported to slightly decrease by volume fraction increase.

Kaloudis *et al.* [64] conducted a CFD simulation to investigate the effect of different volume fractions (0.5-4%) of Al_2O_3 in Al_2O_3/syltherm 800 nanofluid on PTSC performance. They used two different modelings for nanofluid: two-phase simulation and one phase simulation. They concluded that utilizing two-phase modeling of nanofluid had more accuracy in the results. More specifically, the numerical results suggested a maximum of 10% thermal efficiency enhancement at a maximum Al_2O_3 volume fraction of 4%.

Benabderrahmane *et al.* [65] numerically modelled the effect of utilizing nanofluid as well as internal fins in the receiver of PTSC using CFD. In a report [66], conclusions were made that combining both internal fins and nanofluid applications can enhance the heat transfer behaviour of the PTSC greatly. To be specific, the authors utilized four different nanofluids (Al_2O_3/Dowtherm A, Cu/Dowtherm A, SiC/Dowtherm A, C/Dowtherm A) at different volumetric concentration of 1% and found that the metallic nanofluids (Cu/Dowtherm) had better thermal enhancement performance comparing with non-metallic ones (SiC/Dowtherm A, C/Dowtherm A, Al_2O_3/Dowtherm A). Also, the friction factor reported being higher when nanofluid is applied. A 68% increase in convection heat transfer coefficient was reported [67].

Ghasemi *et al.* [68] performed numerical simulations of Al2O3/water and CuO/water as working fluids at different volume fractions, 0.5-3%, in PTC were investigated. They realized that the application of 3% CuO/water nanofluid could lead to a higher convection heat transfer coefficient compared with Al_2O_3/ water and water itself. Although CuO/water nanofluid caused a 7% more increase in heat transfer coefficient compared with Al_2O_3/water nanofluid (35% compared with 28% enhancement), the friction factor caused by the application of Al_2O_3 was lower than the one caused by CuO.

Mwesigye *et al.* [69] conducted another CFD simulation on Cu/Therminol VP-1 nanofluid and the effects of nanofluids with different volume fractions (1-6%) on the thermal and thermodynamic properties of PTSC were studied. At a maximum, 6% volumetric concentration of Cu, convective heat transfer coefficient and thermal efficiency increased to 32% and 12.5%, respectively. Furthermore, the pressure dropsvia the receiver at 400 K and Reynolds number of approximately 650,000 increases roughly by more than 100% (from nearly 3000 to 7734 Pa/m). Finally, the minimum entropy generation rate at a certain Reynolds number was obtained by their model. This point is crucial as it shows the Reynolds

IOP Conf. Series: Materials Science and Engineering **805** (2020) 012015 doi:10.1088/1757-899X/805/1/012015

number in which we have the maximum heat transfer irreversibility and minimum friction irreversibility. Specifically, there is a maximum 30% decrease in the entropy generation rate at 6% volume fraction comparing to the base fluid.

Basbous *et al.* [70] numerically modeled the thermal performance of PTSC utilizing different metallic nanofluids. Syltherm 800 as base fluid and Cu, CuO, and Al_2O_3, and Ag with 5% volumetric concentrations were considered as studied nanofluids. The results suggested that silver is the best nanoparticle to be used to induce the best thermal enhancement followed by copper, cupric oxide and aluminum oxide with convection heat transfer enhancement of 36, 33, 27, and 18%, respectively. They also concluded that as the density of the nanoparticle increases, the thermal performance enhances better. Finally, a 21% decrease in heat loss was observed.

Toghyani *et al.* [71] conducted numerical modeling and optimization of the effect of different nanoparticles on the performance of the PTC used in a Rankin cycle. CuO, SiO_2, TiO_2, and Al_2O_3 were dispersed in Therminol-55 at 2–5.5 vol. %. The results show that the use of nanofluids in solutions can slightly improve thermal efficiency.Finally, the optimization results revealed that using Al_2O_3/Therminol-55, caused the maximum enhancement in the overall exergy efficiency of the system by 11%. Nayak *et al.* [72]modeled Al_2O_3/Synthetic oil nanofluid up to 5% volumetric concentration to investigate and optimize the thermal behavior of the PTSC. They reported an approximately 7% enhancement in heat transfer coefficient when 5% volumetric concentration of Al_2O_3 was used comparing with Synthetic oil only. Ferraro *et al.* [73] used Al_2O_3/Synthetic oil nanofluid to analyse the nanofluid effect on PTSC performance numerically. Their results showed that thermal efficacy and pressure drop almost stays the same. However, the maximum increase of approximately 60% was observed at a 5% volumetric concentration of Al_2O_3 for the convection heat transfer coefficient. Kharkah *et al.* [74] conducted a CFD simulation on the receiver system of PTSC using Al_2O_3/Synthetic oil nanofluid as a working fluid at 5% volumetric concentration. The enhancement of 14.3% was reported in the system thermal efficiency. Table 2 summarizes the details and important notes about numerical studies.

4. Flat plate solar collectors review using nanofluids

Researchers have studied the efficiency and thermal performance of flat-plate solar collectors, and the effects of nanofluids and nanoparticles on them. Extensive investigations on the use of flat plate solar collectors were carried out [85-87]. Yousefi *et al.* [86] experimentally studied the effects of multi-walled carbon nanotubes/water nanofluids as heat transfer fluids on the efficiency of FPSC.The results show that the efficiency of MWCNT nanofluid is lower than when the working fluid is water. To improve the diffusion and reduce the assemblage of nanoparticles in the nanofluid, most researchers add Triton X-100 as a surfactant to the nanofluids [88-90]. Yousefi *et al.* [85] discovered that the addition of Triton X-100 as a surfactant to the Al_2O_3 nanofluid increases the solar collector efficiency.

The efficiency of square FPSC using SiO_2/water nanofluid was experimentally investigated by Noghrehabadi *et al.* [91]. It was found out that the SiO_2/water nanofluid increased the temperature performance and thermal efficiency of the solar collector when compared with water nanofluid. The efficiency of the nanofluid also increased with an increase in the flow rate. In another study Chaji *et al.* [92], two different surfactants (Cetyl trimethylammonium bromide and Triton X-100) were used to enhance the efficiency of the TiO_2/water nanofluid. However, the surfactants produced foam, reducing the efficiency of the collector. In addition, the flow rate of nanofluids also varies and it was discovered that the highest flow rate produced the least efficiency. Meibodi *et al.* [93] studied the thermal efficiency and performance characteristics of SiO_2/ethylene glycol (EG)-water nanofluids with a collector plate surface area of 1.59 m^2. The volume concentrations considered in the experiment were 0.5, 0.75, and 1%, respectively, and the mass flow rates were 0.018, 0.032, and 0.045 kg/s, respectively. When the thermal efficiency increases from 4% to 8% and the concentration increase from 0.5 to 1%, the heat loss parameter is limited to zero.

Alim *et al.* [94] used different types of nanofluids (CuO, Al_2O_3, SiO_2, TiO_2 dispersed in water) to analyze the pressure drops and entropy production in FPSC having an absorption area of 1.51 m^2. The analysis was performed for different volume fractions (1-4%) and volumetric flows (1-4 L/s). Zamzamian *et al.* [95] investigated the effect of synthetic copper nanoparticles/ethylene glycol nanofluids on the efficiency of FPSC. The mean diameter of the nanoparticles was considered in the experiment was 10 nm. Nanofluids with mass fractions of 0.2% and 0.3% were prepared and nanofluids were prepared in a one-step process.The volumetric flow varies from 0.5 to 1.5 L/s and the efficiency of the solar collector is calculated using the ASHRAE standard.Polvongsri and Kiatsiriroat[96] used silver nanofluids as the working fluids to study the performance of flat solar collectors. Three identical closed-loop flat solar collectors were installed having an area of 0.15 m × 1.0 m. The various achievement that has been recorded by researchers using nanofluids is summarized in Table 3.

IOP Conf. Series: Materials Science and Engineering **805** (2020) 012015 doi:10.1088/1757-899X/805/1/012015

5. Evacuated tube solar collector review using nanofluids (Figure 7)

Ghaderin and Sidik [104] used Al_2O_3/distilled water nanofluid as working fluid, and the thermal efficiency of the spherical coil all-glass passive circulating evacuated collector in the horizontal storage tank was studied experimentally. The volume fraction of the nanoparticles was 0.03% and 0.06%, respectively, and the particle size was 40 nm. In this experiment, Triton x-100 was used as a surfactant, and the performance of Al_2O_3 nanofluid and water was compared with the flow in a coil with a flow range of 20-60 l/h. They observed that when the nanofluid volume fraction is 0.06 and the mass flow rate is 60 l/h, the maximum efficiency is 57.63%. With the increase of volume fraction and flow rate of Al_2O_3 nanoparticles, the collector efficiency has been greatly improved. Research results show that alumina nanofluid can be used as the working fluid of ETSC, effectively absorbing solar rays and converting it into thermal energy. Again, Ghaderian *et al.* [105] observed a similar trend studying the effects of the CuO-distilled water nanofluidic ETSC water heater with a built-in coiled thermosiphon system. A volumetric fraction of 0.03-0.06% of nanoparticles with mass flow rates ranging from 20 to 60 l/h was used.When the volume fraction of CuO nanofluids is 0.03%, the efficiency of ESTC is increased to 14% compared to water.

Iranmanesh *et al.* [106] investigated experimentally the thermal performance of evacuated tube solar collectors (ETSC) using graphene nanoplatelets (GNP)/ distilled water nanofluid. Various properties of GNP were studied at four different mass percentage and three volumetric flow rates. From their results, increment in the mass percentage with the use of graphene nanosheets leads to more gain in thermal energy as the higher fluid outlet temperature is reached. Using nanofluid with a mass concentration of 0.01%, the thermal efficiency in the ESTC reached 90.7% and volume flow rate of 1.5 l/min, while ESTC using pure water at the same flow rate attained 35.8%.

IOP Conf. Series: Materials Science and Engineering **805** (2020) 012015 doi:10.1088/1757-899X/805/1/012015

Table 2. Summary of numerical studies conducted on the nanofluid usage in PTSC

Study	Method	Nanofluid	Nanoparticle			Max	Increase	%	Finding(s)
			Type	**ϕmax**	**Size (nm)**	**η_{th}**	**H**	**ΔP**	
Mwesigye *et al.* [75]	CFD	Therminol VP-1	Cu, Ag, Al_2O_3	6 vol. %	<100	13.9	7.9	-	Dispersion of Ag had the highest thermal efficiency enhancement followed by Cu and Al_2O_3. Maximum entropy generation rate decrease of 24% at lower Reynolds number was observed when Al_2O_3/Therminol VP-1 nanofluid is used
Ghasemi *et al.* [68]	CFD	Therminol-66	Al_2O_3	4 vol. %	20	~0.5	-	-	Nusselt number increases considerably by using nanofluid, compared to the base fluid
Bellos *et al.* [76]	Model	Syltherm 800	Al_2O_3, CuO	4 vol. %	-	1.26	41	-	They found that CuO/ syltherm 800 has the maximum thermal efficiency followed by Al_2O_3/syltherm 800.
Bellos *et al.* [75]	Model	Syltherm 800	Al_2O_3, CuO	6 vol. %	-	1.91	-	-	After optimization, the use of CuO nanofluid with 4.35% volumetric concentration of CuO in the solar loop along with Toluene in the organic Rankin cycle was the optimum choice.
Bellos *et al.* [77]	CFD	Syltherm 800 Molten salt	CuO	6 vol. %	-	0.76	31.5	50	CuO/syltherm 800 had better thermal efficiency than CuO/molten salt nanofluid.
Bellos *et al.* [67]	CFD	Syltherm 800	CuO	6 vol. %	-	1.54	130	~163	Combination of both nanofluids and fins can lead to higher efficiency enhancement, convection heat transfer, and pressure drop
Mwesigye *et al.* [78]	CFD	Therminol VP-1	SWCNT	2.5 vol. %	10	4.4	234	-	Higher thermal conductivity doesn't necessarily lead to higher thermal efficiency since the effect of specific heat capacity also should beconsidered
Allouhi *et al.* [79]	Model	Therminol VP-1	TiO_2, CuO, Al_2O_3	5 vol. %	-	~0	83	-	CuO/Therminol VP-1 nanofluid showed the best exergy and energy performance followed by TiO_2 and Al_2O_3
Bellos *et al.* [80]	Model	Syltherm 800	TiO_2, Al_2O_3, and TiO_2/Al_2O_3	3 vol. %	-	0.74	142.1	-	TiO_2/Al_2O_3 based nanofluid had highest performance, followed by TiO_2 and Al_2O_3
Bellos *et al.* [80]	Model	Syltherm 800	Cu, CuO, SiO_2, Al_2O_3, Fe_2O_3, TiO_2	6 vol. %	-	2.2	~24.4	-	Cu had the highest efficiency increase of 2.2% followed by CuO, Fe_2O_3, TiO_2, Al_2O_3, and SiO_2
Kasaeian *et al.* [81]	Model	Thermal oil	MWCNT	6 vol. %	-	-	15	-	Convection heat loss increased by 220% in

IOP Conf. Series: Materials Science and Engineering **805** (2020) 012015 doi:10.1088/1757-899X/805/1/012015

									nanofluid; when the temperature increased from 30 to 100.
Bilal *et al.* [82]	Model	Water	Fe_3O_4	0.6 vol. %	-	1.6	87	-	56% enhancement was observed for Nusselt number at 0.6 vol. % of nanoparticles and highest Reynolds number. A combination of the fin and nanofluid applications can result in a maximum 87% increase in the heat transfer coefficient.
Khular *et al.* [10]	Model	Therminol VP-1	Al_2O_3	0.05 vol. %	5	5–10	-	-	To obtain favorable thermal efficiency and maximum outlet temperature, there should be optimizations on various variables such as nanoparticle shape, size, and material.
Hatami *et al.* [83]	CFD	Water	Cu, Fe_3O_4, Al_2O_3, TiO_2	8 vol. %	-	-	-	-	Cu nanoparticles shows the best thermal enhancement comparing with Fe_3O_4, Al_2O_3, TiO_2
Minea *et al.* [84]	CFD	Water, Water/EG (40%:60%)	Ag/MgO, Al_2O_3/Cu, GO/Co_3O_4	2 vol. %		~5-6	115–125	~5000	Although hybrid nanofluid with hybrid base fluid (water and EG) had better thermal enhancement, it introduces higher pressure drop increase to the system, comparing with hybrid nanofluid with water as the base fluid. The application of hybrid water base nanofluid with acceptable heat transfer enhancement and lower pressure drop would be a better choice over hybrid nanofluid with hybrid base fluid (Water/EG). Convection heat transfer coefficient increase between 115–125% using 0.15% GO/Co_3O_4/W-EG (40%:60%).

IOP Conf. Series: Materials Science and Engineering **805** (2020) 012015 doi:10.1088/1757-899X/805/1/012015

Table 3. Achievements recorded by researchers using various nanofluids

Particle type	Base fluid	Researcher	Particle size(nm)	Findings and remark
TiO_2	Water	Kilic *et al* [97]	30 – 60	The FPSC thermal performance showed an enhancement of 34.43%.
MWCNT		Rastogi *et al* [98]	10 – 30	The absence of surfactant showed that indicates that water is more efficient than MWNCT nanofluids but improved drastically after Triton X-100 was added
Al_2O_3, ZnO, and Fe_2O_3	Water	Colangelo *et al* [99]	45, 60 and 30 respectively	The combination of Al_2O_3/water nanofluid was more stable compared to that of ZnO and Fe_2O_3. Carbon nanostructure-based particles have greater thermal conductivity than Al_2O_3
Ag	Water	Polvongsri and Kiatsiriroat [100]	20	The thermal performance increased when using MWCNT and no surfactant was used
Cu	Water	Jamal-Abad *et al* [101]	35	With little increase in concentration, the efficiency increased to 24%. Surfactant was used but the temperature was high, and the effect on the surfactant was not considered
SiO_2	Water	Mahian *et al.* [102]	12 and 16	The observed fluid flow was turbulent. Solar collector with nanofluid has more efficiency than with water when the viscosity was considered
Cu	Water	He *et al.* [103]	25	As the particle size increases, the FPSC efficiency decreases. In addition, it has been found that the use of Cu-water nanofluids can improve the efficiency of the collector.

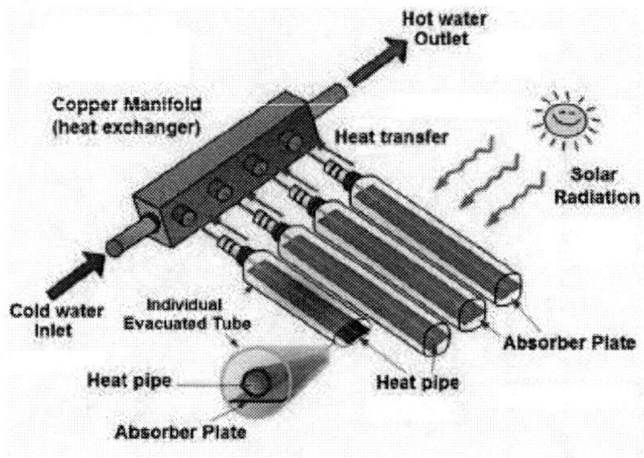

Figure 7. Schematic of Evacuated tube arrangement [4]

In the experiment reported in Ozsoy and Corumlu [107], thermosyphon heat pipe (THP) incorporated into the evacuated tube solar collector and its thermal efficiency were investigated using silver-water nanofluid. Two methods approach was considered by first synthesizing the silver-water nanofluid and also using a cylindrical copper THPs with nanofluid and water both tested. The results show that compared with pure water, THPs mixed with nanofluid improves the efficiency of solar collectorswithin the range of 20.7% and 40%, respectively.

Another approach using thermosyphon was studied by Liu *et al.* [108] considering the thermal performance with nanofluids in an evacuated tube type high-temperature air solar collector. A two-step process was used in the preparation of CuO-water nanofluid having a mean particle size of 50 nm. The results show that the system efficiency and outlet air temperature of the nanofluids are significantly greater than those of the pure water system. In winter, when the volume flow rate is 7.6 m^3/h, the outlet temperature of the nanofluidic system exceeds 170 °C. Mahendran *et al.* [109] conducted thermal performance experiments on ETSC using aqueous titanium oxide nanofluids. The nanofluids were prepared using titanium oxide nanopowders with an average particle size of 30 to 50 nm. A TiO_2-water nanofluid of 0.3% volume was prepared by diluting titanium dioxide nanopowder. The inlet and outlet temperature differences of ETSC using titanium dioxide nanofluids are greater than that of ETSC using water. The results show that the highest outlet temperature of titanium dioxide nanofluidic ETSC is 19% greater than that of water treated ETSC, and the efficiency of titanium dioxide nanofluidic ETSC is 8% higher than that of water treated ETSC.

IOP Conf. Series: Materials Science and Engineering **805** (2020) 012015 doi:10.1088/1757-899X/805/1/012015

Sharafeldin and Grof [110] studied the thermal performance of evacuated tube solar collectors with WO_3/Water Nanofluid.With a diameter of 90 nm, WO_3 nanoparticles are spherical. The volume fractions of three different WO_3 nanoparticles were determined to be 0.014%, 0.028%, and 0.042%, respectively, at different mass flow rates of 0.013 kg/s, 0.015 kg/s, and 0.017 kg/s. The results show that after adding WO_3 nanoparticles, the temperature difference of the liquid can reach 21%. When WO3 nanoparticles were used, the maximum heat gain at 900 W/m^2 solar irradiance was increased to 23%. At the same mass flow rate, the growth ratio of the heat-removable factor of the nanofluid to water is between 1.05 and 1.16. The results show that with the addition of nanoparticles, the efficiency of evacuated tube solar collectors is improved with its efficiency reaching 72.8%.

Saad *et al.* [111] conducted a comprehensive work on the performance of nanofluids with metal oxide as absorbers for vacuum tube solar collectors. In Baghdad, the parameters measured from April 2011 to March 2012 were Al_2O_3/water nanofluid concentrations and tube inclination. They recorded in the research conducted that the optimum tilt angle of the vacuum tube is 41 °per year. When 1%, 0.6%, and 0.3% Al_2O_3/water nanofluids and volume fraction of 28.4, 6.8, and 0.6%, were used, the efficiency of thecollector was also improved.

Hussain *et al.* [112] further studied the evacuated tube solar collector by comparing the thermal efficiencies of metal oxide nanofluids (ZrO_2/water) and metal nanofluids (Ag/water).They noticed when the inlet temperature is high using nanofluids of metal nanoparticles and oxide metals as working fluids can improve the collector thermal performance. Due to the high thermal conductivity of Ag, the efficiency of a 30 nm-sized Ag trap is greater than a 50 nm-sized ZrO_2. Sabiha *et al.* [113] proposed an evacuated tube solar collector using water-based single-walled carbon nanotube nanofluids as an absorption channel to improve energy efficiency. The effects of volume fractions in the range of 0.05-0.2 v % and mass flow rates of 0.008, 0.017 and 0.025 kg/s on energy efficiency were investigated Due to the high thermal performance of SWCNT nanofluids, the capture efficiency of SWCNT nanofluids with a volume ratio of 0.2% is 93.43% higher than water and 71.84% higher than water. At this concentration, it is better to use a nanofluid collector on a cloudy day than to use water on a sunny day.

Tong and his colleagues [114] may be the first to use nanofluids made of multi-walled carbon nanotubes (MWCNT) as working fluids to enhance the thermal performance of closed evacuated u-shape solar collectors.Their calculations show that when using 50 solar collectors of this type, the annual emissions of carbon dioxide and sulfur dioxide can be reduced by 1600kg and 5.3kg, respectively.

IOP Conf. Series: Materials Science and Engineering **805** (2020) 012015 doi:10.1088/1757-899X/805/1/012015

Hayek *et al.* [115] conducted two types of performance tests on glass tubes and thermal design tubes for evacuated collectors. The experiment time was 2 months and various parameters and procedures were considered. For both collectors, at high emissivity, all tilts result in almost the same overall performance. The authors also concluded that the total efficiency of the heat pipe collector is about 15-20% higher than that of the glass water collector. In summary, the heat pipe collector is more efficient than the glass water collector.

Daghigh and Zandi [116] tested and evaluate CuO, TiO_2 and MWCNT nanofluids on an evacuated tube heat pipe solar collector (ETHPSC) and compared its performance with water. Due to the high thermophysical properties of the nanofluids, better performance was observed when used on the collector. In August, the efficiency of collectors increased significantly, reaching 25, 12 and 5% and better results in October with 25, 15, and 7% respectively.

Using another perspective, Kaya and Arslan [117] designed and simulated an evacuated U-tube solar collector using Ag, ZnO, and MgO nanoparticles, ethylene glycol-pure water (EG-PW) and different conditions in analyzing its performance. The highest solar collector efficiency was observed with Ag/EG-PW nanofluid at 68.7% and EG-PW at 26.7%. A more innovative experiment was carried out by Al- Mashat and Hasan [118], studied the efficiency of ESTC using 16 vacuum tubes containing Al_2O_3/water nanofluids as working fluid. The results show that after adding 1% Al_2O_3 and 0.6% A12O3, efficiency is increased by 28.4 and 6.8%, respectively.

Mahendran and Sharma [119] conducted experiments with ETSC using titanium dioxide nanofluids and reported an increase in efficiency of 42.5% compared to 2.0 vol. % of titanium dioxide nanofluidic water.When the volume flow rate of the nanofluid is small, the collector efficiency is greatly improved compared to when base fluid is water. Lu *et al.* [120] used deionized water and water-based CuO nanofluids for evacuated tube solar collectors. The evaporation heat transfer coefficient of nano-copper oxide can be increased by about 30%. With the use of CuO nanofluids, the wall temperature of the open thermosiphon is reduced. Using MWCNT, Al_2O_3, TiO_2, SiO_2, CuO nanoparticles, Kim *et al.* [121] worked on the performance of a U-tube solar collector. The nanoparticles were suspended in a propylene glycol-water base solution at a concentration of 20%. MWCNT nanoparticles achieved the highest thermal performance reaching 62.85% at a volume concentration of 0.2%. Different approaches to enhancing ESTC using nanofluids are highlighted in Table 4.

Cho *et al.* [122] theoretically research on double-layer glass evaporated solar collectors for heat pipes and u-tubes under different working conditions. 0.24% volume of nanofluid made from multi-walled carbon nanotubes is used as a working fluid for the u-tube solar collector. The heat transfer coefficient of nanofluids is about 8% larger than that of water-based fluid.

6. Direct absorption solar collector review using nanofluids

The design of DASC (Figure 8) has been likened to that of a flat plate solar collector and the major difference in the design was identified to be the absence of an absorber tube in the DASC setup [131]. Different types of DASC have been elaborated which has seen different researchers adopting various setup propositions as described in Figures 3 and 4 [31]. Radzi and his group [34] reviewed and reported techniques for DASC including black liquid collectors [132], volume trap collectors [133] as well as small particle collectors [134]. Also, it was previously reported [31, 135, 136] that a distinct concept can convert carbondioxide (CO_2) into hydrocarbons (methane) and other byproducts. The detailed dimensions of components used in the experimental setup with a view to investigating the nanofluid flow and heat transfer efficiency of nanofluid in DASC using Al_2O_3-H_2O as the base fluid [137].

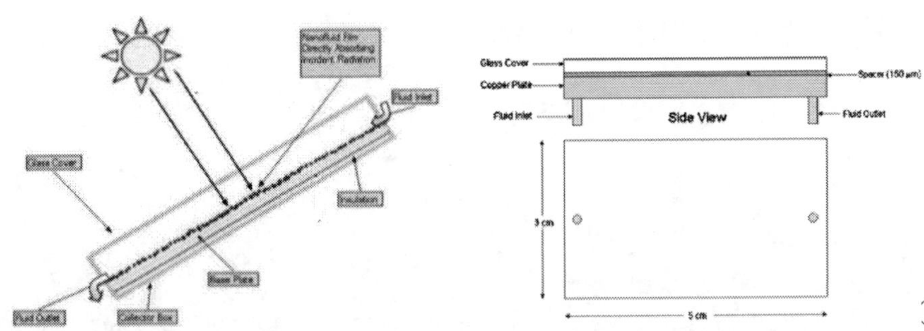

(a) (b)

Figure 8. Experimental setup of a DASC [19,131].

Al_2O_3 and TiO_2 were used as base particles dispersed in water and thermal conductivity was noted to rise up to 32% and 11% respectively [138]. A concise summary of studies with various nanofluids used in DASC is available [33], while in earlier report [34] a compilation on both experimental and patented setups of DASC in various studies also exists as shown in Table 5.

IOP Conf. Series: Materials Science and Engineering **805** (2020) 012015 doi:10.1088/1757-899X/805/1/012015

Table 4. Different approaches to ESTC enhancement using nanofluids

Researchers/Authors	Approach	Nanofluid Used	Volume Concentartion (%)	Mass flow rate (L/h)	Inclination angle (^{0})	Results
Chougule *et al.* [123]	Experimental	CNT (Carbon Nanotubes)	0.15	N/A	31.5^o, 50°	At 31.5^o, an efficiency of 45% was recorded at 50°, 69% efficiency was achieved
Iranmenesh *et al.* [106]	Experimental	GNP (Graphene Nanoplatelets)	0.025, 0.05, 0.075 and 0.1	0.5, 0.1, and 1.5	N/A	At higher concentration, maximum efficiency was obtained.
Kim *et al.* [121]	Theoretical	MWCNT (Multi-walled carbon nanotubes)	0.2	N/A	35, 126	The highest efficiency was gotten using 0.2% vol. of MWCNT reaching 62.8%.
Shahi *et al.* [124]	Numerical	Copper oxide	N/A	N/A	15, 33, 55 & 75	Mass flow rate increases by approximately 40% with enhancement ratio at 5%
He *et al.* [125]	Experimental	Tin Oxide CNT	0.5	N/A	45-60	At 0.5% concentration, CNT performed more better than TiO_2
Eidan *et al.* [126]	Experimental	Al_2O_3 & CuO/acetone	0.25 and 0.5	N/A	30, 45, 60	Optimal performance was observed at 45^0 angle and 70% filling ratio.
Mahbubul *et al.* [127]	Experimental	Single Wall Carbon nanotube	0.2	N/A	N/A	o.2% vol of nanofluids resulted in an efficiency of 56.7% and 66%

		(SWCNT)				respectively.
Kaya *et al.* [128]	Experimental	ZnO/Etylene Glycol-PureWater (ZnO/EG-PW)	1.0, 2.0, 3.0 4.0	0.02, 0.045		There is an enhanced efficiency with an increased volume concentration of 62.87%.
Natividade *et al.* [129]	Experimental	Multilayer graphene	0.00045, 0.00068	6, 24, 42, 60	N/A	Thermal efficiency was upgraded up to 31% and 76% for oth concentrations.
Gan *et al.* [130]	Experimental	Titanium dioxide (TiO_2)	0.5	0.033	3.1180°N, 101.6552°E	The performance of the ETSC increases with mass flow rate and the thermal efficiency reaches up to 16.5%.

IOP Conf. Series: Materials Science and Engineering **805** (2020) 012015 doi:10.1088/1757-899X/805/1/012015

Table 5. Some selected papers and patented DASC at a glance

Study authors	Research outcomes	Developed technique
Kraus and Kraus [139]	Patented	Worked on systems involving particle-fluid absorber
Hunt [140]	Journal paper	A receiver with particle-gas absorber was used in the experiment conducted.
Abdelrahman *et al.* [141]	Journal paper	To achieve ideal absorption, a particle-gas suspensions technique was used.
Bohn [142]	Journal paper	An absorber with pure flowing molten salt was experimented upon.
Parker and Langhoff [143]	Patented	Secondary direct absorber for large magnitude of temperature
Yogev [144]	Patented	The working fluid used was organic with photon concentrated on it.
Hirsch and Steinfeld [145]	Journal paper	A high-temperature reactor with carbon particle having thermochemical properties was studied.
Schunk *et al.* [146]	Journal paper	Researched into the abilities to rotate zinc oxide (ZnO) particle receiver
Tyagi *et al.* [24]	Journal paper	Focused on flat plate solar thermal collector based on nanofluids
Goldman *et al.* [147]	Patented	A novel receiver with semitransparent/translucent tubes
Otanicar et al. [19]	Journal paper	Tested nanofluid-based flat plate collector prototype
Sani *et al.* [148]	Journal paper	Researched into the possibilities of using carbon nanohorns as solar collectors
Taylor *et al.* [30]	Journal paper	A prediction mechanism for the effects of nanofluids on CSP systems was reported
Lu et al. [149]	Journal paper	Copper oxide (CuO) thermosyphon solar collector was used in the prototype.
Lenert and Wang [150]	Journal paper	In other studies, closely related to this research, molten salt has been used. In this case, however, prototype nanofluid receiver was observed and tested for optimum capacities using carbon-coated particles administered in Therminol VP-1.

The adoption of DASC in nanofluids has established a channel to estimate the collector efficiency [19,24, 151, 152,153]. A theoretical numerical study on 2-D models of DASC carried out by Tyagi *et al.* [24] using finite difference method in a bid to solve both radiative transport (i) and energy equations (ii): taking $k_{e\lambda,nanofluid}$ as the spectral extinction coefficient of nanofluids, heat conductivity k, temperature T, Density ρ, specific heat capacity c_p, radiation intensity l_λ, radiative heat flux ∂q_r. In other to prove the solutions, Otanicar *et al.* [19] used varieties of nanofluids in the investigation. Numerical methods that have been applied in by various researchers include finite volume methods (FVM), finite element (FEM), and finite difference method (FDM) [153]. The study also detailed the various numerical studies conducted by researchers to solve equations (i), (ii)

$$\frac{\partial I}{\partial y} = -k_{e\lambda,nanofluid}\ l_\lambda \qquad \text{(i)}$$

$$k\frac{\partial T^2}{\partial y^2} - \frac{\partial q_r}{\partial y} = \rho c_p U \frac{\partial T}{\partial x} \qquad \text{(ii)}$$

Till recent times, researchers have majorly directed their focus on the optical characteristics of water or ethylene glycol suspensions added with varieties of nanoparticles, as well as effective abilities in low-temperature solar collectors in nanofluid-based DASC [25, 32, 154, 155]. It has also been proposed for applications in water heating and purification. The efficiency of DASC has been identified to rely solely on the working fluid. A two-dimensional CFD model for direct absorption of high-temperature molten salt nanofluid concentrating solar receivers to investigate the effects of design and operational variables on receiver performance [156]. The study was able to establish that Carnot efficiency is directly proportional to increasing receiver length, solar concentration, increasing height and decreasing inlet velocity.

Balakin *et al.* [157] also studied a multiphase (Eulerian-Eulerian) CFD model of DASC, validating it with respect to a two (2) independent experimental data. The model is able to reproduce motion and thermal transfer in each phase. As part of the work, they altered the concentration and the nanoparticle sizes as well as the geometry and inclination of the collector. This has resulted in approximately 10% improvement in efficiency while the use of thermomagnetic convection in the collector requiring the use of magnetic nanofluid resulted in a 30% improvement in the efficiency of DASC. Won and Lee [158] investigated the light scattering defect on the thermal performance of DASC using plasmonic nanofluids. The study has shown a link between the type of material the suspended nanoparticles are made of and the absorption cross-section and the reliance on the particle sizes. Conclusions were

IOP Conf. Series: Materials Science and Engineering **805** (2020) 012015 doi:10.1088/1757-899X/805/1/012015

drawn that when light scattering due to nanoparticle suspension can increase the mean optical path length of light packets inside the collector channel, thus enhancing the solar-weighted absorption coefficient.

7. Difficulties and Challenges involved using nanofluids

The new approach of using nanofluids in solar collectors has also come with a lot of setbacks which still needs to be solved;

1) The preparation of the nanofluid is expensive and time-consuming.
2) Due to the toxicity of nanofluids, proper treatment is required.
3) The tendency to form agglomerates after homogenizing and its long requirement for stability is still of great concern.
4) The increase in the fluid circulation rate lead to erosion on the heat transfer walls thus sedimentation reduces.
5) The addition of surfactants at high temperatures can be detrimental.
6) The operating cost of solar collectors coupled with nanofluids is high.

8. Conclusion

This review paper was able to itemize the recent achievement in the use of nanofluids in solar collectors namely PTSC (Parabolic trough solar collector), FPSC (Flat plate solar collector), ETSC (Evacuated tube solar collector) and DASC (Direct absorption solar collector). Nanofluids usage in solar collectors has shown promising efficiencies but the major challenges still lie in the high cost of production, stability, particle agglomeration, pumping power and pressure drop.

The unique rise of thermal conductivity in nanofluids is not yet fully elucidated. However, the long-term stability of nanofluids is still of the paramount need to further achieve better efficiency in solar collectors. The following conclusions can be deduced from this paper;

1) To achieve better efficiency of solar collectors, it is necessary to make sure proper dispersion occurs while mixing the nanoparticle and stability in the long run.
2) Carbon nanotubes both single and multi-walled has shown optimal efficiency when used in solar collectors.
3) Surfactants addition to nanoparticle has helped in achieving stability, but at a higher temperature, it causes corrosion on the wall of solar collectors leading to detrimental

effects in the overall efficiency. Though it contributed to the overall efficiency of solar collectors.

4) Viscosity rise when nanofluids are added to base fluids which also increases the pressure drop therefore a very high pumping power is required.

References

[1] Wada T, Tomiya A, Enomoto M, Sato T, Morishita D, Izumi S, Niizeki K, Suzuki S, Morita T and Kawata G 2016 Radiological impact of the nuclear power plant accident on freshwater fish in Fukushima: an overview of monitoring results. *J. Environ. Radioact.* **151** pp 144-155.

[2] Choi S U S, Zhang ZG, Yu W, Lockwood FE and Grulke E A 2001 Anomalous thermal conductivity enhancement in nanotube suspensions. *Appl. Phys. Lett.* **79** (14) pp 2252-2254.

[3] Kameya Y and Hanamura K 2011 Enhancement of solar radiation absorption using nanoparticle suspension. *Sol. Energ.* **85** (2) pp 299-307.

[4] Ohler A and Fetters I 2014 The causal relationship between renewable electricity generation and GDP growth: a study of energy sources. *Energ. Econ.* **43** pp 125-139.

[5] Tugcu C T, Ozturk I and Aslan A 2012 Renewable and non-renewable energy consumption and economic growth relationship revisited: evidence from G7 countries. *Energ. Econ.* **34** (6) pp 1942-1950.

[6] Alper A and Oguz A 2016 The role of renewable energy consumption in economic growth: evidence from asymmetric causality. *Renew. Sustain. Energ.* Rev. **60** pp 953-959.

[7] Bhutto AW, Bazmib AA, Zahedi G and Klemes J 2014 A review of progress in renewable energy implementation in the Gulf cooperation council countries. *J. Clean Prod.* **71** pp 168-180.

[8] Minardi JE and Chuang HN 1975 Performance of a "Black" liquid flat-plate solar collector. *Sol. Energ.* **17** (3) pp 179-183.

[9] Tian Y and Zhao C Y 2013 A review of solar collectors and thermal energy storage in solar thermal applications. *Appl. Energ.* **104** pp 538-553.

[10] Zarza E 2005 Medium Temperature Solar Concentrators (parabolic Trough Collectors). *Sol. Energ. Convers. Photoenerg. Syst.* **8** 170.

[11] Hussain A and Arif S M 2017 Emerging Renewable and Sustainable Energy Technologies. *Renew. Sustain. Energ.* Rev. **71** pp 12-28.

[12] Rable A 1985 *Active Solar Collectors and Their Application.* Oxford University Press,

USA.

[13] Hakkarainen E, Tähtinen M, Mikkonen H2015 Dynamic model development of linear Fresnel solar field. *ASME 2015 9th Int Conf Energy Sustain* ES 2015, Collocated with ASME 2015 Power Conf ASME 2015 13th Int Conf Fuel Cell Sci Eng Technol ASME 2015 Nucl Forum;1. http://doi.org/10.1115/ES2015-49347.

[14] Sharma K and Kundan L 2014 Nanofluid based concentrating parabolic solar collector (NBCPSC): A new alternative. *Int. J. Res. Mech. Eng. Technol.* **2** pp 2249-5762.

[15] Parabolic Trough Reflector for Solar Thermal System. http://www.alternative-energy-tutorials.com/solar-hot-water/parabolic-trough-reflector.html, 2019 (accessed 13 July 2019).

[16] Florschuetz L W 1979 Extension of the Hottel-Whillier model to the analysis of combined photovoltaic/thermal flat plate collectors. *Sol. Energ.* **22** (4) pp 361-366.

[17] Jamil M M and Sidik N A C 2016 Thermal Performance of Thermosyphon Evacuated Tube Solar Collector using TiO_2/Water Nanofluid. **20** (1) 12-29.

[18] Kabeel A E, Khalil A, Shalaby S M and Zayed M E 2016 Investigation of the thermal performances of flat, finned, and v-corrugated plate solar air heaters. *J. Sol. Energ. Eng.* **138** (5) 051004. https://doi.org/10.1115/1.4034027.

[19] Otanicar TP, Phelan PE, Prasher RS, Rosengarten G and Taylor RA 2010 Nanofluid-based direct absorption solar collector. *J. Renew. Sustain. Energ.* **2** (3) 033102. https://doi.org/10.1063/1.3429737.

[20] Zhu Q Z, Cui Y, Mu L J and Tang L Q 2013. Characterization of thermal radiative properties of nanofluids for selective absorption of solar radiation. *Int. J. Thermophys.* **34** (12) pp 2307-2321.

[21] Taylor RA, Phelan PE, Otanicar TP, Walker CA, Nguyen M, Trimble S and Prasher R 2011 Applicability of nanofluids in high flux solar collectors. *Int. J. Renew. Sustain. Energ.* **3** (2). 023104. https://doi.org/10.1063/1.3571565.

[22] Lee B, Park K, Walsh T and Xu L 2014. Radiative heat transfer analysis in plasmonic nanofluid for direct solar thermal absorption. *J. Sol. Energ. Eng.* **134** (2) 021009. https://doi.org/10.1115/1.4005756.

[23] Otanicar T, Phelan P and Golden J 2009 Optical properties of liquids for direct absorption solar thermal energy systems. *Sol. Energ.* **83** (7) pp 969-977.

[24] Tyagi H, Phelan P and Prasher R 2009 Predicted efficiency of a low-temperature nanofluid-based direct absorption solar collector. *J. Sol. Energ. Eng.* **131** (4) 041004. https://doi.org/10.1115/1.3197562.

[25] Verma S and Tiwari A 2015. Progress of nanofluid application in solar collectors: A review. *Energ. Convers. Manag.* **100** pp 324-346.

[26] Kasaeian A Eshghi A and Sameti M 2015. A review on the applications of nanofluids in solar energy systems. *Renew. Sustain. Energ. Rev.* **43** pp 584-598.

[27] Luo Z, Wang C, Wei W, Xiao G and Ni M 2014 Performance improvement of a nanofluid solar collector based on direct absorption collection (DAC) concepts. *Int. J. Heat Mass Transf.* **75** pp 262-271.

[28] Said Z, Saidur R, Rahim N and Alim M 2014 Analyses of exergy efficiency and pumping power for a conventional flat plate solar collector using SWCNTs based nanofluid. *Energ. Buildings* **78** pp 1-9.

[29] Bandarra Filho EP, Mendoza OS, Beicker CL, Menezes A and Wen D 2014. Experimental investigation of a silver nanoparticle-based direct absorption solar thermal system. *Energy Convers. Manag.* **84** pp 261-267.

[30] Taylor RA, Phelan PE, Otanicar TP, Adrian R and Prasher R 2011 Nanofluid optical property characterization: Towards efficient direct absorption solar collectors. *Nanoscale Res. Lett.* **6** (1) 225. https://doi.org/10.1186/1556-276X-6-225.

[31] Phelan P, Otanicar T, Taylor R and Tyagi H 2013. Trends and opportunities in direct-absorbed solar thermal collectors. *J. Therm. Sci. Eng. Appl.* **5** (2). 021003. https://doi.org/10.1115/1.4023930.

[32] Xu G, Chen W, Deng S, Zhang X and Zhao S 2015 Performance evaluation of a nanofluid-based direct absorption solar collector with parabolic trough concentrator. *Nanomater.* **5** (4) pp 2131-2147.

[33] Raj P and Subudhi S 2018 A review of studies using nanofluids in flat-plate and direct absorption solar collectors. *Renew. Sustain. Energ. Rev.* **84** pp. 54-74.

[34] Radzi A R, Nor A C S and Syahrullail S 2019. Recent progress on concentrating direct absorption solar collector using nanofluids. *J. Therm. Anal. Calorim.* **137** pp 903-922.

[35] Mahmud J M, Isa A M, Nor A C S, Muhammad N A and Muhammad Y 2016 Thermal performance enhancement of flat-plate and evacuated tube solar collectors using nanofluid: A review. *Int. Comm. Heat Mass Transf.* **76** pp 6-15.

[36] Choi S U S 1995 Enhancing thermal conductivity of fluids with nanoparticles, in: D.A. Siginer, H.P. Wang (Eds.), *Developments and Applications of Non-Newtonian Flows*, Fed-V.231/Md-V.66ASME, New York, NY, USA, pp. 99-105.

[37] Bellos E and Tzivanidis C 2018 A review of concentrating solar thermal collectors with and without nanofluids. *J. Therm. Anal. Calorim.* **135** (1) 763-786.

[38] Singh K, Sharma S and Gangacharyulu, D 2013 Experimental study of thermophysical properties of Al_2O_3/water nanofluid. *Int. J. Res. Mech. Eng. Technol.* **3** (2) pp 229-233.

[39] Kasaeian A, Daviran S, Azarian R D and Rashidi A 2015 Performance evaluation and nanofluid using capability study of a solar parabolic trough collector. *Energ. Convers. Manag.* **89** pp 368-375.

[40] Siva R E, Reddy R M and Reddy K K 2015 Experimental study on thermal efficiency of parabolic trough collector (PTC) using Al_2O_3/H_2O nanofluid. *Appl. Mech. Mater.* **787** pp 192-196.

[41] Kandwal S 2019 An experimental investigation into nanofluids (CuO-H_2O and CuO-ethylene glycol) based parabolic solar collector. *Our Heritage* **67** (7) pp 551-571.

[42] Chaudhari K S, Walke P V, Wankhede U S and Shelke RSB 2015 An experimental investigation of a nanofluid (Al_2O_3 + H_2O) based parabolic trough solar collectors. *Bri. J. Appl .Sci. Technol.* **9** pp 551-557.

[43] Coccia G, Di Nicola G, Colla L, Fedele L and Scattolini M 2017 Adoption of nanofluids in low-enthalpy parabolic trough solar collectors: Numerical simulation of the yearly yield. *Energ. Convers. Manag.* **118** pp 306-319.

[44] Menbari A, Alemrajabi A A and Rezaei A 2016 Heat transfer analysis and the effect of CuO/Water nanofluid on direct absorption concentrating solar collector. *Appl. Therm. Eng.* **104** pp 176-183.

[45] Subramani J, Nagarajan P K, Wongwises S, El-Agouz S A and Sathyamurthy R 2017 Experimental study on the thermal performance and heat transfer characteristics of solar parabolic trough collector using Al_2O_3 nanofluids. *Environ. Prog. Sustain. Energ.* **37** (3) pp 1149-1159.

[46] Kasaeian A, Daneshazarian R, Rezaei R, Pourfayaz F and Kasaeian G 2017 Experimental investigation on the thermal behavior of nanofluid direct absorption in a trough collector. *J. Clean. Prod.* **158** pp 276-284.

[47] Sekhar T V, Prakash R, Nandan G and Muthuraman M 2017 Performance enhancement of a renewable thermal energy collector using metallic oxide nanofluids. *Micro Nano Lett.* **13** (2) pp 248-251.

[48] Rehan M A, Ali M, Sheikh N A, Khalil M S, Chaudhary G Q, Rashid T and Shehryar M 2018 Experimental performance analysis of low concentration ratio solar parabolic trough collectors with nanofluids in winter conditions. *Renew. Energ.* **118** pp 742-751.

[49] Subramani J, Nagarajan P K, Mahian O and Sathyamurthy R 2018 Efficiency and heat transfer improvements in a parabolic trough solar collector using TiO_2 nanofluids under turbulent flow regime *Renew. Energ.* **119** pp 19-31.

[50] Marefati M, Mehrpooya M and Shafii M B 2018 Optical and thermal analysis of a parabolic trough solar collector for production of thermal energy in different climates in Iran with comparison between the conventional nanofluids. *J. Clean Prod.* **175** pp 294-313.

[51] Okonkwo E C, Abid M and Ratlamwala T A 2018 Numerical analysis of heat transfer enhancement in a parabolic trough collector based on geometry modifications and working fluid usage. *J. Sol. Energ. Eng.* **140** 5 051009. https://doi.org/10.1115/1.4040076.

[52] Alsaady M, Fu R, Yan Y, Liu Z, Wu S and Boukhanouf R 2018 An experimental investigation on the effect of ferrofluids on the efficiency of novel parabolic trough solar collector under laminar flow conditions. *Heat Transf. Eng.* **40** (9-10) pp.753-761.

[53] De los Rios, Rivera-Solorio M S B and García-Cuéllar A J 2018 Thermal performance of a parabolic trough linear collector using Al_2O_3/H_2O nanofluids. *Renew. Energ.* **122** pp 665-673.

[54] Okonkwo E C, Essien E A, Abid M, Kavaz D and Ratlamwala T A 2018 Thermal performance analysis of a parabolic trough collector using water-based green-synthesized nanofluids. *Sol. Energ.* **170** pp 658-670.

[55] Kasaeian A B, Sokhansefat T, Abbaspour M J and Sokhansefat M 2012 Numerical study of heat transfer enhancement by using Al_2O_3/synthetic oil nanofluid in a parabolic trough collector tube. *World Acad. Sci. Eng. Technol.* **66** pp 1154-1159.

[56] Sokhansefat T, Kasaeian A, Kowsary 2014 Heat transfer enhancement in parabolic trough collector tube using Al_2O_3/synthetic oil nanofluid. *Renew. Sustain. Energ. Rev.* **33** pp 636-644.

[57] Zadeh P M, Sokhansefat T, Kasaeian A B, Kowsary F and Akbarzadeh A 2015 Hybrid optimization algorithm for thermal analysis in a solar parabolic trough collector based on nanofluid. *Energ.* **82** pp 857-864.

[58] Mwesigye A, Huan Z and Meyer J P 2015 Thermal performance of a receiver tube for a high concentration ratio parabolic trough system and potential for improved performance

with Syltherm800-CuO nanofluid. Proc. ASME 2015 Int. Mech. Eng. Congr. Expo. Houston, TX, USA, 13-19 November 2015.

[59] Mwesigye A, Huan Z and Meyer J P 2015 Thermodynamic optimisation of the performance of a parabolic trough receiver using synthetic oil–Al_2O_3 nanofluid. *Appl Energ.* **156** pp 398-412.

[60] Mwesigye A and Huan Z I 2015 Thermodynamic analysis and optimization of fully developed turbulent forced convection in a circular tube with water–Al_2O_3 nanofluid. *J. Heat Mass Transf.* **89** pp 694-706.

[61] Basbous N, Taqi M and Belouaggadia N 2015 Numerical study of a parabolic trough collector using a nanofluid. *Asian J. Curr. Eng. Math.* **4** (3) 40-44.

[62] Bellos E, Tzivanidis C, Antonopoulos K A and Gkinis G 2016 Thermal enhancement of solar parabolic trough collectors by using nanofluids and converging-diverging absorber tube. *Renew. Energ.* **94** pp 213-222.

[63] Wang Y, Xu J, Liu Q, Chen Y and Liu H 2016 Performance analysis of a parabolic trough solar collector using Al_2O_3/synthetic oil nanofluid. *Appl. Therm. Eng.* **107** pp 469-478.

[64] Kaloudis E, Papanicolaou E and Belessiotis V 2016 Numerical simulations of a parabolic trough solar collector with nanofluid using a two-phase model. *Renew. Energ.* **97** pp 218-229.

[65] Benabderrahmane A, Benazza A, Aminallah M and Laouedj S 2016 Heat transfer behaviors in parabolic trough solar collector tube with compound technique. *Int. J. Sci. Res. Eng. Technol.* **5** 11 pp 410-417.

[66] Amina B, Miloud A, Samir L, Abdelylah B and Solano J P 2016 Heat transfer enhancement in a parabolic trough solar receiver using longitudinal fins and nanofluids. *J. Therm. Sci.* **25** (5) pp 410-417.

[67] Bellos E, Tzivanidis C and Tsimpoukis D 2018 Enhancing the performance of parabolic trough collectors using nanofluids and turbulators. *Renew. Sustain. Energ. Rev.* **91** pp 358-375.

[68] Ghasemi S E and Ranjbar A A 2017 Effect of using nanofluids on efficiency of parabolic trough collectors in solar thermal electric power plants. *Int. J. Hydrog. Energ.* **42** (34) pp 21626-21634.

[69] Mwesigye A, Huan Z and Meyer J P 2016 Thermal performance and entropy generation analysis of a high concentration ratio parabolic trough solar collector with Cu-Therminol® VP-1 nanofluid. *Energ. Convers. Manag.* **120** pp 449–465.

[70] Basbous N, Taqi M and Janan M A 2016 Thermal performances analysis of a parabolic trough solar collector using different nanofluids. *Proc. Int. Renew. Sustain. Energ. Conf. (IRSEC),* Marrakech, Morocco. pp 322-326.

[71] Toghyani S, Baniasadi E and Afshari E 2016 Thermodynamic analysis and optimization of an integrated Rankine power cycle and nano-fluid based parabolic trough solar collector. *Energ. Convers. Manag.* **93** pp 93-104.

[72] Nayak Y K, Sharma S P, Sinha U K, Kumar P and Kumar N A 2016 Novel Solution Methodology for the Optimization of Thermal Analysis in the Solar Parabolic Trough Collector. *Ind. J. Sci. Technol.* **9** (48). https://doi.org/10.17485/ijst/2016/v9i48/95013.

[73] Ferraro V, Settino J, Cucumo M A and Kaliakatsos D 2016 Parabolic trough system operating with nanofluids: Comparison with the conventional working fluids and influence on the system performance. *Energ. Procedia* **101** pp 782-789.

[74] Khakrah H, Shamloo A and Hannani S K 2017 Determination of parabolic trough solar collector efficiency using nanofluid: A comprehensive numerical study. *J. Sol. Energ. Eng.* **139** (5) 051006. https://doi.org/10.1115/1.4037092.

[75] Mwesigye A and Meyer J P 2017 Optimal thermal and thermodynamic performance of a solar parabolic trough receiver with different nanofluids and at different concentration ratios. *Appl. Energ.* **193** pp 393-413.

[76] Bellos E and Tzivanidis C 2017 Optimization of a solar-driven trigeneration system with nanofluid-based parabolic trough collectors. *Energies* **10** (7) 848. https://doi.org/10.3390/en10070848.

[77] Bellos E and Tzivanidis C 2017 Parametric investigation of nanofluids utilization in parabolic trough collectors. *Therm. Sci. Eng. Prog.* **2** pp 71-79.

[78] Bellos E, Tzivanidis C and Tsimpoukis D 2018 Thermal, hydraulic and exergetic evaluation of a parabolic trough collector operating with thermal oil and molten salt based nanofluids. *Energ. Convers. Manag.* **156** pp 388-402.

[79] Allouhi A, Amine M B, Saidur R, Kousksou T and Jamil A 2018 Energy and exergy analyses of a parabolic trough collector operated with nanofluids for medium and high temperature applications. *Energ. Convers. Manag.* **155** pp 201-217.

[80] Bellos E and Tzivanidis C 2018 Thermal analysis of parabolic trough collector operating with mono and hybrid nanofluids. *Sustain. Energ. Technol. Assess.* **26** pp 105-115.

[81] Kasaiean A, Sameti M, Daneshazarian R, Noori Z, Adamian A and Ming T 2018 Heat transfer network for a parabolic trough collector as a heat collecting element using nanofluid. *Renew. Energ.* **123** pp 439-449.

[82] Bilal F, Arunachala U and Sandeep H 2018 Experimental validation of energy parameters in parabolic trough collector with plain absorber and analysis of heat transfer enhancement techniques. *J. Phys. Conf. Ser.* **953** (01) 012030. https://doi.org/10.1088/1742-6596/953/1/012030.

[83] Hatami M, Geng J and Jing D 2018 Enhanced efficiency in concentrated parabolic solar collector (CPSC) with a porous absorber tube filled with metal nanoparticle suspension. *Green Energ. Environ.* **3** (2) pp 129-137.

[84 Minea A A and El-Maghlany W M 2018 Influence of hybrid nanofluids on the performance of parabolic trough collectors in solar thermal systems: Recent findings and numerical comparison. *Renew. Energ.* **120** pp 350-364.

[85] Yousefi T, Veysi F, Shojaeizadeh E and Zinadini S 2012 An experimental investigation on the effect of Al_2O_3-H_2O nanofluid on the efficiency of flat-plate solar collectors. *Renew. Energ.* **39** (1) pp 293-298.

[86] Yousefi T, Veysi F, Shojaeizadeh E and Zinadini S 2012 An experimental investigation on the effect of MWCNT-H_2O nanofluid on the efficiency of flat-plate solar collectors. *Exp. Therm. Fluid Sci.* **39** pp 207-212.

[87] Yousefi T, Veysi F, Shojaeizadeh E and Zinadini S 2012 An experimental investigation on the effect of pH variation of MWCNT-H_2O nanofluid on the efficiency of a flat-plate solar collector. *Sol. Energ.* **86** (2) pp 771-779.

[88] Otanicar T P, Phelan P E, Prasher R S, Rosengarten G and Taylor R A 2010 Nanofluid-based direct absorption solar collector. *J. Renew. Sustain. Energ.* **2** (3) 03310. https://doi.org/10.1063/1.3429737.

[89] Liu J, Rinzler AG, Dai H, Hafner JH, Bradley RK, Boul PJ, Lu A, Iverson T, Shelimov K, Huffman CB and Rodriguez-Macias F 1998 Fullerene pipes. *Sci.* **280** (5367) pp 1253-1256.

[90] Li X, Zhu D and Wang X 2007 Evaluation on dispersion behavior of the aqueous copper nano-suspensions. *J. Colloid Interf. Sci.* **310** (2) pp 456-463.

[91] Noghrehabadi A, Hajidavalloo E and Moravej M 2016 Experimental investigation of efficiency of square flat-plate solar collector using SiO_2/water nanofluid. *Case Stud. Therm. Eng.* **8** pp 378-386.

[92] Chaji H, Ajabshirchi Y, Esmaeilzadeh E, Zeinali H S, Hedayatizadeh M and Kahani M 2013 Experimental study on thermal efficiency of flat plate solar collector using TiO_2/water nanofluid. *Mod. Appl. Sci.* **7** (10) 60-69.

[93] Meibodi SS, Kianifar A, Niazmand H, Mahian O and Wongwises S 2015 Experimental investigation on the thermal efficiency and performance characteristics of a flat plate solar collector using SiO_2/EG–water nanofluids. *Int. Comm. Heat Mass Transf.* **65** pp 65-71.

[94] Alim MA, Abdin Z, Saidur R, Hepbasli A, Khairul MA and Rahim N A 2013 Analyses of entropy generation and pressure drop for a conventional flat plate solar collector using different types of metal oxide nanofluids. *Energ. Buildings* **66** pp 289-296.

[95] Zamzamian A, KeyanpourRad M, KianiNeyestani M and Jamal-Abad M T 2014 An experimental study on the effect of Cu-synthesized/EG nanofluid on the efficiency of flat plate solar collectors. *Renew. Energ.* **71** pp 658-664.

[96] Polvongsri S and Kiatsiriroat T 2011 Enhancement of flat plate solar collector thermal performance with silver nano-fluid. In Proceedings of the second TSME international conference on mechanical engineering pp 19-21.

[97] Kiliç F, Menlik T and Sözen A 2018 Effect of titanium dioxide/water nanofluid use on thermal performance of the flat plate solar collector. *Sol. Energ.* **164** pp 101-108.

[98] Rastogi R, Kaushal R, Tripathi, S K, Sharma A L, Kaur I and Bharadwaj L M 2008 Comparative study of carbon nanotube dispersion using surfactants. *J. Colloid Interf. Sci.* **328** (2) pp 421-428.

[99] Colangelo G, Favale E, De Risi A and Laforgia D 2013 A new solution for reduced sedimentation flat panel solar thermal collector using nanofluids. *Appl. Energ.* **111** pp 80-93.

[100] Polvongsri S and Kiatsiriroat T 2014 Performance analysis of flat-plate solar collector having silver nanofluid as a working fluid. *Heat Transf. Eng.* **35** (13) pp 1183-1191.

[101] Jamal-Abad M T, Zamzamian A, Imani E and Mansouri M 2013 Experimental study of the performance of a flat-plate collector using Cu–water nanofluid. *J. Thermophys. Heat Transf.* **27** (4) pp 756-760.

[102] Mahian O, Kianifar A, Sahin A Z, and Wongwises S 2015 Heat transfer, pressure drop, and entropy generation in a solar collector using SiO_2/water nanofluids: Effects of nanoparticle size and pH. *J. Heat Transf.* **137** (6) 061011. https://doi.org/10.1115/1.4029870.

[103] He Q, Zeng S and Wang S 2015 Experimental investigation on the efficiency of flat-plate solar collectors with nanofluids. *Appl. Therm. Eng.* **88** pp 165-171.

[104] Ghaderian J, Sidik N A C,Kasaeian A, Ghaderian S, Okhovat A, Pakzadeh A, Samion S and Yahya W J 2017 Performance of copper oxide/distilled water nanofluid in evacuated tube solar collector (ETSC) water heater with internal coil under thermosyphon system circulations. *Appl. Therm. Eng.* **121** pp 520-536.

[105] Ghaderian J and Sidik N A C 2017 An experimental investigation on the effect of Al_2O_3/distilled water nanofluid on the energy efficiency of evacuated tube solar collector. *Int. J. Heat Mass Transf.* **108** pp 972-987.

[106] Iranmanesh S, Ong H C, Ang B C, Sadeghinezhad E, Esmaeilzadeh A and Mehrali M 2017 Thermal performance enhancement of an evacuated tube solar collector using graphene nanoplatelets nanofluid. *J. Clean. Prod.* **162** pp 121-129.

[107] Ozsoy A and Corumlu V 2018 Thermal performance of a thermosyphon heat pipe evacuated tube solar collector using silver-water nanofluid for commercial applications. *Renew. Energ.* **122** pp 26-34.

[108] Liu Z H, Hu, R L, Lu L, Zhao F and Xiao H 2013 Thermal performance of an open thermosyphon using nanofluid for evacuated tubular high temperature air solar collector. *Energ. Convers. Manag.* **73** pp 135-143.

[109] Mahendran M, LeeG C, Sharma K V and Shahrani A 2012 Performance of evacuated tube solar collector using water-based titanium oxide nanofluid. *J. Mech. Eng. Sci.* **3** pp 301-310.

[110] Sharafeldin M A and Grof, G 2019 Efficiency of evacuated tube solar collector using WO_3/water nanofluid. *Renew. Energ.* **134** pp 453-460.

[111] Saad M, Al-Mashat S and Hasan A A 2013 Evaluation of convective heat transfer and natural circulation in an evacuated tube solar collector. *J. Eng.* **19** (5) pp 613-628.

[112] Hussain H A, Jawad Q and Sultan K F 2014 Experimental analysis on thermal efficiency of evacuated tube solar collector by using nanofluids. *Int. J. Sustain. Green Energ.* **4** pp 19-28.

[113] Sabiha M A, Saidur R, Hassani S, Said Z and Mekhilef S 2015 Energy performance of an evacuated tube solar collector using single walled carbon nanotubes nanofluids. *Energ. Convers. Manag.* **105** pp 1377-1388.

[114] Tong Y, Kim J and Cho H 2015 Effects of thermal performance of enclosed-type evacuated U-tube solar collector with multi-walled carbon nanotube/water nanofluid. *Renew. Energ.* **83** pp 463-473.

[115] Hayek M, Assaf J and Lteif W 2011 Experimental investigation of the performance of evacuated-tube solar collectors under eastern mediterranean climatic conditions. *Energ. Procedia* **6** pp 618-626.

[116] Daghigh R and Zandi P 2019 Improving the performance of heat pipe embedded evacuated tube collector with nanofluids and auxiliary gas system. *Renew. Energ.* **134** pp 888-901.

[117] Kaya H and Arslan K 2019 Numerical investigation of efficiency and economic analysis of an evacuated U-tube solar collector with different nanofluids. *Heat Mass Transf.* **55** (3) pp 581-593.

[118] Al-Mashat SMS and Hasan AA 2013 Evaluation of convective heat transfer and natural circulation in an evacuated tube solar collector. *J. Eng.* **19** (5) pp 613-628.

[119] Mahendran M and Sharma KV 2012 Performance evaluation of evacuated tube solar collector using water-based titanium oxide (TiO_2) nanofluid. *J. Mech. Eng. Sci.* **3** pp 301-10.

[120] Lu L, Liu Z H and Xiao H S 2011 Thermal performance of an open thermosyphon using nanofluids for high-temperature evacuated tubular solar collectors: Part 1: Indoor experiment. *Sol. Energ.* **85** (2) pp 379-387.

[121] Kim H, Ham J, Park C and Cho H 2016 Theoretical investigation of the efficiency of a U-tube solar collector using various nanofluids. *Energ.* **94** pp 497-507.

[122] Cho H and Tong Y 2015 Comparative study on the thermal performance of evacuated solar collectors with U-tubes and heat pipes. *Int. J. Air Cond. Refrig.* **23** (3) 1550019. https://doi.org/10.1142/S2010132515500194.

[123] Chougule S S, Pise A T and Madane P A 2012 Performance of nanofluid-charged solar water heater by solar tracking system. In: Advances in Engineering, Science and Management (ICAESM), International Conference on. IEEE, pp. 247-253.

[124] Shahi M, Mahmoudi A H and Talebi F 2010 Numerical simulation of steady natural convection heat transfer in a 3-dimensional single-ended tube subjected to a nanofluid. *Int. Comm. Heat Mass Transf.* **37** (10) pp 1535-1545.

[125] He Y, Wang S, Tian F and Ren Y2011 Experimental study on the light-heat conversion characteristics of nanofluids. *Nanosci. Nanotechnol.* Lett. **3** (4) pp 494-496.

[126] Eidan A A., Al-Sahlani A, Ahmed A Q, Al-Fahham M and Jalil J M 2018 Improving the performance of heat pipe-evacuated tube solar collector experimentally by using Al_2O_3 and CuO/acetone nanofluids. *Sol. Energ.* **173** pp 780-788.

[127] Mahbubul I M, Khan M M A, Ibrahim N I, Ali H M, Al-Sulaiman F A and Saidur R 2018 Carbon nanotube nanofluid in enhancing the efficiency of evacuated tube solar collector. *Renew. Energ.* **121** pp 36-44.

[128] Kaya H, Arslan K, and Eltugral N 2018 Experimental investigation of thermal performance of an evacuated U-Tube solar collector with ZnO/Etylene glycol-pure water nanofluids. *Renew. Energ.* **122** pp 329-338.

[129] Natividade P S G, Moura G M, Avallone E, Filho E P,Gelamo R V and Gonçalves J C S 2019 Experimental analysis applied to an evacuated tube solar collector equipped with parabolic concentrator using multilayer graphene based nanofluids. *Renew. Energ.* **138** pp 152-160.

[130] Gan Y Y, Ong H C, Ling T C, Zulki N W M, Wang C T and Yang Y C 2018 Thermal conductivity optimization and entropy generation analysis of titanium dioxide nanofluid in evacuated tube solar collector. *Appl. Therm. Eng.* **145** pp 155-164.

[131] Gupta H K Agarwal G D and Mathur J 2014 Experimental evaluation on the effect of nanofluids concentration on the performance of Direct absorption solar collector. *Int. J. Adv. Eng. Nanotechnol.* **1** (12) pp 16-20.

[132] Huang B J, Wung, T Y and Nieh S 1979 Thermal analysis of black liquid cylindrical. *Sol . Energ.* **22** (3) pp 221-224.

[133] Arai N, Itaya Y and Hasatani M 1984 Development of a 'volume heat trap' type solar collector using a fine-particle semitransparent liquid suspension (FPSS) as a heat vehicle and heat storage medium unsteady, 1-D heat transfer in a horizontal FPSS layer heated by thermal radiation. *Sol. Energ.* **32** (1) pp 49-56.

[134] Bertocchi R, Karni J and Kribus A 2004. Experimental evaluation of a non-isothermal high temperature solar particle receiver. *Energ.* **29** (5-6) pp 687-700.

[135] Varghese O K, Paulose M, Latempa T J and Grimes C A 2009. High rate solar photocatalytic conversion of CO_2 and water vapor to hydrocarbon fuels. *Nano Lett.* **9** (2) pp 731-737.

[136] Roy S C, Varghese O K, Paulose M and Grimes C A 2010 Toward solar fuels: Photocatalytic conversion of carbon dioxide to hydrocarbons. *ACS Nano.* **4** (3) pp 1259-1278.

[137] Gupta H K, Agrawal G D and Mathur J 2015 Experimental study of water-based Al_2O_3 nanofluid flow in direct absorption solar collector. *Macromol. Symp.* **357** (1) pp. 30-37.

[138] Mu L, Zhu Q and Si L. Radiative properties of nanofluids and performance of a direct solar absorber using nanofluids. In: ASME 2009 second International conference on micro/nanoscale heat and mass transfer 2009 pp. 549-553.

[139] Kraus R A and Kraus E J 1977. *Solar Thermal-Radiation, Absorption and Conversion System.* USA, Patent No. 4,055,948.

[140] Hunt A J 1979 *A New Solar Thermal Receiver Utilizing a Small Particle Heat Exchanger.* Boston, MA. https://escholarship.org/uc/item/32h764qg.

[141] Abdelrahman M, Fumeaux P and Suter P 1979 Study of solid-gas-suspensions used for direct absorption of concentrated solar radiation. *Sol. Energ.* **22** (1) pp 45-48.

[142] Bohn M S 1987 Experimental investigation of the direct absorption receiver concept. *Energ.* **12** (3-4) pp 227-233.

[143] Parker R Z and Langhoff P W 1993 *Fluid Absorption Receiver for Solar Radiation.* USA, Patent No. 5,182,912.

[144] Yogev A 2004. *Solar Energy System With Direct Absorption of Solar Radiation.* USA, Patent No. 6,776,154.

[145] Hirsc D and Steinfeld A 2004. Solar hydrogen production by thermal decomposition of natural gas using a vortex-flow reactor. *Int. J. Hydro. Energ.* **29** (1) pp 47-55.

[146] Schunk LO, Haeberling P, Wepf S, Wuillemin D, Meier A, Steinfeld A (2008) A receiver-reactor for the solar thermal dissociation of zinc oxide. A receiver-reactor for the solar thermal dissociation of zinc oxide. *J. Sol. Energ. Eng.* **130** (2) 021009. https://doi.org/10.1115/1.2840576.

[147] Goldman A, Meitav A, Yakupov I, Kroizer I, Kokotov Y and Gilon Y 2010. *High Temperature Solar Receiver.* USA, Patent No. 7,690,377.

[148] Sani E, Barison S, Pagura C, Mercatelli L, Sansoni P, Fontani D, Jafrancesco D and Francini F 2010. Carbon nanohorns-based nanofluids as direct sunlight absorbers. *Opt. Express* **18** (5) pp 4613-4616.

[149] Lu L, Liu Z H and Xiao H S 2011 Thermal performance of an open thermosyphon using nanofluids for high-temperature evacuated tubular solar collectors. *Sol. Energ.* **85** (2) pp 379-387.

[150] Lenert A and Wang E N 2012 Optimization of nanofluid volumetric receivers for solar thermal energy conversion. *Sol. Energ.* **86** (1) pp 253-265.

[151] Rehena N, Salma P and Alim M A 2015 Heat transfer and collector efficiency through a direct absorption solar collector with radiative heat flux effect. *Numer. Heat Transf. Part A: Appl.* **68** (8) pp 887-907.

[152] Verma V and Kundan L 2013 Thermal performance evaluation of a direct absorption flat plate solar collector (DASC) using Al_2O_3-H_2O based nanofluids. *IOSR J. Mech. Civil Eng.* **6** (2) pp 29-35.

[153] Rasih R A, Sidik N A C and Samion S 2019 Numerical investigation of direct absorption solar collector using nanofluids: A review. *IOP Conf. Ser.: Mater. Sci. Eng.* **469** (1) 012059. https://doi.org/10.1088/1757-899X/469/1/012059.

[154] Karami M, Bahabadi M, Delfani S and Ghozatloo A 2014 A new application of carbon nanotubes nanofluid as working fluid of low-temperature direct absorption solar collector. *Sol. Energ. Mater. Sol. Cells* **121** pp 114-118.

[155] Verma S and Tiwari A 2015 Progress of nanofluid application in solar collectors: A review. *Energ. Convers. Manag.* **100** pp 324-346.

[156] Karim MA, Arthur O, Yarlagadda PK, Islam M and Mahiuddin M 2019 Performance investigation of high temperature application of molten solar salyt nanofluid in direct absorption solar collector. *Mol.* **24** (2) 285. https://doi.org/10.3390/molecules24020285.

[157] Balakin B V, Zhdaneev O V, Kosinska A and Kutsenko K V 2019. Direct absorption solar collector with magnetic nanofluid: CFD model and parametric analysis. *Renew. Energ.* **136** pp 23-32.

[158] Won K H and Lee B J 2018 Effect of light scattering on the performance of a direct absorption solar collector. *Front. Energ.* **12** (1) pp 169-177.

IOP Conf. Series: Materials Science and Engineering **805** (2020) 012016 doi:10.1088/1757-899X/805/1/012016

Green synthesized silver nanoparticles for cream formulation: its anti-inflammatory and healing activities

OA Adeleye[1]*, AO Badru[1], OE Oyinloye[2], AB Fagbohun[3], LG Bakre[1], OA Bamiro[1], CO Babalola[1], and A Lateef[4]

[1]Department of Pharmaceutics, Olabisi Onabanjo University, Ago-Iwoye, Ogun State, Nigeria; [2]Department of Pharmacology, Olabisi Onabanjo University, Ago-Iwoye, Ogun State, Nigeria; [3]Department of Pharmaceutical Chemistry, Olabisi Onabanjo University, Ago-Iwoye, Ogun State, Nigeria; [4]Department of Pure and Applied Biology, Ladoke Akintola University of Technology, Ogbomoso, Nigeria

*Corresponding Author Email: olutayoadeleye@yahoo.com

Abstract

Among several forms of metallic nanoparticles used in biomedical applications, AgNPs are the most fascinating and widely used. Therefore, this study was carried out to evaluate the anti-inflammatory activity of *Ehretia cymosa*-AgNPs cream. Different extracts of the plant (methanol, ME; n-hexane, NE; ethylacetate, EE) were used to synthesize AgNPs, which were characterized by UV and FTIR spectroscopy. The AgNPs were used to prepare creams, and the physical properties of the creams were evaluated. Anti-inflammatory activities of the creams were evaluated by carrageenan induced rat paw edema method on albino rats. The biosynthesized AgNPs had UV absorption spectra in the range 400-450 nm. The creams had pH of 4.6-5.6 similar to the pH of the skin. Cream formulation NO (no extract) had no significant anti-inflammatory activity within 6 h of treatment while formulations NE, ME and EE had a significant anti-inflammatory activity within 6 h of treatment. The efficacy of the cream formulation of the extracts after 6 h of treatment was in the order EE > ME >NE. There was a significant anti-inflammatory activity with the AgNPs cream formulations. Formulations MS, NS and ES had 87.1, 90 and 100% anti-inflammatory activity, respectively in 4 h of treatment. The ES, MS and NS cream formulations had 100% healing in 4, 5 and 6 h, respectively, while only formulation EE had 100% healing in 6 h. Other cream formulations containing the extract could not achieve 100% healing in 6 h. The study established that *Ehretia cymosa*-synthesized AgNPs cream formulations had a faster healing ability compared with the positive control and *Ehretia cymosa* cream formulations.

Keywords: Silver nanoparticles, *Ehretia cymosa,* cream formulation, anti-inflammatory activity

IOP Conf. Series: Materials Science and Engineering **805** (2020) 012016 doi:10.1088/1757-899X/805/1/012016

1. Introduction

Creams are semi-solid emulsions of oil and water. They are topical preparations which could be medicated or non-medicated. Medicated creams contain active pharmaceutical ingredients (API) most often used as antimicrobial agent, anti-inflammatory agent, anti-acne agent among others, while non-medicated creams do not contain API. They are often used because of their emollient and moisturizing properties. Silver Nanoparticles (SNPs) have gained acceptability and applicability in many sectors like medical, pharmaceutical, chemical, health care, consumer, construction, food, and textile [1]. Nanoparticles are small particles in size range of 1-100 nm which confers an advantage over micro-sized particles in terms of rate of activity [2, 3].

There are several forms of metallic nanoparticles used in biomedical applications; silver nanoparticles (AgNPs) are the most fascinating and most widely used due to their remarkable activities such as antimicrobial, anti-inflammatory and anti-coagulant [4,5,6]. Several studies on the biomedical applications of biosynthesized silver nanoparticle are available in the literature, each with different focus [7, 8]. Also, many researchers have reported the anti-inflammatory potentials of biosynthesized silver nanoparticles [9-12]. Inflammation is the body's way of protecting itself from infection, illness, or injury. Symptoms include pain, swelling, redness, immobility and heat.

Ehretia cymosa is a shrub, which belongs to the family Boraginaceae. It is found abundantly in western, central and eastern Africa. It is commonly called Shekutu in Luhya, Kenya and Jàokè in Yoruba, Nigeria. It is used in traditional folklore medicine as a mild laxative, treatment of muscle stiffness, toothache, inflammation and healing of fractured bone [13]. The stem is used as a chewing-stick for maintaining tooth and gum hygiene. This study was carried out to evaluate the anti-inflammatory activity of a cream formulation incorporated with silver nanoparticles biosynthesized with different extracts of *Ehretia cymosa*.

2 Materials and Methods

2.1 Leaves Collection and Extraction

The leaves of *Ehretia cymosa* plants were collected and identified at the Forest Research Institute of Nigeria (FRIN), Ibadan with voucher No. 112440. The leaves were dried and milled into powder with a blender. The powdered leaves of *Ehretia cymosa* was extracted with three different solvents (N-hexane, ethylacetate, and methanol). The filtrates were concentrated to dryness with a rotary evaporator.

2.2 Synthesis of silver nanoparticles

About 0.1g of the powdered extract was dissolved in 100 ml of distilled water at 60 °C for 1 h; it was allowed to cool and then filtered with Whatman No. 1 paper. The filtrate was centrifuged at 4000 rpm for 15 min. Thereafter, 1 ml of the centrifuged filtrate was transferred into 40 ml of 1 mM aqueous silver nitrate solution and was observed for colour change.

2. 3 Characterization of *Ehretia cymosa*-synthesized silver nanoparticles

The synthesis of *Ehretia cymosa*-silver nanoparticles were monitored by measuring the absorption spectra in the wavelength range of 300-700 nm using T90+ UV/Vis spectrometer and the spectrum was recorded and the maximum absorption wavelength was determined. The silver nanoparticles synthesized were also characterized by Transform infrared (FTIR) spectroscopy which was carried out with PerkinElmer spectrum spectrometer version 10.03.02 to detect the biomolecules responsible for the biosynthesis of the silver nanoparticle.

2.4 Cream preparation

The creams were prepared according to Table 1 with each cream formulation containing 2 g of either the extract from the different solvents or the AgNPs synthesized from the extract of these solvents.Two (2) gram of the extract or the AgNPs was mixed with 0.1 g of chlorocresol and the required quantity of distilled water with the aid of heat to about 70 °C. Nine (9) gram of emulsifying wax, 15 g of white soft paraffin and 6 g of liquid paraffin were melted together at 70 °C. The chlorocresol and extract solution was added to it at same temperature (70 °C) and stirred until it was cold. The pH of the formulations was determined with a digital pH meter. Each gramme (1g) of the formulation was dissolved in 100 ml of distilled water in a beaker and the pH reading was recorded in triplicate [14].

2.5 Evaluation of anti-inflammatory activity of the cream formulations

Carrageenan-induced rat paw edema method according to established protocol [15] was adopted in evaluating the anti-inflammatory activity of the creams. Seven groups of six albino rats were used. Edema was induced by injecting Carrageenan into the sub-planar tissue of the right hind paw in all animal groups. The linear paw circumference was measured before and after induction and also at an hourly interval after the application of the cream formulations.

IOP Conf. Series: Materials Science and Engineering **805** (2020) 012016 doi:10.1088/1757-899X/805/1/012016

Table 1. *Ehretia cymosa* extract and AgNPs cream formulations

Ingredients(g)	NO	NE	EE	ME	NS	ES	MS	Diclo
Extract/AgNPs	-	2.0	2.0	2.0	2.0	2.0	2.0	2.0
Emulsifying wax	9.0	9.0	9.0	9.0	9.0	9.0	9.0	9.0
White soft paraffin	15.0	15.0	15.0	15.0	15.0	15.0	15.0	15.0
Liquid paraffin	6.0	6.0	6.0	6.0	6.0	6.0	6.0	6.0
Chlorocresol	0.1	0.1	0.1	0.1	0.1	0.1	0.1	0.1
Distilled water, to	100.0	100.0	100.0	100.0	100.0	100.0	100.0	100.0

NO, No extract; NE, N-hexane extract; EE, ethylacetate extract; ME, methanol extract; NS, N-hexane extract-AgNPs; ES, Ethylacetate extract-AgNPs; MS, Methanol extract-AgNPs; Diclo, Diclofenac cream

3. Results and Discussion

Peak absorption wavelength for all the three *Ehretia cymosa* extract-mediated silver nanoparticles were observed in the range of 400-450 nm. This confirms that the color change that occurred was due to the conversion of silver nitrate to silver nanoparticles with the extract serving as the a reducing agent. Similar range of absorption wavelength of silver nanoparticle was observed in other studies [7, 16, 17].

Transform infrared (FTIR) spectroscopy was carried out to identify and confirm the biomolecules responsible for the synthesis, reduction, capping and stabilization of the silver nanoparticle. As revealed in Figure 1, the FTIR spectrum shows absorption peaks at 3356.37, 2091.85, 1629.96, 1311.71, and 1038.94 cm^{-1} which is an indication that a capping agent responsible for bio-reduction is present.

The normal human physiological skin pH is in the acidic range [18]. As presented in Table 2, the pH of all the cream formulations was within acidic range (4.6-5.6). This is an indication that the formulations may not cause skin irritation.

Table 2. The pH of cream formulations

Formulation	pH
NO	5.87±0.02
NE	5.25±0.10
ME	5.10±0.07
EE	4.80±0.11
NS	5.20±0.02
MS	5.30±0.02
ES	4.95±0.01
Diclo	5.50±0.06

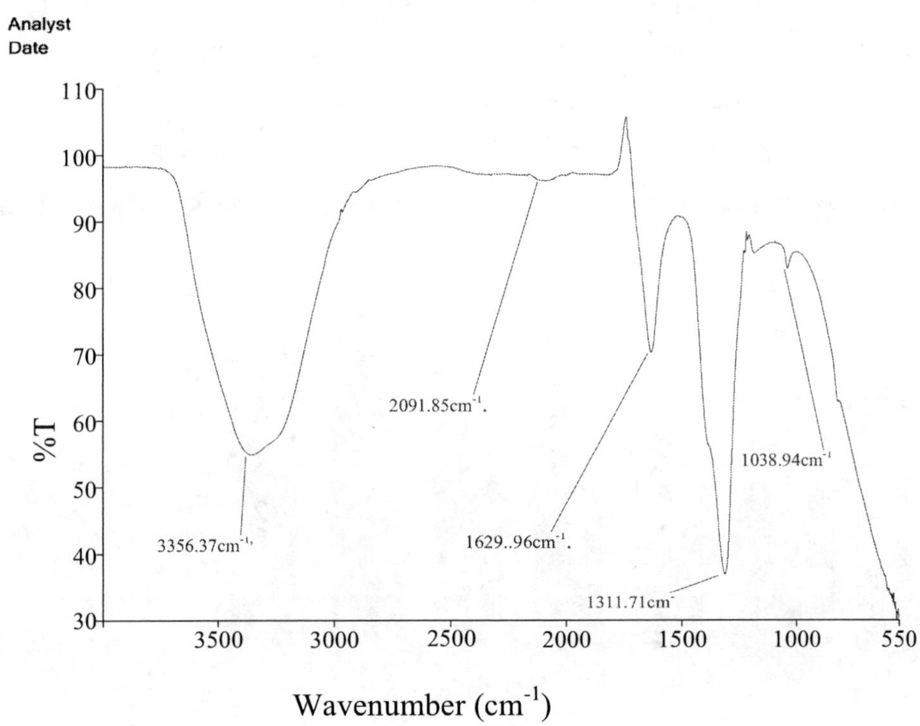

Figure 1. FTIR spectrum of the synthesized AgNPs

The anti-inflammatory activity of the creams is presented in Figures 2, 3 and 4. Formulation NO (cream without any active ingredient) did not have any significant anti-inflammatory activity

IOP Conf. Series: Materials Science and Engineering **805** (2020) 012016 doi:10.1088/1757-899X/805/1/012016

within 6 h of treatment while the N-hexane extract (NE), methanol extract (ME) and ethyl acetate extract (EE) had significant anti-inflammatory activities within 6 h of treatment (Figure 2). The NE, ME and EE creams had 94.5, 96.7 and 100% anti-inflammatory activities, respectively in 6 h. The efficacy of the cream formulations from the extracts after 6 h of treatment was in the order EE > ME > NE.

There was a significant anti-inflammatory activity with the AgNPs cream formulations (Figure 3). Methanol extract silver nanoparticle (MS), N-hexane extract silver nanoparticles (NS) and ethyl acetate extract silver nanoparticle (ES) cream formulations had 87.1, 90 and 100% anti-inflammatory activities, respectively in 4 h of treatment. The ranked order of the anti-inflammatory activity was ES >NS>MS.

The ES, MS and NS cream formulations had 100% healing in 4, 5 and 6 h, respectively while only formulation EE had 100% healing at 6 h (Figure 4), while NE, ME (cream formulations containing the extract) and formulation containing diclofenac, the standard drug could not achieve 100% healing in 6 h. Silver has been used traditionally as a therapy for wound healing especially burns [19]. The formulation of silver sulfadazine cream was based on the knowledge of the traditional use of silver [20]. The wound healing potentials of silver nanoparticle have been reported severally in literature [21-24].

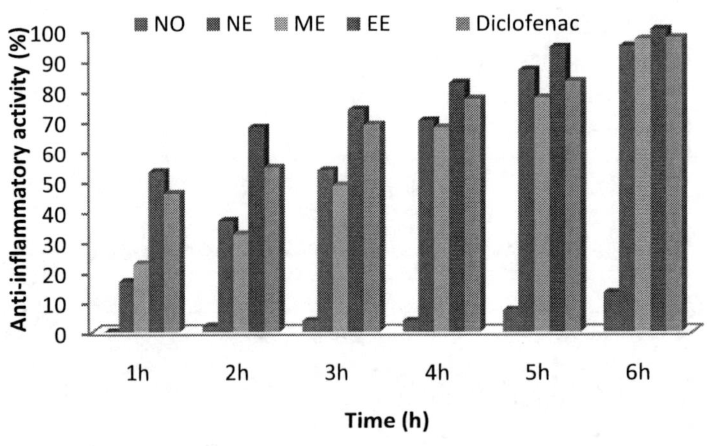

Figure 2. Percentage (%) anti-inflammatory activity of *Ehretia cymosa* cream formulations

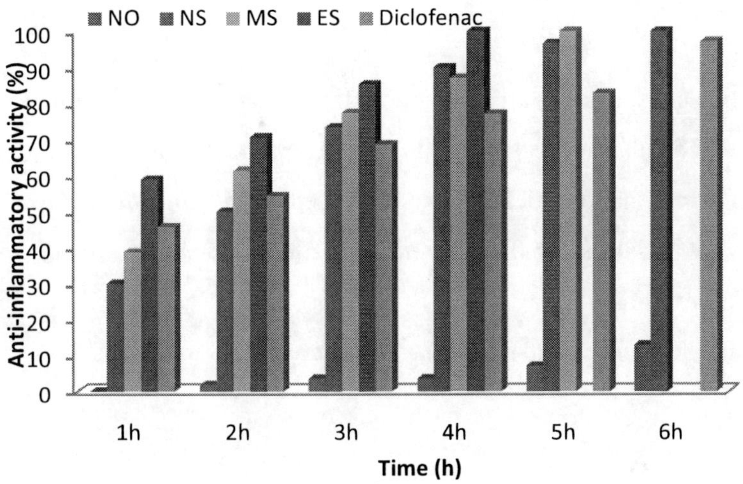

Figure 3. Percentage (%) anti-inflammatory activity of *Ehretia cymosa*-AgNPs cream formulations

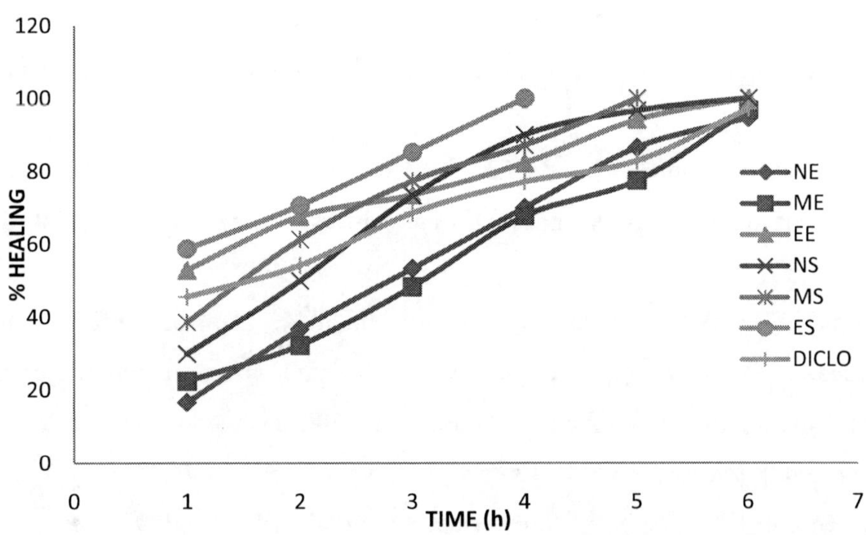

Figure 4. Percentage (%) healing with time on application of the cream formulations

4 Conclusion

Metallic nanoparticles especially silver nanoparticles have widely gained acceptability and applicability in biomedicals. In this study, the anti-inflammatory activity of a cream formulation incorporated with silver nanoparticles biosynthesized with *Ehretia cymosa* leaf extract was evaluated. The study established that *Ehretia cymosa*-synthesized silver nanoparticles cream formulations had a better anti-inflammatory properties and a faster healing ability compared to the positive control (diclofenac) and *Ehretia cymosa* cream formulations.

References

[1] Hungund B S Dhulappanavar G R and Ayachit N A 2015 Comparative evaluation of antibacterial activity of silver nanoparticles biosynthesized using fruit juices. *J. Nanomed. Nanotechnol.* **6** (2) pp 91-97.

[2] Mohamed A and Xing MM 2010 Nanomaterials and nanotechnology for skin tissue engineering. *Int. J. Burns Trauma* **2** (1) pp 29-41.

[3] Tian Y, Chen J, Zahtabi F, Keijzer R and Xing M 2013 Nanomedicine as an innovative therapeutic strategy for pediatric lung diseases. *Pediatric. Pulmonol.* **48** (11) pp 1098-1111.

[4] Sureshkumar M, Siswanto D Y and Lee C K 2010 Magnetic antimicrobial nanocomposite based on bacterial cellulose and silver nanoparticles. *J. Mater. Chem.* **20** (33) pp 6948-6955.

[5] Vijayakumar S, Malaikozhundan B, Saravanakumar K, Durán-Lara EF, Wang MH and Vaseeharan B 2019 Garlic clove extract assisted silver nanoparticle–Antibacterial, antibiofilm, antihelminthic, anti-inflammatory, anticancer and ecotoxicity assessment. *J. Photochem. Photobiol. B: Biol.* **198** pp 111558.

[6] Das P, Ghosal K, Jana NK, Mukherjee A and Basak P 2019 Green synthesis and characterization of silver nanoparticles using belladonna mother tincture and its efficacy as a potential antibacterial and anti-inflammatory agent. *Mater. Chem. Phys.* **228** pp 310-317.

[7] Marslin G, Selvakesavan RK, Franklin G, Sarmento B and Dias AC 2015 Antimicrobial activity of cream incorporated with silver nanoparticles biosynthesized from *Withania somnifera*. *Int. J. Nanomed.* **10** pp 5955.

[8] Gurunathan S, Park J H, Han J W and Kim J H 2015 Comparative assessment of the apoptotic potential of silver nanoparticles synthesized by *Bacillus tequilensis* and *Calocybe indica* in MDA-MB-231 human breast cancer cells: Targeting p53 for anticancer therapy. *Int. J. Nanomed.* **10** pp 4203-4222.

[9] David L, Moldovan B, Vulcu A, Filip A, Olenic L, Perde M, Fischer-Fodor E, Florea A, Crisan M, Chiorean I and Clichici S 2014 Green synthesis, characterization and anti-inflammatory activity of silver nanoparticles using European black elderberry fruits extract. *Colloids Surf. B: Biointerf.* **122** pp 767-777.

[10]Vijayaraj R, Naresh K, Mani P, Senthil J, Kumar G and Jayaseelan T 2016 Green synthesis of silver nanoparticles from ethanolic seed extract of *Acranythes aspera* (Linn.) and its anti-inflammatory activities. *Int. J. Pharm. Ther.* **7** (1) pp 42-48.

[11] Moldovan B, Vulcu A, Filip A, Olenic L, Perde M, Fischer-Fodor E, Florea A, Crisan M, Chiorean I, Clichici S and Baldea I 2017 *In vitro* and *in vivo* anti-inflammatory properties of green synthesized silver nanoparticles using *Viburnum opulus* L. fruits extract. *Mater. Sci. Eng.: C* **79** pp 720-727.

[12] Kumaran NS 2018 *In vitro* anti-inflammatory activity of silver nanoparticle synthesized *Avicennia marina* (Forssk.) Vierh.: A green synthetic approach. *Int. J. Green Pharm.* **12** (3) pp S528.

[13] Lewis W H and Avioli L V 1991 Leaves of *Ehretia cymosa* (Boraginaceae) used to heal fractures in Ghana increase bone remodeling. *Econ. Bot.* **45** (2) pp 281–282.

[14] Adeleye O A, Babalola C O, Femi-Oyewo MN and Balogun G Y 2019 Antimicrobial activity and stability of *Andrographis paniculata* cream containing shea butter. *Nig. J. Pharm. Res.* **15** (1) pp 9-18.

[15] Gupta R and Kaur J 2015 Evaluation of analgesics, antipyretics, and anti-inflammatory activity on *Cordia dichotoma* G Forst leaf. *Pharmacogn. Res.* **7** (1) pp 126-130.

[16] Mulfinger L, Solomon SD, Bahadory M, Jeyarajasingam AV, Rutkowsky SA and Boritz C 2007 Synthesis and study of silver nanoparticles. *J. Chem. Educ.* **84** pp 322.

[17] Adebayo-Tayo B C, Akinsete T O and Odeniyi O A 2016 Phytochemical composition and comparative evaluation of antimicrobial activities of the juice extract of *Citrus aurantifolia* and its silver nanoparticles. *Nig. J. Pharm. Res.* **12** (1) pp 59-67.

[18] Lambers H, Piessens S, BloemAPronk H and Finke P 2006 Natural skin surface pH is on average below 5, which is beneficial for its resident flora. *Int. J. Cosmet. Sci.* **28** (5) pp 359-370.

[19] Pereira R B and Bartolo P J 2013 Traditional therapies for skin wound healing. *Adv. Wound care* **5** (5) pp 208-229.

[20] Fox C L and Modak S 1974 Mechanism of silver sulfadiazine action on burn wound infections. *Antimicrob. Agents Chemother.* **5** (6) pp 582–588.

[21] Neibert K, Gopishetty V, Grigoryev A, Tokarev I, Al-Hajaj N, Vorstenbosch J, Philip A, Minko S and Maysinger D 2012 Wound-healing with mechanically robust and biodegradable hydrogel fibers loaded with silver nanoparticles. *Adv Healthc. Mater.* **1** (5) pp 621-30.

[22] Walker M and Parsons D 2014 The biological fate of silver ions following the use of silver-containing wound care products - a review. *Int Wound J.* **11** (5) pp 496-504.

[23] Lin Y H, Lin J H, Wang S H, Ko T H and Tseng G C 2012 Evaluation of silver-containing activated carbon fiber for wound healing study: *In vitro* and *in vivo*. *J Biomed. Mater. Res. B: Appl. Biomater.* **100** (8) pp 2288-2296.

[24] Liu X, Lee P Y, Ho C M, Lui V C, Chen Y, Che C M, Tam P and Wong K K 2010 Silver nanoparticles mediate differential responses in keratinocytes and fibroblasts during skin wound healing. *Chem. Med. Chem.* **5** (3) pp 468-75.

IOP Conf. Series: Materials Science and Engineering **805** (2020) 012017 doi:10.1088/1757-899X/805/1/012017

Influence of XL-bond nanochemical on strength indices of weak lateritic soil

OS Olaniyan*, VO Ajileye, MO Adewusi, DO Aransiola and JA Ayanlola

Department of Civil Engineering, Ladoke Akintola University of Technology, PMB 4000, Ogbomoso, Nigeria

*Corresponding author, email: osolaniyan@lautech.edu.ng

Abstract

Lateritic soil is among the major construction materials in road pavement. However, obtaining lateritic soil with sufficient strength is difficult and stabilization with cement is expensive. Stabilization with cement and lime has been well reported but there is paucity of information on stabilization with XL-Bond nanochemicals. In this study, the strength characteristics of stabilized lateritic soil with XL-Bond nanochemical were investigated. Lateritic soil samples were taken from Bariga and Mushin, Lagos State, Nigeria borrowpits (A: Latitude 6.5391°, longitude 3.3849°) and (B: Latitude 6.5273°, longitude 3.3414°), respectively. The geotechnical properties which include Moisture Content (MC), Particle Size Analysis (percentage passing sieve no 200), Plastic Index (PI), Optimum Moisture Content (OMC), Maximum Dry Density (MDD), Unconfined Compressive Strength (UCS) and California Bearing Ratio [CBR] (soaked) of lateritic soils were determined at natural state. The lateritic soils were stabilized at 2, 4, 6, and 8% of XL-Bond nanochemical by dry unit weight of the soil samples. The PI, MDD, CBR (soaked and unsoaked) and UCS of stabilized laterite soil samples were determined in line with standard methods.

The MC, percentage passing sieve no 200, PI, OMC, MDD, UCS and CBR (soaked and unsoaked) of soil samples A, B were (45.6%, 38.98%, 5.0%, 10.5%, 2420 g/m^3, 59kPa, 12.0% and 10.0%); (41.1%, 54.52%, 2.0%, 12.50%, 1960g/m^3, 72kPa, 4.0% and 20.0%), respectively. The PI, MDD, CBR (soaked and unsoaked) and UCS of stabilized soil with XL-Bond varied from (3-7%; 2.30-2.35g/cm^3,12-23%, 23-50% and 64-83 kPa) and (3-6%; 1.44-1.50g/cm^3,6-20%, 21-40% and 79-8 kPa), respectively. There is significant effect of varying percentage of XL-Bond on CBR of stabilized samples (P = 0.1004>0.05). In conclusion, stabilization of Laterite soil with XL-Bond nanochemical enhanced its compressive strength. The stabilized lateritic soil can be used as subgrade materials in highway construction.

Keywords: Stabilization, nano-chemical, XL-Bond, lateritic soil, subgrade materials, highway construction

1 Introduction

XL-Bond nanochemicals are emerging nanomaterials that enhance the strength of soil. Nanochemical (XL-Bond) can be added to soils to enhance its workability and load bearing capacity [1]. The nanochemical can significantly improve soil properties such as stability, impermeability and load bearing capacity of soil. Application of Nano chemical (XL- Bond) to subgrades can provide significant improved engineering properties [1]. XL-Bond is water soluble and can be easily applied on the field. It is characterized as heat stable reactive soil stabilizer with excellent waterproof property. It reduces capillary rise and permeability from top of the soil. It also minimizes water permeability of soil bases and expansive properties of soil minerals. XL-Bond is acrylic co-polymer in nature which enhances bonding of soil particles. It is mostly sprayed on compacted soils and in turns reduces drastically the permeability and therefore increases the soil bearing capacity [1]. According to Olaniyan *et al.* [2], the nanochemical converts silanol group (Si–O–H) which makes surface hydrophobic to Alkyl siloxane surface. This implies that it convert water loving to water repellant property of soil.

Lateritic soils are abundant and among the most useable construction materials. The soils are sourced from tropical environment characterized with intense chemical weathering and leached soluble minerals. Lateritic soils are reddish brown, well graded and often located within depth of several tens of metres. They are mostly found in tropics with and useful in most construction work. This makes the study of the soil characteristic important in the areas of consistency limit, grain size distribution, permeability compaction, consolidation and shear strength [2, 3]. The origin of expansive soil is related to a complex combination of condition and processes that bring about formation of clay mineral materials characterized with expansive properties when in contact with water. The expansive soils can be hard and later lose their strength completely when wet [4]. Road pavement constructed on weak subgrades shows distresses of different kinds, which leads to road failure. The poor subgrade will need to be stabilized to enhance its strength if getting a good subgrade materials from borrow pit is unrealistic.

IOP Conf. Series: Materials Science and Engineering **805** (2020) 012017 doi:10.1088/1757-899X/805/1/012017

Most of the asphalt-aggregate based pavement fails as a result of moisture related distress. Failed asphaltic mixture due to moisture ingress is a major problem for State and Federal highway agencies in Nigeria for many years, as is evidenced by increasing budget needs for maintenance and rehabilitation. Water reduces the cohesive strength of the mastic and weakens the pavement's ability to resist damage due to pore pressure and premature cracking. Moisture is an arch enemy of road, which can be tackled by improving design, construction and road maintenance. The aim of the study is to assess the influence of XL–Bond nanochemical on strength indices of weak lateritic soil.

2 Methodology

The geotechnical properties of lateritic soil used were determined from these tests: sieve analysis, Atterberg limit (Liquid Limit, Plastic Limit), Compaction Test, California Bearing Ratio (CBR) and Unconfined compressive strength. The technical specification of the XL-Bond nanochemical used is presented in Table 1.

Table 1. Technical specifications of XL-bond [1]

Property	Description
Appearance	Pale clear to light honey color in concentrate form.
Viscosity at 25 °C	500-800 CPS
Specific gravity	1.05 at 25 °C
Solubility	Mixable with Water, Soluble in Methanol
Flash Point	100 °C (212 F) / non-flammable
XL-Bond	1 kg : 200 liters of water
Penetration	3-7 mm

2.1 Laboratory Test

(i) **Sieve Analysis**

The equipment used for include: set of sieves, mortar and pestle, weighing balance, wash bottle, mechanical sieve shaker, measuring can and porcelain dish. The analysis was done in accordance with the provision of BS1377: Part 2, 1990 [5]. Five hundred (500) g of air-dry representative

IOP Conf. Series: Materials Science and Engineering **805** (2020) 012017 doi:10.1088/1757-899X/805/1/012017

soil sample were placed on the sieve No. 200 and washed through the sieve using distilled water until the water passing through the sieve were clear. Then the dish and the remaining soil-water suspension were dried in oven for at 110 °C. The stack sieves were placed in electrical shaker and the weighed test soil were poured into the topmost sieve. The stacked sieves were clipped into the shaker and shaken for 15 min electronically. The sieves were removed and carefully separated. The particles retained on each sieve were weighed, expressed in gram and percentage weight of the sample. Equations 1, 2 and 3 were used for the calculation.

Weight passing a sieve = Weight of Dry Sample – Weight Retained on the Sieve

$$\%\text{ Weight Retained} = \left(\frac{Weight\,\text{Re}tained}{Total\ dry\ weight\ of\ sample}\right)\times 100\% \quad (1)$$

$$\%\text{ Weight Passing} = 100\% - Percentage\,\text{Re}tained \quad (2)$$

$$\text{Cumulative }\%\text{ Weight Retained} = \left(\frac{Cumulative\ Weight}{Total\ dry\ weight\ of\ sample}\right)\times 100\% \quad (3)$$

(ii) **Consistency Limit Test**

The apparatus used for this test are Liquid- limit device with grooving tool, moisture cans, plastic limit plate, weighing balance, sieve and its pan. The test was done in accordance with the provision of BS1377 (No. 40: BS sieve (0.425). The concentration of XL-Bond nanochemical used was 2.5 g/l.

To determine the Liquid limit test of lateritic soil, 120 g of the lateritic soil were weighed and placed upon the glass plate. The soil was thoroughly mixed using spatula with XL–Bond nanochemical as additive at 2, 4, 6 and 8% of the soil weight, until the consistency of a thick paste was achieved. The apparatus cups were half-filled with the wet soil and levelled. The soil was spread out on the plate to dry. The soil sample was later mixed again, groove was made and the cup was turned at two rotations per second. The cups were dropped until the two parts of the soil along the groove touched. The number of blows at which the closure of groove occurred were recorded and the moisture content was of the soil was determined. The moisture content corresponding to each number of blows represents the Liquid limit.

To determine the Plastic Limit test, 15 g of the lateritic soil were weighed and sieved through a (No. 40: BS sieve (0.425) in accordance to BS1377. The soil were mixed properly on the glass plate and nanochemical XL–Bond as additive in 2, 4, 6 and 8% by weight of soil

IOP Conf. Series: Materials Science and Engineering **805** (2020) 012017 doi:10.1088/1757-899X/805/1/012017

respectively, were added to make it plastic enough to be moulded into small ball. The balls were rolled between the hand and glass plate to form into a thread. As the diameter of the thread decreases, it crumbles and its moisture content was determined as the plastic limit.

(iii) Compaction Test

The apparatus used are BS mould, displaced disk, base plate, a rammer, weighing balance, scraper, moisture cylinder and head pan. The dried lateritic soil samples that passed through the 4.75 mm BS sieve were used in line with the provision of BS1377: Part 2, 1990 [5]. Exactly 800 g of the soil samples were weighed and mixed with a measured volume of water and mixed with XL–Bond as additive in 2, 4, 6 and 8% by weight of the soil sample. The soil mixing was done using scoop and trowel. Five layers of the soil were put into the mould with filter paper covered base plate to remove adhesive force. Thereafter, 62 blows of 4.5 kg metal rammer were applied to each layer, static compaction were employed. After compaction, the excess soils were scrapped with a straight edge. And compacted soil weighed to determine the wet bulk density. A sample of the soil was taken from top and bottom of the compacted soil to determine the moisture content. Maximum Dry Density (MDD) and Optimum Moisture Content (OMC) were determined.

(iv) California Bearing Ratio (CBR) (Soaked and Unsoaked) Test

This method covers the *in situ* of the dry density and optimum moisture content. This test was done by using the established OMC and MDD of the soil determined from the static compaction method. The apparatus used include CBR equipment of 152 mm diameter × 178 mm height, compaction mould with collar and spacer disk 151 mm diameter × 61.4 mm height, compaction rammer, expansion measuring apparatus with dial guage reading to 0.01 mm, compression machine equipped with CBR penetration piston. Precisely 800 g of lateritic soil samples that passed through the 4.75 mm BS sieve were used for compaction. The lateritic soils were broken to avoid reduction of the natural size of the individual particles. The, 150 ml of water was added, and mixed with XL-bond as additive at 2, 4, 6 and 8% by weight of the soil. The mould was arranged and the lateritic soil was compacted. Eight test specimens were compacted at varying moisture content to determine OMC and MDD.

IOP Conf. Series: Materials Science and Engineering **805** (2020) 012017 doi:10.1088/1757-899X/805/1/012017

After compaction, collars were removed and compacted soil trimmed even using straight edge. The weight of the mould and compacted soil were measured and recorded. The mould was placed in the loading frame of the CBR machine and its plunger adjusted by lowering to 1.5 mm of the mould. The surcharge rings were placed and both penetration and load gauge readings were measured and recorded. The loadings and penetrations were also recorded from the bottom by inverting the mould. Moisture content of soil samples taken at the top and bottom of compacted soil were determined.

(v) Unconfined Strength Test

The apparatus used were compression device, evaporating dish, triple beam balance (sensitive to 0.1g), strain gauge and oven. The natural soil samples and the treated samples were compacted in 1000 cm^3 mould at their OMC. The samples were extruded, trimmed and placed centrally on the lower plate of a compression testing machine. A compressive force was applied to the specimen and records of axial deformation and force at regular interval until failure occurred were noted. The unconfined strength of soil sample was determined at the point on the stress strain curve at which failure occurred.

3 Results and Discussion

(a) Grain Size Analysis Test

Particle size analysis of soil determines its engineering properties. The analysis gives the grain size distribution which is used to classify the soil. The sieve analyses of Sample A and B are presented in Tables 2 and 3. Results showed that samples A and B are silty-clay according to AASHTO classification system. The samples are not suitable for subgrade, sub-base and base, because they failed the specific requirement which is, the percentage passing through the sieve No. 200 are greater than 35% and must be stabilized for highway application.

The results of Liquid Limit (LL), Plastic Limit (PL) and Plastic Index (PI) of lateritic samples stabilized with XL-bond at 0, 2, 4, 6 and 8% by weight are presented in Tables 4 and 5. The liquid limit values of the entire samples met the AASHTO classification system requirements which specify material for subgrade course which are greater than 35%. The plastic index of the entire samples collected met the requirement specifies material for subgrade course which is

IOP Conf. Series: Materials Science and Engineering **805** (2020) 012017 doi:10.1088/1757-899X/805/1/012017

below 30%. Also, sample A and sample B met the requirement specification for sub-base and base course which is PI must be below 10%.

Table 2. Sieve analysis result of sample A

Sieve Diameter (mm)	Mass Retained	Percentage Retained (%)	Percentage Passing (%)
20	0	0.00	100.00
8	22.4	4.48	95.52
4	24.4	4.88	90.64
2	33.8	6.76	83.88
1	34.9	6.98	76.90
0.71	30.3	6.06	70.84
0.425	38.9	7.78	63.06
0.355	27.4	5.48	57.58
0.25	31.7	6.34	51.24
0.125	36.5	7.30	43.94
0.075	24.8	4.96	38.98
<0.075	194.90	38.98	0.00

IOP Conf. Series: Materials Science and Engineering **805** (2020) 012017 doi:10.1088/1757-899X/805/1/012017

Table 3. Sieve analysis results for sample B

Sieve Diameter (mm)	Mass Retained (g)	Percentage Retained (%)	Percentage Passing (%)
20	0	0.00	100.00
8	3.5	0.70	99.30
4	8.5	1.70	97.60
2	14.0	2.80	94.80
1	30.0	6.00	88.80
0.71	20.8	4.16	84.64
0.425	35.7	7.14	77.50
0.355	17.6	3.52	73.98
0.25	28.5	5.70	68.28
0.125	50.9	10.18	58.10
0.075	17.9	3.58	54.52
<0.075	272.6	54.52	0.00

Table 4. Summary of Atterberg Limit test for 0-8% XL-bond sample A

Type of Test	0%	2%	4%	6%	8%
Mass of empty can (g)	42.8	42.8	43.0	46.8	45.2
Mass of can + wet soil (g)	51.1	47.6	47.3	55.0	50.1
Mass of can + dry soil (g)	48.5	45.9	45.8	52.1	50.4
Mass of moisture (g)	2.6	1.7	1.5	2.9	2.5
Mass of dry soil (g)	5.7	3.1	2.8	5.3	5.2
Moisture content (%)	45.6	54.8	53.6	54.7	48.1
Number of blows	35	23	21	18	20
Liquid Limit	40	36	38	44	42
Plastic Limit	35	30	35	39	35
Plastic Index	5	6	3	5	7

IOP Conf. Series: Materials Science and Engineering **805** (2020) 012017 doi:10.1088/1757-899X/805/1/012017

Table 5. Summary of Atterberg Limit test for 0-8% XL-bond sample B

Type of Test	0%	2%	4%	6%	8%
Mass of empty can (g)	42.9	47.0	46.7	43.2	42.0
Mass of can + wet soil(g)	50.8	55.1	54.6	53.9	50.4
Mass of can + dry soil(g)	48.5	52.6	52.1	50.3	48.0
Mass of moisture(g)	2.3	2.5	2.5	3.6	3.0
Mass of dry soil(g)	5.6	5.6	5.4	7.1	6.0
Moisture content (%)	41.1	44.6	46.3	50.7	50
Number of blows	41	37	28	15	18
Liquid Limit	40	44	44	42	45
Plastic Limit	37	38	39	39	40
Plastic Index	2	6	5	3	5

(b)Compaction Characteristics of the Soil:

The compaction characteristics, CBR and UCS results of natural and stabilized Samples A and B are presented in Tables 6, 7 and 8. The MDD decreases as percentage of stabilizer (XL-bond) increases. This in agreement with earlier findings [2, 6] which showed decreasing trends of MDD with an increasing OMC and stabilizing agents. For the soaked samples, it falls within the range of 10-25% CBR values specified for subgrade soils by the AASHTO standard and the Nigerian Highway Design Manual Federal Ministry of Works and Housing. The unsoaked CBR ranges from 10-50% for sample A and 20-40% for sample B which met the specification for Federal Ministry of Works and Housing for subgrade, sub base and base materials are <10, <30 and <80%, respectively. However, it was noticed that an increase in the nanochemical additives causes a tremendous increased in compressive strength. The optimum percentage of XL-bond chemical was 6% which produced the highest values of corresponding compressive strength. In order words, the more the addition of XL-Bond, the more the strength gained.

IOP Conf. Series: Materials Science and Engineering **805** (2020) 012017 doi:10.1088/1757-899X/805/1/012017

4 Conclusion and Recommendations

From this study, the following conclusions were deduced: The geotechnical properties of the lateritic soil shows that the percentage passing sieve No 200 for Sample A and B are 38.98 and 54.52%, respectively. The soil falls into A-5 and 6 Soils. The liquid limit values of the entire samples meet the AASHTO classification system requirements, which specify material for subgrade course must be greater than 35%. The MDD decreases with an increase in stabilizer (XL-Bond). The MDD of lateritic soil decreases with an increasing OMC and stabilizing agents. The inclusion of XL-bond increases the strength indices of the weak lateritic soil.

Based on the obtained result from this study, the following recommendations are suggested:

(i) Further research works should be embarked upon in stabilization using other types of nanochemicals such as Zycoprime, Elastobar and Zycosil.

(ii) The Nigeria General Specification [7] did not specify CBR requirement for nanochemical in soil stabilization, hence results from this study can serve as a baseline formulation for the specifications.

Acknowledgement

The authors acknowledge the academic and technical staff of Department of Civil Engineering, LAUTECH, Ogbomoso for creating enabling environment for research.

Table 6. Compaction test for stabilized laterite samples

XL - Bond Solution (%)	Sample A		Sample B	
	OMC (%)	MDD	OMC (%)	MDD (g/cm^3)
0	10.50	2.42	12.50	1.96
2	13.40	2.35	14.00	1.50
4	13.60	2.34	14.20	1.50
6	13.80	2.32	14.40	1.46
8	14.0	2.30	14.40	1.44

Table 7. CBR results of soil samples

XL - Bond Solution (%)	Sample A		Sample B	
	Unsoaked CBR (%)	Soaked CBR (%)	Unsoaked CBR (%)	Soaked CBR(%)
0	10	12	20	4
2	23	18	21	6
4	30	20	26	9
6	38	20	35	14
8	50	23	40	20

IOP Conf. Series: Materials Science and Engineering **805** (2020) 012017 doi:10.1088/1757-899X/805/1/012017

Table 8. Results of unconfined compressive strength for soil samples

XL–Bond Sample (%)	UCS (kPa) Sample A	UCS (kPa) Sample B
0	59	72
2	64	79
4	70	85
6	75	90
8	63	88

References

[1] Zydex 2017 Zydex Industries. http://zydexindustries.com/services/road/soil-stabilization. Accessed 19 January, 2020

[2] Olaniyan OS, Ajileye VO, Adewusi MO and Anifowose OI (2018). Influence of XL-terrasoil on geotechnical properties of weak laterite. *Sci. Focus* **23** (2) pp 122-129.

[3] Butalia TS, Huang J, Kim DG and Croft F 2003 Effect of moisture content and pore water pressure buildup on resilient modulus of cohesive soils in Ohio. Resilient modulus testing for pavement components, ASTM STP 1437, GN Durham, WA Marr and WL De Groff, Eds., ASTM International, West Conshohocken, PA. pp 70-84. https://doi.org/10.1520/STP12523S.

[4] Al-Tabbaa A and Evans WC 2005 Stabilization-solidification treatment and remediation: Part I: Binders and technologies-basic principal. Proceedings of the International Conference on Stabilization/Solidification Treatment and Remediation (pp. 367-385). Cambridge, UK: Balkerma.

[5] Standard B. BS 1377-2: 1990, Methods of test for soils for civil engineering purposes-Part 2: Classification tests. London: UK: British Standard Institution. 1990.

[6] Osinubi KJ 2006 Influence of compactive efforts lime-slag treated tropical black clay. *J. Mater. Civil Eng.* **18** (2) pp 175-181.

[7] The Nigerian General Specification 1997 Nigerian general specification for road works and bridges. Federal Ministry of Works and Housing, Abuja, Nigeria.

IOP Conf. Series: Materials Science and Engineering **805** (2020) 012018 doi:10.1088/1757-899X/805/1/012018

Inhibition efficiency of silver nanoparticles solution on corrosion of mild steel, stainless steel and aluminum in 1.0 M HCl medium

TB Asafa[1*], JK Odusote[2], OS Ibrahim[2], A Lateef[1], MO Durowoju[1], MA Azeez[1], TA Yekeen[1], IC Oladipo[1], EA Adebayo[1], JA Badmus[1], YK Sanusi[1] and O Adedokun[1]

[1]Nanotechnology Research Group, Ladoke Akintola University of Technology Ogbomoso,
[2]Department Materials and Metallurgical Engineering, University of Ilorin, Ilorin.

*Corresponding author: T. B. Asafa (tbasafa@lautech.edu.ng)

Abstract

Research work on corrosion behaviour of metals and their alloys continues to attract considerable attention because of the huge cost required to prevent corrosion and replace failed or contaminated components due to corrosion reactions. Different inhibitors have been used to minimize corrosion rate, but addition of nanoparticles has been reported to enhance inhibition efficiency. In this work, the influence of silver nanoparticles (AgNPs) on inhibition of corrosion of mild steel, stainless steel and aluminum in 1.0 M HCl were investigated. Biosynthesized AgNPs were acquired from Laboratory of Industrial Nanobiotechnology, Ladoke Akintola University of Technology Ogbomoso. The particles were hitherto characterized using FTIR, UV-Vis and TEM. Five concentrations of AgNPs solution (0, 5, 10, 15, and 20 μg/ml) were added to 1.0 M HCl. The corrosion inhibition effects of the AgNPs on the metal samples were analyzed using gravimetric (weight loss), potentiodynamic polarization and gasometric (hydrogen gasevolution) approaches. Gravimetric study was conducted within 2000 h of exposure, and the results showed that weight loss increased with exposure time but decreased with increased concentration of AgNPs solution. In addition, inhibition efficiency was enhanced by the addition of AgNPs solution; 52% for mild steel, 70% for stainless steel and 62% for aluminum. Gasometric results revealed that the volume of hydrogen gas evolution reduced with increased exposure time and concentration of AgNPs solution. Furthermore, potentiodynamic polarization results showed that the presence of AgNPs modified the mechanism of anodic dissolution and cathodic hydrogen gas evolution. These results indicated that AgNPs can be incorporated into the existing inhibitors towards minimizing corrosion rate.

Keywords: silver nanoparticles, gravimetric, gasometric, potentiodynamic polarization

IOP Conf. Series: Materials Science and Engineering **805** (2020) 012018 doi:10.1088/1757-899X/805/1/012018

1. Introduction

Metals and their alloys remain one of the most versatile engineering materials for structural applications. They are used as integral components of machinery, building construction, bridge trusses, screws, military weapons and automobile components among others. These materials are attractive because of their availability, excellent mechanical properties and ease of fabrication. However, metal alloys are susceptible to corrosion leading to degradation of mechanical properties when exposed to corrosive environment [1,2]. Study of corrosion behaviour in metals has continued to attract considerable attention because a large amount of money is expended on corrosion prevention and maintenance or replacement of failed or contaminated components due to corrosion reactions. Over $1.8 trillion or 3-4% of GDP of most industrialized nations is spent on corrosion related activities [3].

Metallic components come in contact with acids during various industrial processes. For instance, acidic solutions are used during acid pickling, chemical cleaning and processing, ore production, industrial cleaning, acid rescaling, oil-well acidification, and petrochemical processes among others [4]. Most industries prefer to use hydrochloric acid as against other acids (such as nitric acid, sulphuric acid and phosphoric acid) because HCl is more efficient and cost effective [5]. This action makes metallic components susceptible to corrosion during industrial processes. To minimize corrosion of industrial equipment and facilities, appropriate corrosion control mechanism must be adopted.

Several approaches have been deployed to minimize corrosion rates in metals. These include cathodic protection, protective coating, heat treatment, alloying, use of inhibitors and environmental modifications [6]. Among these, the use of corrosion inhibitors (CIs) remains one of the most economical [7] and effective approach [8] to corrosion prevention. According to NACE International [9], a corrosion inhibitor, which is usually a chemical substance, effectively minimizes corrosion rate when added in small concentration to an environment. So whenever an inhibitor is present in the corrosion system at a suitable concentration, corrosion rate is significantly decreased. Corrosion inhibitors lead to the formation of a thin film of passive layer on the surface of the material. This hinders the corrosive substance from penetrating into the metal surface, thereby restraining or minimizing either the oxidation or reduction part of the redox corrosion system or render the dissolved oxygen useless [10].

Of the two types of inhibitors-organic and inorganic-the former is synthetic and therefore not eco-friendly because of the toxic products released to the environment after usage [11].

IOP Conf. Series: Materials Science and Engineering **805** (2020) 012018 doi:10.1088/1757-899X/805/1/012018

Consequently, the interest of researchers had shifted to environmentally safe and green inhibitors, which are non-lethal, biodegradable and promptly accessible [7]. In this regard, plant extracts have been used with encouraging results [1,11–16]. Further studies have demonstrated that the use of nanoparticles synthesized from plant extract can significantly improve inhibition performance. The use of plant extracts in synthesizing silver nanoparticles is eco-friendly, reduce energy consumption and can be produced in large scale. Recently, researches have shown that silver nanoparticles can serve as anti-corrosion agent inducing formation of protective layer on the surface of the metals. For instance pitting corrosion in low carbon steel (1010) was significantly reduced by addition of AgNPs [6,17]. AgNPs have been produced from different biological sources through green synthesis. In our previous works [18-24], we produced AgNPs, AuNPs and Ag-AuNPs from diverse materials such as cobweb, cola pod, seeds and seed shell, as well as cocoa pods. Nanoparticles synthesized via green or biological approach have shown to have low toxicity with wide variation in size and shape.

In this study, AgNPs were synthesized from kola nut extract, and were used to inhibit corrosion of mild steel, stainless steel and aluminum in 1.0 M HCl using weight loss, potentiodynamic polarization and hydrogen gas evolution methods.

2. Materials and Methods

2.1 Synthesis and Characteristics of Silver Nanoparticles (AgNPs)

Details of the synthesis and characterization process ofAgNPs used in this study are reported elsewhere [21] while summary of the process is presented herein. AgNPs were produced from silver salt with kolanut pod extract as reducing agent. The synthesized nanoparticles were confirmed by UV-vis which showed a maximum absorbance wavelength of 431.5 nm falling within the range previously reported for AgNPs [21]. Furthermore, FTIR spectra with strong peaks at 3336.85, 2073.48 and 1639.49 cm^{-1} indicate that the 9.5% crude protein in the kolanut was responsible for the stabilization of the synthesized AgNPs (Figure 1a). In addition, silver was the predominant element present in the AgNPs solution based on the EDX pattern (Figure 1b), while the magnesium, sulphur, calcium, sodium and carbon were impurities from the pod extracts and are present on the surface of the biosynthesized AgNPs.The well-dispersed nanoparticles were spherical in shape and poly-disperse in nature, with size ranging from 12 to 80 nm (Figure 1c). The silver is crystalline as revealed by the characteristic ring-like SAED pattern (Figure 1d) which is typical of face-centred cubic crystalline silver [21].

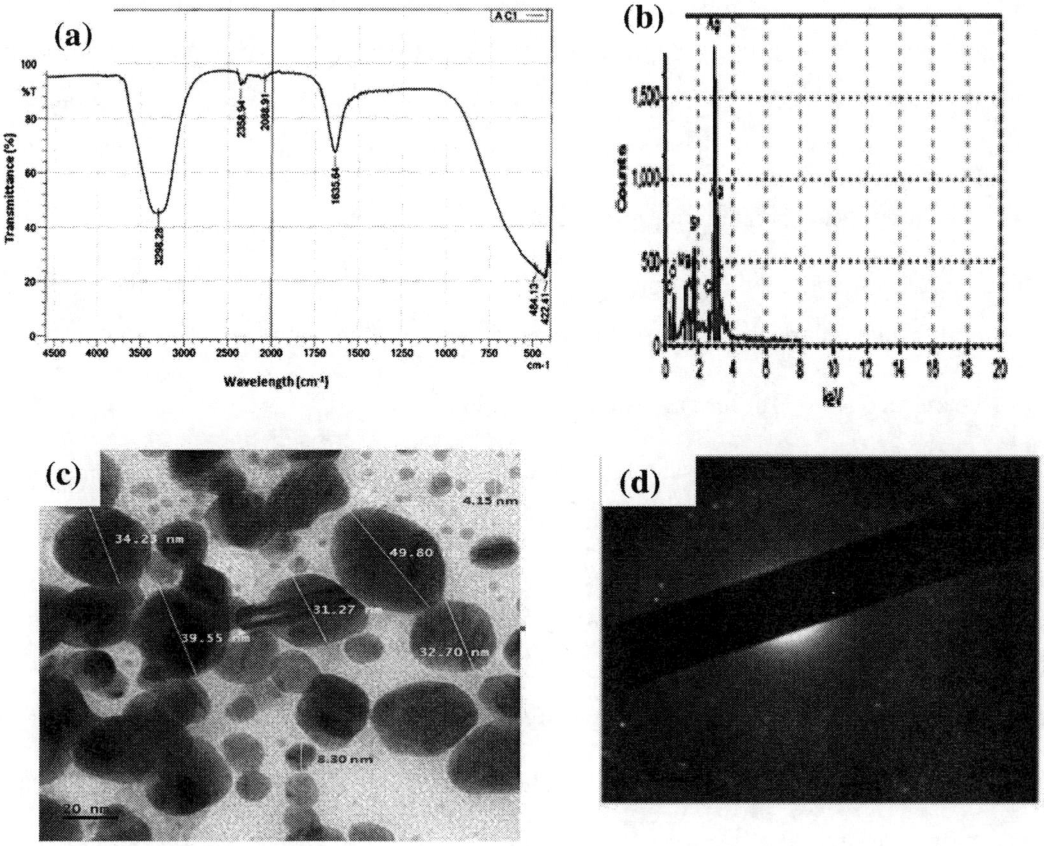

Figure 1. (a) FTIR spectrum (b)energy-dispersive X-ray signal (c)Transmission electron micrograph and (d) selected area electron diffraction pattern of the biosynthesized AgNPs [25]

2.2 Samples Compositions and Preparation

Three different metal alloys (stainless steel, mild steel and aluminum) were selected for this study. The composition of the mild steel was obtained using Olympus Optical Emission Spectroscopy, while those of stainless steel and aluminum were analyzed with Positive Material Identification (PMI) machine following ASTM E1476 standard. The samples were cut into dimensions of 20 mm × 18 mm while 3 mm hole was drilled on each sample to facilitate easy suspension and removal from the corrosion environments. The samples were mechanically polished using emery papers of different grades, degreased in ethanol, cleaned in acetone and then dried and stored in a desiccator to prevent moisture accumulation. Prior to the storage, the weight of each sample was measured using electronic weighing balance.

IOP Conf. Series: Materials Science and Engineering **805** (2020) 012018 doi:10.1088/1757-899X/805/1/012018

2.3 Preparation of corrosion environment

The corrosion environment was prepared by mixing AgNPs solution with 1.0 M HCl. The AgNPs solutions were prepared at different concentrations: 0 µg/ml (control), 5, 10, 15 and 20 µg/ml prepared from the 100 µg/ml stock using distilled water. Then the AgNPs solutions were thoroughly mixed with equal volume of HCl solution in 15 different containers (5 containers for each of the metal alloys). The control solution had no AgNPs added.

2.4 Experimental measurement approaches

Three measurement methods; gravimetric, gasometric (hydrogen gas evolution) and potentiodynamic polarization, were considered for the corrosion study.

2.4.1 Gravimetric/weight loss measurement study

For the gravimetric study, the samples were completely immersed in the corrosion solutions for 400, 800, 1200, 1600 and 2000 h. After immersion for the first 400 h, the specimens were carefully removed, rinsed in distilled water, dried and weighed again. The weight loss was obtained by subtracting the current weight from the original weight. The procedure was repeated after 800, 1200, 1600 and 2000 h of immersion in the corrosion environment. The corrosion rate and inhibition efficiency (IE%) were calculated from the weight loss data according to the following equations [11]:

$$\text{Corrosion rate } (gcm^{-2}h^{-1}) = {87.6\,\Delta W}/{DAT} \quad (1)$$

$$\text{Inhibition efficiency (IE\%)} = \frac{W_o - W}{W_o} \times 100\% \quad (2)$$

where ΔW is the weight loss(g), D is the density of the coupon g/cm^3, Ais the surface area of the mild steel coupon (cm^2), T is the time of exposure (h) [16], W_o and W are the weight loss values in the absence and presence of inhibitor respectively [12].

2.4.2 Electrochemical polarization measurement

Electrochemical study was carried out at room temperature using AUTOLAB potentiostat instrument utilizing a conventional three-electrode cell assembly. The metal samples, each of 20 mm^2, were embedded in resins. The embedded sample, platinum electrode and saturated silver/silver chloride were used as the working electrode, counter electrode and reference electrode, respectively. The surface of the working electrode was made smooth by polishing with emery paper of different grits size. The electrolyte used was 1.0 M HCl solution with varying concentrations of inhibitor (AgNPs). The open circuit corrosion potential was carried

IOP Conf. Series: Materials Science and Engineering **805** (2020) 012018 doi:10.1088/1757-899X/805/1/012018

out for 30 min until a stable potential was attained. Polarization experiment was carried out by changing the potential of the working electrode and monitoring the current produced as a function of time. Tafel polarization was studied between cathodic potential of -250 mV and an anodic potential of +250 mV with a scan rate of 1.0 mVs^{-1} to quantify the influence of inhibitor solution on the corrosion behaviour. The corrosion current densities were estimated from the Tafel curves by extrapolating the linear segments of the anodic and cathodic components to corrosion potential axis. The corrosion rate and inhibition efficiency (IE%) are then obtained from the corrosion current densities. The surface coverage (θ) and inhibitor efficiency (*IE%*) were determined using Eq. (3-4).

$$\text{Surface coverage } \theta = \frac{Io_{corr} - I_{corr}}{Io_{corr}} \quad (3)$$

$$\text{Inhibitor efficiency } IE\ (\%) = \frac{Io_{corr} - I_{corr}}{Io_{corr}} x\ 100\% \quad (4)$$

Where I_{corr} and Io_{corr} are corrosion current density in the presence and absence of the inhibitor, respectively.

2.4.3 Gasometric/ Hydrogen gas evolution measurement approach

The procedure for gasometric measurement of hydrogen gas evolution followed Odusote *et al* [26]. A delivery tube was used to connect the reaction vessel to the burette. Then, 40 ml of HCl and silver nanoparticles solutions of different concentration (5, 10, 15 and 20 μg/ml) were separately introduced into the mylius cell, and the initial volume of air in the burette was recorded. The prepared metal specimens were dropped into the solution and the mylius cell was quickly closed to avoid entry of air. The volume of hydrogen gas evolved due to the corrosion reaction was estimated by the volume of water changed in the burette. The change in volume was recorded every minute in the HCl solution without inhibitor. These procedures were repeated with the HCl solutions diluted with inhibitors [15]. From the volume of hydrogen gas evolved, inhibition efficiency (%IE) was calculated from Eq. (5)[13, 27].

$$\text{Inhibition efficiency (IE\%)} = \left(\frac{V_{H0} - V_{H1}}{V_{H0}}\right) \times 100\% \quad 5$$

where V_{H0} and V_{H1} are volume of hydrogen gas evolved in the absence and presence of the inhibitors respectively.

IOP Conf. Series: Materials Science and Engineering **805** (2020) 012018 doi:10.1088/1757-899X/805/1/012018

3. Results and Discussion

3.1 Elemental compositions of the metal alloys

The elemental compositions of the metal alloy samples are presented in Table 1. The percentage of carbon in the steel sample (Table 1a) indicates that it is a low carbon steel i.e. mild steel, while the stainless steel belongs to the standard ferritic grades containing 14.11% chromium with 0.51% Ni (Table 1b) [2]. The aluminum sample belongs to 1xxx group with some additional minor elements (Table 1c) [28].

Table 1a. Elemental composition of mild steel sample in wt%

Elements	Fe	C	Mn	Ta	P	S	Cr	Cu
Composition	99.00	0.036	0.76	0.07	0.011	0.009	0.03	0.012

Table 1b. Elemental composition of stainless steel sample in wt%

Elements	Fe	Cr	V	Mn	Ni	Zn	Cr	Cu	Mo
Composition	75.10	14.11	0.093	9.51	0.51	0.02	0.03	0.69	0.005

Table 1c. Elemental composition of aluminum sample in wt%

Elements	Al	Si	V	Mn	Fe	Zn	Cr	Cu	Sn
Composition	99.22	0.08	0.024	0.013	0.054	0.02	0.03	0.057	0.01

3.2 Weight loss against time of exposure

The weight losses as a function of time for all the samples are presented in Figure 2. It was observed that the weight loss after the first 400 h of exposure was higher than the subsequent 400 hrs, which may be due to direct contact between the metal surface and environment. However, the rate of materials dissolution reduced after 400 h of exposure possibly due to formation of passive layers on the metal surface, thereby acting as a barrier between the environment and the surface. For all the samples, the weight loss reduced with increased concentration of the inhibitor/AgNPs solution, implying that the presence of AgNPs in the solution enhanced the formation of passive layers on the metal surface and thus improve the inhibition efficiency. This inhibiting property may be due to the presence of some active constituents such as 9.5wt% of crude protein and the richness in alkaloids, caffeine (2.8%), theobromine (0.05%), nicotine and kolatine [29]. Mild steel, stainless-steel and aluminum lost 2.24 g, 1.71 g and 1.45 g, respectively after 2000 h of exposure when no inhibitor was added to the HCl solution. The dissolved mass was reduced to 1.34 g, 1.24 g and 0.77 g, respectively when 5 µg/ml of AgNPs solution was added. At the highest concentration of AgNPs (20 µg/ml), the total dissolved masses were 1.09 g, 0.51 g and 0.58 g, respectively. This indicates that the weight loss strongly dependents on the concentration of the inhibitor and exposure time.

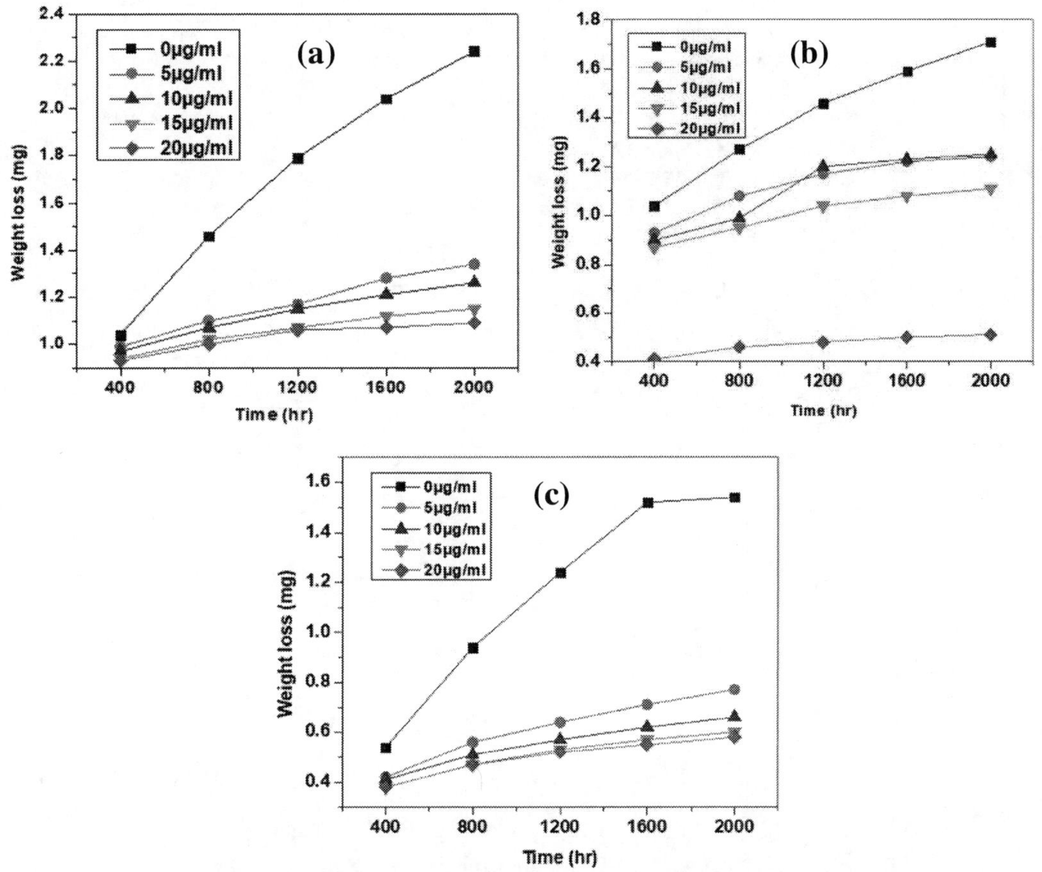

Figure 2. Cumulative weight loss for (a) mild steel (b) stainless steel and (c) aluminum samples against time of exposure in 1.0 M HCl medium with and without AgNPs.

The stability of the inhibition behaviour of the extract was evaluated by the trend of the inhibition efficiency (IE) as a function of time (Figure 3). Inhibition efficiency increased with increased exposure time and concentration of AgNPs solution. The increase in IE may be due to the formation of a protective film attributed to the transition of the metal/solution interface from an active dissolution state to a passive state [26]. This implies that the concentration of the nanoparticle solution reduces dissolution of metallic ions but not sufficient enough to fully inhibit the reactive surface of the different metal alloys from the aggressive environment of the acidic medium. They could serve as a viable inhibitor for metal alloys in HCl solution. However, once adherent film (barrier) was formed on the surface, the effect of the exposure time became less significant because of the difficulty experienced by the corrosion agents in passing through the formed barrier [11].

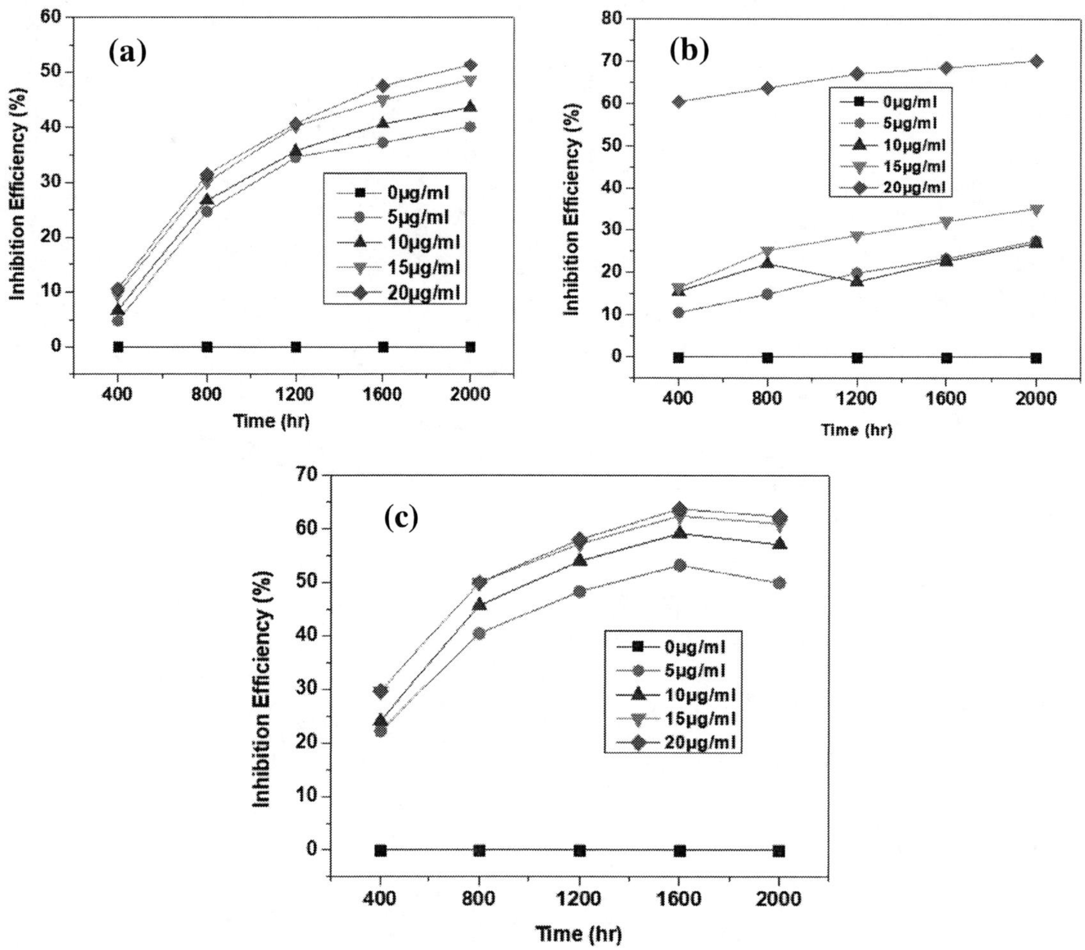

Figure 3. Inhibition efficiency for (a) mild steel (b) stainless steel and (c) aluminum samples against time of exposure in acidic medium with and without AgNPs

The corrosion rates of the alloys exposed to HCl solution are plotted in Figure 4. For all the alloys, the corrosion rates decreased with higher concentration of AgNPs solution. However, inherent corrosion resistance of stainless steel and aluminum also contribute to the decrease in their corrosion rates attributed to the quick formation of passive layer enhanced by the large amount of Cr and Al in the stainless steel and aluminum, respectively. It also connotes adsorption of AgNPs on the surfaces of metal samples which thereby reduce the active surface area [14].

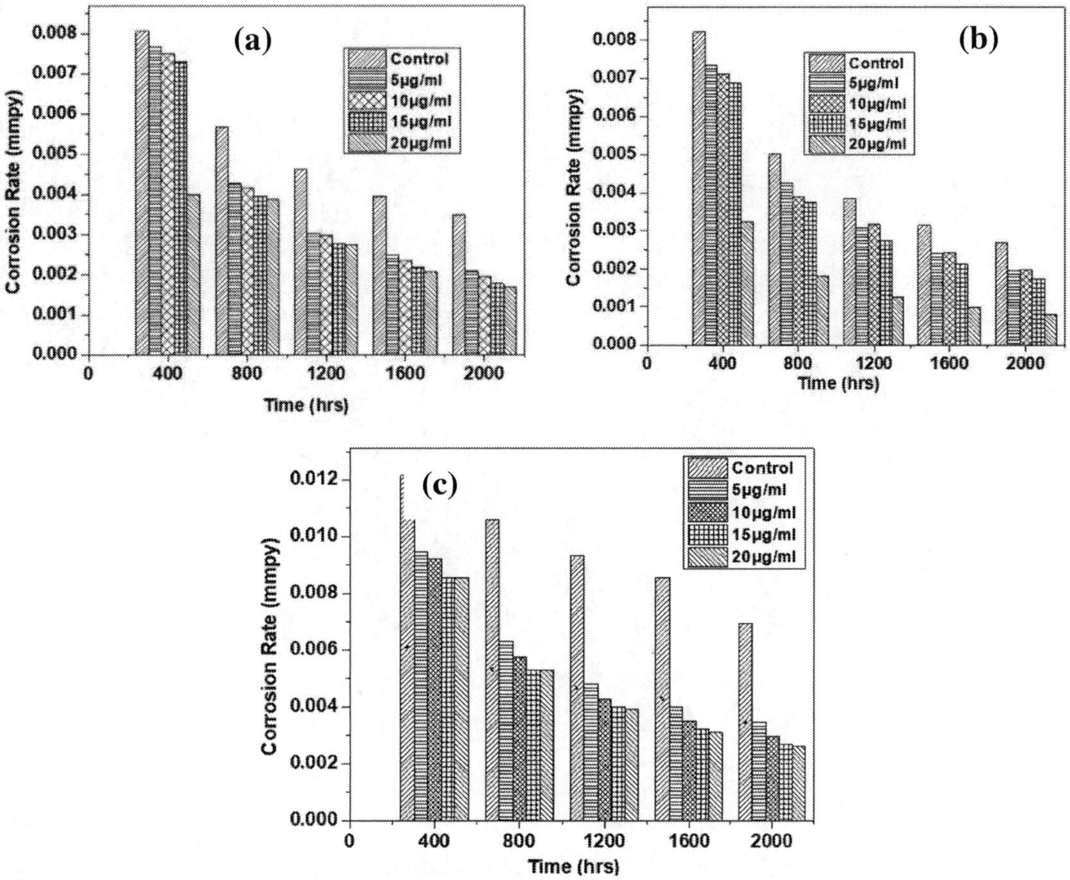

Figure 4. Corrosion rate of (a) mild steel (b) stainless steel and (c) aluminum samples against time of exposure in acidic medium with and without AgNPs.

Figure 5 shows the surface coverage of the samples after immersion in the corrosive medium. The surface area coverage is a measure of the fraction of the surface covered by the adsorbed molecules of AgNPs solution. The adsorption process creates a thin layer separating the metal surface from the corrosive medium thereby acting as a barrier between them. This implies higher inhibition with increased fraction of metal surface area covered by the adsorbed molecules. According to Figure 5, the surface area coverage increases with concentration of AgNPs and time of exposure.

IOP Conf. Series: Materials Science and Engineering **805** (2020) 012018 doi:10.1088/1757-899X/805/1/012018

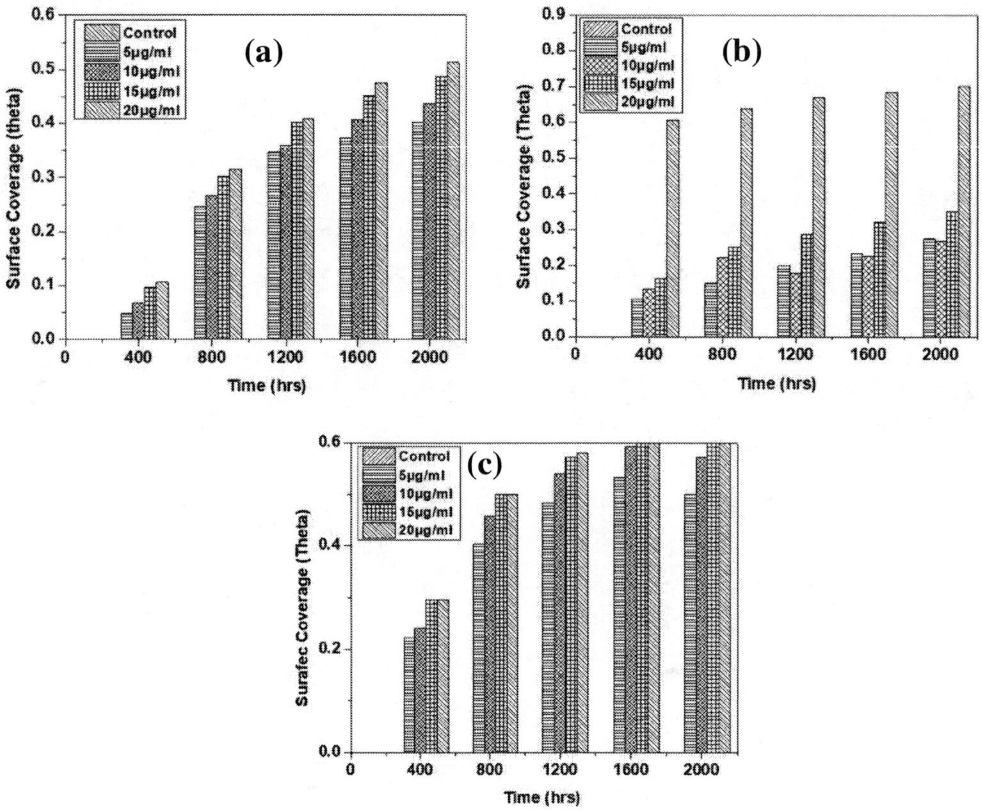

Figure 5. Surface Coverage of (a) mild steel (b) stainless steel and (c) aluminum samples against time of exposure in acidic medium with and without AgNPs

3.3 Potentiodynamic Polarization Studies

The potentiodynamic polarization curves for corrosion behavior of the mild steel, stainless steel and aluminum in 1.0 M HCl are shown in Figure 6. The values of Tafel polarization parameters including the anodic and cathodic Tafel slopes (β_a and β_c respectively) and corrosion current density (i_{corr}) were obtained by extrapolating the linear segments of the curves to the corrosion potential (E_{corr}). The inhibition efficiency along with other parameters are given in Table 2. It is observed that the inhibitors exerted significant effect on both anodic and cathodic reactions indicating that addition of AgNPs retarded both anodic metallic dissolution and cathodic hydrogen gas evolution. This subsequently reduces the value of corrosion current density (i_{corr}).

The decrease in the values of i_{corr} in the presence of inhibitors particularly at higher concentrations is largely due to the fact that inhibitor molecules are adsorbed on the surface of the metals. The direction of displacement in the value of E_{corr} determines the category of the inhibitor (anodic, cathodic or mixed type). If the displacement for the inhibited specimen is more than 85 mV as compared to the E_{corr} value of uninhibited specimen then the inhibitor can be categorized as anodic or cathodic type and mixed type if otherwise. In this study, the values of E_{corr} is generally lower than 85mV for mild steel and aluminum while higher than 85 mV for stainless steel (Table 2). This indicates that the inhibitor acts as mixed type for the former metal samples and anodic/cathodic type for the later. The results depicted in Table 2 showed that the values of β_c are more affected as compared to the values of β_a indicating that the inhibitor acts as cathodic type inhibitors, that is, it has pronounced influence on the cathodic corrosion process for the stainless steel [5].

The values of inhibition efficiency (IE) obtained from the Tafel polarization experiments (Table 2) were higher than those obtained from both weight loss and gasometric experiments (Figures 2 and 7). This behaviour suggests that the IE depends on the method of measurement. Nevertheless, the increase in IE with increased concentration of AgNPs indicates a significant reduction in corrosion rates.

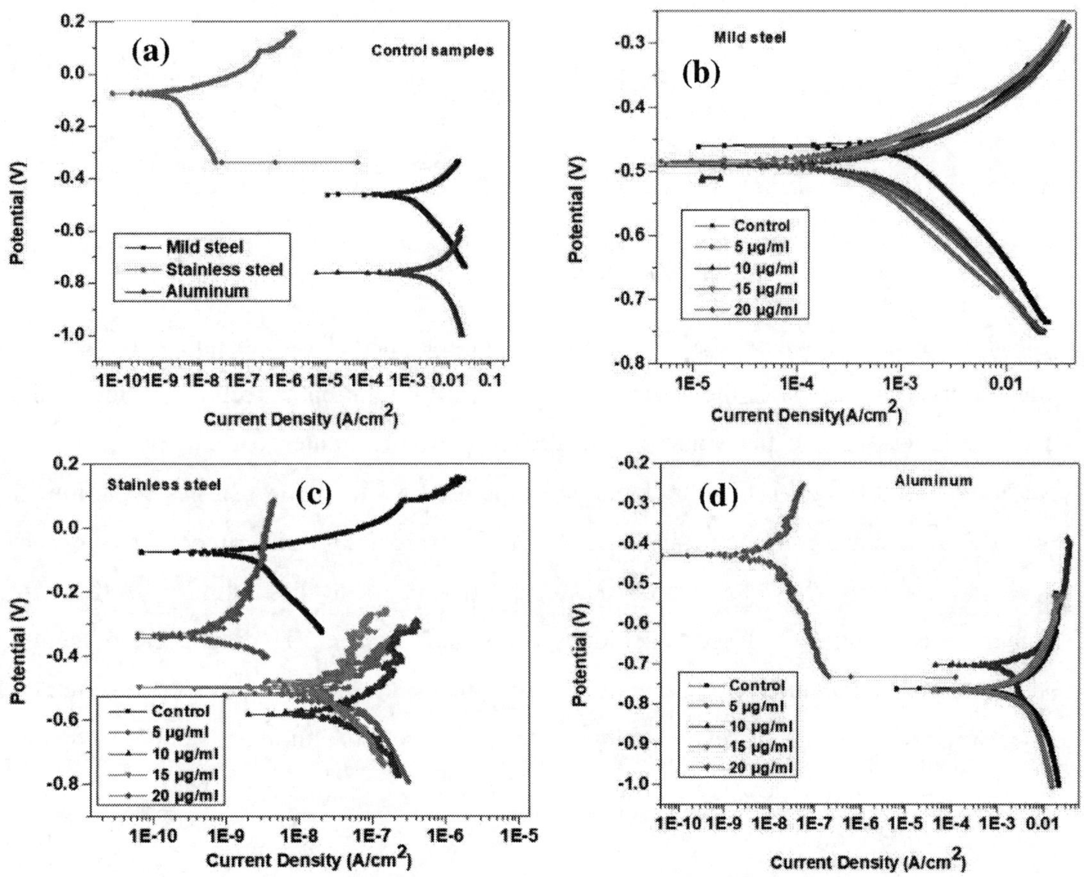

Figure 6. Tafel Polarization curve for (a) control samples (b) mild steel (c) stainless steel and (d) aluminum for immersed in acidic medium with and without AgNPs

IOP Conf. Series: Materials Science and Engineering **805** (2020) 012018 doi:10.1088/1757-899X/805/1/012018

Table 2. Corrosion parameter for the samples in presence and absence of nanoparticles

Metals	Concentration (µg/ml)	E_{corr} (mv)	I_{corr} (mA)	B_a (mV)	β_c (mV)	CR (mmpy)	IE (%)
	Control	-459.51	-2.14	88.88	-2.14	24.89	0
	5	-484.04	-306.35	73.25	131.25	13.55	99.29
Mild Steel	10	-490.65	-819.37	91.57	164.66	9.507	99.73
	15	-487.46	-418.7	88.92	132.38	4.85	99.48
	20	-491.50	-531.92	83.65	132.71	1.98	99.59
	Control	-72.84	-2.43	48.21	316.19	0.000283	0
	5	-548.94	-21.73	221.18	259.68	0.000252	88.78
	10	-581.17	-132.73	505.33	394.93	0.000154	98.16
Stainless	**15**	-497.01	-62.33	131.08	24.24	0.00072	96.08
Steel	**20**	-323.23	-260.22	91.59	93.58	0.0003	99.06
	Control	-760.62	-2.92	124.33	166.73	31.858	0
	5	-766.32	-4.99	192.98	332.647	24.369	41.42
Aluminum	10	-427.72	-31.73	525.35	427.05	20.876	90.78
	15	-763.24	-5.14	245.85	272.32	15.95	93.07
	20	-702.98	-2.79	63.32	806.35	0.345	96.10

3.4 Hydrogen gas evolution measurement

For all samples, the volumes of evolved hydrogen gas increased with time (Figure 7). This was observed to vary with the concentration of AgNPs solution as well as the metal alloys. The continuous rise in the volume of hydrogen gas is attributed to the time required for surface passivation given that the maximum time used for the hydrogen gas evolution study was still below the minimum 400 h used for the weight loss experiment. It is expected, however, that as surface passivation began, the rate of metal dissolution would retard. At higher concentrations, hydrogen evolution was slightly delayed. Hydrogen gas evolution in stainless steel and aluminum commenced a few minutes after the samples were inserted in the corrosion environment. However, evolution was more rapid in aluminum sample compared to stainless steel sample. Generally, the rate of evolution of hydrogen gas decreased with increased concentration as previously observed by Odusote and Ajayi [11]. The highest volume of hydrogen gas evolved in the blank solution suggested that there is no adsorption layer to inhibit both the cathodic and anodic reactions.

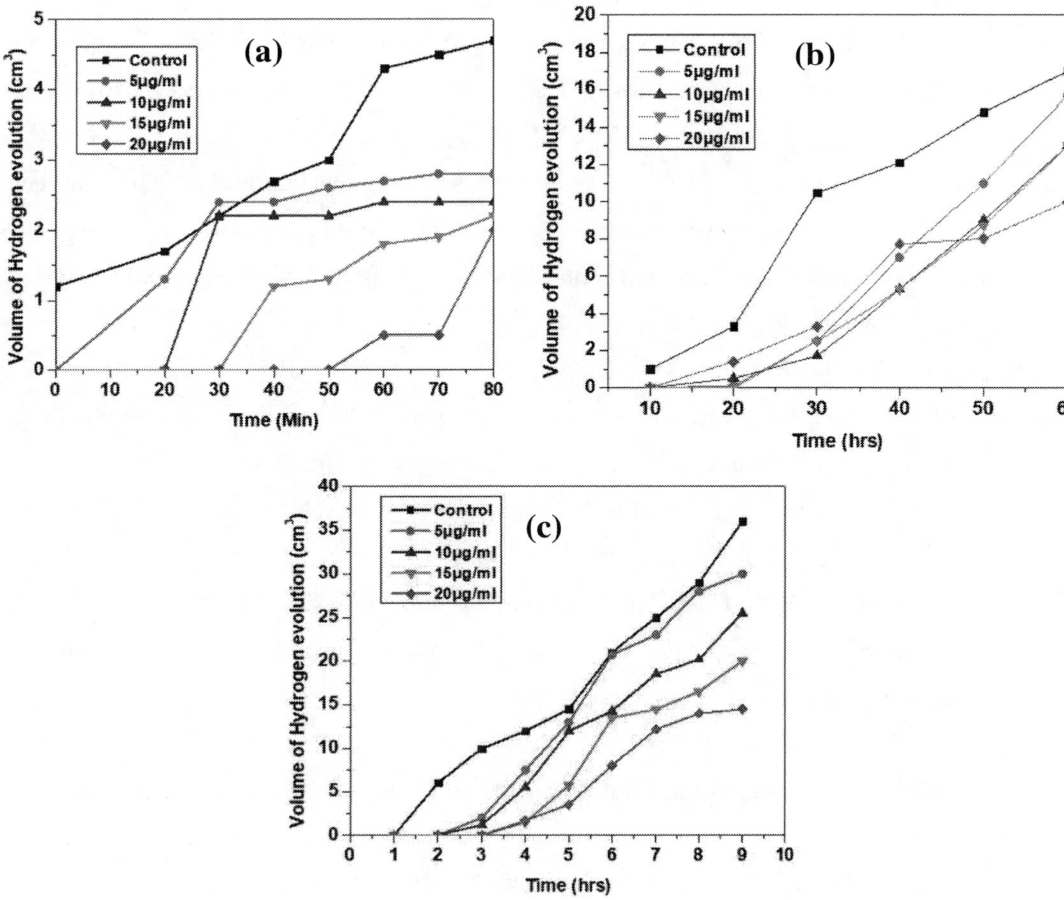

Figure 7. Inhibition efficiency for hydrogen gas evolution study of (a) mild steel (b) stainless steel and (c) aluminum samples against time of exposure in acidic medium with and without AgNPs.

4 Conclusion

Silver nanoparticles solutions, produced via green chemistry approach, have been shown to enhance inhibition of mild steel, stainless steel and aluminum in 1.0 M HCl solution. The results obtained from gravimetric, potentiodynamic polarization and gasometric measurements showed that AgNPs solution possesses significant corrosion inhibition properties for mild steel, stainless steel and aluminum in 1.0 M HCl by influencing the mechanism of anodic dissolution and cathodic hydrogen gas evolution via adsorption which was considered to be spontaneous, exothermic and physisorption. These results indicated that AgNPs can be incorporated into existing inhibitors towards minimizing corrosion rate.

References

[1] Singh A, Ebenso E E, Quraishi M A, Mobin M and Rizvi M 2012 Corrosion inhibition of carbon steel in HCl solution by some plant extracts. *Int. J. Corros.* Article ID 897430. http://dx.doi.org/10.1155/2012/897430.

[2] Wernick S and Pinner R 1972 The Surface treatment and finishing of aluminum and its alloys. Vol. 1-2. Robert Draper Ltd., Teddington, Middx., 1274 p.

[3] Hays GF 2010 Now is the Time. https://www.scientific.net/AMR.95.-2.pdf (Accessed on 17 January, 2020).

[4] lkhathlan HZ, Khan M, Abdullah MM, Al-Mayouf AM, Mousa AA and Al-Othman ZA 2014 *Launaea nudicaulis* as a source of new and efficient green corrosion inhibitor for mild steel in acidic medium: a comparative study of two solvent extracts. *Int. J. Electrochem. Sci.* **9** pp 870-879.

[5] Verma C, Quraishi M A, Kluza K, Makowska-Janusik M, Olasunkanmi L O and Ebenso E E 2017 Corrosion inhibition of mild steel in 1M HCl by D-glucose derivatives of dihydropyrido [2,3-d:6,5-d′] dipyrimidine-2, 4, 6, 8(1H,3H, 5H,7H)-tetraone. *Sci. Rep.*7 Article Number 44432.

[6] Kharissova O V, Rasika Dias H V, Kharisov B I, Olvera Perez B and Jimenez Perez V M 2013 The greener synthesis of nanoparticles *Trends Biotechnol.* **31** (4) pp 240–248.

[7] Benítez LP, Castellar PJ, Percy ED and Bravo MJ 2014 Uso de extractos de plantas como inhibidores de corrosión. *Informador Técnico*. **78** (2) pp155-164.

[8] Sathiyanarayanan S, Marikkannu C and Palaniswamy N 2005 Corrosion inhibition effect of tetramines for mild steel in 1M HCl. *Appl. Surf. Sci.* **241** (3-4) pp 477-484.

[9] Taheri P, Mansouri A, Bachour B, Link A, Ahuja N and Exova M Z 2017 Inspection and mitigation of underground corrosion at anchor shafts of telecommunication. *NACE International Corrosion Conference & Expo,* pp 72-74

[10] Said R M, Salisu A G and Kntapo F M 2017 Corrosion inhibition of mild steel using alkaloids and tanning extracts of *Jatropha curcas* in acidic media. *Bayero J. Pure Appl. Sci.* **10** (1) pp 311–317.

[11] Odusote J K and Ajayi O M 2013 Corrosion inhibition of mild steel in acidic medium by *Jatropha curcas* exract. *J. Electrochem. Sci. Technol.* **4** (2) pp 81–87.

[12] Yang F, Li X, Dai Z, Liu T, Zheng W, Zhao H and Wang L 2017 Corrosion inhibition of polydopamine nanoparticles on mild steel in hydrochloric acid solution. *Int. J. Electrochem. Sci.* **12** (8) pp 7469-7480.

[13] Garcia Inzunza R, Valdez Salas B, Kharshan R, Furman A and Schorr Wienner M

2012 Interesting behavior of *Pachycormus discolor* leaves Ethanol extract as a corrosion inhibitor of carbon steel in 1M HCl: A preliminary study. *Int. J. Corros.* Article ID 980654. https://doi.org/10.1155/2012/980654.

[14] Behzadnasab M, Mirabedini S M, Kabiri K and Jamali S 2011 Corrosion performance of epoxy coatings containing silane treated ZrO_2 nanoparticles on mild steel in 3.5 % NaCl solution. *Corros. Sci.* **53** (1) pp 89-98.

[15] Adeyemi O O, Olubomehin O O and Bello R O 2017 Investigation of *Ageratum conyzoides* aqueous extract as inhibitor for hydrochloric acid corrosion of SX 316 steel. *Acad. J. Sci. Eng.* **8** (1) pp 20-32.

[16] Shuaib-Babata YL, Ibrahim KH, Ambali OI, Yahya AR, Ajao SK, Aremu IN, Pelumi AA 2019 Inhibitive potential of *Prosopis africana* on corrosion of low carbon steel in 1M hydrochloric acid medium. *Int. J. Eng. Mater. Manuf.* **4** (2) pp 66-76.

[17] AbdelHameed R S, Abu-Nawwas A H and Shehata H A 2013 Nano-composite as corrosion inhibitors for steel alloys in different corrosive media. *Adv. Appl. Sci. Res.* **4** pp 126-129.

[18] Lateef A, Azeez MA, Asafa TB, Yekeen TA, Akinboro A, Oladipo IC, Azeez L, Ojo SA, Gueguim-Kana EB and Beukes LS 2016 Cocoa pod husk extract-mediated biosynthesis of silver nanoparticles: its antimicrobial, antioxidant and larvicidal activities. *J. Nanostruct. Chem.* **6** (2) pp 159-169.

[19] Lateef A, Adelere IA, Gueguim-Kana EB, Asafa TB and Beukes LS 2015 Green synthesis of silver nanoparticles using keratinase obtained from a strain of *Bacillus safensis* LAU 13. *Int. Nano Lett.* **5** (1) pp 29-35.

[20] Lateef A, Ojo SA, Azeez MA, Asafa TB, Yekeen TA, Akinboro A, Oladipo IC, Gueguim-Kana EB and Beukes LS 2016 Cobweb as novel biomaterial for the green and eco-friendly synthesis of silver nanoparticles. *Appl. Nanosci.* **6** (6) pp 863-874.

[21] Lateef A, Azeez MA, Asafa TB, Yekeen TA, Akinboro A, Oladipo IC, Ajetomobi FE, Gueguim-Kana EB and Beukes LS 2015 *Cola nitida*-mediated biogenic synthesis of silver nanoparticles using seed and seed shell extracts and evaluation of antibacterial activities. *BioNanoSci.* **5** (4) pp 196-205.

[22] Oladipo I C, Lateef A, Elegbede J A, Azeez M A, Asafa T B, Yekeen T A, Akinboro A, Gueguim-Kana E B, Beukes L S, Oluyide T O and Atanda O R 2017 *Enterococcus* species for the one-pot biofabrication of gold nanoparticles: Characterization and nanobiotechnological applications. *J. Photochem. Photobiol. B: Biol.* **173** pp 250–257.

[23] Ojo S A, Lateef A, Azeez M A, Oladejo S M, Akinwale A S, Asafa T B, Yekeen T A,

Akinboro A, Oladipo I C, Gueguim-Kana E B and Beukes L S 2016 Biomedical and catalytic applications of gold and silver-gold alloy nanoparticles biosynthesized using cell-free extract of *Bacillus safensis* LAU 13: Antifungal, dye degradation, anti-coagulant and thrombolytic activities. *IEEE Trans. Nanobioscience* **15** (5) pp 432-442.

[24] Azeez MA, Lateef A, Asafa TB, Yekeen TA, Akinboro A, Oladipo IC, Gueguim-Kana EB and Beukes LS 2017 Biomedical applications of cocoa bean extract-mediated silver nanoparticles as antimicrobial, larvicidal and anticoagulant agents. *J. Clust. Sci.* **28** (1) pp 149-164.

[25] Lateef A, Azeez M A, Asafa T B, Yekeen T A, Akinboro A, Oladipo I C, Azeez L, Ajibade S E, Ojo S A, Gueguim-kana E B and Beukes L S 2016 Biogenic synthesis of silver nanoparticles using a pod extract of *Cola nitida*: Antibacterial and antioxidant activities and application as a paint additive. *J. Taibah Univ. Sci.* **10** (4) pp 551–562.

[26] Odusote J K, Owalude D O, Olusegun S J and Yahya R A 2006 Inhibition efficiency of *Moringa oleifera* leaf extract on the corrosion of reinforced steel bar in HCl solution. *West Indian J. Eng.* **38** (2) pp 64-70.

[27] Ameh PO, Magaji L, Salihu T 2012 Corrosion inhibition and adsorption behaviour for mild steel by *Ficus glumosa* gum in H_2SO_4 solution. *Afr. J. Pure Appl. Chem.* **6** (7) pp 100-106.

[28] The Aluminum Association 2018 *International Alloy Designations and Chemical Composition Limits for Wrought Aluminum and Wrought Aluminum Alloys.* Crystal Drive, Suite 430, Arlington, VA 22209, 45p. https://www.aluminum.org/sites/default/files/Teal%20Sheet.pdf (Accessed on 17 January, 2020).

[29] Orwa C, Mutua A, Kindt R, Jamnadass R and Anthony S 2009 Agroforestree Database: a tree reference and selection guide version 4.0. World Agroforestry Centre, Kenya.

Recent developments on the photoanodes employed in dye-sensitized solar cell

H A Shittu[1], I T Bello[1,2], M A Kareem[1], M K Awodele, Y K Sanusi[1,3] and O Adedokun[*1,3]

[1]Department of Pure and Applied Physics, Ladoke Akintola University of Technology, Ogbomoso, Nigeria.

[2]Department of Physics, CSET, University of South Africa, Johannesburg 1710, South Africa.

[3]Nanotechnology Research Group, Ladoke Akintola University of Technology, Ogbomoso, Nigeria.

E-mail: oadedokun@lautech.edu.ng

Abstract

The emergence of dye sensitized solar cell (DSSC) as an alternative device for silicon based solar cell has gained a lot of attention from researchers due to its cost-effective, easy fabrication and environmentally friendliness. Photoanodes are semiconductor and as one of the four components of DSSC plays a major role for dye loading and electron conduction. A good photo anode should provide an efficient surface area in dye loading, nanostructure for high light harvesting opportunity, fast electron transport ability and good band gap architecture. Several nanostructures materials have been studied and employed as photoanode in DSSC. They include TiO_2, ZnO, Nb_2O_5, SnO_2, among others. The problem associated with photoanodes used in fabricating DSSC is high recombination rate of electrons that emanate from the number of grains. The dispersed nature of progress reports on developments of photoanodes calls for summary. Hence this review gives a general summary of the progress made in various materials used as photoanode in DSSC and the methods adopted in synthesizing them. In this present review, our attention is not only on synthesis and characterization of the materials alone but also on the effect of different factors influencing photovoltaic characteristics of photoanode for DSSC application.

Keywords: Nanostructure, photoanode, conversion efficiency, DSSC, photovoltaic

IOP Conf. Series: Materials Science and Engineering **805** (2020) 012019 doi:10.1088/1757-899X/805/1/012019

1. Introduction

The continuous depletion of fossil fuel reserves coupled with the associated environmental hazards emanated from combustion calls for more concerted effort for the development of clean, sustainable and renewable energy. Solar energy, with the sun producing over 120,000 TW of energy each year, seems to be the most apparent response [1]. There is no other source of energy as enormous as solar energy radiation. However, it can be considered limitless on the degree of humankind and harnessed at low cost. Profuse energy from the sun is received on earth which is projected to be 3×10^{24} J/yr, which is about 10^4 times more than what the earth uses currently. In other words, it covers only 0.1% of the earth surface with the conversion efficiency of around 10% which would be enough to suit our current needs [2]. Photon energy from the sun can be converted to direct electricity using solid state devices (photovoltaic device) i.e. solar panels, which have undergone three generations with an evolution from the first generation which are silicon based solar cell to the second generation that are based on thin films [3-5] and now to the third generation which are dye-sensitized solar cell (DSSC), Perovskite Solar cell, Quantum-dot Sensitized solar cell (QDSC) and Nanocrystal based solar cell [6-9]. The third generation solar cell shows a relatively low conversion efficiency when they are compared with the silicon based solar cell and thin film based solar cell that have an approximate conversion efficiency of 20-30 % [10, 11].

Ever since the O'regan and Gratzel discovery in 1991, dye sensitized solar cell (DSSC) have captured public view and acquired more research consideration over the following 18 years as shown in Figure 1 [12]. Dye-sensitized solar cell (DSSC) have some advantages over the first two generations which is their simple manufacturing processes and low-cost of production coupled with their other advantages e.g. flexibility, low toxic and good performance in varied light condition [14]. Figure 2 schematically showed components of DSSC, a typical DSSC consists of photoanode made from semiconductor (e.g. TiO_2, ZnO, SnO_2 among others) film deposited on a conductive substrate, a sensitizer (i.e. dye; e.g. Ruthenium dye, organic dyes) by monolayer adsorption or quantum dots (e.g. CdS, CdSe or PbS), an electrolyte (e.g. I^-/I_3^- and Co^{2+}/Co^{3+} redox couples) that is deposited between the counter electrode and sensitizer, and the last component of a DSSC is a counter electrode deposited on another conductive substrate [15].

IOP Conf. Series: Materials Science and Engineering **805** (2020) 012019 doi:10.1088/1757-899X/805/1/012019

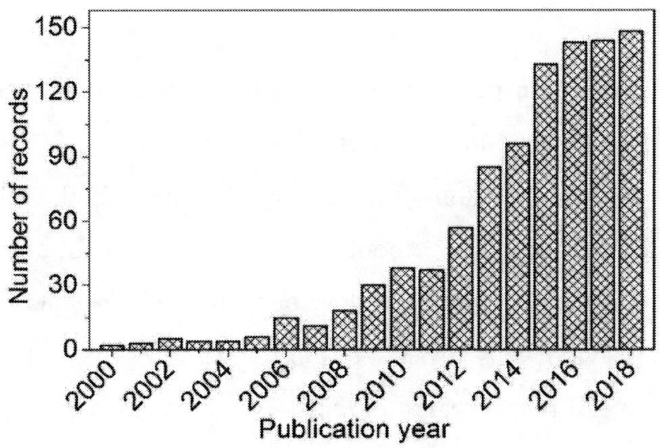

Figure 1. Evolution of the number of publications for Dye-sensitized solar cell [13]

Supply of light by visible light irradiation on the photoanode triggers photo-excitation of the absorbed dye molecules to produce excited electrons that are consequently injected into the wide conduction band of the semiconductor and quickly transferred to the peripheral circuit through the conductive substrate; an electric current is produced in these processes. The initial nature of the dye is consequently returned by electron contribution from the redox electrolyte. Charges are returned by the counter electrode from the external circuit back to the cycling circuit in the cell [15].

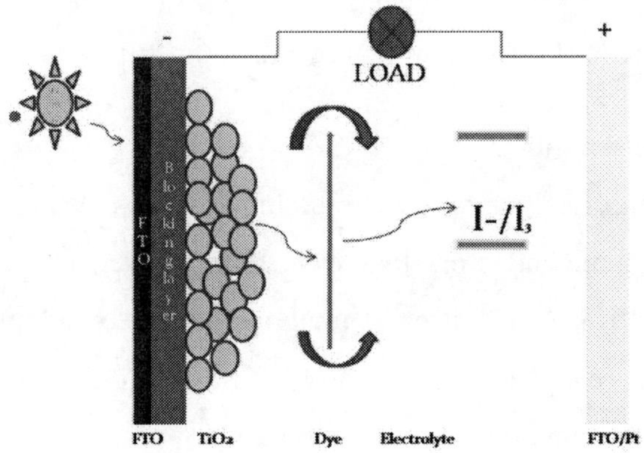

Figure 2. Schematic Diagram of the Working Principle of DSSC

IOP Conf. Series: Materials Science and Engineering **805** (2020) 012019 doi:10.1088/1757-899X/805/1/012019

Production of nanostructured semiconductor photoanodes with effective architectures for high dye loading and fast electron transport have gained a lot of interest by researchers; the utilization of adaptable sensitizers with noticeable light harvesting ability, the use of strong redox electrolytes with favorable compositions for proficient transport of the hole, the optimization of the counter electrode as well as the development of other equivalent alternatives at lower costs [16-19]. Hence, this review highlights latest developments with insights into the photoanodes of DSSC. A general overview of this aspect in DSSC will be given. Detailed information and comprehensive discussion of different issues for DSSC can be found in some special articles [20-24].

2. Recent Developments on Photoanode

A good DSSC photoanode should offer large surface for dye absorption and also transfer the photo-generated electrons from sensitizer to peripheral circuit efficiently. It should have a band gap that matches well with the band structure of the sensitizer for effective injection of photoelectron. Typically, this involves the conduction band of semiconductor 0.2-0.3 eV lower than that of the sensitizer. Owing to these conditions, nanostructured semiconductor materials such as TiO_2, ZnO, graphene, NbO_2 and SnO_2 [25-30] are often employed as photoanode in DSSC.

An efficient surface area for dye loading can be obtained from a nanoscale structure of the semiconductor. Good interconnected semiconductor nanostructure semiconductors give room for effective charge transfer within the hole-transfer and anode to the redox mediator because of the penetration of the electrolyte back of the semiconductor film [26].

3. Materials Used as Photoanode

Nanoparticles, nanorod, nanofiber, nanotube, hierarchical nanostructure, and nanosphere are the different nanostructure of semiconductors that have been studied to improve the photovoltaic performance of the DSSC [31]. Also, different methods have been adopted to prepare effective nanostructured photoanodes (e.g. TiO_2, ZnO, Nb_2O_5 and SnO_2) including nanoparticle, nanorod, nanofiber, nanotube, hierarchical nanostructure, and nanosphere such as hydrothermal/solvothermal processes, sol-gel, electrospining, spray pyrolysis, electrochemical anodization [32-38] and atomic deposition [39].

3.1 Titanium (IV) Oxide (TiO_2)

Anatase, brooktie and rutile are the three forms of TiO_2 with energy band gap 3.23eV, 3.26eV and 3.05eV respectively. Higher electron dispersion coefficient and large surface area for dye-loading are provided by the anatase form of TiO_2 than the rutile form. However, the brooktie form is not considered for DSSC due to difficulty in its production, making the anatase form the most considered DSSC photoanode [40, 41]. Since the inception of researches on DSSC photoanode, TiO_2 nanoparticles gained more attention due to the large surface area provided and dye molecules absorption used for charge production. However, enormous number of grain boundaries and grain size between nanoparticles often cause a high charge recombination rate and inefficient energy conversion behavior. So far, 11.18% conversion efficiency has been obtained using TiO_2 nanoparticles as photoanode [42]. Figure 3 shows the crystal structure of the three forms of TiO_2.

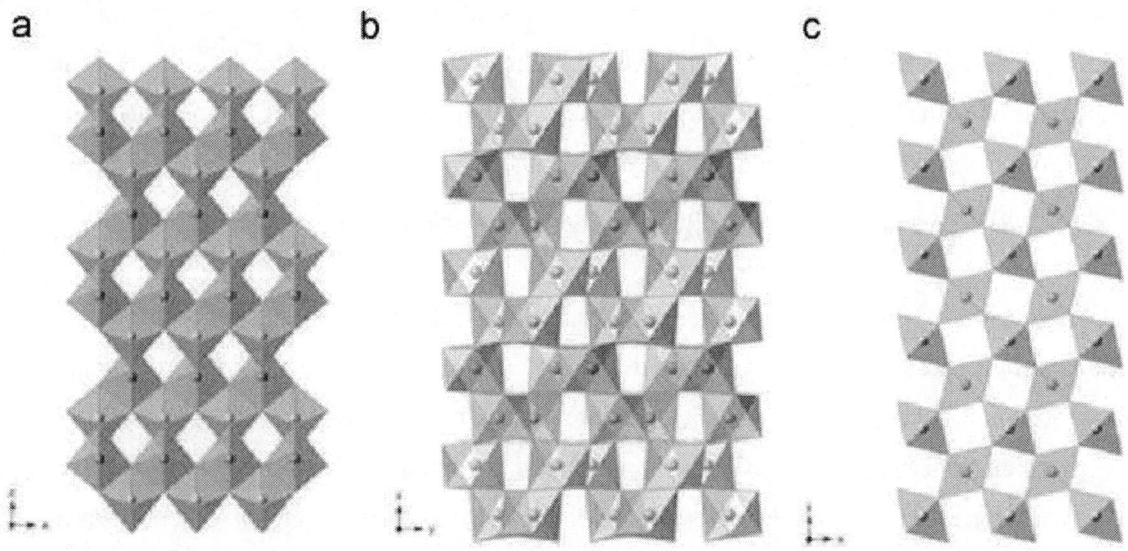

(a) Anatase (b) brooktie (c) Rutile

Figure 3. Crystal structure of the three forms of TiO_2 [43]

Recently, TiO_2 nanoflower (NF) was used in fabrication DSSC by doctor blade method. TiO_2 nanoflower was deposited over the edge of fluorine-doped Tin oxide (FTO) glass substrate with a sensitizer of N719 dye at standard temperature. A thin layer of platinum was used as the counter electrode and I_2 as the electrolyte. The fabricated DSSC have an energy conversion efficiency of approximately 3.64% [44]. Titanium corrosion was used to prepare TiO_2 nanowire (NW) for DSSC fabrication in 2017. Ti foil reaction in alkaline potassium hydroxide at different corrosion time of (6, 12, 24 and 48 h) respectively was used to prepare the TiO_2 nanowire. Back illumination was used to illuminate the fabricated DSSC; metallic substrate used account for the back illumination. The conversion efficiencies obtained at different corrosion time of 6, 12, 24, 48 h were 0.27, 0.42, 0.80, 1.03% respectively. The photovoltaic parameters connected to the efficiency at 1.03% respectively are short circuit current $J_{sc} = 2.08\ mA/cm^2$, open circuit voltage $V_{oc} = 0.69\ V$ and fill factor $FF = 71.5\ \%$ [45].

In 2016, Xia and co-workers fabricated hierarchical anatase form of TiO_2 using one step hydrothermal process. The fabricated film was studied under four different reaction times (3, 6, 9 and 12 h) to achieve different morphology and thickness photoanode. It was discovered that as the reaction time increases, the surface area also increases which gives way to high dye absorption. The fabricated DSSC achieve its highest efficiency at 9 h when the thickness was about 2 μm and the conversion efficiency obtained was 4.11% associated with $J_{sc} = 8.53\ mA/cm^2$, $V_{oc} = 7.4\ V$ and $FF = 66.37\ \%$. The highest efficiency obtained from TiO_2 at 9 h among the four tested samples may be due to its higher Voc and Jsc, which might be credited to large surface area for dye loading [46].

TiO_2 nanotube (NT) boast immense prospect to prevail over the deficiencies of TiO_2 nanoparticles, seeing as their exceptional structure promotes electron transfer and charge parting by forging straight path and as well accelerate charge transport between interfaces [47-49]. The conversion efficiency of DSSC can be improved by TiO_2 nanotube due to their orderly and vertically-oriented characteristic as well as their intrinsic advantage. In other to harness the benefits of TiO_2 nanotube arrays maximally, its structure is to be taken into consideration. TiO_2 nanotube produced from Ti foil of 96% purity via anodization with which three morphologies of nanotube are generated which are grassy tube, open tube with nucleation layer on top and highly ordered nanotube. However, the conversion efficiencies of DSSC are obtained from 4.32%, 4.8% and 5.2% respectively [50]. Slightly higher result can be obtained from TiO_2 NT/NP composite [50, 51]. Figure 4 shows SEM image of highly ordered nanotube array formed on Ti foil for DSSC usage [52]. Electrons in TiO_2 nanotubes have much longer lifetime than those in nanoparticles [53].

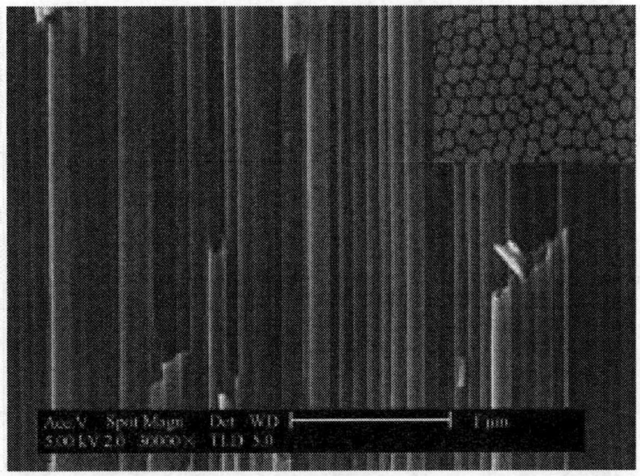

Figure 4. SEM image of highly ordered TiO_2 nanotubes [52]

3.2 Composites of TiO_2 Photoanode

The compositions of TiO_2 semiconductor with other semiconductor have a great impact on DSSC fabrication. Recently, coaxial electrospinning method was employed to prepare Zn doped/TiO_2 hollow fibers. The light absorption was affected by the thickness of the Zn doped/TiO_2, most especially the short circuit current. The conversion efficiency obtained from this composition was 3.12% associated with $J_{sc} = 15.81\ mA/cm^2$, $V_{oc} = 0.566V$ and $FF = 34.91\%$ [54]. DSSC fabricated with TiO_2 nanorod/nanoparticle (NR/NP) via hydrothermal process gives a conversion efficiency of 7.7%, $V_{oc} = 0.732V, J_{sc} = 17.06\ mA/cm^2$ and $FF = 61.7\%$ [55].

TiO_2 nanorod(NR)/nanoflower(NF) based DSSC with some ratio of ($HCl:H_2O$) have improvement in DSSC application. Hierarchical TiO_2 nanorod(NR)/nanoflower (NF) photoanode produced by hydrothermal method composed of ($H_2O:HCl$ = 5:5) shows an largely energy conversion efficiency of 0.85% associated with a $J_{sc} = 2.11\ mA/cm^2$, $V_{oc} = 0.63V$ and $FF = 64\%$, which is much than that of microsphere 0.71% and nanowires of 0.61% efficiency [56]. DSSC with TiO_2 Nanowire (NW)/Nanoparticle (NP) composite with I^-/I^-_3 as electrolyte had an efficiency of 5.5% [57].

ZnO tetrapods was used together with TiO_2 (ZnO/TiO_2), the ZnO/TiO_2 composite was prepared via screen printing method. The structure of ZnO tetrapod/TiO_2 composite obtained via screen

IOP Conf. Series: Materials Science and Engineering **805** (2020) 012019 doi:10.1088/1757-899X/805/1/012019

printing is homogeneously porous and can excite the surface area and improve absorption of dye. It was reported that DSSC based on TiO_2/ZnO/TiO_2, ZnO tetrapod/TiO_2 and ZnO nanorod/TiO_2 photoanode had efficiency of 1.87%, 0.34% and 0.24%, respectively. Figure 5 shows SEM images of different techniques used in obtaining ZnO tetrapod/TiO_2 composites [58]. The photovoltaic parameters of synthesized TiO_2 nanostructures are presented in Table 1.

Figure 5. SEM image of (a) ZnOtetrapods/TiO_2 photoanode fabricated via screen printing (b) ZnOtetrapods/TiO_2photoanode fabricated through colloidal spray coating and (c) ZnOtetrapods/TiO_2 photoanode fabricated via electrophoretic deposition [58].

Table 1. Photovoltaic parameters TiO_2 nanostructures with their synthesis methods

TiO_2 Photoanode	**Method**	**Jsc (mA/cm^2)**	**Voc (V)**	**FF (%)**	**η (%)**	**Ref.**
TiO_2 nanoflower	Doctor Blade	7.80	0.69	67	3.64	[44]
TiO_2 nanowire	Titanium corrosion	2.08	0.69	71	1.0	[45]
TiO_2 nanoparticle	Hydrothermal	8.53	0.74	66.37	4.11	[46]
TiO_2 nanotube	Anodization	13.48	0.67	48	4.32	[50]
TiO_2 nanotube	Anodization	13.45	0.70	55	5.2	[50]
TiO_2 nanotube	Anodization	12.68	0.70	54	4.8	[50]
Zn doped/TiO_2	Electrospinning	15.81	0.56	34.9	13.12	[54]
TiO_2 NR/NP	Hydrothermal	17.06	0.73	61.7	7.7	[55]
TiO_2 NR/NF	Hydrothermal	0.85	2.11	63	0.64	[56]

IOP Conf. Series: Materials Science and Engineering **805** (2020) 012019 doi:10.1088/1757-899X/805/1/012019

3.3 ZnO as Photoanode

The wide-band gap characteristic of ZnO with the energy-band structure and its physical characteristics make ZnO suitable for DSSC application. It also has a higher electronic mobility that makes it encouraging for transportation of electrons which is one of the major characteristics for a proficient photoanode [59]. The high electron mobility of ZnO can be linked to the characterization and syntheses of different nanostructures of ZnO such as nanotubes, nanobelt, nanorods, nanoparticles, nanoflower, nanosheets and nanotetrapods [60]. However, ZnO is unstable with acidic dye and higher charge recombination which account for lower conversion efficiency of ZnO based solar cell than that of TiO_2 [61]. Recently, ZnO nanorods (NR) was synthesized using chemical bath deposition (CBD) alongside electrodeposition; the synthesized ZnO has a large surface area that leads to its large short circuit current. The conversion efficiency of 0.8%, $J_{sc} = 3.47\ mA/cm^2$ and $FF = 51.7\%$ are obtained with high electron transfer and more injected electrons [62]. Hydrothermal process was used recently to grow nanoflowered shape ZnO on an FTO substrate which was employed to DSSC fabrication. The large grain size of the ZnO nanoflower accounts for its great Voc and FF and high recombination rate. The fabricated DSSC has $J_{sc} = 4.2\ mA/cm^2$, $V_{oc} = 0.62V$, $FF = 54\%$ and conversion efficiency of 1.4% [63].

Low temperature hydrothermal process was used to synthesize ZnO flower shaped nanostructure. The axial arrangement of nanoneedles which are accumulated formed the flower shape of the ZnO. The purity of the flowers was high and has good optical characteristics. The light harvesting efficiency result to the reasonable high dye absorption by the photoanode. The fabricated DSSC were reported to have an efficiency of ~1.1% coupled with $J_{sc} = 3.532\ mA/cm^2$, $V_{oc} = 0.611V$ and $FF = 51\%$ [64].

A nanoporous ZnO photo-anode formed from pre-dyed ZnO (Pd-ZnO), the pre-dyed ZnO is made by mixing dyeing procedure with ZnO nanoparticles. The pd-ZnO paste used in fabrication of DSSC which gives $J_{sc} = 1.97\ mA/cm^2$, $V_{oc} = 0.57V$, $FF = 80\%$ and energy conversion efficiency of 0.90%. Using hot-press system on the fabricated DSSC of Pd-ZnO a $J_{sc} = 2.97\ mA/cm^2$, $V_{oc} = 0.59V$, $FF = 75\%$ and efficiency of 1.31% were obtained. However, the pre-dyeing process has insufficient connection with zinc oxide nanoparticles which may serve as the drawback for pd-ZnO [65].

Kumara *et al.* [66], fabricated DSSC with a dense ZnO layer and also mesoporous ZnO layer with spray pyrolysis methods. Higher efficiency of 5.02 % was obtained for dense layer fabricated DSSC and 4.2% for mesoporous layer fabricated DSSC under the same circumstances. High surface area for dye absorption was recorded for the DSSC with dense ZnO layer than that with mesoporous ZnO layer. Other parameters connected to the dense ZnO layer are $J_{sc} = 13.68\ mA/cm^2$, $V_{oc} = 0.537V$ and $FF = 66.3\%$ while those connected to the mesoporous layer hare $J_{sc} = 12.06\ mA/cm^2$, $V_{oc} = 0.565V$ and $FF = 62.4\%$ [66].

Nanosheets ZnO was used to produce Zinc Oxide (ZnO) nanoarchitecture film by a simple two-step synthesis procedure for DSSC photoanode. The ZnO nanoarchitecture was then compared with DSSC with an upright-standing ZnO nanosheet photoanode. ZnO nanoarchitecture was fabricated using dense ZnO nanowire that is grown of primary ZnO nanosheet. Electrodeposition and Aqueous Chemical Growth (ACG) respectively are the two methods used to obtain the ZnO nanoarchitecture. The ZnO nanoarchitecture (nanosheet/nanowire) was dyed on Indium Tin Oxide (ITO) in an ethanol, ruthenium (N719) used as dye, Platinum as the counter electrode and Iodide as the electrolyte [67]. The structure of ZnO nanoarchitecture was analyzed using X-ray diffraction (XRD) which shows that the peak of all diffraction can be indexed to hexagonal ZnO, and the structure contains no impurities. The ZnO (nanosheet/nanowire) based DSSC was compared with conventional ZnO nanosheet array with the same thickness of 5 μm. The conversion efficiency obtained from ZnO nanoarchitecture was 4.8% associated with $J_{sc} = 10.9\ mA/cm^2$, $V_{oc} = 0.68V$ and $FF = 65\%$ as well as the conventional ZnO nanosheet with the efficiency of 2.3% associated with $J_{sc} = 7.2\ mA/cm^2$, $V_{oc} = 0.61V$ and $FF = 52\%$. The efficiency of ZnO nanoarchitecture is as twice as that of ZnO nanosheet. The higher conversion efficiency is said to be obtained due to large internal surface area of the film (Figure 6) without losing the straight conduction pathway for swift collection of electrons [67].

ZnO nanoflower (NF) prepared hydrothermally for DSSC application with a chosen size of 8μm compared with ZnO nanoparticles (NP) has a higher dye-loading and also higher light scattering, due to its large surface area. ZnO nanoflower has a high short-circuit current (Jsc) and higher open-circuit voltage (Voc) because of straight conduction path it possesses. ZnO nanoflower has an efficiency of 5.96% as compared to ZnO nanoparticles that has 4.39% efficiency. However, both structures have nearly the same value of Voc and FF. ZnO nanoflower array has $J_{sc} = 13.5\ mA/cm^2$, $V_{oc} = 0.682V$ and $FF = 64.6\%$ while the ZnO nanoparticles have $J_{sc} = 10.1\ mA/cm^2$, $V_{oc} = 0.680V$ and $FF = 64\%$. Figure 7 showing the current-voltage (I-V) characteristics of the comparison of ZnO nanoflower and ZnO nanoparticles [68].

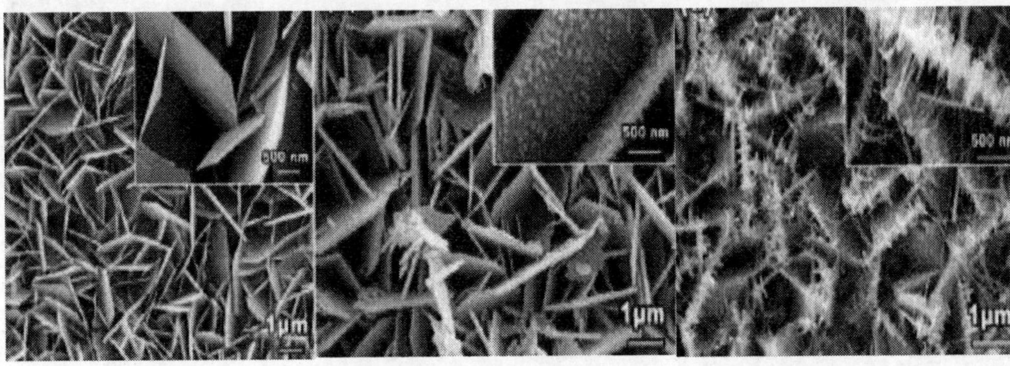

Figure 6. SEM images of the hierarchical ZnO nanowire/nanosheet obtained by aqueous chemical growth after 0, 1, and 4 h, respectively [67]

ZnO nanoparticles doped with Niobium used as photoanode in DSSC has been investigated by Anuntahirunrat and colleagues recently as a means of improving the efficiency of solar cell. Five (5) samples of the dopant were taken in weight percent (wt%) for comparison and to ascertain the most effective one, starting from 1wt% - 5wt% using spin-coating technique. The highest efficiency was recorded at 3wt% with an efficiency of 9.02% associated with $J_{sc} = 16.3\ mA/cm^2$, $V_{oc} = 0.63V$ and $FF = 38.26\ \%$. However, doping of the Niobium with ZnO does not affect the structure of ZnO, and also as the content of Niobium increases light absorption of ZnO [69]. The samples were characterized using scanning electron microscopy (Figure 8).

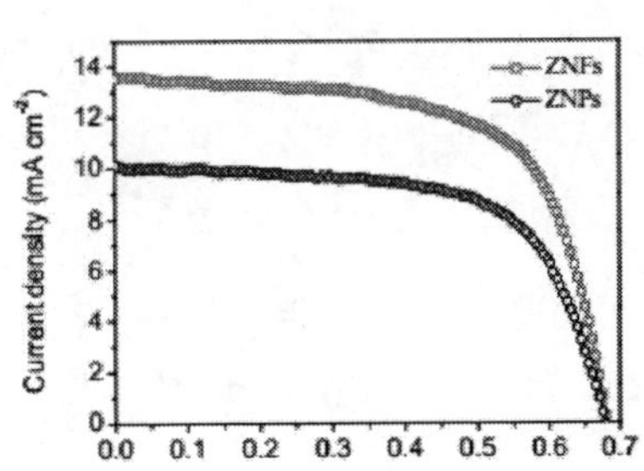

Figure 7. I-V characteristic of ZnO nanoflower and ZnO Nanoparticle photoanode based DSSC [68].

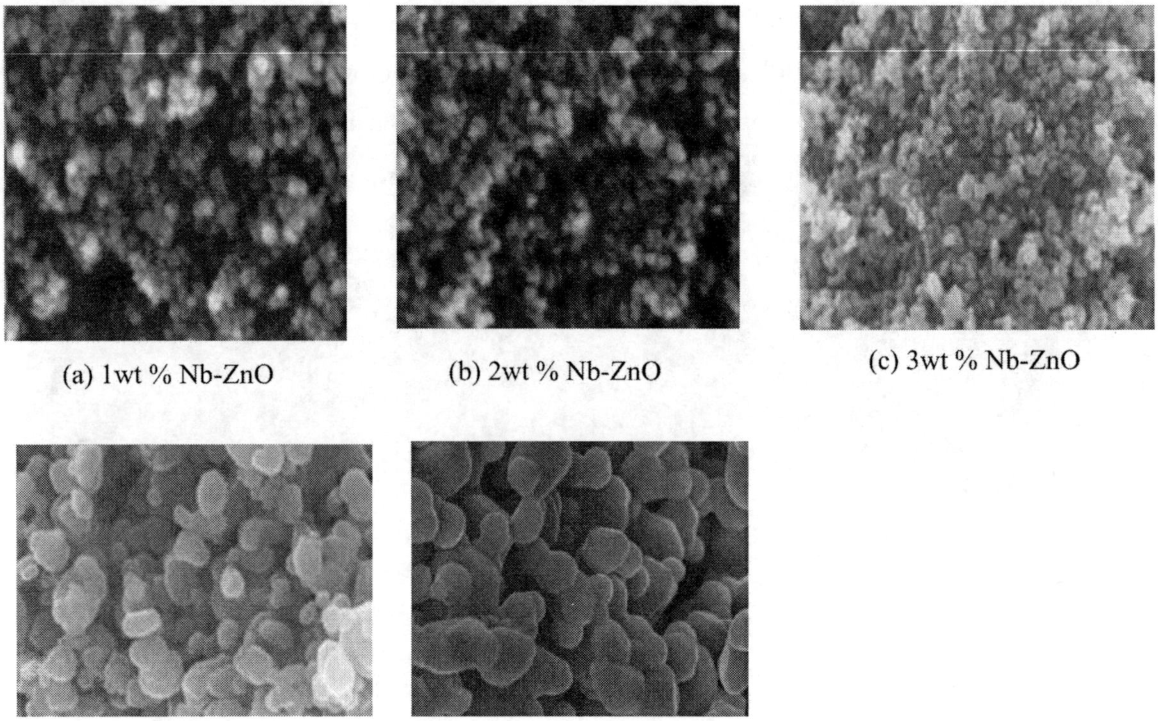

(a) 1wt % Nb-ZnO (b) 2wt % Nb-ZnO (c) 3wt % Nb-ZnO

(d) 4wt % Nb-ZnO (e) 5wt % Nb-ZnO

Figure 8. SEM images different samples of fabricated Nb-doped ZnO nanoparticles [69]

In 2009, Lai *et al.* [70] fabricated DSSC without transparent conducting oxide (TCO) using ZnO film as a replacement for the TCO and ZnO nanorods as the photoanode. Two methods were used to grow ZnO nanorods (NR)/ZnO film which were chemical vapour deposition (CVD) and chemical bath deposition (CBD) respectively. In the first method, one-step chemical vapor deposition (CVD) technique was used to grow ZnO nanorods on ZnO film, while two-step chemical bath deposition (CBD) technique that involved (1) growing of ZnO film on a glass substrate and (2) Deposition of ZnO nanorod array on ZnO substrate. The ZnO film/ZnO nanorods without TCO were compared with ZnO nanorod with TCO which was grown using CBD technique. However, there is uniformity in the thickness of one-step CVD growth technique used than the two-step CBD growth technique. The conversion efficiency obtained from one-step CVD is 0.73% having $J_{sc} = 1.97\ mA/cm^2$, $V_{oc} = 0.80V$ and $FF = 39.4\%$, while the two steps CBD grown sample has an efficiency of 0.66% with $J_{sc} = 3.96\ mA/cm^2$,

$V_{oc} = 0.72V$ and $FF = 20\%$. The efficiency is obtained from the two technique above is low when compared with conventional ZnO based DSSC grown on FTO substrate with an efficiency of 1.37% having $J_{sc} = 5.51\ mA/cm^2$, $V_{oc} = 0.65V$ and $FF = 38.2\%$. It could be noted that the conventional DSSC system has a lower Voc when compared with DSSC without transparent conducting oxide [70].

Recently, ZnO nanoparticles (ZnO NP) were synthesized by Musleh and coworkers [71] using two methods which are sol-gel and hydrothermal technique. Three samples of dye were also adopted in fabrication of DSSC which are Eosin Y (EY), Eosin B (EB) and Rhodamine B (RB). These three dyes were used to replace the conventional dye. X-ray Diffraction (XRD) was used to measure the crystal structure and it size. The precise composition of the structure was done using High Resolution Transmission Electron Microscopy (HRTEM) and Spectroflourimeter was used to check for the photoluminescence properties. It can be deduced from their work that the highest efficiency was obtained from DSSC based on sol-gel technique synthesized with EY dye and the lowest efficiency DSSC was obtained from hydrothermal technique synthesized with EB dye. The factors that may lead to higher efficiency in Sol-gel synthesized with EY dye may be due to the ability of injected electrons, amount of the absorbed dye and the surface structure. The highest efficiency obtained is 1.08% with photovoltaic parameters of $J_{sc} = 4.25\ mA/cm^2$, $V_{oc} = 0.51V$ and $FF = 50.08\%$, while the lowest efficiency obtained may be due to low dye absorption and weak bonding that exist between dye molecules and ZnO nanoparticles with the efficiency of 0.29% with photovoltaic parameters of $J_{sc} = 1.60\ mA/cm^2$, $V_{oc} = 0.37V$ and $FF = 49.76\%$ [71]. The photovoltaic parameters of ZnO nanostructures are presented in Table 2.

3.4 Other materials employed as photoanode

The efficient band gap of Tin Oxide (SnO_2) makes it valuable for DSSC application. Spherical SnO_2 nanoparticles were synthesized using solvothermal method. Two samples of the SnO_2 were prepared at pH 8 and 10; the size of particles in the synthesized SnO_2 increased with the pH value. The two samples have 4.76 and 4.53 band gap respectively. The conversion efficiency of DSSC of SnO_2 samples with pH 8 = 0.0044% and pH 10= 0.0042%. Solid electrolyte used for the fabrication may have accounted for the poor efficiency of the solar cell. The efficiency of the

IOP Conf. Series: Materials Science and Engineering **805** (2020) 012019 doi:10.1088/1757-899X/805/1/012019

DSSC dropped to 0.0013% and 0.0014% respectively after 40 days. The methylene blue dye of poor chemical stability might also account for the reduction in the efficiency of the solar cell [72]. Recently, an architecture of SnO_2 was prepared by combining nanocrystallite aggregate (SnO_2 NA) of SnO_2 synthesized by hydrolysis method at standard room temperature with SnO_2 nanosheet (NS) synthesized via hydrothermal method. A freeze drying treatment was adopted for the SnO_2 synthesized with hydrolysis method.

Table 2. Photovoltaic parameters ZnO nanostructures with their synthesis methods

ZnO Photoanode	Method	Jsc (mA/cm^2)	Voc (V)	FF (%)	η (%)	Ref.
ZnO nanorod	CBD	3.47	0.64	51.7	0.8	[62]
ZnO NF	Hydrothermal	4.2	0.62	54	1.4	[63]
ZnO NF	Hydrothermal	3.53	0.61	51	1.1	[64]
Pd-ZnO	Dyeing	1.97	0.57	80	0.90	[65]
ZnO dense layer	Spray pyrolysis	13.68	0.54	66.3	5.02	[66]
ZnO mesoporous	Spray pyrolysis	12.06	0.57	62.4	4.2	[66]
Zno NF	Hydrothermally	13.5	0.68	64.6	5.96	[68]
ZnO NS	Electrodeposition	7.2	0.61	52	2.3	[67]
ZnO NS-NW	Electrodeposition/ACG	10.9	0.68	65	4.8	[67]
ZnO/Nb	Spin-coating	16.3	0.63	38.26	9.02	[69]
Zno NR	CBD	1.97	0.80	39.4	0.73	[70]
Zno NR	CVD	3.96	0.72	20.0	0.66	[70]
ZnO NP	Sol-gel	4.25	0.51	50.08	1.08	[71]

The SnO_2 architecture was characterized using Transmission Electron Microscopy (TEM), UV-vis spectrometer and X-ray diffraction (XRD). The combined architecture provides an efficiency of 5.59% associated with Jsc of high value and low Voc and FF value was reported. The efficiency obtained from the SnO_2 NA/SnO_2 NS was attributed to large surface area and light scattering ability of the SnO_2 NA [73]. Teh *et al.* [74] synthesized a nanoporous SnO_2 in ammonia aqueous solution and compared with conventional SnO_2 nanoparticles. It was revealed that the nanoporous SnO_2 has low charge transfer and high charge recombination account for the high DSSC performance of the nanoporous SnO_2 based photoanode. The conversion efficiency recorded for nanoporous SnO_2 is 1.66% as against 1.02% for SnO_2 nanoparticles based DSSC [74]. A microsphere SnO_2 structure was hydrothermally synthesized and employed as photoanode in DSSC. Improved light harvesting factor and efficient light scattering ability of the structure propel the good performance of the DSSC. An energy conversion efficiency of 4.55% was obtained from the SnO_2 microsphere based DSSC [75].

In 1999, Burnside and co synthesized a nanocrystal form of strontium titanate ($SrTiO_3$) using hydrothermal technique [76]. The synthesized $SrTiO_3$ was used as photoanode is fabrication of DSSC. However, $SrTiO_3$ has low photocurrent value which could result low dye loading of the material. The $SrTiO_3$ based DSSC have a power conversion efficiency of 1.8% as compared to anatase TiO_2 based DSSC with 6.0% power conversion efficiency [76]. Reduced graphene oxide (r-GO) and titanium dioxide was adopted as photoanode in DSSC recently. The reduced graphene oxide was synthesized by chemical reduction process and mixed with powder of TiO_2. An efficiency of 2.02% was achieved for reduced graphene oxide based solar cell [77]. Nanowire of three forms (monoclinic, pseudo-hexagonal and orthorhombic) of Nb_2O_5 was synthesized via electrospinning of Nb_2O_5 polymeric solution and employed as photoanode in DSSC; the fabricated DSSC with pseudo-hexagonal nanowire form of Nb_2O_5 has greater conversion efficiency. However, monoclinic has a low recombination of charge, efficient electron lifespan and high dye loading ability than orthorhombic and pseudo-hexagonal form. The conversion efficiencies of monoclinic, pseudo-hexagonal and orthorhombic are 1.92%, 3.05% and 2.53% respectively [78]. The photovoltaic parameters of other materials that have been adopted as photoanode in DSSC are shown in Table 3.

IOP Conf. Series: Materials Science and Engineering **805** (2020) 012019 doi:10.1088/1757-899X/805/1/012019

Table 3. Photovoltaic parameters of other materials that have been adopted as photoanode in DSSC

Photoanode	Method	Jsc (mA/cm^2)	Voc (V)	FF (%)	η (%)	Ref.
SnO_2MH	Solvothermal	10.4	7.65	57.3	4.55	[75]
Nb_2O_5	Electrospining	6.68	0.77	59.06	3.05	[78]
$SrTiO_3$	Hydrothermal	3.00	7.89	70.0	1.8	[76]
R-GO	Chemical reduction	10.8	6.24	30.0	2.02	[77]

4. Conclusion

Previous work done and series of attempts to improve the conversion efficiency of DSSC using different nanostructure of semiconductors as photoanodes and with different synthesis method have been summarized in this review. Few of the discussed work can be compiled and applied by researchers to obtain a better efficiency. Combination of two or more nanostructure ranging from nanorods, nanofibers, nanoparticles, nanosphere, nanoflower and nanosheet of photoanodes have proven to enhance the performance efficiency of DSSC. DSSC still has lower conversion efficiency when compared to conventional silicon based solar cell despite the notable improvements. Photoanodes with high electron transfer efficiency with lower recombination electron rate and large surface area are highly anticipated from researchers. Also, composition of TiO_2 nanoparticles as well as 3D nanostructure of Silicon with efficient electron transport and large surface capability could be synthesized via hydrothermal technique and applied as photoanode in DSSC. The counter electrode, electrolytes and sensitizers can also be optimized and the whole structure can enhance the conversion efficiency of DSSC.

References

[1] Lewis NS and Crabtree G 2005 Basic research needs for solar energy utilization: report of the basic energy sciences workshop on solar energy utilization, April 18-21, 2005. https://resolver.caltech.edu/CaltechAUTHORS:LEWsolarenergyrpt05. Accessed on 19 January, 2020.

[2] Grätzel M 2005 Solar energy conversion by dye-sensitized photovoltaic cell. *Inorg. Chem.* **44** (20) pp 6841-6851.

[3] Perlin J 2004 Silicon solar cell turns 50. National Renewable Energy Lab., Golden, CO (USA). https://www.osti.gov/biblio/15009471. Accessed on 19 January, 2020.

[4] Shah AV, Schade H, Vanecek M, Meier J, Vallat-Sauvain E, Wyrsch N, Kroll U, Droz C and Bailat J 2004 Thin-film silicon solar cell technology. *Progr. Photovolt.: Res. Appl.* **12** (2-3) pp 113-142.

[5] Chopra KL, Paulson PD and Dutta V 2004 Thin-film solar cells: an overview. *Progr. Photovolt.: Res. Appl.* **12** (2-3) pp 69-92.

[6] Grätzel M 2003 Dye-sensitized solar cells. *J. Photochem. Photobiol. C: Photochem. Rev.* **4** (2) pp145-153.

[7] Hagfeldt A and Grätzel M 2000 Molecular photovoltaics. *Acc. Chem. Res.* **33** (5) pp 269- 277.

[8] Hoppe H and Sariciftci NS 2004 Organic solar cells: An overview. *J. Mater. Res.* **19** (7) pp1924-1945.

[9] Wöhrle D and Meissner D 1991 Organic solar cells. *Adv. Mater.* **3** (3) pp 129-138.

[10] Green MA, Emery K, Hishikawa Y and Warta W. Solar cell efficiency tables (version 33). *Progr. Photovolt.: Res. Appl.* **17** (1) https://doi.org/10.1002/pip.880.

[11] Grätzel M 2007 Photovoltaic and photoelectrochemical conversion of solar energy *Philos. Trans. Royal Soc. A: Math. Phys. Eng. Sci.* **365** (1853) pp 993-1005.

[12] O'Regan B and Grätzel M 1991 A low-cost, high-efficiency solar cell based on dye-sensitized colloidal TiO_2 films. *Nature* **353** (6346) pp737-740.

[13] García-Salinas MJ and Ariza MJ 2019 Optimizing a simple natural dye production method for dye-sensitized solar cells: examples for betalain (*Bougainvillea* and Beetroot extracts) and anthocyanin dyes. *Appl. Sci.* **9** (12) 2515. https://doi.org/10.3390/app9122515.

[14] Grätzel M 2006 The advent of mesoscopic injection solar cell. *Prog. Photovolt.: Res. Appl.* **14** (5) pp 429-442.

[15] Hagfeldt A, Boschloo G, Sun L, Kloo L and Pettersson H 2010 Dye-sensitized solar cells. *Chem. Rev.* **110** (11) pp 6595-663.

[16] Tétreault N and Grätzel M 2012 Novel nanostructures for next generation dye sensitized solar cell. *Energ. Environ. Sci.* **5** (9) pp 8506-8516.

[17] Zhang S, Yang X, Numata, Y and Han L 2013 Highly efficient dye-sensitized solar cell: progress and future challenges. *Energ. Environ. Sci.* **6** (5) pp1443-1464.

[18] Wang M, Grätzel C, Zakeeruddin S M and Grätzel M 2012 Recent developments in redox electrolytes for dye-sensitized solar cell. *Energ. Environ. Sci.* **5** (11) pp 9394-9405.

[19] Wu M and Ma T 2012 Platinum-free catalysts as counter electrodes in dye-sensitized solar cell. *ChemSusChem* **5** (8) pp 1343-1357.

[20] Wu M, Lin X, Wang Y and Ma T 2015 Counter electrode materials combined with redox couples in dye- and quantum dot-sensitized solar cell. *J. Mater. Chem. A* **3** (39) pp 19638-19656.

[21] Meng K, Chen G and Thampi KR 2015 Metal chalcogenides as counterelectrode materials in quantum dot sensitized solar cell: a perspective. *J. Mater. Chem. A* **3** (46) pp 23074-23089.

[22] Selinsky R S, Ding Q, Faber M S, Wright J C and S J 2013 Quantum dot nanoscale heterostructures for solar energy conversion. *Chem. Soc. Rev.* **42** (7) pp 2963-2985.

[23] Jun H K, Careem M A and Arof A K 2013 Quantum dot-sensitized solar cell-perspective and recent developments: A review of Cd chalcogenide quantum dots as sensitizers. *Renew. Sustain. Energ. Rev.* **22** pp 148-167.

[24] Kramer I J and Sargent E H 2014 The Architecture of Colloidal Quantum Dot Solar Cell: Materials to Devices. *Chem. Rev.* **114** (1) pp 863-882.

[25] Ye M, Chen C, Lv M, Zheng D, Guo W and Lin C 2013 Facile and effective synthesis of hierarchical TiO_2 spheres for efficient dye-sensitized solar cells. *Nanoscale* **5** (14) pp 6577-6583.

[26] Miao X, Pan K, Liao Y, Zhou W, Pan Q, Tian G and Wang G 2013 Controlled synthesis of mesoporous anatase TiO_2 microspheres as a scattering layer to enhance the photoelectrical conversion efficiency. *J. Mater. Chem. A: Mater. Energ. Sustain* **1** (34) 9853-9861.

[27] McCune M, Zhang W and Deng Y 2012 High efficiency dye-sensitized solar cell based on three-dimensional multilayered ZnO nanowire arrays with "caterpillar-like" structure. *Nano Lett.* **12** (7) pp 3656-3662.

[28] Chen L Y and YinY T 2013 Hierarchically assembled ZnO nanoparticles on high diffusion coefficient ZnO nanowire arrays for high efficiency dye-sensitized solar cell *Nanoscale* **5**(5) pp 1777-1780.

[29] Wang H Li B Gao J Tang M Feng H Li J and Guo L 2012 SnO_2 hollow nanospheres enclosed by single crystalline nanoparticles for highly efficient dye-sensitized solar cell. *Cryst. Eng. Comm.* **14** (16) pp 5177-5181.

[30] Chen J, Li C, Xu F, Zhou Y, Lei W, Sun L and Zhang Y 2012 Hollow SnO_2 microspheres for high-efficiency bi-layered dye sensitized solar cell. *RSC Adv.* **2**(19) pp 7384-7387.

[31] Liu X, Fang J, Liu Y and Lin T 2016 Progress in nanostructured photoanodes for dye-sensitized solar cell. *Front. Mater. Sci.* **10** (3) pp 225-237.

[32] Bahramian A 2013 High conversion efficiency of dye-sensitized solar cell based on coral-like TiO_2 nanostructure films: synthesis and characterization. *Ind. Eng. Chem. Res.* **52** (42) pp14837-14846.

[33] Wu HP, Lan CM, Hu JY, Huang WK, Shiu JW, Lan ZJ, Tsai CM, Su CH and Diau EW 2013 Hybrid titania photoanodes with a nanostructured multi-layer configuration for highly efficient dye-sensitized solar cell. *J. Phys. Chem. Lett.* **4** (9) pp 1570-1577.

[34] Bao ZQ, Xie H, Zhu Q, Qian J, Ruan P and Zhou X 2013 Microsphere assembly of TiO_2 with tube-in-tube nanostructures: anisotropic etching and photovoltaic enhancement. *Cryst. Eng. Comm.* **15** (44) pp 8972-8978.

[35] Mir N, Lee K, Paramasivam I and Schmuki P 2012 Optimizing TiO_2 nanotube top geometry for use in dye-sensitized solar cell. *Chem.: Eur. J.* **18** (38) pp 11862-11866.

[36] Chen H Y, Zhang T L, Kuang D B and Su CY 2013 Electrospun hierarchical TiO_2 nanorods with high porosity for efficient dye-sensitized solar cell. *ACS Appl. Mater. Interf.* **5** (18) pp 9205-9211.

[37] Kumar E N, Jose R, Archana P S, Vijila C, Yusoff M M and Ramakrishna S 2012 High performance dye-sensitized solar cell with record open circuit voltage using tin oxide nanoflowers developed by electrospinning. *Energ. Environ. Sci.* **5** (1) pp 5401-5407.

[38] Huo J, Hu Y, Jiang H, Huang W, Li Y, Shao W and Li C 2013 mixed solvents assisted flame spray pyrolysis synthesis of TiO_2 hierarchically porous hollow spheres for dye-sensitized solar cell. *Ind. Eng. Chem. Res.* **52** (32) pp 11029-11035.

[39] Son H, Prasittichai C, Mondloch J E, Luo L, Wu J, Kim D W, Farha O K and Hupp JT 2013 dye stabilization and enhanced photoelectrode wettability in water- based dye-sensitized solar cell through post-assembly atomic layer deposition of TiO_2. *J. Am. Chem. Soc.***135** (31) pp 11529-11532.

[40] Park N G, Schlichthörl G, Van De Lagemaat J, Cheong HM, Mascarenhas A and Frank A J 1999 Dye sensitized TiO_2 solar cell: Structural and photoelectrochemical characterization of nanocrystalline electrodes formed from the hydrolysis of $TiCl_4$. *J. Phys. Chem. B* **103** (17) pp 3308-3314.

[41] Park N G, Van De Lagemaat J and Frank A J 2000 Comparison of dye-sensitized rutile-and anatase based TiO_2 solar cell. *J. Phys. Chem. B* **104** (38) pp 8989-8994.

[42] Nazeeruddin MK, De Angelis F, Fantacci S, Selloni A, Viscardi G, Liska P, Ito S, Takeru B and Grätzel M 2005 Combined experimental and DFT-TDDFT computational study of photoelectrochemical cell ruthenium sensitizers. *J. Am. Chem. Soc.* **127** (48) pp 16835-16847.

[43] Deng D, Kim M G, Lee J Y and Cho J 2009 Green energy storage materials: Nanostructured TiO_2 and Sn-based anodes for lithium-ion batteries. *Energ. Environ. Sci.* **2** (8) pp 818-837.

[44] Akhtar M S, Umar A, Sood S, Jung I, Hegazy H H and Algarni H 2019 Rapid growth of TiO_2 nanoflowers via low-temperature solution process: photovoltaic and sensing applications. *Mater.* **12** (4) 566. https://doi.org/10.3390/ma12040566.

[45] Jin S Shin E and Hong J 2017 TiO_2 Nanowire networks prepared by titanium corrosion and their application to bendable dye-sensitized solar cell. *Nanomater.* **7** (10) pp 315. https://doi.org/10.3390/nano7100315.

[46] Xia R, Wang S, Wang X, Dong W and Fang X 2016 Fabrication of hierarchical anatase TiO_2 nanostructure for dye-sensitized solar cell. *J. Nanosci. Nanotechnol.* **16** (6) pp 5810-5814.

[47] Mor G K, Varghese O K, Paulose M, Shankar K and Grimes C A 2006 A review on highly ordered vertically oriented TiO2 nanotube arrays: Fabrication, material properties and solar energy applications *Sol. Energ. Mater. Sol. Cell* **90** (40) pp 2011-2075.

[48] Shin Y and Lee S 2008 Self-organized regular arrays of anodic TiO_2 nanotubes *Nano Lett.* **8** (10) pp 3171-3173.

[49] Rho W Y, Jeon H, Kim H S, Chung W J, Su J S and Jun B H 2015 Recent progress in dye-sensitized solar cell for improving efficiency: TiO_2 nanotube arrays in active layer. *J. Nanomater.* Article No.: 85. https://doi.org/10.1155/2015/247689.

[50] Roy P, Kim D, Lee K, Spiecker E and Schmuki P 2010 TiO_2 nanotubes and their application in dye-sensitized solar cell. *Nanoscale* **2** (1) pp 45-59.

[51] Roy P, Kim D, Paramasivam I and Schmuki P 2009 Improved efficiency of TiO_2 nanotubes in dye sensitized solar cell by decoration with TiO_2 nanoparticles *Electrochem. Comm.* **11** (5) pp 1001-1004.

[52] Park H, Kim WR, Jeong HT, Lee JJ, Kim HG and Choi WY 2011 Fabrication of dye-sensitized solar cell by transplanting highly ordered TiO_2 nanotube arrays. *Sol. Energ. Mater. Solar Cell* **95** (1) pp 184-189.

[53] Roy P, Albu S P and Schmuki P 2010 TiO_2 nanotubes in dye-sensitized solar cell: Higher efficiencies by well-defined tube tops. *Electrochem. Comm.* **12** (7) pp 949-951.

[54] Arifin Z, Suyitno S, Hadi S and Sutanto B 2018 Improved performance of dye-sensitized solar cell with TiO_2 nanoparticles/Zn-doped TiO_2 hollow fiber Photoanodes. *Energies* **11** (11) 2922. https://doi.org/10.3390/en11112922.

[55] Rajamanickam G, Narendhiran S, Muthu S P, Mukhopadhyay S and Perumalsamy R 2017 Hydrothermally derived nanoporous titanium dioxidenanorods/nanoparticles and their influence in dye-sensitized solar cell as a photoanode. *Chem. Phys. Lett.* **689** pp19-25.

[56] Kim T Y, Kim S J, Han E M, Kim J H, Lee J W and Park K H 2015 Hierarchical nanorod/nanoflower TiO_2 photoanode for natural dye-sensitized solar cell. *Int. J. Electrochem. Sci.***10** pp 9466 -9474.

[57] Heo S Y, Chi W S, Kim J K, Lee C S and Kim J H 2013 Preparation of TiO_2 nanowires/nanoparticles composite photoanodes for dye-sensitized solar cell. *Rapid Comm. Photosci.* **2** (3) pp 82-84.

[58] Yan L, Wu F L, Peng L, Zhang L J, Li P J, Dou S Y and Li T X 2012 Photoanode of dye-sensitized solar cell based on a ZnO/TiO_2 composite film. *Int. J. Photoenerg*. Article ID 613969. https://doi.org/10.1155/2012/613969.

[59] Sharifi N, Tajabadi F and Taghavinia N 2014 Recent developments in dye sensitized solar cell. *ChemPhysChem* **15** (18) pp 3902-3927.

[60] Zhang Q, Dandeneau C, Zhou S X and Cao G 2009 ZnO nanostructures for dye sensitized solar cell. *Adv. Mater.* **21** (41) pp 4087- 4108.

[61] Raj C C and Prasanth R 2016 A critical review of recent developments in nanomaterials for photoelectrodes in dye sensitized solar cell. *J. Power Sour.* **317** pp 120-132.

[62] Boda MA, Çırak BB, Demir Z and Çırak Ç 2019 Facile synthesis of hybrid ZnO nanostructures by combined electrodeposition and chemical bath deposition for improved performance of dye-sensitized solar cell. *Mater. Lett.* **248** pp 143-145.

[63] Umar A, Akhtar M S, Almas T, Ibrahim A A, Al-Assiri M S, Masuda Y, Rahman Q and Baskoutas S 2019 Direct growth of flower-shaped ZnO nanostructures on FTO substrate for dye-sensitized solar cell. *Cryst.* **9** (8) 405. https://doi.org/10.3390/cryst9080405.

[64] Umar A and Ibrahim A A 2018 Fabrication and characterization of dye sensitized solar cell based on flower shaped ZnO nanostructures. *J. Nanosci. Nanotechnol.* **18** (5) pp 3697-3701.

[65] Zhang L and Konno A 2018 Development of flexible dye-sensitized solar cell based on pre dyed Zinc Oxide nanoparticles. *Int. J. Electrochem. Sci.* **13** pp 344-352.

[66] Kumara G R A, Deshapriya U, Ranasinghe C S K, Jayaweera E N and Rajapakse RM G 2018 Efficient dye-sensitized solar cell from mesoporous zinc oxide

nanostructures sensitized by N719 dye. *J. Semicond.* **39** (3) 033005. https://doi.org/10.1088/1674-4926/39/3/033005.

[67] Xu F, Dai M, Lu Y and Sun L 2010 Hierarchical ZnO nanowire-nanosheet architectures for high power conversion efficiency in dye-sensitized solar cell. *J. Phys. Chem. C* **114** (6) pp2776-2782.

[68] Chen X, Tang Y and Liu W 2017 Efficient dye-sensitized solar cell based on nanoflower-like ZnO photoelectrode. *Mol.* **22** (8) 1284. https://doi.org/10.3390/molecules22081284.

[69] Anuntahirunrat J, Sung Y M and Pooyodying P 2017 Efficiency of Nb-doped ZnO nanoparticles electrode for dye sensitized solar cell application. *IOP Conf. Ser.: Mater. Sci. Eng.* **229** (1) 012019. https://doi.org/10.1088/1757-899X/229/1/012019.

[70] Lai M, Tubtimtae A, Lee M and Wang G 2010 ZnO-nanorod dye-sensitized solar cell: new structure without a transparent conducting oxide layer. *Int. J. Photoenerg.* **49** 7095. Article ID 497095. https://doi.org/10.1155/2010/497095.

[71] Musleh H A, Dahoudi N, Zayed H, Shaat S, Tamous H, MShurrab N, Issa A and Asad J 2018 Synthesis and characterization of ZnO nanoparticles using hydrothermal and sol-gel techniques for dye-sensitized solar cell. *J. Univ. Babylon Eng. Sci.* **26** (9) pp 256-267.

[72] Vasanthapriya R, Neelakandeswari N, Rajasekaran N, Uthayarani K, Chitra MS and Kumar S 2018 Synthesis and Characterization of SnO_2 nanostructures for dye-sensitized solar cell. *Mater. Lett.* **220** pp 218-221.

[73] Wang D, Liu S, Shao M, Zhao J, Gu Y, Li Q, Zhang X, Zhao J and Fang Y 2018 Design of SnO_2 aggregate/nanosheet composite structures based on function-matching strategy for enhanced dye-sensitized solar cell performance. *Mater.* **11** (9) 1774. https://doi.org/10.3390/ma11091774.

[74] Teh J J, Guai G H, Wang X, Leong K C, Li M C and Chen P 2013 Nanoporous tin oxide photoelectrode prepared by electrochemical anodization in aqueous ammonia to improve performance of dye sensitized solar cell. *J. Renew. Sustain. Energ.* **5** (2) 023120. https://doi.org/10.1063/1.4798316.

[75] Shang G. Wu J, Tang S, Huang M, Lan Z, Li Y, Zhao J and Zhang X 2012 Preparation of hierarchical tin oxide microspheres and their application in dye-sensitized solar cell. *J. Mater. Chem.* **22** (48) pp 25335-25339.

[76] Burnside S, Moser J, Brooks K and Gra1tzel M 1999 Nanocrystalline mesoporous strontium titanate as photoelectrode material for photosensitized solar devices: increasing photovoltage through flat band potential engineering. *J. Phys. Chem. B* **103** (43) pp 9328-9332.

[77] Ghann W E, Kang H, Uddin J, Chowdhury F, AKhondaker S I, Moniruzzaman M, Kabir M H and Rahman M M 2019 Synthesis and characterization of reduced graphene oxide and their application in dye-sensitized solar cell. *Chem. Eng.* **3** (1) 7. https://doi.org/10.3390/chemengineering3010007.

[78] Viet A L, Jose R, Reddy M V, Chowdari B V R and Ramakrishna S 2010 Nb_2O_5 photoelectrodes for dye-sensitized solar cell: choice of the polymorph. *J. Phys. Chem. C* **114** (49) pp 21795-21800.

Green synthesis of silver nanoparticles (AgNPs) for optical and photocatalytic applications: a review

M A Kareem[1], I T Bello[1,2], H A Shittu[1], M K Awodele[1,3], O Adedokun[*1,3] and Y K Sanusi[1,3]

[1]Department of Pure and Applied Physics, Ladoke Akintola University of Technology, Ogbomoso, Nigeria

[2]Department of Physics, CSET, University of South Africa, Johannesburg 1710, South Africa

[3]Nanotechnology Research Group (*NANO*$^+$), Ladoke Akintola University of Technology, Ogbomoso, Nigeria

*Email: oadedokun@lautech.edu.ng

Abstract

Nano-sized particles of silver (Ag) whose diameter falls within 1-100 nm range possess an exceptional physico-chemical, biological and antimicrobial properties. As a result of their unique properties, silver nanoparticles (AgNPs) have been vigorously investigated. In the last decade, several trials have been made to heighten the green methods of formulating AgNPs to reduce the danger of the by-products from chemical methods. A clear understanding of AgNPs properties is absolutely necessary in order to make the best use of these nanoparticles in various fields, while their effect on man and environment is reduced to the least achievable. This review aims to discuss the green methods of preparing AgNPs and its numerous applications in the area of opto-electronics and environmental remediation. Many natural biomolecules in plants and microorganism were involved in formation, stabilization and bio-reduction of AgNPs. Over the years, several discoveries have reiterated that the catalytic and optical properties of AgNPs are dependent on the size, size-distribution and shape, which show variation by differing their synthetic approaches, stabilizers and reducing agents. In this review, silver nanoparticles have been reported to produce a desired result as a promising photocatalytic material and with a viable application in opto-electronic device. Thus silver nanoparticles are considered useful for having diverse range of applications for the benefits of man.

Keywords: Silver Nanoparticles, Photocatalytic, Degradation, Green synthesis, Stabilizers, Reducing agents

IOP Conf. Series: Materials Science and Engineering **805** (2020) 012020 doi:10.1088/1757-899X/805/1/012020

1. Introduction

Nanotechnology is a very swift growing field that has been used worldwide probably in a large number of different economic product by producing nanoparticles (NPs) and nano products with novel and size-related physico-chemical characteristics that are significantly different from bulk matter [1]. In the last few years, the scientific community has recorded an outstanding progress in the area of nanoscience and nanotechnology [2]. Nanoparticles possess enhanced properties considering particular characteristics, that is, shape, size (1-100 nm) and structure. NPs are of two broad categories, organic and inorganic nanoparticles. Inorganic nanoparticles are semi-conductor nanoparticles (ZnO, ZnS, CdS), metallic NPs (Au, Ag, Cu, Al), and magnetic NPs (Co, Fe, Ni), while organic NPs are carbon nanoparticles such as quantum dots, fullerenes and carbon nanotubes. Several interests for gold (Au) and silver (Ag) NPs increase rapidly as a result of their superior characteristics and flexible properties [3]. In this review, AgNPs is of interest because of its unique and distinctive superior physico-chemical properties which include high thermal and electrical conductivity, surface-enhanced Raman scattering, catalytic activity, chemical stability, biological properties [4] and non-linear optical behaviour [5]. Their superiority is as a result of their shape, size, crystallinity, composition and structure compared to the bulk forms [6, 7]. These novelties have been utilized in a large scope of potential applications such as biomedical devices, medicine, renewable energies, cosmetics, environmental remediation, food, consumer, and industrial purposes [8, 9].

The metal nanoparticles exhibit Surface Plasmon Resonance (SPR) whose strong observation of light is used for producing the optical devices [10]. The SPR is dependent on the free electrons oscillation of the particles and the medium surrounding it and form the conduction band as a result of the electromagnetic field [11]. The spectral and geometrical attributes of silver nanoparticle, such as shape, size, environment and the inter-particle spacing provide important control over the linear and nonlinear optical properties [12], and these properties make them useful in medical imaging, microelectronics, and inks [13]. Besides, silver nanoparticles possess a broad spectrum of bactericidal and fungicidal activities [14] which made them more common in a various order of consumer products such as soaps, plastics, food, pastes and textiles, thereby magnifying their market quality [15, 16].

In other to achieve a desired AgNPs, several methods of synthesis have been improvised including physical, biological and chemical as illustrated in Figures 1 and 2. Generally, a conventional physical and chemical method of synthesis has been shown to be very costly and toxic [17]. There is possibility to achieve desired particle characteristics, nevertheless the

IOP Conf. Series: Materials Science and Engineering **805** (2020) 012020 doi:10.1088/1757-899X/805/1/012020

use of reducing and stabilizing agents which are expensive and toxic in physical and chemical methods of synthesis, makes them unlikely to bring about a favourable result [18, 19]. Among the numerous methods that have been adopted to synthesise nanoparticles, the biologically prepared AgNPs shows a high yield, high stability and solubility. Stable nanoparticles are achieved by reducing silver ions into silver elements by the use of reducing agents and then by nucleation and growing processes [20].

Biological method of production is a very simple, harmless, reliable, and a green approach capable of producing a specified size, morphology under more desirable conditions for translational research in short time. In this regard, green synthesis method of nanoparticles has drawn significant attention of researchers in recent years; extracts from microorganisms, plants and enzymes have been discovered to be a good replacement for reagents used in the synthesis of nanoparticles [21]. The use of green materials has numerous advantages such as low consumption of energy and moderate operation conditions (e.g. temperature and pressure) beyond the use of hazardous chemicals [22].

Therefore, the green method of synthesis utilizing biological organisms such as bacteria, yeast, algae, moulds and extracts from plant have been designed for synthesising nanoparticles [23]. The small size nature and surface configuration of AgNPs increases their surface to volume ratio and make them to exhibit amazing antimicrobial and physico-chemical properties [24]. In comparison with other methods, biosynthesis of nanoparticles using extracts of plant is cost effective, eco-friendly and makes the intended nanoparticles suitable for therapeutic use [25] and as economic and valuable alternative for the mass production of silver nanoparticles.

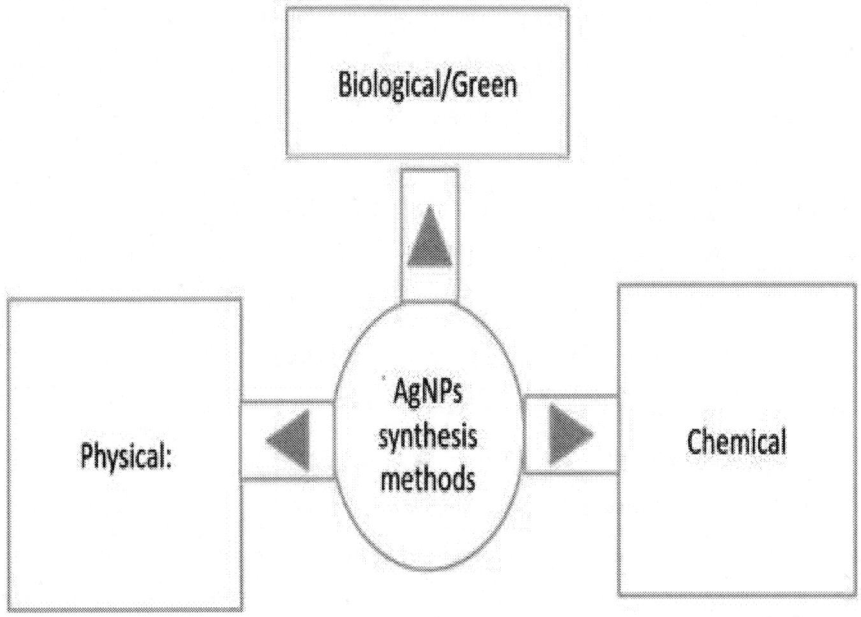

Figure 1. General methods of synthesis of silver nanoparticles

2. Silver nanoparticles (AgNPs)

AgNPs among several other metal nanoparticles have drawn attention as a result of their unique properties which include chemical stability, electrical conductivity, antimicrobial and catalytic activities [26]. Nanosilver portrays properties that are totally different from bulk material as a result of high surface to volume ratio [27].

AgNPs are produced from various chemical, physical and biological approaches that results in different sizes and shapes employed in numerous applications. Hence, the syntheses methods are grouped into two main categories; bottom-up and top-down approaches. In the top-down approach, as shown in Figure 2, the bulk size of silver metal is reduced by mechanical means to nano-scale through the use of sophisticated procedures such as laser ablation and lithography. Bottom-up approach otherwise called self-assembly technique involves dissolution of silver salt into a solvent and reduction of Ag^+ to their element by adding reducing agent and stabilizing agents to stabilize the forming AgNPs to avoid agglomeration [28]. The bottom-up approach produces nanostructures with little or low defects, better long and short range ordering and chemical composition that are more homogenous [29]. AgNPs of desired morphology, size, and shape can be produced by the several physical and chemical methods (such asarc discharge, surface deposition, plasma

IOP Conf. Series: Materials Science and Engineering **805** (2020) 012020 doi:10.1088/1757-899X/805/1/012020

polymerization, emulsion polymerization, laser CVD (Chemical Vapour Deposition), and thermal decomposition in organic solvents, chemical reduction and photo reduction in reverse micelles [30].

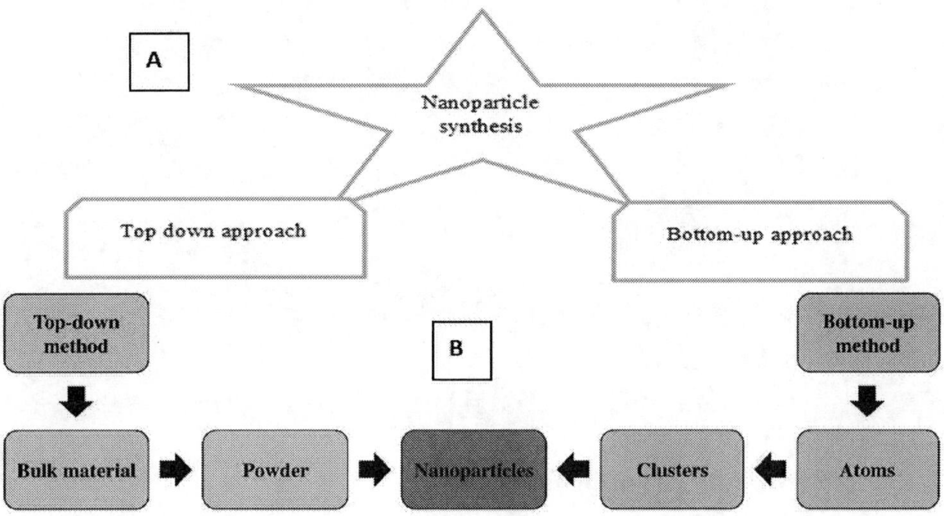

Figure 2. Diverse synthesis approaches of silver nanoparticles (AgNPs) [31]

3. Green synthesis of silver nanoparticles

The conventional synthetic methods of synthesizing NPs are costly, toxic, and not friendly to the environment. In an attempt to bypass these challenges, researchers have found green routes in which natural sources and their outcomes can be employed for synthesising NPs. Figure 3 shows the green synthesis approach of silver nanoparticle utilizing microorganisms (such as yeasts, fungi, actinomycetes and bacteria) and the use of plants and its extracts [32]. It was reported that microorganisms (bacteria and fungi) are used to reform the toxicity of materials by reduction of metal ions, the molecules produced by living organisms are replaced by stabilizer and the reducing agent [33].

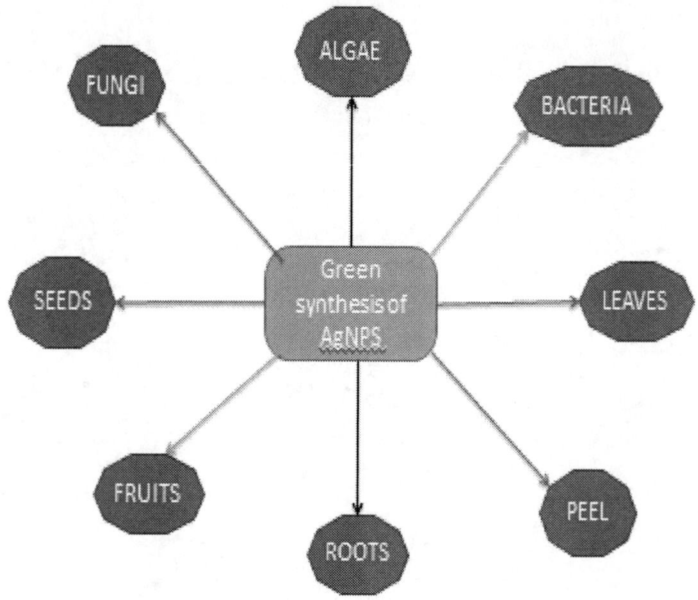

Figure 3. Green synthesis approaches of AgNPs

3.1 Synthesis of AgNPs by using bacteria

The production of inorganic materials using bacteria by extracellular or intracellular makes them a potential bio-factory for producing noble metal NPs whether Au or Ag. Silver nanoparticles have been shown to be biocompatible whereas some bacteria are resistant to silver [34]. Since these bacteria are capable of gathering Ag on the cell walls, they are then recommended industrially for use in recovery of Ag from ore material [35]. There are many bacteria that have been used for synthesising AgNPs (Table 1).

Klaus *et al.* [36] worked on the AgNPs synthesis by the use of *Pseudomonas stutzeri* AG259, Ag resistant bacteria strains. It was reported that the cells accumulate AgNPs in large quantity up to 200 nm [36]. In 2008, the production of AgNPs was also investigated using *Bacillus licheniformis* [37]. From the study, $AgNO_3$ solution was imparted into the biomass of *B. licheniformis*, and appearance of a brownish colour confirmed the formation of AgNPs within the range of 50 nm on the average and stabilized by the use of nitrate enzyme. Nanda and Saravanan [38] also reported AgNPs produced by utilizing culture supernatants from *Staphylococcus aureus*. It was revealed from their study that culture supernatants of any form of bacteria from Enterobacteriaceae make synthesis of AgNPs to be fast [38].

Suresh *et al.* [39] biosynthesized silver nanoparticles by exploiting a bacterium (*Shewanella oneidensis*) capable of reducing metals and incorporated with a solution of silver nitrate. The study revealed a nearly mono-dispersed spherical AgNPs whose size ranges between 2 to 11

IOP Conf. Series: Materials Science and Engineering **805** (2020) 012020 doi:10.1088/1757-899X/805/1/012020

nm and possesses important properties such as being stable, hydrophilic and having a large surface area [39].

In another study [40], stable AgNPs were synthesized by the use of *Bacillus* sp., an airborne bacterium and silver nitrate. The biogenic NP was investigated between the outer and inner cell membranes of the bacterial cells and the obtained diameter was found between 5-15 nm.

According to report [41], AgNPs is produced by reducing aqueous Ag^+ via different culture supernatants of bacteria (*Klebsiella pneumoniae, Enterobacter cloacae,* and *Escherichia coli*). It was concluded that the synthesis rate is much faster, as AgNPs forms within 5 minutes after adding Ag ions to the cell filtrate.

Also, Sintubin *et al.* [42] utilized *Lactobacillus* spp. as capping and reducing agent in the production of AgNPs. The study was conducted with varieties of *Lactobacillus* spp to gather and afterwards reduces Ag^+. The analysis confirmed the lactic acid bacteria to have the ability to produce Ag. It was concluded that, particle localization and distribution relied on *Lactobacillus* species inside the cell and a mean diameter (11.2 nm) of the AgNPs produced.

3.2 Synthesis of AgNPs by using fungi

Fungi possess a unique ability for synthesizing metallic NPs; this is because of their high binding capacity, metal bioaccumulation ability, intracellular uptake and tolerance [43]. Fungi can be used to synthesis NPs whereby it secretes enough enzymes capable of reducing $AgNO_3$ solution through various methods [44]. Kumar *et al.* [45] investigated a rationale of an *in vitro* enzymatic scheme for synthesizing AgNPs. NADPH (nicotinamide adenine dinucleotide phosphate) subjected to phytochelatin and nitrate reductase [45]. The purified nitrate reductase obtained from the fungus, *Fusarium oxysporum* was employed *in vitro* with NADPH. Their study revealed that hydroxyquinoline acts as an electron shuttle to enable Ag formation and a stable silver hydrosol with size range of 10 to 25 nm was produced and stabilized using capping peptide.

IOP Conf. Series: Materials Science and Engineering **805** (2020) 012020 doi:10.1088/1757-899X/805/1/012020

Table 1. Green synthesis of AgNPs using bacteria

Bacteria	Intracellular/ Extracellular	Precursor	Size (nm)	Morphology	Reference
Marine *Ochrobactrum* spp.	Intracellular	$AgNO_3$	38–85	Spherical	[46]
Exiguobacteriumm exicanum	Extracellular	$AgNO_3$	5–40	Spherical and cubic	[47]
Shewanella oneidensis	Extracellular	$AgNO_3$	4 1.5	Spherical	[48]
Lactobacillus spp.	Extracellular and intracellular	$AgNO_3$	2–20	Spherical	[49]
Lactobacillus sp. HG-C3	Extracellular	$AgNO_3$	8–25	Crystalline and spherical	[50]
Serratia nematodiphila	Extracellular	$AgNO_3$	10–31	Spherical and crystalline	[51]
Pseudomonas putida NCIM 2650	Extracellular	$AgNO_3$	70	Spherical	[52]
Bacillus methylotrophicus	Extracellular	$AgNO_3$	10–30	Spherical	[53]
Rhodococcus spp.	Intracellular	$AgNO_3$	5–50	Spherical	[54]
Vibrio alginolyticus	Extracellular and intracellular	$AgNO_3$	50–100	Spherical	[55]

Also, *Trichoderma viride* was extracellularly utilised for the synthesis of AgNPs from aqueous $AgNO_3$. It was reported that *T. viride* has a biological component that is useful for extracellular biosynthesis of Ag and the micrographs revealed that the AgNPs produced have varying morphology with rod-like and spherical NPs; the NPs obtained are in the range of 5 - 40 nm [56]. In another study, Naqvi*et al.,* [57] produced AgNPs and reported the use of *A. flavus* incorporated with antibiotics so as to improve the biocidal efficacy against multidrug-resistant bacteria [57]. Also, Ahmad *et al.* [58] in their studies reported that when aqueous Ag^+ ions are in contact with *Fusarium oxysporum*, they are reduced in solution through an enzymatic process leading to highly stable Ag hydrosol formation. The formed NPs are stabilized in solution by protein-secreted fungus and are found in the range of 5 to 15 nm [58]. Mono-dispersed AgNPs are accomplished by extracellular synthesis from *Aspergillus fumigatus* and the synthesis was reported to be very fast [59].

Furthermore, Li *et al.* [60] synthesized spherical AgNPs from *Aspergillus terreus*, and an average diameter of 1-20 nm was reported [60]. Also, the extracellular production of AgNPs by the use of *F. oxysporum* and the antibacterial effect on textile fabrics was investigated [61]. Vigneshwaran *et al.* [62] reported that mono-disperse AgNPs was achievable using fungus *Aspergillus flavus*. In their study, transmission electron microscopy (TEM) revealed the NPs size to be in the range of 8.92 ± 1.61nm [62]. In addition, the extracellular production of AgNPs by utilizing *Cladosporium cladosporioides* was studied by Balaji *et al.* [63]; the size of the synthesised NPs as measured using TEM are found in the range of 10-100 nm [63].

Summarily, the cell walls of microorganism employed in the biological synthesis plays a vital role in the intracellular production of NPs. Interaction between the positive charged metal ions and negative charged cell wall electrostatically bio-reduces metal ions to NPs [64]. There are numerous fungi that are usable in the synthesis of AgNPs as presented in Table 2.

3.3 Synthesis of AgNPs using plant extracts

Bar *et al.* [65] studied the use of seed extract of *Jatropha curcas* in their attempt to produce AgNPs. The concentration of $AgNO_3$ used varied from10^{-3} to 10^{-2}M and the solutions become reddish after a continuous heating for 15 min at 80 °C temperature. The study revealed the size of AgNPs in the range 15-50 nm at different concentrations with spherical shape. It was concluded that by varying the concentration of $AgNO_3$, AgNPs of any size is achievable [65]. AgNPs synthesis using leaf extract of *Acalypha indica* was investigated by Krishnaraj *et al.* [66] in Erlenmeyer flask, 100 mL of $AgNO_3$ and 12 ml of the extract in aqueous form was combined and the synthesis process was performed under stable conditions in the dark at a temperature of 37°C. The synthesised AgNPs was reported to exhibit a spherical shape of 20-30 nm diameter [66].

IOP Conf. Series: Materials Science and Engineering **805** (2020) 012020 doi:10.1088/1757-899X/805/1/012020

Table 2. Green synthesis of AgNPs using Fungi

Fungi	Intracellular/Extracellular	Precursor	Size (nm)	Morphology	Reference
Aspergillus niger	Extracellular	$AgNO_3$	1-20	Spherical and Polydispersed	[67]
Trichoderma harzianum	Extracellular	$AgNO_3$	34.77	Ellipsoid and Spherical	[68]
Fusarium acuminatum	Extracellular	$AgNO_3$	13	Spherical	[69]
Penicillium fellutanum	Extracellular	$AgNO_3$	5-25	Spherical	[70]
Aspergillus clavatus	Extracellular and intracellular	$AgNO_3$	10-25	Polydispersed and Spherical	[71]
Fusarium oxysporum	Extracellular	$AgNO_3$	10-25	Not reported	[72]
Aspergillus flavus	Extracellular	$AgNO_3$	8.92 ±1.61	Monodispersed	[73]
Schizophyllum commune	Extracellular and intracellular	$AgNO_3$	51-93	Spherical	[74]

Raveendran and colleagues also synthesized AgNPs in a gently heated system using D-glucose as a reducing agent and starch as capping agent. The study discovered that starch and AgNPs have weak binding interactions which can be reversed at a high temperature to allow separation of the particles synthesized [75, 76]. Alternatively, AgNPs produced in a biosynthetic approach using *Mentha piperita* as a bioreducing agent for chloroauric acid ($HAuCl_4$) and silver nitrate. The plant extract (1.5 ml) was mixed with 30 ml of $AgNO_3$ per 1 mM/ml solution. The process was carried out at 28 ^oC for 24 h and the solution was then centrifuged at 6000 rpm for 10 min. The outcome of the investigation is a spherical AgNPs of about 90 nm [77].

Also, *Rumex hymenosepalus* was used by Rodríguez-Leónto *et al.* [78] as reducing agent to synthesize AgNPs using ethanol as stabilizer and $AgNO_3$ as precursor. The research analysis revealed the obtained diameters of the synthesized AgNPs to be 2-40 nm range [78].

IOP Conf. Series: Materials Science and Engineering **805** (2020) 012020 doi:10.1088/1757-899X/805/1/012020

Recently, Dhand *et al.* [79] synthesized AgNPs using tea or coffee extract. In their work, spherical nanoparticles ranging from 20 to 60 nm was formulated and extracts from black tea was reported topossess the ability to break $AgNO_3$ to spherical shaped AgNPs, nano prisms, nanorods and other morphologies in the 20 nm size range [79, 80]. More recently, the green and *Arabica* coffee beans were investigated [81]; the coffees were found to have the characteristics of reducing and stabilizing agents. Both study concluded that the green coffee beans produced spherical AgNPs of 10 to 30 nm range, while the ellipsoidal nanoparticles ranging from 20 to 30 nm were produced by *Arabica* coffee. Also, a biological approach of synthesizing Ag nanoparticles was verified in a study by Ali *et al.* [82]. In their method, silver nitrate of different concentrations from 20 to 0.62 mM was mixed with extract from plant with concentration ranging from 100 to 0.79 mg/ml. Their study confirmed the irregularity in the shapes of Ag particles produced in the range of 5-20 nm and concluded that plant extract with higher concentration will yield larger silver particles [82].

Another researcher synthesized AgNPs by using leaf extract of *Polyalthia longifolia* as reducing agent alongside D-s orbital to improve the stability of nanoparticles [83]. Also, AgNPs produced by Maqdoom and associates [84] through the *Papaya* fruit extract were characterized by FTIR and Absorption spectroscopy. Rout and colleagues [85], also synthesized AgNPs from leaf extract of *Ocimum sanctum*. In their study, it was reported that a spherical-shaped AgNPs was revealed from XRD, UV-vis spectroscopy and SEM analysis. Awwad *et al.*[86] reported that 5 to 40 nm spherical AgNPs was achieved in their study using carob leaf extract and characterized by the use of UV-vis spectroscopy, XRD, SEM and FTIR. The SPR analysis for AgNPs is at 420 nm and XRD showed FCC geometry with crystalline shape [86].

Furthermore, Kasthuri *et al.* [87] synthesized AgNPs from phyllantin as reducing and capping agent. The TEM investigation showed that the AgNPs have an average diameter of 21-39 nm [87]. In 2012, Awwad and Salem discovered mono-dispersed spherical AgNPs as produced by utilizing Mulberry leaves extract to have adiameter of about 20 nm. It is concluded that the SEM history revealed the effective antibacterial activity of silver nanoparticles towards *Shigella* sp and *Staphylococcus aureus* [88].

Moreover, Khalil *et al.* [89] produced AgNPs by the reducing solution of $AgNO_3$ through olive leaf extract. The characterization was done by employing XRD, UV-vis spectroscopy, SEM and TGA. The analysis confirmed that the NPs are almost spherical with size of 20-25 nm in range and exhibits effective antibacterial activity against drug-resistant bacteria isolates [89].

Table 3. Green synthesis of AgNPs using Plant Extracts

Plant Extracts	Reducing/Capping Agent	Precursor	Size (nm)	Morphology	Reference
Catharanthus roseus	Root extract	$AgNO_3$	35-55	Spherical, FCC and crystalline	[90]
Trilobata	Leaf extract	$AgNO_3$	70	Spherical and FCC	[91]
Aloe vera	Leaf extract	$AgNO_3$	70	Spherical, cubical and triangular	[92]
Lens culinary	Seed exudate	$AgNO_3$	13	Crystalline and spherical	[93]
Banana	Peel	$AgNO_3$	23.7	Crystalline and spherical	[94]
Macrotyloma uniflorum	Seed extract	$AgNO_3$	12	FCC and nearly spherical	[95]
Achillea bieberstennii	Flower extract	$AgNO_3$	12 ± 2	Spherical and pentagonal	[96]
Terminalia arjuna	Bark extract	$AgNO_3$	2-100	Spherical	[97]
Ziziphora tenuior	Leaves extract	-----	8-40	Spherical	[98]
Trachy spermumammi	Seed extract	$AgNO_3$	87-99.8	-----	[99]
Solanum lycopersicum	Fruit	-----	10	Spherical	[100]
Honey	honey solution	-----	4	Crystalline and spherical	[101]
Catharanthus roseus	Leaf extract	$AgNO_3$	20	Spherical	[102]

IOP Conf. Series: Materials Science and Engineering **805** (2020) 012020 doi:10.1088/1757-899X/805/1/012020

4. Applications of silver nanoparticles

As a result of the unique properties of silver nanoparticles, they have become vigorously investigated and used extensively in various fields for different application as shown in Figure 4.

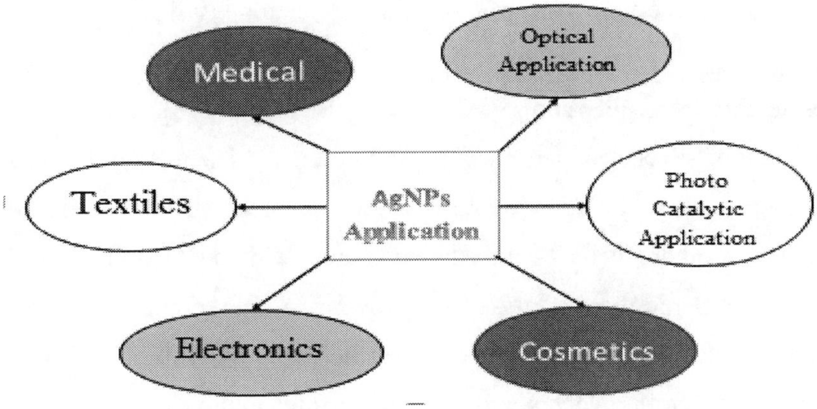

Figure 4. Schematic diagram representing various applications of AgNPs

4.1 Optoelectronic Application

Xie *et al.* [103] worked on substrates of homogeneous silver-coated nanoparticles for enhancement of fluorescence detection. In their study, a monolayer of Fluorescein isothiocyanate (FITC)-conjugated Human Serum Albumin (FITC-HSA) was used to investigate the degree of the fluorescence enhancement and it was tested by the help of laser scanning microscopy at excitation wavelength of 488 nm. The factor for the enhancement was obtained from the spectrofluorometer spectra and it was reported that Au Core-Ag shell nanostructures on the surface of the glass are good substrates with outstanding macroscopic homogeneity when it comes to fluorescence enhancement [103]. Also, Tagad and co-workers investigated the green method of synthesising silver nanoparticles for the fabrication of optical fibre based on hydrogen peroxide sensor (H_2O_2). The characterization of AgNPs was done with UV-vis spectroscopy and atomic force microscopy (AFM). The size of AgNPs was found to vary from 18-51 nm based on LBG and $AgNO_3$ concentration. It was concluded that the fabrication of a sensor based on optical fiber for detection of H_2O_2 was successful through the use of stabilized AgNPs and that the fabricated sensor is a simple, portable and cost effective sensor which is usable in various industrial applications [104].

IOP Conf. Series: Materials Science and Engineering **805** (2020) 012020 doi:10.1088/1757-899X/805/1/012020

In another study, biomediated silver nanoparticles were applied in highly selective copper (II) ion sensor [105]. The consequence of temperature and pH on silver nanoparticles formation was verified and checked by surface plasmon spectra with the help of UV-vis spectrophotometer. The sample characterization was done by TEM and XRD techniques. It was reported from the investigation that the produced AgNPs having no surface modification were successfully utilized in detection of even smallest possible amount of heavy metal copper (II) ion and with efficient specific metal ion detection [105].

4.2 Photocatalytic Application

The effective use of silver nanoparticles in catalytic media depends on the magnitude of the surface energy as well as the surface area. Several researches have established that smaller silver particles are more active catalysts compared with stable colloidal particles. The reduction rate in organic dyes catalysed by growing AgNPs compared to stable and bulk AgNPs is faster [106]. In a study, the results of the photocatalytic investigation of ZnO:Ag composite revealed that the AgNPs and their composites exhibit high catalytic performance in degrading and removing dye [107]. Yao *et al.* [107] in another study formed a visible light photocatalyst Ag_3PO_4/TiO_2 composite by depositing AgNPs onto TiO_2 surface. It was reported that the Ag_3PO_4/TiO_2 hetero-structured photocatalyst have an enhanced activity and is more stable than unsupported Ag_3PO_4.The enhanced activity was ascribed to the effective separation of electron hole and the larger surface area of Ag_3PO_4/TiO_2 composite and also the stability is due to the chemical adsorption of O^- anions and Ag^+ cations in TiO_2 and Ag_3PO_4 respectively [108].

Bi and colleagues reported AgX/Ag_3PO_4 (X=Cl, Br, I) heterocrystals, prepared by *in-situ* ion-exchange method to have some advantages over the single Ag_3PO_4 and was reported to be more hopeful and captivating visible-light-determined photocatalyst than pure Ag_3PO_4 [109].

As reported by Roy *et al.* [110], the photocatalytic reduction of methylene blue dye by biogenic silver nanoparticles produced from yeast (*Saccharomyces cerevisiae*) extract showed that the produced silver nanoparticles were almost spherical in shape with diameter of 10 nm on the average. Also, the photocatalytic study revealed that the biogenic AgNPs exhibit the potential to reduce methylene blue in the presence of solar irradiation and are considered applicable in environmental remediation and textile industry [110].

In addition, Hosseini *et al.* [111] studied the effects of Ag doping on optical, photocatalytic, and structural properties of ZnO nanoparticles. The photocatalytic properties of the samples were examined by decomposing methyl violet in solution and exposed to UV light. It was

IOP Conf. Series: Materials Science and Engineering **805** (2020) 012020 doi:10.1088/1757-899X/805/1/012020

concluded that the XPS analysis showed that the photocatalytic activity of zinc oxide nanoparticles was enhanced by doping ZnO with Ag and also, decolourisation extent of methyl violet is based on the quantity of silver [111]. Recently, the photocatalytic and antibacterial degradation efficacy of nanosilver synthesized from leaf extracts of *Cordia dichotoma* studied by Kumar *et al.* [112], showed that a solution containing methylene blue (MB) was exposed for 6 h and the UV-vis spectra for photocatalytic degradation of MB have a significant reduction in the peak intensity and makes them capable of degrading methylene blue dye and congo red. It was concluded that AgNPs formed from leaf extract of *Cordia dichotoma* have the ability to degrade dye under sunlight [112].

More recently, Guiwei *et al.* [113] developed an easy and green hydrothermal method for the production of Ag_3PO_4 with different morphologies by the use of Ag_3PO_4 as a sacrificial precursor. The analyses of the prepared catalyst were done and the photo-catalytic oxygen production tests showed that a rough-spherical Ag_3PO_4 particles embedded with small particles has the best performance of all the photo-catalytic samples analysed [113]. Moreover, Lateef *et al.* [114] worked on phytosynthesis of silver nanoparticles (AgNPs) utilizing *Synsepalum dulcificum* (miracle fruit plant) for thrombolytic, catalytic, antimicrobial, and anticoagulant applications, the catalytic analysis showed that the malachite green was degraded by approximately 80% in 24 h by the use of silver nanoparticles [114].

5. Conclusion

Naturally occurring sources are capable of converting silver ions ($AgNO_3$) into AgNPs. It is evident from this review that the various compounds which are naturally embedded in plant extracts can be used as reducing as well as stabilizing agents during the synthesis of silver nanoparticles. Green synthesised AgNPs are stable because they possess natural capping agents that do not allow the particles to agglomerate. Green approach of producing AgNPs involving the use of plant extracts have numerous importance such as producing stable products in less time, reduced wastage, serene working environment, ecological friendly and low cost. It is therefore concluded that the possession of these unique characteristics make silver nanoparticles an important factor in advancing nanotechnology processes.

References

[1] Nam J Y and Lead J R 2008 Manufactured nanoparticles: an overview of their chemistry interactions and potential environmental implications. *Sci. Total Environ.* **400** (1-3) pp 396-414.

[2] Rajput K, Raghuvanshi S, Bhatt A, Rai SK and Agrawal PK 2017 A review on synthesis silver nano-particles. *Int. J. Curr. Microbiol. App. Sci.* **6** (7) 1513-28.

[3] Vadlapudi V and Kaladhar D 2014 Review: Green synthesis of silver and gold nanoparticles. *Middle East J. Sci. Res.* **19** (6) pp 834-842.

[4] Gurunathan S, Park J H, Han J W and Kim J H 2015 Comparative assessment of the apoptotic potential of silver nanoparticles synthesized by *Bacillus tequilensis* and *Calocybe indica* in MDA-MB-231 human breast cancer cells: Targeting p53 for anticancer therapy. *Int. J. Nanomed.* **10** pp 4203-4222.

[5] Krutyakov Y A, Kudrynskiy A A, OleninA Y and`Lisichkin G V 2008 Synthesis and properties of silver nanoparticles: advances and prospects. *Russ. Chem. Rev.* **77** (3) pp 233-257.

[6] Syafiuddin A, Salim M R., Kueh A B, Hadibarata T, and Nur H 2017 A Review of silver nanoparticles: Research trends, global consumption, synthesis, properties, and future Challenges. *J. Chin. Chem. Soc.* **64** (7) pp 732-756.

[7] Atwater H A and Polman A 2010 Plasmonics for improved photovoltaic devices. *Mat. Sustain. Energ.* https://doi.org/10.1142/9789814317665_0001.

[8] De M, Ghosh P S and Rotello V M 2008 Applications of nanoparticles in biology. *Adv. Mater.* **20** (22) pp 4225-4241.

[9] Ghosh C R and Paria S 2012 Core /shell nanoparticles: classes, properties, synthesis mechanism, characterization and applications. *Chem. Rev.* **112** (4) pp 2373-2433.

[10] Toudert J, Fernandez H D, Babonneaue S, Camelio T, Girardeau and Solis J 2009 Linear and third order non-linear optical responses of multilayered Ag:Si_3N_4 nanocomposites. *Nanotechnol.* **20** (47) 475705. https://doi.org/10.1088/0957-4484/20/47/475705.

[11] Rao C, Kulkarni G, Thomas P and Edwards P 2002 Size dependent chemistry; properties of nanocrystals. *Chem.: Eur. J.* **8** (1) pp 28-35.

[12] Bruzzone M, Malvaldi G, Arrighini and Guiditti C 2005 Theoretical study of electromagnetic scattering by metal nanoparticles. *J. Phys. Chem. B.* **109** (9) pp 3807-3812.

[13] Monteiro D R, Gorup L F, Takamiya A S, Ruvollo-Filho A C, de Camargo E R and Barbosa D B 2009 The growing importance of materials that prevent microbial adhesion: antimicrobial effect of medical devices containing silver. *Int. J. Antimicrob. Agents* **34** (2) pp 103-110.

[14] Ahamed M, Alsalhi M S and Siddiqui M K 2010 Silver nanoparticle applications and human health. *Clin. Chim. Acta* **411** (23-24) pp 1841-1848.

[15] Fabrega J, Luoma S N, Tyler C R, Galloway T S and Lead J R 2011 Silver nanoparticles: behaviour and effects in the aquatic environment. *Environ. Int.* **37** (2) pp 517-531.

[16] Garcia-Barrasa J, Lopez-de-luzuriaga J M and Monge M 2011 Silver nanoparticles: synthesis through chemical methods in solution and biomedical applications. *Cent. Eur. J. Chem.* **9** (1) pp 7-19.

[17] Gurunathan S, Park J H, Han J W and Kim J H 2015 Comparative assessment of the apoptotic potential of silver nanoparticles synthesized by *Bacillus tequilensis* and *Calocybe indica* in MDA-MB-231 human breast cancer cells: Targeting p53 for anticancer therapy. *Int. J. Nanomed.* **10** pp 4203-4222.

[18] Sharma V K, Yngard R A and Lin Y 2009 Silver nanoparticles: green synthesis and their antimicrobial activities. *Adv. Colloid Interf. Sci.* **145** (1-2) pp 83-96.

[19] Geoprincy G, Srri B V, Poonguzhali U, Gandhi N N and Renganathan S 2013 A review on green synthesis of silver nanoparticles. *Asian J. Pharmaceut. Clin. Res.* **6** (1) pp 8-12.

[20] Chen Y and Yeh C 2002 Laser ablation method: use of surfactants to form the dispersed Ag nanoparticles. *Colloids Surf. A: Physicochem. Eng. Aspects* **197** (1-3) pp 133-139.

[21] Makarov V, Love A, Sinitsyna O, Yaminsky S M, Taliansky M and Kalinina N 2014 Green nanotechnologies: Synthesis of metal nanoparticles using plants. *Acta Naturae* **6** (1) pp 35-44.

[22] Mie R, Samsudin M W, Din L B, Ahmad A, Ibrahim N and Adnan S N 2014 Synthesis of silver nanoparticles with antibacterial activity using the lichen *Parmotrema praesorediosum. Int. J. Nanomed.* **9** pp 121-127.

[23] Kaviya S, Santhanalakshmi J, Viswanathan B, Muthumary J and Srinivasan K 2011 Biosynthesis of silver nanoparticles using *Citrus sinensis* peel extract and its antibacterial activity. *Spectrochim. Acta Part A: Mol. Biomol. Spectr.* **79** (3) pp 594-598.

[24] Thirumalaiarasu V, Prabhu D and Soniya M 2010 Stable silver nanoparticle synthesizing methods and its applications. *Res. J. Biol. Sci.* **1** pp 259-270.

[25] Velayutham K, Rahuman A A, Rajakumar G, Roopan S M, Elango G, Kamaraj C, Marimuthu S, Santhoshkumar T, Iyappan M and Siva C 2013 Larvicidal activity of

green synthesized silver nanoparticles using bark aqueous extract of *Ficus racemosa* against *Culex quinquefasciatus* and *Culex gelidus. Asian Pac. J. Trop. Med.* **6** (2) pp 95-101.

[26] Vidhu V K and Philip D 2014 Catalytic degredation of organic dyes using biosynthesized silver nanoparticles. *Micron* **56** pp 54-62.

[27] Thirunavoukkarasu M, Balaji U, Behera S, Panda P and Mishra B 2013 Biosynthesis of silver nanoparticle from leaf extract of *Desmodium gangeticum* (L.) DC. and its biomedical potential. *Spectrochim. Acta Part A: Mol. Biomol. Spectr.* **116** pp 424-427.

[28] Tolaymat T M, El Badawy A M, Genaidy A, Scheckel K G, Luxton T P and Suidan M 2010 An evidence-based environmental perspective of manufactured silver nanoparticle in syntheses and applications: a systematic review and critical appraisal of peer-reviewed scientific papers. *Sci. Total Environ.* **408** (5) pp 999-1006.

[29] Leela A and Vivekanandan M 2008 Tapping the unexploited plant resources for the synthesis of silver nanoparticles. *Afri. J. Biotechnol.* **7** (17) pp 3162-3165.

[30] Zhang Y, Cheng X, Zhang Y, Xue X and Fu Y 2013 Biosynthesis of silver nanoparticles at room temperature using aqueous *Aloe* leaf extract and antibacterial properties. *Colloids Surf. A: Physicochem. Eng. Aspects* **423** pp 63-68.

[31] Anu M E and Saravanakumar M P 2017 A review on the classification, characterisation, synthesis of nanoparticles and their application. *IOP Conf. Series: Mater. Sci. Eng.* **263** 032019. https://doi.org/10.1088/1757-899X/263/3/032019.

[32] Fayaz AM, Balaji K, Girilal M, Yadav R, Kalaichelvan P T and Venketesan R 2010 Biogenic synthesis of silver nanoparticles and their synergistic effect with antibiotics: a study against gram-positive and gram-negative bacteria. *Nanomater: Nanotechnol. Biol. Med.* **6** (1) pp 103-109.

[33] Arokiyaraj S, Vincent S, Saravanan M, Lee Y, Oh Y K and Kim K H 2017 Green synthesis of silver nanoparticles using *Rheum palmatum* root extract and their antibacterial activity against *Staphylococcus aureus* and *Pseudomonas aeruginosa. Artif. Cells Nanomed. Biotechnol.* **45** (2) pp 372-379.

[34] Slawson R M, Trevors J T and Lee H 1992 Silver accumulation and resistance in *Pseudomonas stutzeri. Arch. Microbiol.* **158** (6) pp 398-404.

[35] Pooley F D 1982 Bacteria accumulated silver during leaching of sulphide ore minerals. *Nature* **296** (5858) pp 642-643.

[36] Klaus T, Joerger R, Olsson E and Granqvist C-G 1999 Silver-based crystalline nanoparticles, microbially fabricated. *Proc. Natl. Acad. Sci.* **96** (24) pp 13611-13614.

[37] Kalimuthu K, Babu R S, Venkataraman D, Bilal M and Gurunathan S 2008 Biosynthesis of silver nanocrystals by *Bacillus licheniformis*. *Colloids Surf. B Biointerf.* **65** (1) pp 150-153.

[38] Nanda A and Saravanan M 2009 Biosynthesis of silver nanoparticles from *Staphylococcus aureus* and its antimicrobial activity against MRSA and MRSE. *Nanomed: Nanotechnol. Biol. Med.* **5** (4) pp 452-456.

[39] Suresh A K, Pelletier D A, Wang W, Moon J W, Gu B and Mortensen N P 2010 Silver nanocrystallites: biofabrication using *Shewanella oneidensis*, and an evaluation of their comparative toxicity on gram-negative and gram-positive bacteria. *Environ. Sci. Technol.* **44** (13) pp 5210-5215.

[40] Pugazhenthiran N, Anandan S, Kathiravan G, Prakash N K U, Crawford S and Ashokkumar M 2009 Microbial synthesis of silver nanoparticles by *Bacillus* sp. *J Nanopart. Res.* **11** (7) pp 1811-1815.

[41] Shahverdi A R, Minaeian S, Shahverdi H R, Jamalifar H and Nohi A 2007 A rapid synthesis of silver nanoparticles using culture supernatants of Enterobacteria: A novel biological approach. *Process Biochem.* **42** (5) pp 919-923.

[42] Sintubin L, De-Windt W, Dick J, Mast J, vanderHa D, Verstraete W and Boon N 2009 Lactic acid bacteria as reducing and capping agent for the fast and efficient production of silver nanoparticles. *Appl. Microbiol. Biotechnol.* **84** (4) pp 741-749.

[43] Sastry M, Ahmad A, Islam Khan M and Kumar R 2003 Biosynthesis of metal nanoparticles using fungi and actinomycete. *Curr. Sci.* **85** (2) pp 162-170.

[44] Mandal D, Bolander M E, Mukhopadhyay D, Sarkar G, and Mukherjee P 2006 The use of microorganisms for the formation of metal nanoparticles and their application. *Appl. Microbiol. Biotechnol.* **69** (5) pp 485-492.

[45] Kumar S A, Abyaneh M K, Gosavi S W, Kulkarni S K, Pasricha R, Ahmad A and Khan M I 2007 Nitrate reductase-mediated synthesis of silver nanoparticles from $AgNO_3$. *Biotechnol. Lett.* **29** (3) pp439-445.

[46] Thomas R, Janardhanan A, Varghese R T, Soniya E, Mathew J and Radhakrishnan E 2014 Antibacterial properties of silver nanoparticles synthesized by marine *Ochrobactrum* sp. *Braz. J. Microbiol.* **45** (4) pp 1221-1227.

[47] Padman A J, Henderson J, Hodgson S and Rahman P K 2014 Biomediated synthesis of silver nanoparticles using *Exiguobacterium mexicanum*. *Biotechnol. Lett.* **36** (10) pp 2079-2084.

[48] Ranganath E, Rathod V and Banu A 2012 Screening of *Lactobacillus* spp, for mediating the biosynthesis of silver nanoparticles from silver nitrate. *IOSR J. Pharm.* **2**(2) pp 237-241.

[49] Ranganath E, Rathod V and Banu A 2012 Screening of *Lactobacillus* spp, for mediating the biosynthesis of silver nanoparticles from silver nitrate. *IOSR J. Pharm.* **2** (2) pp 237-241.

[50] Du J, Singh H and Yi T H 2017 Biosynthesis of silver nanoparticles by *Novosphingobium* sp. THG-C3 and their antimicrobial potential. *Artif. Cells Nanomed. Biotechnol.* **45** (2) pp 211-217

[51] Malarkodi C, Rajeshkumar S, Paulkumar K, Vanaja M, Jobitha G D G and Annadurai G 2013 Bactericidal activity of biomediated silver nanoparticles synthesized by *Serratia nematodiphila*. *Drug Invention Today* **5** (2) pp 119-125.

[52] Thamilselvi V and Radha K 2013 Synthesis of silver nanoparticles from *Pseudomonas putida* NCIM 2650 in silver nitrate supplemented growth medium and optimization using response surface methodology. *Digest J Nanomater. Biostruct.* **8** (3) pp 1101–1111.

[53] Wang C, Kim Y J, Singh P, Mathiyalagan R, Jin Y and Yang D C 2015 Green synthesis of silver nanoparticles by *Bacillus methylotrophicus*, and their antimicrobial activity. *Artif. Cells Nanomed. Biotechnol.* **44** (4) pp 1127-1132.

[54] Otari S, Patil R, Ghosh S, Thorat N and Pawar S 2015 Intracellular synthesis of silver nanoparticle by actinobacteria and its antimicrobial activity. *Spectrochim. Acta Part A Mol. Biomol. Spectr.* **136** pp 1175-1180.

[55] Rajeshkumar S, Malarkodi C, Paulkumar K, Vanaja M, Gnanajobitha G and Annadurai G 2013 Intracellular and extracellular biosynthesis of silver nanoparticles by using marine bacteria *Vibrio alginolyticus*. *Nanosci. Nanotechnol.* **3** (1) pp 21-25.

[56] Ahmad A, Mukherjee P, Senapati S, Mandal D, Khan M I and Kumar R 2003 Extracellular biosynthesis of silver nanoparticles using the fungus *Fusarium oxysporum*. *Colloids Surf. B: Biointerf.* **28** (4) pp 313-318.

[57] Naqvi S Z H, Kiran U, Ali M I, Jamal A, Hameed A and Ahmed S 2013 Combined efficacy of biologically synthesized silver nanoparticles and different antibiotics against multidrug-resistant bacteria. *Int. J. Nanomed.* **8** pp 3187-3195.

[58] Ahmad A, Mukherjee P, Senapati S, Mandal D, Khan M I and Kumar R 2003 Extracellular biosynthesis of silver nanoparticles using the fungus *Fusarium oxysporum*. *Colloids Surf. B: Biointerf.* **28** (4) pp 313-318.

[59] Bhainsa K C and D'Souza S F 2006 Extracellular biosynthesis of silver nanoparticles using the fungus *Aspergillus fumigatus*. *Colloids Surf. B: Biointerf.* **47** (2) pp 160-164.

[60] Li G, He D, Qian Y, Guan B, Gao S and Cui Y 2011 Fungus-mediated green synthesis of silver nanoparticles using *Aspergillus terreus*. *Int J. Mol. Sci.* **13** (1) pp 466-476.

[61] Duran N, Marcarto P, De Souza G, Alves O and Esposito E J 2007Antibacterial effect of silver nanoparticles produced by fungal process on textile fabrics and their effluent treatment. *J. Biomed. Nanotechnol.* **3** (2) pp 203-208.

[62] Vigneshwaran N, Ashtaputre N, Varadarajan P, Nachane R, Paralikar K and Balasubramanya R 2007 Biological synthesis of silver nanoparticles using the fungus *Aspergillus flavus*. *Mater. Lett.* **61** (6) pp 1413-1418.

[63] Balaji D, Basavaraja S, Deshpande R, Mahesh D B, Prabhakar B and Venkataraman A 2009 Extracellular biosynthesis of functionalized silver nanoparticles by strains of *Cladosporium cladosporioides* fungus. *Colloids Surf. B Biointerf.* **68** (1) pp 88-92.

[64] Thakkar K N, Mhatre S S and Parikh R Y 2010 Biological synthesis of metallic nanoparticles. *Nanomed.: Nanotechnol. Biol. Med.* **6** (2) pp 257-262.

[65] Bar, H., Bhui, D. K., Sahoo, G. P., Sarkar, P., Pyne, S. and Misra,A 2009 Green synthesis of silver nanoparticles using seed extract of *Jatropha curcas*. *Colloids Surf. A: Physicochem. Eng. Aspect* **348** (1-3) pp 212-216.

[66] Krishnaraj C, Jagan E G, Rajasekar S, Selvakumar P, Kalaichelvan P T and Mohan N 2010 Synthesis of silver nanoparticles using *Acalypha indica* leaf extracts and its antibacterial activity against water borne pathogens. *Colloids Surf. B: Biointerf.* **76** (1) pp 50-63.

[67] Sagar G, and Ashok B 2012 Green synthesis of silver nanoparticles using *Aspergillus niger* and its efficacy against human pathogens. *Eur. J. Exp. Biol.* **2** (5) pp 1654-1658.

[68] Shelar G B and Chavan A M 2015 Myco-synthesis of silver nanoparticles from *Trichoderma harzianum* and its impact on germination status of oil seed. *Biolife* **3** (1) pp 109-113.

[69] Ingle A, Gade A, Pierrat S, Sonnichsen C and Rai M 2008 Mycosynthesis of silver nanoparticles using the fungus *Fusarium acuminatum* and its activity against some human pathogenic bacteria. *Curr Nanosci.* **4** (2) pp 141-144.

[70] Kathiresan K, Manivannan S, Nabeel M A and Dhivya B 2009 Studies on silver nanoparticles synthesized by a marine fungus, *Penicillium fellutanum* isolated from coastal mangrove sediment. *Colloids Surf. B Biointerf.* **71** (1) pp 133-137.

[71] Verma V C, Kharwar R N and Gange A C 2010. Biosynthesis of antimicrobial silver nanoparticles by the endophytic fungus *Aspergillus clavatus*. *Nanomed.* **5** (1) pp 33-40.

[72] Kumar S A, Abyaneh M K, Gosavi S W, Kulkarni S K, Pasricha R, Ahmad A and Khan M I 2007 Nitrate reductase-mediated synthesis of silver nanoparticles from $AgNO_3$. *Biotechnol. Lett.* **29** (3) pp439-445.

[73] Vigneshwaran N, Ashtaputre N, Varadarajan P, Nachane R, Paralikar K and Balasubramanya R 2007 Biological synthesis of silver nanoparticles using the fungus *Aspergillus flavus*. *Mater. Lett.* **61** (6) pp 1413-1418.

[74] Arun G, Eyini M and Gunasekaran P 2014 Green synthesis of silver nanoparticles usingthe mushroom fungus *Schizophyllum commune* and its biomedical applications. *Biotechnol. Bioprocess Eng.* **19** (6) pp 1083-1090.

[75] Raveendran P, Fu J and Wallen S J 2003 Completely ''green synthesis'' and stabilization of metal nanoparticles. *J. Am. Chem. Soc.* **125** (46) pp 13940-13941.

[76] Amanullah M and Yu L 2005 Environmental friendly fluid loss additives to protect the marine environment from detrimental effect of mud additives. *J. Petrol. Sci. Eng.* **48** (3-4) pp 199-208.

[77] Mubarak AD, Thajuddin N, Jeganathan K and Gunasekaran M 2011 Plant extracts mediated synthesis of silver and gold nanoparticle and its antibacterial activities against clinically associated pathogens. *Colloids Surf. B: Biointerf.* **85** (2) pp 360-365.

[78] Rodríguez-León E, Iñiguez-Palomares R, Navarro R E, Herrera-Urbina R, Tánori J, Iñiguez-Palomares C and Maldonado A 2013 Synthesis of silver nanoparticles using reducing agents obtained from natural sources (*Rumex hymenosepalus* extracts). *Nanoscale Res. Lett.* **8** (1) 318. https://doi.org/10.1186/1556-276X-8-318.

[79] Dhand V, Soumya L, Bharadwaj S, Chakra S, Bhatt D and Sreedhar B 2016 Green synthesis of silver nanoparticles using *Coffea arabica* seed extract and its antibacterial activity. *Mater. Sci. Eng. C* **58** pp 36-43.

[80] Begum N A, Mondal S, Basu S, Laskar S A and Mandal D 2009 Biogenic synthesis of Au and Ag nanoparticles using aqueous solution of black tea leaf extracts. *Colloids Surf. B: Biointerf.* **7** (1) pp 113-118.

[81] Wang M, Zhang W, Zheng X and Zhu P 2017 Antibacterial and catalytic activities of biosynthesized silver nanoparticles prepared using an aqueous extract of green coffee as a reducing agent. *RSC Adv.* **7** (20) pp 12144-12149.

[82] Ali M, Kim B, Belfield K D, Norman D, Brennan M and Ali GS 2016 Green synthesis and characterization of silver nanoparticles using *Artemisia absinthium* aqueous extract-a comprehensive study. *Mater. Sci. Eng. C* **58** pp 359-565.

[83] Kaviya S, Santhanalakshmi J, Viswanathan B, Muthumary J and Srinivasan K 2011 Biosynthesis of silver nanoparticles using *Citrus sinensis* peel extract and its antibacterial activity. *Spectrochim. Acta Part A: Mol. Biomol. Spectr.* **79** (3) pp 594-598.

[84] Maqdoom F, Sabeen H and Zarina S 2013 *Papaya* fruit extract: a potent source for synthesis of bionanoparticle. *J. Environ. Res. Dev.* **7** (4A) pp 1518-1522.

[85] Rout Y, Behera S, Ojha A K and Nayak P 2012 Green synthesis of silver nanoparticles using *Ocimum sanctum* (Tulashi) and study of their antibacterial and antifungal activities. *J. Microbiol. Antimicrob.* **4** (6) pp 103-109.

[86] Awwad A M, Salem N M and Abdeen A O 2013 Green synthesis of silver nanoparticles using carob leaf extract and its antibacterial activity. *Int. J. Ind. Chem.* **4** (1) 29. https://doi.org/10.1186/2228-5547-4-29.

[87] Kasthuri J, Kathiravan K and Rajendiran N 2009 Phyllanthin assisted biosynthesis of silver and gold nanoparticles: a novel biological approach. *J. Nanopart. Res.* **11** (5) pp 1075-1085.

[88] Awwad A M, and Salem N M 2012. Green synthesis of silver nanoparticles by mulberry leaves extract. *Nanosci. Nanotechnol.* **2** (4) pp 125-128.

[89] Khalil M M, Ismail E H, El-Baghdady K Z and Mohamed D 2014 Green synthesis of silver nanoparticles using olive leaf extract and its antibacterial activity. *Arab J. Chem.* **7** (6) pp 1131-1139.

[90] Rajagopal T, Jemimah I A, Ponmanickam P and Ayyanar M 2015 Synthesis of silver nanoparticles using *Catharanthus roseus* root extract and its larvicidal effects. *J. Environ. Biol.* **36** (6) 1283-1289.

[91] Raja S, Ramesh V and Thivaharan V2 017 Green biosynthesis of silver nanoparticles using *Calliandra haematocephala* leaf extract, their antibacterial activity and hydrogen peroxide sensing capability. *Arab J. Chem.* **10** (2) pp 253-261.

[92] Medda S, Hajra A, Dey U, Bose P and Mondal N K 2015 Biosynthesis of silver nanoparticles from *Aloe vera* leaf extract and antifungal activity against *Rhizopus* sp. and *Aspergillus* sp. *Appl. Nanosci.* **5** (7) pp 875-880.

[93] Shams S, Pourseyedi S and Raisi M 2013 Green synthesis of Ag nanoparticles in the present of *Lens culinaris* seed exudates. *Int. J. Agric. Crop Sci.* **5** (23) pp 2812-2815.

[94] Ibrahim H M 2015 Green synthesis and characterization of silver nanoparticles using banana peel extract and their antimicrobial activity against representative microorganisms. *J. Radiat. Res. Appl. Sci.* **8** (3) pp 265-275.

[95] Vidhu V, Aromal S A and Philip D 2011 Green synthesis of silver nanoparticles using *Macrotyloma uniflorum. Spectrochim. Acta Part A: Mol. Biomol. Spectr.* **83** (1) pp 392-397.

[96] Baharara J, Namvar F, Ramezani T, Hosseini N and Mohamad R 2014 Green synthesis of silver nanoparticles using *Achillea biebersteinii* flower extract and its anti-angiogenic properties in the rat aortic ring model. *Mol.* **19** (4) pp 4624-4634.

[97] Ahmed Q, Gupta N, Kumar A and Nimesh S 2016 Antibacterial efficacy of silver nanoparticles synthesized employing *Terminalia arjuna* bark extract. *Artif. Cells Nanomed. Biotechnol.* **45** (6) pp 1192-1200.

[98] Sadeghi B and Gholamhoseinpoor F 2015 A study on the stability and green synthesis of silver nanoparticles using *Ziziphora tenuior* (Zt) extract at room temperature. *Spectrochim Acta Part A: Mol. Biomol. Spectr.* **134** pp 310-315.

[99] Vijayaraghavan K, Nalini S K, Prakash N U and Madhankumar D 2012 One step green synthesis of silver nano/microparticles using extracts of *Trachy spermumammi* and *Papaver somniferum. Colloids Surf. B Biointerf.* **94** pp 114-117.

[100] Umadevi M, Bindhu M and Sathe V 2013 A novel synthesis of malic acid capped silver nanoparticles using *Solanum lycopersicum* fruit extract. *J. Mater. Sci. Technol.* **29** (4) pp 317-322.

[101] Philip D 2010 Honey mediated green synthesis of silver nanoparticles. *Spectrochim. Acta A: Mol. Biomol. Spectr.* **75** (3) pp 1078-1081.

[102] Al-Shmgani HS, Mohammed W H, Sulaiman G M and Saadoon A H 2016 Biosynthesis of silver nanoparticles from *Catharanthus roseus* leaf extract and

assessing their antioxidant, antimicrobial, and wound-healing activities. *Artif. Cells Nanomed. Biotechnol.* **45** (6) pp 1234-1240.

[103] Fang Xie, Mark S. Baker and Ewa M. Goldys 2006 Homogeneous silver-coated nanoparticle substrates for enhanced fluorescence detection. *J. Phys. Chem. B* **110** pp 23085-23091.

[104] Chandrakant KT, Sreekantha R D, Rohini A, Sungha P, Atul K and Sushma S 2013 Green synthesis of silver nanoparticles and their application for the development of optical fiber based hydrogen peroxide sensor. *Sens. Actuat. B*: *Chem.* **183** pp 144-149.

[105] Kirubaharan CJ, Kalpana D, Lee YS, Kim AR, Yoo DJ, Nahm KS, Kumar GG. Biomediated silver nanoparticles for the highly selective copper (II) ion sensor applications. *Ind. Eng. Chem. Res.* **51** (21) pp 7441-7446.

[106] Sharma P and Bhargava M 2013 Applications and characteristics of nanomaterials in industrial environment. *Res. Dev.* (IJCSEIERD) **3** (4) pp 63-72.

[107] Sharma K, Singh G, Singh G, Kumar M and Bhalla V 2015 Silver nanoparticles: facile synthesis and their catalytic application for the degradation of dyes. *RSC Adv.* **5** (33) pp 25781-25788.

[108] Yao W F, Zhang B, Huang C P, Ma C, Song X L and Xu Q 2012 Synthesis and characterization of high efficiency and stable Ag_3PO_4/TiO_2 visible light photocatalyst for the degradation of methylene blue and rhodamine B solutions. *J. Mater. Chem.* **22** (9) pp 4050-4055.

[109] Bi Y P, Ouyang S X, Cao J Y and Ye J H 2011 Facile synthesis of rhombic dodecahedral AgX/Ag_3PO_4 (X = Cl, Br, I) heterocrystals with enhanced photocatalytic properties and stabilities. *Phys. Chem. Chem. Phys.* **13** (21) pp 10071-10075.

[110] Kaushik R, Sarkar C K and Ghosh C K 2015 Photocatalytic activity of biogenic silver nanoparticles synthesized using yeast (*Saccharomyces cerevisiae*) extract. *Appl. Nanosci.* **5** (8) pp 953-959.

[111] Hosseini SM, Abdolhosseini I S, Kameli P and Salamati H 2015. Effect of Ag doping on structural, optical and photocatalytic properties of ZnO nanoparticles. *J. Alloys Compd.* **640** pp 408-415.

[112] Mankamna K, Nikita T, Nidhi G, Kumar A and Surendra N 2016. Antibacterial and photocatalytic degradation efficacy of silver nanoparticles biosynthesized using

Cordia dichotoma leaf extract. *Adv. Nat. Sci.: Nanosci. Nanotechnol.* **7** (4) 045009. https://doi.org/10.1088/2043-6262/7/4/045009.

[113] Guiwei H, Wanliang Y, Wei Z, Li G, Wang X, Yan A and Mengkui T 2019 Facile controlled synthesis of Ag_3PO_4 with various morphologies for enhanced photocatalytic oxygen evolution from water splitting. *RSC Adv.* **9** (32) pp 18222-18231.

[114] Lateef A, Akande MA, Azeez MA, Ojo SA, Folarin BI, Gueguim-Kana EB and Beukes LS 2016 Phytosynthesis of silver nanoparticles (AgNPs) using miracle fruit plant (*Synsepalum dulcificum*) for antimicrobial, catalytic, anticoagulant, and thrombolytic applications. *Nanotechnol Rev.* **5** (6) pp 507-520.

IOP Conf. Series: Materials Science and Engineering **805** (2020) 012021 doi:10.1088/1757-899X/805/1/012021

Characterization and biomedical application of phytosynthesized gold nanoparticles from *Datura stramonium* seed extract

IC Oladipo*[1], A Lateef[2], MA Azeez[2], TB Asafa[3], TA Yekeen[2], SB Ogunsona[1], HM Irshad[4], and SH Abbas[5]

[1]Department of Science Laboratory Technonology, Ladoke Akintola University of Technology, Ogbomoso, Oyo State Nigeria.

[2]Department of Pure and Applied Biology, Ladoke Akintola University of Technology, Ogbomoso, Oyo State Nigeria.

[3]Department of Mechanical Engineering, Ladoke Akintola University of Technology, Ogbomoso, Oyo State Nigeria.

[4]Faculty of Materials and Chemical Engineering, Ghulam Ishaq Khan Institute of Engineering, Science and Technology, Pakistan.

[5]Center for Excellence in Nanotechnology, King Fahd University of Petroleum and Minerals, Saudi Arabia

*corresponding author: icoladipo@lautech.edu.ng

Abstract

This study describes characterization and some biological evaluation of gold nanoparticles synthesized using aqueous seed extract of *Datura stramonium.* The phytosynthesized AuNPs was characterized by UV-Vis spectroscopy, Fourier Transform Infrared Spectroscopy (FTIR), Energy Dispersive X-ray (EDX) and Scanning Electron Microscopy (SEM) and evaluated for antifungal, antioxidant, thrombolytic and anticoagulant activities. The AuNPs were nearly spherical, and agglomerated in nature with size ranges of 75.1-156.5 nm. Gold was the most occurring metal noted on EDX analysis, while the AuNPs showed face-centred crystalline pattern. The UV-visible spectrum of the AuNPs synthesized displayed clear peak at 536 nm. The FTIR peaks at 3423.76, 2920.32, 2360.95 and 1635.69 cm^{-1} attributed to the involvement of proteins in the AuNPs biofabrication and capping. The AuNPs showed potent antifungal activity through inhibitions of mycelia by 57.9, 60.7, 64.5, 69.6 and 80.6 % against *Fusarium solani, Candida albicans, Aspergillus niger, A. flavus* and *A. fumigatus* at 250 μg/ml respectively.

IOP Conf. Series: Materials Science and Engineering **805** (2020) 012021 doi:10.1088/1757-899X/805/1/012021

The AuNPs synthesized showed scavenging properties of 28.9, 37.3, 42.5 and 52.4 % at 250, 500, 750 and 1000 μg/ml against 2,2-diphenyl-1-picrylhydrazyl. Furthermore, the synthesized AuNPs also showed free radical scavenging properties against nitric oxide at 250 (52.20 %), 500 (55.73 %), 750 (56.71 %) and 1000 (59.51 %) μg/ml respectively using sodium nitroprusside as positive control. Blood coagulation was prevented by the AuNPs at 250 (54 %), 500 (55 %), 750 (56 %) and 1000 (59 %) μg/ml. Also, AuNPs showed thrombolytic potential by causing lysis of blood clot thus, showing potential for biomedical applications. An in-expensive and environmental-friendly synthesis of AuNPs from *Datura stramonium* seed has been presented for various nano-biotechnological applications.
Keywords: Antifungal, Antioxidant, Biomedical, Nanoparticles.

1 Introduction

Nanotechnology is the science that involves the manipulation of materials/matters to produce nanoparticles with dimension size between 1 and 100 nanometer [1, 2]. Nanotechnology has shown unprecedented innovations in history as it has opened doors for many feats of discoveries in science and technology. For over two decades, researchers have exploited avenues in nanotechnology that have in turn made the world a better place [3-7]. The indispensability of nanotechnology to man is getting more stronger as many intricate problems in areas of medicine [8,9], biochemistry, microbiology/molecular biology [3, 4, 8], engineering [10] and materials science [10, 11] have been proffered solutions to.

Many methods have been used in the synthesis of nanoparticles and most of these methods have one or two short comings when it comes to the application of the synthesized nanoparticles in biomedical. However, considering the residual and side effects, among all the approaches/methods used in synthesizing nanoparticles, the green method is the cheapest, eco-friendly, safe and easy to carry out. The green synthesis of nanoparticles is commonly used and it has been shown to be useful in biomedical application [3, 4, 13] than any other approach.

Plant/plant materials as a result of their abundance in bio-reductant molecules have been pivotal in the green synthesis of nanoparticles. Phytochemicals present in plants played major roles in reduction of metals, a feature highly useful in nanobiotechnology. Many common and uncommon plants/plants materials have been exploited in the green synthesis of nanoparticles yielding sterling and beneficial applications in all fields especially biomedical [3, 4, 5, 14, 15,

16]. In this study, Jimson weed (*Datura stramonium*); popularly called the Devil's trumpet was evaluated for its nanobiotechnological potentials. It is an annual, herbaceous plant that grows to a height of 1 to almost 2 meters with white/violet trumpet-shaped flowers and spiny fruit [17, 18].

Jimson weed grows indiscriminately around in gardens, road sides and farmlands in Western Africa especially in Nigeria. It is locally called "Gegemu" by the Yoruba tribe and it is popularly known for its intoxicating power and also could be planted explicitly to ward off poisonous snakes. In Nigeria, *D. stramonium* is very important to the locals and also in many countries around the world. It is highly medicinal and it is used for treating asthma/cough, bronchitis, goiter, ulcer and wound dressing [19]. It also possesses abortifacient and antimicrobial activities against *E. coli* and *Pseudomonas aeruginosa* and all these properties could be due to its abundance in phytochemicals such as alkaloids [18].

On the evidence of phytochemical contents and medicinal importance, this study was aimed at establishing the possibility of synthesizing gold nanoparticles using aqueous seed extract of Jimson weed as reducing/capping agents and also evaluating biomedical properties of the synthesized nanoparticles.

2 Materials and Methods

2.1 Sample Collection and Preparation

The Jimson weed fruit was collected from Isale General Area, Ogbomoso, Oyo State by plucking and then conveyed to the laboratory in sterile polythene bag. The seeds were separated from the husk and were examined closely to ensure no sign of infection. The seeds were washed with sterilized distilled water, drained aseptically, air-dried and blended aseptically after two weeks. The seed powder (0.1 g) was weighed and suspended in 10 ml of distilled water, and placed in water bath at 60 °C for 1 h. The extract was filtered and centrifugation was carried out at 4000 rpm for 15 min. The scheme of sample preparation is depicted in Figure 1.

2.2 Biofabrication and Characterization of Gold nanoparticles

Exactly 1 ml of the seed extract was reacted with 40 ml of 1mM $HAuCl_3$ under ambient condition. The synthesis was monitored through the colour change and measurement of the absorption spectrum carried out by UV-vis spectrophotometry (B-UV1800PC spectrophotometer); and also for the determination of biomolecules responsible for the

IOP Conf. Series: Materials Science and Engineering **805** (2020) 012021 doi:10.1088/1757-899X/805/1/012021

biofabrication, capping and stabilization of the AuNPs, the IR spectroscopy spectral was obtained using IR Affinity-1s spectrophotometer (Shimadzu, UK). The AuNPs was also subjected to Scanning Electron Microscope and Energy-Dispersive X-ray spectroscopic analyses for the determination of the elemental composition and size of the AuNPs. The micrograph was taken by using FESEM 6100 Zeiss Ultra Plus (Germany) working at a voltage of 20.0 kV with secondary electrons in low vacuum mode.

2.3 Biomedical Application of the AuNPs

2.3.1Antifungal Activity

The antifungal activities of the AuNPs were determined using mycelial growth inhibition test [21]. AuNPs were included in potato dextrose agar (PDA) at concentration of 200 µg/ml. Agar plug of 6 mm of 48 h-old cultures of *Aspergillus niger, A. fumigatus, Fusarium solani, Candida albicans* and *A. flavus* were used to inoculate the PDA plates at the centre. The control plates contained no nanoparticles. Incubation was done at 28 ± 2°C for 72 h. Measurement of the fungal radial growths in all the plates were taken and the percentage growth inhibitions were calculated using the formula: $\frac{D\ control\ - \text{D test}}{D\ control}\ x\ 100$

Where D is the diameter of fungal growth on the plates.

2.3.2 Antioxidant Activities of Biosynthesized Gold Nanoparticles

The 2, 2-diphenyl-1-picryhydrazyl is widely used for testing preliminary radical scavenging activity of a compound or nanoparticles. Antioxidant activity was measured by the use of the modified DPPH method as reported by Lateef *et al.* [22]. Methanolic solution of DPPH was prepared using 0.04 g of DPPH into 1 liter of methanol and the optical density of the solution was read at 517 nm. To determine the antioxidant activity, 1.0 ml of varied concentrations of the synthesized gold nanoparticles (250, 500, 750 and 1000 µg/ml) were challenged with 4 ml of methanolic DPPH. The reaction was allowed in the dark for 30 min. The absorbance was measured at 517 nm in UV-visible spectrophotometer. Calculation was done using the formula below:

$$\text{DPPH scavenging (\%)} = \frac{Absorbance\ \ control\ - Absorbance\ \ of\ sample}{Absorbance\ \ control} x100$$

2.3.3 Anticoagulant and Thrombolytic Activities of Synthesized Gold Nanoparticles

The anticoagulant activity of the gold nanoparticles was investigated as earlier described by Lateef *et al.* [23]. Exactly 100 µl of the gold nanoparticles (100 µg/ml) was added to 0.5 ml of freely donated human blood, while blood in EDTA served as positive control and blood in clean tube served as negative control. Also, *Datura stramonium* seed extract and aqueous gold solution were used to treat blood samples, which were held at 30 ± 2 °C for 30 min. These were then examined for anticoagulation.

Determination of the thrombolytic activity was done by utilizing the strategies of Azeez *et al.* [24]. Quantification of the thrombolytic activity was carried out in this case [25]. For blood clotting to occur, tubes with 0.5 ml of blood were kept at 37 °C for 30 min and then examined. The weight of clean tube (W1) was subtracted from the weight of tube and blood clot (W2) to obtain the weight of blood clot (W3). Subsequently, 100 µl of the gold nanoparticles, gold chloride solutions, and *Datura stramonium* aqueous seed extract were placed in each blood clot tube. Incubation was done at 37 °C for 90 min, and the tubes were inverted to confirm lysis of blood clot [26, 27]. The tubes were drained, and the weight of the tube left over clot was taken (W4) to obtain the weight of clot that was not lyzed (W5). The percentage thrombolytic activity was obtained as:

$$\frac{W_3 - W_5}{W_3} \times 100$$

3 Results and Discussion

3.1 Biosynthesis and Characterization of AuNPs

The phytosynthesis of the AuNPs using aqueous seed extract of *Datura stramonium* (Plate 1) as the stabilizing /reducing agent brought about change in the colour of the salt solution from pale yellow to light pink within the first 5 min and later stabilized turning to dark pink 22 min later (Figure 2); this similar colour change was reported by Abirami *et al.* [28] and Alaa *et al.* [29]. Furthermore, colours like ruby red, purple, blue-black and light blue have also been observed for gold nanoparticles by many authors [13, 30, 31, 32, 33]. The change in colour is the result of excitation of surface plasmon vibration, which is indicated by the reduction of Au^{3+} ions to Au^0 ions at different time interval [28]. In the interim of each time interval, the peak became

IOP Conf. Series: Materials Science and Engineering **805** (2020) 012021 doi:10.1088/1757-899X/805/1/012021

clear, unique and rising; this unique peak clearly indicates the increase in synthesis of nanoparticles as the time increases till the capping of the reaction, these facts were lent credence to by Reddy *et al.* [34] and Jayaseelan *et al.* [35].

Figure 1. Sample preparation

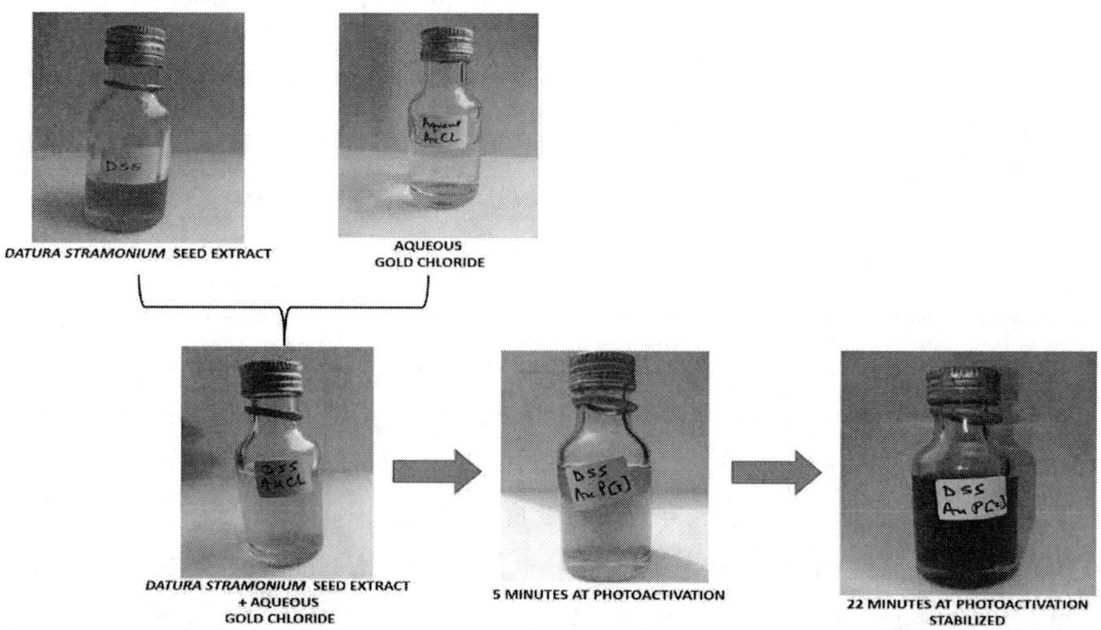

Figure 2. Photoactivation of the gold nanoparticles by *Datura stramonium* seed (DSS) extract

IOP Conf. Series: Materials Science and Engineering **805** (2020) 012021 doi:10.1088/1757-899X/805/1/012021

The UV-vis spectrum of the AuNPs displayed maximum absorbance at the wavelength of 536 nm as shown in Figure 3C and this is within the reported absorption properties of AuNPs i.e surface plasmon resonance in range of 510-614 [13, 26, 36, 37]. The spectra for the aqueous gold chloride which was the precursor (Figure 3A) and the aqueous crude extract of the *Datura stramonium* seed (Figure 3B) that serves as the reducing/capping agents were also presented to establish the occurrence of new reaction in Figure 3C at the mixture of the precursor ($HAuCl_3$) and the reducing/capping agent (aqueous crude extract of the *Datura stramonium* seed) that brought about the reduction of Au^{3+} ions to Au^0.

The AuNPs FTIR spectrum showed prominent peaks at 3423.76, 2920.32, 2360.95 and 1635.69 cm^{-1} (Figure 4). The vibrational peak around 3423 cm^{-1} is assigned to O-H and N-H stretching vibration in amines and amides. Furthermore, the peak at 2920.32 is ascribed to C-H stretch of the alkyl group, the peak around 2360.95 cm^{-1} which is likely to correspond to C≡C from alkyne and the peak at 1635.69 cm^{-1} suggested conjugation effects of C=O stretching of carbonyl groups [33] or esters and N-H binding of proteins. The appearance of these peaks in the AuNPs FTIR spectrum suggested that macromolecules like proteins, carbohydrates and some phytochemicals like phenolic compounds in the *Datura stramonium* seed extract may be responsible for the stabilizing and capping of the AuNPs. Hamed *et al.* [33] reported that phytochemicals from plant extract putatively make a coating that shields metal nanoparticles.

The Energy Dispersive X-ray (EDX) patterns (Figure 5) showed gold as the prominent element showing the yield of 86.00%. Also, the SEM (Figure 6) micrograph shows the fabrication of nearly spherical shaped and agglomerated AuNPs with size range of 75.1-156.5 nm.

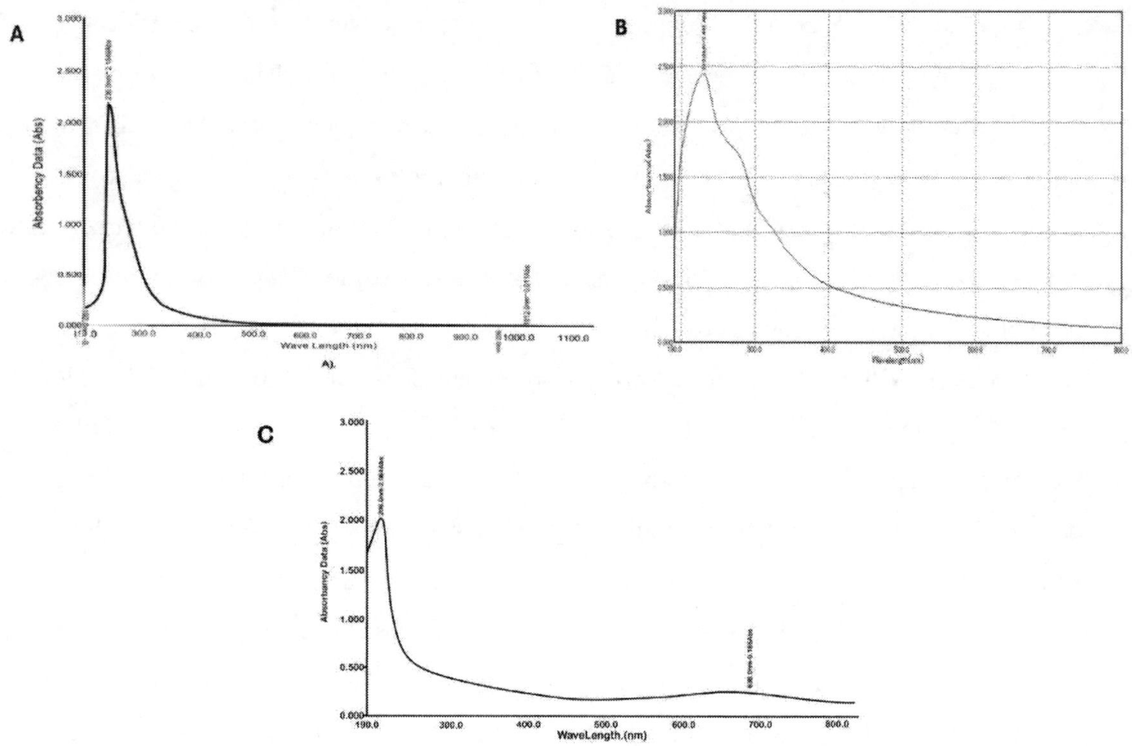

Figure 3. A, UV-vis spectrum of aqueous gold chloride; B, UV-vis spectrum of *Datura stramonium* seed extract; 1C, UV-vis spectrum of biosynthesized gold nanoparticles

IOP Conf. Series: Materials Science and Engineering **805** (2020) 012021 doi:10.1088/1757-899X/805/1/012021

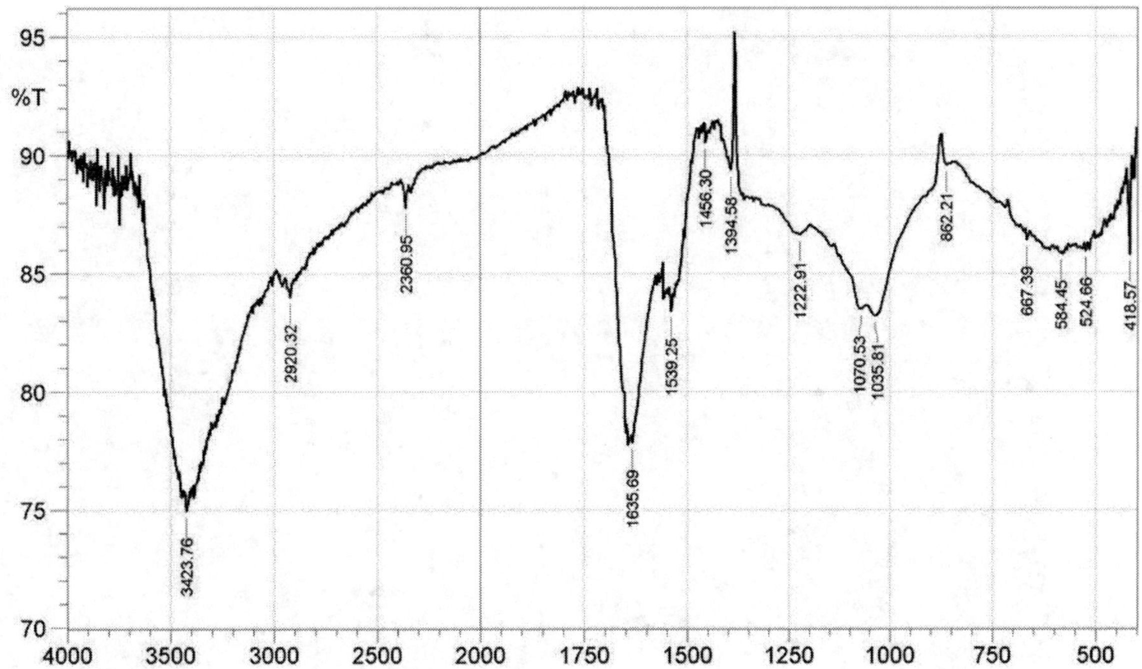

Figure 4. FTIR spectrum of the biosynthesized gold nanoparticles

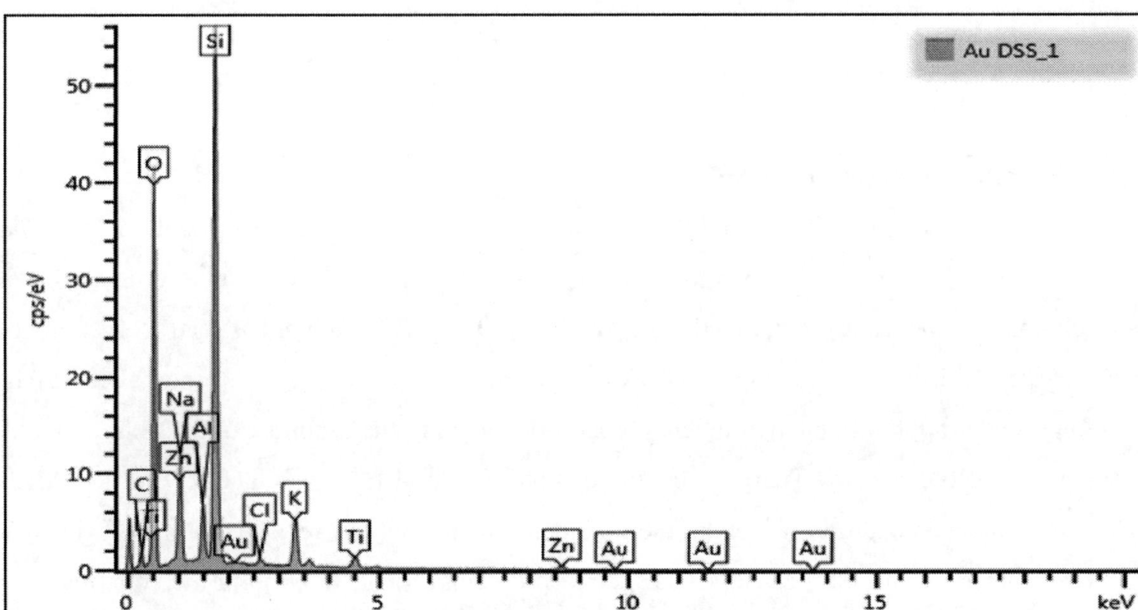

Figure 5. EDX spectrum of the biosynthesized gold nanoparticles

IOP Conf. Series: Materials Science and Engineering **805** (2020) 012021 doi:10.1088/1757-899X/805/1/012021

Figure 6. SEM micrograph of the biosynthesized gold nanoparticles

3. 2Antifungal Activities of AuNPs

The biosynthesized AuNPs showed significant effect on the growth of *C. albicans* (60.7%)*, A. niger* (64.5%), *A. flavus* (69.6%), *A. fumigatus* (80.6%) and *F. solani* (57.9%) in relation the abundant growth of the control and aqueous *D. stramonium* seed extract embedded (Figure 7). Mycotoxins are one out of many to reckon with as they are produced by toxigenic fungi under appropriate environmental conditions making them almost impossible to avoid and are considered the key factor in contamination and foodstuffs spoilage [39, 40]. When the environmental conditions are right, the toxigenic fungi grow and release the mycotoxins into the food materials and later cause severe health problems. Some food-borne mycotoxins are very acute and severe that they start affecting the body system almost immediately, while other food-borne mycotoxins have longer severe accumulative impacts on health, which could even result in cancer or immune deficiency over time [40, 41]. *Fusarium* sp., followed by *Aspergillus* sp. are

IOP Conf. Series: Materials Science and Engineering **805** (2020) 012021 doi:10.1088/1757-899X/805/1/012021

the most common and largest mycOtoxin-producing fungi [42, 43, 44]. Nigeria has suffered huge losses because of the poor standard of farm produce as a result of mycotoxins contamination. Due to the reduction in yield, health and market value of the crops, exportation has been affected drastically over the years [45]. This work might provide solution to the infestation of these fungi as the test organisms used in this study has proven susceptible to the bio-fabricated AuNPs. A green synthesized, eco-friendly AuNPs might be a good replacement to poisonous fumigating chemicals that cause deadly poisoning yearly in Nigeria [46]. *Candida albicans* was also found to be susceptible to the bio-fabricated AuNPs, suggesting that the problem of superficial mycosis caused by this yeast could be possibly mitigated.

3.3 Anticoagulant and Thrombolytic Activity

Clots formation was prevented by the addition of EDTA (which served as positive control) whereas coagulation was noticed in the negative control (blood clot). The synthesized gold nanoparticles on the other hand showed tremendous anticoagulant activities; at 250 µg/ml (54%), 500 µg/ml (55%), 750 µg/ml (56%) and 1000 µg/ml (59%) (Figure 8 and Table 1). The results obtained are in agreement with anticoagulant potentials of metallic nanoparticles synthesized from diverse biomolecules as previously recorded by Azeez *et al.* [24]. Similarly, the gold nanoparticles generated clot dissolution within 5 min of reaction when reacted with previously formed blood clot. The control experiment performed with the aqueous gold chloride and the *D. stramonium* seed extract showed lesser dissolution of blood clot (Figure 9). The AuNPs showed thrombolytic activities of 33%, 46%, 48% and 52% at 250, 500, 750 and 1000 µg/ml respectively (Table 2). Therefore, the thrombolytic activities of the synthesized gold nanoparticles demonstrated fair thrombolytic properties. Though clotting of blood is essential to inhibition of bleeding, yet its dissolution is similarly crucial in inhibiting thrombolysis and sustenance of hemostasis, where nanoparticles assume significant roles in rendering effective lysis of blood clots. The thrombolytic and anticoagulant potentials of gold nanoparticles have been recently reported [13].

IOP Conf. Series: Materials Science and Engineering **805** (2020) 012021 doi:10.1088/1757-899X/805/1/012021

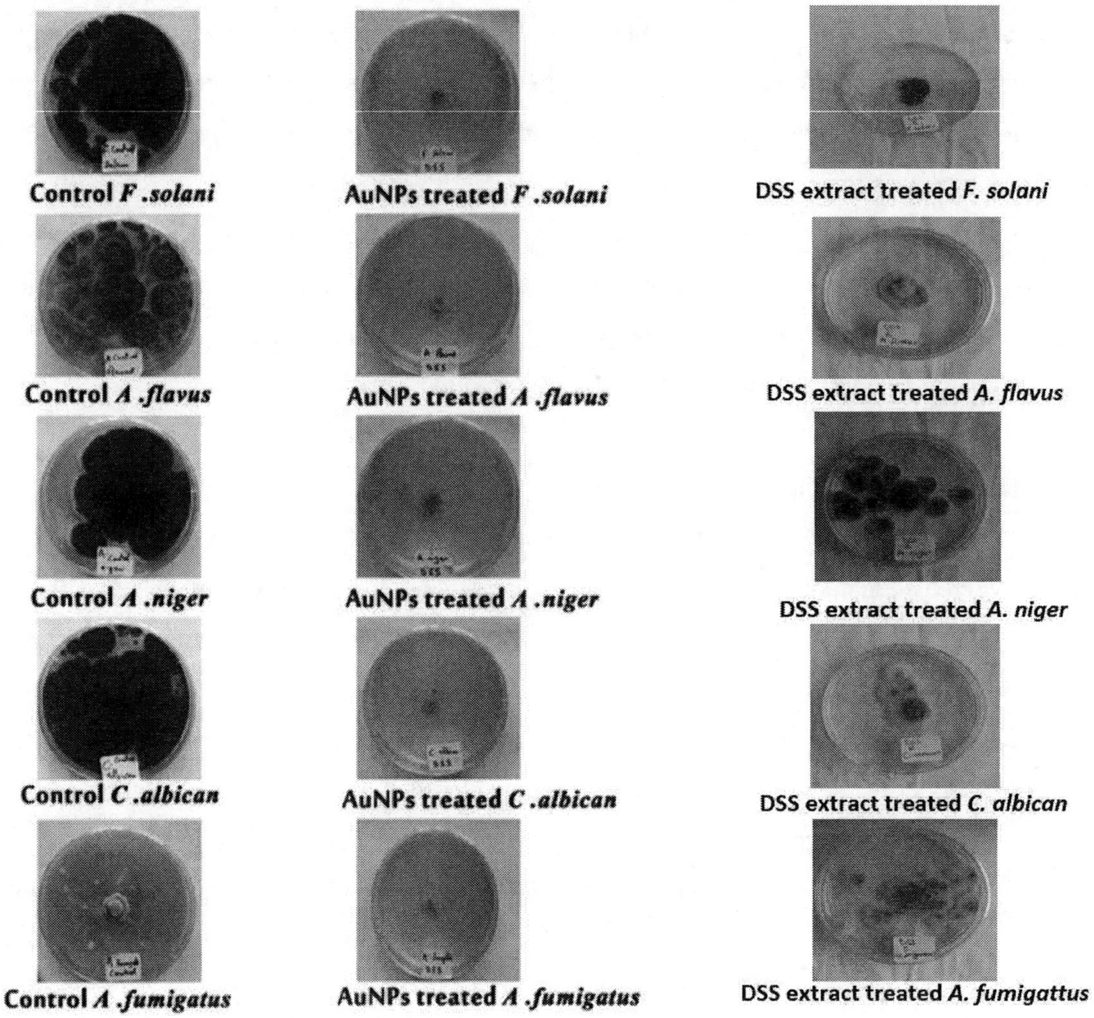

Figure 7. Antifungal activities of the bio-fabricated AuNPs

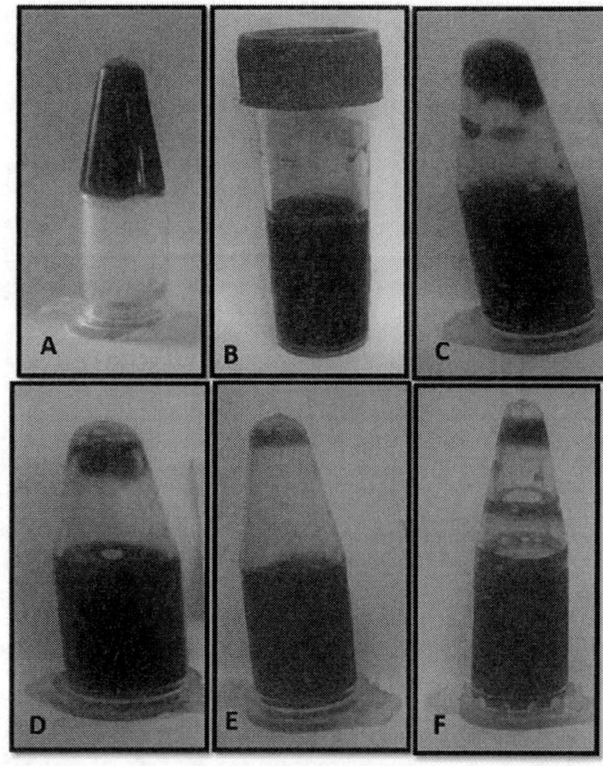

Figure 8. Anticoagulant activities of gold nanoparticles by *Datura stramonium* seed extract {A, control; B, Blood + EDTA; C, Blood + AuNPs (250 µg/ml); D, Blood + AuNPs (500 µg/ml); E, Blood + AuNPs (750 µg/ml); F Blood + AuNPs (1000 µg/ml)}

Table 1. Anticoagulant activities of the biosynthesized AuNPs

Treatments	W_3 (g)	W_5 (g)	$\frac{W_3 - W_5}{W_3} \times 100$ %
Extract	2.32	1.24	46
$HAuCl_3$	2.32	1.10	52
AuNPs (µg/ml)			
250	2.32	1.07	54
500	2.32	1.05	55
750	2.32	1.02	56
1000	2.32	0.95	59

IOP Conf. Series: Materials Science and Engineering **805** (2020) 012021 doi:10.1088/1757-899X/805/1/012021

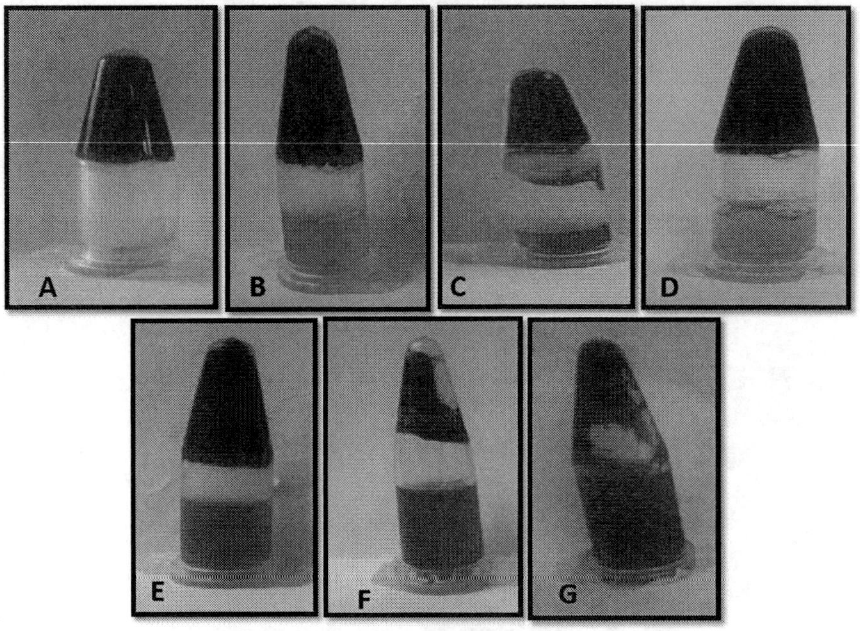

Figure 9. Thrombolytic activities of gold Nanoparticles by *Datura stramonium* seed extract {A, control; B, *D. stramonium* seed extract + blood; C, $AuCl_3$ + blood; D, Blood + AuNPs (250 µg/ml); E, Blood + AuNPs (500 µg/ml); F, Blood + AuNPs (750 µg/ml); G, Blood + AuNPs (1000 µg/ml)}

Table 2. Thrombolytic activities of the biosynthesized AuNPs

Treatments	W_3 (g)	W_5 (g)	$\frac{W_3-W_5}{W_3} \times 100$ % thrombolysis
Extract	2.32	1.61	31
$AuCl_3$	2.32	1.65	29
AuNPs (µg/ml)			
250	2.32	1.56	33
500	2.32	1.24	46
750	2.32	1.20	48
1000	2.32	1.10	52

IOP Conf. Series: Materials Science and Engineering **805** (2020) 012021 doi:10.1088/1757-899X/805/1/012021

The phytosynthesized gold nanoparticles demonstrated significant free radical scavenging properties against 2, 2- diphenyl-1-picryhydrazyl (DPPH) by 28.9, 37.3, 42.5 and 52.4% at 250, 500, 750 and 1000 µg/ml (Table 3). Furthermore, gold nanoparticles demonstrated significant radical scavenging properties against nitric oxide by 52.20, 55.73, 56.71 and 59.51% at 250, 500, 750 and 1000 µg/ml (Table 4). The scavenging activities of gold nanoparticles were compared to those recorded by Oladipo *et al.* [13], in which biosynthesized AuNPs synthesized using the cell-free extract of *Enterococcus* species showed free radical scavenging against DPPH at 33.24-51.47 % at 1- 40 µg/ml. On the evidence of the free radical scavenging properties of the AuNPs in this study, it could be established that the biosynthesized AuNPs has a very potent antioxidant properties. The activity shown by the phytosynthesized gold particles mean that they may be relevant in scavenging for free radicals in the environment as well as in biomedical applications. The free radical scavenging potentials of nanoparticles is because of the functional groups of bioreductant molecules, whose capacity to adhere to the surface of the nanoparticles may bring about amplified surface areas of activity.

Table 3. Antioxidant activities of the AuNPs

Stages	DPPH Free radical scavenging (%)
Control	
Ascorbic acid	82.3
Quercetin	62.7
AuNPs (µg/ml)	
250	28.9
500	37.3
750	42.5
1000	52.4
DSS Extract	58.8

IOP Conf. Series: Materials Science and Engineering **805** (2020) 012021 doi:10.1088/1757-899X/805/1/012021

Table 4. Antioxidant activities of the AuNPs (Nitric Oxide)

Stages	Free radical scavenging (%)
Control	
Sodium nitroprusside	82.0
AuNPs (μg/ml)	
250	52.20
500	55.73
750	56.71
1000	59.51
DSS Extract	59.50

4 Conclusion

This study demonstrated the biofabrication of AuNPs using *Datura stramonium* seed extract as the reducing and capping agent. The deep pink colored AuNPs were slightly agglomerated with sizes between 71.5 and 156.5 nm with near spherical morphology. The particles displayed varied degrees of activities as antifungal, antioxidant, anticoagulant and thrombolytic agents. Supposedly, this is the first documentation of the phytosynthesis of AuNPs from *Datura stramonium* seed extract with potential biomedical applications.

IOP Conf. Series: Materials Science and Engineering **805** (2020) 012021 doi:10.1088/1757-899X/805/1/012021

Reference

[1] Roco MC 2001 From vision to the implementation of the US National Nanotechnology Initiative. *J. Nanopart. Res.* **3** (1) pp 5-11.

[2] Lövestam G, Rauscher H, Roebben G, Klüttgen B S, Gibson N 2010 Considerations on a definition of Nanomaterial for regulatory purposes. Joint Research Centre (JRC) Reference Reports. **80** pp 00-41.

[3] Gupta V K, Gupta M and Sharma S 2001 Process development for the removal of lead and chromium from aqueous solution using red mud – an aluminum industry waste. *Water Res.* **35** (5) pp 1125-1134.

[4] Shanmugavadivu M, Selvam K and Ranjithkumar R 2014 Synthesis of pomegranate peel extract mediated silver nanoparticles and its antibacterial activity. *Am. J. Adv. Drug Deliv.* **2** (2) pp 174-182.

[5] Lateef A, Azeez M A, Asafa T B, Yekeen T A, Akinboro A, Oladipo I C, Ajetomobi F E, Gueguim-Kana E B, Beukes L S 2015 *Cola nitida*-mediated biogenic synthesis of silver nanoparticles using seed and seed shell extracts and evaluation of antibacterial activities. *BioNanoSci.* **5** (4) pp 196-205.

[6] Lateef A, Ojo S A, Folarin B I, Gueguim-Kana E B, Beukes L S 2016 Kolanut (*Cola nitida*) mediated synthesis of silver-gold alloy nanoparticles: antifungal, catalytic, larvicidal and thrombolytic applications. *J. Clust. Sci.* **27** (5) pp 1561-1577.

[7] Oladipo I C and Ogunsona S B 2019 The Utilization of Agro Waste: A nanobiotechnology Point of View. Recent Advances in Biological Research Vol. **5;** Chapter 10 Print ISBN: 978-93-89246-84-1, eBook ISBN: 978-93-89246-85-8.

[8] Salta O V 2004 Applications of nanoparticles in biology and medicine. *J. Nanobiotechnol.* **2** (1) 3. https://doi.org/10.1186/1477-3155-2-3.

[9] Caruthers S D, Wickline S A and Lanza G M 2007 Nanotechnological applications in medicine. *Curr. Opin. Biotechnol.* **18** (1) pp 26-30.

[10] Khorami M and Ganjian E 2011 Comparing flexural behavior of fibre-cement composites reinforced bagasse: Wheat and eucalyptus. *Constr. Build. Mater.* **25** (9) pp 3661-3667.

[11] Karademir A, Aydemir C and Yenidogan S 2011 Sound absorption and print density properties of recycled sheets made from waste paper and agricultural plant fibres. *Afr. J. Agric. Res.* **6** (28) pp 6073-6081.

[12] Morais M G, Vilásia G M, Daniela S, Patricia P and Jorge A V C 2014 Biological applications of nanobiotechnology. *J. Nanosci. Nanotechnol.* **14** (1) pp 1007-1007.

[13] Oladipo I C, Lateef A, Elegbede J A, Azeez M A, Asafa T B, Yekeen T A, Akinboro A, Gueguin-Kana E B, Beukes L S, Oluyide T O and Atanda O R 2017 *Enterococcus* species for the one-pot biofabrication of gold nanoparticles: Characterization and nanobiotechnological applications. *J. Photochem. Photobiol. B: Biol.* **173** pp 250-257.

[14] Van H L, Chi N H T and Huy H T 2013 Synthesis of silica nanoparticles from Vietnamese rice husk by sol–gel method. *Nanoscale Res. Lett.* **8** (1) 58. https://doi.org/10.1186/1556-276X-8-58.

[15] Lee K X, Kamyar S, Mikio M, Noriyuki K, Nurul B A K, Shaza E B and Yen P Y 2016 Green synthesis of gold nanoparticles using aqueous extract of *Garcinia mangostana* fruit peels. *J. Nanomater.* Article ID 8489094, http://dx.doi.org/10.1155/2016/8489094.

[16] Sharma K, Sanket K and Anupam J 2016 Green synthesis of silver nanoparticles by using waste vegetable peel and its antibacterial activities. *J. Pharm. Sci. Res.* **8** (5) pp 313-316.

[17] Dethier M, Cordier Y and Demeyer K 1993. Cultivation of *Datura* species for scopolamine and hyoscyamine production in Burundi. *ISHS Acta Hortic.* **331**. https://doi.org/10.17660/ActaHortic.1993.331.6.

[18] Mairura FS and Setshogo MP 2008 *Datura stramonium* L. Schmelzer GH, Gurib-Fakim A (Editors). pp 221-225.

[19] Burkill HM 2000 The useful plants of West Tropical Africa. 2nd Edition. Volume 5, Families S–Z, Addenda. Royal Botanic Gardens, Kew, Richmond, United Kingdom. 686 pp

[20] Philipov S, Berkov S and Doncheva T S 2007 GC-MS survey of *Datura stramonium* alkaloids. *Comptes Rendus de l'Académie Bulgare des Sciences* **60** (3) pp 239-250.

[21] Lateef A, Folarin BI, Oladejo SM, Akinola PO, Beukes LS and Gueguim-Kana EB 2018 Characterization, antimicrobial, antioxidant and anticoagulant activities of silver nanoparticles synthesized from *Petiveria alliacea* L. leaf extract. *Prep. Biochem. Biotechnol.* **48** (7) pp 646-652.

[22] Lateef A, Azeez MA, Asafa TB, Yekeen TA, Akinboro A, Oladipo IC, Azeez L, Ojo SA, Gueguim-Kana EB and Beukes LS 2016 Cocoa pod extract-mediated biosynthesis of silver nanoparticles: its antimicrobial, antioxidant and larvicidal activities. *J. Nanostruct. Chem.* **6** pp 159-169.

[23] Lateef A, Ojo SA, Elegbede JA, Azeez MA, Yekeen TA and Akinboro A 2017 Evaluation of some biosynthesized silver nanoparticles for biomedical applications: hydrogen peroxide scavenging, anticoagulant and thrombolytic activities. *J. Clust. Sci.* **28** (3) pp 1379-1392.

[24] Azeez M A, Lateef A, Asafa T B, Yekeen T A, Akinboro A, Oladipo I C, Gueguim-Kana EB and Beukes, L.S. 2017 Biomedical applications of cocoa bean extract mediated silver nanoparticles as antimicrobial, larvicidal and anticoagulant agents. *J. Clust. Sci.* **28** (1) pp 149-164.

[25] Lateef A, Akande M A, Ojo S A, Folarin B I, Gueguim-Kana E B and Beukes L S 2016 Paper wasp nest-mediated biosynthesis of silver nanoparticles for antimicrobial, catalytic, anti-coagulant and thrombolytic applications, *3 Biotech.* **6** 140. http://dx.doi.org/10.1007/s13205-016-0459-x.

[26] Ojo S A, Lateef, A Azeez M A, Oladejo S M, Akinwale A S, Asafa T B, Yekeen T A, Akinboro A, Oladipo I C, Gueguim-Kana E B and Beukes L S 2016 Biomedical and catalytic applications of gold and silver-gold alloy nanoparticles biosynthesized using cell-free extract of *Bacillus safensis* LAU 13: antifungal, dye degradation, anticoagulant and thrombolytic activities, *IEEE Trans. NanoBiosci.* **15** (5) pp 433-442

[27] Lateef A, Ojo S A, Oladejo S M 2016 Anti-candida, anti-coagulant and thrombolytic activities of biosynthesized silver nanoparticles using cell-free extract of *Bacillus safensis* LAU 13. *Process Biochem.* **51** (10) pp 1406-1412.

[28] Abirami H, Tajuddin N B, Mohamed H M I, Hussain S J, Premkumar K, Shilu M, Archunan G and Ishtiaq Q 2016 Synthesis of plant mediated gold nanoparticles using *Azima tetracantha* Lam. leaves extract and evaluation of their antimicrobial activities. *Pharmacogn. J.* **8** (5) pp 507-512.

[29] Alaa A A, Aljabali I D, Yazan A, Mazhar S A, Zoubi I D, Khalid M A, Bahaa A I D, Osama A A, Alaaldin M A, Mourad B and David J E 2018 Synthesis of gold nanoparticles using leaf extract of *Ziziphus zizyphus* and their antimicrobial activity. *Nanomater.* **8** (3) 174 https://doi.org/10. 3390/nano8030174.

[30] Inbakandan D, Venkatesan R and Khan S A 2010 Biosynthesis of gold nanoparticles utilizing marine sponge *Acanthella elongata* (Dendy, 1905). *Colloids Surf. B: Biointerf.* **81** (2) pp 634-639.

[31] Ahmad T, Wani I A, Manzoor N, Ahmed J and Asiri A M 2013 Biosynthesis, structural characterization and antimicrobial activity of gold and silver nanoparticles. *Colloids Surf. B: Biointerf.* **107** pp 227-234.

[32] Ramakrishna M, Babu D R, Gengan R M, Chandra S and Rao G N 2016 Green synthesis of gold nanoparticles using marine algae and evaluation of their catalytic activity. *J. Nanostruct. Chem.* **6** (1) pp 1-13.

[33] Hamed A G, Khalid A K, Essam H I and William N S 2019 Synthesis of gold nanoparticles (AuNPs) using *Ricinus communis* leaf ethanol extract, their characterization, and biological applications. *Nanomater.* **9** (5) 765 https://doi.org/10.3390/nano9050765.

[34] Reddy G R, Jayakumar C, Morais A B, Sreenivasn D and Gandhi N N 2012 Green synthesis characterization and *in-vitro* antibacterial activity of polyshaped gold nanoparticles by using *Senna siamea* (Lam.) Plant leaf extract. *Int. J. Green Chem. Bioprocess* **2** (1) pp 1-5.

[35] Jayaseelan C, Ramkumar R, Rahuman A A and Perumal P 2013 Green synthesis of gold nanoparticles using seed aqueous extract of *Abelmoschus esculentus* and its antifungal activity. *Ind. Crops Prod.* **45** pp 423-429.

[36] Mishra A, Kumari M, Pandey S, Chaudhry V, Gupta K C and Nautiyal C S 2014 Biocatalytic and antimicrobial activities of gold nanoparticles synthesized by *Trichoderma* sp. *Bioresour. Technol.* **166** pp 235-242.

[37] Malathi S, Ezhilarasu T, Abiraman T and Balasubramanian S 2014 One pot green synthesis of Ag, Au and Au-Ag alloy nanoparticles using isonicotinic acid hydrazide and starch. *Carbohydr. Polym.* **111** pp 734-743.

[38] Bennett J W and Klich M 2003 Mycotoxins. *Clin. Microbiol. Rev.* **16** pp 497-516.

[39] Bhat RV and Miller JD. Mycotoxins and food supply. Food, Nutrition and Agriculture-Food for the Future., FAO. 1991. https://www.fooddiagnostics.dk/seekings/uploads/FAO-Mycotoxins_and_Food_Supply.pdf. Accessed on 25 January, 2020.

[40] Sokefun E, Ayepola O O and Olasehinde G I 2018 Mycotoxins: Food production and exportation in Nigeria. *Earth Environ. Sci.* **210** (1) 012018. https://doi.org/10.1088/1755-1315/210/1/012018.

[41] Ahmed A I and Jutta P 2015 Mycotoxins: Producing fungi and mechanisms of phytotoxicity. *Agricult.* **5** (3) pp 492-537.

[42] Yazar S and Omurtag G 2008 Fumonisins, Trichothecenes and Zearalenone in Cereals. *Int. J. Mol. Sci.* **9** pp 2062-2090.

[43] Alexandra H H and Lewis E H B 2015 Comparative Ochratoxin toxicity: A review of the available data. *Toxins* **7** pp 4253-4282.

[44] Travis R B and Felicia W 2015 Ochratoxin A and human health risk: A review of the evidence. *Crit. Rev. Food Sci. Nutr.* **55** (13) pp 1860-1869.

[45] Bankole S A and Adebanjo A 2003 Mycotoxins in food in West Africa: current situation and possibilities of controlling it. *Afri. J. Biotechnol.* **2** (9) pp 254-263.

[46] Asogwa E U and Dongo L N 2009 Problems associated with pesticide usage and application in Nigerian cocoa production: a review. *Afr. J. Agric. Res.* **4** (8) pp 675-683.

IOP Conf. Series: Materials Science and Engineering **805** (2020) 012022 doi:10.1088/1757-899X/805/1/012022

Optimization of green synthesis of silver nanoparticles using Response Surface Method (RSM)

M Ibrahim[*1], JB Agboola[2], AS Abdulkareem[3,5], O Adedipe[1] and JO Tijani[4,5]

[1]Department of Mechanical Engineering, Federal University of Technology, Minna.

[2]Materials and Metallurgical Engineering Department, Federal University of Technology, Minna.

[3,]Department of Chemical Engineering, Federal University of Technology, Minna.

[4]Department of chemistry, Federal University of Technology, Minna.

[5]Nanotechnology Research Group, Centre for Biotechnology and Genetic Engineering, Federal University of Technology, Minna.

*Corresponding author: Ibrahim M. (abuhannatu@yahoo.com).

Abstract

This study is aimed at investigating the effect of synthesis parameters such as temperature, pH, volume of precursor and volume of extract on the crystal size of silver nanoparticles (AgNPs) synthesized via plant extract of *Piptadeniastrum africanum* (African Oak) using 2^4 factorial experimental design. The synthesized nanoparticles were characterized by X-ray Diffraction (XRD) and High Resolution Scanning Electron Spectroscopy (HRSEM) fitted with Energy Dispersive Spectroscopy (EDS). The optimum values of the synthesis parameters were found to be: pH 8, reaction temperature 35 ^{o}C, the ratio of the volume of extract to the volume of precursor was kept at 1:10 while silver nitrate $(AgNO_3)_{aq}$ solution was maintained at 1 mM concentration. The HRSEM and XRD analysis showed that the morphology of the nanoparticles is spherical with an average crystal size of 20-25 nm.

Keywords: Synthesis parameters, crystallite size, silver nanoparticles, green synthesis, factorial experimental design, characterization

IOP Conf. Series: Materials Science and Engineering **805** (2020) 012022 doi:10.1088/1757-899X/805/1/012022

1 Introduction

Metallic nanoparticles have recently attracted attentions due to their unique physicochemical properties. These special properties and especially the size give them an edge over other materials. This improves their applications in various human activities [1]. Nanoparticles synthesis basically consist of reduction of metal salt in its oxidation state to zero-valence state called nanoparticles. Nanoparticles can be synthesized using physical methods, chemical methods, and biological methods. The physical and chemical methods of synthesis are complicated because they are carried out using separate stabilizing and capping agents; they are also expensive and they are hazardous to human health due to toxic fumes that are radiated from the chemicals involved. Recently, green synthesis method is preferably used for the synthesis of metallic nanoparticles. Plants and plant extracts play dual role as a reducing, capping and stabilizing agents. In addition, the method is a one pot procedure, economical and more environmentally friendly.

Silver nanoparticles (AgNPs) have wide applications in medical field, water purification, improvement in mechanical properties and corrosion protection of some engineering materials. The green synthesis method is governed by several factors such as temperature, pH, reaction time, volume of metal salt and volume of plant extract. The interaction of these synthesis factors plays a vital role in determining shape and size of synthesized Ag-NPs. Despite numerous studies carried out on synthesis of AgNPs, experimental optimization of operational parameters and factors influencing the crystal size of AgNPs has not been given adequate consideration. Therefore, this study investigates the optimization of the synthesis parameters of AgNPs using Response Surface Method (RSM).

2 Materials and Methods

2.1 Materials

Chemicals used in the study include silver nitrate ($AgNO_3$) purchased from Sigma Aldrich, concentrated hydrochloric acid (HCl) and sodium hydroxide (NaOH) purchased from Kermel England and distilled water. All the chemicals used have percentage purity in the range of 95-99%. Some of the equipment used in the study includes digital weighing balance, pH meter, magnetic stirrer and hot plate, uv-spectrometer, high resolution scanning electron microscopy, X-ray diffractometer, freeze drying machine and high speed refrigerated centrifuge.

2.2 Methodology

2.2.1 Collection and preparation of plant material

Fresh leaves of *Piptadeniastrum africanum* were collected at Bosso area in Minna, Niger State Nigeria. The leaves were thoroughly washed with running tap water and distilled water in order to remove dust and other contaminants and then dried under shade for seven days. The dried leaves were pounded into homogeneous powder using mortar and pestle. The obtained powder was then kept under dry condition for future use.

2.2.2 Preparation of aqueous leaf extracts of *Piptadeniastrum africanum*

Ten grams (10 g) of powdered leaf of *P. africanum* was mixed with 100 ml of distilled water in a 250 ml beaker and the mixture was boiled at 60 ^{o}C for 15 min using magnetic hot plate and stirrer. The extract was obtained by filtration using Whatman no.1 filter paper and stored in brown bottle under room temperature for future use.

2.2.3 Phytochemical analysis of *Piptadeniastrum africanum* leaf

The phytochemical analysis of *Piptadeniastrum africana* leaf extract was carried out in accordance with AOAC methods and reported by Singleton *et al.* [2].

2.2.4 Synthesis of silver nanoparticles using aqueous leaf extracts of *P. africanum* and experimental optimization of synthesis parameters

Silver nanoparticles were synthesized by the mixture of 1 ml of aqueous solution of *P. africanum* leaf extract to 10 ml of 1 mM $AgNO_3$ in a 250 ml beaker while the pH was adjusted to a value of 8 using few drops of 2 M solution of NaOH. The reaction was allowed to take place at room temperature for 60 min. The crystal size of the formed AgNPs was calculated using Derby-Scherrer's equation [3].

$$D = \frac{K\lambda}{\beta Cos\Phi} \qquad (1)$$

Where Φ is the peak width angle, *K* is the shape factor = 0.89, β is the full width at half maximum of the XRD peak, and λ is the wavelength of light used for diffraction = 1.541Å.

IOP Conf. Series: Materials Science and Engineering **805** (2020) 012022 doi:10.1088/1757-899X/805/1/012022

In order to determine the optimum crystal size, four synthesis parameters; reaction temperature, pH, volume of extract and volume of silver nitrate ($AgNO_3$) aqueous solution were varied and the optimization procedure was carried out using 2^4 factorial design of experiment (DOE). Many authors have reported the synthesis of silver nanoparticles within certain range of synthesis conditions. However, the two level factors chosen for this work were based on previous studies and were considered only after preliminary investigations. low level (-) indicate variable values that were considered as low limits and high level (+) indicate variable values that were considered as upper limit as shown in Table 1.

Table 1. High and Low values for 2^4 Factorial Design the for synthesis of AgNPs

Variables	Reaction Temp (°C)	pH	$AgNO_3$ vol. (ml)	Extract vol. (ml)
L/ L	35.00	8.00	10.00	1.00
H/L	60.00	9.00	50.00	3.00

Design Expert software was used to carry out statistical Analysis of Variance (ANOVA), so that the main as well as the combined effects of the parameters considered are represented in empirical models to show the effect of these parameters on the crystal size.

2.2.5 Characterization of Ag-NPs

The synthesized nanoparticles were characterized by X-ray Diffraction (XRD) and High Resolution Scanning Electron Spectroscopy (HRSEM) fitted with Energy Dispersive Spectroscopy (EDS).

3 Results and Discussion

3.1 Phytochemical analysis of *Piptadeniastrum africanum* leaf extract

The result of phytochemical analysis on *P. africanum* leaf extract shows the presence of some prominent biomolecules such as phenols, flavonoids and tannins as presented in Table 2 which are the active ingredients in the plant.

IOP Conf. Series: Materials Science and Engineering **805** (2020) 012022 doi:10.1088/1757-899X/805/1/012022

Table 2. Phytochemical attributes of *P. africanum* leaf extract

S/N	Phytochemicals	Result
1.	Phenol	+
2.	Flavonoids	+
3.	Tannins	+

3.2 Experimental optimization of the synthesis

All the sixteen samples obtained from 2^4 factorial design of experiment were synthesized and characterized by X-ray diffraction (XRD) analysis and the crystal size was calculated by Derby Scherrer's equation. Table 3 shows 2^4 factorial design of experiment with the corresponding responses (crystal size).

Table 3. Experimental matrix for 2^4 factorial design for the synthesis of AgNPs

Run	R.Temp. °C	pH	P/cursor vol. (ml)	Ext.vol. (ml)	Crsyt. Size (nm)
1	35.00	9.00	50.00	1.00	20.2
2	35.00	9.00	10.00	3.00	18.4
3	35.00	9.00	50.00	3.00	22.5
4	60.00	9.00	50.00	1.00	25.0
5	60.00	8.00	50.00	1.00	25.5
6	60.00	9.00	10.00	3.00	25.8
7	35.00	8.00	10.00	1.00	28.0
8	35.00	8.00	50.00	3.00	28.7
9	60.00	9.00	10.00	1.00	30.0
10	60.00	9.00	50.00	3.00	30.2
11	60.00	8.00	10.00	3.00	30.5
12	35.00	8.00	50.00	1.00	34.4
13	60.00	8.00	50.00	3.00	42.6
14	35.00	9.00	10.00	1.00	25.9
15	60.00	8.00	10.00	1.00	20.6
16	35.00	8.00	10.00	1.00	15.4

IOP Conf. Series: Materials Science and Engineering **805** (2020) 012022 doi:10.1088/1757-899X/805/1/012022

3.3 Numerical Optimization

3.3.1 Evaluation

The model terms are temperature (A), pH (B), volume of precursor (C) and extract volume (D). Three factor interaction (3FI) model was utilized and no aliases were found in the model. The degrees of freedom for the model, residuals, lack of fit and pure error are 14, 1, 1 and 0 respectively while the average leverage of the model is 0.9375. The model F-value of 19.55 in Table 4 shows that the model is significant. The model terms (A, B, C, D) are also shown to be significant based on "prob>F" < 0.05. The ANOVA analysis also show that Std. Dev. = 2.05, R-Squared = 0.9572, Mean = 25.63, Adj R-Squared = 0.9082, C.V. % = 8.00, Pred R-Squared = 0.7762, PRESS = 153.59, Adeq Precision = 14.200.

The predicted R-squared value of 0.7762 is in reasonable agreement with the Adjusted R-squared value of 0.9082, while the adequate precision value of 14.200 indicated an adequate signal.

The final regression equations generated in terms of coded factor is given by:

Crystallite size = 25.63 +2.69A + 3.75B + 2.84C + 2.85D + 1.01BC - 0.95BD + 0.44CD - 1.21BCD...(2)

while the final regression equation in terms of actual factor is given by:

Cystallite size = 15.06 + 0.22 temp. - 0.98 pH - 45.95Ext. Vol.- 1.16Salt Vol.+ 5.66 pH * Ext.vol. + 0.15 pH * salt vol. + 1.05 * Ext. vol * salt vol. - 0.12 * pH * Ext.vol. * salt vol......(3)

Table 4. ANOVA for the factorial model

Source Remark	Sum of Square	Df	Mean Square	F-value	P-value Prob>F
Model Significant	656.77	8	82.10	19.55	0.0004
A-Tempt.	115.56	1	115.56	27.52	0.0012
B-pH	225.00	1	225.00	53.38	0.0002
C-Ext. vol.	128.00	1	128.00	30.67	0.0009
D-Salt vol.	129.96	1	129.96	30.75	0.0008
BC	16.40	1	16.40	3.91	0.0887
BD	14.44	1	14.44	3.44	0.1061
CD	3.06	1	3.06	0.73	0.4214
BCD	23. 52	1	23.52	5.63	0.0499
Residual	29.40	7	4.20		
Corr.Total	686.17	15			

3.3.2 Diagnostic Plots

The diagnostic plots are shown in Figure 1, while the overall diagnostic case statistics is presented in Table 5. It can be seen that the actual values and the predicted values of are in close agreement.

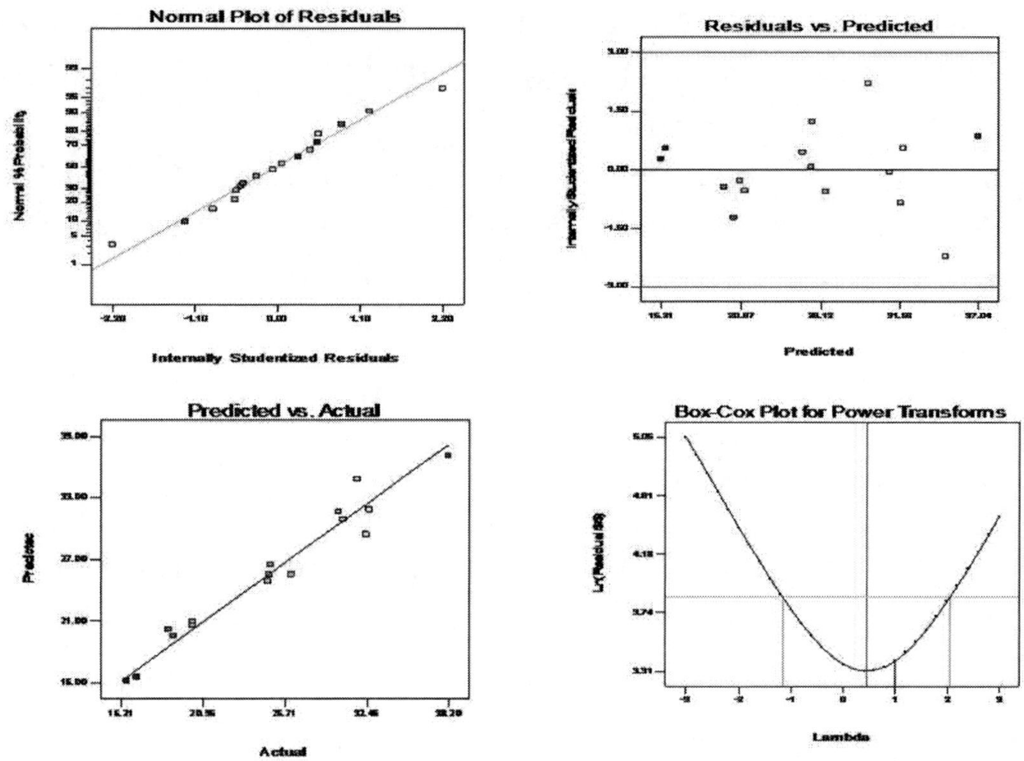

Figure 1. The diagnostic plots of the responses

Table 5. Diagnostic case statistics

Stand Order	Actl Val.	Pred. Val.	Resid	Int. studt Res.	Ext studt Res.	Influence fitted val DFFIT	cooks Dist.	Run Order
1	15.60	15.21	0.39	0.286	0.266	0.302	0.012	13
2	20.20	20.59	-0.39	-0.286	-0.266	-0.302	0.012	11
3	18.50	20.16	-1.66	-1.226	-1.282	-1.453	0.215	12
4	27.20	25.54	1.66	1.226	1.282	1.453	0.215	5
5	16.30	15.5	0.74	0.544	0.515	0.584	0.042	14
6	20.20	20.94	-0.74	-0.544	-0.515	-0.584	0.044	8
7	32.40	29.41	2.99	2.204	3.688	4.18	0.694	3
8	31.80	34.79	-2.99	-2.204	-3.688	-4.18	0.694	16
9	18.90	19.51	- 0.61	-0.452	-0.452	-0.481	0.929	1
10	25.50	24.89	0.61	0.452	0.452	0.481	0.029	4
11	25.80	5.51	0.087	0.065	0.060	0.068	0.001	2
12	30.80	0.89	-0.088	-0.065	-0.060	-0.06	0.001	10
13	25.70	0.46	-0.26	-0.563	-0.533	-0.604	0.045	9
14	32.60	1.84	0.26	0.563	0.533	0.604	0.045	7
15	30.50	1.66	-1.16	-0.858	-0.839	-0.952	0.106	15
16	38.20	37.04	1.16	0.858	0.839	0.9.52	0.105	6

The criteria set for the numerical optimization of the various synthesis factors are detailed in Table 6.

Table 6. Optimization constraints

Name	Goal	Upper Limit	Lower Limit	Upper Weight	Lower Weight	Import
Temp.	in range	60	35	1	1	3
pH	in range	9	8	1	1	3
Ext. vol.	in range	3	1	1	3	3
Salt vol.	in range	50	10	1	1	3
Cryt. Size	Minimize	38.2	15.4			

The result presented in Table 7 indicated that the least crystallite size of 15.43 nm was obtained for the synthesis condition of temperature 35 °C, pH 8, and extract to salt volume ratio of 1:10. These conditions were used for the synthesis of AgNPs in this study.

IOP Conf. Series: Materials Science and Engineering **805** (2020) 012022 doi:10.1088/1757-899X/805/1/012022

Table 7. Final optimized synthesis parameters

No.	Temp.	pH	Ext.vol.	Salt vol.	Cryt. Size	Desblty	
1	35.00	8.00	1.00	10.00	15.43	1.0	Selected
2	35.16	8.01	2.15	10.05	15.55	1.0	
3	35.06	8.02	2.09	10.02	15.58	1.0	
4	36.30	8.01	1.21	10.13	15.58	1.0	
5	35.43	8.00	1.00	11.67	15.49	1.0	
6	35.98	8.00	1.01	10.28	15.46	1.0	
7	35.61	8.01	1.23	10.19	15.48	1.0	
8	35.07	8.00	2.46	10.07	15.35	1.0	
9	35.87	8.01	1.27	10.10	15.53	1.0	
10	35.50	8.00	1.26	10.63	15.46	1.0	
11	35.98	8.00	1.61	10.09	15.55	1.0	
12	35.00	8.00	1.01	13.50	15.60	1.0	
13	35.03	8.00	2.94	10.02	15.59	1.0	
14	35.06	8.00	2.15	10.76	15.59	1.0	
15	35.07	8.00	2.74	10.05	15.57	1.0	
16	35.69	8.00	1.66	10.44	15.56	1.0	
17	36.28	8.01	1.10	10.43	15.60	1.0	
18	36.03	8.00	1.33	10.02	15.52	1.0	
19	35.17	8.00	2.87	10.01	15.59	1.0	
20	35.42	8.00	1.21	10.40	15.40	1.0	
21	35.92	8.01	1.01	10.26	15.51	1.0	
22	35.01	8.01	2.11	10.45	15.59	1.0	
23	37.18	8.00	1.00	11.02	15.79	0.992	
24	35.08	8.10	2.83	10.00	16.83	0.945	
25	35.00	8.00	1.01	38.78	18.33	0.879	
26	35.00	8.00	1.01	44.71	18.96	0.851	
27	51.95	8.03	1.00	10.00	19.00	0.849	
28	35.00	8.00	1.00	45.80	19.08	0.846	
29	35.00	8.01	3.00	36.33	22.80	0.682	
30	35.09	8.44	1.84	50.00	24.93	0.587	

3.4 Characterization of AgNPs

3.4.1 UV-vis spectroscopy

Figure 2 shows UV-vis spectrum of as-synthesized AgNPs. The UV-vis spectrum shows the characteristic surface plasmon resonance band (SPR) at maximum absorption peak corresponding to the wavelength of 400 nm which is within the range previously reported for AgNPs [4].

IOP Conf. Series: Materials Science and Engineering **805** (2020) 012022 doi:10.1088/1757-899X/805/1/012022

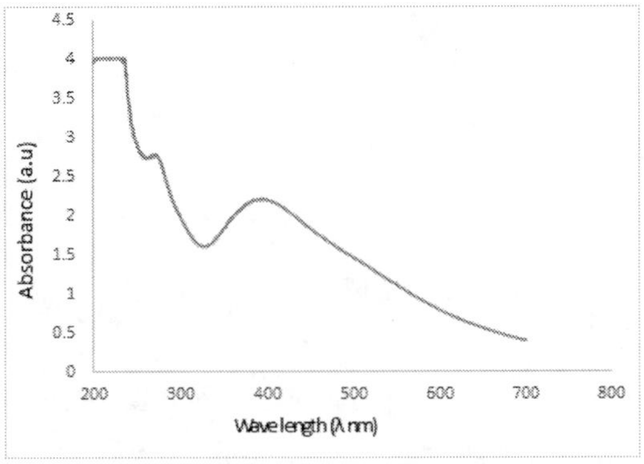

Figure 2. UV-vis spectrum of AgNPs

3.4.2 High Resolution TEM/SEM

The TEM AND SEM images of the as–produced AgNPs are shown in Figure 3. The TEM image shows spherical shaped, mono-dispersed AgNPs, while the SEM image shows spherical shaped and agglomerated AgNPs.

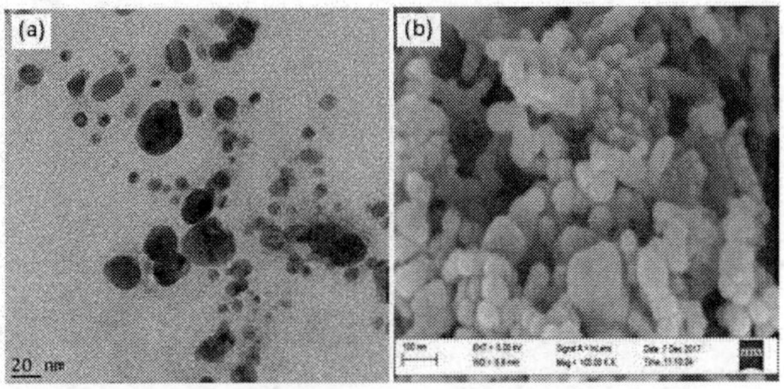

Figure 3. (a)TEM image of AgNPs (b) SEM image of AgNPs

3.4.3 X-ray Diffraction (XRD)

The XRD pattern of the as-produced AgNPs is shown in Figure 4. All reflections correspond to the presence of silver metal with face-centered cubic symmetry. The diffraction intensities were recorded from $2\theta = 20^{o}$ to 100°. Five prominent diffraction peaks at 38.2^{o}, 44.5^{o}, 64.7^{o}, 77.5^{o} and 81.8^{o} shown are in accordance with the JCPDS file corresponds to the set of reflection planes indexed as (111), (200), (220), (311), and (222) respectively. This is similar to published report on XRD of AgNPs [5].

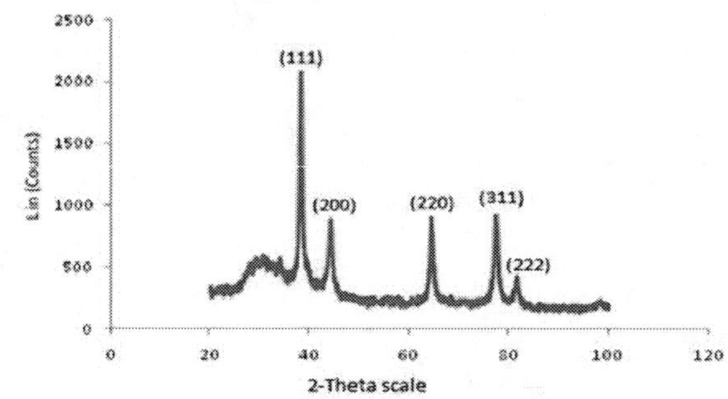

Figure 4. XRD pattern of AgNPs

3.4.4 Energy Dispersive Spectrum (EDS)

Figure 5 shows the EDS spectrum of as-synthesized AgNPs, with the prominence of silver metal and occurrence of impurities like C, O, Na, Pt, and Pb which are from the plant extract.

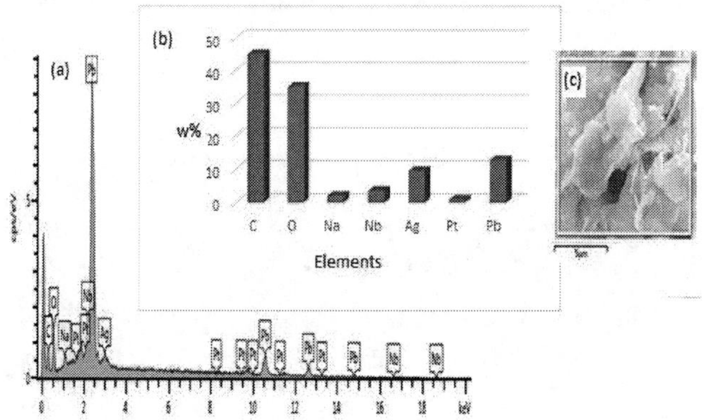

Figure 5. EDS spectrum of as-synthesized AgNPs

3.5 Influence of Synthesis Parameters on the Crystallite Size of Silver Nanoparticles (AgNPs)

In order to investigate the effect of synthesis parameters on the crystallite size of AgNPs, the interaction between the synthesis factors were plotted on 3D surface plots as shown in Figures 6A-C. It was observed that at higher volume of extract, the crystal size of AgNPs is large. This was caused by slowed reduction reaction of Ag^+ to Ag^0 [2]. Oluwaniyi [3] pointed out that the absorption peaks of AgNPs were broader and irregular at higher volume of extract indicating a slow reduction rate of Ag+ to Ag^0 and existence of AgNPs accompanied by increased size distribution.

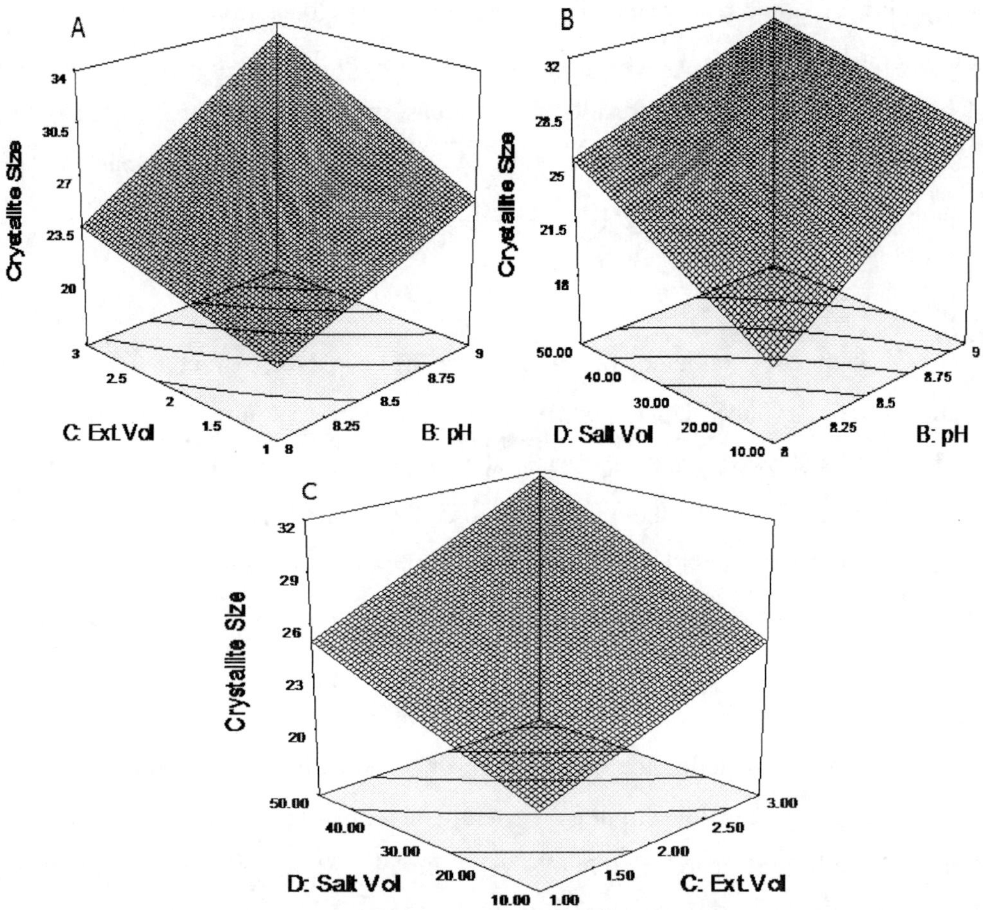

Figure 6. (A) Interaction between pH and extract volume; (B) interaction between pH and salt volume; (C) interaction between extract volume and salt volume

It was also observed that as the volume of Ag^+ solution was increased with respect to the volume of extract, the crystallite size of AgNPs become smaller. This is attributed to existence of more positively charged ions which increased the rate of reaction and hence encouraged the reduction in mean diameter of the NPs. It was reported that as the volume of Ag^+ solution is increased, the absorption peak becomes sharper with excellent enhancement in the absorption band intensity.

As pH of the reaction mixture is increased from lower values to higher values the mean crystal size of AgNPs becomes smaller. However, an increment in reaction temperature was accompanied by increased growth of crystal size of the nanoparticles. This is attributed to the increase movement of free electrons that resulted to rise in the reaction rate. This is in agreement with the opinion of other researchers [6]. However, it has been reported that increment in reaction temperature above 60 ^{o}C usually leads to increase in the intensity of SPR band [7] as a result of bathochromic shift which results in a decrease in the mean diameter of AgNPs. This study has shown the practicability of using RSM to optimize the biosynthesis of AgNPs as earlier established by some authors [8-10]

4 Conclusion

The optimized synthesis parameters of temperature 35°C, pH 8, extract to $AgNO_3$ solution volume ratio of 1:10 produced an optimum crystallite size of 15.4 nm. The characterization of the produced AgNPs showed similar spherical-shaped AgNPs as reported by other researchers. The ANOVA analysis shows that the model as well as the model terms were significant based on P-value <0.05. This means that all the synthesis parameters played a vital role in determining the crystallite size of the synthesized AgNPs.

Acknowledgements

The authors wish to acknowledge the contribution of Nanotechnology Research Group, Centre for Genetic Engineering and Biotechnology, Federal University Technology, Minna, Niger State, Nigeria, Nanotechnology Research Group (*NANO*$^+$), Ladoke Akintola University of Technology, Ogbomoso, Oyo State, Nigeria and all those that contributed technically in this paper.

References

[1] Rao YS, Kotakadi VS, Prasad TN, Reddy AV and Gopal DS 2013 Green synthesis and spectral characterization of silver nanoparticles from Lakshmi tulasi (*Ocimum sanctum*) leaf extract. *Spectrochim. Acta Part A: Mol. Biomol. Spectr.* **103** pp 156-159.

[2] Singleton, V.L., Orthofer, R. and Lamuela-Raventós, R.M 1999 Analysis of total phenols and other oxidation substrates and antioxidants by means of folin-ciocalteu reagent. *Methods Enzymol.* **299** pp. 152-178.

[3] Oluwaniyi OO, Adegoke HI, Adesuji ET, Alabi AB, Bodede SO, Labulo AH and Oseghale CO 2016 Biosynthesis of silver nanoparticles using aqueous leaf extract of *Thevetia peruviana* Juss and its antimicrobial activities. *Appl Nanosci.* **6** (6) pp 903-912.

[4] Lateef A, Folarin BI, Oladejo SM, Akinola PO, Beukes LS, and Gueguim-Kana EB 2018 Characterization, antimicrobial, antioxidant and anticoagulant activities of silver nanoparticles synthesized from *Petiveria alliacea* L. leaf extract. *Prep. Biochem. Biotechnol.* **48** (7) pp 646-652.

[5] Lateef A, Adelere IA, Gueguim-Kana EB, Asafa TB and Beukes LS 2015 Green synthesis of silver nanoparticles using keratinase obtained from a strain of *Bacillus safensis* LAU 13. *Int. Nano Lett.* **5** (1) pp 29-35.

[6] Ahmed S, Saifullah, Ahmad M, Swami BL and Ikram S 2016 Green synthesis of silver nanoparticles using *Azadirachta indica* aqueous leaf extract. *J. Rad. Res. Appl. Sci.* **9** (1) pp 1-7.

[7] Ibrahim HM 2015 Green synthesis and characterization of silver nanoparticles using banana peel extract and their antimicrobial activity against representative microorganisms. *J. Rad. Res. Appl. Sci.* **8** (3) pp 265-275.

[8] Nyakundi EO and Padmanabhan MN 2015 Green chemistry focus on optimization of silver nanoparticles using response surface methodology (RSM) and mosquitocidal activity: *Anopheles stephensi* (Diptera: Culicidae). *Spectrochim. Acta Part A: Mol. Biomol. Spectr.* **149** pp 978-84.

[9] Yıldız N, Ateş Ç, Yılmaz M, Demir D, Yıldız A and Çalımlı A 2014 Investigation of lichen based green synthesis of silver nanoparticles with response surface methodology. *Green Process. Synth.* **3** (4) pp 259-270.

[10] San Chan Y and Don MM 2013 Optimization of process variables for the synthesis of silver nanoparticles by *Pycnoporus sanguineus* using statistical experimental design. *J. Korean Soc. Appl. Biol. Chem.* **56** (1) pp 11-20.

IOP Conf. Series: Materials Science and Engineering **805** (2020) 012023 doi:10.1088/1757-899X/805/1/012023

Preparation and characterization of nanosilica from sorghum husk ash by chemical method

MC Akaleme*, IO Hassan and BJ Olawuyi

Department of Building, School of Environmental Technology, Federal University of Technology, Minna, Nigeria

*Corresponding author: mcakaleme@yahoo.com

Abstract

Extraction of nanosilica from Sorghum Husk Ash (SHA) was performed in this study by acid leaching method. Two methods of calcination was adopted (open and controlled method) in the production of SHA. Amorphous silica was produced by dissolving a quantity of SHA in hydrochloric acid solution which made the removal of heavy metals possible. The extracted/pure silica was rinsed with deionized water which lowers the pH to 7.0. The SHA silica was characterized using XRF, XRD and BET techniques. Silica and mineral contents of SHA were determined by XRF, while the X-ray diffraction patterns revealed the amorphous nature of extracted silica. Pure silica yield from SHA was 89.30%, while moisture content was 1.09%. The average particle size, BET surface area, pore volume, and pore diameter of SHA samples are 77.05 nm, 723.020 m^2/g, 4.448e +01 cc/g, and 2.072e+00 nm respectively. The outcome of the research study reveals that the properties of the extracted silica satisfied all the requirements as established by relevant literature. Consequently, the high volume amorphous nanosilica with minimal mineral contaminants can be produced from SHA by the acid leaching method.

Keywords: Sorghum husk ash; amorphous silica; nanosilica, BET surface area, X-ray diffraction

1 Introduction

Nanotechnology was defined as the science and technology of modifying, examining, and monitoring the performance of material's behaviour at the nanoscale [1, 2]. It was further defined nanoscale materials as a set of substances where at least one measurement is less than approximately 100 nanometers [1]. A nanometer is a million smaller than a millimeter, that is

IOP Conf. Series: Materials Science and Engineering **805** (2020) 012023 doi:10.1088/1757-899X/805/1/012023

smaller than the diameter of a cobweb/human hair. Nanotechnology can also be described as the process of creating a device or material with building blocks at the molecular scale [3].

Silica has wide applications in dental material, ceramics, pharmaceutical products, chromatograph column packing, detergents, adhesives, and electronics [4-6]. Silica has also been adopted as a forerunner for a variety of inorganic and organometallic materials which have wide applications as catalysts and in thin films or coatings for electronic and optical materials [7]. A low energy method to produce pure silica from rice husk ash with 91% extraction yield, has been developed [8, 9], acid leaching [10] and gasification [11].

Guinea corn is a cereal crop locally known as grain sorghum which belongs to the general name of *Sorghum*, and reported to be the most popular cereal grain produced in Africa [12-14]. Guinea corn/sorghum grain is mostly cultivated in northern Nigeria in states like Kaduna, Bauchi, Plateau, and other northern states. When harvested, it is normally processed mechanically by the use of combined harvesters or manually by threshing with sticks leaving a large quantity of residue (husk) constituting environmental waste annually.

Hence, objective of this research work is to extract amorphous silica (using the open and controlled burning) and nanosilica from sorghum husk ash, using the acid (HCl) as leaching agents to remove the heavy metal impurities. The prepared amorphous silica and nanosilica were characterized using XRF, XRD, and BET techniques.

2 Materials and Methods

Sorghum husk was obtained from Bida local government area of Niger state Nigeria, after the harvest seasons. The sorghum husk was washed and sun dried. Samples of sorghum husk were calcinated by open burning and controlled burning at 600 ^{o}C for 3 h at a heating rate of 10 ^{o}C/min in a muffle furnace at National Cereal Research Institute (NCRI) Baddegi, and left over night to cool in the furnace.

2.1 Silica Extraction

After the calcinations, the ashes obtained (from both open and controlled) were collected and subjected to grinding for 10 min in a ball mill. The ground ashes were kept separately in a desiccator for future material characterization tests. Nanosilica was extracted from SHA sample that gives the highest silica by adopting the acid pretreatment method, where 10 g of SHA was weighed and subjected to immersion in the acid solution (hydrochloric acid,) at a concentration of 1N acid solution for 2.5 h with constant stirring at ambient temperature. After the acidic solution was drained off, SHA was rinsed with deionized water until the pH

IOP Conf. Series: Materials Science and Engineering **805** (2020) 012023 doi:10.1088/1757-899X/805/1/012023

rose to 7.0, then it was filtered and oven dried at 50 C for 24 h. In addition, further milling was performed using the Los Angeles abrasion machine.

2.2 Moisture content of silica

Water content of the air dried silica was determined using an air oven method [15]. About 1 g of the sample was heated in aluminum moisture pan at 130 °C for 1 h. The sample was cooled in a desiccator and weighed thereafter. The weight loss (%) was recorded as the water content of sample.

2.3 Characterization

Quantitative chemical analysis of the SHA was executed by X-ray fluorescence. The particle size was assessed by the laser diffraction analyzer Easysize 20 (OMEC, Zhuhai, China). The surface area and pore volume of silica was measured by Brunauer-Emmett-Teller (BET) and Barrett-Joyner-Halenda (BJH) methods, respectively, according to ASTM D3663-03 using Micromeritics Tristar 3000 Surface Area (Micromeritics, Norcross, GA, USA) and Porosity Analyzer (Micromeritics, Norcross, GA, USA).

X-ray diffraction (XRD) of the sorghum husk ash and silica extract were monitored using X-ray minidiffractometer MD-10, at an acceleration voltage of 25 kV and current of 400 μA. The diffraction angle was scanned from 10° to 75° 2θ, at a rate of 3.25° /min.

3.0 Results and Discussion

3.1 Chemical Compositions and Physical Properties of SHA and Silica

Chemical compositions and physical properties of SHA before and after pretreatment are summarized in Tables 1 and 2, respectively. It can be seen that silica oxide forms the main component (76% and 89%) of SHAs with trace elements in the form of composite oxides K_2O, Na_2O, CaO, MgO, Fe_2O_3, and Al_2O_3. In the case of controlled burning SHA, 76.105% of silica was produced by calcining SH at 600 °C for 2 h. Potassium (K) was detected to account for the highest concentration (3.91%) among the metallic trace elements. The SHA after leaching yielded silica content of 89.30% having surface area of 723.020 m^2/g.

The trace elements were reduced to a very low level, especially in the case of potassium which was reduced to 2.3% and aluminum not detected. HCl as leaching agent also have excellent performance in washing off alkali metals and improving the surface area, as well as reducing the carbon residue content of RHA as earlier reported [16]. The average particle size, BET surface area, pore volume, and pore diameter of SHA samples after extraction are given in Table 2.

Table 1. Chemical Composition of SHA and silica extracted (%)

SHA Samples	Si O_2	Al_2O_3	Fe_2O_3	CaO	MgO	K_2O	Na_2O	SO_3	LOI
Open Burning	76.057	2.66	1.48	2.04	1.23	3.90	0.06	0.12	3.1
Controlled Burning	76.105	2.71	1.23	3.32	1.27	3.91	0.07	0.13	3.0
After extraction	89.03	ND	0.36	1.82	0.05	2.30	ND	0.03	2.85

ND, not detected

Table 2. Physical properties of SHA silica

SNPs Tests	Avg part. size	BET surf area	pore Volume	Avg pore dia
After extraction	77.05 nm	723.020 m^2/g	4.448e +01cc/g	2.072e+00 nm

The concentration and water content of the silica was 89.30% and 1.09% respectively. The X-ray diffractogram of amorphous silica concentration in SHA is shown in Figure 1. The broad X-ray diffraction pattern which signifies typical amorphous solids [9] shows that the extracted silica is principally amorphous. Diffraction peak at theta = 22 degree affirm the formation of amorphous silica, however, it has been reported that diffraction broad peak at theta = 22 degree indicates amorphous silica along with some crystalline silica [17].

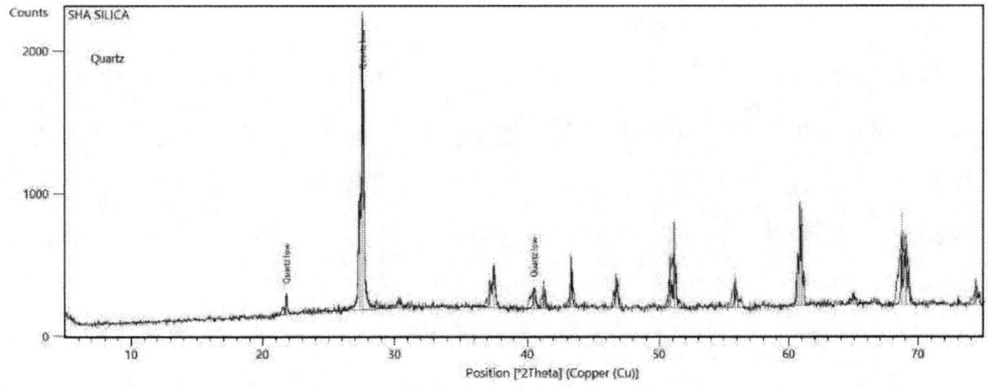

Figure 1. XRD of extracted amorphous silica

4.0 Conclusion

This study affirmed that after treatment, pure silica extracted from SHA was 89.30% with moisture of 1.09%. The average particle size of 77.05 nm falls within the standard range for nanoparticles. In addition, the nanosilica is highly amorphous as revealed by the XRD

IOP Conf. Series: Materials Science and Engineering **805** (2020) 012023 doi:10.1088/1757-899X/805/1/012023

analysis. The results of the research reveals the properties of the extraction silica satisfied all the requirements as established by relevant literature. Consequently, the high volume amorphous nanosilica with minimal mineral contaminants can be produced from SHA by the acid leaching method.

References

[1] Khan MS 2011 Nanotechnology in transportation: Evolution of a revolutionary technology. *TR News* No **277** pp. 3-8.

[2] Birgisson B, Taylor P, Armaghani J and Shah SP 2010 American road map for research for nanotechnology-based concrete materials. *Transp. Res. Rec.* **2142** (1) pp 130-137.

[3] Bakhoum ES, Garas GL, Allam ME and Ezz H 2017 The role of nano-technology in sustainable construction: a case study of using nanogranite waste particles in cement mortar. *Eng. J.* **21** (4) pp 217-227.

[4] Brinker CJ, and Scherer GW 1990 Sol-Gel Science: The Physics and Chemistry of Sol-Gel Processing. *Academic Press, Inc.*, San Diego, CA. 839-880.

[5] Proctor A, Clark PK and Parker CA 1995 Rice hull ash adsorbent performance under commercial soy oil bleaching condition. *J. Am. Oil Chem. Soc.* **72** (4) 459-462.

[6] Sun L and Gong K 2001 Review, silicon-based materials from rice husks and theirapplications. *Ind. Eng. Chem. Res.* **40** (25) pp 5861-5877.

[7] Kalapathy U, Proctor A and Shultz J 2002 An improved method for production of silica from rice hull ash. *Bioresour. Technol.* **85** (3) pp 285-289.

[8] Kalapathy U, Proctor A, and Shultz J 2000 A simple method for production of pure silica from rice hull ash. *Bioresour. Technol.* **73** (3) pp 257-262.

[9] Kamath S R and Proctor A 1998 Silica gel from rice husk ash: preparation and characterization. *Cereal Chem.* **75** (4) pp 484-487.

[10] Chakraverty A and Kaleemullah S 1991 Conversion of rice husk into amorphous silica and combustible gas. *Energ. Conserv. Manag.* **32** (6) pp 565-570.

[11] Luan TC and Chou TC 1990 Recovery of silica from the gasification of rice husk/coal in the presence of apilot fame in a modified fluidized bed. *Ind. Eng. Chem. Res.* **29** (9) pp 1922-1927.

[12] Chukwu O, Orhevba BA and Abubakar I 2011 Determination of some physicochemical properties of guinea corn (*Sorghum vulgare*). *J. Innov. Res. Eng. Sci.* **2** (2) pp 62-66.

[13] Ndububa E E and Nurudeen Y 2015 Effect of guinea corn husk ash as partial replacement for cement in concrete. *IOSR J. Mech. Civil Eng.* **12** (2) 40-45.

[14] Alkamu T, Datok EP and Jambol DD 2017 Guinea corn husk ash as partial replacement of cement in hollow sandcrete block production. *J. Emerg. Trends Eng. Appl. Sci.* **8** (6) pp 263-268.

[15] AOAC 1990 Official Methods of Analysis, 15th ed. *Association of Analytical Chemists*, Washington DC, USA.

[16] Xu W, Wei J, Chen J, Zhang B, Xu P, Ren J and Yu Q 2018 Comparative study of water-leaching and acid-leaching pretreatment on the thermal stability and reactivity of biomass silica for viability as a pozzolanic additive in cement. *Mater.* **11** (9) 1697. https://doi.org/10.3390/ma11091697.

[17] Naqvi J, Shah FH and Mansha M 2011 Extraction of amorphous silica from wheat husk using $KMnO_4$. *J. Faculty Eng. Technol.* **18** (1) pp 39-46.

IOP Conf. Series: Materials Science and Engineering **805** (2020) 012024 doi:10.1088/1757-899X/805/1/012024

Surface sterilization of *Ocimum* seeds and tissues with biosynthesized nanosilver and its effects on callus induction

AA Adebomojo* and AA AbdulRahaman
Department of Plant Biology, University of Ilorin, Ilorin, Nigeria
*Corresponding author: adebomojowaheed@gmail.com

Abstract

Plant tissue culture is a basic and fundamental component of plant biotechnology. Nowadays, nanomaterials especially nanosilver (NS) are being used as an antimicrobial agents for surface sterilization of explants in tissue culture. In this study, biosynthesized nanosilver (BNS) was used for the surface sterilization of *Ocimum* seeds and tissues and its effects on callus induction were evaluated. The seeds and tissues were exposed to different concentrations of BNS (10, 50 and 100 mg/l) as well as 5% Clorox for five exposure times (5, 10, 20, 30 and 60 min) and effects on germination, callus induction and surface sterilization were determined. The BNS was found very effective on surface sterilization as 100% decontamination was achieved with no adverse effect on explant viability and callus formation but rather had stimulating effect on formation of callus. The study concluded that BNS can be used as an antimicrobial agent in surface disinfection of explants therefore extending the frontiers of the potential application of biosynthesized nanosilver in tissue culture.

Keywords: Tissue culture, *Ocimum*, callus, nanosilver, disinfection, plant biotechnology

1. Introduction

Nanotechnology is the study of nano-objects having sizes generally between 1-100 nm [1]. "Nanobiotechnology" is a term that blends nanotechnology and biotechnology terms and is the crossing point of nanotechnology, materials science, molecular biology and biomedical sciences. In both nanobiotechnology and nanotechnology, the principal zone of interest is nanoparticles [2]. Nowadays, plant extracts and microorganisms are presently utilized as bioreductors in nanobiotechnological studies so as to synthesize nanoparticles from metal ions. This technique is termed as "bio-synthesis" or "green synthesis" of nanoparticles [3]. It is an emerging alternative

IOP Conf. Series: Materials Science and Engineering **805** (2020) 012024 doi:10.1088/1757-899X/805/1/012024

technology as it is ecofriendly, compared to conventional physical and chemical synthesis methods, which are energy demanding, expensive, and often generate toxic chemicals and vapours [4].

Silver nanoparticles are the most generally utilized among the other metal nanoparticles. They have antiseptic, antibacterial, antifungal and antiviral properties relying upon their capacity to attack a wide range of organic processes in microorganisms prompting the disruption of cell membrane structure and the breakdown of plasma membrane potential. This subsequently leads to depletion in the intracellular ATP and in this way guaranteeing the cell death. In addition, they are eco-friendly and biocompatible with low lethality towards the earth and human [2].

Presently, plant tissue culture has direct commercial importance, and its applications in basic research such as genetics, biochemistry, cell biology, and biotechnology are evidences for its usefulness. Plant tissue culture does not only provide a method for mass propagation of tissues but also makes possible the production of disease-free and genetically modified plants. It also provides a way for secondary metabolite production [5]. Despite all the advantages of this technique, some methodological obstacles, mainly the contamination of explants, hinders its exploitation as an efficient technique for biotechnological research. Both external and internal contaminations of plant tissues during tissue culture process turn out to be a prevailing problem. This is because microorganisms, mostly fungi, and bacteria which are major contaminants, grow much faster than plant cells in growth media and take up all the nutrients, thus preventing the plants from growing [6].

Surface sterilization is the most important step in plant tissue culture as it is a pre-requisite for tissue culture techniques to be carried out under sterile conditions. Microbial growth rates far exceed plant cell division in culture. Thus, surface sterilization endeavours to remove all microorganisms that will thrive under *in vitro* conditions whilst still maintaining explant viability [7]. Mostly utilized agents for surface sterilization in tissue culture are sodium hypochlorite, ethanol, calcium hypochlorite, mercuric chloride, surfactants, and antibiotics. They are not ecofriendly and are toxic to living organisms. They also have very toxic effects on plant cells and tissues that are cultured. Their effects range from browning of the tissues to complete inhibition of growth in plant cells and tissues under culture conditions [8]. Therefore, an appropriate surface sterilizing agent is very important as the key to successful procedures for plant tissue culture.

IOP Conf. Series: Materials Science and Engineering **805** (2020) 012024 doi:10.1088/1757-899X/805/1/012024

Although, there are a variety of techniques to reduce the possibility of fungal and bacterial contaminations during *in vitro* propagation, examples of such include repetitive sub-culturing and meristem culture (the pathogen-free part of the plant). However, designing a more effective approach to sterilize plant tissues still seems inevitable to eliminate the labour-intensive trial and errors and the time-consuming decontamination approaches in existence. In order to eliminate the tenacious bacterial and fungal contaminations in plant tissue culture, treatment with antibiotics and antifungal agents have been used which greatly lessen the contaminations. However, it has been reported that antibiotics are usually phytotoxic, and have an inhibitory effect on cell multiplication, regeneration, callus induction, and explants' survival [9]. Moreover, lack of an optimal protocol for sterilization of field, greenhouse-derived (*ex vitro*) or orchard explants may result in a scarcity of samples for further tissue culture research [9].

The genus *Ocimum* (family-Lamiaceae), is a versatile aromatic genus well known for its economically important essential oils of enormous medicinal properties [10]. Therefore, in this study, silver nanoparticles synthesized via biological procedure and 5% Clorox were used as surface sterilization agents on the seeds of five *Ocimum* species and tissues of three *Ocimum* species. This was with a view to evaluate the efficiency of BNS as a surface sterilization agent in tissue culture processes compared to 5% Clorox as a standard.

2. Materials and Methods

This research was carried out at the Tissue Culture Laboratory, National Centre for Genetic Resources and Biotechnology (NACGRAB), Moor Plantation, Ibadan, Nigeria. Seeds of five *Ocimum* species (*O. basilicum*, *O. gratissimum*, *O. africanum*, *O. canum* and *O. sanctum*) and tissues (leaves, stems and inflorescence) of three *Ocimum* species (*O. basilicum*, *O. canum* and *O. sanctum*) collected from screen house grown plants were used. The nanosilver (BNS) used was Kola pod (*Cola nitida*) mediated silver nanoparticles [11] obtained from the Nanotechnology Research Group (*NANO*$^+$), Ladoke Akintola University of Technology (LAUTECH), Ogbomoso, Nigeria. The properties of this nanosilver are shown in Table1.

Table 1. Properties of Kola pod (*Cola nitida*) mediated biogenic synthesized silver nanoparticles used [11]

Properties	
Mode of synthesis	By kolanut (*Cola nitida*) pod extract
Max absorbance	431.5 nm
Size	12-80 nm
Stability	Stable, without aggregation
Dispersity	Polydisperse
Shape	Fairly spherical
Phase	Face-centered cubic crystalline
Colour	Dark brown

2.1 *Ocimum* Seeds Sterilization

Prior to the sterilization processes, the seeds were poured in a test tube and filled with water. The floated seeds were regarded as non-viable seeds and were sieved. Prior to the BNS applications, 70% ethanol was first applied for 3 min to the seeds and they were rinsed with sterile distilled water twice. The seeds were later exposed to 100, 50 and 10 mg/l BNS with five exposure times (5, 10, 20, 30 and 60 min). Two control groups were used; negative control group where sterile distilled water was used and the positive control group where 5% commercial Clorox bleach (8.25% active sodium hypochlorite) was used with five exposure times (5, 10, 20, 30 and 60 min) for surface sterilization. They were rinsed in sterile distilled water five times and inoculated on semi solid MS medium (MS basal salts supplemented with 3% sucrose (30 g/L) and 0.1% myoinositol (0.1 g/L), with pH 5.8 ± 0.1) [12]. The cultures were maintained in a growth chamber at 25± 2°C with light intensity of 10-20 μmol m^{-2} s^{-1} provided by white fluorescent bulbs for photoperiod of 16 h of light and 8 h of darkness. The cultures were observed daily for the commencement of seed germination and at the second week, percentage of sterile and germinated seeds were determined by observations of seeds in culture condition. All the surface sterilization processes were conducted under the laminar flow hood to maintain a sterile environment.

2.2 Plant Tissues Sterilization

Plants parts with trichomes (especially with glandular trichomes) such as the leaf, the stem and the inflorescence, were collected from screen house (at flowering stage). These parts were collected in a clean beaker and washed under running tap water for 30 min to wash off adhering dirt coming along with them from the screen house. They were surface sterilized by dipping in concentrations of BNS (100 m, 50 and 10 mg/l) with five exposure times (5, 10, 20, 30 and 60 min). Two control groups were used; negative control group where sterile distilled water was used and the positive control group where 5% commercial Clorox bleach (8.25% active sodium hypochlorite) was used with five exposure times (5, 10, 20, 30 and 60 min) for surface sterilization. They were later rinsed in sterile distilled water five times.

Glandular trichomes were excised from the sterilized plants part (the leaf, the stem and the flower) using a sterile scalpel and forceps with the aid of a dissecting microscope. The glandular trichomes were carefully excised from the epidermal wall without contamination by the epidermal cells and were then inoculated on callus induction media (MS medium supplemented with 1.0 mg/l NAA and 0.5 mg/l BAP for *O. basilicum* and *O.sanctum* and MS medium supplemented with 1.0 mg/l NAA and 1.0 mg/l BAP for *O. canum*). The cultures were maintained in a growth chamber at 25± 2 °C with light intensity of 10-20 μmol m^{-2} s^{-1}provided by white fluorescent bulbs for photoperiod of 16 h of light and 8 h of darkness. The cultures were observed daily for the commencement of callus formation and at the 10th day of culture, percentage of disinfected glandular trichome explants and percentage of explants that induced callus were observed. The intensity of the callus formed was observed after 3 weeks of culture. All the surface sterilization and the culture processes were conducted under the laminar flow hood to maintain a sterile environment.

2.3 Statistical Analysis

Each treatment of 10 samples were replicated thrice, totaling 30 explants per treatment ($n \geq 30$) in order to have an approximately normal distribution of the sample mean [13]. Data were statistically analyzed using one-way Analysis of Variance (ANOVA) with significant means separated using Duncan's Multiple Range Test (DMRT) at $P = 0.05$, on IBM Statistical Package for Social Science (SPSS) statistics version 25 software. Results were presented as mean ± standard deviation (SD).

IOP Conf. Series: Materials Science and Engineering **805** (2020) 012024 doi:10.1088/1757-899X/805/1/012024

3. Results and Discussion

3.1 Effects of Different Concentrations and Exposures of BNS on *Ocimum* Seeds

The percentages of disinfected seeds were observed after 10 days of culture and the control bottles with full contaminated seeds were discarded. Daily observations were made on the remaining cultures to check for any contamination again and no contaminations were recorded after the 10th day in all the treatments. The percentage germinated seeds were observed after 14 days of culture (Table 2).

The seeds of *O. canum* and *O. basilicum* started sprouting after 4 days of culture, while *O. sanctum*, *O. africanum* and *O. gratissimum* started sprouting after the 6th day of culture. No adverse effect was observed on sterilized plantlets compared to the positive control groups. They all started to germinate within the same days with approximately the same height as the controls.

In the negative control group where sterile distilled water was used as disinfectant, the cultures were heavily contaminated and no growth was observed. The contaminants were observed on the seeds after 2 days of culture and grew larger rapidly till they cover the whole surface of the medium as shown in Plate 1C. But in the BNS treated group where full contaminations were not noticed, the contaminants were observed on the seeds after 5 days of culture which was 3 days later than the negative control group. The rate of growth of the contaminants were not as rapid as in the negative control group as shown in the 10 days old culture in plates 1A, D and E. The contaminants were seen mounted on the seeds, thus preventing their growth as seen in Figures 1A, B, D and E where some of the contaminated seeds showed expression of root but no shoot.

The results showed that all used concentrations of BNS were efficient in decontaminating the *Ocimum* plant seeds with varying exposure times and even at higher concentration of 100 mg/l with 60 min exposure time, the BNS seems not to have any harmful effect on the seeds germination.

Nartop [14] also reported successful decontamination (100%) of Lamiaceae seeds with green synthesized silver nanoparticles with no lethal effect on the seeds viability. Although not all disinfected seed showed 100% germination, this does not seem to be associated with the effect of the BNS as there were no significant differences in the percentage germinated where 100% decontaminations were achieved. In some instances, 100% seed germinations were recorded at some high concentration and longer exposure time, as seen in *O. canum* seed. The disparity in percentage germinations could be ascribed to the seeds viability as low seed viability of *Ocimum*

IOP Conf. Series: Materials Science and Engineering **805** (2020) 012024 doi:10.1088/1757-899X/805/1/012024

species had been reported [15, 16]. Also it was reported that *Ocimum* species germination rate varies [15].

When the seeds were treated with very low concentration of BNS (10 mg/l) for 5 min, 100% decontamination was not achieved. However 96% was achieved in *O. canum* being the highest percentage decontamination followed by 83, 66, 63 and 56% achieved in *O. bacilicum, O. sanctum, O. gratissimum* and *O. africanum* respectively. Notwithstanding 100% decontamination was achieved when they were treated with the same concentration for 10 min in *O. bacilicum* and *O. canum* followed by 96% in *O. africanum* and 83% in *O. gratissimum* and *O. sanctum* seeds. At 20, 30 and 60 min exposure time with the same concentration of BNS, 100% decontamination of the seeds were achieved in all the species. This shows that the BNS possessed great antimicrobial properties at a very low concentration which is similar to the report of Matsumura *et al.* [17] that the toxicity of nanosilver to microbial cells is evident even at very low concentration. Although in a related research conducted by Mahna *et al.* [18] on Arabidopsis seeds, 10 mg/l was not effective in decontaminating the seeds. It was reported that plant-derived silver nanoparticles demonstrated better antimicrobial activity than others synthesized through physical and chemical methods [19], which is in consonance with the result obtained in this study.

When the seeds were treated with 100 mg/l BNS for 5 min, 100% decontamination was achieved in all the species with the exception of *O. africanum* (93% decontamination). However, with an increase in the exposure time to 10, 20, 30 and 60 min, 100% decontaminations were achieved in all the species without affecting the seed viability. This differs to the result reported by Mahna *et al.* [18], where treatment of the Arabidopsis seeds with 100 mg/l of non-biosynthesized nanosilver (Nanocid L-2000) for 30 and 60 min resulted in 100% decontamination but none of the seed survived. This shows that BNS is more plant-friendly than other silver nanoparticles synthesized by other methods. Nartop [14] also reported that green synthesized nanosilver showed no negative effect on both germination and morphology of plantlets when used to sterilize Lamiaceae seeds.

In this study, the exposure time was a critical parameter in decontamination of the explants. When BNS was used at higher concentrations of 100 mg/l for 5 min, 100% decontamination was not achieved but as the exposure time increases, full decontaminations were achieved even at the lowest concentration of 10 mg/l. This corroborates the work of Abdi *et al.* [20], which is the first

IOP Conf. Series: Materials Science and Engineering **805** (2020) 012024 doi:10.1088/1757-899X/805/1/012024

time NS was used in tissue culture, where 89% decontamination was achieved. Mahna *et al.* [18] also achieved 100% decontamination with treatment of Arabidopsis seeds in 100 mg/l AgNPs for 1 to 5 min.

In the positive control group where 5% Clorox was used, 100% decontamination was achieved in all the exposure times used in all the species. However, the exposure time of the seeds to the treatment was a critical parameter. When the 5% Clorox was used with short exposure time (5 and 10 min) for the seeds, it functioned as an antimicrobial agent leaving no harm on the seeds and their germination. High germination percentages of up to 86% in *O. africanum*, 96% in *O. basilicum* and *O. canum* and 90% in *O. gratissimum* and *O. sanctum* were observed. However with an increase in the exposure time to 20, 30 and 60 min, there was drastic reduction in the germination percentages (Table 2). Seeds germination of only 10% in *O. africanum*, 6.67% in *O. basilicum*, 3.33% in *O. gratissimum* and 0% in *O. canum* and *O. sanctum* were observed when exposed to the Clorox solution for 60 min. These low germination percentages could be as a result of the prolonged exposure of the explants to sodium hypochlorite (NaOCl). Yildiz and Ekiz [21] opined that direct contact of explant with NaOCl during the sterilization process can have a severe effect on regeneration capacity.

Table 2. Seed germination percentage of surface sterilized *Ocimum* seeds with biosynthesized nanosilver

Sterilization Methods		*O. africanum* seed		*O. basilicum* seed		*O. canum* seed		*O. gratissimum* seed		*O. sanctum* seed	
BNS (mg/l)	Exposure time (min)	(%)D±SD	(%)G±SE	(%)D ± SE	(%)G±SE	(%) D± SE	(%)G ± SE	(%)D ± SE	(%)G ± SE	(%) D± SE	(%)G ± SE
10	5	56.67±5.77b	50.00±10.00c	83.33±5.77b	76.67±5.77b	86.67±5.77b	83.33±5.77b	63.33±5.78c	56.6±5.77c	66.67±5.77c	60.00±0.00b
	10	93.33±5.77a	83.33±5.77b	100.00±0.00a	100.00±0.00a	100.00±0.00a	96.67±5.77a	83.33±5.78b	73.3±15.28b	83.33±5.77b	73.33±15.28b
	20	100.00±0.00a	96.67±5.77a	100.00±0.00a	93.33±5.77a	100.00±0.00a	96.67±5.77a	100.00±0.00a	93.3±5.77a	100.00±0.00a	93.33±5.77a
	30	100.00±0.00a	90.00±0.00ab	100.00±0.00a	100.00±0.00a	100.00±0.00a	100.00±0.00a	100.00±0.00a	96.67±5.77a	100.00±0.00a	96.67±5.77a
	60	100.00±0.00a	93.33±5.77ab	100.00±0.00a	93.33±5.77a	100.00±0.00a	96.67±5.77a	100.00±0.00a	93.33±5.77a	100.00±0.00a	93.33±5.77a
50	5	63.33±5.77b	50.00±10.00c	100.00±0.00a	100.00±0.00a	100.00±0.00a	96.67±5.77a	76.67±5.77b	66.67±11.55b	83.33±5.77b	80.00±0.00b
	10	100.00±0.00a	86.67±5.77b	100.00±0.00a	93.33±5.77a	100.00±0.00a	93.33±5.77a	100.00±0.00a	93.33±5.77a	100.00±0.00a	93.33±5.77a
	20	100.00±0.00a	90.00±0.00ab	100.00±0.00a	96.67±5.77a	100.00±0.00a	100.00±0.00a	100.00±0.00a	90.00±10.00a	100.00±0.00	90.00±10.00ab
	30	100.00±0.00a	100.00±0.00a	100.00±0.00a	96.67±5.77a	100.00±0.00a	100.00±0.00a	100.00±0.00a	90.00±10.00a	100.00±0.00a	93.33±5.77a
	60	100.00±0.00a	96.67±5.77ab	100.00±0.00a	96.67±5.77a	100.00±0.00a	96.67±5.77a	100.00±0.00a	93.33±5.77a	100.00±0.00a	93.33±5.77a
100	5	93.33±5.77b	86.67±5.77a	100.00±0.00	96.67±5.77a	100.00±0.00a	100.00±0.00a	100.00±0.00	93.33±5.77a	100.00±0.00a	90.00±10.00a
	10	100.00±0.00a	100.00±0.00a	100.00±0.00	93.33±5.77a	100.00±0.00a	100.00±0.00a	100.00±0.00	90.00±10.00a	100.00±0.00a	93.33±5.77a
	20	100.00±0.00a	90.00±10.00a	100.00±0.00	96.67±5.77a	100.00±0.00a	96.67±5.77a	100.00±0.00	90.00±10.00a	100.00±0.00a	90.00±10.00a
	30	100.00±0.00a	93.33±11.55a	100.00±0.00	96.67±5.77a	100.00±0.00a	96.67±5.77a	100.00±0.00	96.67±5.77a	100.00±0.00a	96.67±5.77a
	60	100.00±0.00a	93.33±5.77a	100.00±0.00	96.67±5.77a	100.00±0.00a	100.00±0.00a	100.00±0.00	93.33±5.77a	100.00±0.00a	93.33±5.77a
+Control (5% Clorox)	5	100.00±0.00a	86.67±5.77a	100.00±0.00a	93.33±5.77a	100.00±0.00a	96.67±5.77a	100.00±0.00a	86.67±5.77a	100.00±0.00a	83.33±5.77a
	10	100.00±0.00a	86.67±5.77a	100.00±0.00a	96.67±5.77a	100.00±0.00a	96.67±5.77a	100.00±0.00a	90.00±10.00a	100.00±0.00a	90.00±10.00a
	20	100.00±0.00a	56.67±5.77b	100.00±0.00a	56.67±5.77b	100.00±0.00a	53.33±5.77b	100.00±0.00a	63.33±5.77b	100.00±0.00a	56.67±5.77b
	30	100.00±0.00a	53.33±5.77b	100.00±0.00a	23.33±5.77c	100.00±0.00a	16.67±5.77c	100.00±0.00a	30.00±10.00c	100.00±0.00a	6.67±5.77c
	60	100.00±0.00a	10.00±10.00c	100.00±0.00a	6.67±5.77d	100.00±0.00a	0.00±0.00d	100.00±0.00a	3.33±5.77d	100.00±0.00a	0.00±0.00c
-Control		0.00±0.00	0.00±0.00	0.00±0.00	0.00±0.00	0.00±0.00	0.00±0.00	0.00±0.00	0.00±0.00	0.00±0.00	0.00±0.00

Values within column followed by the same letter are not significantly different at P = 0.05; (%) D ± SD is the percentage decontaminated seeds ± standard deviation and, (%) G ± SD is the percentage of germinated seed ± standard deviation

IOP Conf. Series: Materials Science and Engineering **805** (2020) 012024 doi:10.1088/1757-899X/805/1/012024

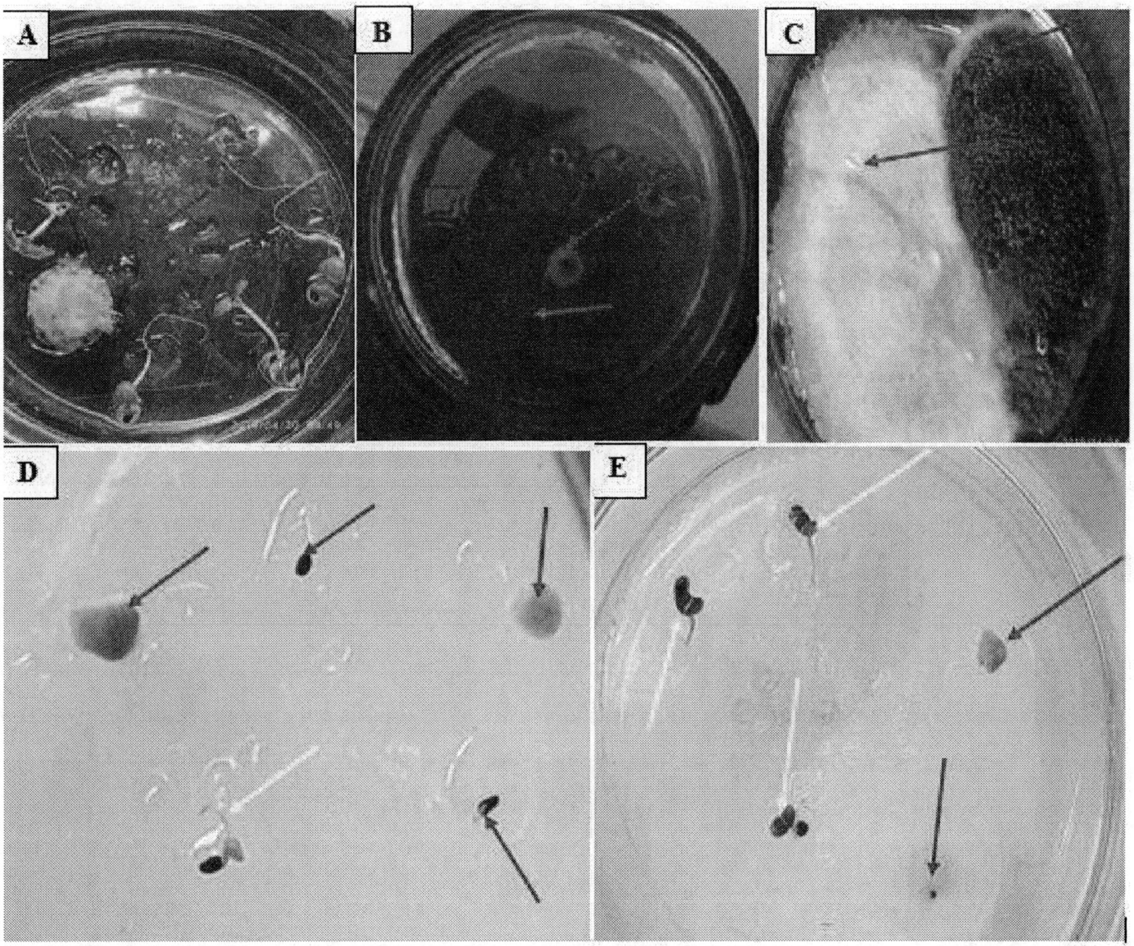

Figure 1. *Ocimum basilicum* culture showing contamination mounted on the seed surface sterilized with 5 mg/l BNS for 5 min (red arrow) (A), *O. canum* contaminated seed showing root but no shoot when treated with 5 mg/l BNS for 5 min (red arrow=contaminant) (green arrow=root) (B),*O. africanum* negative control (seed treated with sterile distilled water) culture showing full contamination (C), *O. gratissimum* culture showing contaminated, non-germinated seeds (red arrows); non-contaminated, non-germinated seeds (blue arrows); non-contaminated, germinated seed (yellow arrow) (D), and *O. sanctum* culture showing contaminated, non-germinated seed (red arrow); non-contaminated, non-germinated seed (blue arrow); non-contaminated, germinated seeds (yellow arrows) (E)

3.2 Effects of Different Concentrations and Exposures to BNS on Glandular Trichome Explants and Callus Culture of *Ocimum* species

The percentage of disinfected glandular trichome explants were calculated after 10 days of culture when all the calli have been formed in all the treatments. The bottles with contaminations were discarded while daily observations were made on the uncontaminated calli for any contamination, and no contaminations were recorded after the 10th day in all the treatments and species. The percentages of explants that induced callus were observed after the 10th day of culture. The intensity of the callus formed was determined after 3 weeks of culture. The data were presented in Table 3 for *O. basilicum*, Table 4 for *O. canum* and Table 5 for *O. sanctum*.

In the negative control where sterile distilled water was used to disinfect the explants, the cultures were heavily contaminated and no growth was observed as the contaminations have covered the surfaces of the media (Figures 2A and 2D). Treatments with BNS for the species where 100% decontaminations were not achieved manifested delay in the onset of contaminations. Contamination was noticed on the 6th day of culture compared with the negative control group which started to show at the 3rd day. In addition, the rate of growth of the contaminations in the treatments with BNS was not as rapid as in the negative control group. This delay observed show that even at very low concentration with little exposure time, the toxicity of BNS to the microbial cells is evident which is in agreement with the report by Matsumura *et al.* [17].

In the positive control group where 5% Clorox was used, 100% decontamination was recorded in all the species with varying exposure time. The highest percentage callus induction was 86% in *O. basilicum* (Table 3) and *O. canum* (Table 4), and 83% in *O. sanctum* (Table 5). The exposure time had significant effect on the callus induction as the degree of callus formation reduces with an increase in the exposure time till no callus was induced at higher exposure time. This showed that prolonged exposure of plant tissues to Clorox even at low concentration is detrimental to them [21].

BNS was effective in decontaminating the explants in all the species with 100% decontamination achieved in all the concentrations cum varying exposure time. Exposure time had significant effect on the explants decontamination as higher decontamination was achieved with an increase in the exposure time. No significant effect was observed for exposure time on the callus induction in treatments where 100% decontamination was achieved. Almost all the disinfected explants induced callus, suggesting that high concentration of BNS (100 mg/l) with longer exposure time (60 min) has no detrimental effect on the explants. This result corroborates that of Mahna *et al.* [18], where nanosilver was used on tomato cotyledon explants and another report [22], where up to 97% decontamination was achieved. In the later, all surface sterilized *Rosmarinus officinalis* stem explants using *Rubia tinctorum* L. cell culture extracts mediated biosynthesized silver nanoparticles induced callus.

The calluses were observed starting to form on the surface of the medium after 8 days of culture in *O. basilicum* and 6 days of culture in *O. canum* and *O. sanctum* in all treatments with BNS. There was evident growth in treatments with higher concentrations of BNS with longer exposure time. This was 2 days earlier than in the positive control where 5% Clorox was used which showed no growth until the 10th day of culture in *O. basilicum* and 8th day of culture in *O. canum* and *O. sanctum.* Moreover the number of explants that induced calluses were higher in treatment with BNS, where 100% decontaminations were achieved compared with the positive control group as shown in Tables 3, 4 and 5 for *O. basilicum, O. canum* and *O. sanctum,* respectively. This result revealed that BNS stimulated the explant to form callus 2 days earlier than Clorox treated explants, which indicates that BNS possess stimulating effect towards callus induction.

In explants exposed to high concentrations of BNS with long exposure times, the degree of callus formation increased in *O. basilicum* and *O. sanctum* as shown in Tables 3 and 5, respectively, while in *O. canum* both the degree of callus formation and root initiation increased as shown in Table 4, Figures 2B and 2C. These results pointed to the possibility of BNS producing a stimulating effect on the induction and intensity of callus formation, which could be as a result of the absorption of the BNS by the explants as observed during the sterilization processes. After rinsing of the plant tissues (explants) three times with sterile distilled water, traces of the BNS colour (brown) could still be seen on them. Stimulating effects of camel milk mediated AgNPs in culture media for callus growth from leaf explants of *Solanum nigrum* (L.) have been reported by Ewais *et al.* [23]. They reported that the frequency of callus formation (89%) and the fresh weight of the callus (4.67 g per leaf explant) were increased on MS medium augmented with 5 mg/l BA, 3 mg/l NAA and 8 mg/l AgNPs. Aghdaei *et al.* [24] and Fazal *et al.* [25] also reported the stimulating effect of AgNPs in culture media.

IOP Conf. Series: Materials Science and Engineering **805** (2020) 012024 doi:10.1088/1757-899X/805/1/012024

Table 3. The effect of different concentrations and exposures of BNS on callus of *Ocimum basilicum* glandular trichome explants cultured on MS medium supplemented with 1.0 mg/l NAA and 0.5 mg/l BAP

Sterilization Methods		Explants		Callus formed		
BNS (mg/l)	Exposure time (min)	Decontaminated (%) ± SD	callus induction (%) ± SD	Intensity of callus formation	Colour	Texture
10	5	0.00±0.00c	0.00±0.00c	-	-	-
	10	86.67±5.77b	83.33±5.77b	+	creamy	Friable and soft
	20	100.00±0.00a	96.67±5.77a	+	creamy	Friable and soft
	30	100.00±0.00a	93.33±5.77a	+	creamy	Friable and soft
	60	100.00±0.00a	93.33±5.77a	+	creamy	Friable and soft
50	5	93.33±5.77b	90.00±10.00a	+	creamy	Friable and soft
	10	100.00±0.00a	93.33±5.77a	+	creamy	Friable and soft
	20	100.00±0.00a	90.00±0.00a	+	creamy	Friable and soft
	30	100.00±0.00a	93.33±5.77a	+	creamy	Friable and soft
	60	100.00±0.00a	100.00±0.00a	++	creamy	Friable and soft
100	5	100.00±0.00a	90.00±0.00c	+	creamy	Friable and soft
	10	100.00±0.00a	93.33±5.77ab	+	creamy	Friable and soft
	20	100.00±0.00a	96.67±5.77ab	+	creamy	Friable and soft
	30	100.00±0.00a	100.00±0.00a	+	creamy	Friable and soft
	60	100.00±0.00a	93.33±5.77ab	++	creamy	Friable and soft
+ve	5	100.00±0.00[a]	83.33±5.77a	+	creamy	Friable and soft
Control	10	100.00±0.00[a]	86.67±5.77a	+	creamy	Friable and soft
	20	100.00±0.00a	83.33±5.77a	+	creamy	Friable and soft
	30	100.00±0.00a	0.00±0.00b	-	-	-
	60	100.00±0.00a	0.00±0.00b	-	-	-
-ve Control		0.00±0.00	0.00±0.00	-	-	-

Values within column followed by the same letter are not significantly different at P = 0.05; ++ = Profuse callus mean + = Moderate callus mean; - = Absent

IOP Conf. Series: Materials Science and Engineering **805** (2020) 012024 doi:10.1088/1757-899X/805/1/012024

Table 4. The effect of different concentrations and exposures of BNS on glandular trichome explants and callus culture of *Ocimum canum* cultured on MS medium supplemented with 1.0 mg/l NAA and 1.0 mg/l BAP

Sterilization Methods		Explants			Callus formed		
		Decontaminated (%) ± SD	callus induction (%) ± SD	Intensity of callus formation	Colour	Texture	Intensity of root formation
BNS (mg/l)	Exposure time						
10	5	0.00±0.00c	0.00±0.00c	-	-	-	-
	10	50.00±0.00b	46.67±5.77b	+	Creamy* and brownish	Hard, Compact and rooty	+
	20	100.00±0.00a	93.33±5.77a	+	Creamy* and brownish	Hard, Compact and rooty	+
	30	100.00±0.00a	93.33±5.77a	+	Creamy* and brownish	Hard, Compact and rooty	+
	60	100.00±0.00a	93.33±5.77a	+	Creamy* and brownish	Hard, Compact and rooty	+
50	5	56.67±5.77b	50.00±10.00b	+	Creamy* and brownish	Hard, Compact and rooty	+
	10	100.00±0.00a	86.67±5.77a	+	Creamy* and brownish	Hard, Compact and rooty	+
	20	100.00±0.00a	90.00±0.00a	+	Creamy* and brownish	Hard, Compact and rooty	+
	30	100.00±0.00a	96.67±5.77a	++	Creamy* and brownish	Hard, Compact and rooty	++
	60	100.00±0.00a	96.67±5.77a	++	Creamy* and brownish	Hard, Compact and rooty	++
100	5	100.00±0.00a	90.00±0.00a	+	Creamy* and brownish	Hard, Compact and rooty	+
	10	100.00±0.00a	93.33±5.77a	+	Creamy* and brownish	Hard, Compact and rooty	+
	20	100.00±0.00a	93.33±5.77a	++	Creamy* and brownish	Hard, Compact and rooty	++
	30	100.00±0.00a	96.67±5.77a	++	Creamy* and brownish	Hard, Compact and rooty	++
	60	100.00±0.00a	96.67±5.77a	++	Creamy* and brownish	Hard, Compact and rooty	++
+ve Control	5	86.67±5.77b	80.00±0.00b	+	Creamy* and brownish	Hard, Compact and rooty	+
	10	100.00±0.00a	86.67±5.77a	+	Creamy* and brownish	Hard, Compact and rooty	+
	20	100.00±0.00a	26.67±5.77c	+	Creamy* and brownish	Hard, Compact and rooty	+
	30	100.00±0.00a	0.00±0.00d	-	-	-	-
	60	100.00±0.00a	0.00±0.00d	-	-	-	-
-ve Control		0.00±0.00	0.00±0.00	-	-	-	-

Values within column followed by the same letter are not significantly different at P = 0.05; Where, ++ = Profuse callus mean; + = moderate callus mean; - = Absent; *= Dominant

IOP Conf. Series: Materials Science and Engineering **805** (2020) 012024 doi:10.1088/1757-899X/805/1/012024

Table 5. The effect of different concentrations and exposures of BNS on callus of *Ocimum sanctum* glandular trichome explants cultured on MS medium supplemented with 1.0 mg/l NAA and 0.5 mg/l BAP

Sterilization Methods		Explants			Callus formed	
BNS (mg/l)	Exposure time	Decontaminated (%) ± SD	callus induction (%) ± SD	Intensity of callus formation	Colour	Texture
10	5	0.00±0.00d	0.00±0.00d	-	-	-
	10	26.67±5.77c	23.33±5.77c	+	Creamy and Greenish**	Hard and Compact
	20	70.00±0.00b	70.00±0.00b	+	Creamy and Greenish**	Hard and Compact
	30	100.00±0.00a	93.33±5.77a	+	Creamy and Greenish**	Hard and Compact
	60	100.00±0.00a	93.33±5.77a	+	Creamy and Greenish**	Hard and Compact
50	5	46.67±5.77b	40.00±0.00c	+	Creamy and Greenish**	Hard and Compact
	10	90.00±10.00a	83.33±5.77b	+	Creamy and Greenish**	Hard and Compact
	20	100.00±0.00a	90.00±0.00ab	+	Creamy and Greenish**	Hard and Compact
	30	100.00±0.00a	93.33±5.77a	+	Creamy and Greenish**	Hard and Compact
	60	100.00±0.00a	96.67±5.77a	++	Creamy and Greenish**	Hard and Compact
100	5	100.00±0.00a	90.00±0.00a	+	Creamy and Greenish**	Hard and Compact
	10	100.00±0.00a	93.33±5.77a	+	Creamy and Greenish**	Hard and Compact
	20	100.00±0.00a	93.33±5.77a	+	Creamy and Greenish**	Hard and Compact
	30	100.00±0.00a	96.67±5.77a	++	Creamy and Greenish**	Hard and Compact
	60	100.00±0.00a	93.33±5.77a	++	Creamy and Greenish**	Hard and Compact
+ve Control	5	73.33±5.77b	70.00±0.00b	+	Creamy and Greenish**	Hard and Compact
	10	100.00±0.00a	83.33±5.77a	+	Creamy and Greenish**	Hard and Compact
	20	100.00±0.00a	76.67±5.77ab	+	Creamy and Greenish**	Hard and Compact
	30	100.00±0.00a	26.67±11.55c	+	Creamy and Greenish**	Hard and Compact
	60	100.00±0.00a	0.00±0.00d	-	-	-
-ve Control		0.00±0.00	0.00±0.00	-	-	-

Values within column followed by the same letter are not significantly different at P = 0.05; Where, ++ = Profuse callus mean; + =Moderate callus mean; - = Absent; **= More dominant

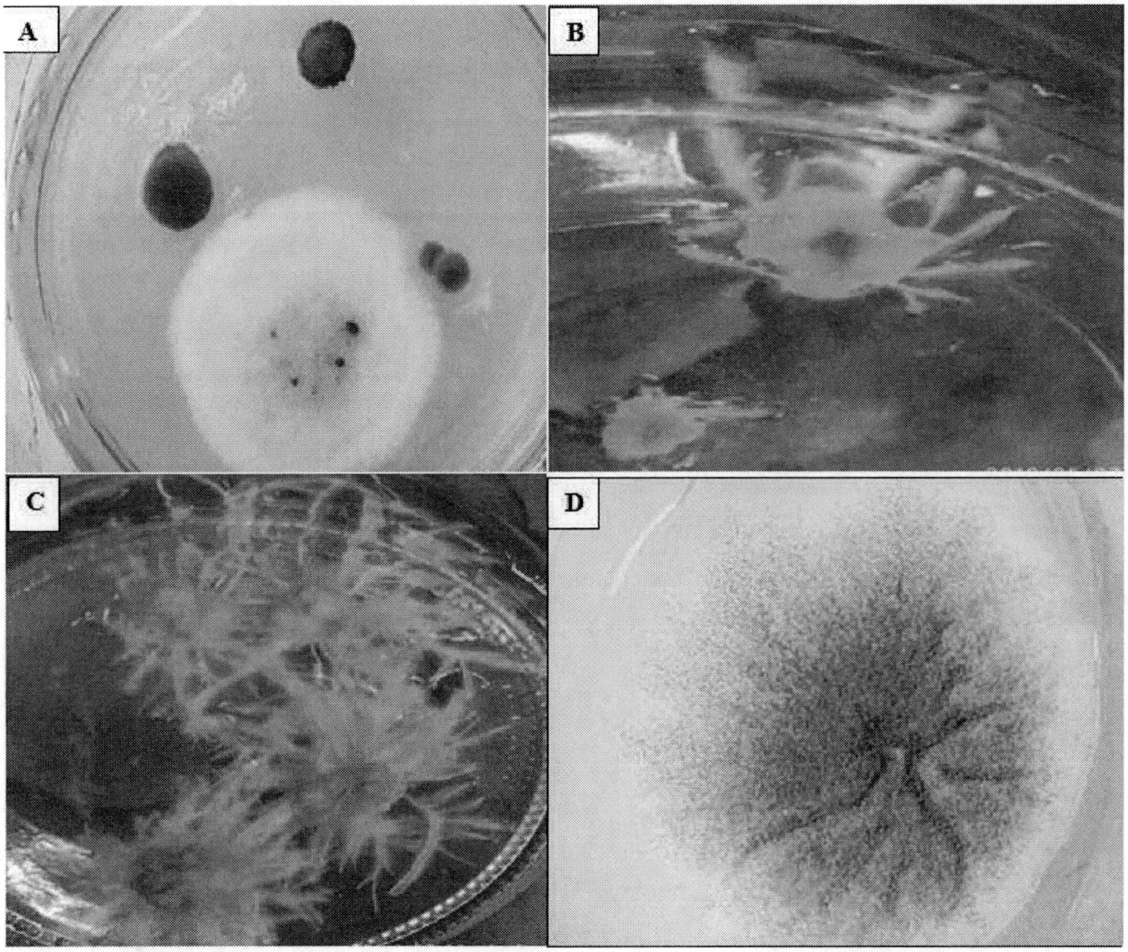

Figure 2.*Ocimum africanum* culture showing full contaminated explants when sterilized with sterile distilled water (A), Formation of moderate rooting callus from glandular trichome explant of *Ocimum canum* surface sterilized with 10 mg/l BNS for 60 min (B), Formation of profuse rooting callus from glandular trichome explant of *Ocimum canum* surface sterilized with 100mg/l BNS for 60 min (C), and *Ocimum gratissimum* culture showing full contaminated explant when sterilized with sterile distilled water (D)

4. Conclusion

The results obtained from this study showed that all used concentrations of BNS were efficient in decontaminating the *Ocimum* plant seeds and tissues with 20 min exposure time. At higher concentration of 100 mg/l with longer expose time (60 min), the BNS posed no harmful effect on the seeds, tissues and callus induction but rather possess stimulating effect on callus formation. This ascertained that biosynthesized nanosilver possess great antimicrobial properties just like chemically synthesized nanosilver and other sterilization agents used in tissue culture. It can be employed as an efficient tool for removing contaminants from plant tissues, at right dose and exposure time. Although, it has not yet become a universal decontamination agent, therefore, to extend its use in *in vitro* culture of other plants, further investigations should be conducted on diverse plant species and different explants.

Acknowledgements
The authors would like to thank the Director, National Centre for Genetic Resources and Biotechnology (NACGRAB), Moor Plantation, Ibadan, Nigeria, for his permission to make use of the Tissue Culture Laboratory and Tissue Culture unit staffs are appreciated for their technical assistance. We are also grateful to Nanotechnology Research Group (*NANO*$^{+}$), Ladoke Akintola University of Technology (LAUTECH), Ogbomoso, Nigeria, for providing the biosynthesized nanosilver used.

References

[1] Liu Y, Tan J, Thomas A, Ou-Yang D and Muzykantov V R 2012 The shape of things to come: importance of design in nanotechnology for drug delivery. *Ther. Deliv.* **3** (2) pp 181-94.

[2] Rudramurthy G R, Swamy M K, Sinniah U R and Ghasemzadeh A 2016 Nanoparticles: alternatives against drug-resistant pathogenic microbes. *Mol.* **21** (7) 836. https://doi.org/10.3390/molecules21070836.

[3] Kaviya S, Santhanalakshmi B, Viswanathan B, Muthumary J and Srinivasan K 2011 Biosynthesis of silver nanoparticles using *Citrus sinensis* peel extract and its antibacterial activity *Spectrochim. Acta A:Mol. Biomol. Spectr.* **79** (3) pp 594-598.

[4] Krishna I M, Reddy G B, Veerabhadram G and Madhusudhan A 2016 Ecofriendly green synthesis of silver nanoparticles using *Salmalia malabarica*: synthesis, characterization, antimicrobial, and catalytic activity studies. *Appl. Nanosci.* **6** (5) pp 681-689.

[5] Khosroushahi A Y, Valizadeh M, Ghasempour A, Khosrowshahli M and Naghdibadi H 2006 Improved Taxol production by combination of inducing factors in suspension cell culture of *Taxus baccata. Cell Biol. Int.* **30** (3) pp 262-269.

[6] Cassells A C 1991 Problems in tissue culture: culture contamination In: *Micropropagation* (Springer: Dordrecht), pp 31-44.

[7] Yildiz M 2012 The pre-requisite of the success in plant tissue culture: high frequency shoot regeneration, recent advances in plant *in vitro* Culture Leva A (Ed.) INTECH *Recent Advances In Plant Tissue Culture* InTech Publication pp 63-90

[8] Mbah E I and Wakil S M 2012 Elimination of bacteria from *in vitro* yam tissue cultures using antibiotics. *J. Plant Pathol.* **94** (1) pp 53-58.

[9] Teixeira da Silva J A, Duong N T, Michi T and Seiichi F 2003 The effect of antibiotic on the *in vitro* growth response of chrysanthemum and tobacco stem transverse thin cell layers (tTcLs) *Sci. Hortic.* **97** (3-4) pp 397-410.

[10] Dolly G, Nidhi S, Sagar B, Raheja S and Agrawal S 2012 *Ocimum kilim* and *scharicum*: A systematic review. *J. Drug Deliv. Ther.* **2** (3) pp 45-52.

[11] Lateef A, Azeez MA, Asafa TB, Yekeen TA, Akinboro A, Oladipo IC, Azeez L, Ajibade SE, Ojo SA, Gueguim-Kana EB and Beukes LS 2016 Biogenic synthesis of silver nanoparticles using a pod extract of *Cola nitida*: Antibacterial, antioxidant activities and application as a paint additive. *J. Taibah Univ. Sci.* **10** (4) pp 551-562.

[12] Murashige T and Skoog F 1962 A revised medium for rapid growth and bio assays with tobacco tissue cultures. *Physiol. Plant.* **15** (3) pp 473-497.

[13] LaMorte W W 2016 Central limit theorem: The Role of Probability Boston, Massachusetts: Boston University School of Public Health. http://sphweb.bumc.bu.edu/otlt/MPH-Modules/BS/BS704_Probability/BS704_ Probability12.html. Acessed on 12 June 2019.

[14] Nartop P 2018 Effects of surface sterilisation with green synthesised silver nanoparticles on Lamiaceae seeds. *IET Nanobiotechnol.* **12** (5) pp 663-668.

[15] Makri O and Kintzios S 2008 *Ocimum* sp. (Basil): Botany, cultivation, pharmaceutical properties, and biotechnology. *J. Herbs Spices Med. Plants* **13** (3) pp 123-150.

[16] Verma K S, Sahin G, Das K A and Gurel E 2016 *In vitro* plant regeneration of *Ocimum basilicum* L. is accelerated by zinc sulfate. *In Vitro Cell Dev. Biol. - Plant* **52** (1) pp 20-27.

[17] Matsumura Y, Yoshikata K, Kunisaki S and Tsuchido T 2003 Mode of bactericidal action of silver zeolite and its comparison with that of silver nitrate. *Appl. Environ. Microbiol.* **69** (7) pp 4278-4281.

[18]Mahna N, Vahed S Z and Khani S 2013 Plant *in vitro* culture goes nano: Nano silver-mediated decontamination of *ex vitro* explants. *J. Nanomed. Nanotechnol.* **4** 2. https://doi.org/10.4172/2157-7439.1000161.

[19] Ghosh S, Patil S, Ahire M, Kitture R, Kale S, Pardesi K, Cameotra S S, Bellare J, Dhavale D D, Jabgunde A and Chopade B A 2012 Synthesis of silver nanoparticles using *Dioscorea bulbifera* tuber extract and evaluation of its synergistic potential in combination with antimicrobial agents. *Int. J. Nanomed.* **7** pp 483-496.

[20] Abdi G, Salehi H and Khosh-Khui M 2008 Nano silver: a novel nanomaterial for removal of bacterial contaminants in valerian (*Valeriana officinalis* L.) tissue culture. *Acta Physiol. Plant.* **30** (5) pp 709-714.

[21] Yildiz M and Erkiz H 2014The effect of sodium hypochlorite solutions on *in vitro* seedling growth and regeneration capacity of sainfoin (*Onobrychis viciifolia* Scop.) hypocotyl explants. *Can. J. Plant Sci.* **94** (7) pp 1161-1164.

[22] Nartop P 2016 Green sterilization of *Rosmarinus officinalis* L. stem surfaces with silver nanoparticles synthesized using *Rubia tinctorum* L. cell culture extracts. *Iranian J. Sci. Technol. Transac. A: Sci.* **42** (2) pp 411-414.

[23] Ewais E A, Desouky S A and Elshazly E H 2015 Evaluation of callus responses of *Solanum nigrum* L. exposed to biologically synthesized silver nanoparticles. *Nanosci. Nanotechnol.* **5** (3) pp 45-56.

[24] Aghdaei M, Salehi H, and Sarmast M K 2012 Effects of silver nanoparticles on *Tecomella undulata* (Roxb·) Seem, micropropagation. *Adv. Hortic. Sci.* **26** (1) pp 21-24.

[25] Fazal H, Abbasi B H, Ahmad N and Ali M 2016 Elicitation of medicinally important antioxidant secondary metabolites with silver and gold nanoparticles in callus cultures of *Prunella vulgaris* L. *Appl. Biochem. Biotechnol.* **180** (6) pp 1076-1092.

Thermophysical properties of nanoparticles in carboxylmethyl cellulose water mixture for heat enhancement applications

F O Akinpelu [1,2], R M Alabison[*1,3] and O A Olaleye[3]

[1]Department of Pure and Applied Mathematics, Ladoke Akintola University of Technology, Ogbomoso, Oyo, Nigeria.

[2]Current address: Department of Mathematics, University of Lagos, Akoka, Yaba, Lagos, Nigeria.

[3]Department of Statistics, The Federal Polytechnic, Ede, Osun State, Nigeria.

*Corresponding Author: allabison1960@gmail.com

Abstract

A successful drilling operation requires among others an effective drilling fluid which is also called drilling mud. Carboxylmethyl cellulose (CMC) which is a derivative of cellulose is commonly used because of its high viscosity. Its performance as a drilling fluid is limited at high temperatures and pressures due to its low thermal conductivity. A colloidal mixture of nanoparticles in CMC enhances the performance of the fluid. In this work, a comparison of the thermophysical properties of three metal oxides namely Titanium Oxide (TiO_2), Aluminium oxide (Al_2O_3) and Copper Oxide (CuO) dispensed in Carboxylmethyl cellulose (CMC) water mixture in drilling operations were investigated at varying temperatures. Maximum of 0.4 volume fraction of the nanoparticles of each of the oxides in less or equal to 0.4 (≤ 0.4) of CMC concentration in water was considered. The governing equations obtained were simplified to a set of Ordinary Differential Equations which were solved numerically using Runge Kutta Scheme (order 4) along with shooting method. Results obtained showed that the metal oxides enhanced the heat transfer capability of the CMC water mixture. Besides, the conductivity enhancement is least with TiO_2 in CMC water mixture and maximum with CuO in CMC water mixture. Viscosity and thermal conductivity increased with increasing volume fraction of the nanoparticles in the dispensing medium. These results were compared with existing literature and found to be in good agreement.

Keywords: Carboxylmethyl cellulose, metal oxide, nanoparticles, Runge Kutta scheme, thermophysical properties, volume fraction

IOP Conf. Series: Materials Science and Engineering **805** (2020) 012025 doi:10.1088/1757-899X/805/1/012025

1 Introduction

The significant of fluids used in drilling operations especially in an environment with high pressure and high temperature condition cannot be overemphasized. These fluids are needed to maintain adequate flow properties of drilled materials and to provide cooling and lubrication of the string and bit of the drill pipe in an environment of such [1]. Carboxylmethyl cellulose (CMC) a highly viscous derivative cellulose is a useful additive capable of maintaining adequate flow properties of drilled materials like clay suspensions in boreholes and rocks crushed into small fragments in telescopic holes of oil and gas [2].

A nanofluid, a potential coolant is a colloidal suspension of nanoparticles inside a base fluid, e.g. ethylene glycol, water and CMC solutions. The existence of nanoparticles of metal carbides, metal oxides, and so on in the convectional fluid enhances its thermal conductivity; thereby improve its heat transfer capability. Nanofluids also exhibit greater viscosity as the volume fraction of the nanoparticles is increased [3].

A number of works abound in literature on formulation of drilling fluids that is nano-based. In their preliminary test investigated the result of the thermophysical properties of iron oxide (Fe_2O_3) nanoparticles on bentonite fluids as base fluid, it was reported that when the volume fraction of iron oxide is raised, nanoparticles in the base fluid increased the viscosity of the base fluid [4]. Similar test conducted showed that the effect of increase in temperature on the properties of nanofluids tested is more significant than that of pressure [5].

Other works include that of Iqbal *et al.* [6] on the comparative investigation of Al_2O_3/H_2O, SiO_2/H_2O and ZrO_2/H_2O nanofluid for heat transfer applications, studies on thermophysical properties of silicon dioxide (SiO_2) in Ethyl Glycol/water mixture for proton exchange membrane fuel cell cooling application [7], and examination of the magnetohydrodynamic radiative flow of Casson fluid on a stretching area with heat source/sink [8].

Greater nanofluid viscosity and its improved capability to transfer heat relative to those of the base fluid like CMC solution ought to be examined for formulating effective drilling fluid with best possible properties within a choice of working conditions. This work therefore presents the comparison of effects of the thermophysical properties of TiO_2, Al_2O_3 and CuO nanoparticles under different temperature conditions. The properties include thermal conductivity and viscosity for drilling operations.

IOP Conf. Series: Materials Science and Engineering **805** (2020) 012025 doi:10.1088/1757-899X/805/1/012025

2. Mathematical Formulation

The fluid considered is electrically conducting, viscid and incompressible induced by buoyancy effect. The flow is two-dimensional and unsteady while the Cartesian coordinate is selected such that the x-axis is along which the fluid flows is perpendicular to the surface and y-axis along it. The fluid properties are taken to be constant and the temperature of the system is negligible. Carboxylmethyl cellulose water (CMC) is the base fluid. Aluminium oxide (Al_2O_3), titanium dioxide (TiO_2) and copper oxide (CuO) are the nanoparticles. Their thermophysical properties used are as follows:

Table 1. Nanoparticles and the base fluids thermophysical properties

Physical properties	Specific heat Capacity $C_p(Jkg^{-1}K^{-1})$	Density $\rho(kgm^{-3})$	Thermal Conductivity $k(Wm^{-1}K^{-1})$	Coefficient of thermal expansion $\beta_T \times 10^5 (K^{-1})$
CMC-water (< 0.4%)	4179	997.1	0.613	21
Al_2O_3 nanoparticles	765	3970	40	0.85
TiO_2 nanoparticles	686.2	4350	8.95	0.72
CuO nanoparticles	531.8	6320	76.5	1.80

Source: [3, 9, 10]

Based on the assumptions, the governing equations for the fluid flow can be modeled as follows;

Continuity Equation

$$\frac{\partial u}{\partial x}+\frac{\partial v}{\partial y}=0 \tag{1}$$

Momentum Equation

$$u\frac{\partial u}{\partial x}+v\frac{\partial v}{\partial y}=\frac{1}{\rho_{nf}}\left(1+\frac{1}{\beta}\right)\frac{\partial \mu_{nf}(T)}{\partial y}\frac{\partial T}{\partial y}\frac{\partial u}{\partial y}+\frac{\mu_{nf}(T)}{\rho_{nf}}\left(1+\frac{1}{\beta}\right)\frac{\partial^2 u}{\partial y^2}$$
$$-\frac{\mu_{nf}(T)}{\rho_{nf}}\left(1+\frac{1}{\beta}\right)\frac{u}{K}+g(\beta_T)_{nf}(T-T_\infty)-\frac{b^*u^2}{K}-\frac{\sigma B_0^2 u}{\rho_{nf}} \tag{2}$$

Energy Equation

$$u\frac{\partial T}{\partial x}+v\frac{\partial T}{\partial y}=\frac{k_{nf}(T)}{(\rho C_p)_{nf}}\frac{\partial^2 T}{\partial y^2}+\frac{1}{(\rho C_p)_{nf}}\frac{\partial k_{nf}(T)}{\partial T}\left(\frac{\partial T}{\partial y}\right)^2+\frac{\mu_{nf}(T)}{(\rho C_p)_{nf}}\left(1+\frac{1}{\beta}\right)\left(\frac{\partial u}{\partial y}\right)^2 +\tau\left[D_B\frac{\partial C}{\partial y}\frac{\partial T}{\partial y}+\frac{D_T}{T_\infty}\left(\frac{\partial T}{\partial y}\right)^2\right] \tag{3}$$

Concentration Equation

$$u\frac{\partial C}{\partial x}+v\frac{\partial C}{\partial y}=D_B\frac{\partial^2 C}{\partial y^2}+\left(\frac{D_T}{T_\infty}\right)\frac{\partial^2 T}{\partial y^2}-k_r(C-C_\infty) \tag{4}$$

With the following boundary conditions:

$$\begin{aligned} &u=Bx \quad v=-Vx \quad T=T_w \quad C=C_w \quad at \quad y=0\\ &u\to 0 \quad T\to T_\infty \quad C\to C_\infty \quad as \quad y\to\infty \end{aligned} \tag{5}$$

$$\mu_{nf}(T)=\mu_{nf}^*[1+a(T_w-T)] \;\; and \;\; k_{nf}(T)=k_{nf}^*[1+b(T-T_\infty)] \tag{6}$$

Using Rosseland approximation, the radiative heat flux, q_r as reported by [11] is given as:

$$q_r=-\frac{4\sigma^*}{3K^*}\frac{\partial T^4}{\partial y} \tag{7}$$

Applying Taylor series and neglecting higher terms

$$q_r=-\frac{16\sigma^* T_\infty^3}{3K^*}\frac{\partial T}{\partial y} \tag{8}$$

Substituting (6) and (8) as well as their derivatives into the governing equations, it became:

Continuity equation

$$\frac{\partial u}{\partial x}+\frac{\partial v}{\partial y}=0 \tag{9}$$

Momentum Equation

$$\begin{aligned} u\frac{\partial u}{\partial x}+v\frac{\partial v}{\partial y}=&-\frac{a\mu_{nf}^*}{\rho_{nf}}\left(1+\frac{1}{\beta}\right)\frac{\partial T}{\partial y}\frac{\partial u}{\partial y}+\frac{\mu_{nf}^*}{\rho_{nf}}[1+a(T_w-T)]\left(1+\frac{1}{\beta}\right)\frac{\partial^2 u}{\partial y^2}\\ &-\frac{\mu_{nf}^*}{\rho_{nf}}[1+a(T_w-T)]\left(1+\frac{1}{\beta}\right)\frac{u^2}{K}+g(\beta_T)_{nf}(T-T_\infty)-\frac{b^*u^2}{K}-\frac{\sigma B_0^2 u}{\rho_{nf}} \end{aligned} \tag{10}$$

Energy Equation

$$u\frac{\partial T}{\partial x}+v\frac{\partial T}{\partial y}=\frac{k_{nf}^{*}(T)}{(\rho C_p)_{nf}}\left[1+b(T-T_\infty)\right]\frac{\partial^2 T}{\partial y^2}+\frac{bk_{nf}^{*}}{(\rho C_p)_{nf}}\left(\frac{\partial T}{\partial y}\right)^2$$
$$+\frac{\mu_{nf}^{*}}{(\rho C_p)_{nf}}\left[1+a(T_w-T)\right]\left(1+\frac{1}{\beta}\right)\left(\frac{\partial T}{\partial y}\right)^2-\frac{16\sigma^{*}T_\infty^3}{3K^{*}(\rho C_p)_{nf}}\frac{\partial^2 T}{\partial y^2}+\tau\left[D_B\frac{\partial C}{\partial y}\frac{\partial T}{\partial y}+\frac{D_T}{T_\infty}\left(\frac{\partial T}{\partial y}\right)^2\right] \quad (11)$$

Concentration Equation

$$u\frac{\partial C}{\partial x}+v\frac{\partial C}{\partial y}=D_B\frac{\partial^2 C}{\partial y^2}+\left(\frac{D_T}{T_\infty}\right)\frac{\partial^2 T}{\partial y^2}-k_r(C-C_\infty) \quad (12)$$

Subject to the following boundary conditions

$$\begin{array}{llllll} u=Bx & v=-Vx & T=T_w & C=C_w & at & y=0 \\ u\to 0 & T\to T_\infty & C\to C_\infty & & as & y\to\infty \end{array} \quad (13)$$

Introducing the following relations

$$u=Bxf'(\eta),\ v=-\sqrt{BV_f}\,f(\eta),\ \eta=\sqrt{\frac{B}{V_f}}y,\ \theta(\eta)=\frac{T-T_\infty}{T_w-T_\infty},\ \phi=\frac{C-C_\infty}{C_w-C_\infty} \quad (14)$$

Thus, $u=Bxf'(\eta)\Rightarrow\frac{\partial u}{\partial x}=Bf'\ and\ \frac{\partial u}{\partial y}=\frac{\partial u}{\partial\eta}\cdot\frac{\partial\eta}{\partial y}=\frac{x\sqrt[3]{B}}{\sqrt{\vartheta_f}}f''$

$$v=-\sqrt{B\vartheta_f}\,f\Rightarrow\frac{\partial v}{\partial y}=\frac{\partial v}{\partial\eta}\cdot\frac{\partial\eta}{\partial y}=-Bf' \quad (15)$$

$$\theta(\eta)=\frac{T-T_\infty}{T_w-T_\infty}\quad\Rightarrow\quad T=T_\infty+\theta(T_w-T_\infty)\quad and\quad \partial T=(T_w-T_\infty)\partial\theta \quad (16)$$

$$\phi(\eta)=\frac{C-C_\infty}{C_w-C_\infty}\quad\Rightarrow\quad C=C_\infty+\theta(C_w-C_\infty)\ and\quad \partial C=(C_w-C_\infty)\partial\theta \quad (17)$$

With the thermophysical properties of nanoparticles expressed theoretically as reported by [10] as follows

$$\mu_{nf}^{*}=\frac{\mu_f}{(1-\psi)^{2.5}},\ \rho_{nf}=(1-\psi)\rho_f+\psi\rho_S,\ (\rho\beta_T)_{nf}=(1-\psi)(\rho\beta_T)_f+\psi(\rho\beta_T)_S$$
$$k_{nf}^{*}=\left[\frac{k_S+2k_f-2\psi(k_f-k_S)}{k_S+2k_f+\psi(k_f-k_S)}\right]k_f,\ (\rho C_p)_{nf}=(1-\psi)(\rho C_p)_f+\psi(\rho C_p)_S,\ V_f=\frac{\mu_f}{\rho_f} \quad (18)$$

Substituting equations (15) – (18) into equations (10) – (12) are re-arranging yielded:

IOP Conf. Series: Materials Science and Engineering **805** (2020) 012025 doi:10.1088/1757-899X/805/1/012025

$$(1+\xi-\theta\xi)\left(1+\frac{1}{\beta}\right)f'''-\xi\left(1+\frac{1}{\beta}\right)\theta' f''+A_2\left[\left(1+\frac{F_S}{D_a}\right)f'^2-ff''\right] \\ +A_3 J_{GR}\xi\theta-\left[\rho_p(1+\xi-\theta\xi)\left(1+\frac{1}{\beta}\right)+A_1 H_a\right]f'=0 \tag{19}$$

$$A_4\left[\frac{4}{3R}+(1+\varepsilon\theta)\right]\theta''+A_4\left[A_6 P_r N_t+\varepsilon\right]\theta'^2 \\ +A_5 P_r\left[f+N_b\phi'\right]\theta'+P_r E_C(1+\xi-\theta\xi)\left(1+\frac{1}{\beta}\right)(f'')^2=0 \tag{20}$$

$$\frac{1}{S_C}\phi''+f\phi''+\frac{1}{S_C}\frac{N_t}{N_b}\theta''-k_r\phi=0 \tag{21}$$

Using the following boundary conditions:

$$\begin{array}{llllll} f'(\eta)=1 & f(\eta)=S & \theta=1 & \phi=1 & at & \eta=0 \\ f'(\eta)\to\infty & \theta\to\infty & \phi\to\infty & & as & \eta\to\infty \end{array} \tag{22}$$

Where,

β is the base fluid parameter

$\xi=a(T_w-T_\infty)$ is the temperature-dependent viscosity parameter

$\varepsilon=b(T_w-T_\infty)$ is the temperature-dependent thermal conductivity parameter

$F_S=\frac{b^*}{x}$ is the Forchheimer parameter $D_a=\frac{K}{x^2}$ is the Darcy number

$P_p=\frac{V_f}{K_B}$ is the porosity parameter $J_{GR}=\frac{g(\beta_T)_f}{axB^2}$, $H_a=\frac{\sigma B_0^2}{B\rho_f}$

$P_r=\frac{\mu_f(C_p)_f}{k_f^*}$ is the Prandtl number $E_C=\frac{x^2B^2}{(C_p)_f(T_w-T_\infty)}$ is the Eckert number

$N_t=\frac{\tau D_T(T_w-T_\infty)}{V_f T_\infty}$ is the thermopheresis parameter $N_b=\frac{\tau D_B(C_w-C_\infty)}{V_f}$

$A_1=\frac{\mu_f}{\mu_{nf}^*}$ $A_2=A_1\left[(1-\psi)+\psi\left(\frac{\rho_S}{\rho_f}\right)\right]$ $A_3=A_1\left[(1-\psi)+\psi\left(\frac{(\rho\beta_T)_S}{(\rho\beta_T)_f}\right)\right]$

$$A_4 = A_1\left[\frac{k^*_{nf}}{k_f}\right] = A_1\left[\frac{k_S + 2k_f - 2\psi(k_f - k_S)}{k_S + 2k_f + \psi(k_f - k_S)}\right]$$

$$A_5 = A_1\left[(1-\psi) + \psi\left(\frac{(\rho C_p)_S}{(\rho C_p)_f}\right)\right] \qquad A_6 = \frac{A_1}{A_4}\left[(1-\psi) + \psi\left(\frac{(\rho C_p)_S}{(\rho C_p)_f}\right)\right]$$

The numerical representation of the solutions was displayed graphically for the velocity and temperature distributions of the fluid flow. A table was also used to illustrate the computational values of the thermophysical properties of the nanofluid in order to enhance the discussion. When $\psi = 0$, the results are reduced to that of the base fluid problem without any nanofluid property. The following values of the dimensionless numbers were used except otherwise to investigate the effects of the fluid parameter β, and the volume fraction of the nanoparticles, ψ.

$\beta = 1.0$, $P_r = 0.7$, $k_r = 0.2$, $S_C = 0.22$, $F_S = 1.0$, $H_a = 0.1$, $P_\rho = 0.3$, $D_a = 1.0$, $E_C = 0.01$

3. Results and Discussion

In Table 2, it is noticed that the effective thermophysical properties of the CMC water mixture and the nanoparticles improved with increase in the volume fraction of the nanoparticles across the temperature. Aluminium oxide has the least density and copper oxide (CuO) has the highest thermal conductivity. The effective viscosity decreased with temperature while the thermal conductivity increased with temperature.

Figure 1 illustrates the velocity distribution of the flow for different values of the base fluid parameter (β) at negligible temperature. It is observed that the velocity of the fluid decreased with increase in β with and without nanoparticles as reported in the work of [12]. This is because increase in β implies decrease in yield stress of the base fluid and the corresponding increase in the dynamic viscosity of the fluid which created internal resistance to the fluid flow. It is also revealed that the flow of the fluid with nanoparticles is faster than without nanoparticles as a result of the enhanced thermal conductivity of the former. The fluid with aluminium oxide (Al_2O_3) nanoparticles is the fastest because it has the least density.

Figure 2 depicts the velocity distribution of the flow for different values of the base fluid parameter (β) at higher temperature. As β increases, the velocity profile of the fluid also increased particularly towards the strain function. This is because the temperature of the fluid is high enough to overpower the resistance of the fluid flow. The fluid with titanium dioxide (TiO_2) nanoparticles has the maximum velocity.

Figure 3 shows the temperature distribution of the fluid flow for different values of the base fluid parameter (β) at negligible temperature. As the β increases, the temperature of the fluid also increased. This is due to the fact that heat is absorbed by the system which implies a rise in the temperature of the fluid with and without nanoparticles. The rise is slower with the fluid with nanoparticles because its heat capacity has been enhanced. The rise in temperature is slowest in the aluminium oxide (Al_2O_3) based drilling fluid because Al_2O_3 is a good insulator.

Figure 4 represents the temperature distribution of the fluid flow for different values of the base fluid parameter (β) at higher temperature. As β increases, the temperature of the fluid decreased, which is in agreement with earlier findings [13]. This is because heat is released which implies reduction in the temperature of the fluid. Heat is released faster in the fluid with nanoparticles because of its enhanced thermal conductivity. It is fastest with copper oxide (CuO) nanoparticles which has the highest thermal conductivity across the temperature.

Table 2. Thermophysical properties of the three nanofluids at various volume fractions of the three nanoparticles

Fluid	Volume Fraction of nanoparticles (ψ)	Density (ρ_{nf})	Heat Capacitance $(\rho C_p)_{nf}$	Viscosity (μ_{nf})		Thermal Conductivity (k_{nf})	
				ξ = 0	ξ = 3.5	ε = 0	ε = 3.5
CMC/Water	0	997.1	4166880.90	0.00890000	0.00323635	0.613	1.68575034
CMC/Water + CuO	0.1	1529.36	4086290.41	0.01158200	0.00421163	0.8119776814	2.8419218849
	0.2	2061.68	4005699.92	0.01554766	0.00565369	1.0592328827	3.7073148949
	0.3	2593.97	3925109.43	0.02170924	0.00789427	1.3747709989	4.8116984948
	0.4	3126.26	3844518.94	0.03191626	0.01160591	1.7914053856	6.2699188496
CMC/Water + TiO_2	0.1	1332.29	4048689.80	0.00890000	0.00421163	0.7711101120	2.7198885392
	0.2	1667.68	3930498.72	0.01158200	0.00565369	0.9733822721	3.4068379521
	0.3	2002.97	3812307.63	0.01554766	0.00789427	1.2122941797	4.2430296290
	0.4	2338.36	3694116.54	0.02170924	0.01160591	1.5094358884	5.2830260940
CMC/Water + Al_2O_3	0.1	1294.39	4053897.80	0.00890000	0.00421163	0.8072556661	2.8253948314
	0.2	1591.68	3940914.72	0.01158200	0.00565369	1.0473972041	3.6658902143
	0.3	1888.97	3827931.63	0.01554766	0.00789427	1.3518603123	4.7315110931
	0.4	2186.26	3714948.54	0.02170924	0.01160591	1.7504837935	6.1266932773

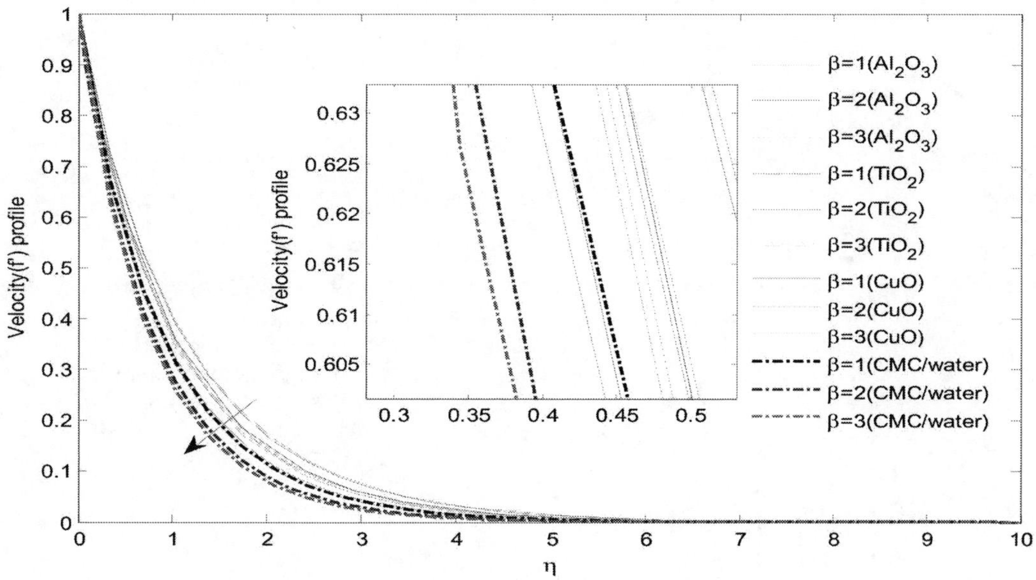

Figure 1. Velocity profile for different values of β when ξ=0 and ε=0

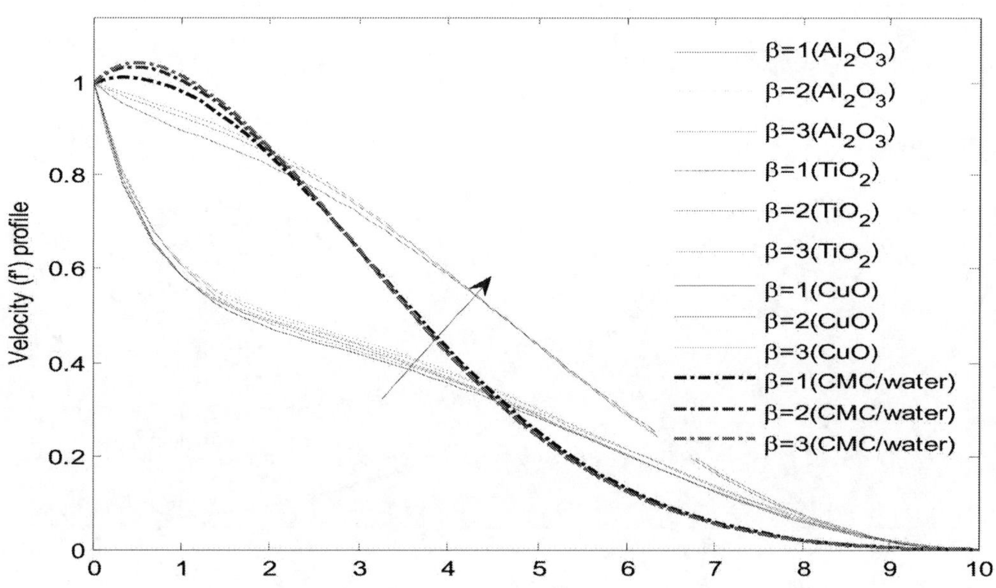

Figure 2. Velocity profile for different values of β when ξ=3.5 and ε=3.5

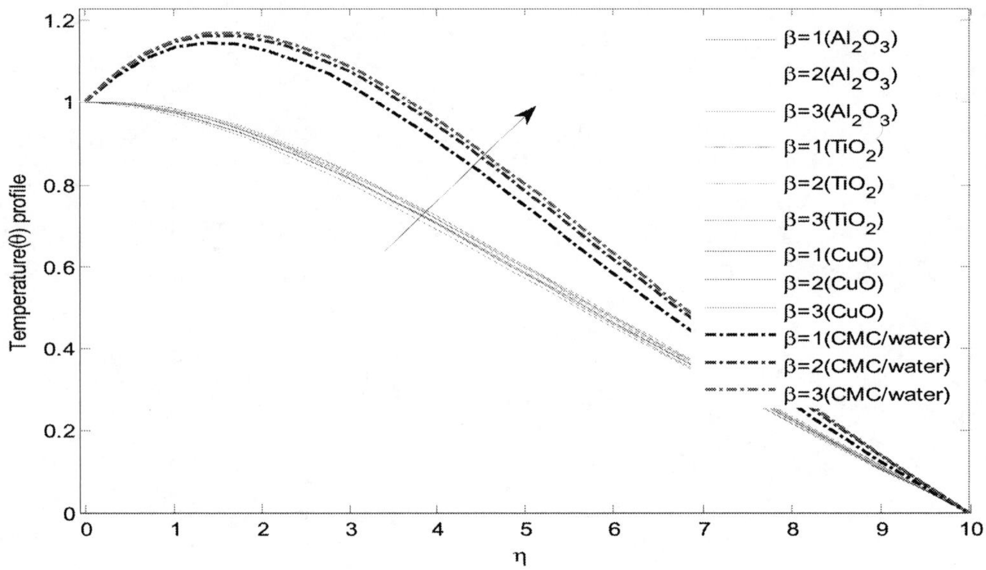

Figure 3. Temperature profile for different values of β when ξ=0 and ε=0

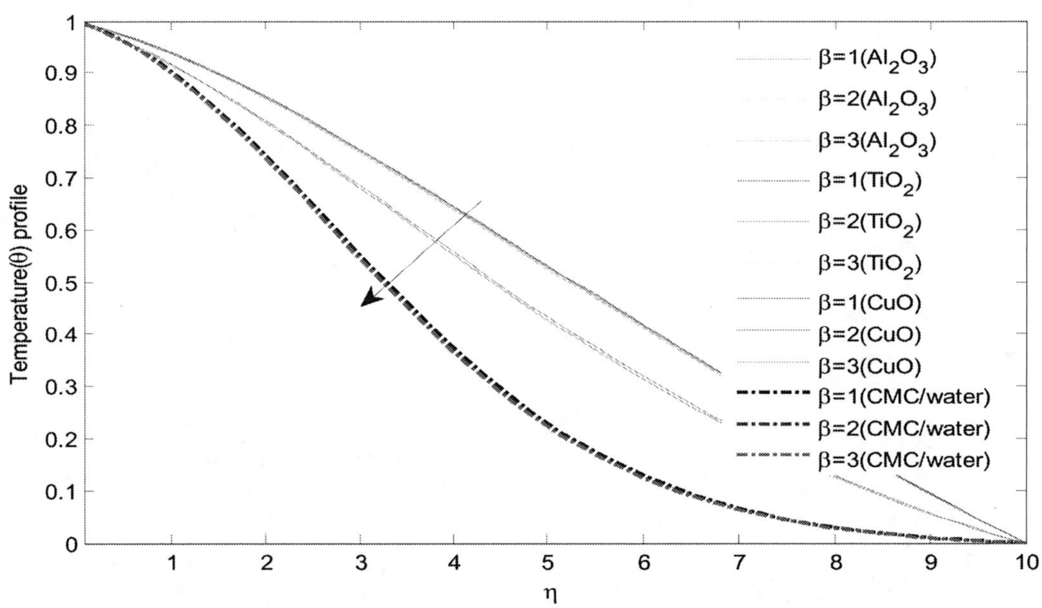

Figure 4. Temperature profile for different values of β when ξ=3.5 and ε=3.5

4 Conclusion

According to the study, it is established that the presence of nanoparticles in CMC solution enhances its viscosity and thermal conductivity across the temperature. Moreover, the flow properties of the fluid with nanoparticles improved as the temperature increased which can bring about reduction in the pumping cost of the fluid. Also, the volume fraction of the nanoparticles improves the heat transfer capability of the drilling fluid while CuO nanoparticles offered the optimal increased cooling effect for the drilling pipe at low and high temperature with highest thermal conductivity and density.

References

[1] Vryzas Z and Kelessidis VC 2017 Nano-based drilling fluids: A review. *Energ.* **10** (4) 540. https://doi.org/10.3390/en10040540.

[2] Benchabane A and Bekkour K 2008 Rheological properties of carboxymethyl cellulose (CMC) solutions. *Colloid. Polym. Sci.* **286** (10) 1173. https://doi.org/10.1007/s00396-008-1882-2.

[3] Rawi NA, Ilias MR, Lim YJ, Isa ZM and Shafie S 2017 Unsteady mixed convection flow of Casson fluid past an inclined stretching sheet in the presence of nanoparticles. *J. Phys.: Conf. Ser.* **890** (1) 012048. https://doi.org/10.1088/1742-6596/890/1/012048.

[4] Amanullah M, AlArfaj MK, Al-abdullatif ZA 2011 Preliminary test results of nano-based drilling fluids for oil and gas field application. *In Proceedings of the SPE/IADC Conference and Exhibition, Amsterdam, Netherlands*. https://doi.org/10.2118/139534-MS.

[5] Barry MM, Jung Y, Lee JK, Phuoc TX and Chyu MK 2015 Fluid filtration and rheological properties of nanoparticles additive and intercalated clay hybrid bentonite drilling fluids. *J. Petrol. Sci. Eng.* **127** pp 338-346.

[6] Iqbal SM, Raj CS, Michael JJ and Irfan AM 2017 A Comparative investigation of Al_2O_3/H_2O, SiO_2/H_2O and ZrO_2/H_2O nanofluid for heat transfer applications. *Digest J. Nanomater. Biostruct.* **12** (2) pp 255-263.

[7] Talib SF, Azmi WH, Zakaria I, Mohamed WA, Mamat AM, Ismail H and Daud WR 2015 Thermophysical properties of silicon dioxide (SiO_2) in ethyl glycol/water mixture for proton exchange membrane fuel cell cooling application. *Energ. Procedia* **79** pp 366-371.

IOP Conf. Series: Materials Science and Engineering **805** (2020) 012025 doi:10.1088/1757-899X/805/1/012025

[8] Vijayaragavan R 2016 Magnetohydrodynamic radiative casson fluid flow over a stretching sheet with heat source/sink. *Adv. Phys. Theor. Applications* **55** pp 13-23.

[9] Tarakaramu N and Narayan PV 2017 Unsteady MHD nanofluid flow over a stretching sheet with chemical reaction. *Mater. Sci. Eng. Conf. Ser.* **263** (6) 062030. https://doi.org/10.1088/1757-899X/263/6/062030.

[10] Kumaresan E, Vijaya Kumar AG and Rushi Kumar B 2017 Chemically reacting on MHD boundary layer flow of CuO-water and MgO-water nanofluids past a stretching sheet in porous media with radiation absorption and heat generation/absorption *Mater. Sci. Eng. Conf. Ser.* **263** (6) 062017. https://doi.org/10.1088/1757-899X/263/6/062017.

[11] Animasaun IL and Oyem AO 2014 Effect of variable viscosity, Dufour, Soret and thermal conductivity on free convective heat and mass transfer of non-Darcian flow past porous flat surface. *Am. J. Comput. Math.* **4** (04) 357. https://doi.org.10.4236/ajcm.2014.44030.

[12] Animasaun IL 2015 Effects of thermophoresis, variable viscosity and thermal conductivity on free convective heat and mass transfer of non-darcian MHD dissipative Casson fluid flow with suction and nth order of chemical reaction. *J. Nig. Math. Soc.* **34** (1) pp 11-31.

[13] Hayat T, Asad S and Alsaedi A 2016 Flow of Casson fluid with nanoparticles. *Appl. Math. Mech* **37** (4) pp 459-470.

IOP Conf. Series: Materials Science and Engineering **805** (2020) 012026 doi:10.1088/1757-899X/805/1/012026

Effect of low temperature synthesis of carbon nanotube nanocomposite on the photovoltaic performance of anode buffer layer in polymer solar cells

H O Oyeshola[1], M A Adisa[2,3], B K Adejumo[1], K K Babalola[1], B A Agboluaje[1], O Adedokun*[1,4] and Y K Sanusi[1,4]

[1]Department of Pure and Applied Physics, Ladoke Akintola University of Technology, Ogbomoso, Oyo State, Nigeria.

[2]Department of Physics and Material Science, Kwara State University, Malete, Kwara State, Nigeria.

[3]Rubber Research Institute of Nigeria, Benin City, Edo State, Nigeria.

[4]Nanotechnology Research Group (*NANO*$^+$), Ladoke Akintola University of Technology, Ogbomoso, Oyo State, Nigeria.

*Email: oadedokun@lautech.edu.ng; adedokunoluwaseun2@gmail.com

Abstract

Today's solar cells are simply not efficient enough and are currently too expensive to manufacture for large-scale electricity generation. However, potential advancements in nanotechnology may open the door to the production of cheaper and slightly more efficient solar cells. This research is based on the study of photovoltaic properties of low temperature synthesized carbon nanotube (CNT) nanocomposite as an anode buffer layer for the PEDOT:PSS based polymer solar cells. CNT was synthesized using simple and cost effective method at low temperature. The structural and optical properties of prepared CNT samples were characterized using X-ray diffraction (XRD) analysis, Fourier Transform Infrared (FTIR) spectroscopy, Scanning Electron Microscopy (SEM) and UV spectroscopy. CNT/PEDOT:PSS nanocomposite solutions was prepared and spin coated on a cleaned glass substrate at different spin coating speed, the fabricated buffer layer thin film devices were annealed from 100 °C to 500 °C, their optical and electrical properties were then analyzed. The XRD of synthesized CNTs nanoparticles show diffraction pattern which exhibit tetragonal structure and FTIR shows functional group of carbon nanotube. The SEM image showed that the obtained sample maintained tubular structure, cluster at 20 nm but properly dispersed at 100 nm. The optical studies of the films show an increase in absorbance as the annealing temperature increases. The photovoltaic performance of the polymer solar cell showed an improved efficiency of 6.44 % for optimized device. It is deduced from this work that low temperature synthesized CNT nanocomposite demonstrated better performance as anode buffer layer for high efficient polymer solar cells.

Keywords: Polymer solar cells; CNT; PEDOT:PSS; nanocomposite, anode buffer layer

1 Introduction

The branch of technology that studies different materials at a nanometric scale is referred to as nanotechnology and it is also defined as science of production, manipulation and use of materials at subatomic level that finds its application in various disciplines [1]. There is increasing global interest in alternative forms of energy, sunlight is a particularly promising clean and readily available source compared with conventional fossil fuels. Solar cells are devices that directly convert solar energy into electricity [2-3]. Nowadays, the processing methods for fabricating devices become increasingly more complex, resulting in higher production costs. Beside the high cost that limits the range of applications in daily life; poor mechanical flexibility also limits use in applications requiring flexible substrates. Organic-based electronic devices have a promising future despite their lower performance as semiconductors because of several unique properties of organic materials. However, the realization of large-scale organic solar cells is slowed because of their poor environmental stability and charge transport across the photoactive medium.

Introduction of carbon nanotubes (CNTs) into the photoactive layers of organic photovoltaic (OPV) cells were intended to mitigate shortcomings of conjugated polymer solar cells. Photovoltaic (PVs) can be rolled out onto a roof or other surfaces, and they are less prone to damage and failure than traditional solar cells. Being organic, polymer PVs are also more environmentally friendly [4-6]. CNTs and conjugated polymers have a similar sp2 type hybridization and exhibit high electrical conductivity [7]. However, even though CNTs have shown potential in the buffer layer, they have not resulted in a solar cell with power conversion efficiency greater than the best tandem organic cells (6.5% efficiency). But, it has been shown in most of the previous investigations that the control over a uniform blending of the electron

IOP Conf. Series: Materials Science and Engineering **805** (2020) 012026 doi:10.1088/1757-899X/805/1/012026

donating conjugated polymer and the electron accepting CNT is one of the most difficult as well as crucial aspects in creating efficient photocurrent collection in CNT-based OPV devices. Therefore, using CNTs as buffer layer in OPV devices is still in the initial research stages and there is still room for novel methods to better take advantage of the beneficial properties of CNTs. In this research, CNT was synthesized at low temperature using simple and cost effective method; the synthesized CNTs were then incorporated into PEDOT:PSS to modify anode buffer layer for the overall efficiency improvement of OPV.

2. Experimental details

2.1 Chemicals and materials

The materials used are graphite powder, nitric acid, sulphuric acid, potassium chlorate, poly (3,4-ethylene dioxythiophene): poly (styrene sulfonate) (PEDOT:PSS), distilled water, ethanol, acetone, isopropanol alcohol (IPA), glass substrates, and FTO coated glass.

2.2 Synthesis of carbon nanotube (CNT)

The graphite powder of 5 g was slowly added to the mixture of 25 ml of nitric acid and 50 ml of sulfuric acid for 30 min. The mixture was later cooled to 5 ^{o}C in ice bath. Then, 25 g of potassium chlorate was added to the above solution under continuous stirring for 30 min leading to the evolvement of heat. After that, the solutions were then placed in air for three days; after the three days, graphite precipitates were at the bottom while the carbon floated. The floating carbon material was transferred into distilled water (1 litre) for purification. After stirring it for one hour, the solution was immediately filtered and the sample was dried. The process was repeated four times.

2.2.1 Characterization of synthesized CNTs

Synthesized CNTs were characterized to investigate the chemical and structural properties using FTIR (BUK, lnfrared spectrophotometer model M530), SEM (ASPEX 3020) and XRD PANalytical system diffractometer (DY-1656), respectively.

IOP Conf. Series: Materials Science and Engineering **805** (2020) 012026 doi:10.1088/1757-899X/805/1/012026

2.3 Substrate preparation

All experimental procedures for the preparation of the thin films were performed under ambient air. Clean rectangular glass slides of dimension 25.4 mm by 76.2 mm were used as substrates. The glass substrates were washed with detergent solution for 10 to 15 min in ultrasonicator and rinsed in distilled water for 15 min at 30°C. The substrate was cleaned with isopropanol alcohol in ultrasonic bath for 15 min at 30 °C and dried in a stream of nitrogen gas (N_2).

2.4 Deposition of anode buffer layer thin film devices

The solutions of CNT/PEDOT:PSS nanocomposites were prepared and deposited on glass substrate at different spin-coating speed using spin-coater machine (Laurel WS-650Hz-23NPP). Spin coating speeds of 3000, 2000 and 1000 rpm were considered respectively for 30 seconds corresponding to thin film layer thicknesses of 32, 35, and 115 nm respectively. Depending on the speed of rotation (rpm) of the spin coater, the desirable thickness of the sample was obtained. It is important to note that the thickness of film being spin -coated depend on both time and speed of rotating stub as specified by equipment manufacturer in absence of surface profilometer.

2.4.1 Optical characterization of anode thin film devices

Optical characterization of thin film sample was carried out using UV-vis spectrophotometer (Avantes Avalight-DH-5-BAL). The photocurrent (current–voltage) of CNT/PEDOT:PSS buffer layer thin films devices were measured under simulated illumination at AM 1.5 G, 100 mW/cm^2 with a Keithley 2400 source meter. The wavelength spectrum for which a material absorbs light and produces photocurrent helps determine whether the material is capable of converting the available light illuminating it into electric energy. Absorption was chosen as the optimization variable because the choice of layer thickness is critical for optimization of the efficiency. UV-vis spectrophotometer was used to record the transmittance and reflectance in percentage (%). Absorbance was calculated using Equation (1) and (2) respectively [12].

$$A + R + T = 1 \quad (1)$$

$$\text{Absorbance (A)} = 2 - \log10(\%T) \quad (2)$$

2.4.2 Photovoltaic properties of thin films devices

The current- voltage characteristics of thin films were carried out under illumination of solar simulator with Keithley source meter (2400 SMU). The short-circuit current (I_{SC}) and open-circuit voltage (V_{OC}) were recorded using equation (3) and (4).

$$FF = \frac{J_m \times V_m}{J_{sc} \times V_{oc}} \quad (3)$$

$$\eta = \frac{P_{out}}{P_{in}} = \frac{J_{sc} . V_{oc} . FF}{P_{in}} \quad (4)$$

3.0 Results and Discussion

3.1 Scanning electron microscope (SEM) images of synthesized CNTs nanoparticles

Studies of the morphology of the synthesized nanoparticles showed that the obtained sample maintained their tubular structure at the origin of their strength. The obtained dried samples at 20 nm are cluster to each other but at 100 nm it shows properly dispersed nanotubes as indicated in Figures 1(a) and 1(b) respectively. The closely contacting interface between them is vital for the separation of electron hole pairs and the transportation of electron and holes, which are important for the hybrid solar cells to obtain high performance in agreement with reports in the literature [13-15].

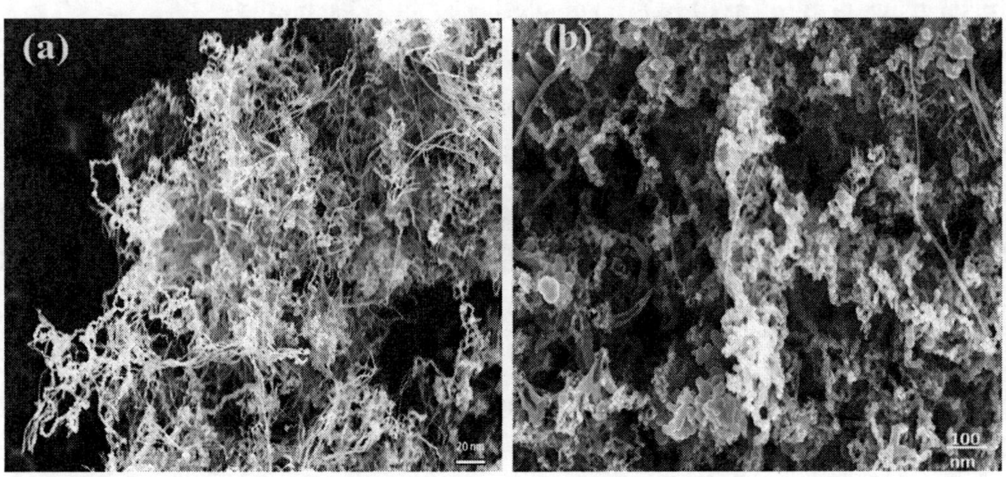

Figure 1. SEM images of (a) CNT at 20 nm (b) CNT at100 nm

3.2 X-ray diffraction (XRD) of synthesized CNTs nanoparticles

Figure 2 shows the XRD analysis of synthesized CNTs. The pattern exhibited the diffraction peaks of (002) and (100) plane of the carbon nanotube. The tetragonal structure of carbon nanotube was confirmed and the value of the lattice constants was in good agreement with the standard JCPDS value as shown in Figure 2

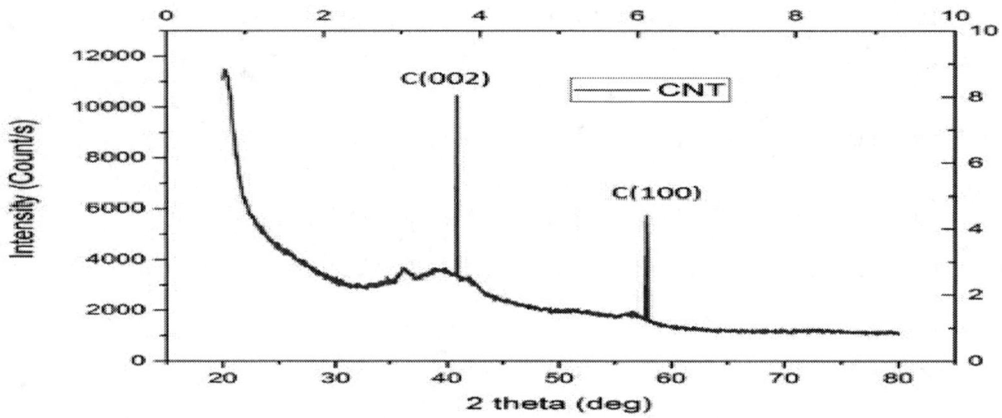

Figure 2. The XRD pattern of synthesized CNTs

3.3 Fourier transform infrared (FTIR) of synthesized CNTs analysis

FTIR spectrum of synthesized CNTs is shown in Figure 3 and was recorded in the range of 4000 to 500 cm^{-1}. It was observed that the band of synthesized CNTs occurs at 3632.98, 3419.89, 1633.79, 1384.27, 1224.71, 1064.69, 1020.41, 963.55, 850.3, 725.25, 623.69, 495.50 and 445.0 cm^{-1}. The strong band at 445 cm^{-1} is ascribed to the N-H, C-H bending mode and the signal at 495.50 cm^{-1} and 963.55cm^{-1} are ascribed to the C-H vibration bands in the synthesized CNT. The adsorption peaks at 1633.97 cm^{-1} and 1020.41 cm^{-1} correspond to the stretching modes of the C=O bond of the carbonic groups. These results show that synthesized CNT have been prepared successfully as reported by earlier workers [7, 11, 16-20].

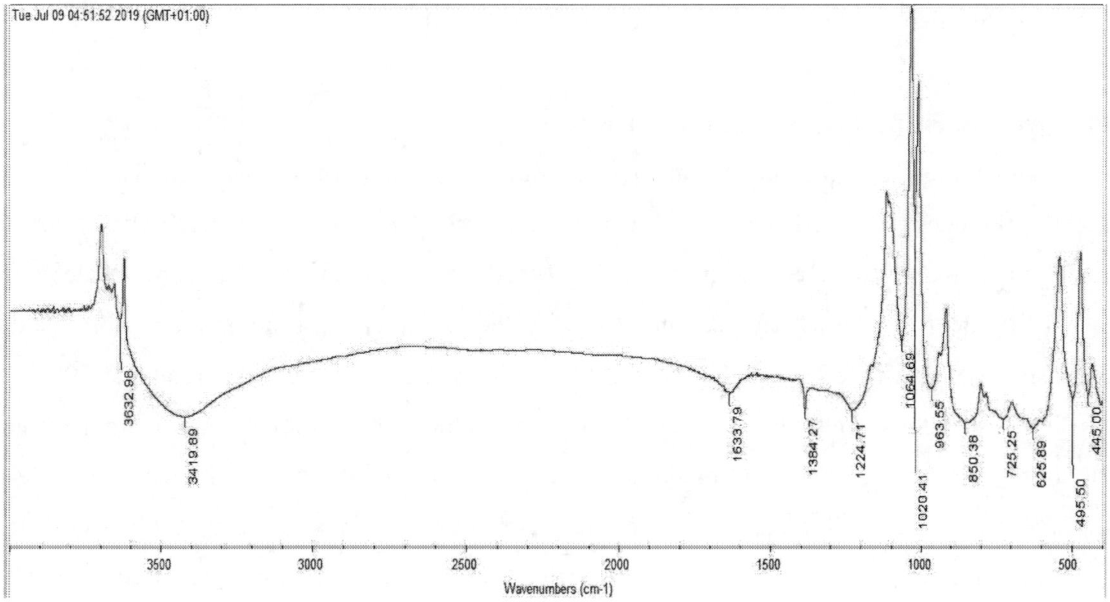

Figure 3. FTIR spectrum of synthesized CNTs

3.4 Optical characteristics of film devices at different speeds

The optical absorbance was calculated using equation 2. Figure 4 shows the absorbance graph of thin film devices spin coated at different spin coating speed using acetone as the solvent. It was observed that as the wavelength increases there is an increase in absorption; this displays the effects of the CNTs nanoparticles on the light absorption and charge separation efficiency of the conjugated polymer buffer layers. UV-visible absorption was measured for 1000, 2000 and 3000 rpm respectively deposited on glass substrates. It was observed from the graph that the highest absorbance value was at 2000 rpm and the least value was at 3000 rpm.

3.5 Optical characteristics of film devices at different annealing temperatures

The absorbance spectra of film devices were annealed at temperature range from 100 to 500 °C with a step of 100 °C. The maximum absorbance was at 200° C and estimated using equation 2, least value was at 400 °C as shown Figure 5. There is an increase in absorbance as the

IOP Conf. Series: Materials Science and Engineering **805** (2020) 012026 doi:10.1088/1757-899X/805/1/012026

wavelength increases; this could be due to structural homogeneity, crystallinity and the thicknesses of films, as reported by other researches [9, 21]. It was observed from the graph that the sample annealed at 400 °C has the highest value of absorbance.

3.6 Photovoltaic Properties of Thin Films Devices

Figure 6 shows the I-V measurement under solar simulation for optimized device. The fill-factor (FF) and efficiency (η) were calculated using equations (3) and (4), respectively. After the optimization of the parameters, the best rotating speed and annealing condition of anode buffer layer thin film device were determined and also the effect of annealing time on the performance of this devices were also investigated. Therefore, the device with annealing temperature of 400 °C at 30 min shows a slightly higher performance compared to others shown in table 1, this improvement can be ascribed to the enhancement of the absorption intensity of the device, increased defects in the hole transport layer that act as recombination centers to improve the conductivity. The optimized fabricated CNT/PEDOT:PSS anode buffer layer thin film device under illumination has the efficiency of 6.44%.

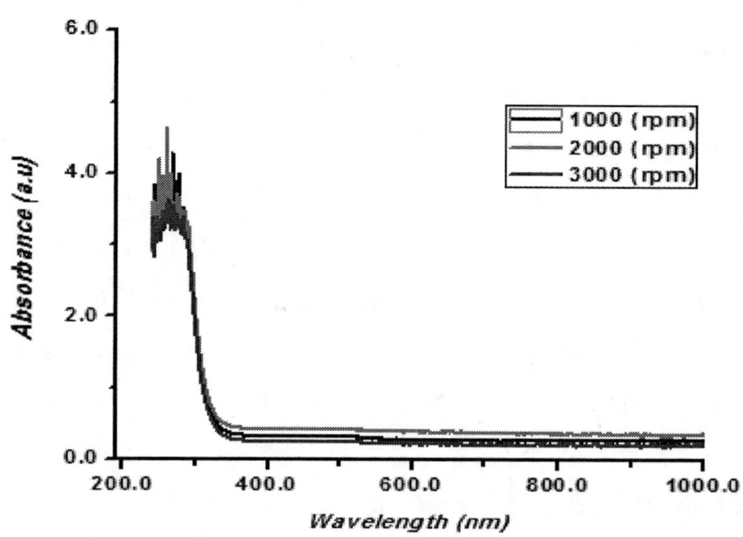

Figure 4. Optical absorbance spectra of film devices at different spin speed (rpm)

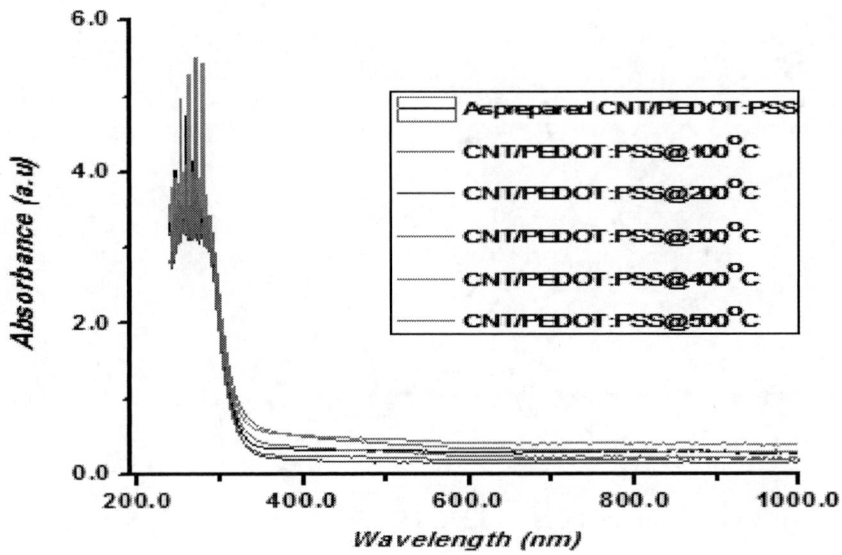

Figure 5. Optical absorbance spectra of film devices at different annealing temperatures

Table 1. Summary of photovoltaic parameter for the optimized anode buffer layer thin film devices

CNT/PEDOT:PSS (rpm)	Annealing (°C)	Time	V_{OC} (V)	J_{SC} (mA/cm^2)	P_{MAX} (W)	FF (%)	Efficiency (%)
2000	400	1 h	0.60	10.54	6.324	66	5.78
2000	400	30 min	0.72	11.02	7.934	72	6.44
2000	400	15 min	0.71	10.68	7.583	67	5.80

IOP Conf. Series: Materials Science and Engineering **805** (2020) 012026 doi:10.1088/1757-899X/805/1/012026

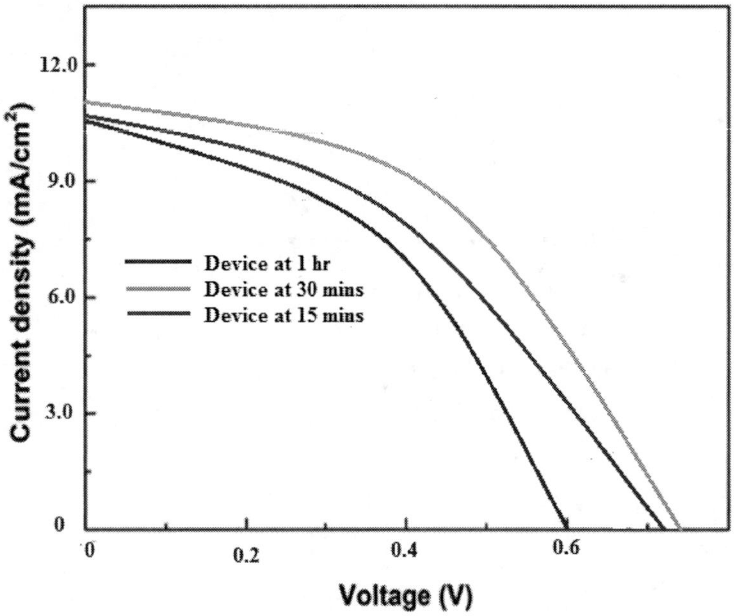

Figure 6. Current voltage curve of optimized CNT/PEDOT:PSS Thin Film Devices

4 Conclusion

This research focused on the synthesis of CNTs using a simple and cost effective method. The structural and electrical characterization was carried-out by using SEM, XRD, FTIR and Kiethley source meter, respectively. Optimization of optical properties was done using UV-vis Spectrophotometer. The synthesized CNTs were incorporated into polymer materials (PEDOT:PSS) anode buffer layer and the impact of post thermal annealing was investigated. It can be concluded that optimized spin coating speed and annealing condition increased the optical properties of the devices so as to enhance the performance of polymer solar cells. The photovoltaic performance evaluation of the polymer solar cells showed that open circuit voltage (Voc) ranged from 0.60 to 0.72 V and short circuit current density (Jsc) was in the range of 10.54 to 11.02 mAcm^{-2}. The improved efficiency of 6.44 % was recorded for device annealed for 30 min which enhance faster travelling of holes to the anode.

References

[1] Siddiki MK, Li J, Galipeau D and Qiao Q 2010 A review of polymer multijunction solar cells. *Energ. Environ. Sci.* **3** (7) pp 867-883.

[2] Baluschev S, Nelles G, Landfester K and Miteva T 2012 Sun-light up conversion in multi-component organic systems: development towards application for solar cells outcome enhancement. *Next Generation (Nano) Photonic and Cell Technologies for Solar Energy Conversion III* **8471** 84710E. https://doi.org/10.1117/12.928929.

[3] Kroon R, Lenes M, Hummelen JC, Blom PW and De Boer B 2008 Small bandgap polymers for organic solar cells (polymer material development in the last 5 years). *Polym. Rev.* **48** (3) pp 531-582.

[4] Said E 2009 Electrolyte Semiconductor Combinations for Organic Electronic Devices. Linköping University: Studies in Science and Technology. Doctoral Dissertations, No. **1228** *ISBN: 978-91-7393-735-1.*

[5] Anokhin DV, Gerasimov KL, Kiriy A and Ivanov DA 2015 Thermal annealing effect on active layer structure in all-polymer organic solar cells. *Appl. Mech. Mater.* **792** pp. 640-644.

[6] Frohne H, Shaheen SE, Brabec CJ, Müller DC, Sariciftci NS and Meerholz K 2002 Influence of the anodic work function on the performance of organic solar cells. *Chem. Phys. Chem.* **3** (9) pp 795-799.

[7] Liang F, Shi F, Fu Y, Wang L, Zhang X, Xie Z and Su Z 2010 Donor–acceptor conjugates-functionalized zinc phthalocyanine: Towards broad absorption and application in organic solar cells. *Sol. Energ. Mater. Sol. Cells* **94** (10) pp 1803-1808.

[8] Shi H, Liu C, Xu J, Song H, Lu B, Jiang F, Zhou W, Zhang G and Jiang Q 2013 Facile fabrication of PEDOT: PSS/polythiophenes bilayered nanofilms on pure organic electrodes and their thermoelectric performance. *ACS Appl. Mater. Interf.* **5** (24) pp 12811-12819.

[9] Alturaif HA, ALOthman ZA, Shapter JG and Wabaidur SM 2014 Use of carbon nanotubes (CNTs) with polymers in solar cells. *Mol.* **19** (11)17329-17344.

[10] Lin JS, Chu WP, Juang FS, Chen NP, Tsai YS, Chen CC, Chen CM and Liu LC 2012 Manufacture of light-trapping (LT) films by ultraviolet (UV) irradiation and their applications for polymer solar cells (PSCs). *Mater. Lett.* **67** (1) pp 42-45.

[11] Nguyen TL, Lee TH, Gautam B, Park SY, Gundogdu K, Kim JY, Woo HY. Single component organic solar cells based on oligothiophene-fullerene conjugate. *Adv. Funct. Mater.* **27** (39)1702474. https://doi.org/10.1002/adfm.201702474.

[12] You J, Dou L, Yoshimura K, Kato T, Ohya K, Moriarty T, Emery K, Chen CC, Gao J, Li G and Yang Y 2013 A polymer tandem solar cell with 10.6% power conversion efficiency. *Nature Comm.* **4** 1446. https://doi.org/10.1038/ncomms2411.

[13] Gusain A, Faria RM and Miranda PB 2019 Polymer Solar Cells—Interfacial Processes Related to Performance Issues. *Front. Chem.* **7** 61. https://doi.org.10.3389/fchem.2019.00061.

[14] Nardes AM, Janssen RA and Kemerink M 2008 A morphological model for the solvent-enhanced conductivity of PEDOT: PSS thin films. *Adv. Funct. Mater.* **18** (6) pp 865-871.

[15] Kan B, Li M, Zhang Q, Liu F, Wan X, Wang Y, Ni W, Long G, Yang X, Feng H and Zuo Y 2015 A series of simple oligomer-like small molecules based on oligothiophenes for solution-processed solar cells with high efficiency. *J. Am. Chem. Soc.* **137** (11) pp 3886-3893.

[16] Awodugba A O, Sanusi Y K and Ajayi I O 2013 Photovoltaic solar cell simulation of Shockley diode parameter in Matlab. *Int. J. Phys. Sci.* **8** (22) pp 1193-1200.

[17] Bakar N A, Supangat K and Sulaiman F 2014 Controlling the morphological structural and optical properties of one-dimensional PCDTBT nanotubes by template wetting *Nanoscale Res. Lett.* **9** 600. https://doi.org/10.1186/1556-276X-9-600.

[18] Bencharrats, F., Zitouni, K., and Gil B 2010 Determination of piezo electric and spontaneous polarization fields in Quantum wells grown along the polar (0001) direction *Superlattices Microst.* **47** (5) pp 592-596.

[19] Bernamache S, Rahal A and Benhaoua B 2014 The effects of solvent nature on spray-deposited ZnO thin film prepared from Zn $(CH_3COO)_2$ $2H_2O$. *Opt.* **125** (2) pp 663-666.

[20] Bertho S, Haeldermans I, Swinnen A, Moons W, Martens T, Lusen L, Vanderzande D, Manca J, Scenes A and Bonfiglio A 2007 Influence of thermal ageing on the stability of polymer bulk heterojunction solar cells. *Sol. Eng. Mat. Sol. Cells* **91** (5) pp 385-389.

[21] Balogun S W, Sanusi Y K and Aina A O 2018 Structural and optical properties of Titanium dioxide thin film deposited by spin coating technique. *Int. J. Dev. Res.* **08** (01) pp 18486-18490.

IOP Conf. Series: Materials Science and Engineering **805** (2020) 012027 doi:10.1088/1757-899X/805/1/012027

Optical studies of titanium dioxide/silver/gold (TiO_2/Ag/Au) nanocomposites as photo anode in dye sensitized solar cells

ST Ogundeji[1], MK Awodele[*], HO Oyeshola and O Adedokun

Department of Pure and Applied Physics, Ladoke Akintola University of Technology, Ogbomoso, Nigeria.

*Corresponding author: M.K. Awodele (mkawodele@lautech.edu.ng)

Abstract

In dye-sensitized solar cells (DSSCs), the performances of the photo anodes depend on the bandgap of semiconducting nanomaterials. Titanium dioxide (TiO_2) is usually used in preparation of photo anode but it absorbs only the ultraviolet light, owing to its large bandgap of about 3.2 eV, and another drawback is that TiO_2 has low electron mobility. In this study, optical studies of (TiO_2/Ag/Au) nanocomposites as photo anode were carried out in order to test the possibility of improving the efficiency of the DSSCs. Dye molecule was extracted from the leave of sensitive plant (*Mimosa pudica*) using ethanol as solvent. TiO_2/Ag/Au were deposited on a glass substrate using doctor blade method; the deposited thin films were annealed in a furnace at 450 ^oC for 1 h after which the annealed thin films were dye loaded for 12 h with *Mimosa pudica* extract. The dye loaded thin films of TiO_2/Ag/Au were then characterized with a UV-vis spectroscopy to get the transmittance. The absorbance and the optical band gap were calculated. The optical absorption spectra and optical bandgap spectra of TiO_2/Ag/Au thin films were examined. The maximum absorption was observed within the range of the visible region when the position of TiO_2/Ag/Au is of volume ratio 1:1:0.8 and optical bandgap of 3.75eV. Anticipatedly, the performance characterization with further review on TiO_2 with Ag/Au nanocomposites using *M. pudica* extract, as a sensitizer will enhance the development of an authentic and competitive dye sensitized solar cell.

Keywords: Dye sensitized solar cells (DSSCs), thin films, *Mimosa pudica*, Photo-anode

IOP Conf. Series: Materials Science and Engineering **805** (2020) 012027 doi:10.1088/1757-899X/805/1/012027

1 Introduction

Fossil fuels are the main source of energy consumed worldwide. It is a non-renewable source of energy. Energy is generated from fossil fuel, and so improves the life condition of mankind [1]. The critical fact is that fossil fuels emit greenhouse gas (carbon dioxide) when it burns which are the major causes of global warming [2] due the long term rise in earth's average surface temperature. As man's population increases, the demand for energy increases, and the more the usage, the more the environmental hazards. Therefore, a viable and renewable energy sources to replace diminishing fossil fuels are required. Various forms of renewable energy sources have been identified including wind, hydro, biomass, biofuel, geothermal and solar energy. Of all the energy sources, solar energy receives the attention of researchers because it was recognised to serve as the primary energy sources for other types of energy [3]. Solar energy provides a clean, cheap and renewable energy for human race, the quest for conversion of solar energy to electricity led to the invention of solar cells.

Solar cells are categorized into three different generations; first generation, second generation and third generation solar cell [4]. The first generation consists of crystalline semiconductor wafers with a thickness of 200-300 nm. The second generation are based on thin film technology having thickness usually in the range 1-2 nm [5]. The third generation technology is mainly focused on the material e.g. wide band gap metal oxide semiconductors that include titanium dioxide (TiO_2), ZnO among others. Dye sensitized solar cells (DSSCs) is an example of third generation technology [6-10].

DSSCs are devices that convert solar to electric energy by light sensitization established on wide energy band semiconductor [11]. It is currently attracting widespread scientific and technological interest because of its minimal production cost, environmental friendliness, simple method of fabrication, lower toxicity manufacturing process, straight forward scale, low weight and potential use in flexible panel as compared to conventional p-n junction devices [12].

A modern DSSC is composed of photo-anode, photosensitizer, electrolyte and a counter electrode. Nanostructure semiconductor metal oxide e.g. TiO_2 are usually deposited over a transparent conducting glass substrate to prepare photo-anode [2, 3, 11].

IOP Conf. Series: Materials Science and Engineering **805** (2020) 012027 doi:10.1088/1757-899X/805/1/012027

The performance of the photo-anodes depends on the band gap, morphology, and composition of metal oxides and thickness of metal oxides. In recent times, a new method has evolved for increasing the light absorption of photo-anode in DSSC; the use of nanoparticles with size ranging between 1-100 nm [13, 14]. Nanostructured metal oxide have emerged as strong candidates for wide range of application on account of its unique physical, chemical, optical, electrical and thermal properties [15-18]. With respect to its unique properties, silver nanoparticles (AgNPs) are being incorporated into photovoltaic application [19-20]. The effects of annealing temperature on the optical properties of AgNPs thin films deposited on glass substrate by spin coating technique has been investigated [5]. The result revealed that light absorption spectral range and thermal stability could be enhanced using silver nanoparticles [5].

The use of inorganic dye has been predominant for the production of the DSSCs [22-26]. The most efficient inorganic dye i.e. ruthenium complexes employ as photo-sensitizers in DSSCs have exhibited high power conversion efficiencies up to 11-13% [27, 28]. The disadvantage of these complexes is high cost and sophisticated preparation techniques. Investigation of low cost, readily available dye as efficient sensitized for DSSCs have been studied. Natural dyes extracted from fruit, leaves, roots and flowers have been proven to be efficient dye as a sensitizer in DSSCs [2, 3, 6].

In this work, natural pigment was extracted from the leaves of *Mimosa pudica*, an organic plant locally known as ''sensitive plant or touch me not plant''. The optical spectral study of the extracted pigment of *M. pudica* using ethanol as solvent was primarily characterized by UV-vis spectrophotometer. Titanium dioxide incorporated with silver and gold nanoparticles (TiO_2:AgNPs:AuNPs) as photo-anode material using doctor blade technique was characterized using a UV-vis spectrophotometer.

2 Materials and Methods

2.1 Materials

The materials utilized for this study included: titanium (iv) oxide nanopowder (< 0.35 nm) of 97% purity supplied by Sigma Aldrich, 100 μg/ml cocoa pod extract-mediated silver nanoparticles (AgNPs) [29] and 340 μg/ml kola pod extract-mediated gold nanoparticles (AuNPs) obtained from the Laboratory of Industrial Microbiology and Nanobiotechnology LAUTECH Ogbomoso.

IOP Conf. Series: Materials Science and Engineering **805** (2020) 012027 doi:10.1088/1757-899X/805/1/012027

2.2 Preparation of natural photo-sensitizers

Mimosa pudica was obtained from a nearby local farm in Ogbomoso, Nigeria. The leaves from the plant were taken and washed severally with distilled water, followed by air-drying. The dried leaves were ground into powder. Two gram of the powder was weighed and poured into a bottle. The photosensitizer was extracted by dissolving 2 g of the fine powder into 50 ml of ethanol; the solution was left undisturbed for 24 h at room temperature. The residue was the filtered out, the resulting dye solution was optically characterized and utilized as natural photosensitizers without any further purifications. To ensure stability before use, the filtrate was kept in a separate clean bottle and protected from direct sunlight.

2.3 Substrate preparation

Microscopic glass slides (25.4 mm × 76.2 mm × 1.0 mm) were used as substrates. The rectangular glass slide was first cleaned in a detergent solution using an ultrasonic bath for 15 min. They were washed with detergent solution for 10-15 min in ultrasonic sonicator and rinsed in distilled water for 15 min at 30 °C. The substrate was cleaned with isopropanol alcohol (IPA) in ultrasonic bath for 15 min at 30 °C and dried in a stream of nitrogen gas (N_2).

2.3.1 Preparation Titanium dioxide (TiO_2) paste

About 30 ml of ethanol was measured and added to 3 g of TiO_2 powder in a beaker. The solution was placed on a magnetic stirrer for 1 h until homogenous mixture is formed.

2.3.2 Preparation of (TiO_2:AgNPs:AuNPs)

The blend solution of (TiO_2: AgNPs: AuNPs) was done by mixing TiO_2 solution with AgNPs and AuNPs solution at different volume ratio (1:1:0.2, 1:1:0.4, 1:1:0.6, 1:1:0.8, 1:1:1) ml and stirred with plastic spatula to obtain homogenous mixture.

2.3.3 Preparation of photo anode (TiO_2:AgNPs:AuNPs)

Blend pastes were deposited on a cleaned glass substrate at different volume ratio using doctor blading technique. The thin film were dried at 100 °C and later annealed at 450 °C in a furnace for 2 h. The temperatures of the films were cooled down to 80

^{o}C before being immersed in to natural dye solution kept at room temperature for 12 h for complete sensitization.

2.3.4 Characterization and measurement

The UV-vis absorption characteristics of natural pigment were analysed using Jenway UV-vis spectrophotometer. It was observed that the natural pigment of *Mimosa pudica* has absorption which wavelength falls within the visible region of the spectrum. The optical absorption spectra of the photo anode were measured using UV-vis spectrophotometer (Avantes, Avalight – DH – 5bal). The absorbance of the samples was determined from transmittance value obtained from the spectrophotometer by using the relation in Eq. (1).

$$A = log\,(1/T) \qquad (1)$$

Where A is the absorbance, $T = I / I_0$, I is the transmitted light and I_0 is the incident light:

The energy band gap of the thin film sample was calculated with the help of the $(\alpha hv)^2$ versus energy band gap value. The theory of the interband absorption shows that at the optical absorption edge, the absorption coefficient α varies with the photon energy hv according to Eq. (2)

$$\alpha(hv) = A\,(hv - E_g)^{1/2} \qquad (2)$$

where A is a constant and E_g is the optical bandgap. Thus a plot of $(\alpha hv)^2$ versus hv is a curve line whose intercept on the energy axis gives the energy gap. The band gap energy of the films have been determined by the extrapolation of the linear regions on the energy axis hv. The absorption coefficient (α) associated with the strong absorption region of the film was calculated from absorbance (A) and the film thickness (t) using the relation.

$$\alpha = 2.303A/t \qquad (3)$$

The thickness of thin films was determined using:

$$t = m/2AD \qquad (4)$$

IOP Conf. Series: Materials Science and Engineering **805** (2020) 012027 doi:10.1088/1757-899X/805/1/012027

Where $m = m_2 - m_1$ and m_1 = mass of substrate before deposition m_2= mass of substrate after deposition, A = Area covered by the films, D = density of TiO_2.

3 Results and Discussion

The UV-vis optical absorption spectra of *Mimosa pudica* using ethanol as solvent are revealed in Figure 1(a). It is observed that the natural pigment showed maximum absorption in the region around at 325, 347 and 430 nm. The absorption of the light falls within the visible region of the electromagnetic spectrum. Figure 1(b) shows the optical transmittance spectra of TiO_2:AgNPs:AuNPs at different volume ratio. The transmittance is generally high between wavelengths of 200-800 nm with a gradual fall near the fundamental absorption region. The sample with the 1:1:0.6 has the highest average transmittance value of about 49.8% and the lowest transmittance of 17% when the volume ratio was 1:1:1. In Figure 1(c), the absorbance spectra of TiO_2:AgNPs:AuNPs at different volume ratio is presented. The spectra revealed that the deposited films have high absorbance in the UV region, whereas the absorbance is low in the visible region. In general, the absorbance of the films decreases with increasing wavelength and decreasing photon energy. For each sample, there is an absorbance level that differs from the other. Peak absorbance was observed at volume ratio1: 1:0.8. In Figure 1 (d), the energy gap spectra of TiO_2:AgNPs:AuNPs was presented. It was observed that the energy gap of the prepared TiO_2:AgNPs:AuNPs was 3.75eV compared to as prepared TiO_2 which was 3.57eV. The inclusion of silver and gold nanoparticles with TiO_2 makes the energy gap to be shifted in positive direction. The shift of conduction band in the positive direction decreases the efficiency of electron–hole separation at the interface and thus increases the electron/ electrolyte recombination rate resulting in reduction in photo-current. The increase in the recombination rate also result lower V_{OC} which is not essential for making an efficient DSSC [30].

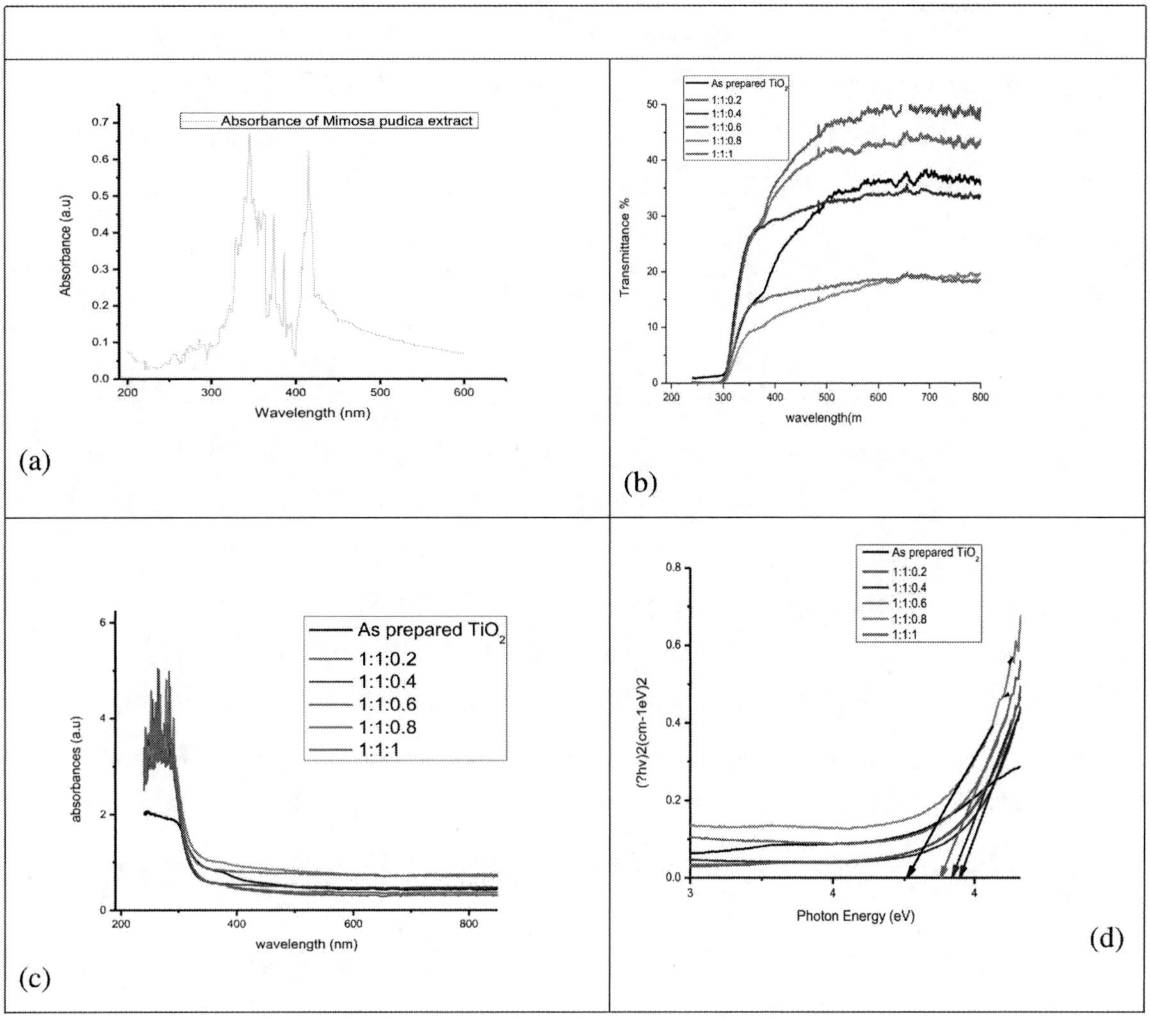

Figure 1. (a) UV-Vis absorption spectrum for light harvesters obtained from leaves of sensitive plant, (b) Transmittance spectra for TiO_2:AgNPs:AuNPs at different volume ratio, (c) Absorbance versus wavelength of TiO_2:AgNPs:AuNPs films blend at different volume ratio, (d) Absorption spectra for TiO_2:AgNPs:AuNPs films blend at different volume ratio

4 Conclusion

The natural photosensitizer from *Mimosa pudica* leaves was successfully appraised as natural photosensitizer in photo-anode of DSSCs. The absorption peak of the extracted pigment falls within the UV-vis region. The optical study using UV-vis spectrophotometry for TiO_2:AgNPs:AuNPs photo-anode at different volume ratio have been investigated. Based on the results, the TiO_2:AgNPs:AuNPs thin film has a good absorbance at doping concentration of 1:1:0.8 in the visible region. The band gap of the as prepared TiO_2 is 3.57eV. With the inclusion of silver and gold nanocomposites, the band gap increases to 3.75eV. On account of its large band gap, it is not advisable to use TiO_2:AgNPs:AuNPs nanocomposites as photo-anode in DSSCs.

References

[1] Edenhofer O, Pichs-Madruga R, Sokona Y, Seyboth K, Kadner S, Zwickel T, Eickemeier P, Hansen G, Schlömer S, von Stechow C and Matschoss P 2011 Renewable energy sources and climate change mitigation: Special report of the intergovernmental panel on climate change. Cambridge University Press, Cambridge, UK.

[2] Sengupta D, Das P, Mondal B and Mukherjee K 2016 Effects of doping, morphology and film thickness of photo-anode materials for dye sensitized solar cell application. *Renew. Sustain. Energ. Rev.* **60** pp 356-376.

[3] Adedokun O, Titilope K and Awodugba A O 2016 Review on natural dye sensitized solar cell (DSSCs). *Int. J. Eng. Technol.* **2** pp 6-13.

[4] Heeger A J 2014 25th anniversary article: bulk heterojunction solar cells: understanding the mechanism of operation. *Adv. Mater.* **26** (1) pp 10-28.

[5] Balogun S W, Abdulhamid A B, Sanusi Y K and Adedokun O 2019 Synthesis and characterization of titanium (IV) oxide loaded with silver nanoparticles thin films. *Am. J. Nano Res. Appl.* **7** (1) pp 1-5

[6] Sanusi Y K, Kazeem A A and Suleman K O 2016 Photo anode thickness and sensitization time effects on overall performance of nanocrystalline TiO_2 based solar sensitized with Roselle flower extract. *Int. J. Sci. Eng. Res.* **7** pp 345-350.

[7] Gardelis S, Nassiopoulou AG, Manousiadis P, Vouroutzis N and Frangis N 2013 A silicon-wafer based pn junction solar cell by aluminum-induced recrystallization and doping. *Appl. Phys. Lett.* **103** (24) 241114. https://doi.org/10.1063/1.4846595.

[8] Jeong W S, Lee J W, Jung S, Yun J H and Park N G 2011 Evaluation of external quantum efficiency of a 12.35% tandem solar cell comprising dye

sensitized and CIGS solar cell. *Sol. Energ. Mater. Sol. Cells* **95** (12) pp 3419-3423.

[9] Li G Shrotriya V Hunang J Yao Y Moriarty T Emery K Yang Y 2011 High efficiency solution processable polymer photovoltanic cells by self-organization of polymer blends. *Mater. Sustain. Energ.* 4: pp 80-84. https://doi.org/10.1142/9789814317665_0012.

[10] Kamat P V 2013 Quantum dot solar cells. The next big thing in photovoltaics. *J. Phys. Chem.* **4** (6) pp 908-918.

[11] Grätzel M 3003 Dye-sensitized solar cells. *J. Photochem. Photobiol C: Photochem.* Rev. **4** (2) pp 145-153.

[12] Gokilamani N, Muthukumarasamy N, Thambidurai M, Ranjitha A, Velauthapillai D, Senthil TS and Balasundaraprabhu R 2013 Dye-sensitized solar cells with natural dyes extracted from rose petals. *J. Mater. Sci.: Mater. Electron.* **24** (9) pp 3394-3402.

[13] Hamidi M, Azadi A and Rafiei P 2008 Hydrogel nanoparticles in drug delivery. *Adv. Drug Deliv. Rev.* **60** (15) pp 1638-1649.

[14] Lawenstam H A 1981 Minerals formed by organisms. *Sci.* **211** (4487) pp 1126-1130.

[15] Lee E J H, Ribeiro C, Longo E and Leite E R 2006 Growth kinetics of tin oxide nanocrystals in colloidal suspension under hydrothermal condition. *Chem. Phys.* **338** (1-3) pp 229-235.

[16] Roy P, Kim D D, Lee K, Spiecker E and Schmuki P 2010 TiO_2 nanotubes and their application in dye-sensitized solar cell. *Nanoscale* **2** (1) pp 45-49.

[17] Linsebigler A Lug L Yates J Y 1995 Photocatalysis on TiO_2 surfaces: Principles, mechanism and selected results. *Chem. Rev.* **95** (3) pp 735-758.

[18] Wang X, Fujimaki M and Awazu 2005 Photonic crystal structures in titanium dioxide (TiO_2) and their optimal design. *Opt. Express* **13** (5) pp 1486-1497.

[19] Saravanan P, Ganapathy M, Charles A, Tamilselvan S, Jeyasekaran R and Vimalan M 2016 Electrical properties of green synthesized TiO_2 nanoparticles . *Adv. Appl. Sci. Res.* **7** (3) pp 158-168.

[20] Sakshum K, Kushagra K, Gauravi X and Prakhar K 2017 Plasmonic study of nanoparticles in organic photovoltic cells: A review. *J. Org. Inorg. Chem.* **3** (1) 2.

IOP Conf. Series: Materials Science and Engineering **805** (2020) 012027 doi:10.1088/1757-899X/805/1/012027

[21] Bardaxoglou G, Rouleau C and Pelletier E 2017 High stability and very slow dissolution of bare and polymer coated silver nanoparticles dispersed in river and coastal water. *J . Aquat. Pollut. Toxicol.* **1** (2) 15.

[22] Shitta M B and Ogedengbe E O B 2016 Performance modelling of *Mimosa pudica* extract as a sensitizer for solar energy conversion. *Energ. Policy Res.* **3** (1) pp 42-49.

[23] Campbell WM, Jolley KW, Wagner P, Wagner K, Walsh PJ, Gordon KC, Schmidt-Mende L, Nazeeruddin MK, Wang Q, Grätzel M and Officer DL 2007 Highly efficient porphyrin sensitizers for dye-sensitized solar cells. *J. Phys. Chem. C* **111** (32) pp 11760-11762.

[24] Choi H, Baik C, Kang SO, Ko J, Kang MS, Nazeeruddin MK and Grätzel M 2008 Highly efficient and thermally stable organic sensitizers for solvent-free dye-sensitized solar cells. *Angew. Chem.*47 (2) pp 327-330

[25] Bessho T, Zakeeruddin SM, Yeh CY, Diau EW and Grätzel M 2010 Highly efficient mesoscopic dye-sensitized solar cells based on donor–acceptor-substituted porphyrins. *Angew. Chem.* **49** (37) pp 6646-6649.

[26] Mathew S, Yella A, Gao P, Humphry-Baker R, Curchod BF, Ashari-Astani N, Tavernelli I, Rothlisberger U, Nazeeruddin MK and Grätzel M 2014 Dye-sensitized solar cells with 13% efficiency achieved through the molecular engineering of porphyrin sensitizers. *Nature Chem.* **6** (3) pp 242-247.

[27] Hao S, Wu J, Huang Y and Lin J 2006 Natural dyes as photosensitizers for dye-sensitized solar cell. *Sol. Energ.* **80** (2) pp 209-214.

[28] Zhou H, Wu L, Gao Y and Ma T 2011 Dye-sensitized solar cells using 20 natural dyes as sensitizers. *J. Photochem. Photobiol. A: Chem.* **219** (2-3) pp 188-194.

[29] Lateef A, Azeez MA, Asafa TB, Yekeen TA, Akinboro A, Oladipo IC, Azeez L, Ojo SA, Gueguim-Kana EB and Beukes LS 2016 Cocoa pod extract-mediated biosynthesis of silver nanoparticles: Its antimicrobial, antioxidant and larvicidal activities. *J. Nanostruct. Chem.* **6** (2) pp 159-169.

[30] Watson DF and Meyer GJ 2004 Cation effects in nanocrystalline solar cells. *Coord. Chem. Rev.* **248** (13-14) pp 1391-406.

IOP Conf. Series: Materials Science and Engineering **805** (2020) 012028 doi:10.1088/1757-899X/805/1/012028

Nanobiosensors: applications in biomedical technology

AT Banigo[1]*, TO Azeez[1], KO Ejeta[1], A Lateef[2] and E Ajuogu[1]

[1]Department of Biomedical Technology, Federal University of Technology, Owerri, Imo State, Nigeria.

[2]Nanotechnology Research Group (*NANO*[+]), Department of Pure and Applied Biology, Ladoke Akintola University of Technology, Ogbomoso, Oyo State, Nigeria.

*Corresponding author: almabanigo2@gmail.com

Abstract

Biosensor devices are composed of bioreceptor, transducer and detector that detect and aid in measuring parameters of some primary metabolites, immunological molecules and many more materials. These devices are of various types including piezoelectric which exhibit high efficiency based on sensitivity, response time, selectivity and linearity. Currently, newly developed nanobiosensors help in transduction and are employed to sense biomolecules bearing high sensitivity. Nanobiosensors also could be homogeneous or heterogeneous in nature and equally function in sensing mechanism of the biosensing technology. Thus, different nanobiosensors are greatly utilized to reduce poison in products, disease diagnostics and in many biomedical applications. Based on all these factors and the positive impact of using these devices; nanobiosensor types, applications, challenges and preferred solution in biomedical technology were considered and discussed in this work.

Keywords: Nanobiosensor, fabrication, challenges, application, biomedical technology

1. Introduction

Biosensor consists of three essential components such as transducer, bioreceptor and detector, which is used to detect some primary metabolites, immunological molecules and many more materials as shown in Figure 1 [1, 2]. The biosensor acts as a tool for the production of a form of signal from a biological response [1]. The two major high points of this device are the successful merging of biological and electronic components to generate signal during any analysis for utmost performance and the ability of the bioreceptor to remain stable in a non-biological environment. Biosensors are presented in different types because of their transduction and biological recognition elements such as piezoelectric [3] and potentiometric for transduction, and antibodies, organelles and enzymes for biorecognition element [1, 4, 5]. Also, the characters of the biosensor required for its high efficiency may include sensitivity, response time, selectivity and linearity as highlighted in Table 1 [6].

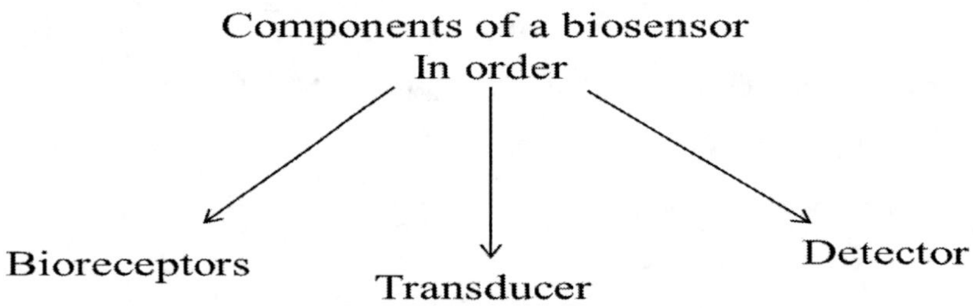

Figure 1. The basic components of a biosensor

Table 1. Characteristics of biosensor and their descriptions [6]

S/N	Characteristics	Description
1	Sensitivity	Electrode response per substrate concentration value
2	Response time	Time taken for about 95% response to be obtained
3	Selectivity	Proper reduction of chemicals interference is necessary
4	Linearity	Aid in high substrate concentration detection

The production of the premier oxygen biosensor by Lel and Clark in 1960s paved better opportunities for the development of many biosensors in Medicine, Biomedical Engineering and Technology (BMET) and Nanotechnology (NT) applications. Therefore, these devices pose potential benefits in fields of BMET, Medicine, Tissue Engineering (TE) and NT to enhance healthcare delivery and preserve human lives [7]. Specifically, some biosensors are applied in checking blood glucose level [8, 9] and *in vivo* studies, to detect disease-specific biomarkers. At cellular level, a biosensor of micro or nanoscale is required and the biosensors with nano-dimensions including nanowires and nanotubes have been introduced and produced for proper efficient diagnostic and Biosensing matters. This device may detect the pH level or other factors

IOP Conf. Series: Materials Science and Engineering **805** (2020) 012028 doi:10.1088/1757-899X/805/1/012028

[10]. Chan and Nie [11] and Tomasulo *et al.* [12] reported that "nanocantilevers had been employed to check the amount of serum protein markers and identify the parts of the specific DNA moieties". A typical example of nanodevices, quantum dot is a unique nanocrystal that had been and could still be used to identify particular DNA or protein [13]. This is applied in an area of specialization termed as "nanotechnology".

Nanotechnology is broadly referred to as the multidisciplinary study area that applies science, engineering and technology to efficiently manipulate and also manufacture materials and instruments on nanoscale ranging from 1 to 100 nm [14, 15]. It encompasses several technologies from various educational sectors including medicine, electronics, biomaterials, chemical engineering and biotechnology to synthesize well deserved materials for mankind. Advancement in this fast developing field has resulted to the introduction of biosensors at nanoscale for drug target, discovery and delivery, and many more biomedical applications bearing dynamic properties. Other application areas include the environment where nanotechnology is used to manufacture products for cleaning up environmental pollution. Most recently, biosensors best known to analyse the tiniest of biological interactions at an insignificant scale and with high precision, versatility and sensitivity have been reported [16, 2, 17]. A typical example is the nanobiosensor synthesized from the synergy of nanotechnology and biosensors.

A nanotechnology structured biosensor, consisting of a nanomaterial ranging from 1 to 100 nm becomes so unique, bearing several atomic constituents with are placed in or near the surface and have the important physical, chemical and physicochemical properties much more different from the same materials at the bulk level. Nanobiosensors may be homogeneous or heterogeneous in nature and they equally act in sensing mechanism of the biosensing technology [2, 18]. The former occurs in solution without a phase separation and later concentrates on a solid material such as capture material embedded with a chosen receptor including antigen that aids in the identification and capture of the analyte (antibody) to be detected.

A combination of electrical systems and the devices of the nanomaterials generate nanoelectomechanical system (NEMS) that are so active in the event of electrical transduction [2]. Most electronic and mechanical properties of some nanomaterials including nanotubes, nanorods, nanowires, nanoparticles and thin films consisting of crystalline matter are well known and utilized in improving transduction mechanisms and biological signalling [2]. Nanoparticles amongst other nanomaterials are well studied and analysed up to this moment. This review

IOP Conf. Series: Materials Science and Engineering **805** (2020) 012028 doi:10.1088/1757-899X/805/1/012028

focuses on nanobiosensor types, applications, challenges and preferred solution in Biomedical Technology.

2. Types of nanobiosensors

Nanobiosensors are grouped into different types with great considerations on the nanomaterials and biosensing operation. These nanomaterials possess properties that enhance the sensing mechanism. They include nanotubes-based sensors, nanoparticles-based biosesnsors, nanowire-based sensors, quantum dots-based sensors among others [2].

2.1. Nanotube-based sensors

Nanotube-based sensors (NTBSs) are a type of nanobiosensors with carbon nanotubes to improve the specificity of reaction. The NTBSs aid in monitoring glucose and detecting biomolecules. The development of glucose biosensors was concentrated on the utilization of nanotubes as immobilizing surfaces specifically for glucose oxidase which apparently quantifies glucose present in different bodily fluids including tears and saliva. Carbon nanotubes (CNTs) also known as hollow cylindrical tubes are sensing materials present in nanoelectromechanical sensors (NEMS) and are hollow cylinders-like that are produced in the rolled graphite sheet [19]. Its carbon atoms arrangement in the graphene sheet may be chiral, armchair or zigzag [20]. The different chirality, graphene wall number and tube diameter are points to consider for CNT conductivity [20, 21]. According to Bareket-Keren and Hanein [20], 'They have special structures, are chemically inert, electrically and mechanically stable and have high tensile strength, high thermal conductivity, high surface to volume ratio and great electrocatalytic activity'[6]. Their structures are also special in their form.

The CNTs are produced using more than two (2) dynamic techniques including chemical vapour deposition (CVD). The later method most times yields CNTs in mass production and then purified for avoidance of structural damage and elimination of unwanted metals required as catalysts as CNT are produced [22]. The CNTs concentrate on monitoring glucose in urine and blood. They are clearly applied in material science and optoelectronics with unique properties including physical geometry, mechanical strength, asbestos-like characteristics and physicomechanical properties [2]. In recent past, CNTs have been extensively used in tissue engineering, biosensors, molecules as well as drug delivery and for diagnostics [6]. Based on the

IOP Conf. Series: Materials Science and Engineering **805** (2020) 012028 doi:10.1088/1757-899X/805/1/012028

number of layers of the concentric graphite covered by a type of hemispheres known as fullerene, CNTs are categorized into two namely; single walled CNTs (SWCNTs) and multi walled CNTs (MWCNTs).

The SWCNTs are single objects between 0.4-2.5 nm in diameter and lengths usually up to a few millimeters, whereas MWCNTs are usually SWCNTs arranged in a coaxial manner, 2-100 nm in diameter and lengths ranging from 1-100 μm or more [20]. In addition, SWCNTs are sometimes metallic or semiconducting with MWCNTs bearing the metallic behavior [22]. Based on all these qualities, SWCNT and MWCNT have greatly been considered in developing biosensors for utmost functionality and output. Single walled (CNT) has been well known for enzymatic detection of glucose and its increment in enzymatic activity [23] and are also largely applied for recording and simulating neuronal activity [20]. Figures 2 and 3 depict the typical structures of SWCNT and MWCNT.

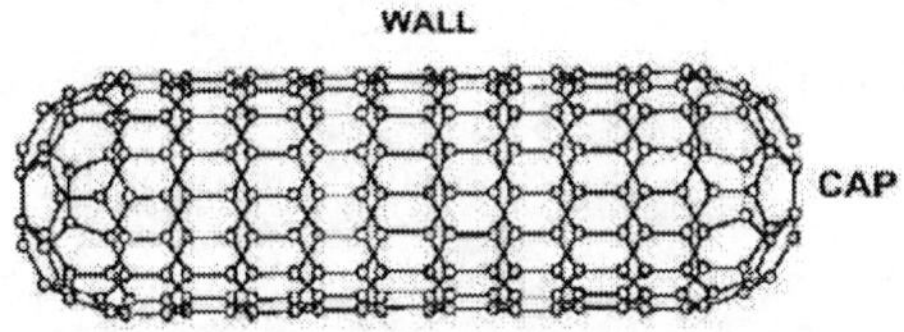

Figure 2. SWCNT structure displaying cap

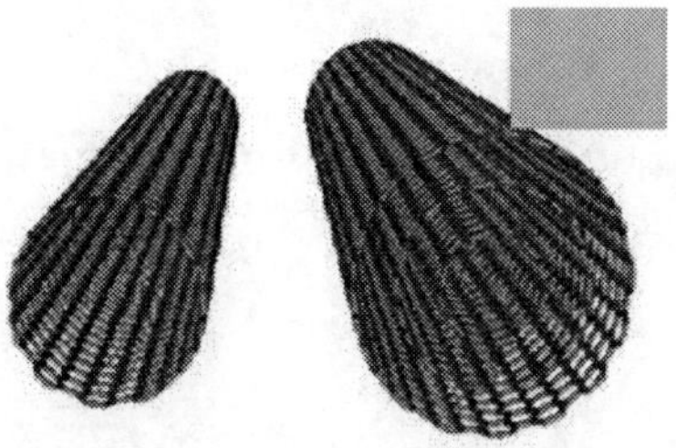

Figure 3. SWCNT and MWCNT structures and wall [6] with a difference [24]

IOP Conf. Series: Materials Science and Engineering **805** (2020) 012028 doi:10.1088/1757-899X/805/1/012028

2.1.1. CNTs in Biomedical Applications

More advances have been made in both fields of CNT-based technology and Biomedical Engineering Technology (BMET), where CNTs perform as biomaterials in the production of bionanosensors, vaccine and drug delivery vehicles, scaffolds, smart materials and biomolecule nano self-assembly carriers in an emerging and sophisticated subfield of BMET termed as Tissue Engineering (TE) (Figures 4-7). The CNTs help in some tissues' growth and proliferation [25], but have deficient solubility in aqueous media [22].This deficiency can be corrected by increasing the chemical functional groups to parts of CNTs wall, including the tips [22]. According to Shi *et al.* [26], "CNTs are suitable materials for the synthesis of scaffolds in reproduction of bones bearing basic properties such as biocompatibility, biodegradability, cell proliferation and adhesion, absorption and many others".

A typical example is the ligation of the CNTs and a biomaterial, collagen which yielded high mechanical properties obtained from CNTs, thereby replacing the lost features of the already produced scaffolds [27, 22, 28-30]. Another example is the output realized from the ligation of CNTs and synthetic polymers resulting in a material so biocompatible for regeneration of tissues. Furthermore, the use of MWCNT and poly (L-lactide) together yielded high crystallization, direct-current conductivity and many other properties as studied by Zhang and colleagues [31] and electrospun nanocomposite scaffolds (ENSs) were produced from the encapsulation of MWCNTs in poly (lactic acid) (PLA) nanofibers by McCullen and partners [32]. The ENSs had fewer diameters and more conductivity than PLA only. Therefore, conductivity greatly seen as a physical property became more present as CNTs were added.

IOP Conf. Series: Materials Science and Engineering **805** (2020) 012028 doi:10.1088/1757-899X/805/1/012028

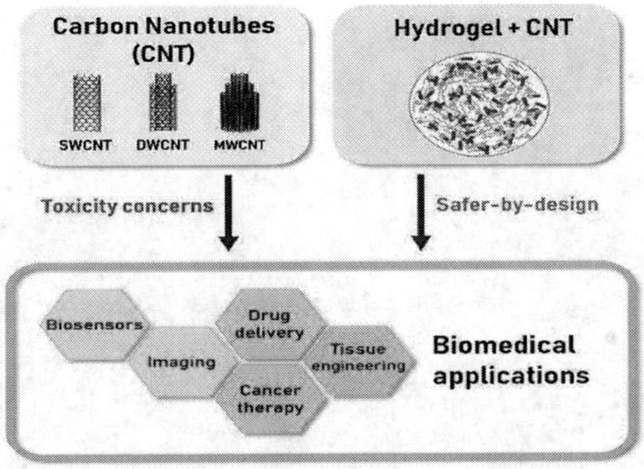

Figure 4. A schematic diagram displaying the application of CNT in biomedical engineering and technology [33]

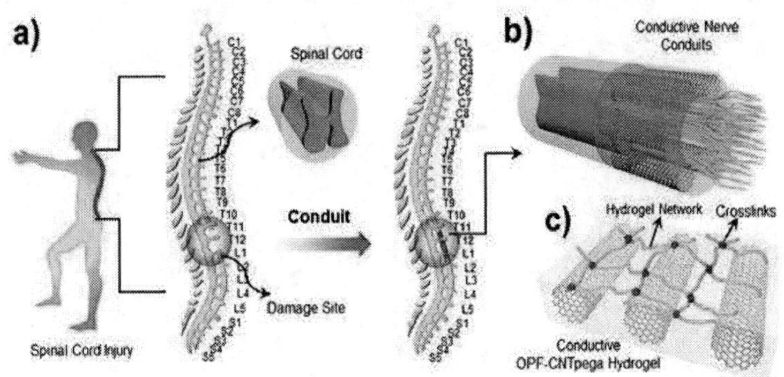

Figure 5. (a) A schematic representation of the human spinal cord. (b) Conductive nerve conduits for the treatment of spinal cord injury (c) Conductive oligo (poly (ethylene glycol) fumarate) (OPF)-CNT pega hydrogel construct [33]

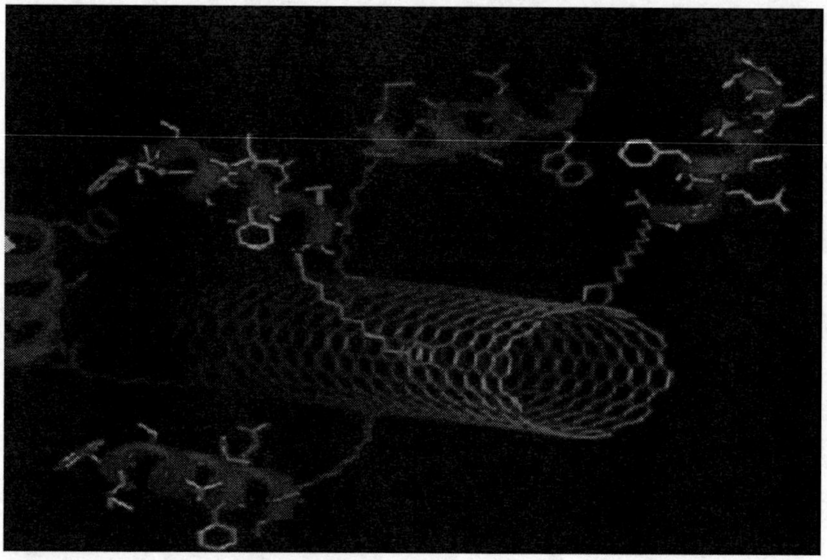

Figure 6. A pictorial diagram of drug loaded carbon nanotubes [24]

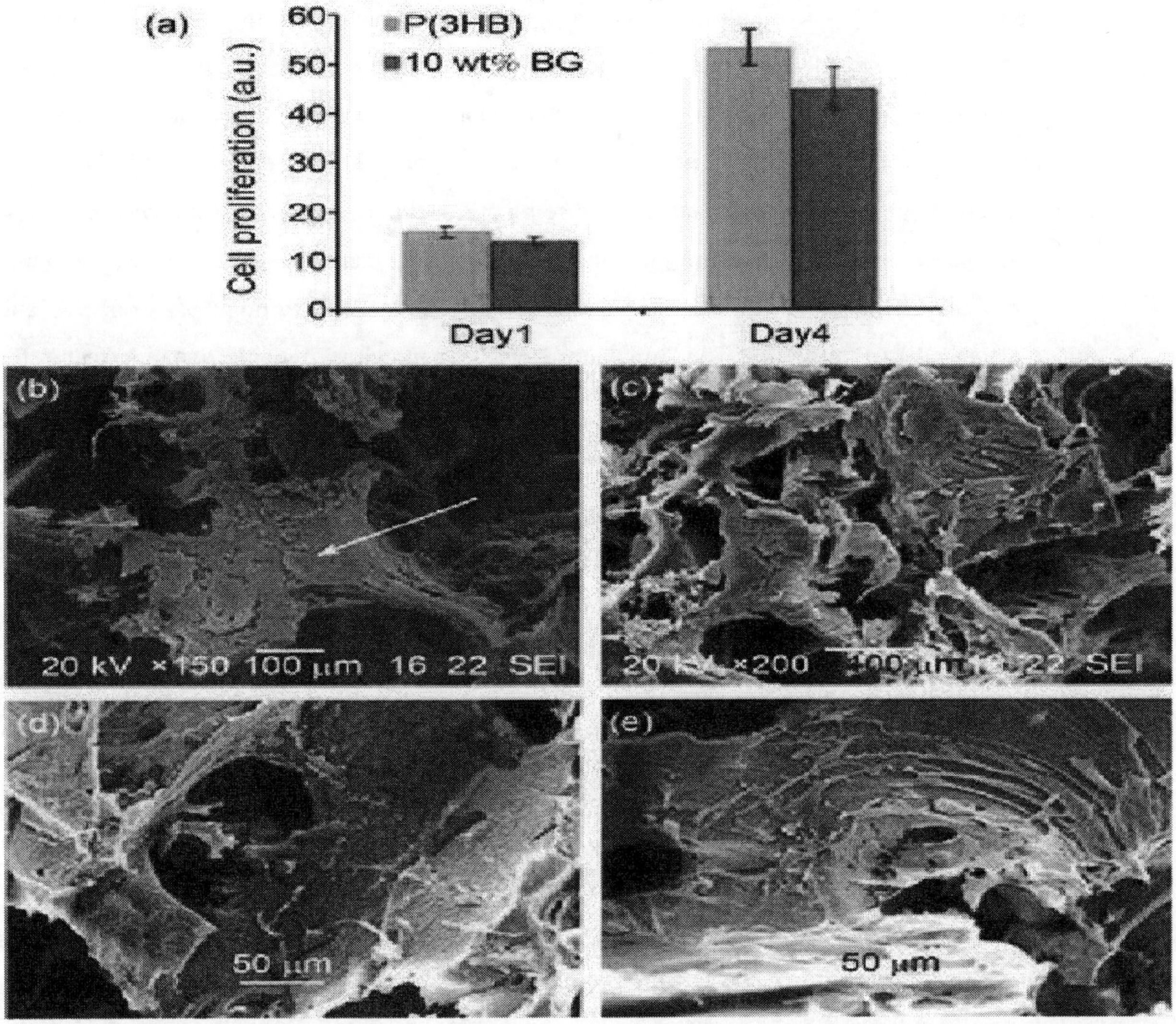

Figure 7. One CNT-based tissue engineering scaffold (a) cell proliferation for P(3HB) and P(3HB)/m-BG composite scaffolds on the first and fourth days (b) P(3HB) foams (c) P(3HB)/m-BG foams, (d) P(3HB) foams displaying the surrounding cells and bridging the pores and, (e) P(3HB)/m-BG foams having the properties of the cells to bridge the pores and also to consume the substrate contours [34]

2.2. Nanowire-based sensors (NWBSs)

Nanowire-based sensors are nanobiosensing devices specially made up of nanowires or fibres at nanoscopic scale covered by many macromolecules (bionanowires) including fibrin proteins and

DNA molecules [35]. According to Touhami [17], 'A bionanowire resembles one dimensional (1D) fibril like nanomade structure bearing diameter ranging from 1 nm to tens of nanometers and appreciable length'. Based on the uniquely modified properties, nanowires become independently analyte. They may be applied in biomedical engineering, medicine, bionanotechnology, life sciences and other disciplines. Devices containing nanowires are best used for optical excitation, electrons transport and carriers, means to find species, including biological and chemical ones (proteins, viruses or drug molecules) through ultrasensitive and direct electrical methods [36]. The thin nature of the nanowires remains a major factor for developing the remote nanobiosensors to study the environment and also in-body sensing [17].
Nanowires, nanobelts and others are best used for biosensing research because of the special properties and ability to be made into high density nanoscale instruments. Nanowires also help in electron transport and excitation of optics. Based on these factors, they are involved in nanoscale device integration and performance. The NWBSs in Figure 8 bear special range of properties including semiconduction, metallic, biocompatibility, insulation and high surface to volume ratio [37, 38, 39],which enables the materials to function better. Also, the attached material supports nanowire detection of biomolecules while employing working mechanism in Figure 9. Nanotechnology methods applied in developing NWBSs effectively aid in the conversion of biological and chemical sources into signals electronically or digitally in Figure 9 [37]. For instance, nanowires were used to check for antibodies (proteins) in Figure 10.

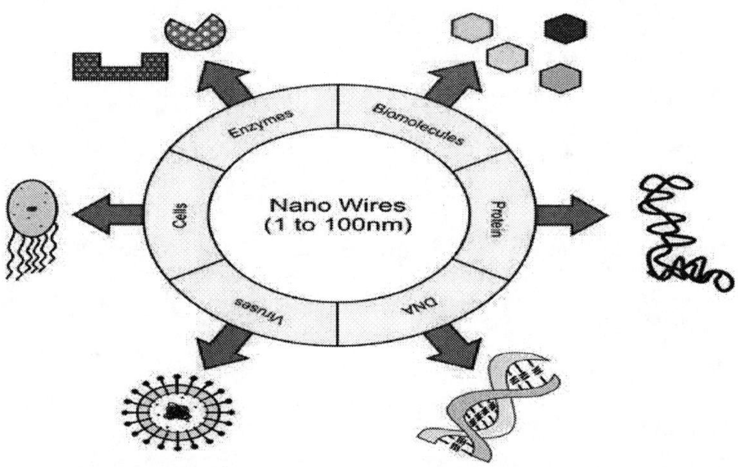

Figure 8. NWBSs required to check biomolecules [37]

IOP Conf. Series: Materials Science and Engineering **805** (2020) 012028 doi:10.1088/1757-899X/805/1/012028

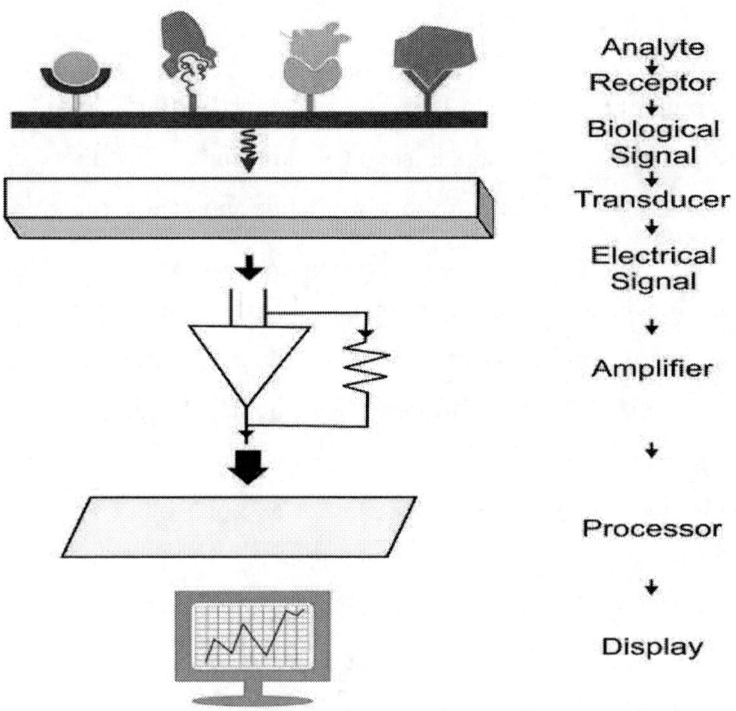

Figure 9. Detection of nanowire working procedure in a descending order [37]

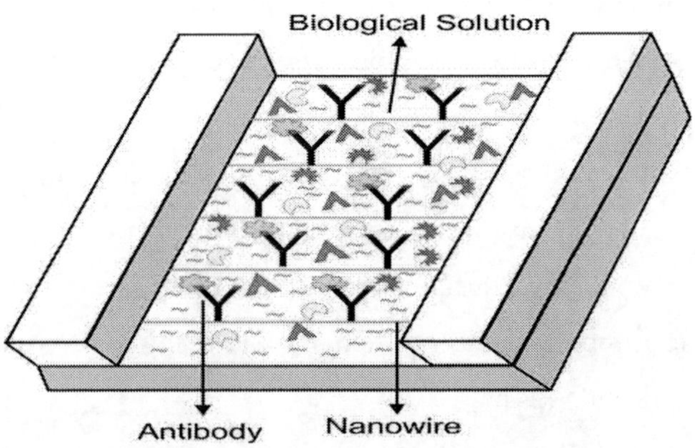

Figure 10. Nanowires used to identify antibodies. The antibody (Y) was glued to the nanowire in orange colour and the active sites in red colour ligate to the molecule in green colour due to specificity [37]

IOP Conf. Series: Materials Science and Engineering **805** (2020) 012028 doi:10.1088/1757-899X/805/1/012028

2.3.Quantum dots nanobased sensors (QDNSs)

Quantum dots (QDs) are types of nanoparticles that are inorganic fluorescent semiconductor in nature and are made up of 9 or more to less than 51 atoms bearing a diameter between 3-9 nm [40,41]. There is a size and a particular shape controlling ability that help some important features. They hardly degrade or alter in any form but apparently functions in terms of light imaging in human systems [40]. The QDs are said to be useful vehicles popularly used for delivering genes for prevention of any harm that could affect the cell membrane [24]. They can also be used as self-trafficking transfection tool for Multiple Drug Resistant (ABCB1) siRNA [42] and as radio sensitizers or near-infrared (NIR) optical imaging devices for viewing and treatment of cancer and other tumours [24]. Sequence variations including point mutation, deletions and single nucleotide substitutions can act as point of identification in diagnosis of cancer. The QDNSs have widely been used to diagnose point mutation when it comes in synergy with oligonucleotide ligation reaction assay (OLA) as reported by some researchers [43, 44]. For instance, the quantum dot based mutation detection assay has gracefully been employed to diagnose two types of point mutation (Kras and Braf) in victims having a special type of tumor [43, 44].

Semiconductor quantum dots have been considered to be extensively used as the beginning tools for the synthesis of good sensitive, simple, selective and rapid growing biosensors for macromolecules (DNA, proteins and RNA) and human cells detection. These nanobiosensors possess various dynamic optical and electronic properties including distinctive photo electrochemical features, broad absorption spectrum and others [45]. Unfortunately, they are quite toxic in nature when applied in the presence of deadly heavy metals harmful to human. Therefore, it is of utmost importance to run toxicity and many other tests before they are made available for human application [24].

3. Biosensor/bioelectrode materials/materials for nanobiosensors

Several materials are considered suitable for the production of nanobiosensors. The biosensing materials should be biocompatible, affordable, and reproducible for them to be developed in large quantity. Most of the recently produced materials exhibit and enhance the stability, sensitivity, mechanical and chemical strength of the sensor for any practical application [46].

IOP Conf. Series: Materials Science and Engineering **805** (2020) 012028 doi:10.1088/1757-899X/805/1/012028

These include but are not limited to carbon based materials (graphene-based materials, carbon nanotubes and carbon nanoparticles) in Figure 11, polymers (conductive and insulating) in Figures 12 and 13, hybrids in Figure 14, and organic and inorganic nanomaterials including magnetic and metal nanoparticles. The characteristics of conducting polymeric nanobiosensors (CPNBSs) such as morphology, size, shape and many others determine their activities and CPNBSs are widely used in biomedical, food, agricultural, nuclear and nuclear applications. For instance, cationic polymers are deemed worthy to be used as a tool for non-viral vectors for delivering genes because of their unique properties including flexibility and facile synthesis in Figure 15. When in contact with macromolecules such as DNA through electrostatic force attraction at physiological pH, there tends to be quick gene delivery and their molecular weight, structure and other aspects are well needed for the efficiency of gene transfection. Some examples of the polymers used in this process are poly (L-ornithine), poly (amidoamine) dendrimers and poly (L-lysine) as synthetic and dextran, chitosan and gelatin as natural ones [47].

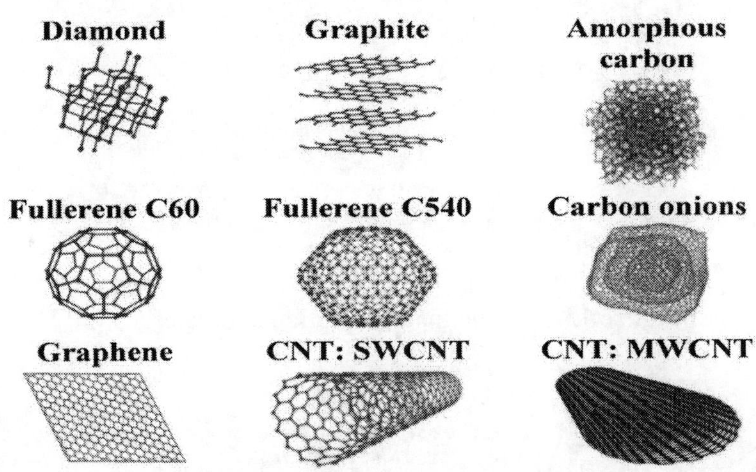

Figure 11. Carbon based nanomaterials [48]

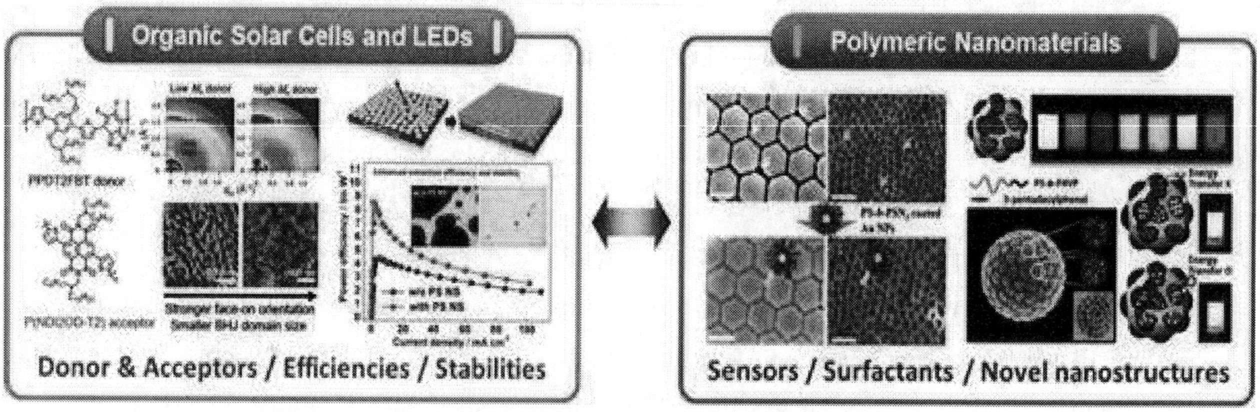

Figure 12. Polymer based nanomaterials [49]

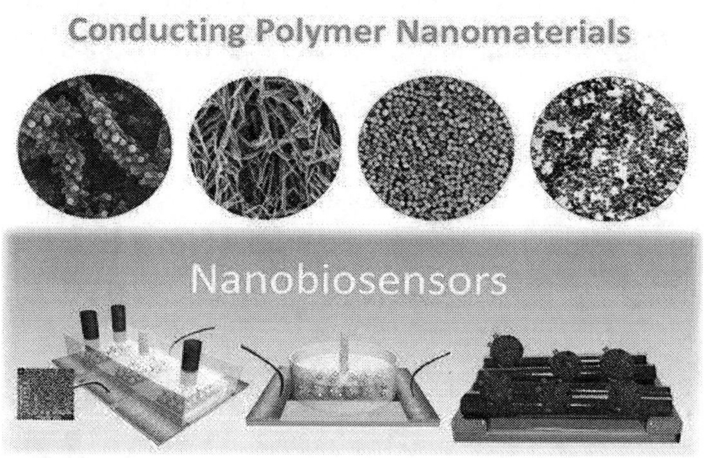

Figure 13. Conducting polymeric nanomaterials [50]

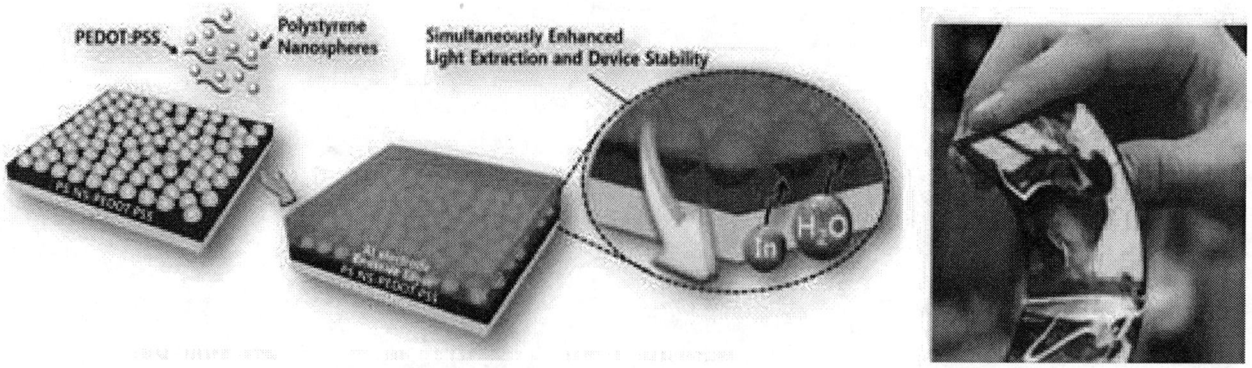

Figure 14. Hybrid nanoparticles with much efficiency and stability [49]

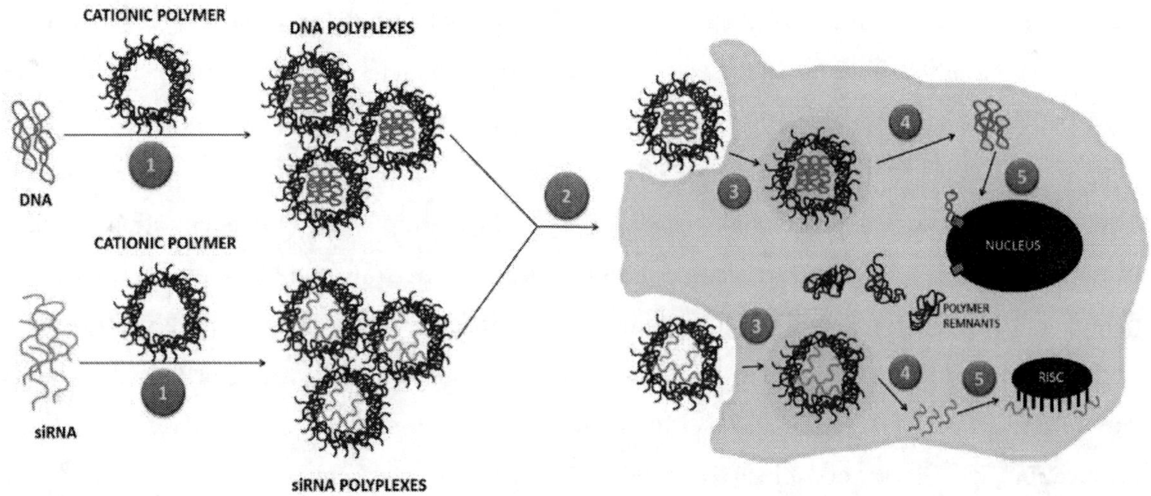

Figure 15. The intracellular delivery of DNA and siRNA using polymeric nanoparticles; A) The cationic polymers and anionic DNA and siRNA in a complex situation to form polyplexes; B) The cells absorbs the polyplexes through several endocytic routes C) Proper releasing of polyplexes from some endo-lysosomal areas; D) Removal of DNA and siRNA from polyplexes and keeping behind polymer material; E) DNA transfer to nucleus for expression by a type of protein and ligation of siRNA by RNA-induced silencing complex [47]

4. Design and optimization of nanobiosensors

The use of nanbiosensors for detection of biomaterials, biological and chemical products (such as drugs, cholesterol, enzymes, blood glucose, RNA, DNA, protein to mention a few) involves transformation of chemical or biological response to eletrical signal. This indicates concentration or composition, pH, rheological properties, energy, polarization, amplitude, decay time, and many more based on sensitivity and stability within a shortest time or seconds [6, 51, 52]. Silver, gold, platinum, and palladium nanoparticles are used in design and fabrication of biosensors because of their versatile specific behaviours.The design of nanobiosensors is in line with the chart presented in Figure 16. It requires:(1) recognition element that can identify biological and chemical products; (2) suitable and sensitive manomaterials called transducers that senses information from recognition element from those products and transmit signals including wavelength, wave propagation, time and distribution of the spectrum. The transducers may be solid articles, ceramics, metals, glasses or plastics that transform substrate into measurable

IOP Conf. Series: Materials Science and Engineering **805** (2020) 012028 doi:10.1088/1757-899X/805/1/012028

signals like electrochemical, piezoelectric, thermal, opto-electronic and mass type; (3) contact surface technique or binding specificity and transducer contact surface probe which should be governed by thin film model; (4) detector for transfer of electrical signal from transducer to amplifier typed microprocessor for a specific task.

The optimization parameters include the senstivity and dynamic range of concentration, quantity of the substrate or sensory materials, pH, temperature, sensing time and wave phase, rheological parameters and many more [53, 54]. In glucose nanobiosensor design as reported by Ridhuan *et al.* [54], the optimization was also based on selectivity as a feature of high performance, which can be influenced by reaction and measurement of substrate, specificity, effectiveness and cost implication. ZnO is a good electrode used to fabricate electrochemical glucose biosensors. It possesses high isoelectric point (IEP) with large surface area which makes it an effective and reliable surface for immobilisation of Glucose oxidase (GOx) enzyme because of its biocompatibility and other unique properties like low toxicity, high degree of electron mobility and easy to construct at low cost. ZnO nanorods grow on a seeded substrate by hydrothermal method which make them useful to modify electrode to detect glucose level. Sol-gel technique is used to produce seeded substrates in which ITO (indium tin oxide) substrates is coated by seed layer. This method is simple, economical, effective and suitable for producing a high-quality ZnO seed layer for growing nanostructures by the above mentioned method. The use of ITO glass substrate is attributed to low cost and resistivity compared with gold (Au)and carbon glass electrodes and electron mediator.

GOx immobilisation step increased the reliability and performance of glucose detection. In transcription factor (TF) based nanobiosensor, sensory proteins that regulate the physiology of cell gene expression are known as transcription factors, which have been reported to be used for sensing antibiotics, amino acids, vitamins, succinate, sugars, butanol, triacetic acid lactone and malonyl-CoA by Liu *et al.* [53]. The ChnR was controlled by constitutive promoter called *Acinetobacter* sp. Chnb promoter (Pb) with reduction in metabolic burden of replicated and produced plasmid by low copy number of plasmid.

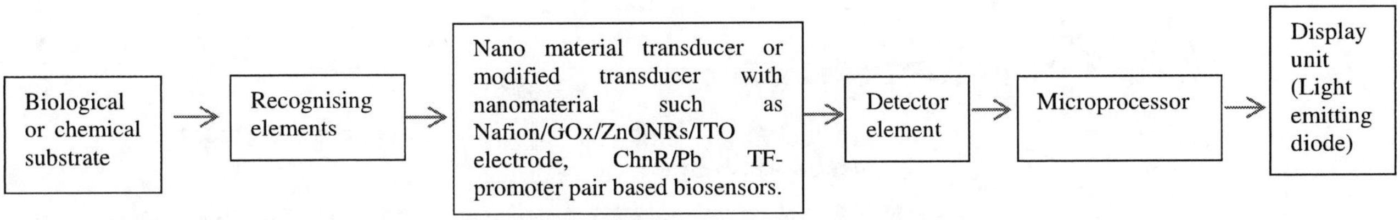

Figure 16. Design flow chart of nanobiosensors

The developmental procedures usually focus on the principle and application of the sensing devices including printing techniques for many purposes [52], heat and pull method, and chemical etching for optical nanobiosensors [17]. The use of printing techniques may be attributed to high resolution, large-area devices, roll-to-roll compatibility, energy utilization, temperature requirement, stress and impact on substrate or biomolecules, crystallization, carrier mobility fastness efficiency and damage to the substrate which are needed for design and optimization of non-biosensors development [55, 56]. Printing techniques employ image carrying medium (ICM) that stores information based on deposition of substrate or analyte from recognition element to be printed in ink and show on display unit. Printing techniques are of two (2) classes: Convectional or master technique uses screen printing technology, and non-impact printing technology which operates on drop-on-demand mode [48].

4.1 Heat and pull method

Heat pulling devices like laser-based micropipette puller are usually employed to fabricate nanobiosensor where in an optical fibre is placed in the micropipette puller and heated by carbon VI oxide (CO_2) laser. The fibre is stretched along tension axis in Figure 17. When optical fibre is pulled, tapering of heated region occurred and the fibre pulled into two, each tapered optical fibre have one large end and one end with nanometer dimensions. It was reported that temperature varies with applied tension for pulling process with change in tip diameters of fibre of a range of less than 20 nm and greater than 1000 nm [2, 51] which was determined by scanning electron microscopy (SEM).

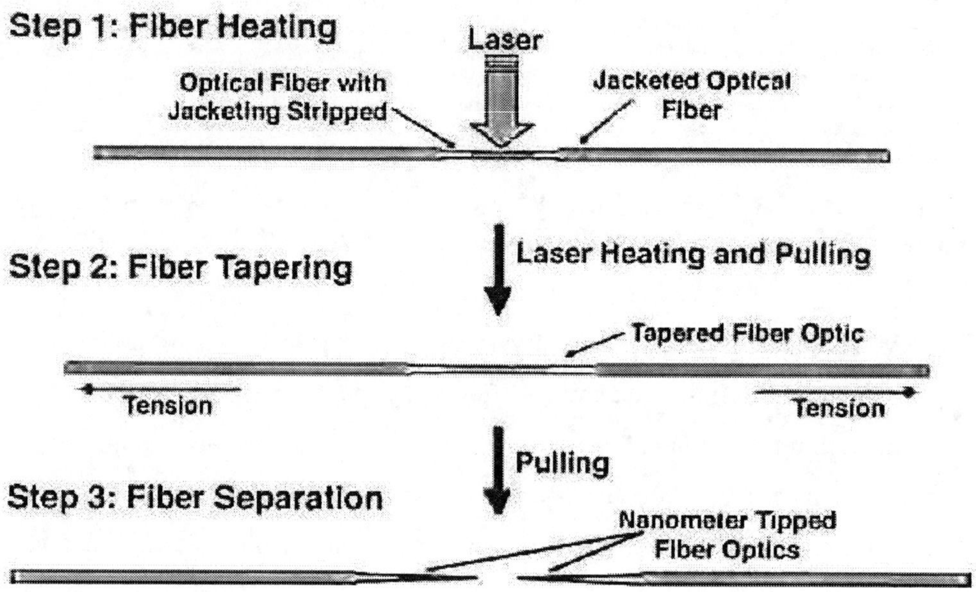

Figure 17. Cartoon depicting the heated pulling process used for nanofiber fabrication [31]

4.2 Chemical etching

This is of two types, namely; Turner etching and tube etching [6]. Hydrofluoric acid (HF) is used in chemical etching process for removal of silica core of an optical fibre into a smooth tip at a large taper angle after removal of polymer cladding. In the Turner method, the fibre is placed in the meniscus between HF and an organic over layer, causing the HF to etch the silica and the organic layer formed to protect it. This process creates fibres with larger taper angles than those produced via the heated pulling method. These larger taper angles allow excitation light to travel closer to the edge of the fibre before being trapped, providing a better excitation process.

5. Application of nanobiosensors to biomedical technology

Nanobiosensors, small and smart technology are widely used in broad fields of biomedical technology such as tissue engineering, 3D printing, biotechnology, biomaterials, sport, medical technology and many more. They are employed for several purposes including diagnosing, preventing, clinical testing, treating, monitoring of human diseases as illustrated in Figures 18-20

IOP Conf. Series: Materials Science and Engineering **805** (2020) 012028 doi:10.1088/1757-899X/805/1/012028

[57, 58]. A typical example is the transistor-based biosensor bearing nanowire characteristics and protein-nanoparticles biosensors aimed at detecting the ultra sensitivity of DNA and proteins respectively without performing PCR or curing with bisulfite in the case of DNA. These sensors may further serve to check plant diseases and some unparallel issues associated with biomarkers and lack of minerals. Also, the SPR biosensors plays a significant role in vitamin analysis, while other types could check for the availability and efficiency of antibiotics and then, immunobiosensors with other factors find a particular type of *E. coli*.

According to Cavalcanti *et al.* [59],'Nanorobot sensor device (NRSD) has potential in medicine. These medical NRSDs are produced to solve the common inconvenient procedure of collecting blood samples from diabetic patients to manage and control glucose level by activating the proteomic-based result to check the biochemical alterations related to hyperglycemia'. The outer part of NRSD contains carbon metal nanocomposites with properties including hardness and atomic smoothness [58]. The NRSD with radiofrquency idetification device (RFID) CMOS transponder machine help *in vivo* positioning.The NRSD decreases the patients lost time from hyperglycemia mishap and also give a unique meidcal attention and type of medication of the client [58].

Localized surface plasmon resonance (LSPR) nanobiosensor, an optical type, has a potential of carrying out prototypical immunoassay on biotin (B) and anti-biotin (AB). Rai *et al.* [58] reported that 'this type of sensor checks for any form of difference in local refractive index (LRI) by proper monitoring of the sensor extinction maximum and UV-visible spectroscopy. According to Nagrath *et al.* [60], a microfluidic nanobiosensor could be designed and fabricated to identify and isolate peripheral blood tumor cells in most cancer patients. Also, an optical fiber nanobiosensor (OFNBS) was developed for cancer marker check-up, telomerase at the level of one cell having a nanoscale tip [61]. Some other nanobiosensors are produced to detect harmful bacteria, its metabolic work and its reproduction in an antibiotic environment. A typical example is the superparamagnetic iron oxide nanoparticles which have been applied in the diagnosis of *Mycobacterium avium* spp. *paratuberculosis* (MAP) via magnetic resting [62]. In a nutshell, these nanobiosensors are applied extensively in monitoring glucose, sense pH, identify asthma, DNA, proteins or microorganisms, diagnose cancer, and facilitate drug delivery [63].

IOP Conf. Series: Materials Science and Engineering **805** (2020) 012028 doi:10.1088/1757-899X/805/1/012028

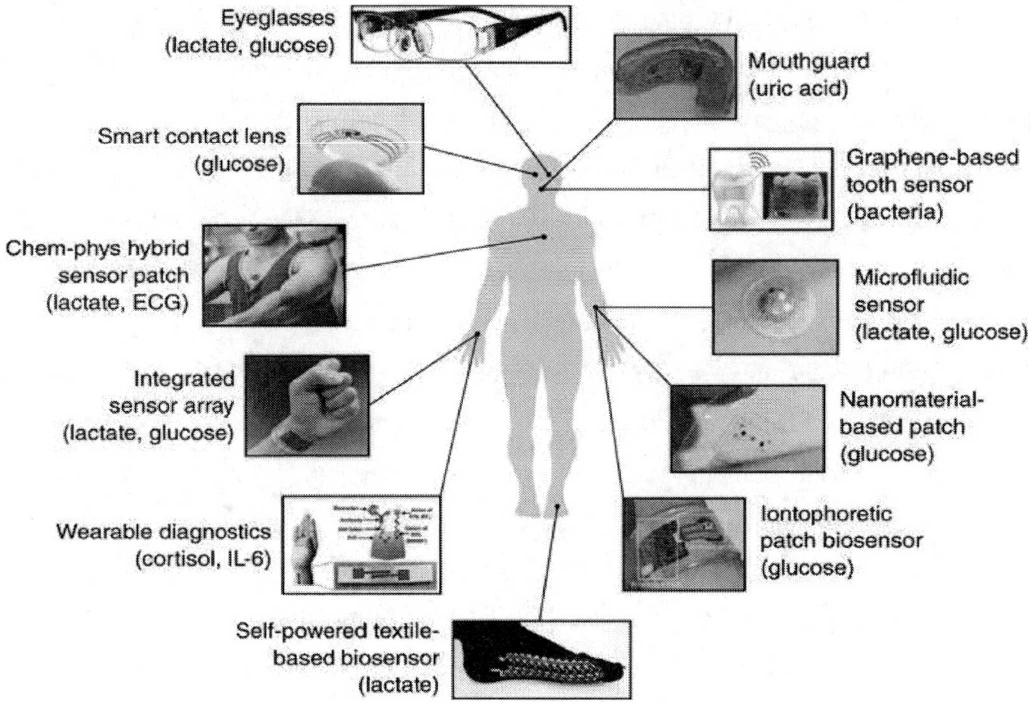

Figure 18. Utilization of nanobiosensors in biomedical technology [64]

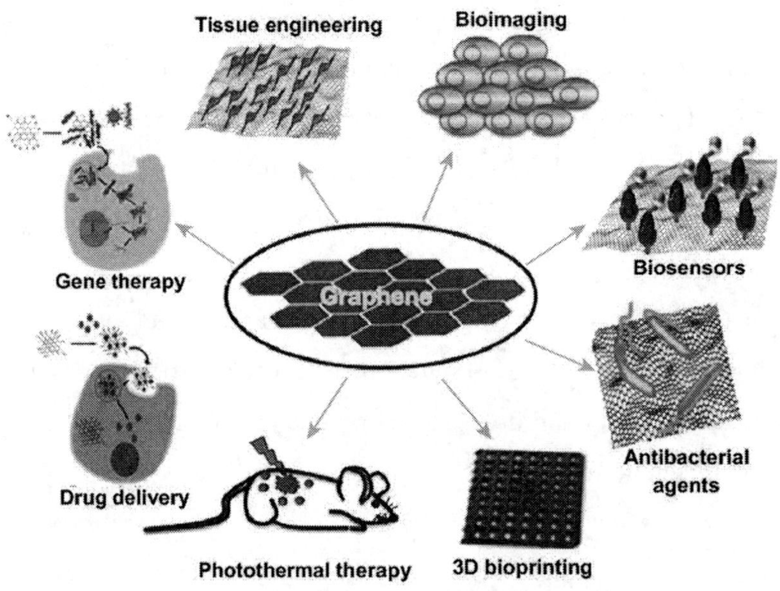

Figure 19. The application of graphene in different subdisciplines of biomedical technology [65]

IOP Conf. Series: Materials Science and Engineering **805** (2020) 012028 doi:10.1088/1757-899X/805/1/012028

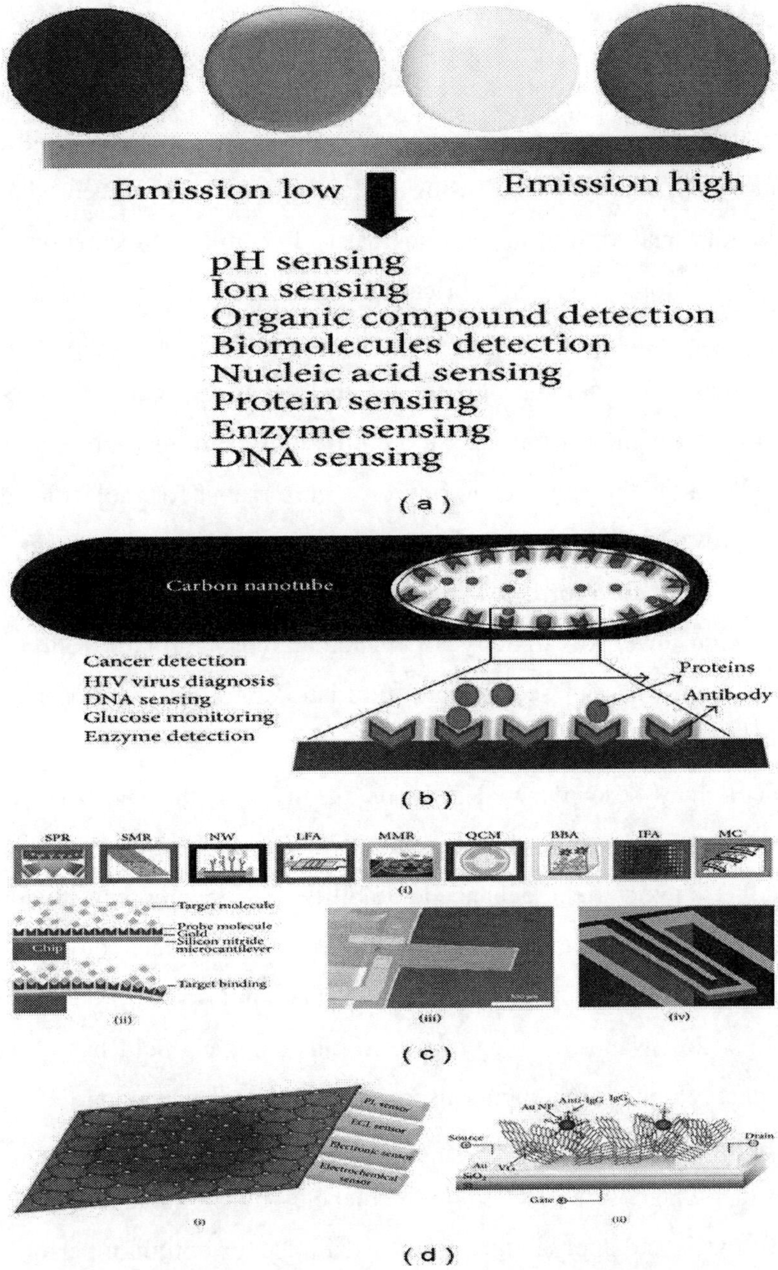

Figure 20. Application of nanobiosensors in tissue engineering (a) Change in colour of quantum dots may occur based on their emission wavelength; (b) CNT based biosensor is used for the identification biomolecules produced by the cell; (c) MEMS based biosensors (surface plasmon resonance, lateral flow assay, microring resonator, microcantilever etc.); d) Graphene and its derivatives and vertically-oriented grapheme based field effect transistor-sensor [7].

IOP Conf. Series: Materials Science and Engineering **805** (2020) 012028 doi:10.1088/1757-899X/805/1/012028

6. Challenges and future research

The possible challenges that might occur with technology of nanobiosensing includes, but not limited to scalability emanating from the nanowire or nanoparticles synthesis bearing specific properties including shape or size (using bottom-up procedures), reproducible fabrication and failure to quantify analytically, both biological and chemical information in a quick and cheap manner [66]. Therefore, modified fabrication methods mainly applied for multiplexing and to fabricate reproducible devices could ultimately assist in the synergy of technology and diagnostic tools, thus increasing the amount of different analytes with specific sensors on one platform as well as creating great impact on disease identification, and also introducing more reliable and efficient tools such as the nanotechnology on a chip meant for molecular diagnosis which will eventually improve healthcare delivery. A typical example of nanotechnology biochips is the nanofluidic type developed for DNA and protein isolation and analysis [66]. Others may include presence of high level of toxicity of nanomaterials; miniaturization and integration of microfluidic systems; lack of security and futuristic effect of nanobiosensors in terms of risks; less biocompatible and biostable.

These problems may pose potential risks on the lives of the end users, patients or even the laboratory research analysts. Therefore, more effective and efficient therapeutic solutions with little or no adverse toxic effect, enhanced durability and biocompatibility could greatly influence and improve healthcare delivery. These may thereby make the technology to gain more weight in biomedical as well as clinical applications and industrial productions in the area of nanobiosensors [24]. Nanotechnology as a multidisciplinary field has come to stay and cause positive impact globally. Nanobiosensors with magnetic nanoparticles might serve as a significant tool for telomerase activity screening in living experiments. Nanorobot sensor, a unique type, can diagnose as well monitor diabetes by producing a form of proteomic-based output for diagnosis and apply its increased sensitivity for optimum performance, thus diagnose and treat different types of cancer. They could also identify some microbial pathogens and its toxins in clients too. In the nearest future, the production of wearable biosensors for real-time sensing of most physiological features and development of more biosensors for other aspects of medicine will be used for proper understanding of human diseases. Development of these novel nanobiosensor tools and solutions will ultimately be used to positively change the medical conditions in the world.

7. Conclusion

Many nanobiosensosrs have been and are still being produced in the world, and yet the likes of multiplexed, affordable and high throughput devices have not been fully developed. More works should be done to achieve this goal and many others in the healthcare system as future nanobiosensors will aid in accurate as well as precise medical diagnosis and treatment than the old and current nanobiosenosors.

References

[1]Malhotra S, Verma A, Tyagi N and Kumar V 2017 Biosensors: principle, types and applications. *Int. J. Adv. Res. Innov. Ideas Educ.* **3** (2) pp 3639-3644.

[2]Malik P, Katyal V, Malik V, Asatkar A, Inwati G and Mukherjee T K 2013 Nanobiosensors: concepts and variations. Article ID 327435. http://dx.doi.org/10.1155/2013/327435.

[3] Darsanaki R K, Azizzadeh A, Nourbakhsh M, Raeisi G and Aliabadi A M 2013 Biosensors: functions and applications. *J. Biol. Today's World* **2** (1) pp 53-61.

[4] Schubert FL, Wollenberger UL, Scheller FW, Muller HG 1991 Artificially coupled reactions with immobilized enzymes: biological analogs and technical consequences. *Bioinstrumentation and Biosensors* (New York: Marcel Dekker). pp 19.

[5] Wilson J S 2005 *Sensor Technology Handbook* (Amsterdam, Boston: Elsevier).

[6] Pandit S, Dasgupta D, Dewan N and Ahmed P 2016 Nanotechnology based biosensors and its application. *Pharma. Innov. J.* **5** (6) pp 18-25.

[7] Hasan A, Nurunnabi Md, Morshed M, Paul A, Polini A, Kuila T, Hariri M Al, Lee Y and Jaffa A A 2014 Recent Advances in Application of Biosensors in Tissue engineering. *BioMed. Res. Int.* Article ID 307519. https://doi.org/10.1155/2014/307519.

[8] Khatun Z, Nurunnabi M, Cho K J and Lee Y 2012 Imaging of the GI tract by QDs loaded heparin-deoxycholic acid (DOCA) nanoparticles. *Carbohy. Polym.* **90** (4) pp 1461-1468.

[9] Khatun Z, Nurunnabi M, Cho K J and Lee Y 2012 Oral delivery of near-infrared quantum dot loaded micelles for noninvasive biomedical imaging. *ACS Appl. Mater. Interf.* **4** (8) pp 3880-3887.

[10] Giepmans B N G, Adams S R, Ellisman M H and Tsien R Y 2006 The fluorescent toolbox for assessing protein location and function. *Sci.* **312** (5771) pp 217-224.

[11] Chan W C and Nie S 1998 Quantum dot bioconjugates for ultrasensitive nonisotopic

detection. *Sci.* **281** (5385) pp 2016-2018.

[12] Tomasulo M, Yildiz I, Kaanumalle S L and Raymo F M 2006 pH-sensitive ligand for luminescent quantum dots. *Langmuir* **22** (24) pp 10284-10290.

[13] Xie H Y, Liang J G, Zhang Z L, Liu Y, He Z K and Pang D W 2014 Luminescent CdSe-ZnS quantum dots as selective Cu^{2+} probe. *Spectrochim. Acta A: Mol. Biomol. Spectr.* **60** (11) pp 2527-2530.

[14] Niemeyer C M 2002 The developments of semisynthetic DNA-protein conjugates. *Trends Biotechnol.* **20** (9) pp 395-401.

[15] Whitesides GM, Mathias J P and Seto C T 1991 Molecular self-assembly and nanochemistry: a chemical strategy for the synthesis of nanostructures. *Sci.* **254** (5036) pp 1312-1319.

[16] Dzyadevych SV, Arkhypova VN, Soldatkin AP, El'Skaya AV, Martelet C and Jaffrezic-Renault N 2008 Amperometric enzyme biosensors: Past, present and future. *Innov. Res. BioMed. Eng.* **29** (2-3) pp 171-180.

[17] Touhami A 2014 Biosensors and nanobiosensors: design and applications. *Nanomed.* **15** pp 374-403.

[18] Wang G 2018 Nanotechnology: The new features. arXiv preprint arXiv:1812.04939. https://arxiv.org/abs/1812.04939. Accessed on 31 January, 2020.

]19] Mahar B, Laslau C, Yip R and Sun Y 2007 Development of carbon nanotube-based sensors-a review. *IEEE Sens. J.* **7** (2) pp 266-284.

[20] Hanein Y and Bareket-Keren L 2013 Carbon nanotube-based multi electrode arrays for neuronal interfacing: progress and prospects. *Front. Neural Circuit.* **6** pp 122. https://doi.org/10.3389/fncir.2012.00122.

[21] Charlier J C, Blase X and Roche S 2007 Charlier JC, Blase X, Roche S. Electronic and transport properties of nanotubes. *Rev.Mod. Phys.* **79** (2) pp 677-732.

[22] Haniu H, Saito N, Matsuda Y, Tsukahara T, Usui Y, Narita N, Hara K, Aoki K, Shimizu M, Ogihara N, Takanashi S, Okamoto M, Kobayashi S, Ishigaki N, Nakamura K and Kato H 2012 Basic potential of carbon nanotubes in tissue engineering applications. *J. Nanomat.* Article ID 343747. https://doi.org/10.1155/2012/343747.

[23] Azamian B R, Davis J J, Coleman K S, Bagshaw C B and Green M L H 2002 Bioelectrochemical single-walled carbon nanotubes. *J. Am. Soc.* **124** (43) pp 12664-12665.

[24] Qiao W, Wang B, Wang Y, Yang L, Zhang Y and Shao P 2010 Cancer therapy based on nanomaterials and nanocarrier systems. *J. Nanomat.* Article ID 796303. https://doi.org/10.1155/2010/796303.

[25] Bosi S, Ballerini L and Prato M 2014 Carbon nanotubes in tissue engineering. *Topics Curr. Chem.* **348** pp 181-204.

[26] Shi X, Hudson J L, Spicer P P, Tour J M, Krishnamoorti M and Mikos A G 2006 Injectable nanocomposites of single-walled carbon nanotubes and biodegradable polymers for bone tissue engineering. *Biomacromol.***7** (7) pp 2237-2242.

[27] Cao Y, Zhou Y M, Shan Y, Ju H X and Xue X J 2007 Preparation and characterization of grafted collagen-multiwalled carbon nanotubes composites. *J. Nanosci. Nanotechnol.*7 (2) pp 447-451.

[28] MacDonald A R, Laurenzi B F, Viswanathan G, Ajayan M P and Stegemann J P 2005 Collagen-carbon nanotube composite materials as scaffolds in tissue engineering *.J. Biomed. Mater. Res. A* **74** (3) pp 489-496.

[29] Tan W, Twomey J, Guo D, Madhavan K and Li M 2010 Evaluation of nanostructural, mechanical, and biological properties of collagennanotube composites. *IEEE Trans. Nanobiosci.* **9** (2) pp 111-120.

[30] Tosun Z and McFetridge P S 2010 A composite SWNT-collagen matrix: characterization and preliminary assessment as a conductive peripheral nerve regeneration matrix. *J. Neural Eng.* **7** (6) 066002. https://doi.org/10.1088/1741-2560/7/6/066002.

[31] Zhang D, Kandadai M A, Cech J, Roth S and Curran S A 2006 Poly(L-lactide) (PLLA)/multiwalled carbon nanotube (MWCNT) composite: characterization and biocompatibility evaluation. *J. Phys. Chem. B* **110** (26) pp 12910-12915.

[32] McCullen S D, Stevens D R, Roberts W A, Clarke L I, Bernacki S H, Gorga R E and Loboa E G 2007 Characterization of electrospun nanocomposite scaffolds and biocompatibility with adipose-derived human mesenchymal stem cells. *Int. J. Nanomed.* **2** (2) pp 253-263.

[33] Simon J, Flahaut E and Golzio M 2019 Overview of carbon nanotubes for biomedical applications. *Mater.* **12** (4) 624. https://doi.org/10.3390/ma12040624.

[34] Wang X J and Liu Z 2012 Carbon nanotubes in biology and medicine: An overview. *Chin. Sci. Bull.* **57** (2–3) pp 167-180.

[35] Patolsky F, Zheng G and Lieber CM 2006 Nanowire-based biosensors. *Anal. Chem.* **78** (13) pp 4260-4269.

[36] Patolsky F, Zheng G and Liebert C M 2006 Nanowire sensors for medicine and the life sciences. *Future Nanomed.* **1** (1). https://doi.org/10.2217/17435889.1.1.51.

[37] Arora P, Sindhu A, Dilbaghi N and Chaudhury A 2013 Engineered multifunctional nanowires as novel biosensing tools for highly sensitive detection. *Appli. Nanosci.* **3** (5) pp 363-372.

[38] Sung W J and Bae Y H 2003 A glucose oxidase electrode based on polypyrrole with polyanion/PEG/enzyme conjugate dopant. *Biosens. Bioelectr.* **18** (10) pp 1231-1234.

[39] Yin A J, Li J, Jian W, Bennett A J, Xu J M 2001 Fabrication of highly ordered metallic nanowire arrays by electrodeposition. *Appi. Phys. Lett.***79** (7) pp 1039-1041.

[40] Cai W and Chen X 2007 Nanoplatforms for targeted molecular imaging in living subjects *Small* **3** (11) pp 1840-1854.

[41] Kim K Y 2007 Nanotechnology platforms and physiological challenges for cancer therapeutics. *Nanomed.: Nanotechnol. Biol. Med.* **3**(2) pp 103-110.

[42] Klein S, Zolk O, Fromm F, Schrödl F, Neuhuber W and Kryschi C 2009 Functionalized silicon quantum dots tailored for targeted siRNA delivery. *Biochem. Biophys. Res. Comm.* **387** (1) pp 164-168.

[43] Yeh H, Ho Y, Shih L and Wang T 2006 Homogeneous point mutation detection by quantum dot-mediated two-color fluorescence coincidence analysis. *Nucl. Acids Res.* **34** (5) pp e35. https://doi.org/10.1093/nar/gkl021.

[44] Zhang C Y, Yeh H C, Kuroki M T and Wang T H 2005 Single-quantum-dot-based DNA nanosensor. *Nature Mater.* **4** (11) pp 826-831.

[45] Ma F, Li C and Zhang C 2018 Development of quantum dot-based biosensors: principles and applications. *J. Mater. Chem. B* **6**(39) pp 6173-6190.

[46] Andreescu D, Andreescu S and Sadik O 2005 New materials for biosensors, biochips and molecular bioelectronics. *Compr. Anal. Chem.* **44** pp 285-327.

[47] Rai R, Alwani S and Badea I 2019 Polymeric nanoparticles in gene therapy: new avenues of design and optimization for delivery applications. *Polym.* **11** (4) 745. https://doi.org/10.3390/polym11040745.

[48] Sandulescu R, Tertis M, Cristea C and Bodoki E 2015 *New Materials for the Construction*

of Electrochemical Biosensors (London: IntechOpen). http://dx.doi.org/10.5772/60510.

[49] Kim B 2015 P*olymer Nano-Electronics Laboratory* (Korea: Advanced Institute of Science and Technology).

[50] Park C S, Lee C and Kwon O S 2016 Conducting polymer based nanobiosensors. *Polym.* **8** (7) 249. https://doi.org/10.3390/polym8070249.

[51] Atta N F, Galal A and Ali S M 2011 *Nanobiosensor for Health Care* (Rijeka, Croatia: InTech Open). Pp 87-126.

[52] Satvekar R, Tiwale B and Pawar S 2014 Emerging trends in medical diagnosis : a thrust on nanotechnology *Med. Chem.***4** (4) pp 407-416.

[53] Liu Y, Liu Y and Wang M 2017 Design, optimization and application of small molecule biosensor in metabolic engineering. *Front. Microbiol.* **8** https://doi.org/10.3389/fmicb.2017.02012.

[54] Ridhuan N S, Razak K A and Lockman Z 2018 Fabrication and characterization of glucose biosensors by using hydrothermally grown ZnO nanorods. *Sci. Repor.* **8** pp 1-12.

[55]Matzeu G 2015 *Development , Optimisation and Applications of Screen-Printed Electrochemical Sensors* (Dublin: City University). http://doras.dcu.ie/20762/ (Accessed on 01 February, 2020).

[56] Pavinatto F J 2015 Biosensors with printed active layer. *Disp. Imag.* **2** (1) pp 69-91.

[57] Ahangar P, Cooke M E, Weber M H and Rosenzweig D H 2019 Current biomedical applications of 3D printing and additive manufacturing. *Appl. Sci.* **9** (8) 1713. https://doi.org/10.3390/app9081713.

[58] Rai M, Gade A, Gaikwad S, Marcato P D and Durán N 2012 Biomedical applications of nanobiosensors: the state-of-the-art. *J. Braz. Chem. Soc.* **23** (1) pp 14-24.

[59] Cavalcanti A, Shirinzadeh B and Kretly L C 2008 Medical nanorobotics for diabetes control. *Nanomed.: Nanotechnol. Biol. Med.* **4** (2) pp 127-138.

[60] Nagrath S, Sequist L V, Maheswaran S, Bell D W, Irimia D, Ulkus L, Smith M R, Kwak E L, Digumarthy S, Muzikansky A, Ryan P, Balis U J, Tompkins R G, Haber D A and Toner M 2007 Isolation of rare circulating tumour cells in cancer patients by microchip technology. *Nature* **450** (7173) pp 1235-1239.

[61] Zheng X T and Li C M 2010 Single living cell detection of telomerase over-expression for cancer detection by an optical fiber nanobiosensor. *Biosens. Bioelectron.* **25** (6) pp 1548-

1552.

[62] Kaittanis C, Naser S A and Perez J M 2007 One-step, nanoparticle-mediated bacterial detection with magnetic relaxation. *Nano Lett.* **7** (2) pp 380-383.

[63]Agrawal S and Prajapati R 2012 Nanosensors and their pharmaceutical applications: a review. *Int. J. Pharm. Sci. Nanotechnol.* **4** (4) pp 1528-1535.

[64] Kim J, Campbell A S, De Avila B E and Wang J 2019 Wearable biosensors for healthcare monitoring. *Nature Biotechnol.* **37** (4) pp 389-406.

[65] Syama S and Mohanan P V 2019 Comprehensive application of graphene: emphasis on biomedical concerns. *Nano-Micro. Lett.* **11** 6. https://doi.org/10.1007/s40820-019-0237-5.

[66]Bellan L M, Wu D and Langer R S 2011 Current trends in nanobiosensor technology. *Wiley Interdiscip. Rev. Nanomed. Nanobiotechnol.* **3** (3) pp 229-246.

IOP Conf. Series: Materials Science and Engineering **805** (2020) 012029 doi:10.1088/1757-899X/805/1/012029

Experimental study, phenomenological and adaptive neuron analysis of creep property of *Momordica augustisepala* L nanofiber reinforced polyvinyl alcohol polymer film composite

A J Adeyi[1*], M O Durowoju[1] and O Adeyi[2]

[1]Department of Mechanical Engineering, Ladoke Akintola University of Technology Ogbomoso, Oyo State, Nigeria

[2]Department of Chemical Engineering, Michael Okpara University of Agriculture, Umudike, Abia State, Nigeria

*Corresponding author: adeyi.abiola@yahoo.com

Abstract

The viscoelastic property of packaging material affects foods' postharvest management and biophysical quality preservation of food. This study investigated the creep viscoelastic property of neat polyvinyl alcohol film and polyvinyl alcohol film reinforced *Momordica augustisepala* nanofiber composite. The applicability and comparison of Burger phenomenological and Adaptive neuron model to analyze the creep data were investigated. The instantaneous modulus and creeping speed of neat and reinforced polyvinyl alcohol were 7.53710 MPa, 10.27580 MPa and 3.00590 MPa.s, 2.85391 MPa.s, respectively. The coefficient of determination of Burger model prediction was 0.99889 and 0.99972 for neat and composite polyvinyl alcohol film, while the coefficient of determination of Adaptive neuron model was 0.99986 for both neat and reinforced polyvinyl alcohol film. The *Momordica augustisepala* nanofiber improved the creep viscoelastic property of polyvinyl alcohol film and Adaptive neuron model showed superiority over Burger model in prediction of experimental data. The results are vita in material design, life prediction and usage prescription.

Keywords: *Momordica augustisepala*, viscoelastic property, polyvinyl alcohol film, adaptive neuron

1 Introduction

Postharvest or post processed handling of foods and fruits contributes to the physical quality and acceptability of the product. The biophysical quality and hygiene of food products can be

IOP Conf. Series: Materials Science and Engineering **805** (2020) 012029 doi:10.1088/1757-899X/805/1/012029

preserved or protected using active packaging material that can withstand mechanical abuse during display, packing, shelving, transportation and storage; and creation of barrier resistance for the packed food product. For a packaging material to be this active, the investigation and understanding of its behavior necessitates not only the static responses but also the time dependent or viscoelastic response, because knowledge of viscoelastic properties of foods and agricultural materials are important when considering harvesting and post harvesting processes [1]. In addition, the data on viscoelastic properties are a basic requirement as input for mathematical models that describe and predict the internal stresses and cracking during different post harvesting processes.

Mathematical and phenomenological or physical model exist for the analysis, understanding of the structure, evaluation and forecasting of viscoelastic (creep and relaxation) data. The phenomenological models are particularly useful because a larger number of rheological parameters could be estimated from creep data and elastic, while viscoelastic and viscous flow characteristics can be predicted separately [2]. Mathematical model like Findley power law is also important in this regard. However, power law analysis shows not much detail as phenomenological model does and power law shows deficiency when the effect of temperature changes is introduced. The Burger phenomenological model is represented in Figure 1 with its four elements for the creep analysis and modeling. It is a systematic combination of springs and dash pots to mimic or analyze both elastic and viscous responses of a typical material.

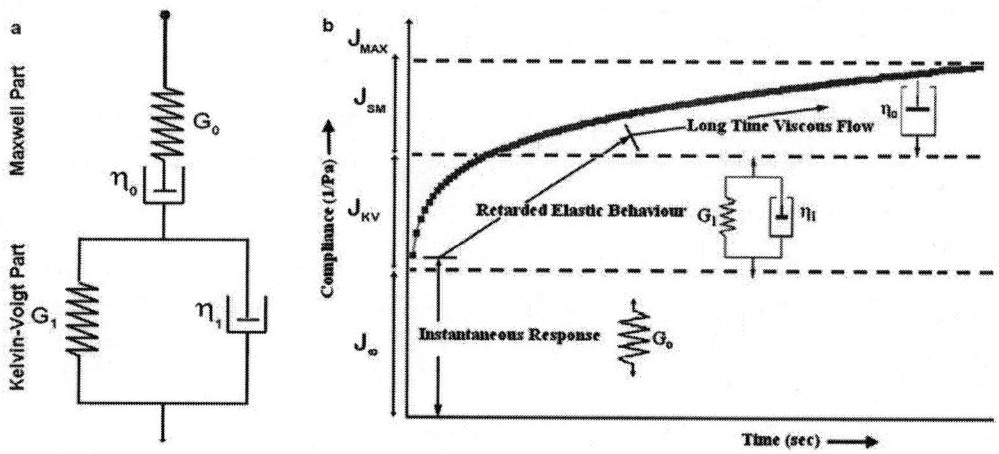

Figure 1. Burger model [3]

IOP Conf. Series: Materials Science and Engineering **805** (2020) 012029 doi:10.1088/1757-899X/805/1/012029

In a typical creep compliance experiment, samples are loaded with constant stress in linear material deformation region and deformation or strain is measured with respect to time. The creep compliance is the ratio of strain to stress or the inverse of Young Modulus. Burger model for creep investigation is the connection of Maxwell unit and a Kelvin unit in series and the model separates the creep strain into the instantaneous deformation (Maxwell spring), viscoelastic deformation (Kelvin unit), and the viscous deformation (Maxwell dash-pot) [4]. The Burger model equation is depicted in equation 1.

$$\varepsilon = \frac{\sigma}{E1} + \frac{\sigma}{E2}\left[1 - exp\left(-\frac{tE2}{\eta 2}\right)\right] + \frac{\sigma}{\eta 1}t \qquad 1$$

where *E1* and *E2* are elastic moduli, η_2 and η_1 are viscosities, σ is the imposed stress and t is the creep time. This model has only one retardation time, $\tau 2 = \eta 2/E2$ which is the time required to generate 63.2% deformation in the Kelvin unit [4] or the creeping speed.

Adaptive network based fuzzy inference system (ANFIS) is a soft computing or machine-learning methods which has the capability to understand the dynamics of experimental data and creation of accurate models for simulation and control. It is the fusion between the neural network and the fuzzy inference system, characterized by transparency of fuzzy systems and learning ability as neural networks [5]. It is based on Takagi-Sugeno fuzzy and it uses a hybrid learning algorithm. The ANFIS has layers which can be called inputs, if part, rules and normalization, then part and output [6]. A typical structure of ANFIS is represented in Figure 2, which is composed of 6 layers.

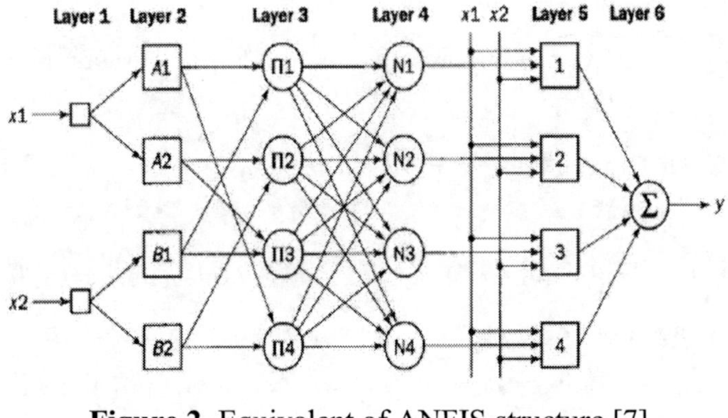

Figure 2. Equivalent of ANFIS structure [7]

The two inputs (x, y) of the system are presented (ANFIS supports multiple input, single output systems). Square nodes (adaptive nodes) show that parameters in these nodes are adjustable, to

IOP Conf. Series: Materials Science and Engineering **805** (2020) 012029 doi:10.1088/1757-899X/805/1/012029

be learned, while the circle nodes (fixed nodes) demonstrate that they are fixed parameters. A common rule set with two fuzzy if-then rules is as follows:

Rule 1: If x is $AA1$ and y is $BB1$, then $ff1 = pp1x + qq1x + rr1$ 2

Rule 2: If x is $AA2$ and y is $BB2$, then $ff2 = pp2x + qq2x + rr2$ 3

Where A, B are linguistic terms that are user-defined and representing a range of values. The sequence and functions of the layers are as follow: Layer 1: Square node equipped with node function

$$O_i^L = \mu_{Ai}(x) \quad 4$$

Assuming x and y are the two typical input values fed at the two input nodes, which then transform those values to the membership functions (triangle, generalized bell-shaped, Gaussian membership). O_i^L is the membership function of Ai and x is the input parameter to the node. Ai is the linguistic label connected with the node function.

Layer 2: This node multiplies the incoming signal and sends the product out. Each node output is the firing strength of a rule.

$$wi = (x) \times (y), i=1,2 \quad 5$$

Layer 3: circle node. Node computes the ratio of i-th rule's firing strength to the sum of all rules' firing strengths:

$$w' = wi/w1+w2, i = 1, 2 \quad 6$$

Layer 4: Square node with node function:

$$O_i^4 = w_i' f_i = w'_i(p_i x + q_i y + r_i) \quad 7$$

p, q, r – parameter set (consequent, linear, parameters)

Layer 5: circle node. This node computes the overall output as summation of all incoming signals.

$$O^5_i = \text{overall output} = \sum_i w'_i f = \sum_i w_i f_i / \sum_i w_i \quad 8$$

In the present day, the applicability of active packaging material to food and postharvest materials keeps growing. Polymers are functionalized with suitable reinforcement or fillers to create material that will not only protect the physical properties of the wrapped or sealed food products but also keeps its biological qualities. Fillers or reinforcements are capable of creation of quality barrier against oxygen and water vapor, which can affect the food quality, improvement of mechanical resistance, tearing, busting and moisture absorption of polymer

IOP Conf. Series: Materials Science and Engineering **805** (2020) 012029 doi:10.1088/1757-899X/805/1/012029

material. Polyvinyl alcohol (PVA) is a class of thermoplastic polymer with moderate strength and high barrier resistance property. It is highly soluble in water and completely biodegradable when disposed. The objectives of this study are to investigate the effect of *Momordica augustisepala* nanofiber reinforcement on the creep property of polyvinyl alcohol film using Burger Phenomenological or physical model and to compare the Burger model efficiency with Adaptive Neuron system model.

2 Materials and Methods

2.1 Materials

The materials used include alkalized *Momordica augustisepala* L fiber, sulphuric acid, acetic acid, polyvinyl alcohol; weighing balance (0.0000 g) was used for fiber measurements. Matlab 2010 b and Microsoft excel 2010 software were used for data analysis.

2.2 Sample Development

Momordica augustisepala fiber was processed in accordance with the method of Adeyi *et al.* [9]. Nanoscale fiber was further extracted using the method of Nascimento *et al.* [10]. Two creep samples, neat polyvinyl alcohol film and 3 % volume fraction reinforced polyvinyl alcohol film, were prepared by surface casting to test the effect of the nanofiber on the viscoelastic property of polyvinyl alcohol film.

2.3 Sample Characterization

Dynamic Mechanical Analysis (DMA) 242E machine was used to study the creep property of the samples. Each sample was loaded with a constant stress of 5 MPa for 30 min at room temperature and strain data were collated. Nonlinear fitting in Microsoft excel was used to establish the rheological parameter and prediction according to Burger model while Matlab 2010b was used for development of Adaptive neuro fuzzy artificial intelligent model.

IOP Conf. Series: Materials Science and Engineering **805** (2020) 012029 doi:10.1088/1757-899X/805/1/012029

3 Results and Discussion

3.1 Burger Model and Prediction

The summary of the data used in this study is given in Table 1.

Table 1. Summary of creep experimental data

Statistical Parameter	Time (min)	% strain of Neat PVA	% strain of Reinforced PVA
Min	0.8460	0.6721	0.4939
Max	28.9091	1.0363	0.8460
Standard Deviation	8.3819	0.0814	0.0864

The result of the creep test carried out on neat polyvinyl alcohol and reinforced polyvinyl alcohol film is represented in Figure 3.

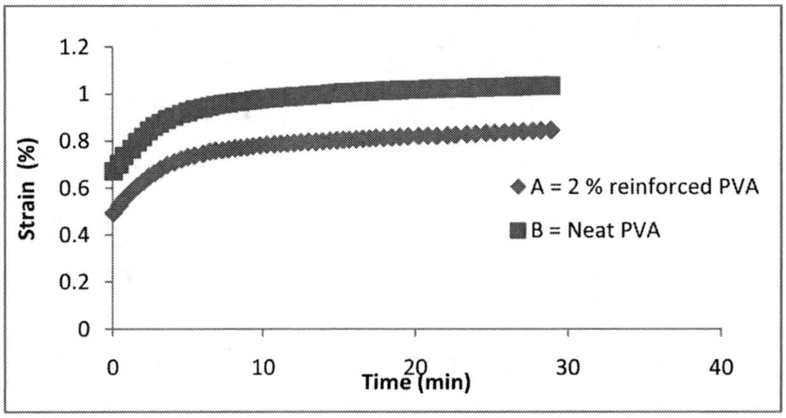

Figure 3. Experimental creep data of neat and reinforced PVA

It can be observed from the figure that the two samples passed through instantaneous deformation, primary, secondary and tertiary creep without rupture due to the quantity of load applied and the time of load application. However, the reinforced polyvinyl alcohol shows more resistance to creeping compared to the neat polyvinyl alcohol film.

The rheological parameters established according to Burger mode is summarized in Table 2.

Table 2. Summary of rheological parameter of both neat and reinforced PVA

Sample	$E1$ (MPa)	$E2$ (MPa)	$\eta 2$(MPa.s)	$\eta 1$(MPa.s)	$\tau 2$(s)
Neat PVA film	7.5371	16.5781	1915.3102	48.8437	2.85391
Reinforced PVA film	10.2758	18.2734	1711.3935	52.1502	3.00590

The table shows that nanofiber reinforcement improved the property of the PVA film. For instance, the stiffness value of reinforced PVA in the elastic ($E1$) and transition ($E2$) region is greater than that of neat PVA. The viscous term ($\eta 1$) in the PVA is also smaller than the viscous term in the reinforced PVA; this means a lower viscous flow in reinforced PVA and shows ability to better resist time dependent deformation than neat PVA. Furthermore, the transition region of neat film is shorter than the transition region of reinforced film judging from the value of $\eta 2$, which means that the reinforced PVA sample did not lose elastic property easily as neat PVA film did. Finally, with the value of $\tau 2$, it is obvious that neat PVA film sample creeps faster than reinforced PVA film. The overall result shows that reinforced PVA film has improved rheological properties. The result here conforms to the result of Plaseied and Fatemi [10] in the study of tensile creep and deformation modeling of vinyl ester polymer and its nanocomposite. The prediction of Burger model is represented in Figure 4.

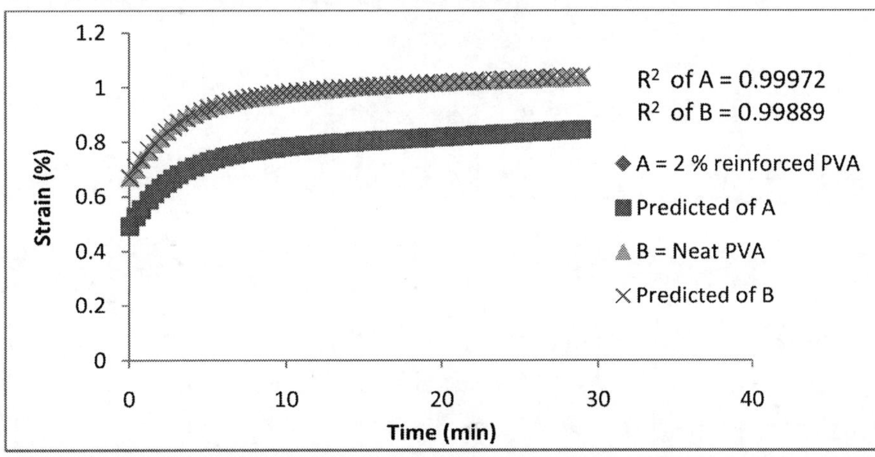

Figure 4. Comparison of experimental and Burger model predicted data

The figure shows that Burger model has capability to predict the experimental data with high accuracy. The coefficient of determination (R^2) for neat PVA creep response is 0.99889, while the coefficient of determination (R^2) of reinforced PVA is 0.99972.

3.2 ANFIS model and Prediction

In this study, the effect of membership function type on speed of error convergence or simulation was initially determined. The indicator for speed of error convergence was the epoch number.

The minimum epoch number is also important to prediction speed and computer memory allocation. Therefore, the membership function with minimum epoch number represents the optimum ANFIS structure. Several trials with constant four membership functions were done to search for the optimum membership function type. Figure 5 represents some of the membership functions tried.

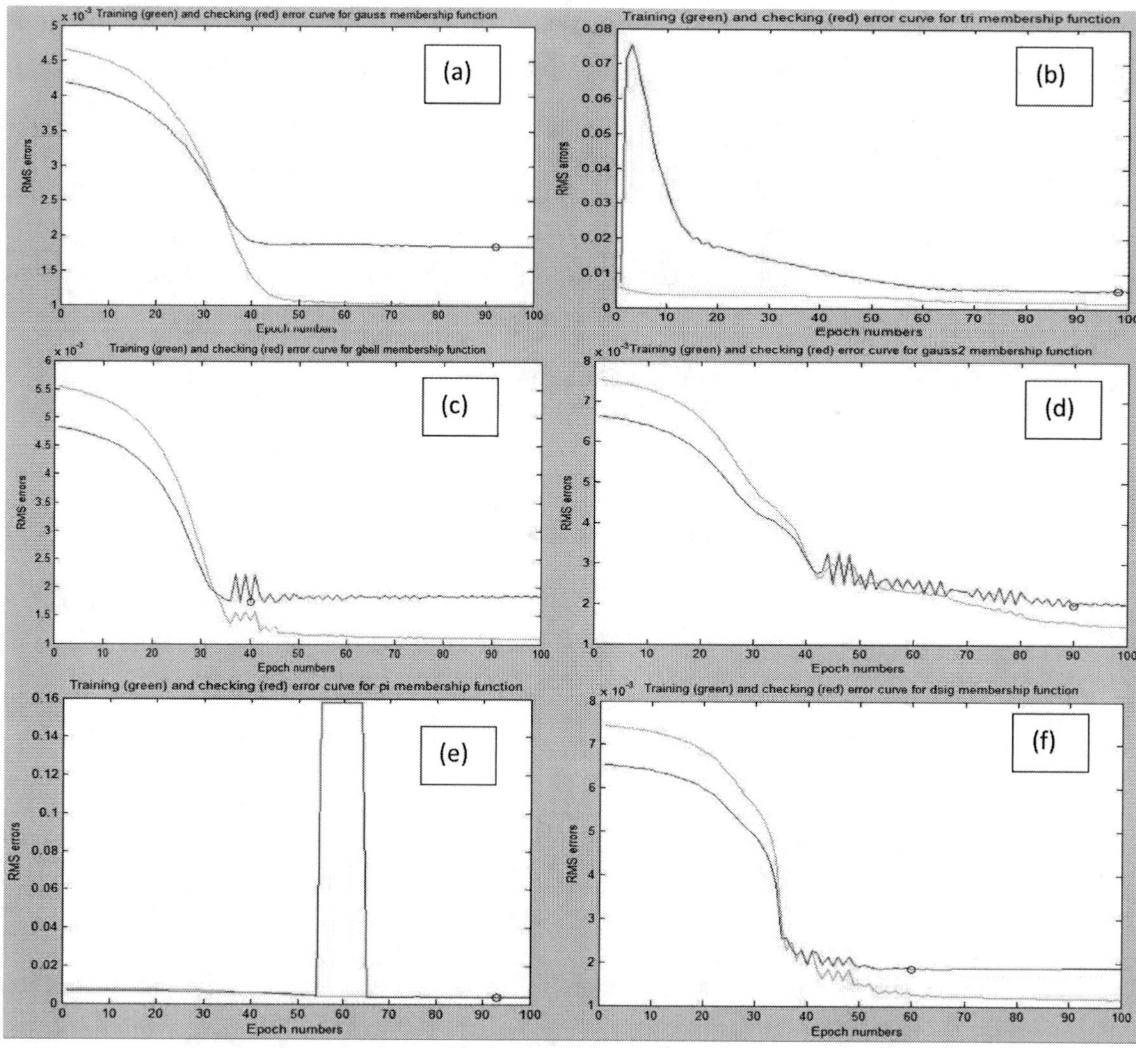

Figure 5. Effect of membership function type (a) gaussmf (b) trimf (c) gbellmf (d) gauss2mf (e) pimf and (g) dsigmf on the speed of error convergence or simulation

It could be observed from Figure 5 that gbell membership function type has the highest speed of error convergence or simulation with forty epoch number. The closest to gbell membership function is the dsig membership function, which converged at sixty epoch number. Tri membership function did not show any sign of convergence even after the exhaustion of the allocated forty epoch number. Gauss2 and gauss membership functions converged at ninety and about ninety three epoch number, respectively. The Pi membership function converged at about ninety four epoch number after initial upsurge between fifty-five to sixty-five epoch number. Therefore, gbell membership function represents the optimum membership function for the development of ANFIS model structure.

The gbell input membership function and output membership function are represented in Figures 6 (a) and (b). The membership function is a fuzzy reasoning parameter used in data approximation and prediction. During the course of modeling and prediction, the membership function is expected to undergo some refinement for optimum data prediction as a result of adaptiveness of ANFIS.

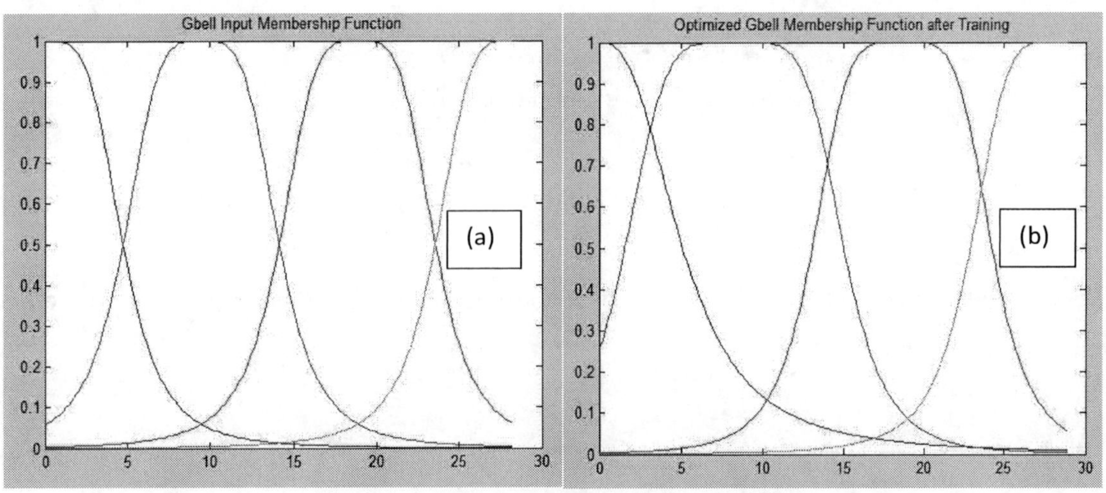

Figure 6. Input gbell membership function

Furthermore, comparing the input and optimized output membership function; it could be observed that the crossing point for the membership function line cross jumped from 0.50 y-axis to 0.80, 0.75 and 0.65 y-axis, respectively. The correlation coefficient signifies the prediction efficiency. Correlation coefficient is the square root of coefficient of determination and statistically, the closer the value of correlation coefficient to unity, the better the prediction.

IOP Conf. Series: Materials Science and Engineering **805** (2020) 012029 doi:10.1088/1757-899X/805/1/012029

Figure 7 depicts the correlation coefficient of the developed ANFIS structure and from the figure, the correlation coefficient is 0.99993; therefore, square of correlation coefficient is 0.99986 which is the coefficient of determination of ANFIS prediction. Comparing the two applied models, it is established that ANFIS performed better than Burger model. ANFIS has been previously used as a reliable tool for modelling and prediction of powder/water extraction process behaviour [11].

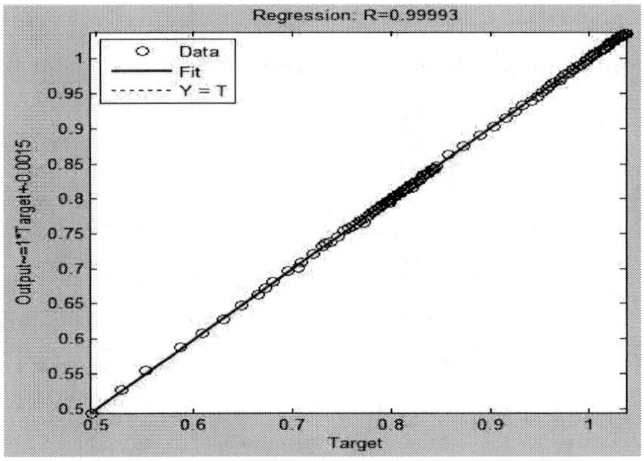

Figure 7. Regression plot of the ANFIS

In addition, the ANFIS generated rules for the modeling purpose is represented in Figure 8. The figure shows that 16 fuzzy rules were used for the modeling and prediction. The fewer number of rules could be related to the observed speed of the gbell membership function over the other tested membership function.

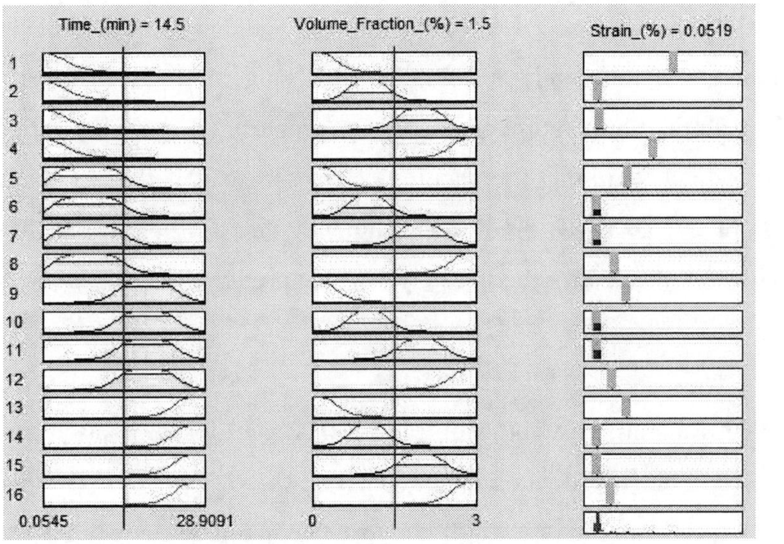

Figure 8. ANFIS generated rules

The qualitative representation of the prediction is represented in Figure 9. It could be observed that ANFIS had superiority in modeling and prediction of the experimental data. The ability of ANFIS to model and predict the experimental data for neat and reinforced PVA together at once also show robustness in its operation as seen in Figure 9 over Burger model that models the experimental data for each sample separately.

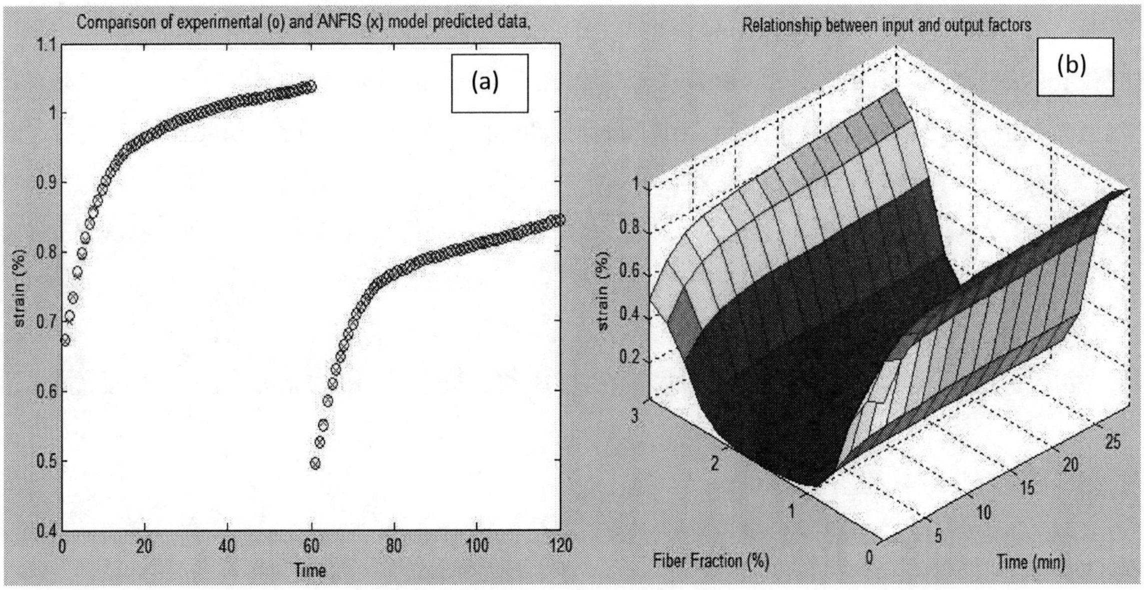

Figure 9. Prediction efficiency of ANFIS and data relationship

IOP Conf. Series: Materials Science and Engineering **805** (2020) 012029 doi:10.1088/1757-899X/805/1/012029

4.0 Conclusion

The results obtained in this study enable the following vital conclusions; that *Momordica augustisepala* L natural fiber improved the creep property of polyvinyl alcohol film, Burger model is adequate in determining the rheological parameters of the investigated samples and Adaptive neuron fuzzy has higher prediction efficiency than Burger model. Therefore, Adaptive neuron model is recommended for the life performance prediction of the investigated samples.

References

[1] Chakespari AG, Rajabipour A and Mobli H 2010 Anisotropic relaxation and creep properties of apple (cv. Shafi Abadi and Golab Kohanz). *Adv. J. Food Sci. Technol.* **2** pp 200-205.

[2] Xu Y, Wu Q, Lei Y and Yao F 2010 Creep behavior of bagasse fiber reinforced polymer composites. *Bioresour. Technol.* **101** (9) pp 3280-3286.

[3] Yilmaz MT 2012 Comparison of effectiveness of adaptive neuro-fuzzy inference system and artificial neural networks for estimation of linear creep and recovery properties of model meat emulsions. *J. Texture Stud.* **43** (5) pp 384-399.

[4] Buragohain M (2008) *Adaptive network based fuzzy inference system (ANFIS) as a tool for system identification with special emphasis on training data minimization* (Doctoral dissertation). Department of Electronics and Communication Engineering, Indian Institute of Technology, Guwahati. Guwahati - 781 039, India.

[5] Didekovà Z and Kajan S 2007 Applications of intelligent hybrid systems in Matlab. http://dsp.vscht.cz/konference_matlab/MATLAB09/prispevky/021_didekova.pdf. Accessed on 26 January, 2020.

[6] Pusat S, Akkoyunlu MT, Pekel E, Akkoyunlu MC, Özkan C and Kara SS 2016 Estimation of coal moisture content in convective drying process using ANFIS. *Fuel Process. Technol.* **147** pp 12-17.

[7] Sathiyamurthy S, Thaheer A and Jayabal S 2013 Prediction and optimization of mechanical properties of particles filled coir-polyester composites using ANN and RSM algorithms. *Ind. J. Fiber Text. Res.* **38** pp 81-86.

[8] Adeyi A, Durowoju M and Adeyi O 2018 Experimental studies and artificial neural networks (ANN) modeling of moisture absorption characteristics of polyester/*Momordica* fibre reinforced composite. *Int. J. Mech. Eng. Technol.* **9** (11) pp 1453-1467.

[9] Nascimento P, Marim R, Carvalho G and Mali S 2016 Nanocellulose produced from rice hulls and its effect on the properties of biodegradable starch films. *Mater. Res.* **19** (1) pp 167-174.

[10] Plaseied A and Fatemi A 2009 Tensile creep and deformation modeling of vinyl ester polymer and its nanocomposite. *J. Reinf. Plast. Compos.* **28** (14) pp 1775-1788.

[11] Oke EO, Adeyi O, Adeyi AJ and Adekunle KF 2018 Modelling of *Grewia mollis* stem bark gum extraction yield using neuro-fuzzy technique. *Int. J. Eng. Res. Afr.* **34** pp 70-80.

Evaluation of microalgal-based nanoparticles in the adsorption of heavy metals from wastewater

VO Adenigba[1]*, IO Omomowo[2], JK Oloke[2], BA Fatukasi[1], MA Odeniyi[1], and AA Adedayo[2]

[1]Department of Science Laboratory Technology, Ladoke Akintola University of Technology, Ogbomoso, P.M.B 4000 Oyo State, Nigeria.

[2]Department of Pure and Applied Biology, Ladoke Akintola University of Technology, Ogbomoso, P.M.B 4000, Oyo State, Nigeria.

*E-mail address: moadesiyan@lautech.edu.ng

Abstract

Bioremediation remains an eco-friendly and inexpensive method employed in getting rid of the toxic compounds like heavy metals present in wastewater. Different microorganisms are being used in either removing or reducing the concentration of heavy metals in wastewater but microalgae have a high affinity for heavy metals. In this study, nanoparticles from microalgae were evaluated for the potential to absorb lead and zinc from pharmaceutical effluent. Silver (AgNPs) and gold nanoparticles (AgNPs) were synthesized from both *Nannochloropsis* sp (NN) and *Chlorella vulgaris* (CV). The nanoparticles were introduced into a pharmaceutical effluent and the mixture was allowed to interact for 48 h after which the concentration of zinc and lead was determined using atomic absorption spectrophotometer. The AgNPs of *Nannochloropsis* sp (NN-AgNPs) and *C. vulgaris* (CV-AgNPs) had percentage reduction of 70.35 % and 74.62% respectively for zinc. Also gold nanoparticles of *Nannochloropsis* sp (NN-AuNPs) and *C. vulgaris* (CV-AuNPs) had percentage reduction of 60.32% and 66.83% respectively for zinc. The concentration of lead was reduced by CV-AgNPs (66.10 %), NN-AgNPs (68.86%), CV-AuNPs (57.41%), and NN-AuNPs (66.53 %). This study has established algal based nanoparticles as adsorbent with affinity for heavy metals and can be adopted for bioremediation of wastewater.

Keywords: Heavy metals, microalgae, nanoparticles, pharmaceutical effluent, adsorption

1.0 Introduction

All living beings all over the world depend on water for survival and for the various developmental processes which has the potential to enhance the economic status and growth of a community. Water is very important because it is highly required for industrial activities and by both rural and urban dwellers. The availability of water has therefore pose a major challenge in the economic development of any country. The contamination of the aquatic environ as a result of industrial activities has resulted in serious environmental concerns. Contaminated water contains complex materials especially heavy metals that have been implicated for their high toxicity, non-biodegradability and ability to bioaccumulate [1]. Heavy metals are released into the environment from various sources; these are agricultural activities, mining industries, metallurgical industries, pharmaceutical industries, chemical plants and domestic wastewater. Some metals (Pb, Zn, Cu, Hg) that are heavy in nature could pose a severe threat to human health when they can be accumulated biologically in the food chain [2]. Heavy metals are known to cause damages to the central nervous functions, mental, lungs, kidneys and other organs [3-5]. Moreover, heavy metals on the contrary do accumulate through the food chain because they are not degradable by microorganisms. Once they are released into the environment, they exert adverse effects on the receiving aquatic environment and other ecological receptors.

Heavy metals are responsible for causing several diseases [6]. It is therefore very important to either reduce or completely remove these unfriendly pollutants from the environment since they are toxic and cannot be degraded by microorganisms [7]. Various industrial activities has caused the issue of heavy metals discharged directly or indirectly into the water bodies thereby increasing the rate of waste pollution, especially in developing countries [8]. Wastewaters obtained from both petrochemical industries and refineries contain heavy metals; their toxic nature has presented them as threat to the environment and human. When emptied into water bodies, it imparts taste and odor to drinking water with even minute concentrations of their chlorinated derivatives and creates an oxygen demand in the river [9]. Several precipitation techniques that are used to remove these metals from industrial effluents include membrane filtration [10-13], ion exchange [14-16], and adsorption [17-20].

Algal biomass as an adsorbent in removing heavy metals from contaminated water bodies has attracted attention [21-23]. They are rich sources of several metabolites and they contain functional groups which take a great part in the attachment of heavy metals to the wall of the

IOP Conf. Series: Materials Science and Engineering **805** (2020) 012030 doi:10.1088/1757-899X/805/1/012030

algae [24-26]. Improving the adsorptive potential of algae in the removal of heavy metals can be achieved by employing algal nanoparticles which can enhance the adsorption of heavy metals through a larger surface area.

Chemical, physical and biological techniques are methods through which preparation of metal nanoparticles is usually carried out [8]. Chemical and physical approaches produce toxic materials and harmful to the environment, more expensive and required high pressure [9]. In contrast, cost-effective, sustainable, non-toxic, ability to use without massive equipment and chemical materials are advantages of biological method over former methods [27]. *Kappaphycus alvarezi* [28] and *Diclyota indica* [21] have been used as bioagent to synthesize nanoparticles which were used as adsorbent for heavy metal removal. Marine algae such as microalgae used as reducing agent in biosynthesizing nanoparticles can produce various metabolites which include carbonyl groups, phenolics, polysaccharides, proteins, amines, amides, flavonones, alkaloids, and other reducing-agents in the biological extracts [29].

In this study, nanoparticles were synthesized from both *Nannochloropsis* sp and *Chlorella vulgaris* and they were evaluated as adsorbents in removing heavy metals (zinc and lead) from wastewater.

2 Materials and methods

2.1 Maintenance and harvesting of microalgal cultures

Nannochloropsis sp and *Chlorella vulgaris* which were previously isolated from waste water in our laboratory was cultured in BG 11 medium. The algal biomass was harvested on day twenty one through centrifugation at 5000 rpm for 20 min. The biomass of the microalgae was washed with sterile distilled water to remove the culture medium that was present in the cell [30].

2.2 Silver and gold nanoparticles synthesis from microalgae

One gram of algal powder was suspended in 100 ml of distilled water, boiled at 100 °C for 20 min [30]. The cell-free extract of each microalgae was separated from the mixture in a centrifuge at 5000 rpm for 10 min. In order to synthesize both silver and gold nanoparticles from the extract, 90 ml each of 1 mM $AgNO_3$ and 1 mM gold chloride was added to 10 ml of the extract in separate bottles. The mixtures were incubated and the colour change was monitored until it stabilized [31].

IOP Conf. Series: Materials Science and Engineering **805** (2020) 012030 doi:10.1088/1757-899X/805/1/012030

2.3 Characterization of the nanoparticles

UV-visible spectra of samples were recorded using Ultraviolet-visible spectroscopy from 300-500 nm. Morphology of nanoparticles was characterized by both Scanning (SEM) and transmission Electron Microscopes (TEM) to obtain the shape and size of nanoparticles. Fourier-transform infrared spectroscopy (FTIR, Nicolet Avatar, Thermo, US) in the range of 4000-400 cm^{-1} was used to determine the presence of the various functional groups on the samples.

2.4 Bioadsorption of zinc and lead from pharmaceutical effluent

Twenty millilitres of each nanoparticle (CV-AgNPs, CV-AuNPs, NN-AgNPs, and NN-AuNPs) and the algal biomass of *Chlorella vulgaris* (CV) and *Nannochloropsis* sp (NN) was added to 200 ml of the industrial effluent. The mixture was subjected to manual agitation during the experiment. The ability of the nanoparticles and algal biomass to absorb both zinc and lead was determined after 12 h using Atomic Absorption Spectrophotometer. The percentage reduction in concentration of each metal by the nanoparticles was calculated:

$$\text{Reduction in metal concentration} = \frac{\textit{Initial concentr ation} - \textit{Final concentration}}{\textit{Initial concentration}} \times 100$$

3 Results and Discussion

The colour change which is an indicator of nanosynthesis was observed after 20 min of adding silver nitrate to the cell-free extracts of both *C. vulgaris* and *Nannochloropsis* sp. There was colour change from pale green to brown which is indicative of silver nanoparticles. The surface plasmon resonance for the AgNPs was obtained with broad absorbance at a wavelength of 426 nm for *C. vulgaris* (Figure 1a) and 430 nm for *Nannochloropsis* sp (Figure 1b). These absorption spectra are similar to those reported for AgNPs of *Turbinaria conoides*, a brown marine alga which had a broad spectrum at 420 nm [32]. The synthesis of AgNPs from both *C. vulgaris* and *C. calcitrans* with a strong Plasmon vibration at 436 nm and 420 nm respectively has been reported [33]. Sudha *et al.* [31] reported a broad peak of 420 nm from AgNPs synthesized from *Scenedesmus abundans*.

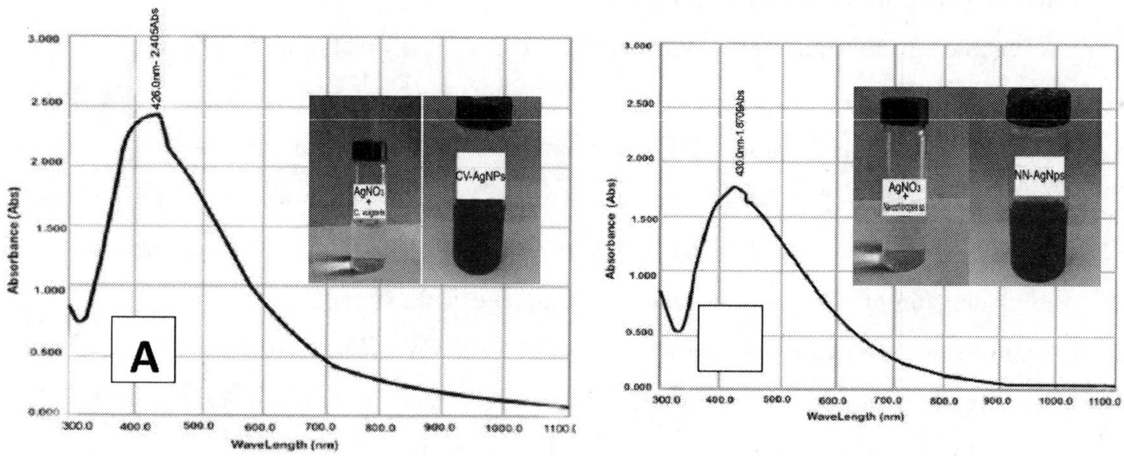

Figure 1. UV-visible spectra of silver nanoparticles synthesized from A (*C. vulgaris*) and B (*Nannochloropsis* sp) with colour change in reaction bottles after incubation.

The gold nanoparticles of both *C. vulgaris* and *Nannochloropsis* sp was synthesized with change in colour after 2 h of incubating the extract of each microlaga with gold chloride. A purple colour was seen in both bottles and synthesis of gold chloride was further confirmed with UV-visible spectra at 520 nm and 535 nm for *C. vulgaris* and *Nannochloropsis* sp respectively (Figures 2a and b). This result is in agreement with previous reports on synthesis of gold nanoparticles [34] where AuNPs was synthesized using *Camellia sinensis* and *Juniperus communis* and had absorbance spectra within ranges around 530-540 nm.

The micrographs for silver and gold nanoparticles of *C. vulgaris* and *Nannochloropsis* were taken using both scanning electron microscope for silver nanoparticles(CV-AgNPs and NN-AgNPs) and transmission electron microscopy for gold nanoparticles (CV-AuNPs and NN-AuNPs) in other to obtain the shape and size of each nanoparticles. The CV-AgNPs and NN-AgNPs were polydispersed and spherical in shape with size ranging from 27 to 90 nm and 41 to 88 nm respectively (Figures 3a and b). Also CV-AuNPs and NN-AuNPs had spherical polydispersed particles with particle range from 7 to 20 nm and 7 to 26 nm (Figures 3c and d).

IOP Conf. Series: Materials Science and Engineering **805** (2020) 012030 doi:10.1088/1757-899X/805/1/012030

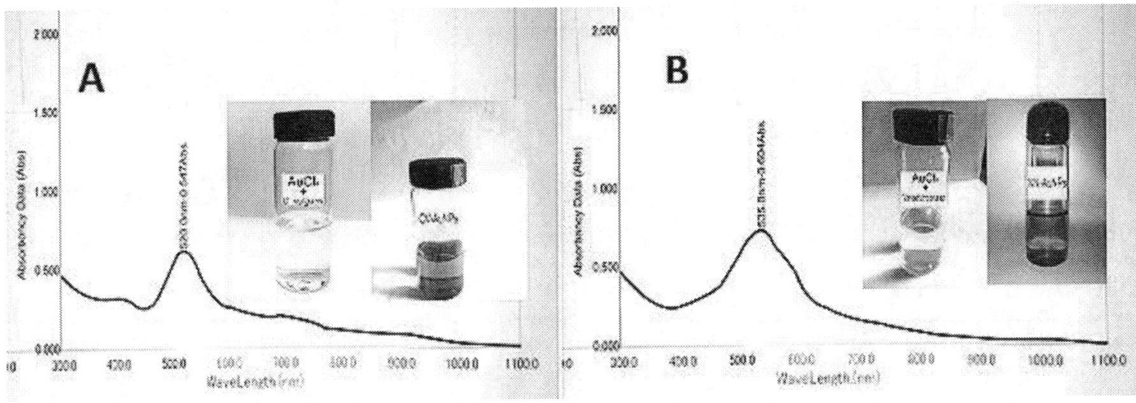

Figure 2. UV-visible spectra of gold nanoparticles synthesized from A (*C.vulgaris*) and B (*Nannochloropsis* sp) with colour change in reaction bottles after incubation.

The micrographs for silver and gold nanoparticles of *C.vulgaris* and *Nannochloropsis* was taken using both scanning electron microscope for silver nanoparticles(CV-AgNPs and NN-AgNPs) and transmission electron microscopy for gold nanoparticles (CV-AuNPs and NN-AuNPs) in other to obtain the shape and size of the nanoparticles. The CV-AgNPs and NN-AgNPs were polydispersed and spherical in shape with size ranging from 27 to 90 nm and 41 to 88 nm respectively (Fig. 3a and b). Also CV-AuNPs and NN-AuNPs had spherical polydispersed particles with particle range from 7 to 20 nm and 7 to 26 nm (Fig 3c and d).

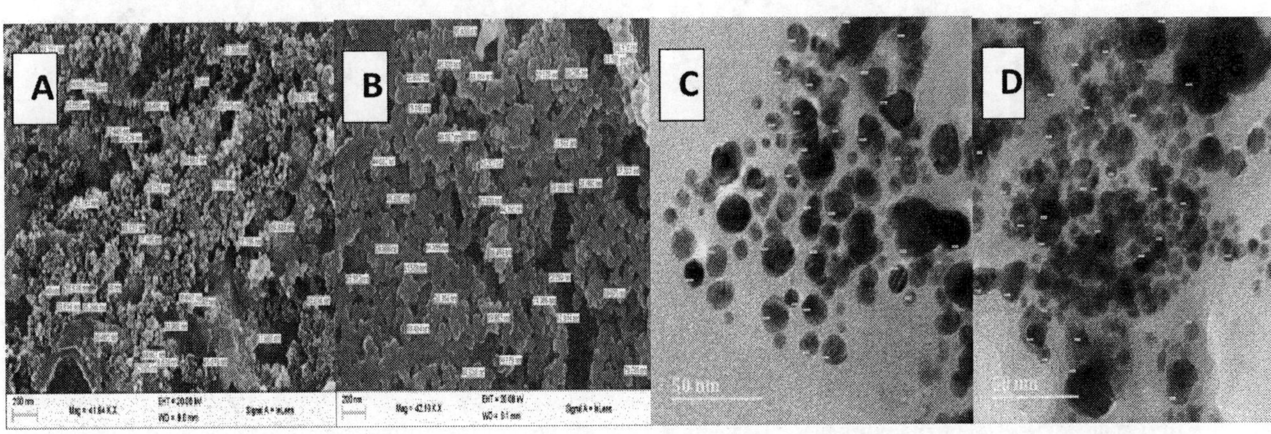

Figure 3. Electron micrographs of A (CV-AgNPs), B (NN-AgNPs), C(CV-AuNPs) and D (NN-AuNPs).

IOP Conf. Series: Materials Science and Engineering **805** (2020) 012030 doi:10.1088/1757-899X/805/1/012030

The EDX of the nanoparticles revealed the different elemental composition. The silver nanoparticles had silver as the element with the highest weight composition of 47.82 % for CV-AgNPs and 54 % for NN-AuNPs with a strong signal strength at 3 keV (Figures 4a and b). The weight percentage of gold in the gold nanoparticles was higher than that of the other elements present: CV-AuNPs had gold composition of 96.62%, while 79 % gold was present in NN-AuNPs (Figures 4c and d).

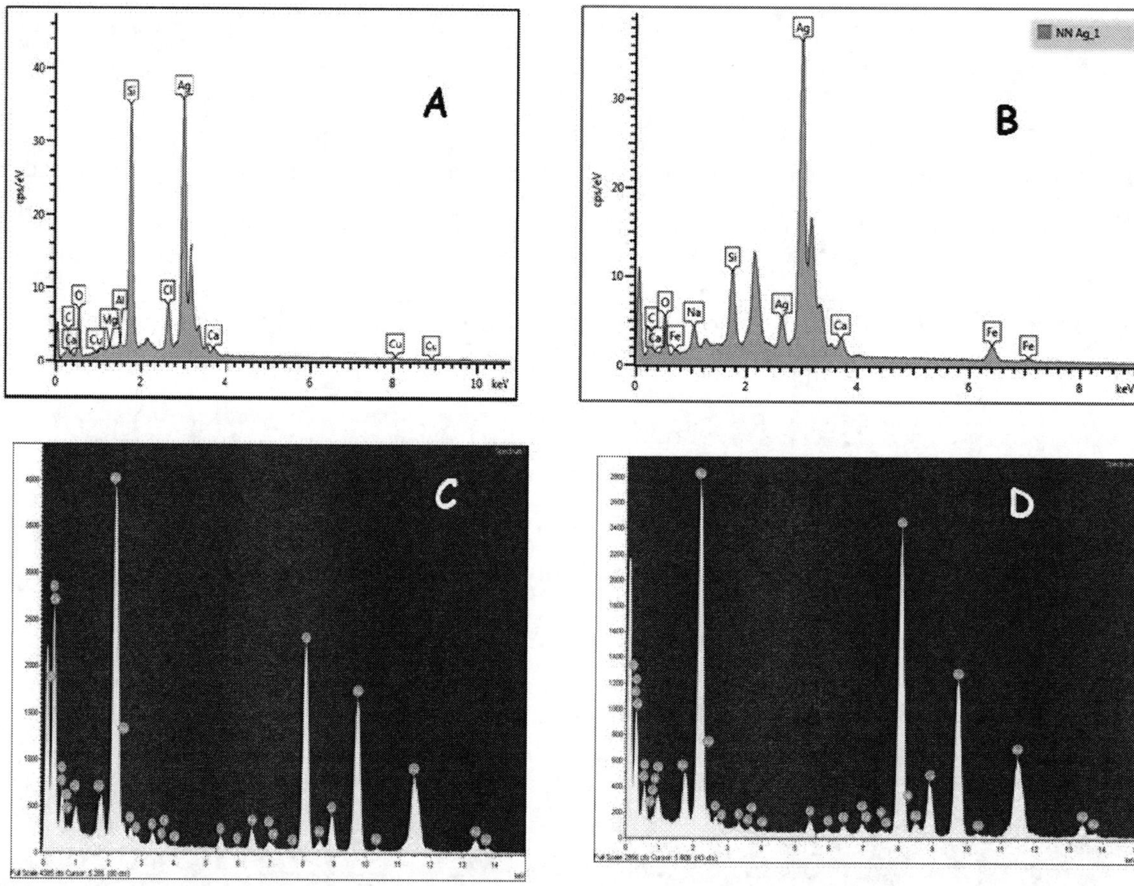

Figure 4. EDX of (A) CV-AgNPs, (B) NN-AgNPs, EDX of (C) CV-AuNPs, (D) CV-AuNPs

The FTIR spectra showed absorbane for CV-AgNPs at 3272 (O-H of alcohol), 2103(broad C≡ N stretching vibration of amine), and 1636 cm^{-1} (C=C of olefin) (Figure 5a), CV-AuNPs at 3295 (O-H of alcohol), 2103 (broad C≡ N stretching vibration of amine), and 1636 cm^{-1} (C=C of olefin) (Figure 5b), NN-AgNPs at 3294 (O-H of alcohol), 2104 (broad C≡ N stretching vibration of amine), and 1638 cm^{-1}(C=C of olefin) (Figure 5c), and NN-AuNPs at

3280 (O-H of alcohol), 2106 (broad C≡ N stretching vibration of amine), and 1636 cm^{-1} (C=C of olefin) (Figure 5d). The function groups reperesented on each nanoparticles include O-H, C≡ N and C=C. FTIR analysis showed the involvement of hydoxyl and unsaturated hydrocarbons in the reduction, stabilization and capping of the metal nanoparticles [35-36].

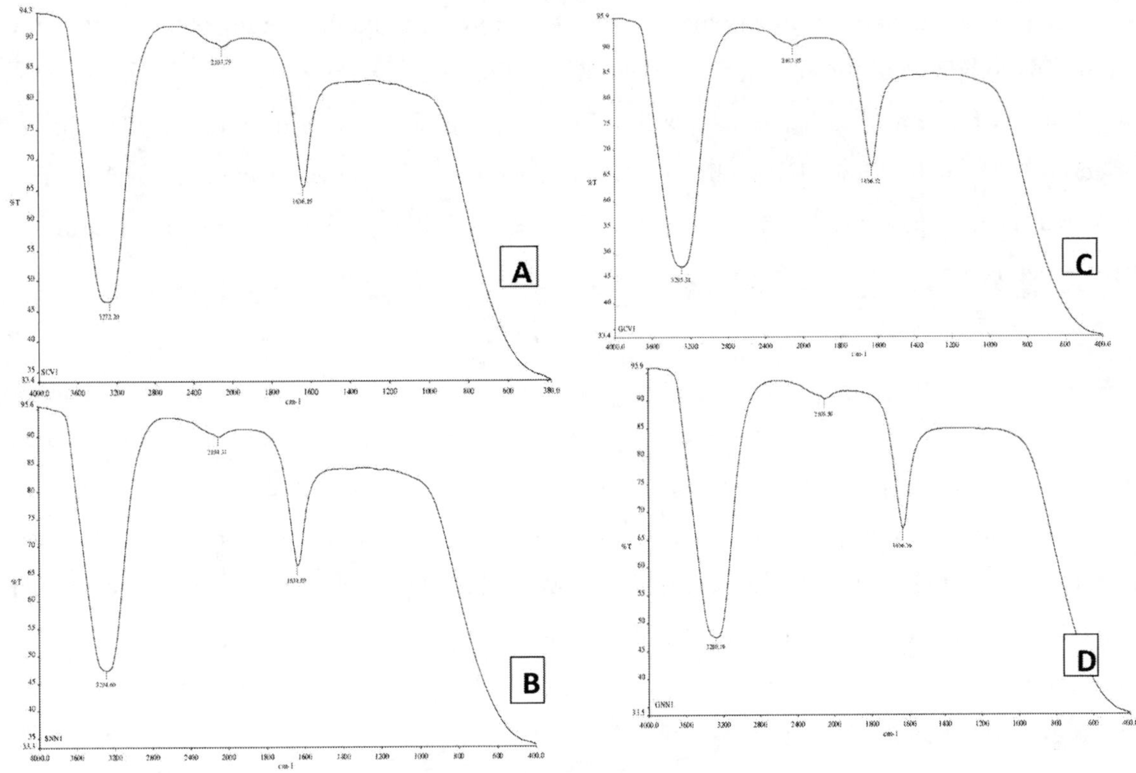

Figure 5. FTIR spectra of (A) CV-AgNPs, (B) NN-AgNPs, (C) CV-AuNPs, (D) NN-AuNPs

The silver and gold nanoparticles of *C. vulgaris* and *Nannochloropsis* sp were introduced into pharmaceutical effluent and the concentration of each metal was obtained after 12 h of adsorption. The percentage reduction in concentration of zinc by CV-AgNPs,NN-AgNPs, CV and NN were 74.62, 70.35, 46.98 and 48.49 % (Figure 6A) while 66.83, 60.32, 46.98 and 48.49 % were achieved by CV-AuNPs, NN-AuNPs, CV and NN respectively (Figure 6B). Similarly, the adsorption potential of CV-AgNPs (66.10 %), NN-AgNPs (68.86%), CV (30.29%) and NN (55.51%) for the adsorption of lead were obtained (Figure 6C) while CV-AuNPs (57.41%), NN-AgNPs (66.53 %), CV (30.29%) and NN (55.51%) were also recorded (Figure 6D).

IOP Conf. Series: Materials Science and Engineering **805** (2020) 012030 doi:10.1088/1757-899X/805/1/012030

The two microalgae (CV and NN) and their nanoparticles were efficient in adsorbing both zinc and gold from the effluent, however, the nanoparticles outperformed the algal biomass with nanoparticles of *C. vulgaris* (74.62 and 66.83 %) having the maximum efficiency for zinc and *Nannochlopsis* sp (68.86 and 66.53%) with the maximum efficiency for lead. The improved performnace of both silver and gold nanoparticles compared to their biomass is as a result of the increased surface area of reaction which consequentially increased the binding sites available for the metals to adhere to the cells [37]. The application of both microalgae and their nanoparticles as adsorbent for removing heavy metals from effluent is a prospective technology which is cost effective and environmental friendly. Different strains of microalgae and nanoparticles from different metal oxides have been documented for their activities in reducing the concentration of heavy metals present in effluent [6, 37].

4 Conclusion

Heavy metal remains a major component of industrial effluent which puts both man and its environment in great risk due to their persistence, non-biogeradability and ability to bioaccumulate. Various technologies are being adopted to either remove or reduce the heavy metal concentration to become less toxic. Two microalgae and their nanoparticles were employed in this work as bioadsorbent for the removal of both zinc and lead. The nanoparticles out performed the microalgal biomass as adsorbent owing to increased surface of reaction which led to increased metal adsorption.

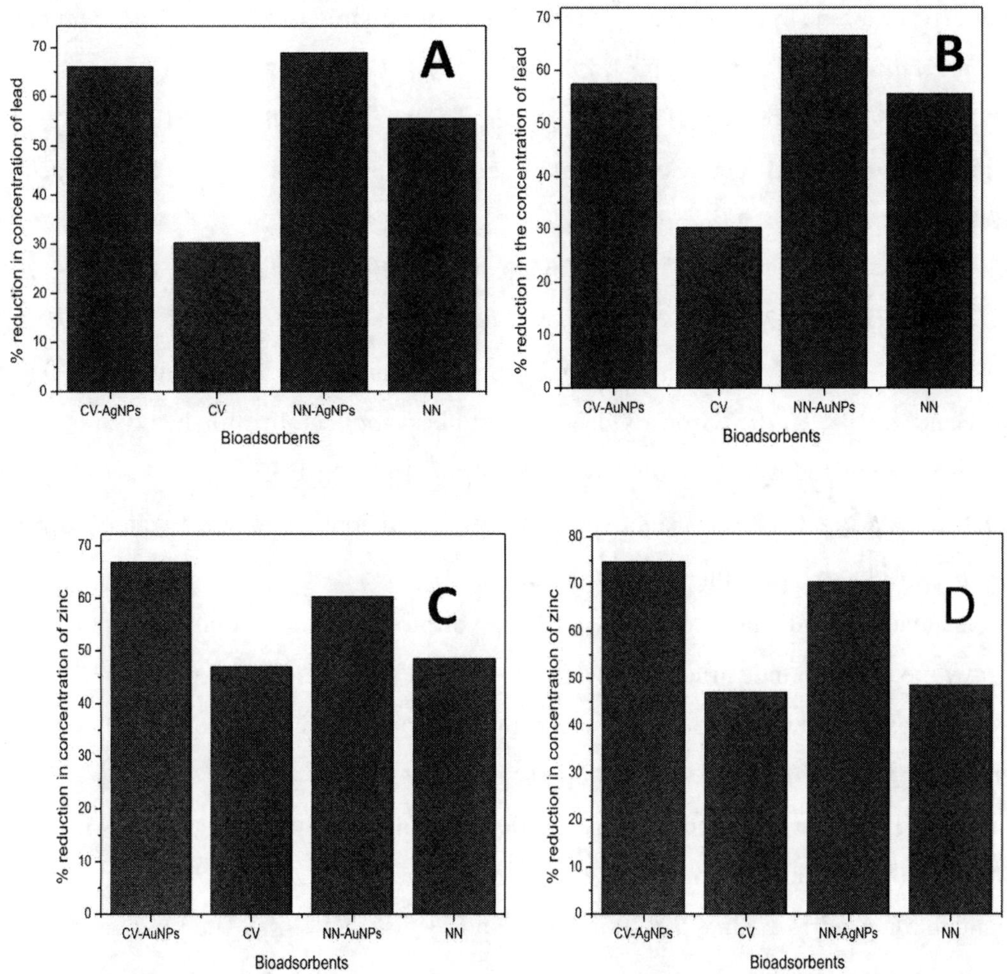

Figure 6. Percentage reduction in concentration of lead by (A) CV-AgNPs and NN-AgNPs, (B) CV-AuNPs and NN-AuNPs; zinc by (C) CV-AuNPs and NN-AuNPs, (D) CV-AgNPs and NN-AgNPs.

References

[1] Nazal MK 2019 Marine algae bioadsorbent for adsorptive removal. *In: Advanced Sorption Process Applications.* http://dx.doi.org/10.5772/intechopen.80850.

[2] Cheraghi M, Lorestani B and Yousefi N 2009 Effect of waste water on heavy metal accumulation in Hamedan Province vegetables. *Int. J. Bot.* **5** (2) pp 190-193.

[3] Reglero M M, Taggart M A, Lidia M G and Rafael M 2009 Heavy metal exposure in large game from a lead mining area: Effects on oxidative stress and fatty acid composition in liver. *Environ. Pollut.* **157** (4) pp 1388-1395.

[4] Gybina AA and Prohaska JR 2009 Copper deficiency results in AMP-activated protein kinase activation and acetylCoA carboxylase phosphorylation in rat cerebellum. *Brain Res.* **1204** pp 69-76.

[5] Kampa M and Castanas E 2008 Human health effects of air pollution 2008. *Environ. Pollut.* **151**(2) pp 362-367.

[6] Al-Saad KA, Amr MA, Hadi DT, Arar RS, Al Sulaiti MM, Abdulmalik TA, Alsahansy NM and Kwak JC 2012 Iron oxide nanoparticles: applicability for heavy metal removal from contaminated waters. *Arab J. Nucl. Sci Appl.* **45** (2) pp 335-246.

[7] Fu F and Wang Q (2011) Removal of heavy metal ions from wastewaters: a review. *J Environ. Manag.* **92** (3) pp 407-418.

[8] Srivastava NK and Majumder CB 2008 Novel biofiltration methods for the treatment of heavy metals from industrial wastewater. *J. Hazard. Mater.* **151** (1) pp 1-8.

[9] Viraraghavan T and Alfaro F 1998 Adsorption of phenol from wastewater by peat, fly ash and bentonite. *J. Hazard. Mater.* **57** (1-3) pp 59-70.

[10] Labanda J Khaidar M S and Llorens J 2009 Feasibility study on the recovery of chromium (III) by polymer enhanced ultrafiltration. *Desalination* **249** (2) pp 577-581.

[11] Landaburu-Aguirre J, García V, Pongrácz E and Keiski RL 2009 The removal of zinc from synthetic wastewaters by micellar-enhanced ultrafiltration: statistical design of experiments. *Desalination* **240** (1-3) pp 262-269.

[12] Landaburu-Aguirre J, Pongrácz E, Perämäki P and Keiski RL 2010 Micellar enhanced ultrafiltration for the removal of cadmium and zinc: Use of response surface methodology to improve understanding of process performance and optimisation. *J. Hazard. Mater.* **180** (1-3) pp 524-534.

[13] Acero JL, Benítez FJ, Leal AI and Real FJ 2005 Removal of phenolic compounds in water by ultrafiltration membrane treatments. *J. Environ. Sci. Health A: Toxicol. Hazard Subst. Environ. Eng.* **40** (8) pp 1585-1603.

[14] Kang SY, Lee JU, Moon SH and Kim KW 2004 Competitive adsorption characteristics of Co^{+2}, Ni^{+2}, and Cr^{+3} by IRN-77 cation exchange resin in synthesized wastewater. *Chemosphere* **56** (2) pp 141-147.

[15] Alyüz B and Veli S 2009 Kinetics and equilibrium studies for the removal of nickel and zinc from aqueous solutions by ion exchange resins. *J. Hazard. Mater.* **167** (1-3) pp 482-488.

[16] Caetano M, Valderrama C, Farran A and Cortina JL 2009. Phenol removal from aqueous solution by adsorption and ion exchange mechanisms onto polymeric resins. *J. Colloid Interf. Sci.* **338** (2) pp 402-409.

[17] Al-Jlil SA and Alsewailem FD 2009 Saudi Arabian clays for lead removal in wastewater. *Appl. Clay Sci.* **42** (3-4) pp 671-674.

[18] Hima KA, Srinivasa RR, Vijaya SS, Suryanarayana SB and Venkateshwar V 2007 Biosorption: An eco-friendly alternative for heavy metal removal. *Afr. J. Biotechnol.* **6** (25) pp 2924-2931.

[19] Kour J, Homagai PL, Cagnin M, Masi A and Pokhrel MR 2013 Adsorption of Cd (II), Cu (II), and Zn (II) from aqueous solution onto nitrogen-functionalized *Desmostachya bipinnata*. *J. Chem.* Article ID 649142. http://dx.doi.org/10.1155/2013/649142.

[20] Ma Y, Gao N, Chu W and Li C 2013 Removal of phenol by powdered activated carbon adsorption. *Front. Environ. Sci. Eng.* **7** (2) pp 158-165.

[21] Shargh AZ, Sayadi MH and Heidari A 2018 Green biosynthesis of palladium oxide nanoparticles using *Dictyota* indica seaweed and its application for adsorption. *J. Water Environ. Nanotechnol.* **3** (4) pp 337-347.

[22] Vijayakumar S and Manoharan C 2012 Treatment of dye industry effluent using free and immobilized cyanobacteria. *Bioremed. Biodeg.* **3** (10) pp 1-6.

[23] Vijayaraghavan K and Yun YS 2008 Bacterial biosorbents and biosorption. *Biotechnol. Adv.* **26** (3) pp 266-291.

[24] Davis T A, Volwsky B and Mucci A 2003 A review on Biochemistry of heavy metals biosorption by brown algae. *Water Res.* **37** (18) pp 4311 4330.

[25] Shankar S, Ahmad A and Sastry M 2003 Geranium leaf assisted biosynthesis of silver nanoparticles. *Biotechnol. Prog.* **19** (6) pp 1627-1631.

[26] Huang J, Li Q, Sun D, Lu Y, Su Y, Yang X, Wang H, Wang Y, Shao W, He N, Hong NJ and Chen C 2007 Biosynthesis of silver and gold nanoparticles by novel sundried *Cinnamomum camphora* leaves. *Nanotechnol.* **18** (10) 105104. https://doi.org/10.1088/0957-4484/18/10/105104.

[27] González-Muñoz MJ, Rodríguez MA, Luquea S and Álvareza JR 2006 Recovery of heavy metals from metal industry wastewaters by chemical precipitation and nanofiltration. *Desalination* **200** (1-3) pp 742-744.

[28] Radihika RSR, Gayathri S and Karthih MG 2018 Comparative studies on biosorption of zinc oxide and iron oxide nanoparticles using marine macroalgae *Kappaphyaus alvarezii* *Ind. J. Geo. Mar. Sci.* **47** (01) pp 219-227.

[29] Asmathunisha N and Kathiresan K 2013 A review on biosynthesis of nanoparticles by marine organisms. *Colloids Surf. B: Biointerf.* **103** pp 283-287.

[30] Azizi SAhmad MB, Namvar F and Mohamad R 2014 Green biosynthesis and characterization of zinc oxide nanoparticles using brown marine macroalga *Sargassum muticum* aqueous extract. *Mater. Lett.* **116** pp 275-277.

[31] Kathiraven T, Sundaramanickam A, Shanmugam N and Balasubramanian T 2015 Green synthesis of silver nanoparticles using marine algae *Caulerpa racemosa* and their antibacterial activity against some human pathogens. *Appl. Nanosci.* **5** (4) pp 499-504.

[32] Rajeshkumar S, Chellapadian K and Gurusamy A 2012 Green synthesis of silver nanoparticles using marine brown algae *Turbinaria conoides* and its antibacterial activity. *Int. J. Pharm. Biol. Sci.* **3** (4) pp 502-510.

[33] Karthikeyan P Mohan D Abishek G and Priya R 2015 Synthesis of silver nanoparticles using phytoplankton and its characteristics. *Int. J. Fish. Aquat. Stud.* **2** (6) pp 398-401.

[34] Geraldes AN, da Silva AA, Leal J, Estrada-Villegas GM, Lincopan N, Katti KV and Lugão AB 2016 Green nanotechnology from plant extracts: synthesis and characterization of gold nanoparticles. *Adv. Nanopart.* **5** (03) Article ID 69898. https://doi.org/10.4236/anp.2016.53019.

[35] Sheng PY, Ting YP, Chen JP and Hong L 2004 Sorption of lead, copper, cadmium, and nickel by marine algal biomass: characterization of biosorptive capacity and investigation of mechanisms. *J. Colloid Interf. Sci.* **275** (1) pp 131-141.

[36] Devi JS and Bhimba BV 2012 Anticancer activity of silver nanoparticles synthesized by the seaweed *Ulva lactuca in vitro. Sci. Rep.* **1** (4) 242. https://doi.org/10.4172/scientificreports.242.

[37] Taman R, Ossman ME, Mansour MS and Fanag HA 2015 Metal oxide nanoparticles as an adsorbent for removal of heavy metals. *J. Adv. Chem. Eng.* **5** (3) pp 125-132.

IOP Conf. Series: Materials Science and Engineering **805** (2020) 012031 doi:10.1088/1757-899X/805/1/012031

Setting time and standard consistency of Portland cement binders blended with rice husk ash, calcium carbide and metakaolin admixtures

M Babako and JA Apeh

Department of Building, Federal University of Technology, Minna.

apehjoe@futminna.edu.ng

Abstract

The use of pozzolanic materials to reduce carbon dioxide (CO_2) emissions and enhanced properties of mortar and concrete has received increased interests in the last decades. In this study, admixtures such as metakaolin (MK), rice husk ash (RHA) and calcium carbide waste (CCW) partially replaced Portland cement (PC) at 5, 10, 20 % in multiple combinations of aforementioned admixtures as binders and the effect of setting times (initial and final) and standard consistency on the binders were evaluated. The Department of Environment (DOE) approach for mix proportions was employed. The mixes are in six groups consisting of control PC, each admixture replaced PC at 5, 10 and 20 % replacement level in binary mixes, then two of the admixtures replaced PC at same replacement level in ternary and the three admixtures replaced PC in quaternary mixes respectively. Results indicated that RHA significantly influenced the standard consistency of the binders incorporating the three admixtures as compared with CCW and MK. The requirement of water in the binder increased with increase in percentage replacement level of RHA within the binder due to its porous nature. At 5 % replacement level, it was 9 % above control value, 21 % and 39 % at 10 and 20 % replacement level subsequently. Similarly, setting times at the aforementioned replacement levels are 27 % and 6 % above the control value for initial and final setting time. For 20 % replacement level, it was 24 % and 38 % above the control value for the initial and final setting time respectively. This is due to reduction in PC content leading to less content of Tricalcium silicate aluminate (C_3A) and amount of carbon content within the pozzolanic materials which retard the rate of reactions. By and large, the admixtures retard the setting times of quaternary binder, and increased its water of consistency which can be overcome by activation. However, they are useful in the construction of concrete structural elements or structures that requires longer time

IOP Conf. Series: Materials Science and Engineering **805** (2020) 012031 doi:10.1088/1757-899X/805/1/012031

for placement. Furthermore, it will be worthwhile to investigate same properties at the nanoscale since these have been established at the micro level.

Keywords: Portland cement, pozzolans, rice husk ash, calcium carbide waste, metakaolin, consistency, setting times

1 Introduction

The use of concrete is over centuries and has retained its popularity because of its properties. Setting time as one of the fresh properties of concrete is vital in concrete construction [1]. In concreting, this property (setting time) find application in its development in terms of transportation, placing, compacting and finishing. The placement of concrete in formwork depends on the setting time for it to harden [2, 3]. With the advent of new types of concrete such as high strength concrete (HSC), self-compacting concrete (SCC) and high performance concrete (HPC), application of Pozzolans is imperative so as improve their performance, achieve better engineering properties and most especially reduce cost and lower the carbon foot print. Mineral admixtures such as rice husk ash (RHA), calcium carbide waste (CCW), metakaolin (MK), fly-ash (FA) and others are applied to replace cement partially for more efficacy of blended concrete.

Since these mineral admixtures differ in chemical, mineralogical composition and particle characteristics, their effects on concrete properties including setting time characteristics is also likely to differ. It is not out of place to say that knowledge of the setting characteristics of an admixture is imperative as this will enhance the scheduling of stages in concrete construction operations vis-a-vis transporting, placing, compacting and finishing of concrete work. Such vital knowledge cannot be over emphasized in the choice for a retarder or an accelerator admixture.

The mixing of water with cement triggers-off hydration products formation. Through the rigid behavior of the matrix, its setting times is determined. The beginning of the hardening of mortar or concrete is termed the initial setting time, while its sufficient hardness is termed final setting time [3]. Previous work [4] on cementing material from calcium carbide residue-rice husk ash showed that the product of the mixture is pozzolanic. Also, similar work on activated CCW indicated that it has prolonged setting times compared with control values [5].

When calcium carbide (CaC_2) reacts with water, CCW and acetylene gas is obtained (equation 1).

IOP Conf. Series: Materials Science and Engineering **805** (2020) 012031 doi:10.1088/1757-899X/805/1/012031

$$CaC_2 + 2H_2O \longrightarrow C_2H_2 + Ca(OH)_2 \quad (1)$$

It is a by-product from acetylene gas (C_2H_2) production [6]. Acetylene gas find application in ripening of fruits, a veritable choice for heating processes such as flame heating, flame gouging, welding, flame hardening, thermal spraying and other heating applications [7]. However, in Nigeria, acetylene gas finds application in oxy-acetylene gas welding. The carbide residue (CCW) is disposed as waste which is becoming a health hazard [8]. Also, CCW contains $Ca(OH)_2$ and is alkaline with pH greater than twelve [9]. To mitigate this negative effect on the environment, attempts were made to use CCW positively in both building and construction industry. When CCW is mixed with rice husk ash (RHA), pozzolanic reaction occurs and mortar produced from this mix has a compressive strength of 15 MPa at 28 days [10].

It has been shown that cement paste and concrete containing CCW at 1 % addition of CCW had initial and final setting times of 78 % and 57 % of control values, but increased in consistency by 14 % [11]. Also 5 % addition of CCW to PC mix, increased consistency but decreased the setting times [12]. From these results, CCW can influence properties of PC blended cement such as consistency and setting times. Consistency can be said to be the extent of wetness or dryness; an indication that a concrete sample is workable or not through the entire process of concreting operations (transportation, placement, finishing) without segregation. RHA mortar paste has higher consistency compared with control paste values. The consistency increases with increase in replacement level of PC with RHA [13-15]. This is not unconnected with the natural structure of RHA with large surface area [16]. The adsorption capacity, high fineness and specific surface area require more water [17, 18] which could be as high as 100 % [19]. The use of RHA of high fineness also leads to increased water demand as opined in earlier studies [20-22. From the effect of RHA and RHA-CCW in concrete, the setting time of RHA mortar decreased compared with that for RHA-CCW mortar [23].

Rice husk ash (RHA) is a pozzolanic material which is produced by incineration of (RH) under controlled temperature (550- 800 ^{o}C), time and rate of burning. When it is incinerated under these conditions, highly reactive amorphous silica (SiO_2) is obtained. At incineration temperature above 700 ^{o}C, the amorphous SiO_2 changes to a crystalline SiO_2 (Cristoballite, Quartz and Tridymite) and below 500 ^{o}C, the product has high LOI which affects rate of reaction. In the aforementioned cases, the reactivity of RHA is greatly decreased. It has been shown to have low specific gravity [24], hence finds application in the making of light weight building materials. It has been reported that when PC is replaced at 10 % level, optimum workability and strength is obtained [25]. Its usage improves concrete properties; mitigate global warming and waste disposal problems [26, 27]. In a study [28], the physico-chemical and hardened properties of blended cement binders containing RHA and (MK), there was an increase in water demand and setting times. MK is a highly reactive alumino silicate and alumina. These oxides combine with slake lime ($Ca(OH)_2$ at ambient temperature with moisture to form compounds that were virtually identical to the compounds in hydrated Portland cement. MK is widely used for its highly pozzolanic properties and considered to have twice the reactivity of most other pozzolans and considered a very viable admixture. It is not like other SCMs like fly ash, silica fume which are by-products of an industrial process. When kaolin clay is calcined within a temperature range of 650-800 °C, metakaolin is obtained; a relatively new material in the concrete industry, effective in strength development and mitigates sulphate attack. The pozzolanic reactions changed the microstructure of concrete and chemistry of hydration products by consuming the released CH to form more calcium silicate hydrate (C-S-H), thus enhancing properties of mortar and concrete [29]. When PC is replaced with MK, up to 20 % increase in strength and good results in other variables were obtained when compared with control values. The use of MK as a supplement in concrete up to 20 % can reduce cost of cement and also mitigates negative environmental pollution due to cement production.

Results so far analyzed indicated prolonged consistency and setting times. Since pozzolans enhance both fresh and hardened properties and ensure the production of concrete at minimum cost, there has been increasing use of pozzolans both in the building and construction industry. Although CCW as a pozzolan improves strength of concrete, it prolongs setting due to its low pozzolanicity. However, (MK) though improves early strength development, lowers workability due to high heat of hydration which is detrimental to hardened properties of mortar and concrete. It is also weak in magnesium sulphate environment as well as at high temperatures. But when RHA replaced PC, it accelerates the early hydration of Tri-calcium silicate (C_3S) and increase in the early hydration rate of C_3S which is attributed to its high fineness [30]; but due to its porous nature, absorbs more water. Thus, extent of scope of application of binary blends of CCW, MK, RHA and the ternary blends of CCW-MK, and MK-RHA in building and construction industry may be limited. However, the multiple applications of these admixtures in the form of blended binders have the potential to compensate for the differences due to their synergistic interactions.

IOP Conf. Series: Materials Science and Engineering **805** (2020) 012031 doi:10.1088/1757-899X/805/1/012031

Hence, this study was set out to evaluate the fresh properties of blended binders containing RHA, CCW and MK with a view to determine its suitability as a retarder or accelerator in concrete production.

2 Materials and experimental investigation

2.1 Materials

PC, CCW, RHA and MK were used in the production of quaternary paste binders. Tests were carried out on the physical and chemical properties of PC, CCW, RHA and MK to determine their suitability (Table 1). The mix proportions of the binder were divided into six groups: Group 1a (100 % PC), Group 1b BM (PC + MK), Group 2 BM (PC + RHA), Group 3 BM (PC + CCW), Group 4 TM (PC + CCW+ MK), Group 5 QM (PC + CCW + RHA + MK) as indicated in Table 2. A local dealer of Dangote Cement Company, Nigeria provided the PC (CEM 1-42.5 N). RH was incinerated in a local furnace for 3 h at 600 ^{o}C, and cooled to room temperature in a desiccator and then ground using a local mill and sieved through a 75 μm sieve to obtain RHA. Kaolin was obtained from Madagali, Yobe state, Nigeria and prepared and calcined at 700 ^{o}C for 2 h at a rate of 10 ^{o}C/min at the Central Services Laboratory, National Cereals Research Institute, Badeggi, Niger State, Nigeria to obtain MK that was used for the study. It was then cooled to room temperature in a desiccator, ground with a local mill and sieved through a 75 μm sieve. CCW was obtained from a local panel beater at Katerin-Gwari, Minna, Nigeria. The CCW was air dried, sieved through a 75 μm sieve. Preliminary tests were carried out on the materials.

2.2 Preparation of samples

Six groups of paste samples were prepared using the admixtures RHA, CCW and MK which replaced PC at 5, 10 and 20 %. The constituents were thoroughly mixed on a glass sheet. The mixtures include 1 control sample of plain PC, 3 binary, 1 ternary and 1 quaternary mixture. PC was partially replaced by RHA, CCW, and MK at replacement levels of 0, 5, 10 and 20 %. Maximum replacement of PC by supplementary materials was 20 % by mass for all mixtures (Table 2).

IOP Conf. Series: Materials Science and Engineering **805** (2020) 012031 doi:10.1088/1757-899X/805/1/012031

2.3 Methods

For the determination of consistency and setting times of PC, PC-RHA, PC –CCW and PC-MK pastes, six groups of samples were tested. The normal consistency and setting time tests were conducted in accordance with BS EN 196-3 provisions [31]. The normal consistency, setting times (initial and final) of PC pastes were determined using Vicat apparatus in accordance with ASTM C 191 [32]. The water of consistency and setting times of the prepared pastes were directly determined from the relation in equation (2)

$$WC\ (\%) = L/W \times 100 \qquad (2)$$

Where WC is consistency water, L is water required to produce a suitable paste and W is the mass of cement sample (300g).

The initial setting time is the time taken to reach the initial set, while the final setting time is the time taken to attain final set of paste. Using the Vicat apparatus, a needle of 1 mm square penetrated the paste at every 10 min interval till the index scale showed 5 + 0.5 mm from bottom of the mould. For the final setting time, the needle was replaced with an annular attachment. The needle was released every 30 min interval until it makes no visible indentation on the test specimen surface. For each mixed sample, three tests were carried out and the mean result was recorded.

3 Results and discussion

3.1 Physical and chemical properties of constituent materials

3.1.1 Physical Properties

Table 1 showed the physical and chemical properties of PC, CCW and MK respectively. The average specific gravity of RHA, CCW and MK showed that CCW with a value of 2.19 is very light and fine compared with other constituents like PC with a value of 3.15. RHA and MK are both lighter than PC with a value of 3.15. The specific surface area of CCW, RHA and MK are higher compared with PC value which means that they are much finer than PC particles and thus occupy more space than PC. Generally, the specific gravity of mineral admixture is less than that of PC therefore more volume is obtained when it replaces PC.

IOP Conf. Series: Materials Science and Engineering **805** (2020) 012031 doi:10.1088/1757-899X/805/1/012031

Table 1. Physical and Chemical Properties of Materials

Physical Properties	**Materials**			
	PC (wt %)	**RHA (wt %)**	**CCW (wt %)**	**MK (wt %)**
Specific gravity (g/cm^3)	3.16	2.25	2.19	2.52
Blaine Fineness (cm^3/g)	3045	-	-	7483
Loss on ignition (LOI)	2.98	15.76	31.70	1.22
Chemical composition				
Calcium Oxide, CaO %	64.19	1.30	93.69	0.39
Silicon dioxide, SiO_2 %	19.57	73.60	6.49	54.30
Aluminum Oxide, Al_2O_3 %	5.47	0.68	2.00	38.30
Ferric Oxide, Fe_2O_3 %	3.24	0.93	1.87	4.28
Magnesium Oxide, MgO %	2.01	0.35	0.70	0.08
Manganese Oxide, MnO %	1.25	-	0.47	-
Sodium Oxide, Na2O %	0.26	0.12	0.10	0.12
Potassium Oxide, K_2o %	0.43	2.37	0.18	0.50
SO_3	2.74	-	0.36	0.22

3.1.2 Chemical Properties

From Table 1, the sum of SiO_2 + Al_2O_3 +Fe_2O_3 (SAF) exceeds 70 % and thus met provisions for pozzolanic materials [33, 34]. Amount of SiO_2 and Al_2O_3 in RHA and MK depicts their reactive capability with the primary hydrate of cement. It is interesting to note that CCW with less SiO_2 and Al_2O_3 has the highest CaO content of 95.69 % ensuring that the Ca/Si ratio is maintained (kept at not less than pH 12) for reaction to continue.

3.2 Standard consistency and setting times

3.2.1 Standard Consistency of PC-CCW-RHA-MK

The result of standard consistency tests for the quaternary binder, PC-CCW-RHA-MK at different percentage replacement levels is shown in Figure 1. In a study on ternary binder with OPC-FA-SF at 20FA 10SF and 30FA 10 SF as replacement of OPC, it was shown that requirement of water increased with increase in percentage replacement level due to its fineness [35]. From this study, it was found that as CCW, RHA and MK replaced PC to make quaternary binder; the normal consistency became higher as replacement level increased. This can be attributed to the high fineness of RHA and MK compared with PC value. Their specific surface value is much higher compared with PC value (Table1). At 5 % replacement level of PC with CCW-RHA-MK binder, its water demand increased by 9 %, and 21 %, 39 % at 10 % and 20 % replacement levels, when compared to PC value (Figure 1). For the binary binders (PC-CCW,

IOP Conf. Series: Materials Science and Engineering **805** (2020) 012031 doi:10.1088/1757-899X/805/1/012031

PC-RHA, PC-MK), consistency is same with that of PC (control) value (Table 3) at 5 % and 10 % replacement levels but increased by about 5 % at 20 % replacement level except for PC-CCW binder whose consistency is same with that of PC at all replacement levels. This showed that CCW does not demand much water due to its low pozzolanicity compared with RHA and MK values. The increase in consistency for the quaternary binder is fruitful as it facilitates more reactions between constituents and the primary hydrate of PC.

Table 2. Mix proportion of the materials

S/NO	Mix ID	Materials (gm) PC	RHA	CCW	MK
Group 1a	100 % PC	300	-	-	-
Group 1b, BM	95PC- 5MK	285	-	-	15
	90PC- 10MK	270	-	-	30
	80PC- 20MK	240	-	-	60
Group 2, BM	95PC- 5RHA	285	15	-	-
	90PC- 10RHA	270	30	-	-
	80PC- 20RHA	240	60	-	-
Group 3, BM	95PC- 5CCW	285	-	15	-
	90PC- 10CCW	270	-	30	-
	80PC- 20CCW	240	-	60	-
Group 4, TM	95PC- 5CCW-5MK	285	-	15	15
	90PC- 10CCW-10MK	270	-	15	15
	80PC- 20CCW-20MK	240	-	30	30
Group 5, QM	95PC-5CCW-5RHA-5MK	255	15	15	15
	90PC-10CCW-10RHA-10MK	210	30	30	30
	80PC-20CCW-20RHA-20MK	120	60	60	60

BM, binary mix; TM, ternary mix; QM, quaternary mix; PC, Portland cement

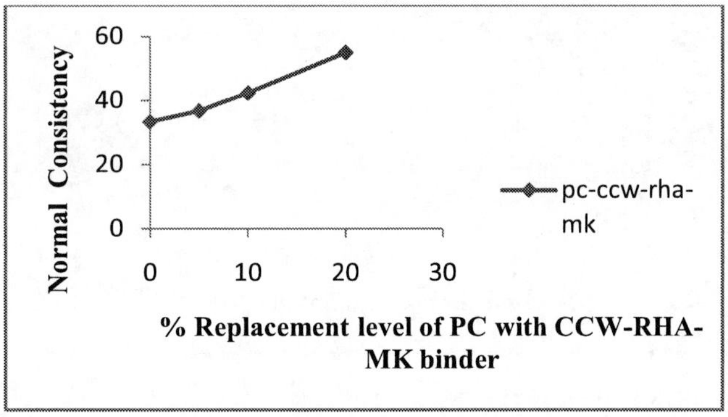

Figure 1. Consistency of PC-CCW-RHA-MK

3.2.2 The initial and final setting times of PC-CCW-RHA-MK binders

For quaternary binders containing PC, CCW, RHA and MK, the setting times, initial (IS) and final (FS) are shown in Figures 2 and 3. It can be observed that the effect of the admixtures prolonged the setting times of the quaternary binders when compared with control value. The observed retardation in setting times is mainly due to the effect of lower cement content. It can also be observed that the setting times for binary and ternary binders increased as percentage replacement level of PC with the admixtures increased. This shows that the admixtures used in this study retards the setting times of binary, ternary and quaternary binders [36, 37]. If quaternary binder (PC-CCW-RHA-MK) is compared with ternary binder (PC-CCW-MK), (Table 3 and Figure 3) at 5 % replacement level, IS decreased by 27 %, at 10 % replacement level, increased by 8 % and then decreased again at 20 % replacement level by 24 %.

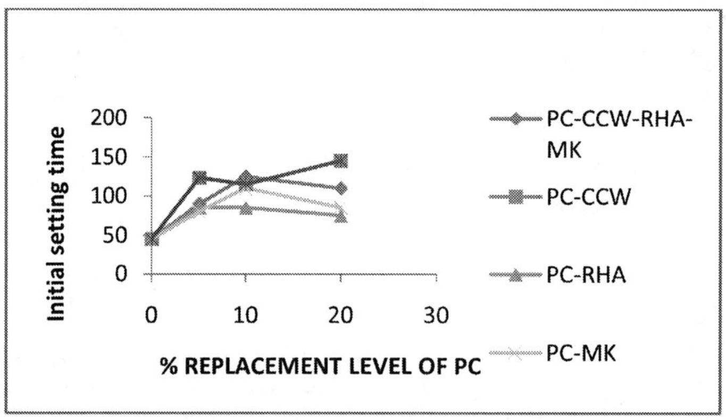

Figure 2. Initial setting times of binders

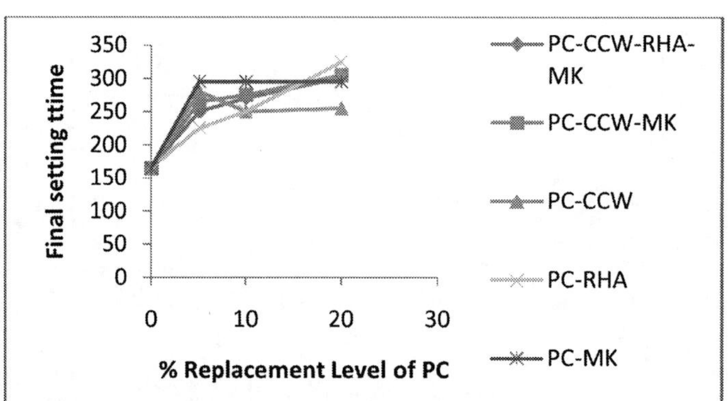

Figure 3. Final setting times of binders

IOP Conf. Series: Materials Science and Engineering **805** (2020) 012031 doi:10.1088/1757-899X/805/1/012031

Table 3. Consistency, initial and final setting time results

S/NO	Mix ID	PC (g)	Materials RHA	CCW	MK	Std Consistency	IS (min)	FS (min)
Group 1a	100 % PC	300	-	-	-	33.33	45	165
Group 1b, BM	95PC- 5MK	285	-	-	15	33.00	80	295
	90PC- 10MK	270	-	-	30	34.00	110	295
	80PC- 20MK	240	-	-	60	34.33	85	295
Group 2, BM	95PC- 5RHA	285	15	-	-	32.67	85	225
	90PC- 10RHA	270	30	-	-	33.67	85	250
	80PC- 20RHA	240	60	-	-	35.00	75	325
Group 3, BM	95PC- 5CCW	285	-	15	-	31.67	105	280
	90PC- 10CCW	270	-	30	-	32.00	60	250
	80PC- 20CCW	240	-	60	-	32.33	70	255
Group 4, TM	95PC- 5CCW-5MK	285	-	15	15	33.00	123	265
	90PC- 10CCW-10MK	270	-	15	15	34.67	115	275
	80PC- 20CCW-20MK	240	-	30	30	39.00	145	305
Group 5, QM	95PC-5CCW-5RHA-5MK	255	15	15	15	36.67	90	250
	90PC-10CCW-10RHA-10MK	210	30	30	30	42.33	125	270
	80PC-20CCW-20RHA-20MK	120	60	60	60	55.00	110	301

This effect of RHA on IS depict disorder; but on FS, at 5 % level of replacement, it decreased by 6 %; at 10 % level by 2 % and by 1 % at 20 % replacement level which showed that RHA progressively decreased FS of binders studied. This can be attributed to pozzolanic reaction of RHA at later stage. Similarly, when PC-CCW-MK is compared with PC-CCW, at 5 % level of replacement, decreased FS by 5 %, at 10 % level of replacement decreased by 1 % and then increased by 26 %. The effect of MK decreased FS by 5 % and 1 % at 5 and 10 % replacement levels but increased it by 26 % at 20 % replacement level. Furthermore, comparison between PC-CCW-MK with PC-MK revealed that CCW prolonged IS by 35 %, 4 % and 41 % but decreased FS by 10 %, 7 % and 3 % at 5, 10 and 20 % replacement levels respectively. These results

IOP Conf. Series: Materials Science and Engineering **805** (2020) 012031 doi:10.1088/1757-899X/805/1/012031

showed that as % replacement level increased, IS is increased and then reduced FS which implies that reaction becomes faster at later stage of hydration.

4 Conclusion

From the analysis and discussion, results showed that the effect of the admixtures on standard consistency and setting times of the binders differs. The following conclusions are hereby drawn from the study:

i. At 5 % replacement of PC with RHA, CCW, MK, influence on consistency is limited (< 9 %) but becomes significant (21 %) at 10 % and (39%) at 20 % replacement levels which are still within standard limits.

ii. Standard consistency in quaternary binder (PC-CCW-RHA-MK) increased with increase in replacement level of PC with RHA due to high demand for water and its high specific surface area. This suggests that if percentage level of RHA is checked in binder, standard consistency requirements and water issues can be addressed.

iii. Replacement of PC by CCW, RHA, MK at 5 % level has negligible influence on consistency but at 20 % replacement level increased consistency which is at limits and similar to that of PC. This suggests that absorption and adsorption properties of CCW and MK are similar to that of PC.

iv. When ternary (Group 4) binder, PC-CCW-MK is compared with quaternary, PC-CCW-RHA-MK, with the addition of RHA, setting properties increased, suggesting that increase of pozzolanic materials increase hydration process leading to increase in time period. Delayed setting times is advantageous in that concrete can be worked for longer periods resulting in time delays in transport and between mixing and using concrete becomes less critical. This type of binder finds application in reduction of cold joints in larger concrete pours.

v. Setting time of quaternary binder increases at lower replacement level (5 %) of RHA but decreases at higher replacement level (20%) of RHA in the binder. This is due to the high specific surface area of RHA and MK which facilitates pozzolanic reaction with $Ca(OH)_2$ to speed up the hydration process.

vi. Retardation of setting time is the overall effect of CCW, RHA and MK on standard consistency and setting times (initial and final) of the quaternary binder. Since there is less tri-calcium-aluminate (C_3A) due to lower cement content, the influence of increasing levels of RHA is to provide greater retardation in the setting time. This type of binder (PC-CCW-RHA-MK) can be used in the manufacture of special types of concrete such as reinforced marine concrete (RMC) and also in construction works in marine areas with RMC where longer period is required for its placing.

References

[1] Brooks JJ, Johari MM and Mazloom M 2000 Effect of admixtures on the setting times of high-strength concrete. *Cem. Concr. Compos.* **22** (4) pp 293-301.

[2] Gardner NJ, 1986. Concrete pressure on formwork. Proc. Inst. Civil Eng. **80** (1) pp 145-159.

[3] Naik TR, Singh SS and Ramme BW 2001 Time of setting influenced by inclusion of fly ash and chemical admixtures. *Am. Concr. Inst.* (Special Publication) **199** pp 393-414.

[4] Jaturapitakkul C and Roongreung B 2003 Cementing material from calcium carbide residue-rice husk ash. *J. Mater. Civil Eng.* **15** (5) pp 470-475.

[5] Sun H, Li Z, Bai J, Memon SA, Dong B, Fang Y, Xu W and Xing F 2015 Properties of chemically combusted calcium carbide residue and its influence on cement properties. *Mater.* **8** (2) pp 638-651.

[6] Makaratat N, Jaturapitakkul C, Namarak C and Sata V 2011 Effects of binder and $CaCl_2$ contents on the strength of calcium carbide residue-fly ash concrete. *Cem. Concr.Compos.* **33** (3) pp 436-443.

[7] Liu X, Zhu B, Zhou W, Hu S, Chen D and Griffy-Brown C 2011 CO_2 emissions in calcium carbide industry: An analysis of China's mitigation potential. *Int. J Greenhouse Gas Control* **5** (5) pp 1240-1249.

[8] Abiya SE, Odiyi BO, Salau IO 2015 Effect of calcium carbide waste on the growth and biomass of Okra (*Abelmoschus esculentus* l. Moench). *IOSR J. Environ. Sci. Toxicol. Food Technol.* **9** (1) pp 68-71.

[9] Walker R and Pavia S 2010 Behaviour and properties of Lime-Pozzolan Pastes. In: 8th International Mansonry Conference (eds: Dresden. W. Jäger, B. Haseltine, and A. Fried) Dresden University of Technology, Germany, pp 353-364.

[10] Zhang MH, Lastra R and Malhotra VM 1996 Rice-husk ash paste and concrete: some aspects of hydration and the microstructure of the interfacial zone between the aggregate and paste. *Cem. Concr. Res.* **26** (6) pp 963-977.

[11] Ogork EN and Ayuba S 2014 Influence of sawdust ash (SDA) as admixture in cement paste and concrete. *Int. J. Innov. Sci. Eng. Technol.* **1** (10) pp 736-743.

[12] Abdurra'uf MG, Ogork EN, and Haruna SI 2017 Effect of calcium carbide wastes as admixture in mortar. *Sch. J. Eng. Technol.* **5** (11) pp 655-660.

[13] Calica Jr MG 2008 Influence of rice husk ash as supplementary material in cement paste and concrete. *NLR J.* **2** pp 80-92.

[14] Kartini K, Mahmud HB and Hamidah MS 2010 Absorption and permeability performance of Selangor rice husk ash blended grade 30 concrete. *J. Eng. Sci. Technol.* **5** (1) pp 1-6.

[15] Marthong C 2012 Effect of rice husk ash (RHA) as partial replacement of cement on concrete properties. *Int. J. Eng. Res. Technol.* **1** (6) pp 1-9.

[16] Le HT 2015 Behaviour of rice husk ash in self- compacting high performance concrete. PhD Thesis, Institute of Building Materials Science, Bauthaus University, Weimar, Germany.

[17] Vinothan MK and Baskar G 2015 Study of structural behaviour on pozzolanic material (rice husk). *Int. J. Civ. Eng. Technol.* **6** (9) pp 31-46.

[18] Ganesan K, Rajagopal K and Thangavel K 2008 Rice husk ash blended cement: assessment of optimal level of replacement for strength and permeability properties of concrete. *Construc. Build. Mater.* **22** (8) pp 1675-1683.

[19] Rashid MH 2016 Strength behavior of cement mortars assimilating rice husk ash. *Int. J. Adv. Agric. Environ. Eng.* **3** (3) pp 288-293.

[20] Chindaprasirt P, Jaturapitakkul C and Rattanasak U 2009 Influence of fineness of rice husk ash and additives on the properties of lightweight aggregate. *Fuel* **88** (1) pp158-162.

[21] Van Tuan N, Ye G, Van Breugel K, Fraaij AL and Dai Bui D 2011 The study of using rice husk ash to produce ultra high performance concrete. *Constr. Build. Mater.* **25** (4) pp 2030-2035.

[22] Van Tuan N, Ye G, Van Breugel K and Copuroglu O 2011 Hydration and microstructure of ultra high performance concrete incorporating rice husk ash. *Cem. Concr. Res.* **41** (11) pp 1104-1111.

[23] Zahemen TL, Alao J and John W 2019 Performance of rice husk ash-calcium carbide waste in concrete. *Adv. Mater. Res.* **1155** pp 41-53.

[24] Givi AN, Rashid SA, Aziz FN and Salleh MA 2010 Contribution of rice husk ash to the properties of mortar and concrete: a review. *J. Am. Sci.* **6** (3) pp 157-165.

[25] Krishna NK, Sandeep S and Mini KM 2016 Study on concrete with partial replacement of cement by rice husk ash. *IOP Conf. Ser.: Materi. Sci. Eng.* 149 (1) 012109. https://doi.org/10.1088/1757-899X/149/1/012109.

[26] Kumar PC, Palli MR and Patnaikuni I 2011 Replacement of cement with rice husk ash in concrete. *Adv. Mater. Res.* **295** pp 481-486.

[27] Brown DK 2012 *Unprocessed rice husk ash as a partial replacement of cement for low-cost concrete* (Doctoral dissertation, Massachusetts Institute of Technology, USA).

[28] Abo-El-Enein SA, Hammad HM, EL-Sokary TM, Mekky SD and Mustafa ME 2014 Physico-chemical and mechanical properties of blended cement pastes containing rice husk ash and metakaolin. *Al-Azhar Bull.Sci.* 2014 **25** (1-A) pp 7-14.

[29] Dinakar P, Sahoo PK and Sriram G 2013 Effect of metakaolin content on the properties of high strength concrete. *Int. J. Concr. Struct. Mater.* **7** (3) pp 215-223.

[30] Della VP, Kühn I and Hotza D 2002 Rice husk ash as an alternate source for active silica production. *Mater.* Lett. **57** (4) pp 818-21.

[31] BS EN 196 1995 Part 3: Methods of testing cement; determination of setting time and Soundness. British standard institution. London, 1995.

[32] ASTM C 191 2008 Standard test method for normal consistency and setting of hydraulic cement. *Annual book of ASTM standards* pp 172-174.

[33] ASTM C 618-00 2000 Class F: Specification for coal fly ash and raw of calcined natural pozzolans for use as a mineral admixture in Portland cement concrete. Annual Book of ASTM Standards, Philadelphia, USA.

[34] ASTM C618 2000: Requirements for Pozzolanic additives. ASTM International West Conshocken Philadelphia, USA.

[35] Nochaiya T, Wongkeo W and Chaipanich A. Utilization of fly ash with silica fume and properties of Portland cement–fly ash–silica fume concrete. *Fuel* **89** (3) pp 768-774.

[36] Bendapudi SC and Saha P 2011 Contribution of fly ash to the properties of mortar and concrete. *Int. J. Earth Sci. Eng.* **4** (6) pp 1017-1023.

IOP Conf. Series: Materials Science and Engineering **805** (2020) 012031 doi:10.1088/1757-899X/805/1/012031

[37] Rao GA 2003 Investigations on the performance of silica fume-incorporated cement pastes and mortars. *Cem. Concr. Res.* **33** (11) pp 1765-1770.

IOP Conf. Series: Materials Science and Engineering **805** (2020) 012032 doi:10.1088/1757-899X/805/1/012032

Green synthesis of silver nanoparticles from leaf extract of *Tetrapleura tetraptera* and its antimicrobial activity

B O Ogunsile*[1], D O Seyinde[2] and B A Salako[1]
[1]Department of Chemistry, University of Ibadan, Ibadan, Nigeria
[2]Department of Chemical Sciences, Anchor University, Lagos, Nigeria
*E-mail: ogunsile@yahoo.com.

Abstract
Plants have very high potential as alternative source of bio-reducing and capping agents in nanoparticles (NPs) synthesis. Silver nanoparticles (AgNPs) were synthesized from *Tetrapleura tetraptera* leaf extract. The UV-vis spectrophotometer was used to observe the formation of the NPs at various time intervals and ratios of leaf extract to $AgNO_3$ solution. The properties of the AgNPs synthesized were characterized by FTIR and Scanning Electron Microscope (SEM) equipped with Energy Dispersive X-ray Spectroscopy (EDX). Antimicrobial activities of the AgNPs synthesized were evaluated against ten human pathogens using agar well diffusion method. The rate of formation of the AgNPs was shown to increase with respect to time and ratio of leaf extract to $AgNO_3$ solution. The EDX result showed signal energy peak for silver atom at 3.0-3.8 KeV. The AgNPs was characterized with FTIR peaks around 3280, 1620, 1400 and 1040 cm^{-1} assigned to O-H, C=O, C–O–H and C-C groups respectively, as possible stabilizing and capping groups. The AgNPs showed maximum inhibitory activity against *Staphylococcus aureus, Escherichia coli* and *Salmonella* sp.

1 Introduction
The synthesis of nanoparticles using environmentally friendly biosynthetic methods is fast emerging [1]. Biosynthetic methods is currently being used in the production of large quantities of nanoparticles that are free from contamination with well defined size and morphology while reducing the chance of failure with greater ease of characterization [2]. Green synthesis of nanoparticles is an alternative approach to the more complex chemosynthetic procedures. Most significantly is its advantage that conditions of high temperature, pressure, energy and toxic chemicals are not required in the biosynthesis [3]. Biologically synthesized methods employ

IOP Conf. Series: Materials Science and Engineering **805** (2020) 012032 doi:10.1088/1757-899X/805/1/012032

microorganisms, marine algae, live plants, plant tissues, fruits and plant extracts for the production of various nanoparticles [4-6].

Although nanoparticles can be synthesized using different biogenic routes, their production using simple and less expensive plant extract to a rather complex and expensive methods based on plant tissue and microbial processes is an increasing focus of attention [3,7-16]. The synthesis of nanoparticles using plant extract is advantageous in that it is easily scalable and can be used as both a bio-reducing and capping agents. Plants are known to contain different concentrations and compositions of bioreducing agents [17]. The quality and quantity of the nanoparticles produced are affected by the source and nature of the plant extract and other operational variables like pH, reaction time, and volume ratio of plant extract to the metallic precursor [18].

Silver nanoparticles are of great interests to nanotechnologist because of their unique physiochemical properties when compared to their nano-scaled counterparts [19], especially in its role in nanomedicine as therapeutic and diagnostic agents [20-23]. Silver nanoparticles have been produced from various plant extracts such as tea extract [24], beet juice [25], *Mussaenda glabrata* leaf extract [26], *Thymbra spicata* extract [27], *Pedalium murex* leaf extract [22], aqueous extract from geranium leaf [28], blackberry, blueberry, pomegranate, and turmeric extracts [29].

Tetrapleura tetraptera (known locally as Aridan) is a species of flowering plant in the pea family normally found in Western Africa. It contains high amount of essential phytochemicals and nutrients needed for proper body functioning [30]. Due to its high vitamin content, it is used as a dietary supplement, a spice and as medicine. Studies have been done to discover its molluscicidal [31], anticonvulsant [32], hypotensive, anti-inflammatory [33] and antipathogenic properties [34] but its silver nanoparticles potential had not be reported. This paper is aimed at synthesizing silver nanoparticles from *T. tetraptera* leaf extract and evaluation of its antimicrobial activity against ten selected human pathogens.

2 Materials and methods

2.1 Selection and collection of plant material

The leaves of *T. tetrapleura* were collected from the botanical garden at the University of Ibadan campus, Ibadan, Nigeria. The leaves were sliced and washed thoroughly with de-ionized water.

2.2 Biosynthesis of silver nanoparticles

About 60 g of the sliced leaves was boiled in 600 ml of de-ionized water on the hot plate to obtain the leaf extract. The extract obtained was filtered into a 250 ml Erlenmeyer flask and kept in the refrigerator. Equal volume (20 ml) of 1 mM $AgNO_3$ solution was added to the extract in a conical flask and the mixture was incubated on a hot plate at 70 °C under agitation for 10-45 min. The change in colour as a function of the bioreduction of silver ion to metallic silver nanoparticles was monitored.

2.3 Effect of time

The effect of time on the bioreduction of silver ion by the leaf extract was investigated. Exactly 30 ml of *T. tetraptera* leaf extract was added to 90 ml of 1 mM $AgNO_3$. The absorbance of the reaction mixture was read on UV-visible spectrophotometer at 15 min intervals for 45 min.

2.4. Effect of volume ratio

The effect of varying the volume of 1 mM $AgNO_3$ solution used to a fixed volume of the leaf extract in the bioreduction process was monitored using UV-visible spectrophotometer. Approximately 5 ml of 1 mM $AgNO_3$ solution was added to 5 ml of leaf extract and incubated on a hot plate at 70 °C for 10 min, after which a color change was observed. Ratios 1:2 and 1:3 of $AgNO_3$ solution to leaf extract were also carried out.

2.5 Characterization of the synthesized silver nanoparticles

Four characterization techniques were used for the synthesized AgNPs. The formation of the NPs with respect to time and ratios of leaf extract to $AgNO_3$ solution was monitored using the UV-vis spectrophotometer (Perkin Elmer, Lamda 25, Germany). The bioreduction of silver ions was observed by measuring the UV-visible spectrum of the solutions at wavelength range of 300-900 nm. Organic functional groups present in the leaf extract and AgNPs were detected using FTIR (Perkin Elmer LS-55-Luminescence spectrometer). The elemental composition of metal nanoparticles was established using energy dispersive x-ray spectroscopy (EDX), while scanning electron microscope (SEM) was employed for morphological characterization.

2.6. Antimicrobial activity by well diffusion method

The synthesized AgNPs from *T. tetrapleura* was screened for its antimicrobial activity by well diffusion method against ten (10) pathogenic organisms (six bacteria and four fungi). The pure cultures of organism were sub-cultured on sterile nutrient agar at 45 °C. Each strain was swabbed uniformly on the individual plates using sterile cotton swabs. Then, 8 mm well size was made on sterile Petri dishes using sterile cork borer. Using micropipette, the nanoparticles solution was released into the wells on all plates. The negative and positive control used for bacteria and for fungi are solvent, gentamycin and ticonazole. After incubation at 37 °C for 24 h, the levels of zones of inhibition were measured.

3 Results and discussion

The aqueous silver ions were reduced to zero valent silver nanoparticles when added to natural leaf extract of *T. tetrapleura* as observed by the color change from faint yellow to reddish brown. The color change has been attributed to the surface plasmon resonance phenomenon (SPR) [35]. The synthesized nanoparticles showed a spectrum range around 390 to 450 nm. It is reported in literature that AgNPs show spectrophotometric absorption measurements in the wavelength ranges of 400-450 nm [36]. The aggregation of the silver nanoparticles in the solution is indicated by the presence of broad resonance (Figure 1).

The bioreduction of the silver ions occurs fairly rapidly with more than 90% of reduction of Ag^+ ions completed less than 1 h after its addition to the leaf extract. Higher SPR peaks intensities were observed at the higher time intervals which indicate the increase in the production of smaller silver nanoparticles (Figure 2). After addition of the leaf extract to 1 mM $AgNO_3$ solution at different ratios, the curve shows increased absorbance at higher volume ratio. This apparently indicates the extracellular bioreduction of the Ag^+ ions [35].

FTIR spectroscopic studies of both the leaf extract and the AgNPs produced were carried out to investigate the possible bioreducing and capping agents present in the leaf extract. The spectra of the *T. tetraptera*-AgNPS showed FTIR peaks around 3280, 1620, 1400 and 1040 cm^{-1} assigned to O-H, C=O, C–O–H and C-C groups absorption bands as the possible stabilizing and capping groups (Figure 4). The result revealed that the silver nanoparticles were synthesized due to the bioreducing property of the leaf extract.

IOP Conf. Series: Materials Science and Engineering **805** (2020) 012032 doi:10.1088/1757-899X/805/1/012032

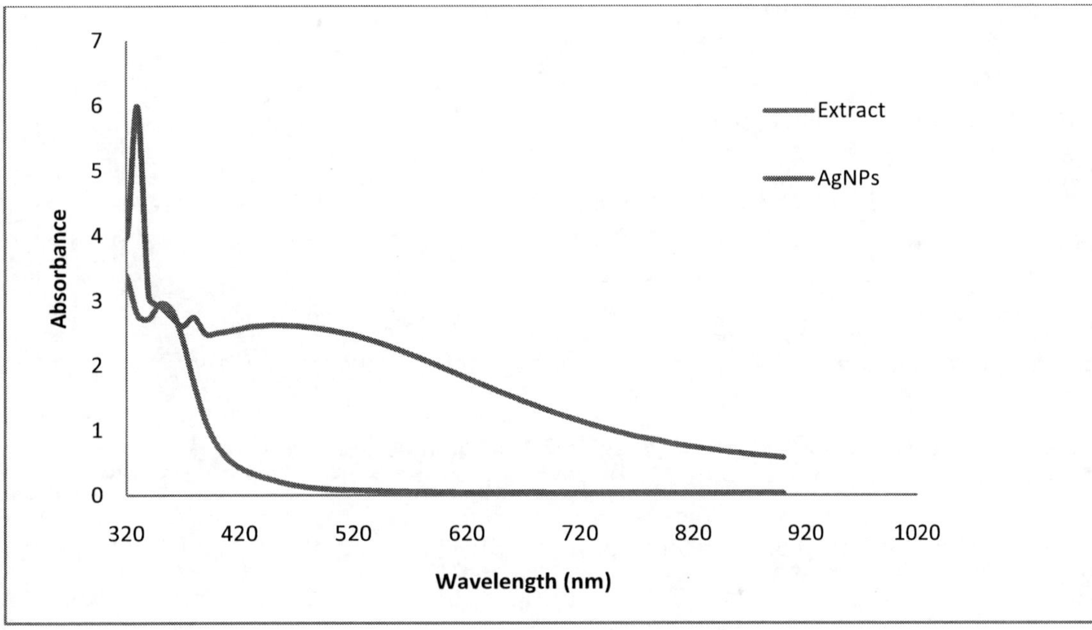

Figure 1. UV-vis spectra of *T. tetraptera* extract and its AgNPs

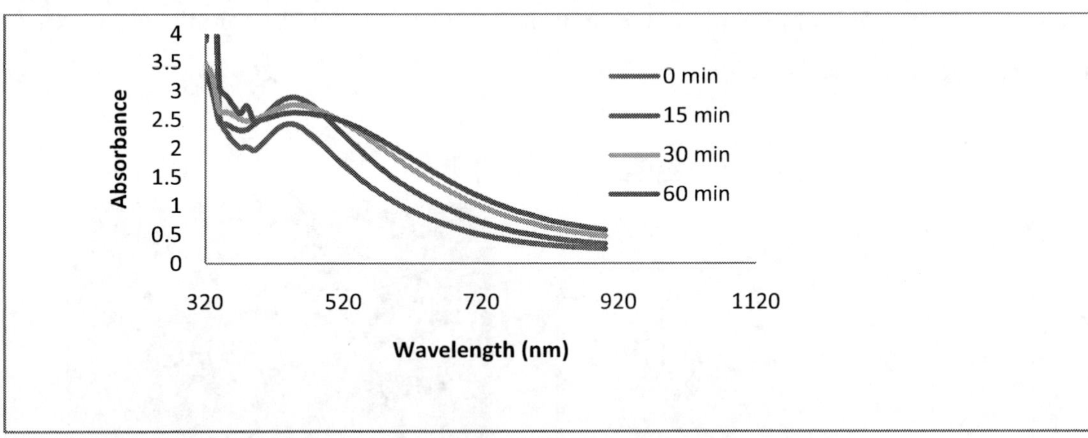

Figure 2.UV-vis spectra of *T. tetraptera* AgNPs at various time intervals

IOP Conf. Series: Materials Science and Engineering **805** (2020) 012032 doi:10.1088/1757-899X/805/1/012032

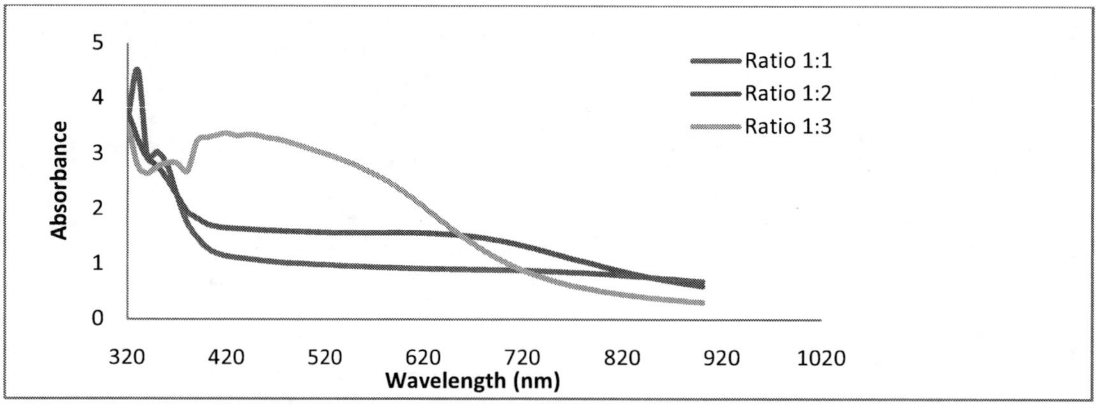

Figure 3. UV-vis spectra of *T. tetraptera* AgNPs at different volume ratio

The morphology of the AgNPs as revealed by SEM showed the formation of spherical, crystalline AgNPs (Figure 5). The EDX result (Figure 6) confirmed the elemental silver signal energy peak at 3.0-3.8 KeV. The presence of signals from other trace element like carbon and oxygen is likely as a result of the direct usage of the synthesized AgNPs from the native reaction solution prior to further purification.

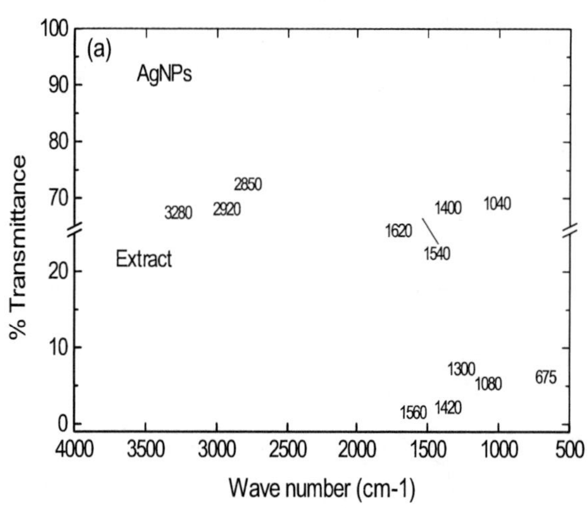

Figure 4. FTIR spectra of *T. tetraptera* plant and AgNPs obtained from *T. tetraptera* extract

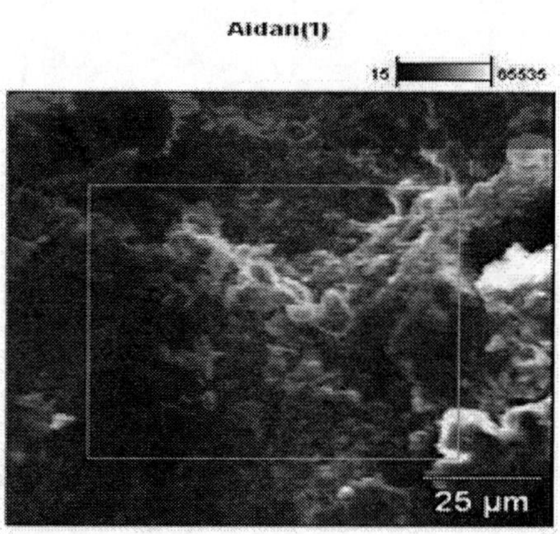

Figure 5. SEM images of the AgNPs synthesized by *T. tetraptera* extract

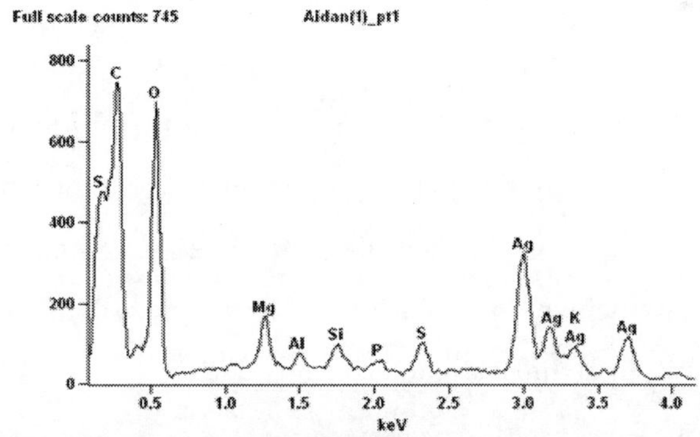

Compound %

	C	O	Mg	Al	Si	P	S	K	Ag
Aidan(1)_pt1	0.60	46.93	3.93	0.84	1.74	0.88	2.46	4.33	38.27

Figure 6. EDX signal of AgNPs synthesized by *T. tetraptera* extract

The AgNPs showed the greatest antimicrobial activity against *S. aureus* (14 mm), *E. coli* (14 mm) and *Salmonella* sp (14 mm) and no inhibitory effect against *B. subtilis, Pseudomonas* sp, *Rhizopus* sp and *P. aeruginosa* (Table 1). The synthesized AgNPs showed greater antimicrobial property towards bacteria pathogens compared to fungi pathogens under this study [36].

Table 1. Zone of inhibition of the synthesized silver nanoparticles from *T. tetraptera* plant extract against various microbes

	Zone of inhibition (mm)			
Organism/concentration	100mg/ml	50 mg/ml	25mg/ml	12.5mg/ml
S. aureus	14	12	10	-
E. coli	14	10	-	-
B. subtilis	-	-	-	-
Pseudomonas sp	-	-	-	-
Salmonella sp	14	12	10	
Klebsiella pneumoniae	12	10	-	
Candida albicans	12	10	-	-
Aspergillus niger	12	10	-	-
Rhizopus sp	-	-	-	-
P. aeruginosa	-	-	-	-

4 Conclusion

The study presents an easy, economical and efficient route for the production of AgNPs using *T. tetraptera* leaf extract. The synthesized AgNPs were characterized and confirmed using UV-vis spectroscopy, FTIR and SEM equipped with EDX. The silver ions were believed to be reduced to metallic AgNPs by the phytochemicals present in *T. tetraptera* leaf extract. The synthesized AgNPs showed efficient antifungal and antibacterial activities.

References

[1] Gour A, Jain N K 2019 Advances in green synthesis of nanoparticles. *Artif. Cells Nanomed. Biotechnol.* **47** (1) pp 844-851.

[2] Abdelghany T M, Al-Rajhi A M H, Al Abboud M A, Alawlaqi M M, Ganash Magdah A, Helmy E A M and Mabrouk A S 2018 Recent advances in green synthesis of silver nanoparticles and their applications: about future directions- A review. *BioNanoSci.* **8** (1) pp 5-16.

[3] Singh A, Jain D, Upadhyay M, Khandelwal N and Verma H 2010 Green synthesis of silver nanoparticles using *Argemone mexicana* leaf extract and evaluation of their antimicrobial activities. *Dig. J. Nanomater. Biostruct.* **5** (2) pp 483-489.

[4] Chandran S P, Chaudhary M, Pasricha R, Ahmad A and Sastry M 2006 Synthesis of gold nanotriangles and silver nanoparticles using *Aloe vera* plant extract. *Biotechnol. Prog.* **22** (2) pp 577-583.

[5] Luangpipat T, Beattie I R, Chisti Y and Haverkamp R G 2011 Gold nanoparticles produced in a microalga. *J. Nanopart. Res.* **13** (12) pp 6439-6445.

[6] Rajesh S, Raja D P, Rathi J M and Sahayaraj K 2012 Biosynthesis of silver nanoparticles using *Ulva fasciata* (Delile) ethyl acetate extract and its activity against *Xanthomonas campestris* pv. *malvacearum*. *J. Biopest.* **5** pp 119-128.

[7] Dhillon G S, Brar S K, Kaur S and Verma M 2012 Green approach for nanoparticle biosynthesis by fungi: current trends and applications. *Crit. Rev. Biotechnol.* **32** (1) pp 49-73.

[8] Li X, Xu H, Chen Z S and Chen G 2011 Biosynthesis of nanoparticles by microorganisms and their applications. *J. Nanomater.* Article ID 270974. https://doi.org/10.1155/2011/270974.

[9] Ali D M, Thajuddin N, Jeganathan K and Gunasekaran M 2011 Plant extract mediated synthesis of silver and gold nanoparticles and its antibacterial activity against clinically isolated pathogens. *Colloids Surf. B Biointerf.* **85** (2) pp 360-365.

[10] Bankar A, Joshi B, Kumar A R and Zinjarde S 2010 Banana peel extract mediated novel route for the synthesis of silver nanoparticles. *Colloids Surf. A: Physicochem.* Eng. *Aspects* **368** (1-3) pp 58-63.

[11] Baskaralingam V, Sargunar C G, Lin Y C and Chen J C 2012 Green synthesis of silver nanoparticles through *Calotropis gigantea* leaf extracts and evaluation of antibacterial activity against *Vibrio alginolyticus. Nanotechnol. Dev.* **2** (1) e3. https://doi.org/10.4081/nd.2012.e3.

[12] Castro L, Blázquez M L, Muñoz J A, González F, García-Balboa C and Ballester A Biosynthesis of gold nanowires using sugar beet pulp. *Process Biochem.* **46** (5) pp 1076-1082.

[13] Daisy P and Saipriya K 2012 Biochemical analysis of *Cassia fistula* aqueous extract and phytochemically synthesized gold nanoparticles as hypoglycemic treatment for diabetes mellitus. *Int. J. Nanomed.* **7** pp 1189-1202.

[14] Kaler A Nankar R Bhattacharyya M S and Banerjee U C 2011 Extracellular biosynthesis of silver nanoparticles using aqueous extract of *Candida viswanathii. J. Bionanosci.* **5** (1) pp 53-58.

[15] Lee H J, Lee G, Jang N R, Yun J H, Song J Y and Kim B S 2011 Biological synthesis of copper nanoparticles using plant extract. *Nanotechnol.* **1** (1) pp 371-374.

[16] Park Y, Hong Y N, Weyers A, Kim Y S and Linhardt R J 2011 Polysaccharides and phytochemicals: a natural reservoir for the green synthesis of gold and silver nanoparticles. *IET Nanobiotechnol.* **5** (3) pp 69-78.

[17] Mukunthan K and Balaji S 2012 Cashew apple juice (*Anacardium occidentale* L.) speeds up the synthesis of silver nanoparticles. *Int. J. Green Nanotechnol.* **4** (2) pp 71-79.

[18] Dwivedi A D and Gopal K 2010 Biosynthesis of silver and gold nanoparticles using *Chenopodium album* leaf extract. *Colloids Surf A: Physicochem. Eng. Aspects* **369** (1-3) pp 27-33.

[19] Gong P, Li H, He X, Wang K, Hu J and Tan W 2007 Preparation and antibacterial activity of Fe_3O_4@Ag nanoparticles. *Nanotechnol.* **18** (28) pp 604-611.

[20] Ghiută I, Cristea D, Croitoru C, Kost J, Wenkert R, Vyrides I, Anayiotos A and Munteanu D 2018 Characterization and antimicrobial activity of silver nanoparticles, biosynthesized using *Bacillus* species. *Appl. Surf. Sci.* **438** pp 66-73.

[21] Khatami M, Varma R S, Zafarnia N, Yaghoobi H, Sarani M and Kumar V G 2018 Applications of green synthesized Ag, ZnO and Ag/ZnO nanoparticles for making clinical antimicrobial wound-healing bandages. *Sustain. Chem. Pharm.* **10** pp 9-15.

[22] Anandalakshmi K, Venugobal J and Ramasamy V 2016 Characterization of silver nanoparticles by green synthesis method using *Pedalium murex* leaf extract and their antibacterial activity. *Appl. Nanosci.* **6** (3) pp 399-408.

[23] Shankar S and Rhim J-W 2015 Amino acid mediated synthesis of silver nanoparticles and preparation of antimicrobial agar/silver nanoparticles composite films. *Carbohydr. Polym.* **130** pp 353-363.

[24] Nadagouda M N and Varma R S 2008 Green synthesis of silver and palladium nanoparticles at room temperature using coffee and tea extract. *Green Chem.* **10** (8) pp 859-682.

[25] Kou J and Varma R S 2012 Beet juice – induced green fabrication of plasmonic AgCl/Ag nanoparticles. *ChemSusChem.* **5** (12) pp 2435-2441.

[26] Francis S, Joseph S, Koshy E P and Matthew B 2017 Green synthesis and characterization of gold and silver nanoparticles using *Mussaenda glabrata* leaf extract and their environmental applications to dye degradation. *Environ. Sci. Pollut. Res. Int.* **24** (21) pp 17347-17357.

[27] Veisi H, Azizi S and Mohammadi P 2018 Green synthesis of the silver nanoparticles mediated by *Thymbra spicata* extract and its application as a heterogeneous and recyclable nanocatalyst for catalytic reduction of a variety of dyes in water. *J. Clean Prod.* **170** pp 1536-1543.

[28] Rivera-Rangel R D, Gonzalez-Munoz M P, Avila-Rodriguez M, Razo-Lazcano T A and Solans C 2018 Green synthesis of silver nanoparticles in oil-in-water microemulsion and nano-emulsion using geranium leaf aqueous extract as a reducing agent. *Colloids Surf. A: Physicochem. Eng. Aspects* **536** pp 60-67.

[29] Nadagouda M N, Iyanna N, Lalley J, Han C, Dionysiou D D and Varma R S 2014 Synthesis of silver and gold nanoparticles using antioxidants from blackberry, blueberry, pomegranate, and turmeric extracts. *ACS Sustain. Chem. Eng.* **2** **(7)** pp 1717-1723.

[30] Enema O J, Umoh U F, Thomas P S, Adesina S K and Eseyin O A 2019 Phytochemical and antioxidant studies of leaf of *Tetrapleura tetraptera* (Schumand Thon) Taubert (Mimosaceae). *Br J. Pharm. Med. Res.* **04** (03) pp 1865-1875.

[31] Akin-Idowu P E, Ibitoye D O, Ademoyegum O T and Adeniyi O T 2011 Chemical composition of the dry fruit of *T. tetraptera* and its impact on human health. *J. Herbs Spices Med. Plants* **17** (2) pp 52-61.

[32] Akintola O O, Bodede A I and Ogunbanjo O R 2015 Nutritional and Medicinal Importance of *Tetrapleura tetraptera* Fruits (Aridan) *Afr. J. Sci. Res.* **6** (4) pp 33-38.

[33] Aderibigbe A O, Iwalewa E O, Adesina S K, Adebanjo A O and Ukponmwan O E 2007 Anti-convulsant, analgesic, and hypothermic effects of aridanin, isolated from *Tetrapleura tetraptera* fruit in mice. *J. Biol Sci.* **7** (11) pp 1520-1524.

[34] Awofisayo S O, Udoh I E and Mbagwu H O C 2010 Antibacterial effects of the aqueous and ethanolic extracts of *Tetrapleura tetraptera* pods on the pathogens in nosocomial wound infections. *J. Pharmacogn. Herbal Formul.* **1** (2) pp 18-23.

[35] Logeswari P, Sivagnanam S and Jayanthi A 2012 Synthesis of silver nanoparticles using plants extract and analysis of their antimicrobial property. *J. Saudi Chem. Soc.* **19** (3) pp 311-317.

[36] Mittal A K, Kaler A and Banerjee U C 2012 Free radical scavenging and antioxidant activity of silver nanoparticles synthesized from flower extract of *Rhododendron dauricum. Nano Biomed. Eng.* **4** (3) pp 118-124.

IOP Conf. Series: Materials Science and Engineering **805** (2020) 012033 doi:10.1088/1757-899X/805/1/012033

Biosynthesis and bactericidal activity of silver nanoparticles (AgNPs) using cell-free bioflocculant from domestic wastewater bacterial consortium

E A Adebayo*[1], E O Kumi[1], V O Ajibade[1], A O Akinjo[1], O S Abiola[1], M A Raji[1], Y Adepoju[1], and O T Oduoye[2]

[1]Biotechnology Laboratory, Department of Pure and Applied Biology, Ladoke Akintola University of Technology, PMB 4000, Ogbomoso Nigeria

[2]National Centre for Genetic Resources and Biotechnology, Ibadan, Oyo State, Nigeria.

*Email: eaadebayo@lautech.edu.ng; egokeadebayo@gmail.com

Abstract

Water bodies are becoming contaminated by chemicals and harmful microorganisms like bacteria, which often cause serious illnesses and at times death. It is therefore pertinent to find a mean of purifying wastewaters and larger polluted water bodies. This research aimed at isolating bioflocculant–producing bacteria to synthesis silver nanoparticles (AgNPs) for water purification and antimicrobial activities. Wastewater was collected from eateries (sites A and B) around LAUTECH Ogbomoso; the bioflocculant–producing bacteria were isolated and characterized to species level using molecular tools and registered in the GenBank database. Kaolin (5 g/L) suspension was used to measure the flocculating activity of isolated organisms at an optical density (OD) of 550 nm. The cell-free bioflocculant was used to synthesize AgNPs and its potentials to reduce the microbial load of polluted water were evaluated. The isolated strains were characterized and identified as **(A)** *Bacillus polymyxa* (KY705398), *Bacillus* species (KY705399), *Bacillus alvei* (KY705400) and *Bacillus subtilis* (KY705401); **(B)** *Klebsiella pneumoniae* (KY706100), *Providencia stuartii* (KY706101), and *Acinetobacter* species (KY706102). The consortium organisms gave flocculating activity of 92.46% (A) and 60% (B). The UV-vis spectroscopy showed AgNPs with surface plasmon resonance at 570 nm and 460 nm for sites A and B respectively. FTIR peaks of AgNPs (A) (3525, 3284, and 1179 cm^{-1}) and (B) (3611, 3307 and 2492 Cm^{-1}), pointed to protein as both capping and stabilizing agents for the synthesized nanoparticles. Synthesized AgNPs exerted total reduction of microbial load in faecal polluted wastewater of 146 CFU/ml at a 10^{-5} factor at a concentration of 20 µg. The results obtained revealed that cell-free bioflocculant AgNPs has bactericidal activity and it could be used as efficient water purifying agent.

Keywords: Silver nanoparticles, bioflocculant, wastewater, antibacterial activity

1. Introduction

The world we live in today becomes more civilized with every passing second; more houses, more cars, more plastics, paper and fabric manufacture and more consumable products but all these do not come without a price. With the establishment of industries that make these products, pollution of air, water and land have also increased and the ill effects that come with them. Indiscriminate dumping and release of these wastes into water bodies have led to an increase in the rate of water pollution. Wastes from slaughter houses, industrial wastes before, during and after production, agricultural runoffs, domestic wastes and so on have adverse effects on our water quality [1]. The alleviated levels of pollutants have led to an increase in biological oxygen demand (BOD), chemical oxygen demand (COD), total dissolved solids (TDS), and total suspended solids (TSS), not leaving out toxic metals such as Cd, Cr, Ni and Pb which in unison/synergy and individually, make water unsuitable for drinking, irrigation, cleaning, aquatic life and pose great threats to human health [1]. This research work provides an insight to new, dynamic and safe way of saving these polluted waters by using biosynthesized AgNPs from cell-free bioflocculant from domestic wastewater.

Flocculants are basically of three different types which include bioflocculants, inorganic flocculants and synthetic flocculants [2]. Bioflocculants are biopolymers secreted by microbes like bacteria and fungi. They are capable of causing a phenomenon called; bioflocculation which is the clumping together of fine, dispersed and organic particles [3]. They are biodegradable, safe, and eco-friendly with major components such as extracellular polymeric macromolecules such as polysaccharides, proteins and nucleic acid [2].

The eco-friendliness and potency of green synthesized AgNPs has increased its considerable attention in nanomedicine and nanobiotechnology. Silver nanoparticles produced from bioflocculants were reported to have strong antibacterial activity in sewage water and wastewater treatment [4]. Also, bioflocculant stabilized-AgNPs with capacity to reduce bacterial population in sewage water have been reported [2].

Several efforts have been made to combat the pollution and bring water bodies to a semblance of their original state with the use of chemicals such as chlorine (Cl), potassium aluminium sulphate $KAl(SO_4)_2$ and the likes, but the problem with these chemicals is that they can be persistent in nature and in the body system, causing irritation and harmful side effects to the body. Depending on the route of administration, harmful side effects could include; airway irritation, skin irritation, hives, eye irritation and life threatening allergic reactions especially when a person had prolonged exposure to them [5].

This is why there is need for safer alternatives ways of water purification; one that will not persist after use, cause no harm and be just as effective as conventional chemical water purifiers. This study looks into these possibilities with the production of bioflocculant synthesized AgNPs that can reduce the microbial load in polluted water.

2 Materials and Methods

2.1 Sample location and collection

Samples were collected from two different sources namely; "A" eatery inside campus and "B" eatery under-G, at LAUTECH, Ogbomoso, Oyo state. Using a sterile glass container, the samples were collected at the waste water point of the two eateries.

2.2 Isolation of the micro-organisms

After collecting the samples of waste water, serial dilution was performed and aliquot of appropriate dilutions were plated unto nutrient agar and incubated at 37 ^{o}C for 24 h. On observation, the morphologically distinct colonies were sub-cultured at 37 ^{o}C for 24 h on nutrient agar slant.

2.3 Screening of isolates for bioflocculant production

The isolates were screened for bioflocculant production using the adapted protocol of Xia [6] which involves a bioflocculant producing broth (BPB), consisting of 10 g glucose, 2 g KH_2PO_4, 0.2 g $MgSO_4.7H_2O$, 0.1g NaCl, 0.5 g $CaCO_3$ and 0.5 g yeast extract, all dissolved in 1 litre of de-ionized water. The pH of the broth was adjusted to 7.0 and the medium sterilised in an autoclave at 121 ^{o}C for 15 min. After cooling, it was inoculated with pure culture of the isolates and incubated in a rotary shaker at 37 ^{o}C for 3 days at a rotating speed of 120 rpm.

2.4 Determination of the flocculating activity

The flocculating activity was determined according to the methods developed by Kurane and Matsuyama [7] and modified by Gao *et al.* [8] which uses a suspension of kaolin clay as test material for determining the flocculating activity. The kaolin clay was suspended in distilled water at a concentration of 5 g/l and pH was adjusted to 7; 1% $CaCl_2$ was also prepared. To check the flocculating activity, 0.1 ml of cell-free supernatant of each isolate was mixed with 9 ml of kaolin clay suspension and 0.25 ml $CaCl_2$ solution. A control was also made containing the same components except the 0.1 ml of bacterial supernatant that was replaced with 0.1 ml of ordinary BPB solution. After mixing gently, the mixture was allowed to settle for 5 min at room temperature; then optical density of the upper clarified phase for the control and isolates was measured using a spectrophotometer set at a wavelength of 550nm. The flocculating activity of the isolates was calculated using the following formula:

$$\text{Flocculating activity } (\%) = \left[\frac{B-A}{B}\right] X\, 100$$

Where A is the optical density of the control and B is the optical density of the isolate.

2.5 Identification of bioflocculant producing organism

The organisms with the highest flocculating activities were morphologically and biochemically identified according to Bergey's manual of determinative bacteriology. Molecular characterization was carried out using the methods of Akinola and Adebayo [9] with isolation of genomic DNA, PCR amplification and 16S rRNA sequencing. PCR analysis was performed with 16S rRNA bacterial primers:forward primer 27F (AGAGTTTGATCMTGGCTCAG) and 1525R (AAGGAGGTGWTCCARCCGCA) reverse primer were used for the amplification and amplified sequence reaction was purified. The purified amplified product loaded on the 3131xl gene analyzer from Applied Bio-systems to give the sequences. BigDye cycle terminating sequencing kit method was used and BioEdit sequence alignment editor was used for product analysis. The 16S rRNA gene sequences obtained in this study were compared and registered in GenBank database.

2.6 Synthesis and characterization of silver nanoparticles

The cell-free bioflocculant (1 ml) was added separately to the reaction vessel containing 25 ml of 1 mM silver nitrate ($AgNO_3$) and photoactivated under the sunlight for AgNPs synthesis. Visual observation of change in colour was used to monitor formation of AgNPs and absorbance was measured in the range of 300-800 nm using a UV-visible spectrophotometer (Genesys 10 UV Thermoelectron Corporation, UK). Fourier transform-infrared (FTIR) spectroscopic analysis was done, where the nanoparticles were mixed with potassium bromide salts at 25 ^{o}C and pressed to a pellet for analysis over a wave number range of 4000 to 600 cm^{-1}.

2.7 Bacterial activity of synthesized silver nanoparticles

Waste water from a butchers market was collected in a sterile container and was brought to the laboratory for serial dilution, pour plated on nutrient agar and incubated at 37 ^{o}C for 24 h. After careful observation and counting, the 10^{-5} factor was selected as the standard, having a CFU of 146 colonies per ml. After 24 h, the selected dilution factor was again plated on nutrient agar along with 10, 20, 40, 60, 80 and 100 mg dose of the synthesized AgNPs. Plates

IOP Conf. Series: Materials Science and Engineering **805** (2020) 012033 doi:10.1088/1757-899X/805/1/012033

without the concentrations were also prepared as controls. The plates were incubated at 37 °C for 24 h. After the incubation time, the antibacterial activities of the AgNPs were evaluated based on the number of colony reduction that had occurred.

3. Results and Discussion

The identity of the isolated organisms showed bacteria from four genera of *Bacillus, Klebsiella, Acinetobacter* and *Providencia*. The accession number of each species, quantity and purity of the isolated genomic DNA are recorded in Table 1. Concentration of DNAs varied from 253.2 ng/μl (KY706102) to 33.3 ng/μl (KY705401). The A260 to A280 DNA concentration ratio was between 1.5 (least) in KY705401 and 2.02 (highest) in KY705399. Most isolated bioflocculant producing bacteria in this study fall within the genus *Bacillus,* such as *Bacillus polymyxa, Bacillus alvei, Bacillus subtilis* and *Bacillus* species. Several reports available revealed *Bacillus* species as good bioflocculant-producing bacteria. Akinola and Adebayo [9] reported the bioflocculant producing bacteria isolated such as *Bacillus licheniformis, Bacillus thuringiensis* and *Bacillus cereus* from eatery wastewater, while Sathiyanarayanan *et al.* [2] gave a report of *Bacillus subtilis* MSBN17 as a good producer of bioflocculant. *Klebsiella pneumoniae* isolated from soil sample with high bioflocculant production has been reported by Nakata and Kurane [10]. Nie *et al.* [11] reported the production of bioflocculant by *Klebsiella pneumoniae* isolated from municipal wastewater and its application in municipal wastewater treatment. Optimization of bioflocculant production using consortium bacteria has been previously conducted by Harun *et al.* [12]. The idea of using consortium organisms is to optimize the production of bioflocculant, increase the activity and performance of its application.

The A260/A280 ratio obtained for most isolates fell below the acceptable optimum range (1.80), except for a *Bacillus* species (2.02). Amount and quality of extracted DNA was good enough for PCR amplification and other molecular assays [13]. Figures 1a and b showed the amplicon length of amplified DNA of isolated organisms at 1500 bp on 1Kb DNA marker.

The same amplicon lengths and band pattern shown by all isolated organisms using 16S rRNA indicated homogeneity in DNA component [9]. The amplicon length reported in this study agreed with previous works of Rajashekhar *et al.* [14], that reported an amplicon length of about 1500 bp for *Bacillus* species, while Singh *et al.* [15] recorded an amplicon length of 1271 bp for different species of *Bacillus.* Senesi *et al.* [16]r eported lower amplicon length on *Bacillus* species ranging from 800 to 150 bp, while Akinola and Adebayo [9] documented 700 bp amplicon lengths for *Bacillus* species isolated from eatery wastewater. It is clearly

IOP Conf. Series: Materials Science and Engineering **805** (2020) 012033 doi:10.1088/1757-899X/805/1/012033

shown from literature that using 16S rRNA for identification of *Bacillus* species gave different amplicon lengths that varied from one species to the other [9].

Table 1. The identity of the isolates with DNA quantity and purity determined using Nanodrop-2000 spectrophotometer

Code	Name of the Organism	Accession number	Nucleic acid (ng/µl)	A_{260}/A_{280}
A	*Klebsiella pneumoniae*	KY706100	223.7	1.73
B	*Acinetobacter* species	KY706102	253.2	1.53
C	*Providencia stuartii*	KY706101	130.3	1.66
D	*Bacillus polymyxa*	KY705398	33.9	1.54
E	*Bacillus* species	KY705399	85.3	2.02
F	*Bacillus alvei*	KY705400	90.1	1.69
G	*Bacillus subtilis*	KY705401	33.3	1.5

* organisms A, B and D classified as consortium B and organisms; E, F, G and H classified as consortium A.

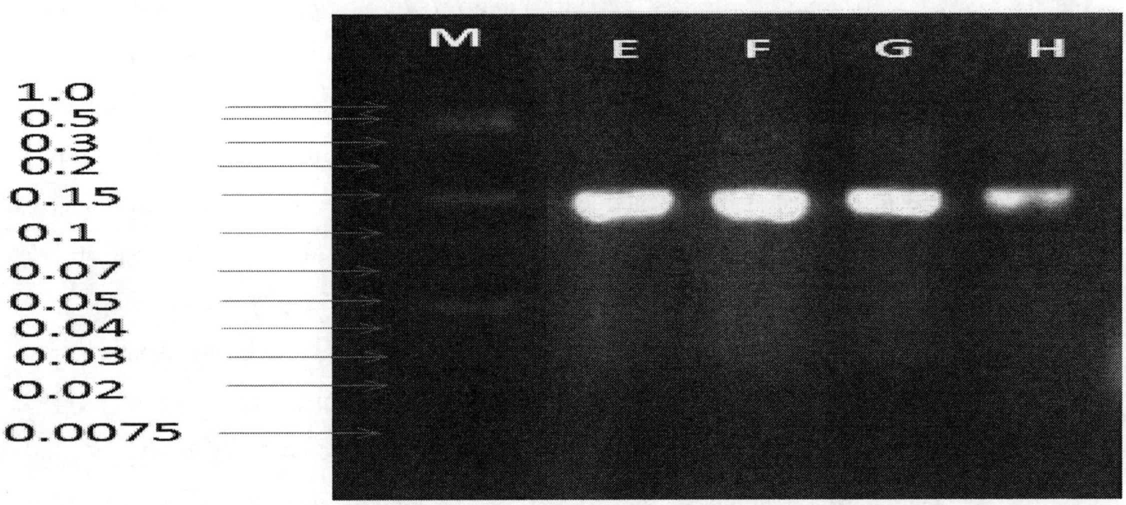

Figure 1a. 16S rRNA fingerprint of species labelled E, F, G, H, isolated from "A" eatery. Amplified on 1.5% agarose gel Lane M: Marker (1 Kb DNA Marker; GeneRuler)

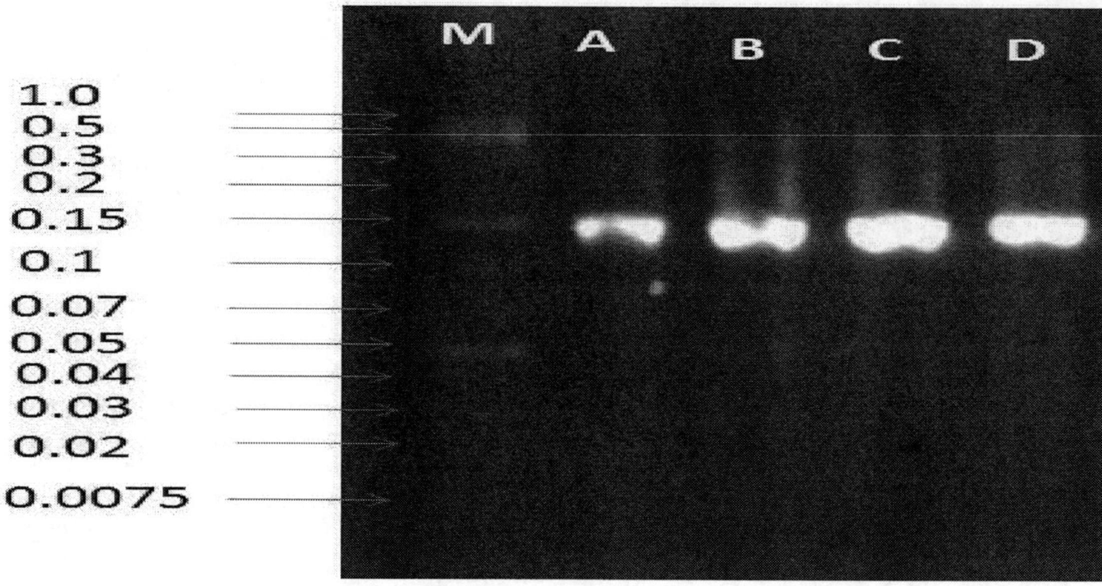

Figure 1b. 16S rRNA fingerprint of species labelled A, B, & D, isolated from "B" eatery. Amplified on 1.2% agarose gel Lane M: Marker (1KB DNA Marker; GeneRuler)

The consortium organisms used in this study [A; *Bacillus polymyxa* (KY705398), *Bacillus* species (KY705399), *Bacillus alvei* (KY705400) and *Bacillus subtilis* (KY705401]; and [B; *Klebsiella pneumoniae* (KY706100), *Providencia stuartii* (KY706101), and *Acinetobacter* species (KY706102)] gave flocculating activities 92.46% and 60% for A and B respectively (Table 2). The highest flocculating activities of 92.46% obtained in consortium A which are all *Bacillus* species indicated better interaction and combination among the same species compared to consortium B, which are of different species. Salehizadeh and Shojaosadati [17] reported their findings which showed that *Bacillus* species significantly secrete biopolymer flocculant during growth. Bajlan *et al.* [18] reported *Bacillus coagulans* isolated from domestic waste water with flocculating activity of 62.7%, while Simphiwe *et al.* [19] observed flocculating activity of 89% from wastewater treatment plant and Piyo *et al.* [20] reported organisms which produced at 72% from sediment of algal bay. The differences obtained in the level of bioflocculants production in this study and previous works may be as a result of wastewater sources, waste water compositions and environmental factors.

Synthesized AgNPs using cell-free bioflocculant as reducing and stabilizing agent was carried out, with evidence of a colour change observed after 45 min of synthesis (Figure 2). The appearance of the brown colour was an indication of the formation of colloidal AgNPs in the medium. The change of brown colour due to the excitation of surface plasmon resonance is typical of AgNPs [2, 21]. The synthesis of AgNPs nanoparticles was further confirmed spectrophotometry. Maximum absorbance was displayed by the biosynthesized AgNPs at the wavelengths of 540 and 460 nm for A and B respectively (Figure 3).

Table 2. Flocculating activities (%) of the bacterial consortium A and B

Sample	12 h	24 h	36 h	48 h
A	35.02	53.01	66.03	92.46
B	32.09	47.80	53.00	60.00

The absorption spectra of the consortium B fell is similar to 435 nm reported by Sathiyanarayanan *et al.* [2] for AgNPs synthesized from bioflocculant of *Bacillus subtilis* MSBN17. Several reports available on AgNPs are within the range of obtained values in the current study [22, 23, 24].

The FTIR analysis was carried out to identify the possible interaction between silver salts and bioflocculant, which accounted for the reduction of Ag^+ ions and stabilization of the nanoparticles. The FTIR spectral analysis revealed the presence of absorption peaks of 3525, 3284, and 1179 cm^{-1} (Figure 4) which correspond to alcohol groups, amine group and alcohol and 3611, 3307, 2492 cm^{-1} (Figure 5) which correspond to aliphatic primary amine, carboxylic acid and amine groups for consortium A and B respectively. The biomolecules obtained have shown to be rich in amine groups which accounted for the reduction in the metal ions (Ag^+), capping and stabilization of AgNPs [25].

The current study is a novel approach by using bioflocculant to synthesize AgNPs for waste water treatment. Wastewater, drinking water, and industrial downstream processing have been widely known to be treated by bioflocculants. The faecal polluted water collected from meat butchery market that had a CFU of 146 colonies per ml at the 10^{-5} dilution factor was totally sanitized by synthesized AgNPs. The total viable organisms present in the polluted water were totally killed and eliminated at 20 (μg/ml) of synthesized AgNPs (Table 3). Sathiyanarayanan *et al.* [2] reported significant reduction of microbial load of initial (average) 250 CFU/ml (in 10^{-5} dilution) in sewage water to 7 CFU/ml by AgNPs (0.50 mg/l) synthesized from bioflocculant treatment. The synthesized AgNPs from this study has exerted bactericidal effect at low concentration, against faecal polluted organisms. The results from the current study revealed that bioflocculant cell-free mediated AgNPs has bactericidal activity and it could be used as efficient water purifying agent.

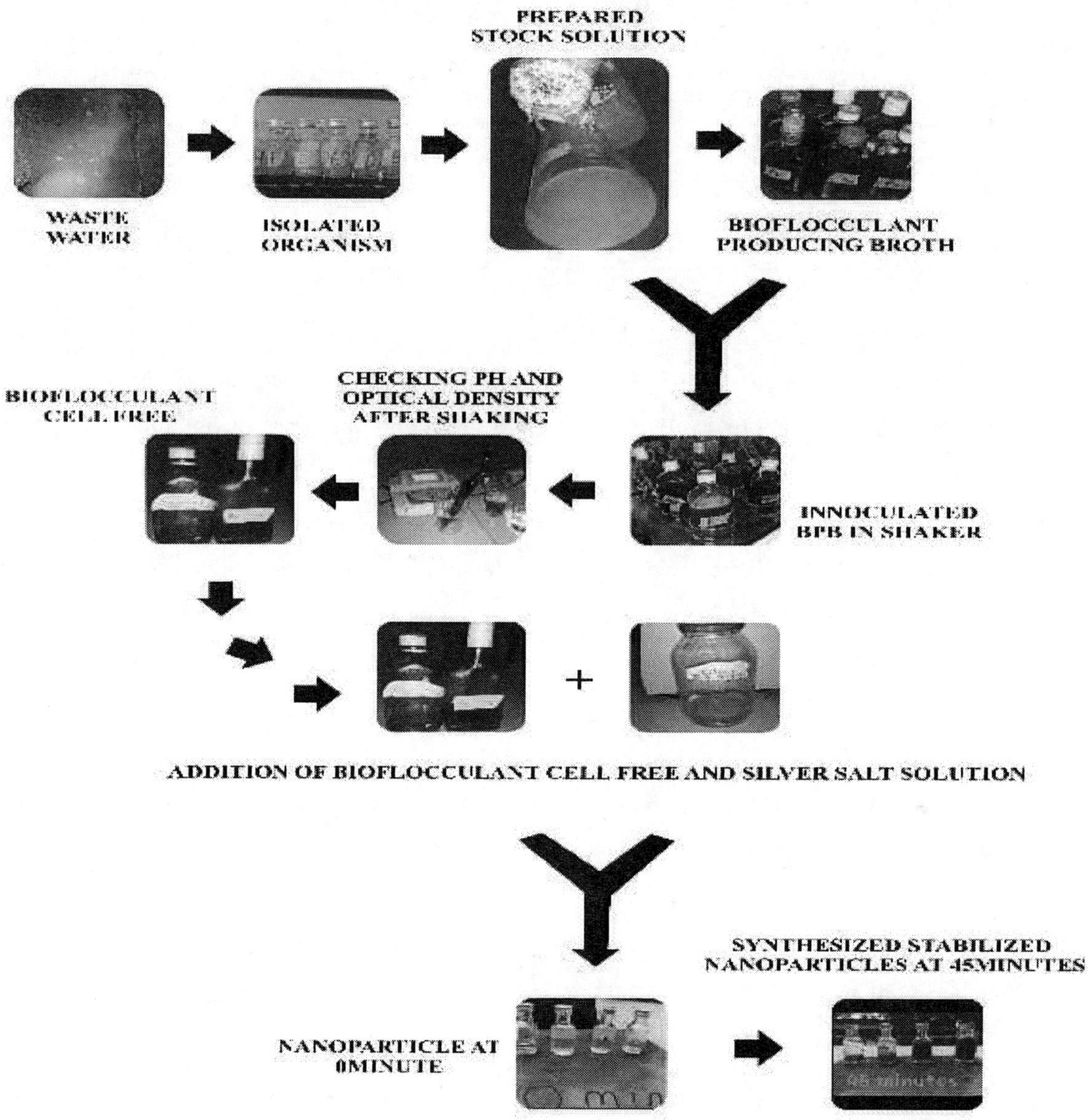

Figure 2. Flow chart of the biosynthesis of silver nanoparticles

IOP Conf. Series: Materials Science and Engineering **805** (2020) 012033 doi:10.1088/1757-899X/805/1/012033

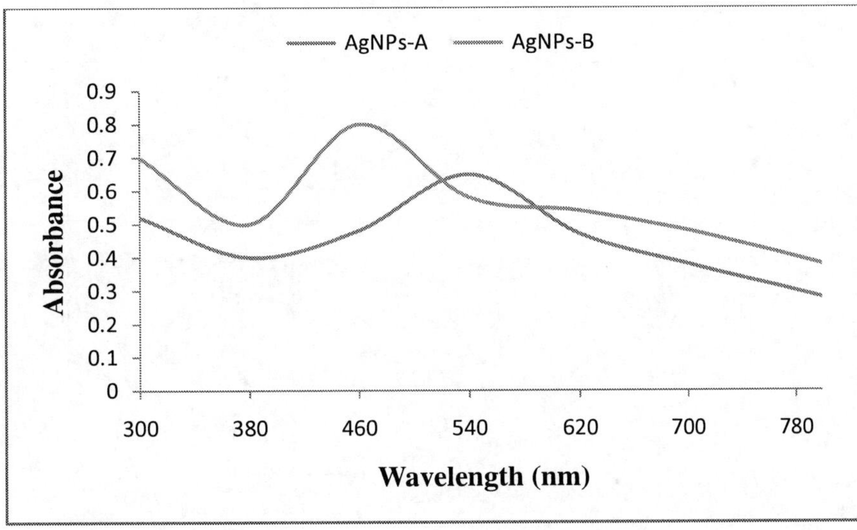

Figure 3. Absorbance spectra of synthesized AgNPs (A and B)

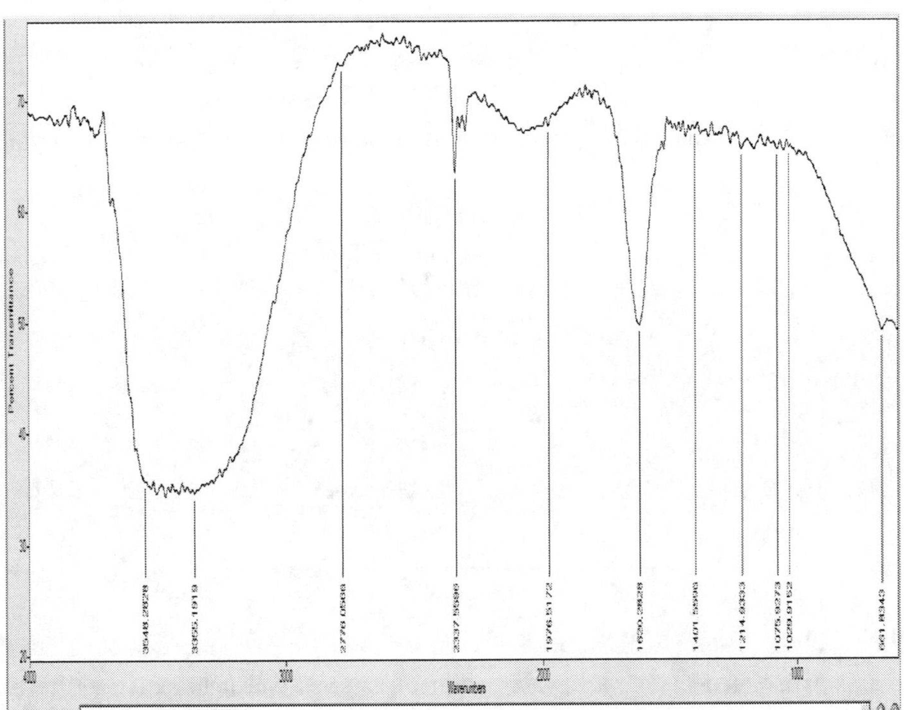

Figure 4: FTIR spectrum of bioflocculant-based synthesized AgNPs using consortium A

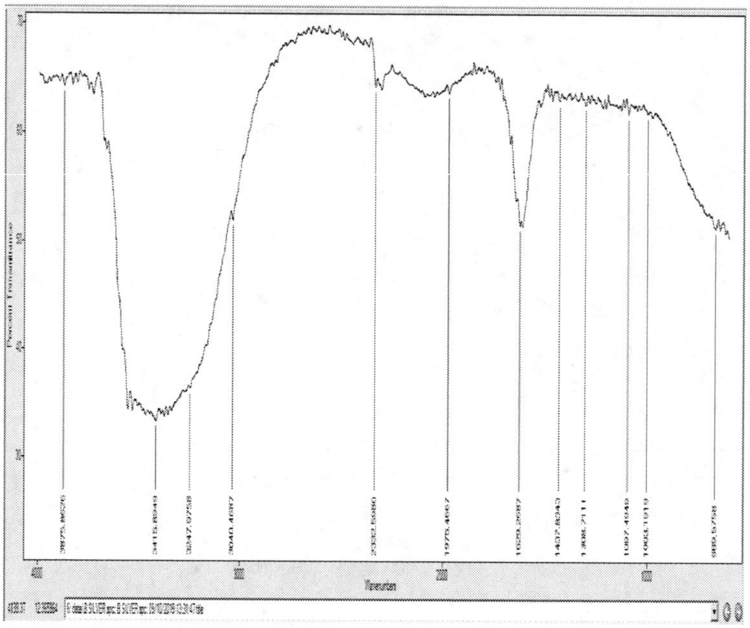

Figure 5. FTIR spectrum of bioflocculant-based synthesized AgNPs using consortium B

Table 3. Total colony reduction of microbial loads of waste water by synthesized AgNPs

Concentrations used (µg/ml)	Total colony reduction (%) A	Total colony reduction (%) B
10	98	92
20	100	100
40	100	100
60	100	100
80	100	100
100	100	100

4. Conclusion

The organisms isolated from wastewater point of eateries were characterized to the species level and registered in the GenBank database. The isolated bacteria were established to be bioflocculant-producing bacteria, with higher flocculating activities at 48 h of incubation. Cell-free bioflocculants obtained was used to synthesize AgNPs which exhibited a colour

IOP Conf. Series: Materials Science and Engineering **805** (2020) 012033 doi:10.1088/1757-899X/805/1/012033

change and stabilized after 45 min of exposure to sunlight. The AgNPs were able to exert total bactericidal effect on faecal polluted water at a concentration of 20 µg/ml. Hence, from the results obtained, the bioflocculant mediated-AgNPs exhibited bactericidal capabilities and thus could be used as efficient water purifying agent.

References

[1] Odigie J O 2015 Harmful effects of wastewater disposal into water bodies: a case review of the Ikpoba river, Benin city, Nigeria. *Trop. Freshwater Biol.* **23** pp 87-101.

[2] Sathiyanarayanan G, Kiran GS and Selvin J 2013 Synthesis of silver nanoparticles by polysaccharide bioflocculant produced from marine *Bacillus subtilis* MSBN17. *Colloids Surf. B: Biointerf.* **102** pp 13-20.

[3] Adebayo-Tayo B and Adebami G E 2014 Production and characterization of bioflocculant produced by *Bacillus clausii* NB2. *Innov. Rom. Food Biotechnol.* **14** pp 13-25.

[4]Manivasagan P, Kang K, Kim D G and Kim S-K 2015 Production of polysaccharide-based bioflocculant for the synthesis of silver nanoparticles by *Streptomyces* sp. *Int. J. Biol. Macromol.* **77** pp159-167.

[5] Weisel CP, Richardson SD, Nemery B, Aggazzotti G, Baraldi E, Blatchley III ER, Blount BC, Carlsen KH, Eggleston PA, Frimmel FH and Goodman M 2009 Childhood asthma and environmental exposures at swimming pools: state of the science and research recommendations. *Environ. Health Perspect.* **117** (4) pp 500-507.

[6] Xia S, Zhang Z, Wang X, Yang A, Chen L, Zhao J, Leonard D and Jaffrezic-Renault N 2008 Production and characterization of a bioflocculant by *Proteus mirabilis* TJ-1. *Bioresour. Technol.* **99** (14) pp 6520-6527.

[7] Kurane R and Matsuyama H 1994 Production of a bioflocculant by mixed culture. *Biosci. Biotechnol. Biochem.* **58** (9) pp 1589-1594.

[8] Gao J, Bao H Y, Xin M X, Liu Y X. Li Q and Zhang Y F 2006 Characterization of a bioflocculant from a newly isolated *Vagococcus* sp. W31. *J. Zhejiang Univ. Sci.* B. 7 (3) pp 186-192.

[9] Akinola O and Adebayo E A 2018 Screening and characterization of bioflocculant-producing bacteria isolated from domestic waste water in Bowen University, Iwo, Osun State Nigeria. *Nig. J. Biotechnol.* **35** (2) pp 159-166.

[10] Nakata K and Kurane R 1999 Production of an extracellular polysaccharide bioflocculant by *Klebsiella pneumoniae. Biosci. Biotechnol. Biochem.* **63** (12) pp 2064-2068.

[11] Nie M, Yin X, Jia J, Wang Y, Liu S, Shen Q, Li P and Wang Z 2011 Production of a novel bioflocculant MNXY1 by *Klebsiella pneumoniae* strain NY1 and application in precipitation of cyanobacteria and municipal wastewater treatment *J. Appl. Microbiol.* **111** (3) pp 547-558.

[12] Harun AA, Mohammad NA, Ikhwanuddin M, Ismail N, Ibrahim Z, Kasan NA, Feng B, Peng J, Zhu X, Huang W and Nasir NM 2017 Factors influencing the production of extracellular polymeric substances by *Rhodopseudomonas acidophila. J. Fisheries Aquat. Sci.* **12** (4) pp 71-76.

[13] Plaza GA, Upchurch R, Brigmon RL, Whitman WB and Ulfig K 2004 Rapid DNA Extraction for Screening Soil Filamentous Fungi Using PCR Amplification. *Pol. J. Environ. Stud.* **13** (3) pp 315-318.

[14] Rajashekhar M, Shahanaz E and Vinay K 2017 Biochemical and molecular characterization of *Bacillus* spp. isolated from insects. *J. Entomol. Zool. Stud.* **5** (5) pp 581-588.

[15] Singh P, Barolia SK, Sharma DK and Patni A 2016 Molecular characterization and identification of *Bacillus* spp. MLSU-PSSKB isolated from fecal contents of *Pteropus giganteus* in Udaipur, Rajasthan, India. *J. Appl. Life Sci. Int.* **7** pp 1-6.

[16] Senesi S, Celandroni F, Tavanti A and Ghelardi E 2001 Molecular characterization and identification of *Bacillus clausii* strains marketed for use in oral bacteriotherapy. *Appl. Environ. Microbiol.* **67** (2) pp 834-839.

[17] Salehizadeh H and Shojaosadati SA 2001 Extracellular biopolymeric flocculants: recent trends and biotechnological importance. *Biotechnol. Adv*. **19** (5) pp 371-385.

[18] Bajlan JS, Alden SB and Al-Saffar OB 2013 Characterizations of bioflocculant produced by *Bacillus coagulans* 8B and application in domestic wastewater treatment. *J. Madenat Alelem Univ. College* **5** (1) pp 47-55.

[19] Buthelezi SP, Olaniran AO and Pillay B 2010 Production and characterization of bioflocculants from bacteria isolated from wastewater treatment plant in South Africa. *Biotechnol. Bioprocess Eng.* **15** (5) pp 874-881.

[20] Nontembiso P, Sekelwa C, Leonard MV and Anthony OI 2011 Assessment of bioflocculant production by *Bacillus* sp. Gilbert, a marine bacterium isolated from the bottom sediment of Algoa bay. *Mar. Drugs* **9** (7) pp 1232-1242.

[21] Ahmad A, Mukherjee P, Senapati S, Mandal D, Khan MI, Kumar R and Sastry M 2003 Extracellular biosynthesis of silver nanoparticles using the fungus *Fusarium oxysporum*. *Colloids surf. B: Biointerf.* **28** (4) pp 313-318.

[22] Adebayo E A, Oke M A, Lateef A, Oyatokun A A, Abisoye O D, Adiji I P, Fagbenro D O, Amusan T V, Badmus J A, Asafa T B, Beukes LS, Gueguim-Kana E B and Abbas S H 2019 Biosynthesis of silver, gold and silver–gold alloy nanoparticles using *Persea americana* fruit peel aqueous extract for their biomedical properties. *Nanotechnol. Environ. Eng.* **4** (1) 13. https://doi.org/10.1007/s41204-019-0060-8.

[23] Lateef A, Ojo S A, Akinwale A S, Azeez L, Gueguim-Kana E B and Beukes L S 2015 Biogenic synthesis of silver nanoparticles using cell-free extract of *Bacillus safensis* LAU 13: antimicrobial, free radical scavenging and larvicidal activities. *Biologia* **70** (10) pp 1295-1306.

[24] Malathi S, Ezhilarasu T, Abiraman T and Balasubramanian S 2014 One pot green synthesis of Ag, Au and Au-Ag alloy nanoparticles using isonicotinic acid hydrazide and starch. *Carbohydr. Polym.* **111** pp 734-743.

[25] Adebayo EA, Ibikunle J B, Oke A M, Lateef A, Azeez M A, Oluwatoyin A O, Ajala V A, Olowoporoku T B, Okunlola O C, Ogundele O A, Badmus J A, Asafa T B, Beukes L S, Gueguim-Kana E B, and Abbas S H 2019 Antimicrobial and antioxidant activity of silver, gold and silver-gold alloy nanoparticles phytosynthesized using extract of *Opuntia ficus-indica*. *Rev. Adv. Mater. Sci.* **58** (1) pp 313-326.

IOP Conf. Series: Materials Science and Engineering **805** (2020) 012034 doi:10.1088/1757-899X/805/1/012034

Synthesis and characterization of silver nanoparticles-doped titanium (IV) oxide thin films

SW Balogun*[1], AB Abdulhamid[2], YK Sanusi[1, 3], and O Adedokun[3]

[1]Department of Physics and Materials Science, Kwara State University Malete, Ilorin, Nigeria

[2]Department of Physics, School of Science, Kwara State College of Education, Ilorin, Nigeria.

[3]Department of Pure and Applied Physics, Ladoke Akintola University of Technology, Ogbomoso, Nigeria

*Corresponding author: Sunday Wilson Balogun (sbalogun94@gmail.com)

Abstract

Current advancements in nanotechnology appear to open a more efficient, stable, and cheaper way of producing solar cells to replace the too expensive means of inorganic solar cells production. The creation of nanoscale materials for advanced structures has led to a growing research interest in the area of photovoltaic energy conversion using photovoltaic devices. This research investigates the impact of annealing temperature on the optical properties of silver nanoparticles-doped titanium (IV) oxide thin film (TiO_2:AgNPs) deposited by spin-coating method on glass substrate. AgNPs were prepared using the leaf extract of *Gliricidia sepium* as a reducing agent for silver nitrate. Deposition of TiO_2:AgNPs composite solution was performed in different volume ratio. Deposition of TiO_2:AgNPs volume ratio (1: 0.2) was performed with different spin-coating speed for 30 seconds at 7 different thicknesses. Characterization of the optical properties of thin films was carried out by using a UV-vis single beam spectrophotometer; this was used to calculate the absorbance and the bandgap energy. Sixteen 16 samples of TiO_2:AgNPs deposited at 1000 rpm on the glass substrate were annealed at temperature range of 50 to 425 °C with step of 10 °C interval in a tubular furnace. It was observed from the results that the peak absorption of photon energy occurred at 375 °C in the visible spectrum of the wavelength band. The optimal thickness for peak absorbance of the TiO_2:AgNPs blend layer occurred at 115 nm in the visible spectrum and at the corresponding spin speed of 1000 rpm. The optimized fabrication process with blend layer thickness of 115 nm yielded the best absorbance at an annealed temperature of 375 °C in the visible spectrum. The volume ratio of (1:0.2) gave the peak absorption at 0.75 au. At 1000 rpm with the corresponding thickness of 115 nm, TiO_2:AgNPs blend has the peak absorbance. The energy band gap of the blend thin film is 3.58 eV at 375 °C in the visible spectrum of wavelength band. The result obtained showed that AgNPs enhanced light absorption, broadened absorption spectral range and thermal stability of titanium (IV) oxide film. The result can be used as a guideline in the design and fabrication of solar cells.

Keywords: Organic thin film; silver nanoparticles; titanium dioxide; annealing; blend AgNPs:TiO_2; optical properties

1. Introduction

Demand for eco-friendly and clean energy has increased in recent times leading to increase in research towards alternative source of energy and development of solar energy. Researchers have focused attention on nanotechnology which is an arm of science and technology that deals with different materials at a nanoscale. Nanotechnology involves production, manipulation and use of materials at subatomic level that finds its application in materials science, engineering, electronics, photovoltaics, biological, and chemical sensors [1-23]. Silver nanoparticles uniqueness is found in its optical, electrical, and thermal stability. Silver nanoparticles characteristics include low sintering temperatures, high electrical conductivity and stability. Silver nanoparticles can be prepared using the following methods: ultrasound, anodization, co-precipitation, sol-gel method, chemical vapor deposition and mechanochemical-thermal synthesis [24-27]. This research investigates effect of annealing temperature on the optical properties of silver nanoparticles doped titanium (IV) oxide (TiO_2:AgNPs) thin films deposited by spin-coating method on glass substrate.

2. Materials and Methods

2.1. Preparation of thin films of TiO_2:AgNPs

Titanium (IV) oxide nanopowder (< 35nm), 97% purity was purchased from Sigma-Aldrich Company (USA), while silver nitrate and other chemical reagents were purchased locally. The glass substrates of dimension 25.4 mm by 76.2 mm were washed with liquid detergent in ultrasonic sonicator for 10 to 15 min and then rinsed in distilled water for 15 min at 27 °C. The

IOP Conf. Series: Materials Science and Engineering **805** (2020) 012034 doi:10.1088/1757-899X/805/1/012034

substrates were further cleaned with isopropanol alcohol (IPA) in ultrasonic bath for 15 min at 27 ^{o}C and dried in a stream of nitrogen gas (N_2). To prepare TiO_2, 25ml of ethanol was added into 3 g of TiO_2 powder. The solution then underwent ageing process for 3 h at room temperature without heat to allow homogeneous mixture and TiO_2 powder to fully diffuse into the solvent. Ethanol was chosen as a solvent because it has no characteristic absorption and emission in the visible spectrum of wavelength band. Then, 0.168 g of silver nitrate was dissolved in 100 ml deionized water which gives colorless clear solution.

Then, 3 g of leaves of *Gliricidia sepium* were thoroughly washed, rinsed and cut into pieces. Thereafter, the pieces of leaves were immersed in 300 ml of boiling distilled water for 30 min and followed by filtration. The filtered extract was dissolved into 100 ml of 0.01M of aqueous silver nitrate solution and stirred for homogenous mixture. The product was stored in a brown bottle shielded from the ray of light. Blend solution (TiO_2:AgNPs) was done by mixing TiO_2 solution with AgNPs solution at different composite ratio (1:1, 1: 0.8, 1: 0.6, 1: 0.4, 1: 0.2) and stirred using magnetic stirrer for homogeneous mixture. TiO_2 solution was deposited on glass substrate as reference (control sample), while the blend solution of TiO_2:AgNPs were deposited at 5 different volume ratio (1:1, 1: 0.8, 1: 0.6, 1: 0.4, 1: 0.2) ml using spin-coater (Laurel WS-650Hz-23NPP).

The blend solution with volume ratio of (1:0.2) ml has highest absorption value after optical properties characterization. This optimized absorption TiO_2:AgNPs (1: 0.2) volume ratio was used to deposit another seven films of layer thicknesses of 30 nm, < 35nm, 35 nm, 87 nm, 98 nm, 115 nm, and 146 nm corresponding to spin coating speeds of 4000, 3000, 2000, 1500, 1250, 1000, and 750 rpm, respectively for 30 seconds on glass substrate. Depending on the speed of rotation of the spin coater, the desirable thickness of the sample was obtained. It is important to note that the thickness of film being spin-coated depend on both time and speed of rotating stub as specified by equipment manufacturer. Another method of determination of thickness is by using Equation (1) in the absence of surface profilometer.

$$t = \frac{M_2 - M_1}{AD} \quad (1)$$

Where t = thickness (nm), M_1 = mass of substrate before deposition, M_2 = mass of substrate after deposition, A = Area covered by the film, D = Density of the material.

$$D = \frac{M}{V}$$

IOP Conf. Series: Materials Science and Engineering **805** (2020) 012034 doi:10.1088/1757-899X/805/1/012034

The fabricated TiO_2:AgNPS at 1000 rpm with corresponding thickness of 115 nm, has best absorption value after optical properties characterization. At the deposition rate of 1000 rpm seventeen sample films were prepared and annealed at temperature range of 50 to 425 °C with 10 °C interval in a carbolite tubular furnace (model Srw 21-501042 Type-CT17) for 30 min. One sample was used as a reference.

2.2 Characterization

The TiO_2 solution and synthesized AgNPs thin films prepared on clean glass substrates by spin-coating method were characterized using UV-vis spectrophotometer (Avantes, Avalight-DH-5-BAL) to determine the absorption values in arbitrary unit (au) and spectrum using the recorded transmittance and reflectance (%). Absorbance was calculated using Equation (2)

$$A = 2 - log(\%T)\text{au} \qquad (2)$$

Where A is the absorbance, %T is the percentage transmittance.

The wavelength band for which a material absorbs photon energy and convert it to electricity helps determine whether the material is capable of converting the available light illuminating it into electric energy. Absorption was chosen as the optimization variable because the choice of layer thickness is a function for optimization of the efficiency.

(i) **Graph of absorption against photonenergy**

Photon energy values were determined from the inverse relationship between energy in electron volts (eV) and wavelength of the UV-visible spectrum using equation (3).

$$E = \frac{hc}{q\lambda}(\text{eV}) \qquad (3)$$

Where E=Photon energy, h = planck's constant, C = speed of light, λ = wavelength in nanometres (nm), q = electronic charge, 1 eV = 1.6×10^{-19} J

3 Results and Discussion

The absorbance readings of TiO_2 (control) and TiO_2 loaded with AgNPs prepared at different composite ratio are as shown in Figure 1a. The blend solution with composite ratio of (1:0.2) ml has the highest absorbance (a u) and this optimized volume ratio was then deposited at different spin-coating speeds of 4000, 3000, 2000, 1500 , 1250 ,1000, and 750 rpm in Figure 1b. Thermal annealing was carried out on all the samples as shown in Figures 1c and 1d. The optimized film thickness for peak absorbance of the TiO_2:AgNPs blend layer occurred at 115 nm in the visible band of spectral and at the corresponding spin speed of 1000 rpm. The optimized fabrication process with blend layer thickness of 115 nm yielded the best absorbance at annealed temperature of 375 °C in the visible spectrum as shown in Figures 1c and 1d. The graph of the plot of absorbance versus photon energy in eV of TiO_2:AgNPs is shown in Figures 1e and 1f. The energy band gap of the blend thin film is 3.58 eV at 375 °C in the wavelength range of visible light. The result also revealed that heat treatment or thermal annealing is a tool that can be used to tune the optical properties of TiO2:AgNPs blend thin film.

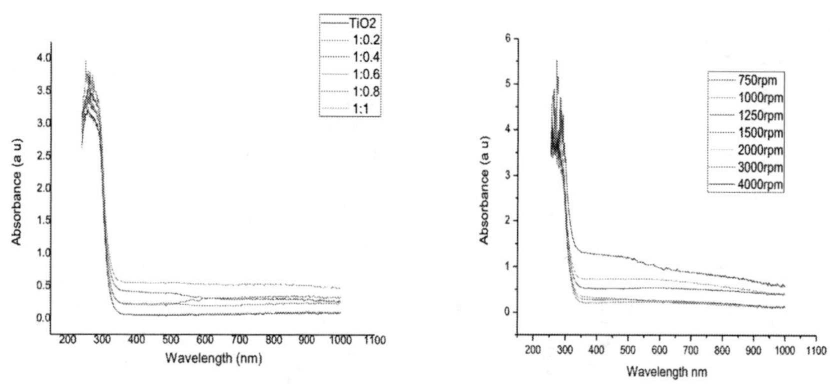

(a) (b)

Figure 1. Absorbance of TiO_2 and blend TiO_2:AgNPs (a), TiO_2:AgNPs (1:0.2) at different layer thicknesses (b)

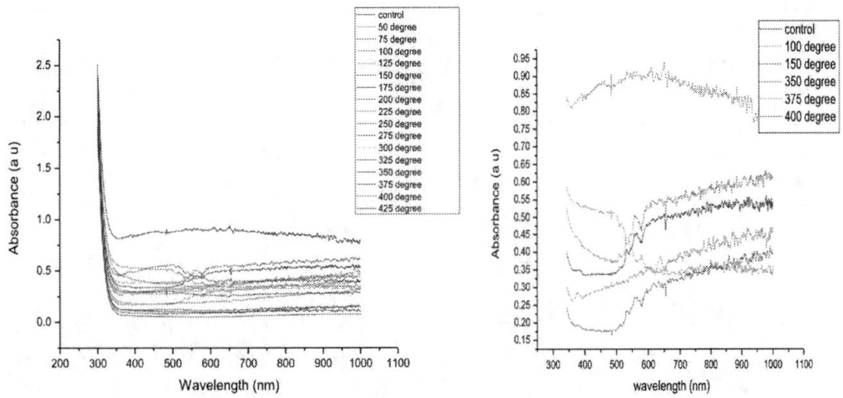

(c) (d)

Figure 1. TiO_2:AgNPs at different annealing temperature (c), TiO_2:AgNPs (1:0.2,1000 rpm) at selected annealing temperature (d)

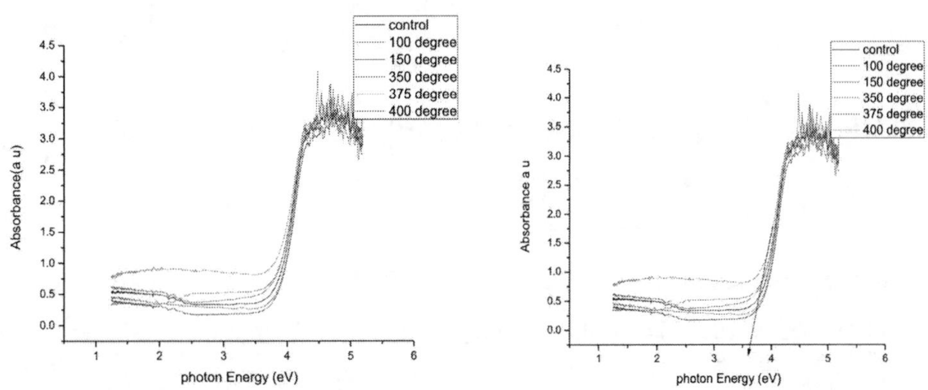

(e) (f)

Figure 1. Graph of absorption vs. photon energy (e), the slanting line at 375 °C (f)

4. Conclusion

The result obtained showed that gradual thermal annealing, in a controlled manner revealed a more stable and efficient control in tuning the TiO_2:AgNPs thin films. The results can form a guideline to design and fabricate solar cells. Results obtained could lead to a breakthrough in performance improvement and stability of solar cells.

Acknowledgement

The authors extend appreciation to technologists at Materials Science and Engineering Laboratory, Kwara State University, Malete-Ilorin, Nigeria for their assistance with the laboratory work.

References

[1] Atwater H A and Polman A 2010 Plasmonics for improved photovoltaic devices *Mater. Sustain. Energ.* https://doi.org/10.1142/9789814317665_0001.

[2] Heeger A J 2014 Heeger AJ. 25th anniversary article: bulk heterojunction solar cells: understanding the mechanism of operation. *Adv. Mater.* **26** (1) pp 10-28.

[3] Hamidi M, Azadi A and Rafiei P 2008 Hydrogel nanoparticles in drug delivery. *Adv. Drug Deliver. Rev.* **60** (15) pp 1638-1649.

[4] Lowenstam H A 1981 Minerals formed by organisms. *Sci.* **211** (4487) pp 1126-1130.

[5] Lee EJ H, Ribeiro C, Longo E and Leite E R 2006 Growth Kinetic of tin oxide nanocrystals in colloidal suspensions under hydrothermal conditions *J. Chem. Phys.* **328** (1-3) pp 229-235.

[6] Roy P, Kim D, Lee K, Spiecker E and Schmuki P 2010 TiO_2 nanotubes and their application in dye-sensitized solar cells. *Nanoscale* **2** (1) pp 45-59.

[7] Linsebigler A L, LuG, J T and Yates J T 1995 Photocatalysis on TiO_2 surfaces: principles, mechanisms, and selected results. *Chem. Rev.* **95** (3) pp 735-758.

[8] Fujishima A and Honda K 1972 TiO_2 photoelectrochemistry and photocatalysis *Nature* **238** (5358) pp 37-38.

[9] Wang X, Fujimaki M and Awazu K 2005 Photonic crystal structures in titanium dioxide (TiO_2) and their optimal design. *Opt.* Express **13** (5) pp 1486-1497.

[10] Anpo CM 1997 Photocatalysis on titanium oxide catalysts: Approaches in achieving highly efficient reactions and realizing the use of visible light. *Catal. Surv. Asia.* **1** (2) pp 169-179.

[11] Kim H M, Kokubo T, Miyaji F and Nakamura T 1996 Preparation of bioactive Ti and its alloys via simple chemical surface treatment. *J. Biomed. Mater. Res.* 32 (3) pp 409-417.

[12] Mann S 2001 Biomineralization: Principles and Concepts in Bioinorganic Materials Chemistry (*Oxford Univ. Press, Oxford*)

[13] Abhishek G, Sharad V, Pramod KS and Nitin K 2011 Formulation, characterization and application on nanoparticle: a review. *Der. Pharmacia Sin.* **2** (2) pp 17-26.

[14] Saravanana P, Ganapathyb M, Charlesc A, Tamilselvana S, Jeyasekarand R and Vimalane M 2016 Electrical properties of green synthesized TiO_2 nanoparticles *Adv. Appl. Sci. Res.* **7** (3) pp 158-168.

[15] Sakshum K, Kushagra K, Gauravi X and Prakhar K 2016 Plasmonic study of nanoparticles in organic photovoltic cells: a review. *J. Org. Inorg. Chem.* **3** (1) 2.

[16] Bardaxoglou G, Rouleau C and Pelletier E 2017 High stability and very slow dissolution of bare and polymer coated silver nanoparticles dispersed in river and coastal waters *J. Aquat. Pollut. Toxicol.* **1** (2) 15.

[17] Piccinno F, Gottschalk F, Seeger S and Nowack B 2012 Industrial production quantities and uses of ten engineered nanomaterials in Europe and the world. *J. Nanopart. Res.* **14** 1109. https://doi.org/10.1007/s11051-012-1109-9.

[18] Echegoyen Yand Nerín C 2013 Nanoparticle release from nano-silver antimicrobial food containers. *Food Chem. Toxicol.* **62** pp 16-22.

[19] Mackevica A, Olsson M E and Hansen S F 2017 The release of silver nanoparticles from commercial toothbrushes. *J. Hazard. Mater.* **322** pp 270-275.

[20] Matsuhisa N, Inoue D, Zalar P, Jin H, Matsuba Y, Itoh A, Yokota T, Hashizume D and Someya T 2017 Printable elastic conductors by *in situ* formation of silver nanoparticles from silver flakes. *Nat. Mater.* **16** (8) pp 834-840.

[21] Reidy B, Haase A, Luch A, Dawson K A and Lynch I 2015 Mechanisms of silver nanoparticles release, transformation and toxicity: A critical review of current knowledge and recommendations for future studies and applications. *Mater.* **6** (6) pp 2295- 2350.

[22] Huang L Dai T, Xuan Y, Tegos GP and Hamblin M R 2011 Synergistic combination of chitosan acetate with nanoparticle silver as a topical antimicrobial: efficacy against bacterial burn infections. *Antimicrob. Agents Chemother.* **55** (7) pp 3432-3438.

[23] Le Ouay B and Stellacci F 2015 Antibacterial activity of silver nanoparticles: A surface science insight. *Nano Today* **10** (3) pp 339-354.

[24] Kumar S S, Venkateswarlu P, Rao V R and Rao G N 2013 Synthesis, characterization and optical properties of zinc oxide nanoparticles. *Int. Nano Lett.* **3** (1) 30. https://doi.org/10.1186/2228-5326-3-30.

IOP Conf. Series: Materials Science and Engineering **805** (2020) 012034 doi:10.1088/1757-899X/805/1/012034

[25] Meulenkamp EA 1998 Synthesis and growth of ZnO nanoparticles. *J. Phys. Chem. B* **102** (29) pp 5566-5572.

[26] Prasad K and Anal K J 2009 ZnO nanoparticles: synthesis and adsorption study. *Nat. Sci.* **1** (2) pp 129-135.

[27] Sadraei R 2016 A simple method for preparation of nano-sized ZnO. *Res. Rev.: J. Chem.* **5** (2) pp 45-49.

IOP Conf. Series: Materials Science and Engineering **805** (2020) 012035 doi:10.1088/1757-899X/805/1/012035

Antidiabetic properties of phytosynthesized gold nanoparticles (AuNPs) from *Datura stramonium* seed

*I C Oladipo[1], A Lateef[2], M A Azeez[2], T B Asafa[3], T A Yekeen[2], S B Ogunsona[1], H M Irshad[4], and S H Abbas[5]

[1]Department of Science Laboratory Technology, [2]Department of Pure and Applied Biology, [3]Department of Mechanical Engineering, Ladoke Akintola University of Technology, Ogbomoso, Oyo State Nigeria.

[4]Faculty of Materials and Chemical Engineering, Ghulam Ishaq Khan Institute of Engineering, Science and Technology, Pakistan.

[5]Center for Excellence in Nanotechnology, King Fahd University of Petroleum and Minerals, Saudi Arabia

*Corresponding author: icoladipo@lautech.edu.ng

Abstract

The gold nanoparticles synthesized from *Datura stramonium* seed was investigated for its antidiabetic properties in the present study. A total number of 42 healthy male albino rats with body weight ranges of 150±20 g were purchased and 7 days acclimatization was carried out after which they were randomly divided into 7 groups (A-G) of 6 animals each. Group A, the control was not induced while the other six groups (B-G) were induced with 150 mg/kg of alloxan by intraperitoneal administration. After 72 h, measurement of the tail vein blood glucose was done to confirm hyperglycemia. Experimental rats with fasting glucose levels (5.6-6.9 mmol/l) were considered diabetic. Group B received no treatment, Group C received 1000 μg/kg of *Datura stramonium* seed extract, Group D received 500 μg/ml of synthesized AuNPs, Group E received 750 μg/ml of synthesized AuNPs, Group F received 1000 μg/ml of synthesized AuNPs and Group G received 1000 μg/ml of standard drug (glibenclamide). This treatment was administered every day for 21 days. The result showed that the weight of the rats reduced after diabetes induction but after 21 days of treatment, the weight of the treated groups increased when compared to the untreated group. Furthermore, it was observed that after the treatment with 500 and 750 μg/ml AuNPs, the blood sugar was down regulated unlike the untreated groups. Due to the alloxan induction, there were increases in aspartate aminotransferase (AST), alanine aminotransferase (ALT), alkaline phosphatase (ALP), urea and creatinine levels in the serum of experimental rats, which were significantly reduced after the treatment with *Datura stramonium* seed extract and the synthesized nanoparticles at 500, 750 and 1000 µg/ ml. The histopathology of the pancreas of alloxan induced rats showed severe fibrotic interlobular connective tissues with severely diffused, degenerated, and necrotized Islets of langerhans which were also normalized by treatments with AuNPs. It was confirmed in this research that AuNPs could be embedded in drugs or folk medicine aimed at reducing blood sugar levels.

Keywords: Gold nanoparticles, antidiabetic, alloxan, blood sugar, diabetes, histology.

1 Introduction

Plants are indispensable to man and of course have profound importance in so many branches of Science and Technology. Plants contain abundance of biological materials that are plentiful in bio-reductant molecules [1] and thus made plant materials good reducing and capping agents. One of the branches of science where plant materials play a major role in the properties of the finished product is Nanotechnology [2]. The word Nanotechnology was used first in 1974 by Professor Norio Taniguchi [3, 4] and since then has been the gate way to many innovations beyond the confines of human's imagination, spreading its tentacles to all fields and disciplines [5]. Nanotechnology manages the control of materials within the nanometric scale in order to produce particles that exhibit better characteristics which the parent materials do not possess. According to the National Nanotechnology Initiative (NNI), Nanotechnology centers on the generation of nanoparticles with at least one dimension sized from 1 to 100 nanometers [6, 7].

The role of plant materials in the eco-friendly production of nanoparticles are significant due to great deal of macromolecules they constitute [8]. Numerous researchers have revealed diverse plant materials whose extracts have been used as capping and reducing agents in the biological route production of nanoparticles [2], such as zinc oxide [9], while Shanmugavadivu *et al.* [4] reported the synthesis of AgNPs using pomegranate peel extract as the reducing agent. In another research endeavour, Lateef *et al.* [8] reported the biological route production of

AgNPs from cocoa pod husk, while Adelere *et al.* [10] reported the production of AgNPs from extract of wonderful kola (*Buchholzia coriacea*) to mention a few.

Datura stramonium, otherwise called Jimson weed, Devil's trumpet, and locally called Gegemu by Yoruba tribe in Western Africa is an annual, herbaceous plant that grows to a height of 2 meters with white or violet trumpet-shaped flowers and spiny fruit [11, 12]. *Datura stramonium* has medicinal uses in numerous nations around the globe. It is used to alleviate bronchitis, dressing of fresh wound [13], abortifacient [12], goiter treatment, ulcer treatment and has antimicrobial properties. More than 70 alkaloids have been reported to be identified from various plant parts [14]. These medicinal properties that *Datura stramonium* possesses triggered this research work at establishing the plausibility of assessing antidiabetic properties of the phytosynthesized AuNPs from *Datura stramonium* seed extract.

2 Materials and Methods

2.1 Synthesis and Characterization of Gold nanoparticles

Gold nanoparticles (AuNPs) were synthesized using extract of *Datura stramonium* seeds. Exactly 1 ml of the seed extract was reacted with 40 ml of 1mM $HAuCl_3$ under ambient condition. The extract was obtained through hot water (60 ^oC) extraction of dried seed powder (0.1 g in 100 ml of distilled water). The gold nanoparticles were characterized by UV-vis spectroscopy, Fourier transform infrared spectroscopy (FTIR) and scanning electron microscopy (SEM).

2.2 Anti-diabetic Activity of Gold Nanoparticles Synthesized from *Datura stramonium* Seed Extract on Alloxan Induced Diabetic Rats

2.2.1 Animal management

A total number of 42 healthy male albino rats weighing between 150-200 g were purchased at the Anatomy Department of Adeleke University, Ede, Osun State. The animals were acclimatized for seven days prior to the experimental research and the beddings were changed every 48 h. The experimental was carried out under standard conditions and the animals had free access to food and water.

IOP Conf. Series: Materials Science and Engineering **805** (2020) 012035 doi:10.1088/1757-899X/805/1/012035

2.2.2 Diabetes Induction and Treatment in experimental rats

Measurement of weights and blood glucose levels of all the rats were carried out prior to induction and all rats were confirmed to be healthy. The rats were divided randomly into 7 groups comprising of 6 rats each. The animals in Group A, the control were not induced while the other six groups were induced with 150 mg/kg of alloxan by intraperitoneal administration. Measurement of the tail vein blood glucose with the aid of an Accu-Check Sensor Comfort glucometer was carried out after 72 h in order to confirm hyperglycemia. The animals with fasting glucose levels between 5.6 and 6.9 mmol/l were considered diabetic. Group B - D was induced with diabetes while Group B received no treatment, Group C received 1500 μg/kg of jimson seed extract, Group D received 500 μg/ml of the phytosynthesized AuNPs, Group E received 750 μg/ml of synthesized AuNPs, Group F received 1000 μg/ml of AuNPs and Group G received 1000 μg/ml of standard drug (glibenclamide). This treatment was administered every day for 21 days.

All rats had free access to clean water and food, while their beddings were changed every 48 h. The rats were placed on 24 h fasting prior to the sacrificing.

2.2.3 Clinical examination of the experimental animals

The weight and fasting blood sugar of the rats were taken four times prior to animal sacrifice. The animals were sacrificed by cervical dislocation after 21 days of administration, [15].

2.2.4 Serum sample collection

Blood was drawn from the heart of all the animals and then centrifugation was done at 2000 revolution per minute for 10 min to separate the components of the blood into different layers. The upper layer which is the serum was obtained and collected into a plain sample bottles. The bottles were well labeled accordingly and recorded.

2.2.5 Pancreas sample collection

The pancreas was excised and washed free of the blood with 0.25 M sucrose solution. The pancreas was divided into two parts each; the first part was fixed in a separate container containing 10% formal saline. The second part of the pancreas was cut into small pieces, mashed in mortar and pestle, then homogenized in an equal volume of chilled phosphate buffer pH 7.0 and the homogenate was then kept in clean bottle.

2.3 Biochemical Assays

2.3.1 Liver Function Biochemical Assay

2.3.1.1 Determination of Aspartate Aminotransferase (AST) activity

Assay of aspartate aminotransferase (AST) catalyses the exchange of amino group from L-aspartate to α a-ketoglutarate with the production of oxaloacetate and glutamate. The measure of oxaloacetate is estimated by changing over it into pyruvate through treatment with aniline citrate which is trailed by responding the pyruvate with 2,4-dinitrophenylhydrazine to frame 2,4-dinitrophenyl hydrazone derivative, which is darker hued in basic medium. The absorbance of this hydrazone derivative is related to AST activity. Buffered substrate was mixed with 0.1 ml of serum and incubated in a water bath at 37 °C for 30 min. Exactly 0.1 ml distilled water was included to the blank tubes, instead of serum. Also, Dinitrophenylhydrazine (DNPH) reagent was included and incubated at room temperature for 20 min. To cap it all, 5.0 ml of 0.4 N sodium hydroxide was included. Standards were prepared and treated in the same manner and read at 546 nm after 10 min.

2.3.1.2 Determination of Alanine Aminotransferase (ALT)

Alanine aminotransferase catalyses the exchange of amino group from L-alanine to α-ketoglutarate with the development of pyruvate and glutamate. The pyruvate produced is permitted to react with 2,4-dinitrophenylhydrazine to from 2,4-dinitrophenylhydrazone derivative which is brown colored in basic medium. The absorbance of this hydrazone derivative is related to ALT activity. Buffered substrate was mixed with 0.1 ml of serum and incubated in a water bath at 37 °C for 30 min. 0.1 mL distilled water was then added to the blank tubes instead of serum. Also, Dinitrophenylhydrazine (DNPH) reagent was included and incubated at room temperature for 20 min. Then, 5.0 ml of 0.4N sodium hydroxide was included and standards were also treated in the same manner were prepared and read at 520 nm after 10 min.

2.3.1.3 Determination of Alkaline Phosphatase (ALP) activity

Serum alkaline phosphatase hydrolyses a colorless substrate of phenolphthalein monophospahte giving rise to phosphoric acid and phenolphthalein, which at alkaline pH values turns into a pink colour that can be spectrophotometrically determined [16]. An aliquot of 1.0 ml of water was pipetted into two test tubes (test tube SA and ST), to each of the test tube was added 1 drop of the chromogen substrate and was mixed and then incubated at 37 ^{o}C for 15 min. Then to the test tube SA, 0.1 ml of the sample was added and to the test tube ST, 0.1 ml of the

IOP Conf. Series: Materials Science and Engineering **805** (2020) 012035 doi:10.1088/1757-899X/805/1/012035

standard alkaline phosphatase was added. The test tubes were mixed and incubated at 37 °C for 20 min. To each of the test tube, 5 ml of the colour developer was added. Absorbance was read against a water blank at wavelength of 550 nm. The activity of ALP in the sample was calculated using the formula:

UL of Alkaline phosphatase = SA Absorbance / ST Absorbance × 100

2.4 Renal function biochemical Assay

2.4.1 Determination of creatinine level

Creatinine in alkaline solution reacts with picric acid to form a coloured complex. The amount of the complex formed is directly proportional to the creatinine concentration. Picric acid solution (0.5 ml of 25 mol/l) was added to 0.5 ml of 0.4 Mol/l sodium hydroxide solution and the reaction was initiated by 0.1 ml of serum. The changed absorbance at 530 nm was recorded after 30 seconds and exactly 2 min later. Creatinine activity was then calculated from change in absorbance per min.

2.4.2 Determination of Urea

Urea in the sample originates from the coupled reaction described below. The colour complex formed is measured in the spectrometer.

$$Urea + H_2O \xrightarrow{\text{(Urease).}} 2NH_4^+ + CO_2$$

$$NH_4^+ + Salicylate + \xrightarrow{\text{NaCl (nitroprusside)}} Indophenol$$

Sample (10 μl) was pipetted in to the cuvette and 1m l of Reagent A (62 mmol/l Sodium salicylate, 3.4 mmol/l sodium nitropruside, 20 mmol/l Phosphate buffer pH 6.9 and 500 U urease) was added, then thoroughly mixed and incubated for 10 min at room temperature. After incubation, 1.0 ml of Reagent B (7 mmol/l sodium hypochloride and sodium hydroxide) was also added, thoroughly mixed and also incubated at room temperature for 10 min. The absorbance was read at 600 nm against reagent blank.

2.5 Oxidative Stress Biochemical Assay

2.5.1 Determination of Total Protein

Protein concentration was determined based on the principle of Folin-Lowry method [17]. Protein reacts with the Folin- Ciocalteau reagents to give a coloured complex. The intense blue colour so formed is as a result of the reaction of the alkaline copper with the proteins in Buiret test and the reduction of phosphomolybdate by tyrosine tryptophan present in the protein. Exactly 6 ml of alkaline copper reagent was added to 1.2 ml (0.300 μg of BSA) and thoroughly

IOP Conf. Series: Materials Science and Engineering **805** (2020) 012035 doi:10.1088/1757-899X/805/1/012035

mixed. The solution was kept for 10 min. Folin-Ciocalteau reagent (0.3 ml) was added to 0.2 ml of the sample and thoroughly mixed, after which the solution was kept for 30 min. Absorbance was read at 500 nm against the blank reagent. The result of the test was read from the protein standard curve.

2.5.2 Determination of glutathione activity

Glutathione reduces the hydrogen peroxide (H_2O_2) into water (H_2O). 5,5`-dithiobis (2-nitrobenzoic acid) (DTNB), a water soluble aromatic disulphide react with thiol (SH) compound at pH 8.0 and produces one mole coloured complex per mole thiol. The colour intensity of this complex anion is measure at 412 nm to determine the thiol (SH) concentration. Glutathione level was expressed as μmol/ml blood. Whole blood (0.4 ml) mixed with 1.6 ml of phosphate buffer was added in 2.0 ml of TCA and centrifuged at 3000 rpm for 15 min. The supernatant (1.0 ml) was added to 1.5 ml of Na_2HPO_4 solution and 0.25 ml of DTNB, after which the colour was read at 412 nm against blank reagent.

2.5.3 Determination of superoxide dismutase activity

The ability of Superoxide dismutase to inhibit the auto oxidation of epinephrine at pH 10.2 makes this reaction a base for simple assay of dismutase. Then 1 ml of the sample was added to 9 ml of distilled water to make a 1 in 10 dilution. An aliquot of the diluted sample was added to 2.5 ml of 0.05 M carbonate buffer of pH 10.2 to equilibrium in the spectrophotometer and the reaction started by addition of 0.3 ml of adrenaline [18]. Absorbance was read at 480 nm against the blank and the calculation is as follows;

$$\text{Increase in absorbance per minute} = A_3 - A_0 / 2.5$$

Where A_0 = absorbance at time (t= 0)

A_3 = absorbance at time (t = 150 seconds)

$$\%\ \text{inhibition} = 100 - 100\frac{(\Delta A480\ \text{inhibited})}{(\Delta A480\ \text{uninhibited})}$$

$$\text{Activity} = \frac{\%\ \text{inhibition}\ \ \text{x d.f}}{50\%\ \text{inhibition}}$$

$$\text{Specific Activity} = \frac{\text{Activity}}{\text{Protein concentration}} = \frac{\text{Units of enzymes}}{\text{mg}/ml\ \text{protein}}$$

2.5.4 Determination of malondialdehyde

Small amount of malondialdehyde (MDA) are produced during lipid peroxidation and these are able to react with thiobarbituric acid (TBA) to generate a pink coloured complex which is acidic solution to absorb light at 532 nm. A portion of the sample (0.1 ml) was diluted in 0.9 ml of distilled water to make a dilution of 1 in 10. An aliquot of 0.4 ml of the sample was mixed with 1.6 ml tris KCl buffer to which 0.5 ml of 30% Trichloro-acetic acid (TCA) was added. A portion of TBA (0.5 ml) was then added and kept in water bath for 45 min at 80 °C. Centrifugation of the produced pink colored reaction mixture was done at 14000g for 15 min. The absorbance of the pink supernatant was then read against reference blank at 532 nm in spectrophotometer

$$\text{MDA (units/ mg protein)} = \frac{\text{absorbance} \times \text{volume of mixture}}{\text{E 532} \times \text{volume of sample} \times \text{mg protein}}$$

2.6 Histological Analysis of Pancreas Using Haematoxylin and Eosin Techniques

The pancreas tissue was dehydrated to remove water present in the sample; removal of the water was done by immersing the tissue in increasing concentration of ethanol until 100% which was the final stage of dehydration. The tissue was immersed in different xylene immersions. In this stage, the ethanol was replaced with xylene gradually and the tissue was embedded by replacing the xylene with molten paraffin wax. After the final infiltration, the tissue cassettes was transferred to an embedding station, which served as a reservoir of molten wax, hot plates and cold plate for setting blocks. The infiltrated tissue was removed from the cassettes within a suitable sized metal mould, and the whole mould was placed on the cold plate to finally set.

Cutting of the tissue was done by using a microtome to sizes ranging from 2 to 50 μm. Once sections were cut, they were floated on a warm water bath that helped to remove wrinkles. Then, glass slides were then transferred to a warm oven for about 15 min to help the section adhere to the slide. The slides were deparaffinized by passing them through xylene, alcohol and water prior staining. The slides were then kept in the solution for an extended period of time and then taken back through acid-alcohol to remove part of the stain during regressive staining. While with a progressive staining, the slide was dipped in the haematoxylin until the desired intensity was achieved. The slide was then covered a glass cover slip in order to prevent scratching of the tissue, and to provide good optical quality under the microscope.

2.7 Statistical Analysis

Data were analyzed using Graph pad prism 6. Data were expressed as mean ± Standard error of mean (Mean ± SEM). Mean values were compared using one-way analysis of variance and differences between mean were determined using Tukey tests. P-value less than 0.05 ($P < 0.05$) were taken to be statistically significant.

3 Results and Discussion

3.1 Biosynthesis and Characterization of AuNPs

The seed extract of Jimson weed facilitated the formation of AuNPs as shown in Figure 1.As previously reported from our laboratory, the AuNPs were nearly spherical, and agglomerated in nature with size of 75.1-156.5 nm. The EDX analysis displayed gold (Au) as the prominent metal present, while the SAED pattern of the AuNPs showed the face-centred crystalline nature of AuNPs. The UV-visible spectrum of the AuNPs synthesized displayed clear peak at 536.0 nm. The FTIR analysis showed prominent peaks at 3423.76, 2920.32, 2360.95 and 1635.69 cm^{-1} attributed to the involvement of proteins in the biofabrication and capping of AuNPs (Figure 2).

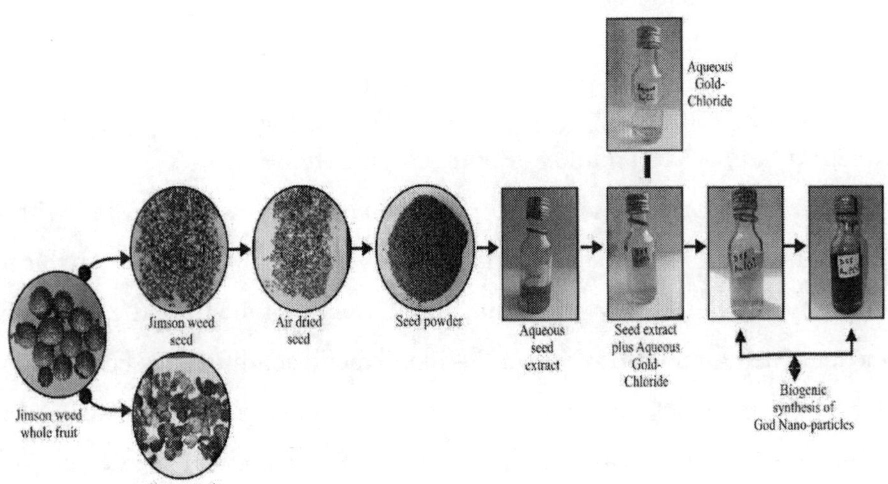

Figure 1. Sample preparation and synthesis of AuNPs by seed extract of Jimson

IOP Conf. Series: Materials Science and Engineering **805** (2020) 012035 doi:10.1088/1757-899X/805/1/012035

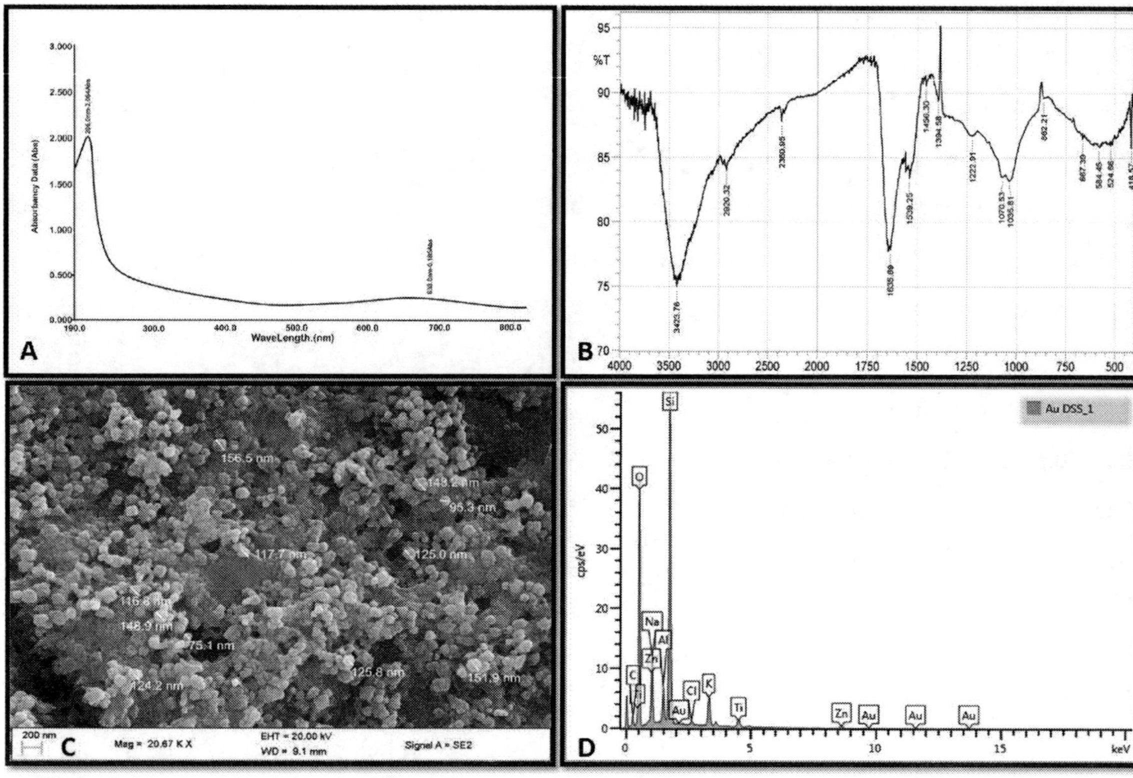

Figure 2. The characteristics of AuNPs; A, UV-vis spectrum; B, FTIR spectrum; C, FESEM micrograph; D, EDX spectrum

3.2 Antidiabetic activity of AuNPs on alloxan induced diabetic rats

Table 1 showed the weight of diabetic rats treated with AuNPs synthesized with jimson seed extract. According to the result, the weight of the experimental animals reduced after diabetes induction whereas, after 21 days of treatment, the weight of the treated groups increased compared with the untreated group. Body weight loss in diabetic condition has been attributed to increased protein catabolism, which is secondary to insulin deficiency [19, 20]. This observation is in agreement with previous report on Streptozotocin (STZ) induced diabetic rats whereas, weight gain was observed upon treatment with seed extract of *D. stramonium* [21]. In addition, literature has it that various plant extracts improved body weight loss in STZ-induced diabetic animals [22, 23, 24]. Similarly, the improved body weight observed suggests that *D. stramonium*

IOP Conf. Series: Materials Science and Engineering **805** (2020) 012035 doi:10.1088/1757-899X/805/1/012035

seed extract and the phytosynthesized AuNPs might have protective effects against protein catabolism and muscle wasting possibly by enhancing insulin secretion and action [21].

Table 1. Weight of experimental rats before and after alloxan induction

	Group A	Group B	Group C	Group D	Group E	Group F	Group G
IW	159±6.8	192±8.1	221.6±12.4	182.5±4.0	193.7±3.2	195±3.8	184.0±3.00
WAI	166±7.7	154±4.5	174.8±4.6	174.8±2.77	182.2±4.5	180±5.4	171.7±2.6
FW	186±8.9	169±7.3	185.8±5.9	185.3±3.75	198.3±2.85	197.8±4.6	181.8±5.8

IW, Initial weight; WAI, Weight After Induction; FW, Final weight; Group A, Control ; Group B, Induced with Diabetes only; Group C, Induced with Diabetes + extract ; Group D, Induced with Diabetes +500 μg/ml of AuNPs; Group E, Induced with Diabetes + 750 μg/ml of AuNPs; Group F, Induced with Diabetes +1000 μg/ml of AuNPs; Group G, Diabetes +1000 μg/ml of Standard drug

The fasting blood sugar of alloxan induced diabetic rats is shown in Table 2. Rats with blood glucose above 5.6 mmol/L were considered diabetic. Treatment with the extract and synthesized AuNPs however normalized the increase in fasting blood sugar of the rats within 500 and 750 μg/ml after 21 days of treatment. Highest percentage glucose reduction was observed in-group D, subsequently Group G, E, and F. Observation shows that the synthesized AuNPs was effective against hyperglycemia. Effective blood glucose control is the key to managing diabetes, preventing its complications and improving the quality of life in patients with diabetes. In order words, a sustained reduction in hyperglycaemia will decrease the risk of developing diabetic complications [25].

The potentials of AuNPs in treating inflammatory and autoimmune diseases [26] have prompted investigation into its anti-oxidative and anti-hyperglycemic activities [27, 28]. According to Selvaraj *et al.* [27] and Edrees *et al.* [28], AuNPs was effective in lowering blood glucose by increasing insulin level and sensitivity, and preventing insulin resistance in rats. In addition, plant phytochemicals modulate glucose metabolism by increasing insulin secretion and action, decreasing gluconeogenesis, protecting pancreatic β-cells against oxidative stress and inflammation, inhibiting α-glucosidase and α-amylase, promoting pancreatic β-cell proliferation,

IOP Conf. Series: Materials Science and Engineering **805** (2020) 012035 doi:10.1088/1757-899X/805/1/012035

enhancing expression and translocation of glucose transporters, and inhibiting apoptosis [29, 30, 31, 32, 33]. Phytochemical studies on the root and leaves of *D. stramonium* revealed important compounds that serve as good capping agent and have good sugar lowering effects [21, 34]. In this study, similar compounds in the seed extract might have contributed to the improved activity observed in the synthesized gold nanoparticles.

Table 2. Effects of treatment with AuNPs on the fasting blood glucose of diabetic rats

	Group A	Group B	Group C	Group D	Group E	Group F	Group G
Initial blood glucose	5.2±0.2	6.9±0.2	7.0±0.7	15.3±1.3	11.5±1.1	9.4±1.4	17.6±2.3
After Treatment	5.2±0.2	6.3±0.4	4.7±0.4	4.1±0.3	4.2±0.5	4.5±0.3	6.3±0.2
Percentage glucose reduction (%)	0.00	8.70	32.86	73.20	63.48	52.13	64.20

FBS, Fasting blood sugar; Groups A-G are as defined in Table 1

3.3 Effects of AuNPs on Aspartate Amino Transferase (AST), Alkaline Phosphatase (ALP) and Alanine Amino Transferase (ALT) Activities of alloxan induced diabetic rats

The result of the liver function markers of diabetic rats treated with the extract and synthesized AuNPs are shown in Figures 3-5. Significant ($p < 0.05$) increase was noted in the AST and ALT activities of diabetic untreated group (Group B), Group E and F when comparison was done with control group, whereas treatment with extract (Group C), 500 µg/ml, 750 µg/ ml and 1000 µg/ ml of the phytosynthesized AuNPs particles (Group D, E and F) and standard drug (Group G) caused significant ($p < 0.05$) decline when contrasted with Group B. No significant difference ($p > 0.05$) was noted in the AST and ALT activities of Group D, E and F when comparison was done with Group C.

Significant increment ($p < 0.05$) was noted in the ALP activity of the diabetic control group (Group B), Group E and F when compared with the control group. However, ALP activity decreased significantly ($p < 0.05$) after treatment with extract (Group C), 500 µg/ ml, 750 µg /ml and 1000 µg /ml of synthesized AuNPs (Group D, E and F) and standard drug (Group G) when

IOP Conf. Series: Materials Science and Engineering **805** (2020) 012035 doi:10.1088/1757-899X/805/1/012035

compared with Group B. The increased ALP activity noted in Group D, E and F when contrasted with Group C were statistically insignificant.

The liver function biomarkers such as AST, ALT and ALP are metabolic enzymes with ability to escape into the blood stream upon liver damage due to hyperglycemic conditions [28]. In present study, diabetic induction with alloxan caused damage to the liver. Conversely, the ability of *Datura stramonium* seed extract and AuNPs to control these enzymes affirms it potentials to prevent liver damage. Although the biochemical parameters in the AuNPs treated group differ significantly with the untreated diabetic group, however, the activity of the in the extract treated group showed weak organ protective activity. Previous studies have reported the protective role of AuNPs [28] on the liver in diabetic rats. In present study, both the extract and AuNPs were able to restore the liver integrity of diabetic rats to near normal.

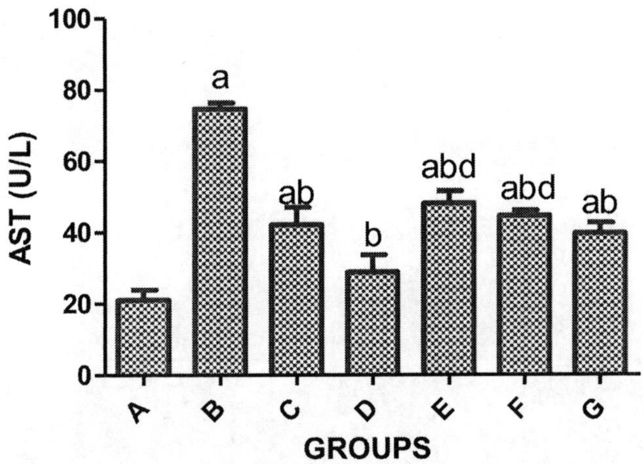

Figure 3. Effect of AuNPs on aspartate amino transferase (AST) activities of experimental rat serum (Groups A-G are as defined in Table 1)

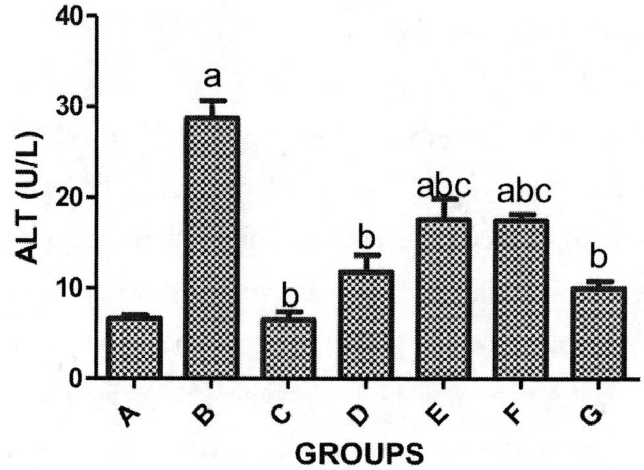

Figure 4. Effect of AuNPs on alanine amino transferase (ALT) activities of experimental rat serum (Groups A-G are as defined in Table 1)

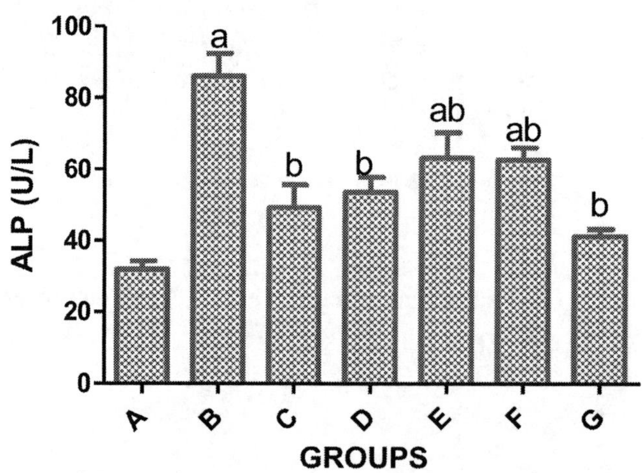

Figure 5. Effect of AuNPs on alkaline phosphatase (ALP) activities of experimental rat serum (Groups A-G are as defined in Table 1)

3.4 Effects of extracts and AuNPs on total protein, urea and creatinine levels of alloxan induced diabetic rats

Figures 6-8 present the effects of treatment of diabetic rats with extracts and AuNPs on renal function of rats. Total protein level decreased significantly ($p < 0.05$) in all the groups (diabetic untreated (Group B), Group C, D, E, F and G) when contrasted with the control group. Significant ($p < 0.05$) increment noted in the urea and creatinine levels of the untreated diabetic control declined significantly ($p < 0.05$) upon treatment with extract (Group C), 500 µg/ ml, 750 µg /ml and 1000 µg /ml of synthesized AuNPs (Group D, E and F) and standard drug (Group G). Significant ($p < 0.05$) increase was also noted in the urea and creatinine levels of Group D, E and F when contrasted with the normal control group, whereas no significant ($p > 0.05$) difference was noted between group G and the control.

Assessment of urea and serum creatinine levels are established indicators for diabetes related renal dysfunction. Similar to the observation in present study, Virk [35] reported increase in these parameters of diabetic rats, which was subsequently brought down upon treatment with AuNPs [28]. According to Shanmugasundaram *et al.* [36], increase in blood urea and serum creatinine level in rats induced with Alloxan for 14 days indicates an abnormal glomerular filtration. Furthermore, increased blood urea levels can be attributed to a high rate of protein breakdown during diabetes and the failure of excretion due to an abnormal renal filtration [37]. In the present study, the extracts and AuNPs demonstrated some renal protective activities in diabetic rats.

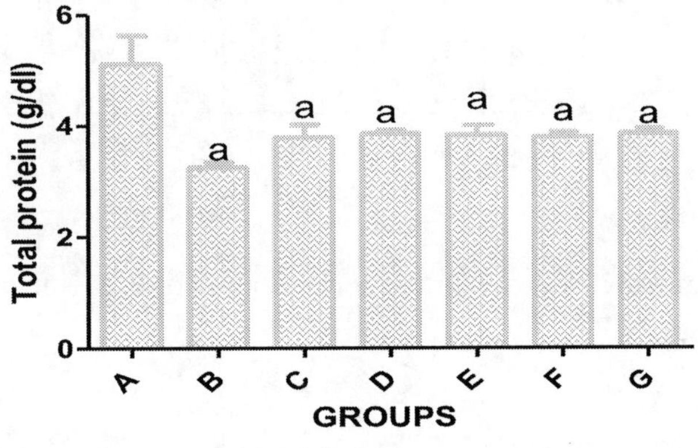

Figure 6. Evaluation of AuNPs on total protein concentration of experimental rat pancreas (Groups A-G are defined in Table 1)

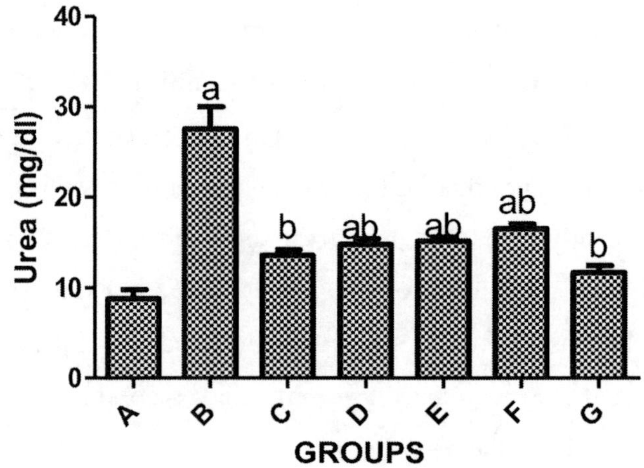

Figure 7. Evaluation of AuNPs on urea concentration of experimental rat pancreas (Groups A-G are defined in Table 1)

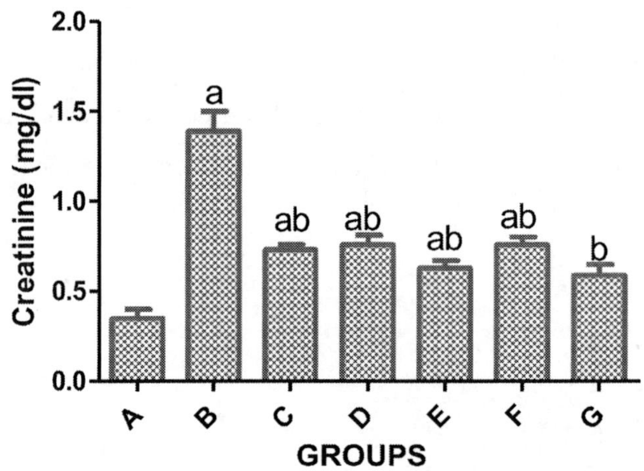

Figure 8. Evaluation of AuNPs on creatinine concentration of experimental rat pancreas (Groups A-G are defined in Table 1)

IOP Conf. Series: Materials Science and Engineering **805** (2020) 012035 doi:10.1088/1757-899X/805/1/012035

3.5 Effects of extracts and AuNPs on Glutathione (GSH) and Malodialdehyde (MDA) Concentrations and Superoxide Dismutase (SOD) activity of alloxan induced diabetic rats

Table 3 shows the antioxidant status of alloxan induced diabetic rats treated with the extracts and synthesized AuNPs. The result showed significant ($p<0.05$) decline in GSH concentration and SOD activities of Group B, Group C, D, E and F when comparison was made with control group. Conversely, there was significant increment ($p < 0.05$) in GSH concentration of Group G when contrasted with group B whereas no significant difference was observed when compared with the normal control. The SOD activity of Group G showed insignificant ($p > 0.05$) difference with the untreated diabetic group (Group B). Significant increase ($p < 0.05$) observed for the MDA levels of the untreated diabetic rats (Group B) declined significantly ($p < 0.05$) after treatment with extract (Group C), 500 µg /ml, 750 µg /ml and 1000 µg /ml of synthesized AuNPs (Group D, E and F) and standard drug (Group G). No significant difference ($p > 0.05$) was noted when the control was contrasted with the treated groups. Hyperglycemia is said to enhance oxidant production and destabilizes the homeostasis balance provided by the antioxidant defense system of the body [38]. Treatment of diabetic rats reversed reactive oxygen species (ROS) induced lipid peroxidation under hyperglycemic conditions, which is in concurrence with past report in diabetic rodents treated with AuNPs [27]. As previously revealed by Belayneh *et al.* [21], the antioxidant activity of the root extract of *D. stramonium* suggested that the β-cell protective effect secondary to the antioxidant activity of the root extract might have contributed to the antihyperglycemic activity observed in STZ-`induced diabetic mice

Oxidative stress as one of the accompanying complications of hyperglycemia is linked with changes in liver function markers [39, 40, 41]. Consequently, increase in ALT, ALP and AST activities observed in diabetes is the effects of ROS induced organ damage. In this study, reversal of the imbalances in the antioxidants and liver enzymes responsible for the cell dysfunction and destruction, leading to tissue injury observed in hyperglycemic conditions by *Datura stramonium* seed extracts and AuNPs could be attributed to lipid peroxidation inhibiting potentials.

Table 3. Effects of extracts and Gold Nanoparticles on Glutathione (GSH) and Malodialdehyde (MDA) Concentrations and Superoxide Dismutase (SOD) activity of alloxan induced diabetic rats

Groups	GSH (μmol/mg protein)	SOD (μmol/mg protein)	MDA (μmol/mg protein)
A	33.13 ± 3.90	2.55± 0.45	0.52± 0.10
B	8.53 ± 0.58 [a]	0.32± 0.05 [a]	2.54 ± 0.69[a]
C	18.05 ± 0.79[a]	0.95 ± 0.09 [a]	0.75 ± 0.09 [b]
D	11.98 ± 2.37 [a]	1.07 ± 0.16[a]	0.51 ± 0.11[b]
E	11.32 ± 2.68[a]	1.02 ± 0.35 [a]	0.63 ± 0.11 [b]
F	12.88 ± 3.17 [a]	0.96 ± 0.11[a]	0.76± 0.11[b]
G	26.58 ± 3.36 [b]	0.86 ± 0.28 [a]	0.69 ±0.13[b]

Values are expressed as Mean ± S.E.M of GSH, MDA concentration and SOD activities using student t-test, Value having "a" superscript is significantly different from the control, Values having "b" superscript is significantly different from diabetic untreated control group; groups A-G are as defined in Table 1

3.6 Histopathological analysis of the Pancreas

The histopathology of the pancreas of diabetic rats showed very poor architecture, severe fibrotic interlobular connective tissues (blue arrow) and septa were observed (red arrow). Mild vascular congestion, with severely diffused, degenerated, and necrotized Islets of langerhans (white arrow) was also seen but this was normalized by treatments with AuNPs and moderately normalized by *Datura stramonium* seed extract (Figure 9). Destruction of insulin secreting β cells of islets of Langerhans by alloxan is achieved via damage to β-cell DNA [42], mitochondria [43], lysosomes [44], and plasma membrane [45] of the pancreas. This reduction in insulin secreting cells lowered insulin production and consequently resulted in hyperglycemia noted. The histopathological damage noted are consistent with the damage earlier reported by Hadi *et al.* [46]. *Datura stramonium* seed extract and AuNPs restored the insulin secreting potentials of the pancreas as reflected in the blood glucose reduction. Treatment might have reversed lipid peroxidation to restore the pancreatic integrity [47, 48].

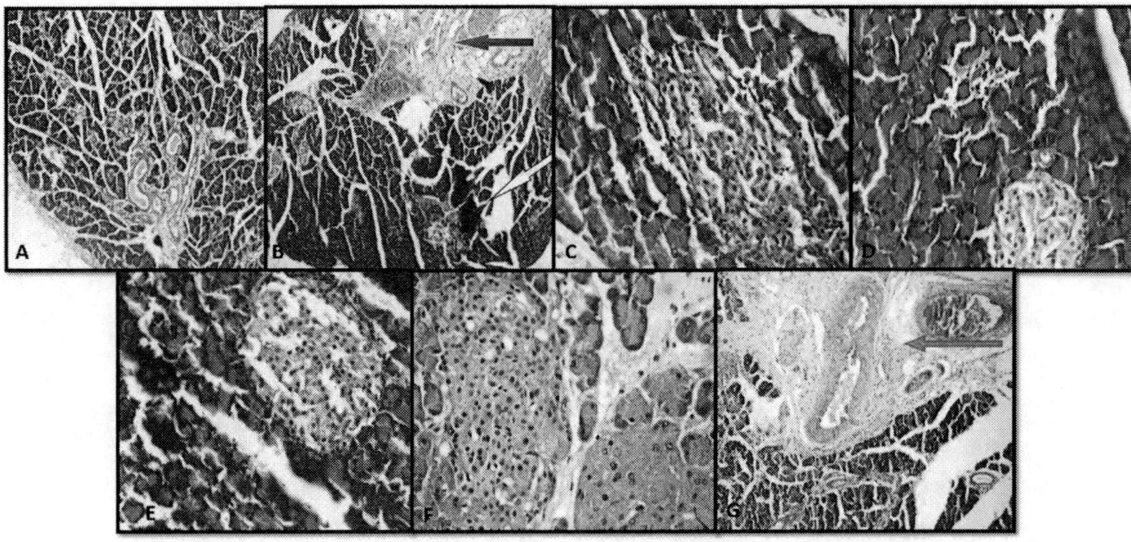

Figure 9. Histological study of the pancreas of experimental rats (Groups A-G are defined in Table 1)

4 Conclusion

The findings of this study indicated that the AuNPs phytosynthesized from *Datura stramonium* seed extract were effective in lowering blood glucose by increasing insulin level in rats. Also, diabetic induction with alloxan caused damage to the liver but both the extract and AuNPs were able to restore the liver integrity of diabetic rats to near normal thus, affirming its potentials to prevent liver damage. The extracts and AuNPs demonstrated some renal protective effects in diabetic rats by reducing the increment in urea and creatinine levels. The severe damage to the pancreas was also normalized by treatments with AuNPs. Thus, AuNPs could be embedded in drugs or folk medicine as anti-diabetic agent.

References

[1] Kumar R, Roopan SM, Prabhakarn A, Khanna VG and Chakroborty S 2012 Agricultural waste *Annona squamosa* peel extract: biosynthesis of silver nanoparticles. *Spectrochim. Acta Part A: Mol. Biomol. Spectr.* **90** pp 173-176.

[2] Oladipo I C and Ogunsona S B 2019 The Utilization of Agro Waste: A nanobiotechnology Point of View. Recent Advances in Biological Research

Vol. **5;** Chapter 10 Print ISBN: 978-93-89246-84-1, eBook ISBN: 978-93-89246-85-8.

[3] Taniguchi N 1983 Current status in, and future trends of, ultraprecision machining and ultrafine materials processing. *CIRP Ann.* **32** (2) pp 573-582.

[4] Shanmugavadivu M, Selvam K and Ranjithkumar R 2014 Synthesis of Pomegranate Peel Extract Mediated Silver Nanoparticles and its Antibacterial Activity. *Am. J. Adv. Drug Deliv.* **2** (2) pp 174-182.

[5] Baker S and Satish S 2012 Endophytes: toward a vision in synthesis of nanoparticle for future therapeutic agents. *Int. J. Bio-Inorg. Hybd. Nanomat.* **1** (2) pp 67-77.

[6] Roco MC 2001 From vision to the implementation of the US National Nanotechnology Initiative. *J. Nanopart. Res.* **3** (1) pp 5-11.

[7] Lövestam G, Rauscher H, Roebben G, Klüttgen B S, Gibson N 2010 Considerations on a definition of Nanomaterial for regulatory purposes. Joint Research Centre (JRC) Reference Reports. **80** pp 00-41.

[8] Lateef A, Ojo S A, Folarin B I, Gueguim-Kana E B, Beukes L S 2016 Kolanut (*Cola nitida*) mediated synthesis of silver-gold alloy nanoparticles: antifungal, catalytic, larvicidal and thrombolytic applications. *J. Clust. Sci.* **27** (5) pp 1561-1577.

[9] Fatimah I 2018 Biosynthesis and characterization of ZnO nanoparticles using rice bran extract as low-cost templating agent. *J. Eng. Sci. Technol.* **13** (2) pp 409- 420.

[10] Adelere IA, Lateef A, Aboyeji DO, Abdulsalam R, Adabara NU and Bala JD 2017. Biosynthesis of silver nanoparticles using aqueous extract of *Buchholzia coriacea* (wonderful kola) seeds and their antimicrobial activities. *Ann. Food Sci. Technol.* **18** pp 671-679.

[11] Dethier M, Cordier Y and Demeyer K 1993. Cultivation of *Datura* species for scopolamine and hyoscyamine production in Burundi. *ISHS Acta Hortic.* **331**. https://doi.org/10.17660/ActaHortic.1993.331.6.

[12] Mairura FS and Setshogo MP 2008 *Datura stramonium* L. Schmelzer GH, Gurib-Fakim A (Editors). pp 221-225.

[13] Burkill HM 2000 The useful plants of West Tropical Africa. 2nd Edition. Volume 5, Families S–Z, Addenda. Royal Botanic Gardens, Kew, Richmond, United Kingdom. 686 pp

[14] Philipov S, Berkov S and Doncheva T S 2007 GC-MS survey of *Datura stramonium* alkaloids. *Comptes Rendus de l'Académie Bulgare des Sciences* **60** (3) pp 239-250.

[15] Oladipo IC, Adebayo EA and Kuye OM 2015 Effect of monosodium glutamate in ovaries of female Sprague-dawley rats. *Int. J. Curr. Microbiol. Appl. Sci.* **4** (5) pp 737-745.

[16] Klein B, Read PA and Babson AL 1960 Rapid method for the quantitative determination of serum alkaline phosphatase. *Clin. Chem.* **6** (3) pp 269-275.

[17] Lowry OH, Rosebrough NJ, Farr AL and Randall RJ 1951 Protein measurement with the Folin phenol reagent. *J. Biol. Chem.* **193** pp 265-275.

[18] Banting FG 1926 An address on diabetes and insulin: being the Nobel lecture delivered at Stockholm on September 15th, 1925. *Can. Med. Assoc. J.* **16** (3) pp 221-232.

[19] Eleazu CO, Iroaganachi M, Okafor PN, Ijeh II and Eleazu KC 2013 Ameliorative potentials of ginger (*Z. officinale* Roscoe) on relative organ weights in streptozotocin induced diabetic rats. *Int. J. Biomed. Sci.* **9** (2) pp 82-90.

[20] Stirban A, Gawlowski T and Roden M 2014 Vascular effects of advanced glycation endproducts: clinical effects and molecular mechanisms. *Mol. Metab.* **3** (2) pp 94-108.

[21] Belayneh YM, Birhanu Z, Birru EM and Getenet G 2019 Evaluation of *in vivo* antidiabetic, antidyslipidemic, and *in vitro* antioxidant activities of hydromethanolic root extract of *Datura stramonium* L. (Solanaceae). *J. Exp. Pharmacol.* **11** pp 29-38.

[22] Sharma S, Choudhary M, Bhardwaj S, Choudhary N and Rana AC 2014 Hypoglycemic potential of alcoholic root extract of *Cassia occidentalis* Linn. in streptozotocin induced diabetes in albino mice. *Bull. Faculty Pharm. Cairo Univ.* **52** (2) pp 211-217.

[23] Sulyman AO, Akolade JO, Sabiu SA, Aladodo RA and Muritala HF 2016 Antidiabetic potentials of ethanolic extract of *Aristolochia ringens* (Vahl.) roots. *J. Ethnopharmacol.* **182** pp 122-128.

[24] Zhang Y, Feng F, Chen T, Li Z and Shen QW 2016 Antidiabetic and antihyperlipidemic activities of *Forsythia suspensa* (Thunb.) Vahl (fruit) in streptozotocin-induced diabetes mice. *J. Ethnopharmacol.* **192** pp 256-263.

[25] Attele AS, Zhou YP, Xie JT, Wu JA, Zhang L, Dey L, Pugh W, Rue PA, Polonsky KS and Yuan CS 2002 Antidiabetic effects of *Panax ginseng* berry extract and the identification of an effective component. *Diabetes* **51** (6) pp 1851-1858.

[26] Mukherjee P, Bhattacharya R, Wang P, Wang L, Basu S, Nagy JA, Atala A, Mukhopadhyay D and Soker S 2005 Antiangiogenic properties of gold nanoparticles. *Clin. Cancer Res.* **11** (9) pp 3530-3534.

[27] BarathManiKanth S, Kalishwaralal K, Sriram M, Pandian SR, Youn HS, Eom S and Gurunathan S 2010 Anti-oxidant effect of gold nanoparticles restrains hyperglycemic conditions in diabetic mice. *J. Nanobiotechnol.* **8** 16. https://doi.org/10.1186/1477-3155-8-16.

[28] Edrees HM, Elbehiry A and Elmosaad YM 2017 Hypoglycemic and anti-inflammatory effect of gold nanoparticles in streptozotocin-induced type 1 diabetes in experimental rats. *Int. J. Diabetes Res.* **6** (1): pp 16-23.

[29] Banshidhar B and Yadav D 2013 Current researches on plants having antidiabetic potential: an overview. *Res. Rev.* **2** (2) pp 4-17.

[30] Jiang CS, Liang LF and Guo YW 2012 Natural products possessing protein tyrosine phosphatase 1B (PTP1B) inhibitory activity found in the last decades. *Acta Pharmacol. Sin.* **33** (10) pp 1217-1245.

[31] Romano B, Pagano E, Montanaro V, Fortunato AL, Milic N and Borrelli F 2013 Novel insights into the pharmacology of flavonoids. *Phytother. Res.* **27** (11) pp 1588-1596.

[32] Kennedy DO 2014 Polyphenols and the human brain: plant "secondary metabolite" ecologic roles and endogenous signaling functions drive benefits. *Adv. Nutr.* **5** (5) pp 515-533.

[33] Sayem AS, Arya A, Karimian H, Krishnasamy N, Ashok Hasamnis A and Hossain CF 2018 Action of phytochemicals on insulin signaling pathways accelerating glucose transporter (GLUT4) protein translocation. *Mol.* **23** (2) 258. https://doi.org/10.3390/molecules23020258.

[34] Shobha G, Soumya C, Shashidhara KS and Moses V 2014 Phytochemical profile, antibacterial and antidiabetic effects of crude aqueous leaf extract of *Datura stramonium*. *Pharmacophore* **5** (2) pp 273-278.

[35] Virk P 2018 Antidiabetic activity of green gold-silver nanocomposite with *Trigonella foenum graecum* L. seeds extract on streptozotocin-induced diabetic rats. *Pak. J. Zool.* **50** (2) pp 711-718.

[36] Shanmugasundaram R, Devi KV, Soris TP, Maruthupandian A and Mohan VR 2011 Antidiabetic, antihyperlipidaemic and antioxidant activity of *Senna auriculata* (L.) Roxb. leaves in alloxan induced diabetic rats. *Int. J. Pharm. Tech. Res.* **3** (2) pp 747-756.

[37] Palanivel R, Thangavel M, Selvendiran K and Sakthisekaran D 2001 Insulinomimetric effect of ammonium paratungstate on protein metabolism in streptozotocin induced diabetic rats. *Biomed.-.Trivandrum Then Taramani* **21** (2/3), pp 23-30.

[38] King GL and Loeken MR 2004 Hyperglycemia-induced oxidative stress in diabetic complications. *Histochem. Cell Biol.* **122** (4) pp 333-338.

[39] Giugliano D, Ceriello A and Paolisso G 1996 Oxidative stress and diabetic vascular complications. *Diabetes Care* **19** (3) pp 257-267.

[40] Johnson PC, Doll SC and Cromey DW 1986 Pathogenesis of diabetic neuropathy. *Ann. Neurol.* **19** (5) pp 450-457.

[41] Ruggiero D, Lecomte M, Michoud E, Lagarde M and Wiernsperger N 1997 Involvement of cell-cell interactions in the pathogenesis of diabetic retinopathy. *Diabetes Metab.* **23** (1) pp 30-42.

[42] Yamamoto H, Uchigata Y and Okamoto H 1981 DNA strand breaks in pancreatic islets by *in vivo* administration of alloxan or streptozotocin. *Biochem. Biophys. Res. Comm.* **103** (3) pp 1014-1020.

[43] Boquist L and Ericsson I 1984 Inhibition by alloxan of mitochondrial aconitase and other enzymes associated with the citric acid cycle. *FEBS Lett.* **178** (2) pp 245-248.

[44] Zhang H, Zdolsek JM and Brunk UT 1992 Alloxan cytotoxicity involves lysosomal damage. *Acta Pathol. Microbiol. Immunol. Scand.* **100** (1-6) pp 309-316.

[45] Watkins D, Cooperstein SJ and Lazarow A 1964 Effect of alloxan on permeability of pancreatic islet tissue *in vitro*. *Am. J. Physiol.* **207** (2) pp 436-440.

[46] Adil Hadi M, Zaidan HK, Al-Saadi AH. Histopathological changes of pancreatic tissues in hyperglycemic male rats treated with mixture of plants extracts. *Int. J. ChemTech. Res.* **9** pp 501-513.

[47] Usha K and Saroja S 2000 Antitubercular potential of selected plant materials. *J. Med. Arom. Plant Sci.* **22** pp182-184.

[48] Bhandari U and Pillai KK 2005 Effect of ethanolic extract of *Zingiber officinale* on dyslipidaemia in diabetic rats. *J. Ethnopharmacol.* **97** (2) pp 227-230.

IOP Conf. Series: Materials Science and Engineering **805** (2020) 012036 doi:10.1088/1757-899X/805/1/012036

Effects of pH, temperature and agitation on the biosynthesis of iron nanoparticles produced by *Trichoderma* species

*SO Kareem, TM Adeleye and RO Ojo
Department of Microbiology, Federal Univeristy of Agriculture, Abeokuta, Ogun State, Nigeria.
*email: sharafkareem@yahoo.co.uk

Abstract

There is an increasing global search for commercially and environmentally clean synthesis of metallic nanoparticles from bioresources. However, there is insufficient reported information on the production of iron nanoparticles (FeNPs) from fungi such as *Trichoderma* species which has potentials for greater for yield of FeNPs as compared with bacterial sources. Filtrates obtained from biomass of pure cultures grown (72 h) in Potato Dextrose Broth were reacted with 1 M ferric chloride solution. A change in coloration monitored by ultraviolet-visible spectrophotometer (200-600 nm) as compared with control (ferric chloride solution) indicated a positive result. Nanoparticle synthesized in the reactive filtrate was characterized using utltraviolet-visible spectroscopy and Fourier transform infrared spectroscopy (FTIR). The effects of some physical parameters such as agitation, pH and temperature were monitored. The UV-vis spectrum revealed the peak of absorbance of synthesized nanoparticle by selected *Trichoderma* species at 275 nm. Optimum conditions for nanoparticle biosynthesis were observed at pH 4.5, 35 °C and when cells were agitated. Peaks of transmittance were observed at 950, 1800, 2250, 3000 and 3500 cm^{-1}. These peaks represent the C-H, C=O, C≡N, C=H and the OH functional groups respectively. The presence of the alkene, carboxyl and phenol groups suggests a capping of the NPs by the organism after redox reaction. The properties of extracellular iron nanoparticles synthesized by *Trichoderma* species from this study, presents the fungi as a bioresource for synthesis of stable NPs.

1 Introduction

Nanotechnology is a multidisciplinary branch of science encompassing numerous disciplines in science and technology [1]. Nanoparticles have become important due to their size compatibility with cells (10-100 nm), viruses (20-450 nm), proteins (5-50 nm) and genes (2 nm wide by 10-

IOP Conf. Series: Materials Science and Engineering **805** (2020) 012036 doi:10.1088/1757-899X/805/1/012036

100 nm long). They can therefore, move within the body without interruption of normal functions and also access places that are rather unreachable by other materials [2]. Magnetic nanoparticles (MNPs) are one of the most commonly used materials within the nanoscale which possess a type of core/shell structure that consists of a magnetic core trapped within an organic polymeric coating [1]. MNPs naturally possess a hydrophobic surface which has a large surface-to-volume ratio and an agglomerating potential [3]. MNPs, such as iron nanoparticles (FeNPs) which exhibit special magnetic properties different from the bulk of other nanomaterials, are beginning to attract significant interests in applications, such as a storage medium in magnetic memory devices to probes and vectors in the biomedical sciences [4]. The large surface-to-volume ratio of MNPs makes available in abundance, a chemical site of activity for biomolecules to bind, as a result allowing for precision in their engineering and design to suit specific functions, such as prolonged circulation within the blood-stream, target specificity to tissues, ease of optical detection and drug delivery [4, 5]. In addition, superparamagnetic FeNPs have the ability to escape mononuclear phagocytes of the immune system [1].

Currently, the biological approach to the synthesis of nanoparticles is gaining more interest over the physico-chemical methods, which have numerous disadvantages [6]. Biological systems have unique features for the synthesis of nanomaterials with well-defined and predetermined properties. Certain microorganisms such as filamentous fungi play an important role in bioremediation through the reduction of toxic metals. These fungal isolates have become the favorites for the nanotechnologist due to the wide variety of advantages they offer over bacteria, yeast, actinomycetes, plants, and other physico-chemical properties they possess [7, 8]. These features include ease of handling, simple nutrient requirements, high wall-binding capacity and an intracellular metal uptake capability. There are reports on the use of *Trichoderma* species for the synthesis of nanoparticles [9, 10]. However, there is paucity of information on the synthesis of iron nanoparticles from this group of organisms. This research focuses on biological synthesis of FeNPs from *Trichoderma* species isolated from soil samples. We also report the effects of some physical parameters on the synthesis of FeNPs and characterization of the synthesized material using simple techniques.

IOP Conf. Series: Materials Science and Engineering **805** (2020) 012036 doi:10.1088/1757-899X/805/1/012036

2 Materials and Methods

2.1 Sample collection

Soil samples were collected from farm area of the Federal University of Agriculture, Abeokuta, into sterile polyethylene bags and taken to the Microbiology laboratory.

2.2. Isolation and identification of pure cultures

Serial dilution of the soil sample was done and aliquot (1 ml) of appropriate diluent was inoculated on acidified Potato Dextrose Agar (PDA) plates using pour plate method. Plates were incubated for 3 days at 25 °C. Colonial and microscopic examination characteristics of isolates were compared with reported standards [11]. Identified pure cultures of *Trichoderma* species were stored on PDA slants at 4 ºC and sub-cultured bi-monthly.

2.3 Production of fungal biomass

Biomass was prepared by introducing spores of *Trichoderma* species (1 x 10^7 spores /ml) into 250 ml flasks containing 100 ml of sterile Potato Dextrose Broth. The flasks were cotton-plugged and incubated at 25 °C and agitated at 120 rpm for 72 h. The biomass of each isolate was harvested and then centrifuged at 5000 rpm for 20 min. The filtrates obtained were discarded and the biomass (residue) was washed twice with sterile distilled water.

2.4 Iron nanoparticle biosynthesis

Harvested biomass was washed with after 24 h incubation in sterile distilled water. Cell filtrates (50 ml) obtained were reacted with 50 ml of 1M Ferric chloride solution. The flasks were incubated on an orbital shaker at for 72 h [12]. A conical flask containing only Ferric chloride solution without the filtrates was set up as a control experiment. The flasks were retrieved from the orbital shaker and observed for presence and intensity of color changes. Synthesis and formation of iron nanoparticle in the reaction mixture was monitored using UV-visible spectrophotometry between the wavelengths of 200-600 nm.

2.5 Characterization of synthesized FeNPs

The synthesized iron nanoparticles were characterized using ultraviolet-visible spectroscopy and Fourier transform infra-red spectroscopy (FTIR).

2.6 Effects of pH, temperature and agitation on synthesis of FeNPs

The effects of parameters such as pH (4.5-6.5) and temperature (25-40 °C) on the synthesis of the nanoparticle by the selected *Trichoderma* species were monitored. Effect of agitation at optimum pH and temperature on synthesis was also studied.

IOP Conf. Series: Materials Science and Engineering **805** (2020) 012036 doi:10.1088/1757-899X/805/1/012036

3 Results and Discussion

3.1 Screening for FeNPs production in isolated *Trichoderma* species

About five (5) fungal isolates identified as *Trichoderma* species were obtained from soil sample. Isolate with the most reactive filtrate was selected. Colonial characteristic of the isolate is shown on Figure 1. Synthesis of FeNPs indicated by development of brownish coloration as compared with the control in the reaction mixture of the filtrates with Ferric chloride is shown on Figure 2. As observed in this study, fungi such as *Trichoderma* have been reported to extracellularly produce reducing enzymes in simple culture medium a medium [13,14] Nitrate-dependent reductases have been identified as the main mediators of bioreduction progression along with a shuttle quinine extracellular process leading to the reduction of $FeCl_3$ to Fe_2O_3 [15].

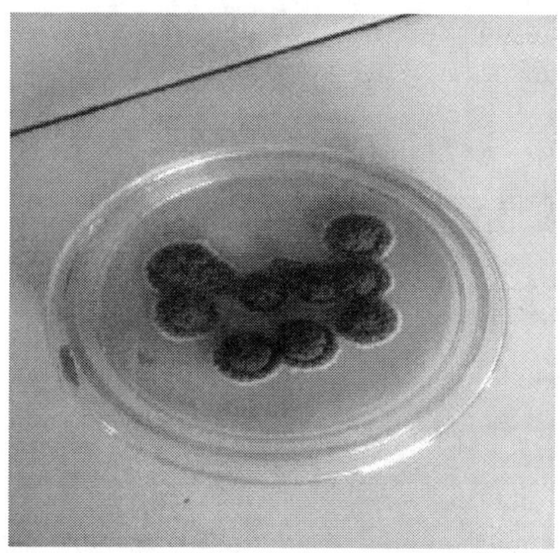

Figure 1. Plate showing growth of selected *Trichoderma* species on PDA

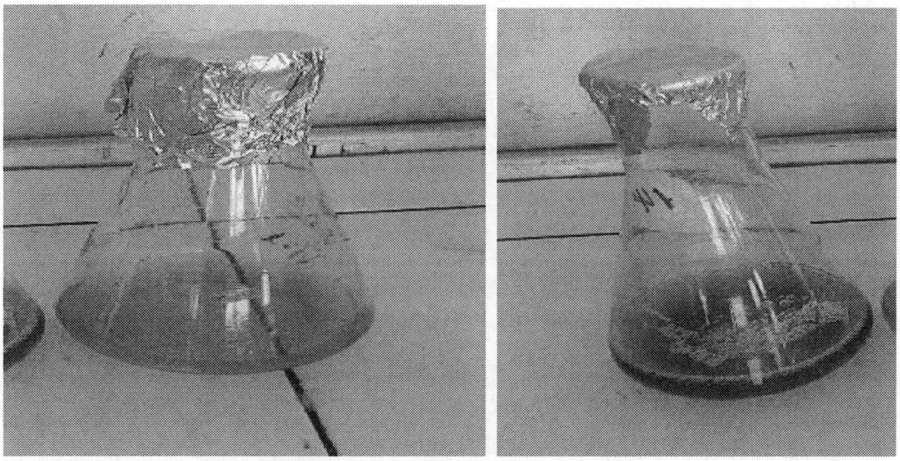

Figure 2. Contol solution of $FeCl_3$ (Left); reaction of fungal biomass with $FeCl_3$ (right)

3.2 Characterization of FeNPs using ultraviolet-visible spectrophotometry

The absorption spectrum of the nanoparticle synthesized by the selected *Trichoderma* species was observed to peak at 275 nm as shown on Figure 3. This result is in the range reported in previous study on the synthesis of iron nanoparticles from *Pleurotus* species with absorbance at 265 nm wavelength [16]. Absorbance at the range has been explained to be due to electronic excitation in tryptophan and tyrosine residue in the protein present [17]. It has also been reported that solutions of metals nanoparticles tend to have distinct color characteristics and corresponding peaks. Further analysis of these peaks helps to differentiate the nanomaterials synthesized by their bond structures and functional groups in a given spectrum [18].

3.3 Optimization studies

The effects of pH, temperature and agitation on the synthesis of FeNPs from selected *Trichoderma* species studied at 275 nm were observed to be optimal at pH 4.5, temperature of 35 °C with agitation of the cells as shown on Figures 4, 5 and 6 respectively. A further increase of both pH and temperature led to decrease in biosynthesis. This reduction is similar to reports on inactivation of activity at increased physical conditions such as pH and temperature [19]. Characterization of physical parameters has been considered important and required along with others *in vitro* and *in vivo* studies [20].

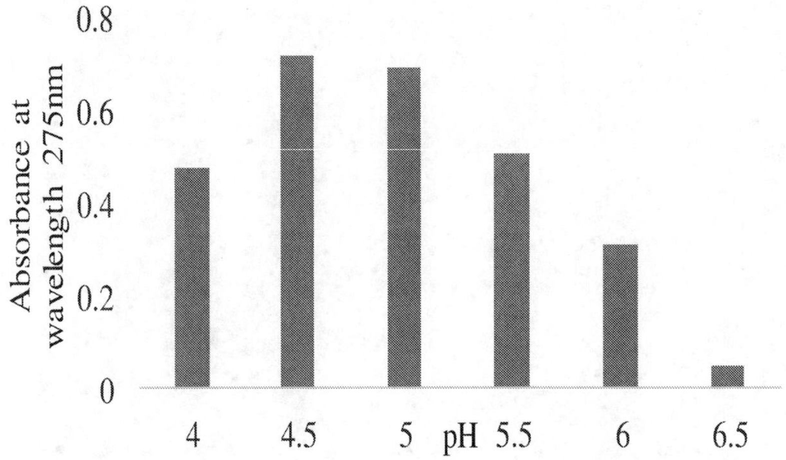

Figure 3. Effect of change in pH on iron nanoparticles produced by *Trichoderma* sp. at 275nm

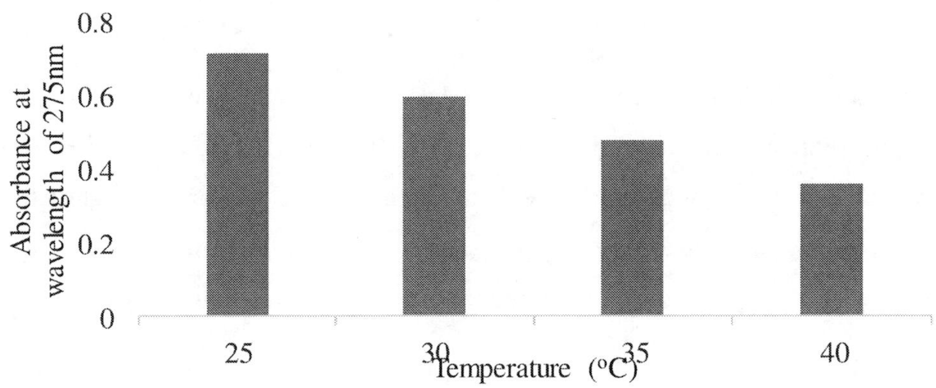

Figure 4. Effect of change in temperature on iron nanoparticles produced by *Trichoderma* sp. at 275 nm

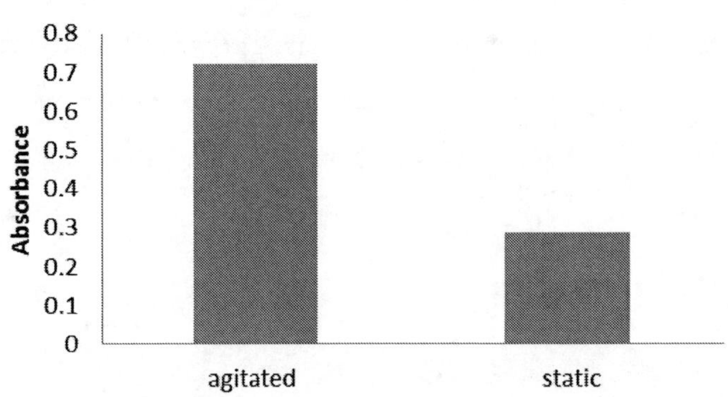

Figure 5. Effect of agitation on iron nanoparticles produced by *Trichoderma* sp. at 275 nm

IOP Conf. Series: Materials Science and Engineering **805** (2020) 012036 doi:10.1088/1757-899X/805/1/012036

3.4. FTIR analysis of iron nanoparticles produced by *Trichoderma* species

Characterization of synthesized FeNPs using FTIR revealed the percentage transmittance of the biosynthesized iron nanoparticle showing peaks in the range of 950-500 cm^{-1} and is represented on Figure 6. The result reveals the surface composition and the ligand banding pattern of nanosized compound. It identifies specific chemical bonds or functional groups present based on their absorption signatures which is measured by absorption energy dissipated by the stretch and bending of the chemical bonds through infrared spectroscopy [14]. The peak occurring at 950 cm^{-1} is the finger print region which represents the CH (aromatic and out of plane band). The functional groups responsible for the nanosizing are found in the 1800, 2200, 3000 and 3500 cm^{-1} wavelengths representing the C=O (carboxylic acid), C≡N (nitrile), C=H (alkene) and the OH (phenol) groups respectively. C-H and OH groups can be associated with capping of the nanoparticle after reduction. Biological synthesis of Fe nanoparticles is a redox reaction a which have been reported to be significantly influenced by the presence of functional groups such as the carboxyl, and mercaptans (phenols and alcohols) as observed to be present in this study [21]. The Fourier transform infrared spectroscopy provides information on the binding properties of biosynthesized NPs.

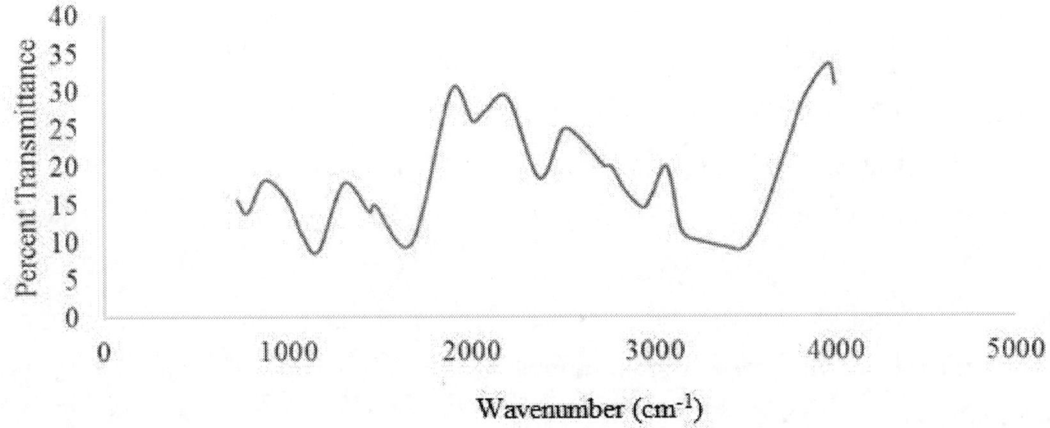

Figure 6. FTIR spectrum of iron nanoparticles produced by *Trichoderma* sp.

IOP Conf. Series: Materials Science and Engineering **805** (2020) 012036 doi:10.1088/1757-899X/805/1/012036

4 Conclusion

The results of this study showed the potential of synthesis of FeNPs from *Trichoderma* species using simple techniques. Furthermore, mild conditions of pH and temperature for optimum synthesis of the nanoparticle by the fungus, as a better alternative to chemical synthesis which is usually characterized by harsh conditions. The study also confirms the significant role of agitation during the green synthesis of nanoparticles. The study moreover, had addressed the paucity of information reported on the synthesis of FeNPs from filamentous fungi such as *Trichoderma* species.

References

[1] Estelrich J, Escribano E, Queralt J and Busquets MA 2015 Iron oxide nanoparticles for magnetically-guided and magnetically-responsive drug delivery. *Int. J. Mol. Sci.* **16** (4) pp 8070-8101.

[2] Medeiros SF, Santos AM, Fessi H and Elaissari A 2011 Stimuli-responsive magnetic particles for biomedical applications. *Int. J. Pharmaceut.* **403** (1-2) pp 139-161.

[3] Lu AH, Salabas EE and Schüth F 2007 Magnetic nanoparticles: synthesis, protection, functionalization, and application. *Angew. Chemie. Int. Ed.* **46** (8) pp1222-1244.

[4] Craik D 1995 Magnetism: Principles and Applications; Wiley: Chichester, NY, USA.

[5] Fang C and Zhang M 2009 Multifunctional magnetic nanoparticles for medical imaging applications. *J. Mater. Chem* **19** (35) pp 6258-6266.

[6] Devi TP, Kulanthaivel S, Kamil D, Borah JL, Prabhakaran N and Srinivasa N 2013 Biosynthesis of silver nanoparticle from *Trichoderma* species. *Ind. J. Exp. Biol.* **51** pp 543-547.

[7] Dias MA, Lacerda IC, Pimentel PF, De Castro HF and Rosa CA 2002 Removal of heavy metals by an *Aspergillus terreus* strain immobilized in a polyurethane matrix. *Lett. Appl. Microbiol.* **34** (1) pp 46-50.

[8] Sanghi R and Verma P 2009 Biomimetic synthesis and characterisation of protein capped silver nanoparticles. *Bioresour. Technol.* **100** (1) pp 501-504.

[9] Vahabi K, Mansoori GA and Karimi S 2011 Biosynthesis of silver nanoparticles by fungus *Trichoderma reesei* (a route for large-scale production of AgNPs). *Insciences J.* **1** (1) pp 65-79.

[10] Fayaz AM, Balaji K, Girilal M, Yadav R, Kalaichelvan PT and Venketesan R 2010 Biogenic synthesis of silver nanoparticles and their synergistic effect with antibiotics: a study against gram-positive and gram-negative bacteria. *Nanomed.: Nanotechnol. Biol. Med.* **6** (1) pp 103-109.

[11] Barnett JA 2003 Beginnings of microbiology and biochemistry: the contribution of yeast research. *Microbiol.* **149** (3) pp 557-567.

[12] Rahim KA, Mahmoud SY, Ali AM, Almaary KS, Mustafa AE and Husseiny 2017 Extracellular biosynthesis of silver nanoparticles using *Rhizopus stolonifer* (mould). *Saudi J. Biol. Sci.* **24** (1) pp 208-216.

[13] Ahmad A, Senapati S, Khan MI, Kumar R and Sastry M 2005 Extra-/intracellular biosynthesis of gold nanoparticles by an alkalotolerant fungus, *Trichothecium* sp. *J. Biomed. Nanotechnol.* **1** (1) pp 47-53.

[14] Rautio JJ, Smit BA, Wiebe M, Penttilä M and Saloheimo M 2006 Transcriptional monitoring of steady state and effects of anaerobic phases in chemostat cultures of the filamentous fungus *Trichoderma reesei.* BMC genomics. **7** 247. https://doi.org/10.1186/1471-2164-7-247.

[15] Bakshi M, Mahanty S and Chaudhuri P 2017 Fungi-mediated biosynthesis of nanoparticles and application in metal sequestration. In: Surajit D, and Dash HR (eds.) Handbook of metal-microbe interactions and bioremediation. CRC Press: Boca Raton, USA. Pp 423-434.

[16] Mazumdar H and Haloi N 2011 A study on biosynthesis of iron nanoparticles by *Pleurotus* sp. *J Microbiol Biotechnol. Res.* **1** (3) pp 39-49.

[17] Durán N, Marcato PD, Alves OL, De Souza GI and Esposito E 2005 Mechanistic aspects of biosynthesis of silver nanoparticles by several *Fusarium oxysporum* strains. *J. Nanobiotechnol.* **3** 8. https://doi.org/10.1186/1477-3155-3-8.

[18] Carter N 2015 Physical properties of iron oxide nanoparticles. Honors College 210. https://digitalcommons.library.umaine.edu/honors/210/ (Accessed on 3 February, 2020).

[19] Jha AK and Prasad K 2010 Understanding biosynthesis of metallic/oxide nanoparticles: a biochemical perspective. In: Kumar SA, Thiagarajan S, and Wang SF (eds.) Biocompatible nanomaterials synthesis, characterization and applications. NOVA Science Publishers Inc., Hauppauge, NY, USA.

IOP Conf. Series: Materials Science and Engineering **805** (2020) 012036 doi:10.1088/1757-899X/805/1/012036

[20] Hall JB, Dobrovolskaia MA, Patri AK and McNeil SE 2007 Characterization of nanoparticles for therapeutics. *Future Med.* **2** (6). https://doi.org/10.2217/17435889.2.6.789.

[21] Vitta Y, Figueroa M, Calderon M and Ciangherotti C 2020 Synthesis of iron nanoparticles from aqueous extract of *Eucalyptus robusta* Sm and evaluation of antioxidant and antimicrobial activity. *Mater. Sci. Energ. Technol.* **3** pp 97-103.

Optimization studies on biosynthesis of iron nanoparticles using *Rhizopus stolonifer*

*TM Adeleye, SO Kareem and AA Kekere-Ekun
Department of Microbiology, Federal University of Agriculture, Abeokuta, Ogun State, Nigeria.
*email: adeleyetm@funaab.edu.ng

Abstract

Isolates from pure culture of *Rhizopus stolonifer* were aseptically grown in potato dextrose broth to early exponential phase and centrifuged. The biological synthesis of Iron nanoparticles from the fungus was carried out using simple techniques. Filtrates obtained by simple methods were reacted with 1 M Ferric chloride solution. Characterization of the synthesized iron nanoparticles was monitored using Ultraviolet-visible spectrophotometer and Fourier transform infrared spectroscopy (FTIR). An increase in synthesis of the iron nanoparticles by over 200% occurred when culture of selected isolate was agitated. At 325 nm, peaks of absorbance (3.5) were read at pH 4.5 and 6, while maximum production of the iron nanoparticles was reached at 35 ºC. Peaks of transmittance of biosynthesized iron nanoparticles from selected isolate as shown by the FTIR spectrum were located at 750, 1100, 1700, 2500 and 3750 cm^{-1}, representing the CH, C-O, C=O, SH and OH groups respectively. The varying degree of transmittance and pH reported presents the selected *R. stolonifer* isolate as a biological entity for the synthesis of stable iron nanoparticles.

Keywords: *Rhizopus stolonifer*, iron nanoparticles, optimization, UV-vis spectrophotometry, FTIR

1 Introduction

Nanotechnology, a study of materials at the nanoscale (that is less than 100 nm), is fast becoming one of the most active research fields in Biology, Chemistry, Physics, Mathematics, Technology and Engineering [1]. The study of nanosized particles has attracted increasing attention due to their unique size-dependent properties which has various applications [2, 3]. Industrial demands of these particles continue to increase as the number of potential applications for the materials grows [4]. Over the decades, applications of soft magnetic nanoparticles are being studied because of their impact on the delivery of drugs, magnetic resonance imaging, cell separation, antimicrobial activities, and the treatment for cancer, [5]. For biochemical applications, there is the need for ecofriendly and nontoxic synthesis of nanoparticles [6]. Iron NPs have been considered more toxicologically desirable because iron is ubiquitous in the body and can be handled by the body even at higher doses than other metals [7]. FeNPs have strong magnetic properties which as a result aid the ease of separation of the particles from the medium after synthesis [8, 9]. Use of chemical and physical methods for the synthesis of nanoparticles has been regarded as not suitable for biochemical applications due to the generation of toxic chemicals [4]. Biological methods on the other hand, are considered as green technology and are being manipulated through the use of bacteria, fungi, algae, actinomycetes, and plants [10]. The use of magnetotactic bacteria has become one of the most well-known biological methods for production of magnetic nanoparticles. Magnetotactic bacteria are known to synthesize magnetite (Fe_3O_4) through the mechanism of direct mineralization. There is however, a need to cheaply mass produce magnetic nanoparticles and fungal synthesis of iron magnetic nanoparticles rather than other microorganisms is preferred due to the ease of handling, low maintenance cost and ease of downstream processing [4, 11]. *Rhizopus oryzae* has been reported as a natural 'nanofactory' with a simplified process scale-up and reduced production costs for the synthesis of gold nanoparticles from fungal protein extracts [12]. The aim of this study was to screen and optimize selected parameters for synthesis of nanoparticles from *R. stolonifer* isolated from soil samples.

2 Materials and Methods

2.1 Isolation and identification of pure cultures from soil samples

Soil samples were randomly collected from different locations in Abeokuta in sterile bottles and immediately conveyed to the laboratory. Standard procedures of serial dilution and pour plate method on sterilized Potato Dextrose Agar (PDA) plates acidified with acetic acid were employed for the isolation. Pure forms of filamentous fungi with light black spores were subcultured into PDA plates. Microscopic examination was done by viewing stained slides (x40) under the light microscope. Structures observed under the microscope and compared with reported standards [12]. Identified pure cultures of *R. stolonifer* were stored on PDA slants at 4 °C and sub-cultured bi-monthly.

2.2 Production of fungal biomass

Biomass was prepared by introducing fungal spores (1 x 10^6 spores /ml) into Erlenmeyer flasks containing 100 ml of sterile Potato Dextrose Broth. The conical flasks containing fungal spores were cotton plugged and placed in an Orbital Shaker, incubated at 25 °C and agitated at 100 rpm for 72 h. The biomass was harvested and then centrifuged at 5000 rpm for 20 min. The filtrates obtained were discarded and the biomass (residue) was washed twice with sterile distilled water.

2.3 Mycosynthesis of Iron nanoparticles

Ferric chloride solution (1 mM) was prepared by dissolving 16.22 g of anhydrous ferric chloride into 1000 ml of distilled water. Biomass (10 g) was inoculated into conical flasks containing 100 ml of distilled water and placed in the orbital shaker again at 100 rpm for 72 h. Mixture of the biomass and the distilled water was then centrifuged again after which the resulting cell filtrate (50 ml) was mixed with 50 ml of 1 M Ferric chloride solution. This flask was then finally agitated in the orbital shaker for 72 h. A conical flask containing only Ferric chloride solution without the filtrate was also placed in an orbital shaker to serve as a control sample. The development of color change and stability as compared with the control indicated synthesis of FeNPs. Fungal isolate with the most reactive filtrate was selected for further studies.

2.4 Characterization and Optimization studies of synthesized FeNPs

The synthesized iron nanoparticles were characterized using ultraviolet-visible spectroscopy and Fourier transform infra-red spectroscopy (FTIR). Effects of pH (4.5-6.5), temperature (25-40 °C) and agitation on the biosynthesis of iron nanoparticles by selected fungal isolates were monitored.

3.0 Results and Discussion

3.1 Screening for FeNPs biosynthesis from fungal isolates

Out of twenty (20) fungi isolated from soil samples, ten (10) were identified as *R. stolonifer* as shown on Figure 1. Four isolates showed positive results indicated by a color change from the initial brown color to a black color indicating a production of iron nanoparticles after incubation, while no color change was observed in control conical flask containing only ferric chloride solution. Filtrates from fungal biomass reacting the most with the ferric chloride solution as indicated with the intensity of color formation within 72 h of incubation, was selected for further studies. The development of the black color is as a result of the extracellular secretion of the fungal enzymes or metabolites which led to decomposition of $FeCl_3$ to Fe_2O_3. Observation from this study agrees with earlier reports [13] that fungi are able to synthesize high amounts of proteins and other enzymes which are used for bioreduction of metals which triggers the synthesis of biologically compatible nanoparticles, without much sophistication of instruments as used in this study. The reducing enzymes secreted from fungi have been reported to play the vital role in the biosynthesis of nanoparticles. These metabolites are secreted extracellularly into the solution in both a static and agitated cultures [14, 15].

3.2 Ultraviolet-visible spectrophotometry

The ultraviolet-visible spectrophotometry of the solution showed the peak of absorbance (surface plasma resonance absorption spectrum) of the nanoparticles produced by *Rhizopus stolonifer* at 325 nm as shown in Figure 3. This is in agreement with previous studies that the band gap of absorption of FeNPs detected by UV-vis exists between 200-900 nm. The UV-visible spectrophotometry has been described as a reliable technique used in the primary characterization of synthesized metal nanoparticles. The technique has also been used in the monitoring the synthesis and stability of metal nanoparticles, as they have high extinction co-efficient and bright color which is visible by naked eye. Results obtained therefore have qualitatively reliable information of the nanoparticles [16].

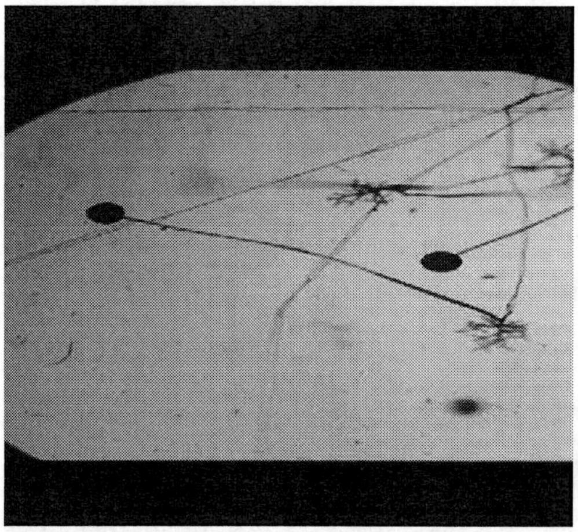

Figure 1. Selected isolate of *Rhizopus stolonifer* (X40)

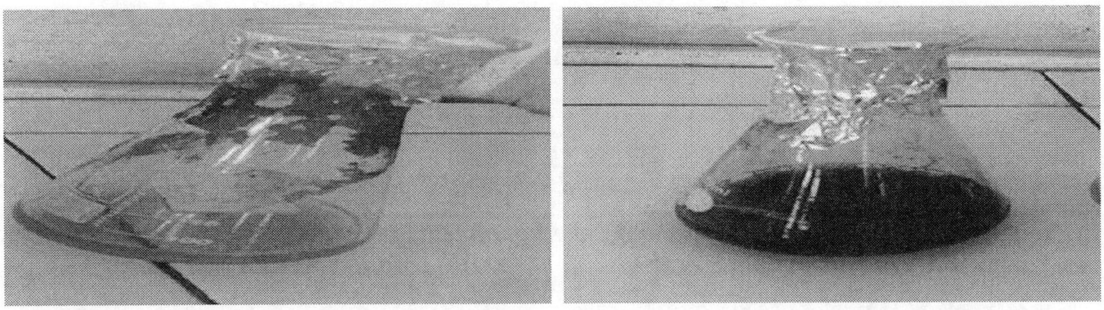

Figure 2. Control solution containing $FeCl_3$ (left); eeaction of fungal biomass with $FeCl_3$ (right)

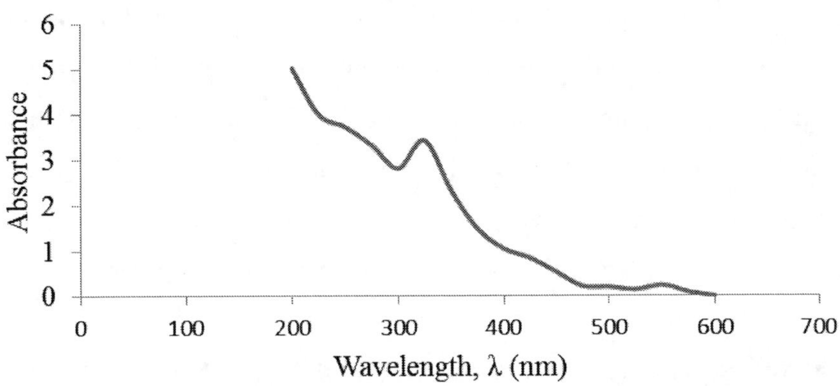

Figure 3. UV-visible spectrum of iron nanoparticles produced by *Rhizopus stolonifer*

3.3 Optimization studies

The effects of pH, temperature and agitation on the synthesis of FeNPs from selected *R. stolonifer* were studied at 325 nm. Synthesis was observed to be optimal at two peaks of pH (4.5 and 6.0) with absorbance values of 3.430 and 4.311 respectively as shown in Figure 4. Study on the effect of temperature as shown on Figure 5 shows that optimum biosynthesis was observed at 35 °C with an absorbance value of 4.311. The biosynthetic process was carried out with and without agitation. There was increase in absorbance values from 1.783 (without agitation) to 3.430 (with agitation) during biosynthesis of FeNPs by selected fungal isolate at optimum pH and temperature studied. This study agrees with previous studies on the effect of physical parameters such as pH and temperature on the activity of membrane bound oxidoreductases in the synthesis of metal nanoparticles [17]. Peak of FeNPs biosynthesis observed at low pH of 4.5 in this study is similar to the previous research on fungal oxidoreductases enzymes which were

IOP Conf. Series: Materials Science and Engineering **805** (2020) 012037 doi:10.1088/1757-899X/805/1/012037

observed to be sensitive to pH. The enzyme activity was in alternative manner being inactivated at a high pH while at low pH, get activated [18].

3.4 FTIR analysis of iron nanoparticles produced by *Rhizopus stolonifer*

FTIR spectrum showed peaks located within the region of 617.4 cm^{-1} to 3908.6 cm^{-1}. The FTIR gives information on surface composition and the ligand banding pattern of compound. It is used to identify and specify the chemical bonds or functional groups in a compound based on their specific unique absorption signatures which measure absorption energy dissipated by the stretch and bending of the chemical bonds through infrared spectroscopy [16]. This pattern is shown in Figure 6. The peaks at 750 cm^{-1} and 1100 cm^{-1} are the fingerprint region which representing the CH (alkane and out of plane band) and the C-O (carboxylic) groups respectively. The functional groups responsible for the nanosizing are the carboxylic acid (C=O), mercaptans (S-H) and the free hydroxyl (OH), which were observed at 1700 cm^{-1}, 2500 cm^{-1} and 3750 cm^{-1}, respectively. The C=OH and S-H groups can be associated with the fungus used and therefore playing the role of capping the nanoparticles after reduction. Green synthesis of metallic nanoparticles is a redox reaction and the presence of the carboxyl, sulfhydryl and mercaptans (phenols and alcohols) as observed in this study, have been reported to be responsible for the transformation of metal ions to nanoparticles earlier studies [19].

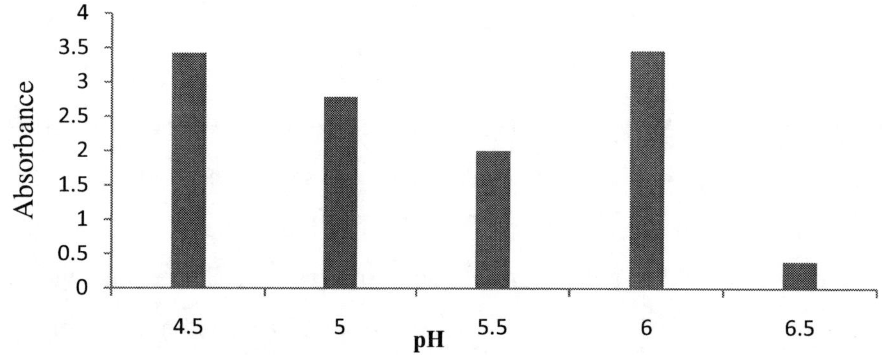

Figure 4. Effect of pH on biosynthesis of FeNPs from *Rhizopus stolonifer* extract

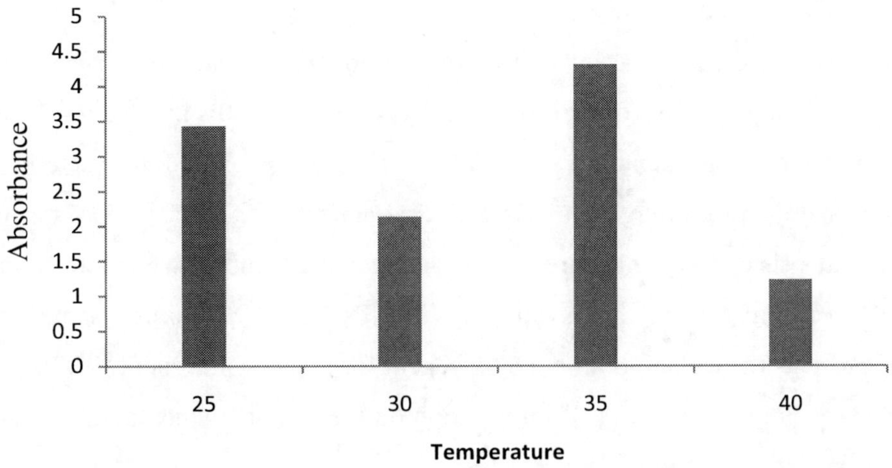

Figure 5. Effect of temperature on the biosynthesis of FeNPs from *Rhizopus stolonifer* extract

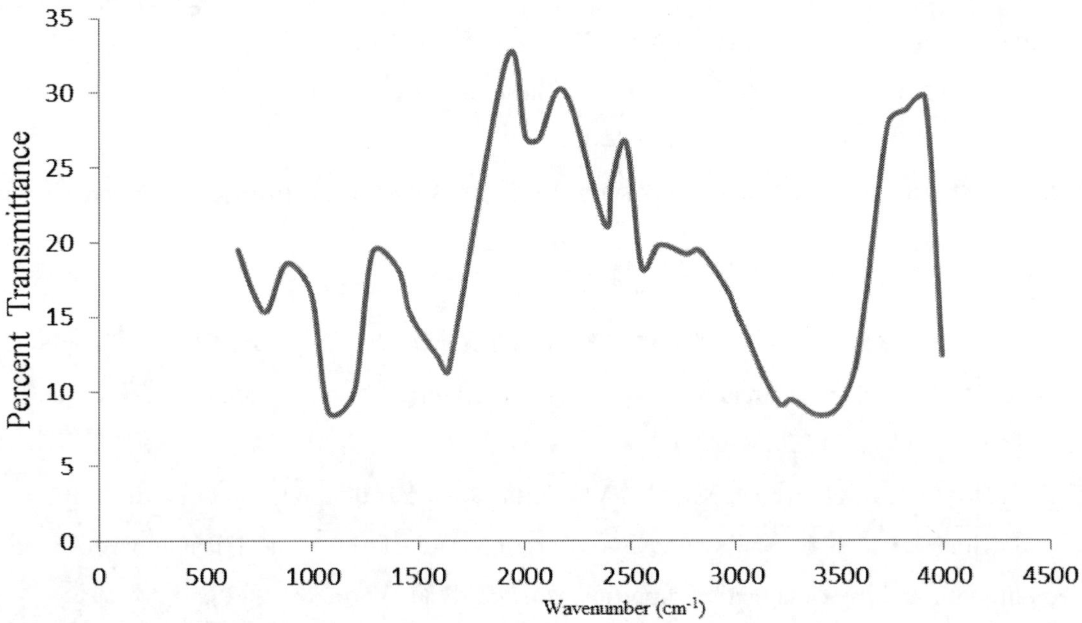

Figure 6. FTIR spectrum of FeNPs produced by *Rhizopus stolonifer* extract

IOP Conf. Series: Materials Science and Engineering **805** (2020) 012037 doi:10.1088/1757-899X/805/1/012037

4 Conclusion

The present study presents the potentials of cell extract of *Rhizopus stolonifer* to synthesize FeNPs through simple techniques. Most of reports on the synthesis of FeNPs involved the use of magnetotactic bacteria. Fungi are generally regarded as safe as compared to bacteria for industrial applications and therefore fungal isolates with desirable characteristics as presented in this study should be synthesis of NPs which are increasingly gaining significance in a number of industries. The mild conditions for optimum synthesis of FeNPs from *R. stolonifer,* as observed in this study further presents the isolate as an attractive agent for green synthesis of FeNPs. This study therefore, helps to address the dearth of information on the use of filamentous fungi such as *R. stolonifer* in green synthesis of iron nanoparticles.

References

[1]Koopmans RJ and Aggeli A 2010 Nanobiotechnology—quo vadis?: Ecology and Industrial Microbiology. *Curr. Opin. Microbiol.* **13** (3) pp 327-334.

[2] Hulkoti NI and Taranath TC 2014 Biosynthesis of nanoparticles using microbes—a review. *Colloids Surf. B: Biointerf.* **121** pp 474-83.

[3] Sastry M, Ahmad A, Khan MI and Kumar R 2003 Biosynthesis of metal nanoparticles using fungi and actinomycete. *Curr. Sci.* **85** (2) pp 162-170.

[4] Abdeen M, Sabry S, Ghozlan H, El-Gendy AA, Carpenter EE. Microbial-physical synthesis of Fe and Fe3O4 magnetic nanoparticles using *Aspergillus niger* YESM1 and supercritical condition of ethanol. *J. Nanomater.* Article ID 9174891. https://doi.org/10.1155/2016/9174891.

[5] El-Gendy AA, Ibrahim EM, Khavrus VO, Krupskaya Y, Hampel S, Leonhardt A, Büchner B and Klingeler R 2009 The synthesis of carbon coated Fe, Co and Ni nanoparticles and an examination of their magnetic properties. *Carbon* **47** (12) pp 2821-2828.

[6] ain N, Bhargava A, Tarafdar JC, Singh SK and Panwar J 2013 A biomimetic approach towards synthesis of zinc oxide nanoparticles. *Appl. Microbiol. Biotechnol.* **97** (2) pp 859-869.

[7] Vogel CF, Charrier JG, Wu D, McFall AS, Li W, Abid A, Kennedy IM and Anastasio C 2016 Physicochemical properties of iron oxide nanoparticles that contribute to cellular ROS-

dependent signaling and acellular production of hydroxyl radical. *Free Radic. Res.* **50** (11) pp 1153-1164.

[8] Estelrich J, Escribano E, Queralt J and Busquets MA 2015 Iron oxide nanoparticles for magnetically-guided and magnetically-responsive drug delivery. *Int. J. Mol. Sci.* **16** (4) pp 8070-8101.

[9] Lemine OM 2009 Microstructural characterisation of α-Fe2O3 nanoparticles using, XRD line profiles analysis, FE-SEM and FT-IR. *Superlattices Microstruct.* **45** (6) pp 576-582.

[10] Rai M, Yadav A and Gade A 2008 CRC 675—current trends in phytosynthesis of metal nanoparticles. *Crit. Rev. Biotechnol.* **28** (4) pp 277-284.

[11] Gade A, Ingle A, Whiteley C and Rai M 2010 Mycogenic metal nanoparticles: progress and applications. *Biotechnol.Lett.* **32** (5) pp 593-600.

[12]Barnett JA 2003 Beginnings of microbiology and biochemistry: the contribution of yeast research. *Microbiol.* **149** (3) pp 557-567.

[13] Das SK, Dickinson C, Lafir F, Brougham DF and Marsili E 2012 Synthesis, characterization and catalytic activity of gold nanoparticles biosynthesized with *Rhizopus oryzae* protein extract. *Green Chem.* **14** (5) pp 1322-1334.

[14]. Bakshi M, Mahanty S and Chaudhuri P 2017 Fungi-mediated biosynthesis of nanoparticles and application in metal sequestration. In: Surajit D, and Dash HR (eds.) Handbook of metal-microbe interactions and bioremediation. CRC Press: Boca Raton, USA. Pp 423-434.

[15] Ahmad A, Senapati S, Khan MI, Kumar R, Ramani R, Srinivas V and Sastry M 2003 Intracellular synthesis of gold nanoparticles by a novel alkalotolerant actinomycete, *Rhodococcus* species. *Nanotechnol.* **14** (7) 824. https://doi.org/10.1088/0957-4484/14/7/323.

[16]. Rautio JJ, Smit BA, Wiebe M, Penttilä M and Saloheimo M 2006 Transcriptional monitoring of steady state and effects of anaerobic phases in chemostat cultures of the filamentous fungus *Trichoderma reesei.* BMC genomics. **7** 247. https://doi.org/10.1186/1471-2164-7-247.

[17]. Khandel P and Shahi SK 2018 Mycogenic nanoparticles and their bio-prospective applications: current status and future challenges. *J. Nanostruct. Chem.* **8** (4) pp 369-391.

[18]. Jha AK and Prasad K 2010 Understanding biosynthesis of metallic/oxide nanoparticles: a biochemical perspective. In: Kumar SA, Thiagarajan S, and Wang SF (eds.) Biocompatible

nanomaterials synthesis, characterization and applications. NOVA Science Publishers Inc., Hauppauge, NY, USA.

[19]. Vitta Y, Figueroa M, Calderon M and Ciangherotti C 2020 Synthesis of iron nanoparticles from aqueous extract of *Eucalyptus robusta* Sm and evaluation of antioxidant and antimicrobial activity. *Mater. Sci. Energ. Technol.* **3** pp 97-103.

Ecofriendly production of silver nanoparticles from the seeds of *Carica papaya* and its larvicidal and antibacterial efficacy against some selected bacterial pathogens

AD Aina[a]*, O Owolo[a], M Adeoye-Isijola[a], O D Olukanni[b], A Lateef[c,d], T Egbe[a], FO Aina[e], TB Asafa[d,f] and SH Abbas[g]

[a]Department of Microbiology, Babcock University PMB 21244, Ikeja, Lagos, Nigeria

[b]Department of Chemical Sciences, Redeemer's University, Gbogan-Osogbo Road, PMB 230, Ede 232001

[c]Department of Pure and Applied Biology, Ladoke Akintola University of Technology PMB 4000, Ogbomoso, Nigeria

[d]Nanotechnology Research Group (*NANO*$^+$), LAUTECH, Ogbomoso, Nigeria

[e]Department of Maternal and Child Health, Babcock University, PMB 21244, Ikeja, Lagos, Nigeria

[f]Department of Mechanical Engineering, Ladoke Akintola University of Technology, PMB 4000, Ogbomoso, Nigeria

[g]Center of Excellence in Nanotechnology (CENT), King Fahd University of Petroleum and Minerals, Dhahran 31261 Saudi Arabia

Abstract

Carica papaya seed extract (CPE) was used in the synthesis of silver nanoparticles (AgNPs) in this study. For the characterization of biosynthesized particles, UV-vis spectroscopy, FTIR, FESEM and EDX were used. Antimicrobial and larvicidal efficacies of the synthesized AgNPs were assessed in the fight against certain pathogens and the *Aedes aegypti* 4th instar larvae. The absorption spectrum of AgNPs peaked at 433 nm with a broad peak of 3000 to 3800 cm^{-1} for different functional groups as presented in the FTIR analysis. A FESEM evaluation revealed a number of spherical particle structures with an average of 20-69 nm. With zones of inhibition between 10-24 mm, the AgNPs synthesized inhibited selected microorganisms. After 12 h of exposure, the nanoparticles had LC_{50} and LC_{90} lethal concentration on the *Aedes aegypti* larva at 14.56 and 33.89 µg/ml respectively. This study demonstrates possibility of using *Carica papaya* seeds in AgNPs synthesis.

Keywords: Antibacterial, *Carica papaya,* dengue fever, larvicidal, pathogens, silver nanoparticles

1 Introduction

The development of microbial resistance to antimicrobial agents is one of the biggest challenges facing bio-scientists. Zika and dengue fever both transmitted by *Aedes aegypti* mosquito are another major public health problem that bio-scientists are looking to tackle. One way to control these mosquito-borne diseases is to interfere with the transmission of diseases through either the eradication of mosquitoes, the preservation of mosquitoes against bitten persons or through a high-scale mortality in mosquito larvae at vector breeding sites [1]. The application of chemical larvicides is detrimental to the environment because they are not biodegradable and thereforecan persist for a long time in the environment [2]. In addition, synthetic larvicidal also disrupt biological frameworks of natural control which at times promote resistance [3]. Scientists are therefore looking for alternative natural products that can serve as both larvicidal and antimicrobial agents.

Silver has been known for a long time to prevent microbial contamination. It protects against quite a number of microorganisms such as *Escherichia coli* and *Pseudomonas aeruginosa* [4]. However, the metal has received much attention recently and is being used in the synthesis of silver nanoparticles. *Azadirachta indica* leaf extract used in silver nanoparticles synthesis (AgNPs) was reported by Benakashani *et al.* [5]. Numerous researchers in silver nanoparticles synthesis with remarkable biomedical applications also used several other species including *Cola nitida*, *Theobroma cacao, Buchholzia coriacea* and *Synsepalum dulcificum* [6-14].

In biomedical research, the different applications for certain parts of the *Carica papaya* plant have been used a number of times. Okeniyi *et al.* [15] documented the anti-moebic and antihelmintic potentials in *Carica papaya.* Bamisaye *et al.* [16] also reported on the plant's ethnobotanical perspectives. The antimicrobial efficacy of its root against certain infectious microorganisms was also tested by Tiwari *et al.* [17]. In a previous paper, the flavonoid content,total phenolic content andtotal antioxidant activity of the various plant parts of the papaya tree were determined and compared [18].

Several portions of the *Carica papaya* plant have also been utilised for green synthesis of nanoparticles. Jain *et al.* [19] reported the usage of *Carica* fruit extract in AgNPs synthesis. It's antimicrobial efficacy against some microorganisms such as *Escherichia coli* and *Pseudomonas aeruginosa* were likewise reported [19]. In addition, Kokila and others in 2013 also reported the usage of the AgNPs synthesized from its peel extract as antioxidant and antimicrobial agents [20]. Mahanty *et al.* [21] evaluated the possible application of silver nanoparticles in aquaculture as alternatives to antibiotics. Banala *et al.* [22] also evaluated the

IOP Conf. Series: Materials Science and Engineering **805** (2020) 012038 doi:10.1088/1757-899X/805/1/012038

bactericidal activity of silver nanoparticles synthesized from the leaf extract of this plant against human pathogenic strains like *Staphylococcus aureus, Bacillus subtilis, Microbacterium luteus, Klebsiella pneumoniae* and *E. coli*. In the synthesis of AgNPs, Mude *et al.* [23] also used the callus extract to synthesize AgNPs, while Sankar *et al.* [24] took further steps in the synthesis and application of colloidal copper oxide in photocatalytic degradation with the extracts from this plant. It is worthy of note in the field of nanotechnology that nanoparticles from the seeds of *Carica papaya* have not been synthesized or investigated for its biological or antimicrobial properties. Therefore, the study discussed here aims to analyze and apply *Carica papaya* seed as antimicrobial, larvicidal and bio-reducing agents in the green synthesis of AgNPs.

2 Materials and Methods

2.1 Sample collection

Fresh *Carica papaya* was purchased from Ilishan-Remo market in Ogun state and was stored in the Babcock University Microbiology Laboratory. The fruit was cat open and the seeds were removed and dried for 5 days (30 ± 2 ° C) under shade. The dried seeds were pulverised and stored in air-tight bottles at room temperature for further use in water/air resistant bottles. For antimicrobial studies, *Staphylococcus aureus* was the selected Gram positive organism used, while *Pseudomonas aeruginosa, Escherichia coli* and *Klebsiella pneumoniae* served as the selected Gram negative organisms. These bacterial strains were collected from the Teaching Hospital of Ladoke Akintola University Ogbomoso, Nigeria.

2.2 Collection and Identification of Mosquito Larva

The mosquito larvae were collected using a net, and kept in sealed containers from polluted body of water in Babcock University, Nigeria. In this study, the target mosquito larvae were the 4th instar larva *Aedes aegypti* mosquito. Identified *Aedes aegypti* mosquito larvae (identified by a medical entomologist) were isolated from the other mosquito species and placed separately in a container full of water.

2.3 Preparation of extract

The *Carica papaya* seed aqueous extract used for this analysis was done following the method of Lateef *et al.* [25]. One gram of the seed was soaked in 100 ml distilled water. This was accompanied by heating in the water bath for 1 h at 60 ^{o}C temperature. The extract was filtered and centrifuged at 4000 rpm for 20 min and the resultant supernatant was obtained.

2.4 Biogenic synthesis of the AgNPs

The AgNPs was biogenically synthesised by introducing 1 ml of the prepared extract into 40 ml of 1 mM $AgNO_3$ and the colour change was observed. The entire reaction which was about 2 h was conducted in ambient condition [25].

2.5 Characterization of the AgNPs

The synthesized AgNPs were characterized by a UV-visible spectrophotometer (Cecil USA) that measured its absorbance spectrum at 190-1100 nm. FTIR spectroscopy identified biomolecules participating in the green synthesis. The FTIR model 8400S (Shimadzu Japan) developed vibrational frequencies in NPs from FTIR spectrum using KBr process. To assess the identity and ratio of elemental particles, EDX analysis (Silicon-Drift Detector X-Max daN Oxford Instruments UK) was carried out. FESEM (Lyra 3 Tescan Czech Republic) was used to examine the presence, size and structural morphology of synthesized silver nanoparticles. Image J was used to analyze the image attributes [26].

2.6 Antimicrobial Activity

The agar well-diffusion method was used to assess the efficacy of biosynthesized *Carica papaya* extract mediated silver nanoparticles (CPE-AgNPs) against certain selected clinical isolates. An 18 h-old culture of each species grown in peptone broth was used to inoculate an already prepared Mueller Hinton agar plate. Holes were drilled using a 6 mm sterile cork borer on the inoculated plate and labelled 10, 20, 40, 60, 80 and 100 μg/ml respectively. The resulting holes were then filled with 100 μl of different concentrations from the synthesized CPE-AgNPssilver nitrate (positive control) and water (negative control). The plates were incubated at 37 ^{o}C for 18 h after which the zone of inhibition was then determined [27]. The analysis was done in triplicates.

2.7 Larvicidal Efficacy

The WHO recommended guideline was used to test the efficacy of the AgNPs with the 4th *Aedes aegypti* instar larva [28]. A microtitre plate of 96-hole was used. The 5 holes of 20, 40, 60, 80 and 100 μg / ml respectively were loaded with five larvae. In each respective hollow, 300 μl of each AgNP concentration was then dispensed as labelled. A control experiment was conducted in which the larvae were exposed to sterile distilled water. Triplicate tests were performed and after 12 h of treatment readings were obtained. This was followed by statistical assessment of mortality percentage and Probit analysis for the LC_{50} and LC_{90} computation.

3 Results and Discussion

3.1 Synthesis and Characterization of the Synthesized AgNPs

Upon the addition of the plant extract to the silver nitrate, the solution still remained colourless. After 10 min, a gradual change in colour was observed; initially starting faintly and in 30 min, a stable deep brown colouration was formed in the reaction vessel (Figure 1), which is indicative of the formation of AgNPs as it has been observed by several other researchers [8-12]. The colour change of the solution of metallic salt is a sign of nanoparticles development. Earlier researchers reported extensively the advances of different color shades including light yellow to darker colloidal AgNPs [4, 7-9, 29-32]. A light brown coloration was observed by Mahanty *et al.* [21 during the synthesis of silver nanoparticles from *Carica papaya* leaves. It was believed that this colour shift during the formation of silver nanoparticles from *Carica papaya* is induced by the different biomolecules present in the plant and are of responsible for the synthesis and stability of silver nanoparticles in its seed extract [8].

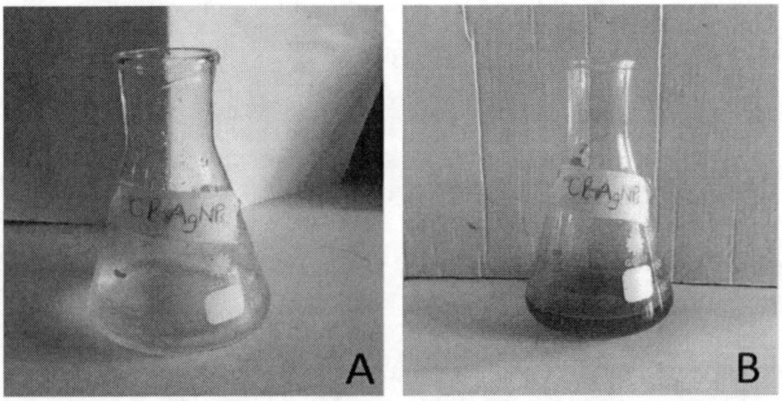

Figure 1. Synthesis of the CPE-AgNPs (a) 0 min after the addition of the plant extract to the silver nitrate (b) Formation of deep brown colouration after 30 min

The CPE-AgNPs formed were examined with the UV-vis spectrophotometer as shown in Figure 2. The absorption range of the synthesized CPE-AgNPs displayed a peak magnitude at 433 nm. The peak is in line with previously reported investigations on silver nanoparticles [4, 8, 25, 33-35]. Jain *et al.* [19] recorded a maximum absorption of 450 nm which was due to surface plasmon excitation when using *Carica papaya* fruit extract for silver nanoparticles synthesis.

IOP Conf. Series: Materials Science and Engineering **805** (2020) 012038 doi:10.1088/1757-899X/805/1/012038

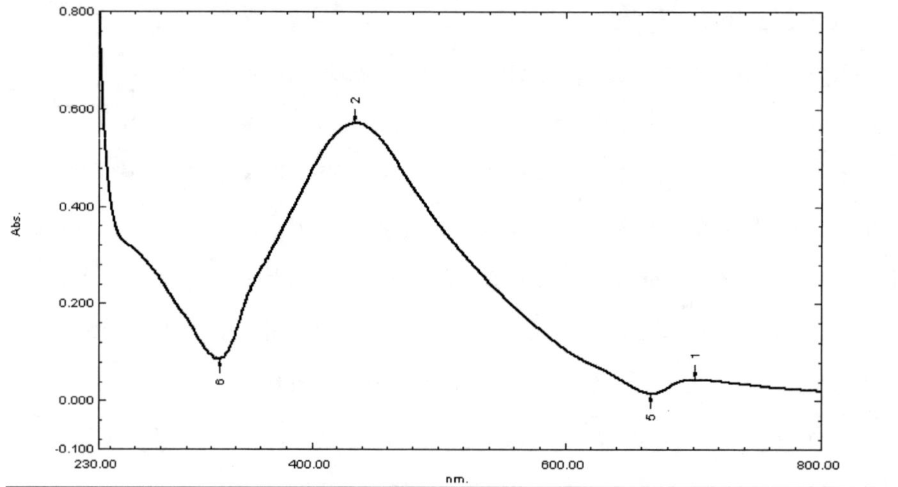

Figure 2. UV-vis spectra of AgNPs synthesized from *Carica papaya* seed extract

Specific functional classes on the surface of the AgNPs were shown in the FTIR study of the CPE-AgNPs (Figure 3). It has a large range from 3000 to 3800 cm^{-1} defined as the ones with vibration of O-H and/or N-H correlated with n-substituted amide [36] at 2426 cm^{-1} and 2360 cm^{-1}, and the ones with ambient CO_2 absorption [37-38]. The N-H bending of amides of proteins is reported at 1635 cm^{-1}.

The EDX ranges of the *Carica papaya* controlled AgNPs is shown in Figure 4. Powerful signals at 1.5 keV and 3.0 keV were recorded from Ag atoms while weak signals at 2.7, 2.9 and 3.2 keV were noted. Figure 4 also displayed certain components such as C and O which are known as impurities from the sample. Several other scientists [39-41] reported similar findings of heavy silver signalling varying from 1.5 to 5.0 keV.

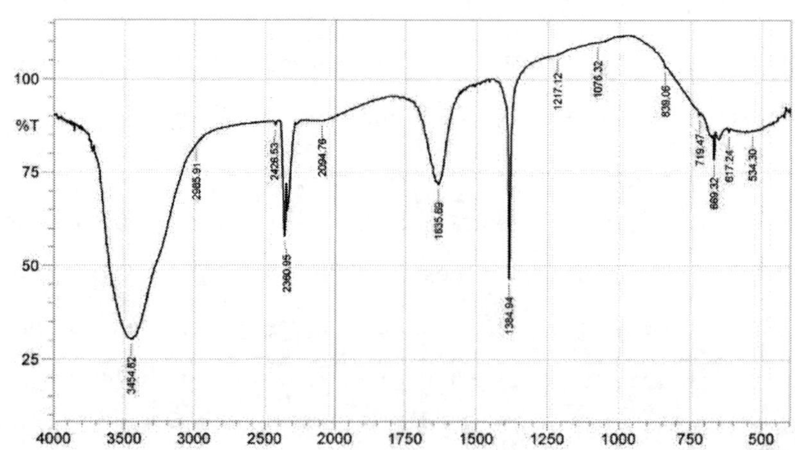

Figure 3. The FTIR spectrum of the synthesized AgNPs

IOP Conf. Series: Materials Science and Engineering **805** (2020) 012038 doi:10.1088/1757-899X/805/1/012038

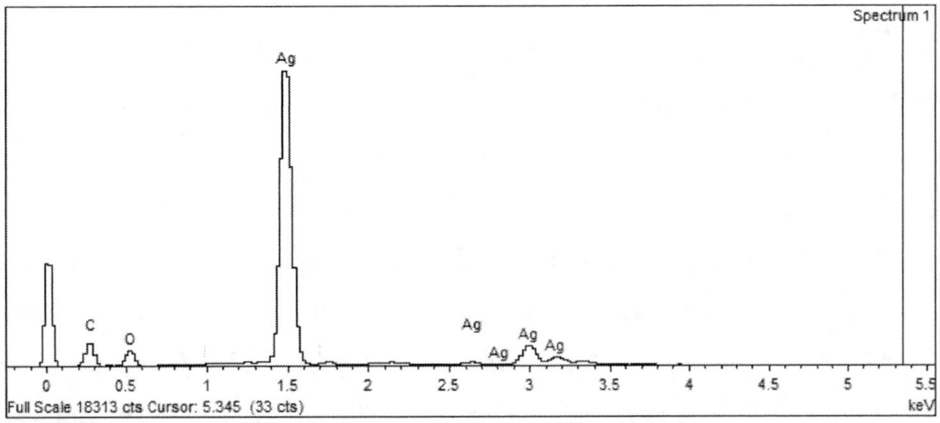

Figure 4. Energy dispersive spectrum of the synthesized AgNPs

In Figure 5, FESEM analysis on CPE-AgNPs provided additional details on the size and morphology of the synthesized AgNPs. As a consequence of its high surface ratio, the particles produced formed clusters or aggregates. There are also many spherical nanoparticles from 20 to 69 nm. This picture shows a large amount. The viability of *Carica papaya* seed for the biosynthesis of silver nanoparticles in this study has been verified by similar FESEM findings from synthesized silver nanoparticles from other plants [8].

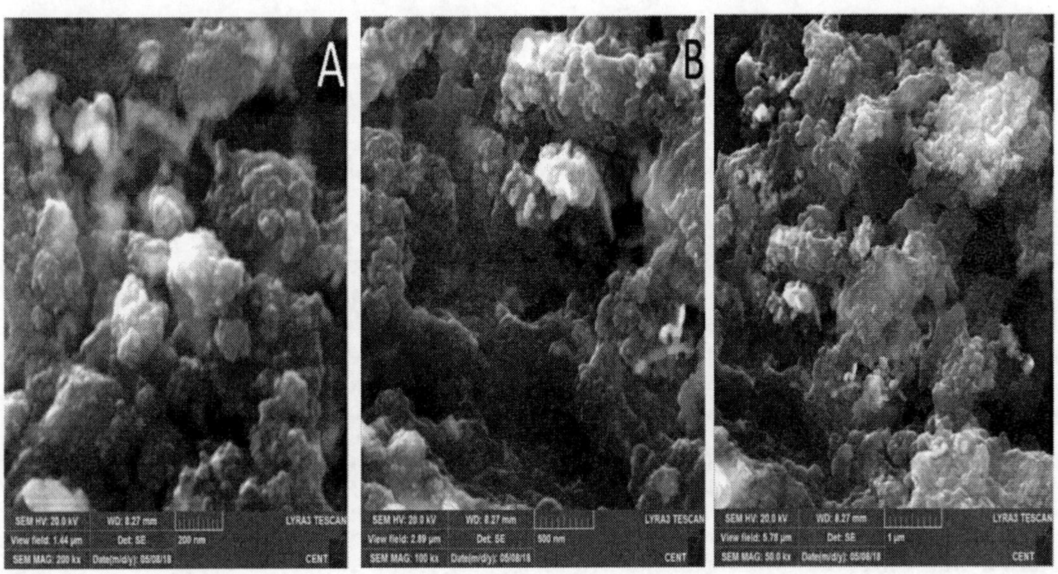

Figure 5. Field emission scanning electron micrograph of the synthesized AgNPs at different magnifications (a) 200kx (b) 100kx (c) 50kx

3.2 Antimicrobial activity

Of all the four selected pathogens tested, the synthesized AgNPs inhibited the growth of *Klebsiella pneumoniae*, *Pseudomonas aeruginosa* and *Escherichia coli*, all of which were

IOP Conf. Series: Materials Science and Engineering **805** (2020) 012038 doi:10.1088/1757-899X/805/1/012038

inhibited distinctly with zones ranging from 10-24 mm. A wide range of pathogenic organisms have been inhibited by biosynthesized AgNPs effectively [42-46]. At 100 µg of CPE-AgNPs *Staphylococcus aureus* was inhibited by 16.0 mm, while other concentrations (10-80 µg) did not have any inhibitory effect. Panacek *et al.* [47] reported similar resistance trends to AgNPs. It is proposed that flagellin present in large amounts in bacteria may cluster the silver nanoparticles and thus decreases its antibacterial effectiveness. Ultimately, the results of this study showed a significant antibacterial impact of the nano-sized silver synthesized from *Carica papaya* seed. The above details are described in both Table 1 and Figure 6.

Table 1. Zone of inhibition of the synthesized AgNPs against some selected pathogens

Mean Zone of Inhibition (mm) ± standard deviation (SD)									
	AgNPs						$AgNO_3$	CPE	Distilled H_2O
	10µg/ml	20µg/ml	40µg/ml	60µg/ml	80µg/ml	100µg/ml			
Isolate									
taphylococcus aureus	NZ	NZ	NZ	NZ	NZ	16.0±01	6.2±02	NZ	NZ
lebsiella pneumoniae	NZ	11±02	17±02	24±02	20±02	15.0±01	6.4±01	NZ	NZ
seudomonas aeruginosa	10±01	12±02	13±01	NZ	NZ	NZ	6.1±01	NZ	NZ
scherichia coli	13±01	10±01	NZ	NZ	12±02	20.0±01	6.3±01	NZ	NZ

NZ, No zone

3.3 Larvicidal Activity

The biosynthesized AgNPs were found to be highly toxic to the 4th instar larvae *Aedes aegypti*. Mortality percentages in the 20, 40 and 60 µg/ml were found to be 80%, 86.6% and 93.3% respectively, while the 80 and 100 µg/ml showed 100 % mortality as shown in Figure 7. However, no mortality was recorded in the control group. Furthermore, the AgNPs exhibited a concentration dependent activity against mosquito larvae since the percentage mortality was observed to increase with increasing concentrations of the biosynthesized nanoparticles. The 50 % lethal concentration (LC_{50}) of the nanoparticles was 14.54 µg/ml, while the LC_{90} was 33.9 µg/ml. Similar results were recorded also in the previous study by Lateef *et al.* [25], where AgNPs synthesized from cell-free extract of *Bacillus safensis* exhibited larvicidal potential against Anopheline larvae. The larvicidal activity of the AgNPs could be because of the penetration of the particles to debilitate cell metabolism because of their attachment to enzymes and DNA. CPE-AgNPs larvicidal activities were similar to

IOP Conf. Series: Materials Science and Engineering **805** (2020) 012038 doi:10.1088/1757-899X/805/1/012038

reported larvicidal activity of some AgNPs of bacterial extract origin and plants against *Aedes aegypti* larvae [48].

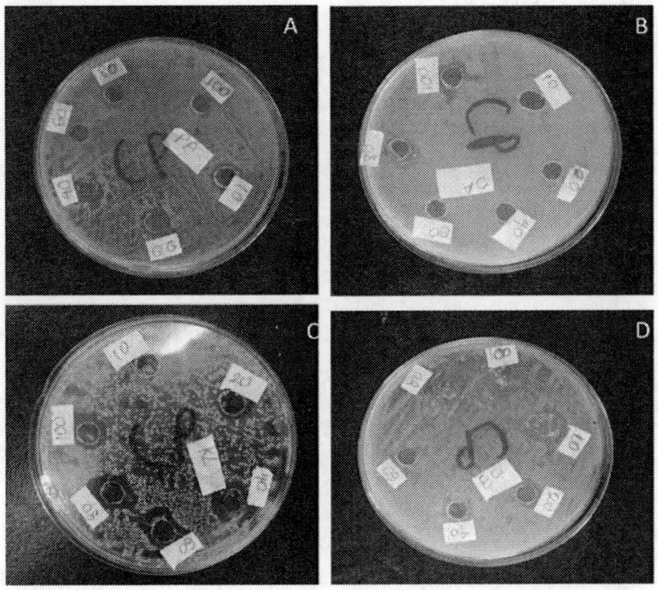

Figure 6. The antibacterial activities of the CPE-AgNPs against some clinical bacterial isolates (a) *Pseudomonas aeruginosa* (b) *Staphylococcus aureus* (c) *Klebsiella pneumoniae* (d) *Escherichia coli*

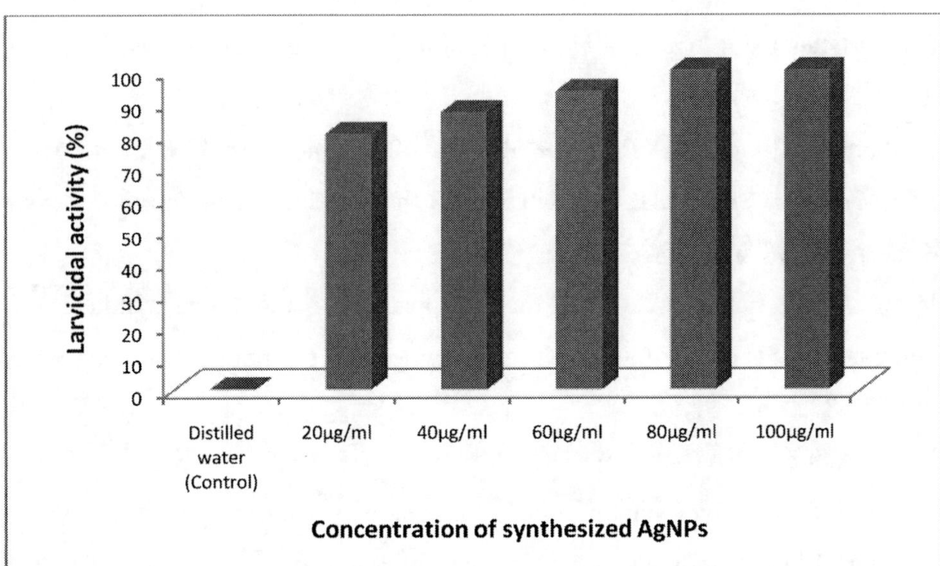

Figure 7. Larvicidal activity of the biosynthesized CPE-AgNPs on *Aedes* mosquito larvae

IOP Conf. Series: Materials Science and Engineering **805** (2020) 012038 doi:10.1088/1757-899X/805/1/012038

4 Conclusion

This study has revealed a cheaper and eco-friendly biosynthesis of AgNPs using seeds of *Carica papaya.* It is proof that the seeds of the plant possess the biomolecules and phytochemicals necessary for the reduction and stabilization agents in the synthesis of silver nanoparticles.The CPE-AgNPs had remarkable antibacterial and larvicidal activities, making available a new agent for combat against multi-drug resistant bacteria and the Zika and Dengue fever vector. More investigation into the cytotoxicity of these biosynthesized silver nanoparticles is however required to further clarify issues on its application.

References

[1] Mohan DR and Ramaswamy M 2007 Evaluation of larvicidal activity of the leaf extract of a weed plant *Ageratina adenophora* against two important species of mosquito *Aedes aegypti* and *Culex quinqufaciatus. Afr. J. Biotechnol.* **6** (5) pp 631-638.

[2] Tiwary M, Naik S, Tewary D K, Mittal P K and Yadav S 2007Chemical composition and larvicidal activities of the essential oil of *Zanthoxylum armatum* DC (Rutaceae) against three mosquito vectors. *J. Vect. Borne Dis.* **44** (3) pp 198-204.

[3] Mathivanan T, Govindarajan K, Elumalai K and Ananthan A 2010 Mosquito larvicidal and phytochemical properties of *Ervantania coronaria* Stap f (Family Apocynaceae). *J. Vect. Borne Dis.* **47** (3) pp 178-180.

[4] Ahmed S, Saifullah A, Ahmad M, Swami B L and Ikram S 2016 Green synthesis of silver nanoparticles using *Azadirachta indica* aqueous leaf extract. *J. Rad. Res. Appl. Sci.* **9** (1) pp 1–7.

[5] Benakashani F, Allafchian A R and Jalali S A H 2016 Biosynthesis of silver nanoparticles using *Capparis spinosa* L leaf extract and their antibacterial activity. *Karbala Int. J. Modern Sci.* **2** (4) pp 251-258.

[6] Adelere I A and Lateef A 2016 A novel approach to the green synthesis of metallic nanoparticles: The use of agro-wastes enzymes and pigments. *Nanotechnol. Rev.* **5** (6) pp 567-587.

[7] Adelere I A, Lateef A, Aboyeji DO, Abdulsalam R, Adabara NU and Bala JD 2017 Biosynthesis of silver nanoparticles using aqueous extract of *Buchholzia coriacea* (wonderful kola) seeds and their antimicrobial activities. *Ann. Food Sci. Technol.* **18** (4) pp 671-679.

[8] Lateef A, Azeez M A, Asafa T B, Yekeen T A, Akinboro A, Oladipo I C and Beukes L S 2015 *Cola nitida*-mediated biogenic synthesis of silver nanoparticles using seed and

seed shell extracts and evaluation of antibacterial activities. *BioNanoSci.* **5** (4) pp 196-205.

[9] Lateef A, Azeez M A, Asafa T B, Yekeen T A, Akinboro A, Oladipo I C and Beukes L S 2016 Biogenic synthesis of silver nanoparticles using a pod extract of *Cola nitida*: Antibacterial and antioxidant activities and application as a paint additive. *J. Taibah Univ. Sci.* **10** (4) pp 551-562.

[10] Lateef A, Azeez M A, Asafa T B, Yekeen T A, Akinboro A, Oladipo I C and Beukes L S 2016 Cocoa pod husk extract-mediated biosynthesis of silver nanoparticles: its antimicrobial antioxidant and larvicidal activities. *J. Nanostruct. Chem.* **6** (2) pp 159-169.

[11] Lateef A, Akande M A, Azeez M A, Ojo S A, Folarin B I, Gueguim-Kana E B and Beukes L S 2016 Phytosynthesis of silver nanoparticles (AgNPs) using miracle fruit plant (*Synsepalum dulcificum*) for antimicrobial catalytic anticoagulant and thrombolytic applications. *Nanotechnol. Rev.* **5** (6) pp 507-520.

[12] Azeez M A, Lateef A, Asafa T B, Yekeen T A, Akinboro A, Oladipo I C, Gueguim-Kana E B and Beukes L S 2017 Biomedical applications of cocoa bean extract-mediated silver nanoparticles as antimicrobial larvicidal and anticoagulant agents. *J. Clust. Sci.* **28** (1) pp 149-164.

[13] Yekeen T A, Azeez M A, Lateef A, Asafa T B, Oladipo I C, Badmus J A, Adejumo SA and Ajibola A A 2017 Cytogenotoxicity potentials of cocoa pod and bean-mediated green synthesized silver nanoparticles on *Allium cepa* cells. *Caryologia: Int. J. Cytol. Cytosystem. Cytogenet.* **70** (4) pp 366-377.

[14] Yekeen T A, Azeez M A, Akinboro A, Lateef A, Asafa T B, Oladipo I C, Oladokun SO and Ajibola A A 2017 Safety evaluation of green synthesized *Cola nitida* pod seed and seed shell extract-mediated silver nanoparticles (AgNPs) using an *Allium cepa* assay. *J. Taibah Univ. Sci.* **11** (6) pp 895-909.

[15] Okeniyi JA, Ogunlesi TA, Oyelami OA and Adeyemi LA 2007 Effectiveness of dried *Carica papaya* seeds against human intestinal parasitosis: A pilot study. *J. Med. Food* **10** (1) pp 194-196.

[16] Bamisaye FA, Ajani EO and Minari JB 2013 Prospects of ethnobotanical uses of pawpaw (*Carica papaya*). *J. Med. Plants Stud.* **1** (4) pp 171-177.

[17] Tiwari P, Kumar K, Panik R, Pandey A, Pandey A, Sahu P K and Raipur C G 2011 Antimicrobial activity evaluation of the root of *Carica papaya* Linn. *Int. J. PharmTech Res.* **3** (3) pp 1641-1648.

IOP Conf. Series: Materials Science and Engineering **805** (2020) 012038 doi:10.1088/1757-899X/805/1/012038

[18] Maisarah A, Nurul Amira B, Asmah R and Fauziah O 2013 Antioxidant analysis of different parts of *Carica papaya. Int. Food Res. J.* **20** (3) pp 1043-1048.

[19] Jain D, Daima H, Kachhwaha S and Kothari S 2009 Synthesis of plant-mediated silver nanoparticles using *Carica papaya* fruit extract and evaluation of their antimicrobial activities. *Digest J. Nanomater. Biostruct.* **4** (3) pp 557-563.

[20] Kokila T, Ramesh P S and Geetha D 2016 Ecotoxicology and Environmental Safety Biosynthesis of AgNPs using *Carica Papaya* peel extract and evaluation of its antioxidant and antimicrobial activities. *Ecotoxicol. Environ. Saf.* **134** pp 467-473.

[21] Mahanty A, Mishra S, Bosu R, Maurya U K, Netam S P and Sarkar B 2013 Phytoextracts-synthesized silver nanoparticles inhibit bacterial fish pathogen *Aeromonas hydrophila. Ind. J. Microbiol.* **53** (4) pp 438-446.

[22] Banala R, Nagati V and Karnati P 2015 Green synthesis and characterization of *Carica papaya* leaf extract coated silver nanoparticles through X-ray diffraction electron microscopy and evaluation of bactericidal properties. *Saudi J. Biol. Sci.* **22** (5) pp 637-644.

[23] Mude N, Ingle A, Gade A and Rai M 2009 Synthesis of silver nanoparticles using callus extract of *Carica papaya*-a first report. *J. Plant Biochem. Biotechnol.* **18** (1) pp 83-86.

[24] Sankar R, Manikandan P and Malarvizhi V 2014 Green synthesis of colloidal copper oxide nanoparticles using *Carica papaya* and its application in photocatalytic dye degradation. *Spectrochim. Acta Part A: Mol. Biomol. Spectr.* **121** pp 746-750.

[25] Lateef A, Ojo S A, Akinwale A S, Azeez L, Gueguim-Kana E B and Beukes L S 2015 Biogenic synthesis of silver nanoparticles using cell-free extract of *Bacillus safensis* LAU 13: Antimicrobial free radical scavenging and larvicidal activities. *Biologia* **70** (10) pp 1295-1306.

[26] Defante A P, Vreeland W N, Benkstein K D and Ripple D C 2018 Using image attributes to assure accurate particle size and count using nanoparticle tracking analysis. *J. Pharmaceut. Sci.* **107** (5) pp 1383-1391.

[26] Perez C, Paul M and Bazerque P 1990 Antibiotic assay by agar well diffusion method. *Acta Biol. Med. Exp.* **15** pp 113-115.

[27] WHO 2005 Guidelines for laboratory and field testing of mosquito larvicides. *World Health Organization* (No. WHO/CDS/WHOPES/GCDPP/2005.13). https://apps.who.int/iris/bitstream/handle/10665/69101/WHO_CDS_WHOPES_GCDPP_2005.13.pdf (Accessed on 2 February, 2020).

[28] Prabu H and Johnson I 2015 Plant-mediated biosynthesis and characterization of silver nanoparticles by leaf extracts of *Tragia involucrata, Cymbopogon citronella, Solanum verbascifolium* and *Tylophora ovata. Karbala Int. J. Modern Sci.* **1** (4) pp 237-246.

[29] Bindhu MR and Umadevi M 2013 Synthesis of monodispersed silver nanoparticles using *Hibiscus cannabinus* leaf extract and its antimicrobial activity. *Spectrochim. Acta Part A: Mol. Biomol. Spectr.* **101** pp 184-190.

[30] Bindhu MR and Umadevi M 2015 Antibacterial and catalytic activities of green synthesized silver nanoparticles. *Spectrochim. Acta Part A: Mol. Biomol. Spectr.* **135** pp 373-378.

[31] Thirumurugan A, Tomy N A, Ganesh R J and Gobikrishnan S 2010 Biological reduction of silver nanoparticles using plant leaf extracts and its effect on increased antimicrobial activity against clinically isolated organisms. *Der Pharm. Chem.* **2** (6) pp 279-284

[32] Lateef A, Adelere I A, Gueguim-Kana E B, Asafa T B and Beukes L S 2015 Green synthesis of silver nanoparticles using keratinase obtained from a strain of *Bacillus safensis* LAU 13. *Int. Nano Lett.* **5** (1) pp 29-35.

[33] Krishnaraj C, Jagan E G, Rajasekar S, Selvakumar P, Kalaichelvan P T and Mohan N 2010 Synthesis of silver nanoparticles using *Acalypha indica* leaf extracts and its antibacterial activity against water borne pathogens. *Colloids Surf. B: Biointerf.* **76** (1) pp 50-56.

[34] Basavegowda N and Rok Lee Y 2013 Synthesis of silver nanoparticles using Satsuma mandarin (*Citrus unshiu*) peel extract: A novel approach towards waste utilization. *Mater. Lett.* **109** pp 31-33.

[35] Unuabonah EI, Adie GU, Onah LO and Adeyemi O G 2009 Multistage optimization of the adsorption of methylene blue dye onto defatted *Carica papaya* seeds. *Chem. Eng. J.* **155** (3) pp 567-579.

[36] Zhao Y, Frost RL, Yang J and Martens WN 2008 Size and morphology control of gallium oxide hydroxide GaO(OH) nano- to micro-sized particles by soft-chemistry route without surfactant. *J. Phys. Chem. C* **112** (10) pp 3568-3579.

[37] Gopalakrishnan M, Purushothaman V, Ramakrishnan V, Bhalerao G M and Jeganathan K 2014 The effect of nitridation temperature on the structural optical and electrical properties of GaN nanoparticles. *CrystEngComm.* **16** (17) pp 3584-3591.

[38] Islam N U, Amin R, Shahid M, Amin M, Zaib S and Iqbal J 2017 A multi-target therapeutic potential of *Prunus domestica* gum stabilized nanoparticles exhibited

prospective anticancer antibacterial urease-inhibition anti-inflammatory and analgesic properties. *BMC Complem. Altern. Med.*. **17** 276. https://doi.org/10.1186/s12906-017-1791-3.

[39] Puchalski P, Dabrowski P, Olejnikzac W, Krukowski P, Polanski K and Kowalczyk K 2007 The study of silver nanoparticles by scanning electron microscopy energy dispersive X-ray analysis and scanning tunnelling microscopy. *Mater. Sci.-Poland* **25** (2) pp 473-478.

[40] Lateef A, Folarin BI, Oladejo SM, Akinola PO, Beukes LS and Gueguim-Kana EB 2018 Characterization, antimicrobial, antioxidant, and anticoagulant activities of silver nanoparticles synthesized from *Petiveria alliacea* L. leaf extract. *Prep. Biochem. Biotechnol.* **48** (7) pp 646-652.

[41] Rai M, Deshmukh S, Ingle A and Gade A 2012 Silver nanoparticles: the powerful nanoweapon against multidrug-resistant bacteria. *J. Appl. Microbiolo.* **112** (5) pp 841-852.

[42] Lara HH, Ayala-Núñez NV, Turrent L and Padilla CR 2010 Bactericidal effect of silver nanoparticles against multidrugresistant bacteria. *World J. Microbiol. Biotechnol.* **26** (4) pp 43-56.

[43] Sondi B and Salopek-Sondi A 2004 Silver nanoparticles as antimicrobial agent: a case study on *E coli* as a model for Gram-negative bacteria. *J. Colloid Interf. Sci.* **275** (1) pp 177-182.

[44] Salem W M, Haridy M, Sayed W F and Hassan N H 2014 Antibacterial activity of silver nanoparticles synthesized from latex and leaf extract of *Ficus sycomorus. Ind. Crops Prod.* **62** (4) pp 228-234.

[45] Shankar S, Jaiswal L,Aparna R S L and Prasad V 2014 Synthesis characterization *in vitro* biocompatibility and antimicrobial activity of gold silver and gold silver alloy nanoparticles prepared from *Lansium domesticum* fruit peel extract. *Mater. Lett.* **137** pp 75-78.

[46] Panáček A, Kvítek L, Smékalová M, Večeřová R, Kolář M, Röderová M and Zbořil R 2018 Bacterial resistance to silver nanoparticles and how to overcome it. *Nature Nanotechnol.* **13** (1) pp 65-71.

[47] Benelli G 2016 Plant-mediated biosynthesis of nanoparticles as an emerging tool against mosquitoes of medical and veterinary importance : a review. *Parasitol. Res.* **115** (1) pp 23-34.

IOP Conf. Series: Materials Science and Engineering **805** (2020) 012038 doi:10.1088/1757-899X/805/1/012038

[48] Aina DA, Owolo O, Lateef A, Aina FO, Hakeem AS, Adeoye-Isijola M, Okon V, Asafa TB, Elegbede JA, Olukanni OD and Adediji I. Biomedical applications of *Chasmanthera dependens* stem extract mediated silver nanoparticles as antimicrobial, antioxidant, anticoagulant, thrombolytic, and larvicidal agents. *Karbala Int. J. Modern Sci.* **5** (2) pp 71-80.

IOP Conf. Series: Materials Science and Engineering **805** (2020) 012039 doi:10.1088/1757-899X/805/1/012039

Engaging science diplomacy for nanotechnology development in Africa

I P Ezekiel

Science Policy and Innovation Studies, National Centre for Technology Management (NACETEM), North West Zone, Kano, Kano State, Nigeria

E-mail: itegbeyogene@gmail.com

Abstract

The concept of science diplomacy is gaining ground as a global strategy in addressing global concerns such as global peace, insecurity, climate change and environmental impact. This study reduces science diplomacy to an effective means for the strategic development of nanotechnology in Africa. African nations are currently not encouragingly close to the leading nations in nanotechnology, yet there seem to be extant diplomatic relationship with many of these forefront nations. African diplomats are state actors in science diplomacy to propose foreign policies that will meet the domestic demand for science and technology development in Africa, especially for emerging technologies like nanotechnology. The necessity of inclusion of competent scientists with diplomatic skills as members of the diplomatic corps is recommended here as one of the ways to develop nanotechnology in Africa. The scientist diplomats will function to foster international scientific collaborations, drive platforms for national research facility development and for non-state actors to thrive in their domestic nano-research. Scientifically informed foreign policies are presented here to have potentials to significantly assist Africa in developing nanotechnology and provide pathways for overcoming the numerous constraints to nanotechnology development in Africa. Critical areas of intervention include human capacity development, national nano-research laboratory facilities, platforms for institutional collaborations, post-graduate student training and robust exchange programs. These machineries also benefit independent individual researchers by leveraging on the international networks created by science attachés through science diplomacy.

Keywords: Nanotechnology, science diplomacy, Africa, human capacity development, Science and Technology, international collaboration

IOP Conf. Series: Materials Science and Engineering **805** (2020) 012039 doi:10.1088/1757-899X/805/1/012039

1. Introduction

It is non-controversial that science and technology (S&T) is the economic driver of the 21st century [1-3]. This fact is not likely to change in the future because the innovative ecosystem is constantly improving on the dividends of S&T. Science and technology is dynamic, and innovation (I) is developing very rapidly. The absence of a nation's consciousness and commitment to ST&I could propel different levels of technological ignorance and inefficiencies that will be enough to cripple such a nation's economy. The world economic powers pride on their knowledge-based economy that is anchored on ST&I. One of such technologies that are revolutionizing innovations in the world today is nanotechnology. Nanotechnology has miniaturized devices and gadgets, enabled compatibility and flexibility with improved service delivery [3-5].

The proposed single statistical definition of nanotechnology by the Organization for Economic Co-operation and Development (OECD) provides a comprehensive overview. The OECD proposes that, "It is the understanding of processes and phenomena and the application of science and technology to organisms, organic and inorganic materials, as well as parts, products and models thereof, at the nanometre-scale (but not exclusively below 100 nanometres) in one or more dimensions, where the onset of size-dependent phenomena usually enables novel applications" [6]. The unique properties exhibited by nanomaterials which are different from their bulk sizes, have engineered novel researches and applications in diverse areas such as health, agriculture, environmental remediation, food industry, automobile industry, academia, space technology, manufacturing industry, electronics, energy, oil and gas industry [4, 5, 7]. It is a convergence of different fields with a high propensity of transforming nations' approach to poverty eradication and guaranteeing sustainable economic development [2].

The continent of Africa is in apt need of technologies that can reduce poverty, improve the living standard of the people, ensure a cleaner environment and provide a platform for technological advancement. Unfortunately, the continent has not done much to show priority to the development of nanotechnology, except for South Africa's contribution which is yet to be identified among the frontline nations [8]. Recent data on the indicators of nanotechnology development under three main categories of Science, Innovations and Industry shows the leading nations in the development of nanotechnology [9]. It did not reflect significant changes for any of the Africa nations to gain global audience.

However, one encouraging and interesting hope is that the frontline nations in nano-articles publications, patenting and national standards shown in Figures 1, 2 and 3 respectively, have diplomatic relationships with most of the African countries. One pertinent question to ask is how effective are these diplomatic relationships and foreign policies helping to develop required technologies in Africa such as nanotechnology? The answer to this question will be strongly dependent on how African nations especially the two economic giants; Nigeria and South Africa, use the instrumentality of diplomacy to prioritize their technology needs. Important to note is that it takes knowledgeable scientists to be able to engage in scientific diplomatic negotiations that will yield meaningful benefits.

This study therefore, discussed the importance of engaging science diplomacy by African nations to aid foreign policies that will attract the development of nanotechnology to their home countries. It reviewed the offices and the diplomatic corps staffing of Nigeria and South Africa embassies and their high commissions abroad as case studies. It also looked at the role of tertiary institutions and stakeholders research-based public and private establishment in the development of nanotechnology.

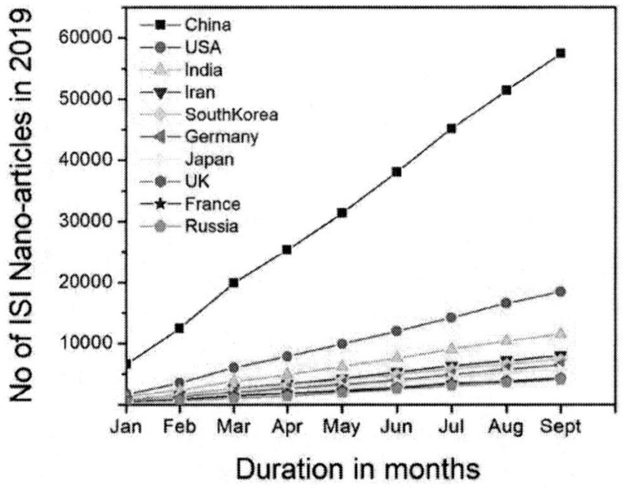

Figure 1. Number of ISI nano-articles published in 2019 by 10 leading countries in nanotechnology/nanoscience development. Data from Web of Science (ISI Web of Knowledge) [9]

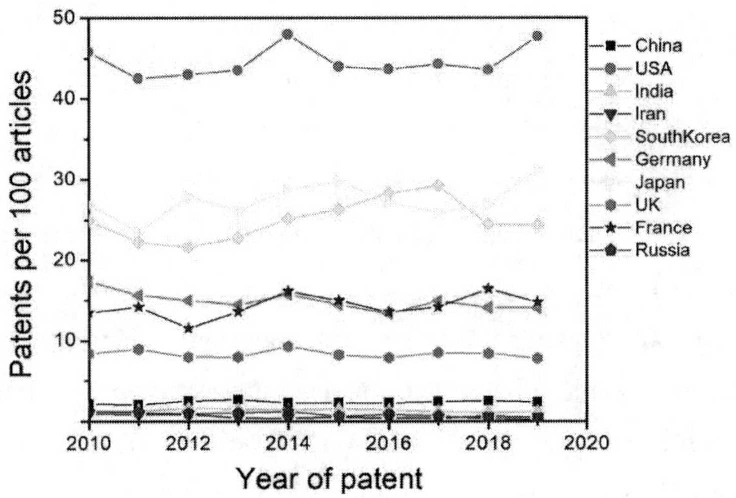

Figure 2. Ratio of nanotechnology patents to nano-articles in nanotechnology/nanoscience development. Data from Web of Science (ISI Web of Knowledge) [9]

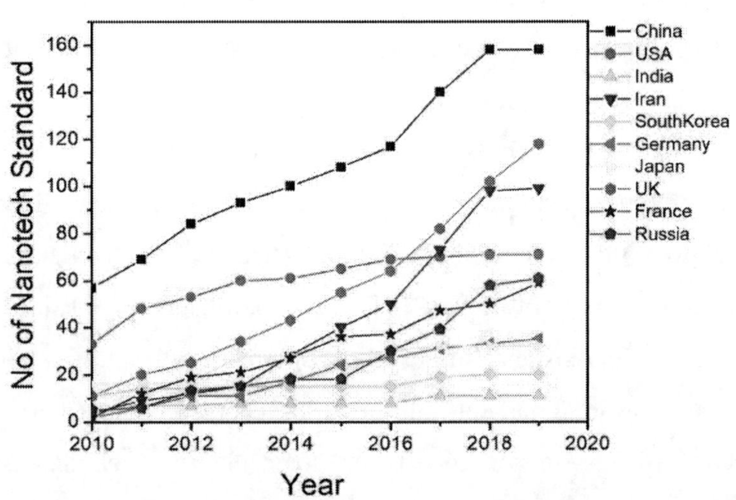

Figure 3. Number of national nanotechnology standards. Data from Statnano Databank [10]

2. Science Diplomacy and National ST&I Development

Although, science diplomacy is a relatively new area of diplomacy study, the concept is however not entirely new. Science diplomacy is the deployment of scientific knowledge to aid

IOP Conf. Series: Materials Science and Engineering **805** (2020) 012039 doi:10.1088/1757-899X/805/1/012039

diplomacy for sound foreign policies that also contribute to improve global issues such as peace, political stability, economic ties, climate change, health and environmental management [11-13]. There are three dimensions of science diplomacy: diplomacy for science, science in diplomacy and science for diplomacy [14]. Diplomacy for science takes place when state actors such as government units facilitate international cooperation to advance scientific goals. Science in diplomacy plays out when science is the bedrock upon which the frameworks for foreign policies or international agreements are formulated.

Science for diplomacy is an attribute of using scientific assets to help improve diplomatic relationship between a strong and a weak nation. These three dimensions often overlap in operation especially for the technologically advanced nations. In any case, scientific and diplomatic skills are highly required. This implies training a scientist in diplomacy, referred to as "diplomat scientist" or training a diplomat in deep scientific knowledge, referred to as "scientist diplomat" [13]. The flexibility in bridging this gap by the choice of training will be dependent on which of the trainings is easier to achieve. Arguably, it might be easier for a scientist to develop diplomatic skills than for a diplomat to engage in deep scientific knowledge because the rout to becoming a renowned scientist entails fundamental understanding of scientific principles and applications, rigorous laboratory experiments or in-depth theoretical analysis, hands-on experiences and publication records of repute.

One of the most referenced examples of the origin of the practice of science diplomacy concept dated back to 1300 BC regarding a formal peace treaty agreement between Ramesses II of Egypt and Hattusili III of Hittite [12, 15]. This relationship gave Hattusili an opportunity to request for the scientific assistance of a medical doctor from Ramesses Kingdom to help his sister get pregnant. Towards the end of the 18th century, the American Colonies sent Benjamin Franklin as a Minister to France to seek for technical support for their decolonization from the British. His achievement as an accomplished scientist and an innovator gave him the capacity to reach a common ground with the French for a successful mission [12]. The work of Turekian [12] shows further examples of the relationship between science and diplomacy in strengthening diplomatic ties and advancing science from the 18th to 19th century.

Since the 19th century, the United States of America developed a strategic model for the dispatch of scientific envoys to nations with ST&I capacity to keep abreast with global development of ST&I and remain a leading nation in technology matters [16]. However, this suffered some setbacks at some stage due to difficulties in recruiting suitable personnel and funding challenges [16, 17]. The ST&I interest in the regime of USA former President Barrack Obama, reintroduced the practice of scientific envoys [16]. The numerous efforts before, during and after his regime have contributed to strengthening the capacity of the American Association for the Advancement of Science (AAAS) in advancing the course of science diplomacy. The AAAS currently has the only open access journal on science diplomacy launched in 2012. Some other notable developed countries that have shown similar interest in using science diplomacy for ST&I development include France, Germany, Switzerland, United Kingdom and Japan [18]. They show different areas of interest where they require the application of science diplomacy in improving national ST&I and addressing common global challenges.

3. The instrumentality of science diplomacy in developing nanotechnology in Africa

African nations, especially Nigeria and South Africa must prioritize the development of nanotechnology because of its multidisciplinary nature and the potential to transform ST&I landscape, enhance socio-economic thrive, public health and sustainable development. Such priority will guide science and technology foreign policy formulation. Science diplomacy is central to the achievement of this fate and it is reduced here to an instrument that African countries can deploy for nanotechnology development. Of the three dimensions of science diplomacy, diplomacy for science and science in diplomacy will be most instrumental for Africa. This is because science for diplomacy requires that a nation's science and technology assets be deployed as soft powers in persuading the weaker nations into bilateral agreements, especially in the recent numerous economic sanctions around the world. Africa does not currently have the sufficient capacity to engage science for diplomacy effectively. Some of the vital envisaged areas of engagement for nanotechnology development in Africa include human capacity development, national nano-research laboratory facilities, platforms for institutional collaborations, post-graduate student training and robust exchange programs that have direct bearing on nanotechnology training.

IOP Conf. Series: Materials Science and Engineering **805** (2020) 012039 doi:10.1088/1757-899X/805/1/012039

3.1 The need for Science Attachés from Africa

Science attachés are critical human assets in achieving effective science diplomacy in the 21st century for nanotechnology development in Africa. They are members of a nation's diplomatic corps whose main functions are to give advice to the ambassadors of their home country on science and technical matters, report on scientific development in their host countries, and represent the home country in scientific and technical meetings. Every country can redefine these roles to suit their national ST&I priorities. For example, when USA introduced the idea of science attachés in 1950, their functions among the main functions stated above were such as, facilitating exchanging of scientists, providing assistance in the procurement of scientific apparatus, chemical and biological, arrange for collaborative research projects and cooperate with all USA scientific groups abroad. Africa can leverage on the soft powers of their natural and young human resources, using diplomacy for science and science in diplomacy to influence foreign policies that will consciously develop nanotechnology. It is not out of place for a country to formulate foreign policies that will directly aid the development of a particular technological need [19].

Very shocking to reveal is that most of the leading nations in nanotechnology like China and USA have a department for science and technology in their foreign embassies in many of the developed nations, while developing countries in Africa like Nigeria and South Africa do not have S&T departments in their embassies. These leading nations carefully select the countries with ST&I capabilities for diplomatic scientific representation. USA is one of the developed leading countries in nanotechnology. A case study of China [20], India [21] and France [22, 23] shows that they have S&T departments at their embassies in USA, while South Africa [24] and Nigeria [25], as the two leading economies in Africa do not have S&T department. There is the need for African countries to consider the establishment of S&T department in their embassies, especially in the countries that are leading in nanotechnology. This will facilitate opportunities for stronger and sustainable scientific collaborations. The S&T department usually have science envoys, science attachés, or well-informed technical personnel as members of staff.

Countries like Brazil and Egypt are not among the top twenty countries in basic research on nano-science and nanotechnology, however, Figure 4 shows that they stand out among the developing nations compared to South Africa and Nigeria in ISI nano-articles publication.

IOP Conf. Series: Materials Science and Engineering **805** (2020) 012039 doi:10.1088/1757-899X/805/1/012039

These results might not be strange because Brazil has a unit for Education, Science, Technology and Innovation [26], while Egypt has an academic and cultural collaboration unit in their foreign embassies [27]. Such international representation could facilitate research opportunities and collaborations in nano-science between the scientists from the host and home countries. Therefore, a significant impact on research outputs is expected. The benefits derived from individual scientist efforts and collaborations might not have national and global impacts compared to research collaborations that are established based on national platforms resulting from science diplomacy. Certainly, developing nations of African can benefit from developed countries in the training of personnel, curriculum development and acquisition of equipment to drive nanotechnology agenda of the continent [28].

Funding is necessary to sustain research until results are evident. Individual researchers might lack the capacity for sustenance when there are no guaranteed immediate results unless government or well-established private organizations commit their funds. Science diplomacy can provide the opportunity for joint research and funding that will have a global impact.

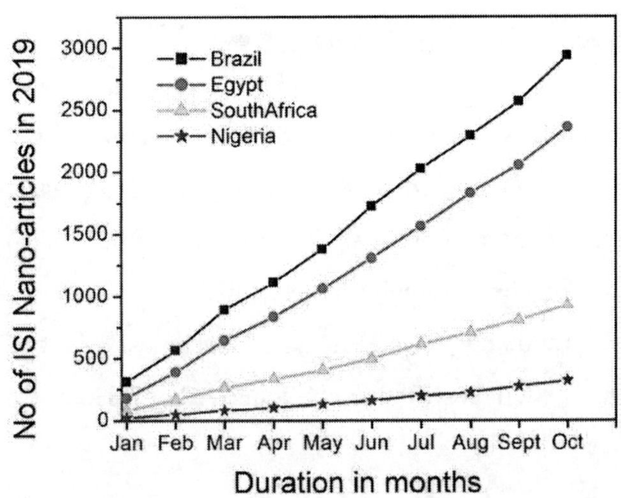

Figure 4. Number of ISI nano-articles published in 2019 by some developing countries. Data from Web of Science (ISI Web of Knowledge) [9]

IOP Conf. Series: Materials Science and Engineering **805** (2020) 012039 doi:10.1088/1757-899X/805/1/012039

3.2 Recommendations and defining functions for Science attachés from Africa

There is the need for the government of African countries to understand the impact that nanotechnology can have in all the sectors of the economy for improved standard of living of the populace. This invites a fundamental role for the tertiary research institutions and national research centres. They should act as consistent sensitization agents to the government through seminars, workshops, short courses and conferences that involves active participation of stakeholders from government. This will raise government confidence and support commitment in prioritizing nanotechnology. Such priority will naturally give credence to tailored foreign policy drive, human capacity development such as training of concerned personnel, post-graduate student scholarships, exchange programmes and facility development. Adequate and appropriate research facilities are vital to sustaining trained personnel and graduates, absence of which could lead to brain drain or diversion of research interest.

African nations should consider introducing the department of science, technology and innovation in their foreign embassies and high commissions in developed nations with excellent nanotechnology development and applications such as the first ten leading countries. The staff composition should consist of an informed science and policy envoy supported by science attaché(s). In addition to the roles defined for USA science attachés when the concept started in 1950, we recommend that African science attachés engage in an annual stakeholder's forum in their home countries for interface with domestic scientists to discuss latest development in nanotechnology with a view to strategise on how Africa can feature for global competitiveness. An annual review of the effectiveness of the collaboration platforms initiated by science attachés will ensure that challenges to efficiency are managed both for the home and host scientists.

4. Conclusion

The potential in using nanotechnology to develop African ST&I is enormous. Nanotechnology has the capacity to cause an economic revolution and improve the living standard of the people. This study has shown the importance of engaging science diplomacy in developing nanotechnology in Africa. Sensitizing the government and policy makers on the need for nanotechnology development is crucial through the instrumentality of workshops, seminars, conferences and short courses. Tertiary research institutions are fundamental to guiding African governments in prioritizing nanotechnology development. Government commitment will reflect in their budgeting, inclusion of the department of science, technology and innovation at various embassies and high commissions in the leading countries. Since science diplomacy involves the combination of scientific and diplomatic skills, science envoy and science attachés with both skills are strong considerations to form the core staff composition in the ST&I department.

Acknowledgment

The author is thankful to the National Centre for Technology Management, NACETEM.

References

[1] Michalski W 1999 21st century technologies: a future of promise. *The OECD Observer* **217** 50.

[2] Council I 2004 Inventing a better future *A strategy for building worldwide capacities in science and technology.* http://goob.free.fr/iup/Biologie_Moleculaire/Rapport%20Interacademic%20Council.pdf (Accessed on 2 February, 2020).

[3] Kuhlmeier D, Sandetskaya N and Allelein S 2012 Application of nanotechnology in miniaturized systems and its use in medical and food analysis. *Recent Pat Food, Nutr. Agric.* **4(3)** pp 187-199.

[4] Khan K and Rehman S 2016 Wonders of Nanotechnology *Development and Prospective Applications of Nanoscience and Nanotechnology: Nanomaterials and their Fascinating Attributes* **1** pp 3-49. (Sharjah, UAE: Bentham Science Publishers).

[5] Thiruvengadam M, Rajakumar G and Chung I M 2018 Nanotechnology: current uses and future applications in the food industry. *3 Biotech* **8** (1) 74. https://doi.org/10.1007/s13205-018-1104-7.

[6] Friedrichs S and Beuzekom B 2018 Revised proposal for the revision of the statistical definitions of biotechnology and nanotechnology. *OECD Science Technology and Industry Working Papers.* https://doi.org/10.1787/085e0151-en (Accessed on 2 February, 2020).

IOP Conf. Series: Materials Science and Engineering **805** (2020) 012039 doi:10.1088/1757-899X/805/1/012039

[7] Khan I, Saeed K and Khan I 2019 Nanoparticles: Properties, applications and toxicities. *Arabian J. Chem.* **12** (7) pp 908-931

[8] Itegbeyogene P E, Adebudo G O and Zayyanu M M 2018 Sustainable nanotechnology development in Nigeria – Hazards and policy direction. *Science Focus* **23** (2) pp 66-80.

[9] Web of Science (ISI Web of Knowledge Data) 2019 https://statnano.com/report/s88 (Accessed on 13 October, 2019).

[10] Statnano Databank 2019 https://statnano.com/report/s105 (Accessed 13 November 2019).

[11] Su P and Mayer M 2018 Science diplomacy and trust building: ‘Science China ‘in the Arctic. *Global Policy* **9** pp 23-28.

[12] Turekian V 2018 The Evolution of Science Diplomacy. *Global Policy* **9** pp 5-7.

[13] Moomaw W R 2018 Scientist diplomats or diplomat scientists: Who makes science diplomacy effective? *Global Policy* **9** pp 78-80.

[14] Koppelman B, Day N, Davison N, Elliott T and Wilsdon J 2010 New frontiers in science diplomacy: Navigating the changing balance of power. http://sro.sussex.ac.uk/id/eprint/45416/ (Accessed on 2 February, 2020).

[15] Gluckman P D, Turekian V C, Grimes R W and Kishi T 2017 Science diplomacy: A pragmatic perspective from the inside. *Sci. Diplom.* **6** (4) pp 1-13.

[16] El-Baz F 2010 Science attachés in embassies. *Sci.* **329** (5987) pp 13. https://doi.org.10.1126/science.1189621.

[17] Lexow W 1966 The science attaché program. *CIA Historical Review Program Washington DC.* https://www.cia.gov/library/center-for-the-study-of-intelligence/kent-csi/vol10no2/pdf/v10i2a02p.pdf (Accessed on 2 February, 2020).

[18] Flink T and Schreiterer U 2010 Science diplomacy at the intersection of S&T policies and foreign affairs: toward a typology of national approaches. *Sci. Public Policy* **37** (9) pp 665-677.

[19] Zolfagharzadeh M M, Aslani A, Sadabadi A A, Sanaei M, Lesan Toosi F and Hajari, M 2017 Science and technology diplomacy: a framework at the national level. *J. Sci. Technol. Policy Manag.* **8** (2) pp 98-128.

[20] http://us.chineseembassy.org/eng/sgxx/bmxx/kjc/. Accessed on 10 November, 2019.

[21] https://www.indianembassyusa.gov.in/pages?id=2&subid=2. Accessed on 10 November, 2019.

[22] https://franceintheus.org/spip.php?article415. Accessed on 10 November, 2019.

[23] https://www.france-science.org/-Organization-.html. Accessed on 10November, 2019.

[24] http://www.saembassy.org/. Accessed on 10 November, 2019.

[25] http://www.nigeriaembassyusa.org/index.php?page=the-embassy. Accessed on 10 November, 2019.

[26] http://washington.itamaraty.gov.br/en-us/our_staff.xml. Accessed on 10 November, 2019.

[27] https://eecous.net/services/service-2/. Accessed on 10 November, 2019.

[28] Elegbede JA and Lateef A 2019 Green nanotechnology in Nigeria: The research landscape, challenges and prospects. *Ann. Sci. Technol* **4** (2) pp 6-38.

Isolation and characterization of cellulose nanofibres from three common Nigerian grasses

KO Ejeta*, TO Azeez, AT Banigo, KI Nkuma-Udah and E Ajuogu

Department of Biomedical Technology, Federal University of Technology, PMB 1526, Owerri, Imo state, Nigeria.

*Correspondence author: koejeta@yahoo.co.uk

Abstract

Nanocellulose may be in nanocrystal or nanofibre form which may be extracted from purified cellulose by different methods. These include methods like dual-opposite-spinneret electrospinning, mechanical methods, a combination of chemical and mechanical methods, cryocrushing and enzymatic approaches. Nitrocelluloses have been extracted from various plants but not much has been reported on the yield from grasses. In this study, cellulose nanofibres (CNFs) were extracted from three common grasses; *Pennisetum purpureum*, *Cynodon dactylon* and *Axonopus compressus* by formic acid hydrolysis with the micrograph of CNF for each grass obtained. The CNFs are web-like long fibrous structure with diameter ranging between 3 to 5 nm and yield of over 75%. The crystallinity index averaged 76% and the onset temperature of thermal decomposition was 199 °C. Thus, *Pennisetum purpureum*, *Cynodon dactylon* and *Axonopus compressus* are good eco-friendly sources of CNF for potential application as new source of nanofillers for reinforcement of nanocomposite films.

Keywords: Nanocellulose, yield, grasses, crystallinity, nanofiller

1 Introduction

Nanocellulose may be crystal called nanocrystal cellulose (CNC) or fibre called cellulose nanofibres (CNF) [1, 2]. Cellulose material provides a source for preparation of CNC or CNF [3]. Moreso, CNC are chiral or rod like and contains only crystalline domain, while CNF are web-shaped fibres which consists of both crystalline and amorphous domains [4, 5]. The promising nature of nanocellulose to improve material properties for many applications such as optical filters, sensor sand reinforcement of polymer resins in composites have been motivating factors for research communities. Other factors that have steered tremendous interest in isolation and characterization of nanocellulose include the aspect ratio, crystallinity index, surface area and mechanical properties which are very high and its variable ability to absorb and reflect visible light as well as low cost production [2, 6-8]. These properties make nanocellulose to be regarded as the new generation of nanomaterials for the development of high performance composite materials [7-9].

Isolation approaches of cellulose nanofibre and cellulose nanocrystal differs in many ways. For instance, cellulose nanocrystal is mostly isolated by acid using H_2SO_4, HCl, HBr or H_3PO_4 and enzyme hydrolysis whereas cellulose nanofibre is isolated by mechanical, chemomechanical and combination of enzyme and mechanical processes [10-14]. Reports from investigations have shown that the properties of nanocellulose depend on the source, pretreatment methods and experimental conditions [14-17]. The morphology of nanocellulose is usually studied by Transmission Electron Microscopy (TEM), Atomic Force Microscopy (AFM) and polarized light scattering [18-21]. Studies on cellulose plant fibres for various numerous applications have generated great interest worldwide because they are sustainable, green and environmental friendly [6-10, 22]. In the present study, isolation and characterization of CNF from three Nigeria grasses; Elephant grass (*Pennisetum purpureum*), Bermuda grass (*Cynodon dactylon*) and carpet grass (*Axonopus compressus*) were investigated.

2. Materials and methods

2.1 Isolation of cellulose from the grass samples

Three grass samples used in this study were sun dried for 3 days, washed thoroughly with distilled water and oven dried at 60 °C for 24 h. Dried grass (10 g) was de-lignified using 300 ml of 1 M NaOH for 1 h and 30 min at 80 °C, then washed with distilled water and centrifuged. This was followed with addition of 150 ml of 5% NaOH for 20 min at 80 °C, and then washed with distilled water and centrifuged. The cellulose obtained was dried in an oven at 70 °C for 3 h and milled with mortar and pestle to obtain the powder.

2.2 Formic acid Hydrolysis of Cellulose to obtain CNF

The procedure described by Hamid *et al.* [20] was used for acid hydrolysis of isolated cellulose. Exactly 90 ml of 50% of formic acid was added to 3 g of cellulose powder in two neck flasks followed by the addition of 1g $FeCl_3$ at 90 °C and in a rotator flask at 400 rpm for 4 h. The CNF was obtained, washed with distilled water, centrifuged 5 to 7 times at 1400 rpm for 30 min, dried in an oven at 70 °C for 3 h and powdered using mortar and pestle. The yield of CNF was determined using equation (1)

$$Y(in\ \%) = \frac{M_1 - M_3}{M_2 - M_0} x100 \quad (1)$$

2.3 Scanning Electron Microscopy and Transmission Electron Microscopy of extracted CNFs

Microstructure surface of extracted CNFs was determined by Scanning Electron Microscope (SEM) with model EVO MA10 (Carl Zeiss SMT Germany) at a set accelerated voltage of 15 kV after being coated with gold. Transmission Electron Microscope (TEM) with image analysis system model Image BioPlusVer 701 (Media Cybernetics Inc. USA) was used to determine the morphology of CNF and its diameter.

2.4 Crystallinity Index

Crystallinity index and crystal size of samples were determined using X-ray diffraction; X-ray diffractometer with measurement condition of 40 kV and current 30 mA. The range of scattering angle (2θ) was from 10-80° with the scan rate of 10°/min. The crystallinity index (CrI) was calculated using equation (2) and crystal size determined using equation (3).

$$CrI\ (\%) = (1 - I_m / I_{200}) \times 100 \quad (2)$$

Where I_m =intensity minimum between the 200 and 110 peaks and I_{200} = maximum peak intensity at lattice diffraction 200.

Or

$$CrI\ (\%) = (I_{crystalline} / I_{total}) \times 100$$

$$\tau = \frac{K\lambda}{\beta cos\theta} \quad (3)$$

Where k = Scherer constant, λ = wavelength of X-ray radiation and θ = Bragg angle and β is the line broadening at half the maximum intensity.

IOP Conf. Series: Materials Science and Engineering **805** (2020) 012040 doi:10.1088/1757-899X/805/1/012040

3. Results and Discussion

Figure 1 shows the micrograph of CNF obtained; they are long fibrous structure consisting of both crystalline and amorphous domain. They have average width of 3.5 nm and length ranged from 1.5 to 2.0 microns. The CrI and crystalline size increased up to 78.23% and 3.65 nm respectively. This result is in agreement with works reported in literature [4, 5].

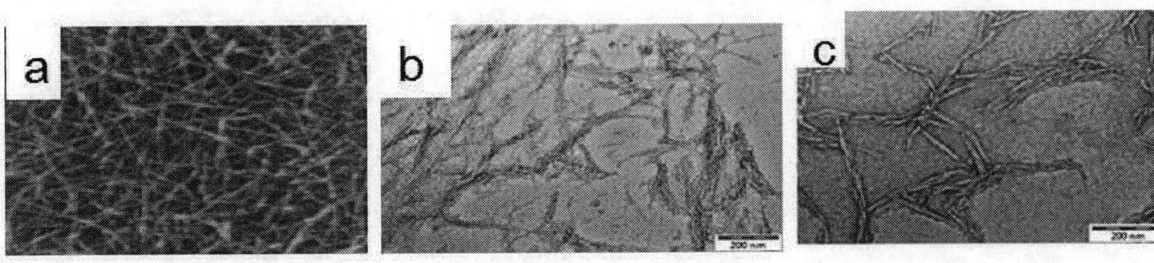

Figure 1. TEM for CNFs obtained from (a) elephant (b) bermuda (c) carpet grass

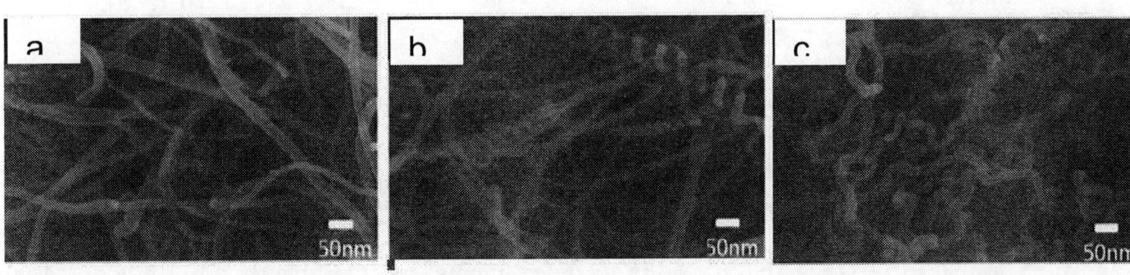

Figure 2. SEM for CNFs obtained from (a) elephant (b) bermuda (c) carpet grass

The electron micrograph revealed that before ultrasonication, the cellulose filaments presented a diameter in the scale of tens of micrometres.

Table 1. Yield of cellulose and CNF from the grass Samples

Samples	Yield (%)	Crystallinity %	Crystalline Size (%)
Cellulose	-	42.10	2.00
1 (CNF)	79.23	76.30	3.45
2 (CNF)	81.05	78.23	3.65
3 (CNF)	76.42	73.04	3.13

Table 2. Chemical composition of lignocelluloses materials from grass samples

Grass	Cellulose (mm)	Hemicellulose	Lignin
Elephant	45	28	20
Bermuda	43	25	16
Carpet	47	30	24

Table 1 and 2 show the yield of cellulose and CNF from the grass samples analyzed. The yield of CNF was over 75% for all samples of grasses, CrI and crystalline size increased up to 78% and 3.65 nm, respectively. The crystallinity index increased significantly from 42% in crude cellulose to 78% in CNF. The higher crystallinity index values indicate the removal of amorphous non-cellulosic compounds induced by the purification treatment (pulping and bleaching treatments) performed to purify cellulose stem.

The yield of cellulose obtained from cellulose nanofibres of elephant, bermuda and carpet grasses were found to be more than that of Walnut shell as reported by Zheng *et al.* [24] by 64.42, 56.93 and 71.53% [24]. This indicates that elephant, bermuda and carpet grasses possess low amorphous constituents (lignin, hemicelluloses, ash and extractive soluble materials) compared with Walnut shell. This corroborates morphology and network structure in which the more cellulose or crystalline cellulose, the less the irregular pore formation as shown in SEM and TEM (Figures 1 and 2). It can also be observed that the crystallinity index of cellulose nanofibre of elephant, bermuda and carpet grasses are greater than that of walnut fibre and SantaFe Kraft pulp mill as reported [23- 25]. Futhermore, crytalline size of cellulose nanofibre of used grasses were also more than that of Santa Fe Kraft pulp mill as reported [25].

4. Conclusion

Cellulose nanofibres (CNF) were successfully isolated from all the three samples of Nigerian grasses investigated using formic acid hydrolysis with yield of over 75%. Isolated CNF exhibited web-like long fibrous structure with diameter of 3 to 5 nm. The crystalllinity index averaged 78% and the onset temperature of thermal decomposition was 199 ^{o}C. The nanocellulose isolated from the grass samples have high potential to be used as new source of cellulose based nanofillers for reinforcement of bionanocomposite films.

References

[1] Peng BL, Dhar N, Liu HL and Tam KC 2011 Chemistry and applications of nanocrystalline cellulose and its derivatives: a nanotechnology perspective. *Can. J. Chem. Eng.* **89** (5) pp 1191-206.

[2] Turbak AF, Snyder FW and Sandberg KR 1983 Microfibrillated cellulose, a new cellulose product: properties, uses, and commercial potential. *J. Appl. Polym. Sci. Symp.* **37**, No. CONF-8205234.

[3] Turbak AF, Snyder FW and Sandberg KR 1984 Microfibrillated cellulose—a new composition of commercial significance. Atlanta, GA., USA. Chapter 16 pp 115-124.

[4] Herrick FW, Casebier RL, Hamilton JK and Sandberg KR 1983 Microfibrillated cellulose: morphology and accessibility. InJ. Appl. Polym. Sci.: Appl. Polym. Symp. **37**, No. CONF-8205234.

[5] Berglund L 2005 Cellulose-based nanocomposites In: AK Mohanty, M Misra, and L Drzal (eds.). *Natural fibers, biopolymers and biocomposites.* Boca Raton, Florida: CRC Press, 807-832.

[6] He W, Jiang S, Zhang Q and Pan M 2013 Isolation and characterization of cellulose nanofibers from *Bambusa rigida. BioResour.* **8** (4) pp 5678-5689.

[7] Lavoine N, Desloges I, Dufresne A and Bras J 2012 Microfibrillated cellulose–Its barrier properties and applications in cellulosic materials: A review. *Carbohydr. Polym.* **90** (2) pp 735-764.

[8] Missoum K, Martoïa F, Belgacem MN and Bras J 2013 Effect of chemically modified nanofibrillated cellulose addition on the properties of fiber-based materials. *Ind. Crops Prod.* **48** pp 98-105.

[9] Effendi DB, Rosyid NH, Nandiyanto AB and Mudzakir A 2015 Sintesis Nanoselulosa. *Jurnal Integrasi Proses.* **5** (2). http://dx.doi.org/10.36055/jip.v5i2.199.

[10] Jonoobi M, Harun J, Mishra M and Oksman K 2009 Chemical composition, crystallinity and thermal degradation of bleached and unbleached kenaf bast (*Hibiscus cannabinus*) pulp and nanofiber. *BioResour.* **4** (2) pp 626-639.

[11] Chan HC, Chia CH, Zakaria S, Ahmad I and Dufresne A 2013 Production and characterisation of cellulose and nano-crystalline cellulose from kenaf core wood. *BioResour.* **8** (1) pp 785-794.

[12] Jonoobi M, Oladi R, Davoudpour Y, Oksman K, Dufresne A, Hamzeh Y and Davoodi R 2015 Different preparation methods and properties of nanostructured cellulose from various natural resources and residues: a review. *Cellulose* **22** (2) pp 935-969.

[13] Wulandari WT, Rochliadi A and Arcana IM 2016 Nanocellulose prepared by acid hydrolysis of isolated cellulose from sugarcane bagasse. *IOP Conf. Ser.: Mater. Sci. Eng.* **107** 012045. https://doi.org/10.1088/1757-899X/107/1/012045.

[14] Aprilia NS, Davoudpour Y, Zulqarnain W, Khalil HA, Hazwan CC, Hossain MS, Dungani R, Fizree HM, Zaidon A and Haafiz MM 2016 Physicochemical characterization of microcrystalline cellulose extracted from kenaf bast. *BioResour.* **11** (2) pp 3875-3889.

[15] Haafiz MM, Eichhorn SJ, Hassan A and Jawaid M 2013 Isolation and characterization of microcrystalline cellulose from oil palm biomass residue. *Carbohydr. Polym.* **93** (2) pp 628-634.

[16] Ahmad Z, Roziaizan NN, Rahman R, Mohamad AF and Ismail WI 2016 Isolation and characterization of microcrystalline cellulose (MCC) from rice husk (RH). *MATEC Web Conf.* **47** 05013. https://doi.org/10.1051/matecconf/20164705013.

[17] Nuruddin M, Chowdhury A, Haque SA, Rahman M, Farhad S, Jahan MS and Quaiyyum A 2011 Extraction and characterization of cellulose microfibrils from agricultural wastes in an integrated biorefinery initiative. *Cellulose Chem. Technol.* **45** (5-6) pp 347-354.

[18] Siqueira G, Bras J and Dufresne A 2010 *Luffa cylindrica* as a lignocellulosic source of fiber, microfibrillated cellulose and cellulose nanocrystals. *BioResour.* **5** (2) pp 727-740.

[19] Wicaksono R, Syamsu K, Yuliasih I, Nasir M and Street K 2013 Cellulose nanofibers from cassava bagasse: Characterization and application on tapioca-film. *Chem. Mater. Res.* **3** (13) pp 79-87.

[20] Hamid SB, Chowdhury ZZ, Karim MZ and Ali ME 2016 Catalytic isolation and physicochemical properties of nanocrystalline cellulose (NCC) using HCl-$FeCl_3$ system combined with ultrasonication. *BioResour.* **11** (2) pp 3840-3855.

[21] Saurabh CK, Dungani R, Owolabi AF, Atiqah NS, Zaidon A, Aprilia NS, Sarker ZM and Khalil HA 2016 Effect of hydrolysis treatment on cellulose nanowhiskers from oil palm (*Elaeis guineesis*) fronds: Morphology, chemical, crystallinity, and thermal characteristics. *BioResour.* **11** (3) pp 6742-6755.

[22] Börjesson M and Westman G 2015 Crystalline nanocellulose—preparation, modification, and properties. Cellulose-fundamental aspects and current trends. 159-191. https://dx.doi.org/10.5772/61899.

[23] Zheng D, Zhang Y, Guo Y and Yue J 2019 Isolation and characterization of nanocellulose with a novel shape from walnut (*Juglans regia* L.) shell agricultural waste. *Polym.* **11** (7) 1130. https://doi.org/10.3390/polym11071130.

[24] Wang QQ, Zhu JY, Reiner RS, Verrill SP, Baxa U and McNeil SE 2012 Approaching zero cellulose loss in cellulose nanocrystal (CNC) production: recovery and characterization of cellulosic solid residues (CSR) and CNC. *Cellulose* **19** (6) pp 2033-2047.

[25] Aguayo MG, Fernández Pérez A, Reyes G, Oviedo C, Gacitúa W, Gonzalez R and Uyarte O 2018 Isolation and characterization of cellulose nanocrystals from rejected fibers originated in the kraft pulping process. *Polym.* **10** (10) 1145. https://doi.org/10.3390/polym10101145.

IOP Conf. Series: Materials Science and Engineering **805** (2020) 012041 doi:10.1088/1757-899X/805/1/012041

Synthesis, bioactivities and cytogenotoxicity of animal fur-mediated silver nanoparticles

G O Akintayo[1], A Lateef[2,8], M A Azeez[1,8], T B Asafa[3,8], I C Oladipo[4,8], J A Badmus[5,8], S A Ojo[2], J A Elegbede[2], E B Gueguim-Kana[6], L S Beukes[7] and T A Yekeen[1,8*]

[1]Laboratory of Ecotoxicology, Genetics and Nanobiotechnology, [2]Laboratory of Industrial Microbiology and Nanobiotechnology, Department of Pure and Applied Biology, [3]Department of Mechanical Engineering, [4]Department of Science Laboratory Technology, [5]Department of Biochemistry, Ladoke Akintola University of Technology, Ogbomoso, Nigeria

[6]Department of Microbiology, [7]Microscopy and Microanalysis Unit, School of Life Sciences,University of KwaZulu-Natal, Private Bag X01, Scottsville, Pietermaritzburg 3209, South Africa

[8]Nanotechnology Research Group (*NANO*+) (www.nanotech.lautech.edu.ng)

*Corresponding author: tayekeen@lautech.edu.ng

Abstract

Animal waste materials are rarely used in the synthesis of nanoparticles compared to microorganisms and plant materials. The use of animal fur (goat) in synthesis could assist in turning waste to wealth. Thus, potentials of animal fur in the synthesis of silver nanoparticles (AF-AgNPs), its biological activities and safety through cytogenotoxicity were investigated. Animal fur (1 g) was hydrolyzed with 100 ml of 0.1 M NaOH at 90 °C for 1 h, cooled and centrifuged at 4000 rpm for 30 min. The extract (1 ml) was added to 1 mM $AgNO_3$ (40 ml) to reduce Ag^+ to its nanoparticles. The AF-AgNPs was characterized using UV–vis spectroscopy, Fourier-transform-infrared spectroscopy (FTIR), transmission electron microscopy (TEM) and Energy Dispersive X-ray (EDX) analysis. Larvicidal, antioxidant, anticoagulant and thrombolytic potentials of AF-AgNPs were studied. Onion bulbs (20) were exposed to 0.01, 0.10, 1.0, 10.0 and 100.0 µg/ml of AF-AgNPs solution for its cytogenotoxicity study with $AgNO_3$ solution and distilled water as controls. Microscopic (24, 48 and 72 h) assessment of the onion cells and macroscopic (72 h) evaluation of the roots were also studied. The AF-AgNPs solution was brownish with surface plasmon resonance at 419 nm. Evaluation of FTIR spectra showed that protein molecules were used as capping and stabilization agents. The AF-AgNPs had size range of 11.67-31.47 nm, caused 60-100% mortality of exposed *Anopheles* mosquito larvae in 12 h, and scavenged DPPH (40-59%) and hydrogen peroxide (75-94%). The nanoparticles also exhibited anticoagulant and thrombolytic potentials on human blood with 25% lysis compared to 13% observed for only extract. Various chromosomal aberrations and growth inhibition were induced by AF-AgNPs especially at 72 h of 100 µg/ml. Extract from animal fur was explored in biogenic synthesis of nanoparticles and found to have high potentials as antioxidant, anticoagulant, thrombolytic agents. Inhibition of cell growth observed especially at highest concentration can be explored in anticancer drugs though with caution due to AF-AgNPs potential to induce chromosomal aberrations.

1 Introduction

Nanoparticles are found useful in different facet of human life with applications in pharmaceutics, catalysis, cosmetics, food packaging, medical devices, textiles, electronics, optics, fuel cells, biosensors, water treatment technology and in agriculture [1-4]. Green synthesized nanoparticles are mostly from micro-organisms and plant materials. Microorganisms including bacteria [5, 6] and fungi [7-9] as well as various plant materials [10-13] had been reported to mediate the synthesis of various nanoparticles. The extracts from various materials used in the synthesis of nanoparticles have been found to possess biomolecules that serve as reducing and capping agents.

In the biosynthesis of nanoparticles, animal materials and their metabolites are seldom used despite the fact that they possess biomolecules that can serve as reducing and capping agents. Reported cases of biosynthesis involving the use of animal materials include that of cobweb, paper wasp net, and cockroaches wings [14-16]. Further more, honey from bees (insect) had been reported to mediate green synthesis of gold, silver, carbon, platinum, and palladium nanoparticles by acting as both stabilizing and reducing agents as well as a precursor in nanoparticle synthesis [17]. Cow milk had also been reported in the synthesis of AgNPs and with potency as antifungal agent [18]. The proteins present in these biological materials are involved in the reduction and stabilization of nanoparticles so produced [19].

IOP Conf. Series: Materials Science and Engineering **805** (2020) 012041 doi:10.1088/1757-899X/805/1/012041

Quite a lot of animal waste products are found in the environment that constitute nuisance. These include animal fur, human hair, and chicken feathers among others. Employing these materials in the green synthesis will assist in turning waste to wealth and while the benefit that could be derived from the nanoparticles will still be achieved. This study sought to use animal fur (goat fur) in the synthesis of AgNPs and evaluate its bioactivities and cytogenotoxicity.

2 Materials and Methods

2.1 Experimental materials

Sample of animal fur (goat) were obtained from a local farmer in Ogbomoso, while the larvae of *Anopheles* mosquito were collected from stagnant water and were prepared for the experiment.

2.2 Green Synthesis of silver nanoparticles

The fur extract was made by adopting the modified method of Tszydel *et al.* [20] by hydrolyzing 1 g in 100 ml of 0.1M NaOH at 90 °C for 1 h. The hydrolyzed fur was cooled and centrifuged at 4000 rpm for 30 min and supernatant was obtained. This extract was used to synthesize AgNPs as previously described by Lateef *et al.* [21] where 1 ml of the extract was added to the reaction vessel containing 40 ml of 1 mM silver nitrate ($AgNO_3$) solution for the reduction of silver ion to nano size. The reaction was carried out for 15 min under bright condition for photoactivation. The formation of AgNPs was monitored through visual observation of change in colour (Figure 1) as described by Morones *et al.* [22].

2.3 Characterizations of the synthesized AgNPs

The AgNPs synthesized through the fur extract was characterized using UV-visible spectroscopy, Fourier transform infrared (FTIR) spectroscopy, Scanning electron microscopy, Transmission electron microscopy (TEM), and Energy Dispersive X-ray (EDX).

2.4 Applications of synthesized nanoparticles

2.4.1Antioxidant activities

DPPH free radicals scavenging assay was carried out using the modified method of Williams *et al.* [23] where 1, 1-diphenyl-2-2picrylhydrazyl (DPPH) free radical scavenging potential of the AF-AgNPs was determined with different concentrations (1, 2, 5, 10, 20 and 40 µg/ml) of the nanoparticles. The antioxidant activity was determined by adding 1 ml of AF-AgNPs to 3.0 ml of

DPPH solution, and then kept in a dark room for 30 min before finally read at 517 nm with spectrophotometer (Shimadzu, UK).

$$\%\text{DPPH scavenging effect} = \frac{AC-AS}{AC} \times 100$$

Where: AC= absorbance of negative control and AS= absorbance of sample.

Hydrogen peroxide scavenging assay was evaluated using the modified method of Nabavi *et al.* [24, 25]. A solution of H_2O_2 (40 mM) was prepared in phosphate buffer of pH 7.4. The concentration of H_2O_2 was determined by absorption using spectrophotometer. A 4 ml of different concentrations of 1, 2, 5, 10, 20, 40, 60 and 80 µg/ml of AF-AgNPs and distilled water was added to 0.6 ml, 40 mM of H_2O_2 solution. The absorbance of H_2O_2 at 610 nm was determined after 20 min against a blank solution (phosphate buffer with H_2O_2).

$$\%\ H_2O_2 \text{ scavenging effect} = \frac{AC-AS}{AC} \times 100$$

Where: AC= absorbance of negative control and AS= absorbance of sample

2.4.2 Anticoagulant activity

The anticoagulant activity was performed by mixing 0.5 ml of 150 µg/ml AF-AgNPs with 5 ml of freshly collected human blood from an adult volunteer. This was held at ambient condition to observe coagulation of blood. A control experiment was set up for comparison [26].

2.4.3 Thrombolytic activity

The thrombolytic assay was performed by recording the weight of Eppendorf tube (X_1), dispensed 0.5 ml of blood into the tube then allowed to stand for 30 to 45 min for clot formation at room temperature of 37 °C (X_2). The weight of the clot was calculated as $A=X_2-X_1$. The AF-AgNPs (100µl) was added to the blood clot formed and incubated at room temperature of 37 °C for 90 min. The free fluid generated was removed and the weight was recorded again (X_3) while the weight of clot left was calculated as $B=X_3-X_1$. Percentage clot lysis was calculated as:

$$\frac{A-B}{A} \times 100$$

Finally, the microscopic analysis of the released fluid was performed using x40 objective lens [12, 27].

2.4.4 Larvicidal activity of AF-AgNPs

The AF-AgNPs was subjected to a dose-response bioassay for larvicidal activity against *Anopheles* mosquito's larvae using the method of Bansal *et al.* [28]. Equal volume (10 ml) of different AgNPs concentrations (10, 20, 40, 60, 80 and 100 µg/ml) was used for the toxicity test

IOP Conf. Series: Materials Science and Engineering **805** (2020) 012041 doi:10.1088/1757-899X/805/1/012041

against 10 *Anopheles* mosquito larvae in triplicates. The percentage mortality rates were determined at 12 h interval for 36 h.

2.5 *Allium cepa* assay

Two hundred and fifty (250) onion bulbs (*Allium cepa*) were purchased at Wazo market in Ogbomoso and sun dried for two weeks. The outer scales of the bulbs and dried roots were carefully removed leaving the root primordial intact. Five concentrations i.e. 0.01, 0.1, 1.0, 10 and 100 µg/ml of AF-AgNPs, salt solution of silver and extract alone were prepared from the stock solutions and distilled water was used as diluent as well as negative control. Twenty onion bulbs were planted per concentration and were grown in dark cupboard. The test solutions were changed every 24 h to supply fresh nutrient to the growing bulbs. At 24, 48 and 72 h, 5 out of the onion bulbs planted for each concentration and controls were harvested and thoroughly rinsed in distilled water and fixed in ethanol: glacial acetic acid (3:1 v/v) in sample bottles per onion bulb, and samples were kept in the refrigerator. Already fixed and refrigerated root tips (2 or 3) were picked and hydrolyzed in 1N HCl at 65 °C for 3 min. The meristematic region of the root tip of two roots was cut on a clean glass slide and macerated using needle. A drop of aceto-orcein stain was added and the tissue on the slide was allowed to stain for 10 min, followed by scoring under light microscope as previously described [29-32]. At 72 h, roots of 5 bulbs per concentration and controls were harvested to evaluate growth inhibition.

2.6 Statistical analysis

Data were analyzed using SPSS 20.0. The mean root lengths obtained for the treated bulbs at different concentrations were compared using one way analysis of variance and Duncan multiple range test as post hoc test. Mitotic index, mitotic inhibition and percentage aberrant cells were calculated as follow:

$$\text{Mitotic Index MI} = \frac{number\ \ of\ dividing\ \ cells\ \ in\ treated\ \ A.cepa}{total\ number\ \ of\ cells\ counted} \text{x } 100$$

$$\text{Mitotic Inhibition} = \frac{MI\ of\ negative\ \ control\ -MI\ of\ treated}{MI\ of\ negative\ \ control} \text{x } 100$$

$$\%\ \text{aberration per total cells scored} = \frac{number\ \ of\ abberant\ \ cells}{total\ number\ \ of\ cells\ scored} \text{x } 100$$

IOP Conf. Series: Materials Science and Engineering **805** (2020) 012041 doi:10.1088/1757-899X/805/1/012041

3 Results and Discussion

3.1 Green synthesized AgNPs

The animal fur mediated the synthesis of AgNPs within 8-10 min of reaction in the vessel under ambient conditions of room temperature (30 ± 2 °C) with a brown colouration, which further intensified with time but stabilized within 15 min of exposure to sunlight (Figure 1). A number of authors have reported variations in the colour of AgNPs colloidal solutions due to the composition of bioreductant responsible for the synthesis of the nanoparticles. Lateef *et al.* [21] reported the formation of dark brown AgNPs solution using the crude extracellular keratinase of strain of *B. safensis* LAU 13, Jeevan *et al.* [33] reported formation of yellowish brown AgNPs using cultured supernatant of *Pseudomonas aeruginosa,* while Oladipo *et al.* [6] reported dark brown AgNPs using cell free-extracts of *Enterococcus* species.

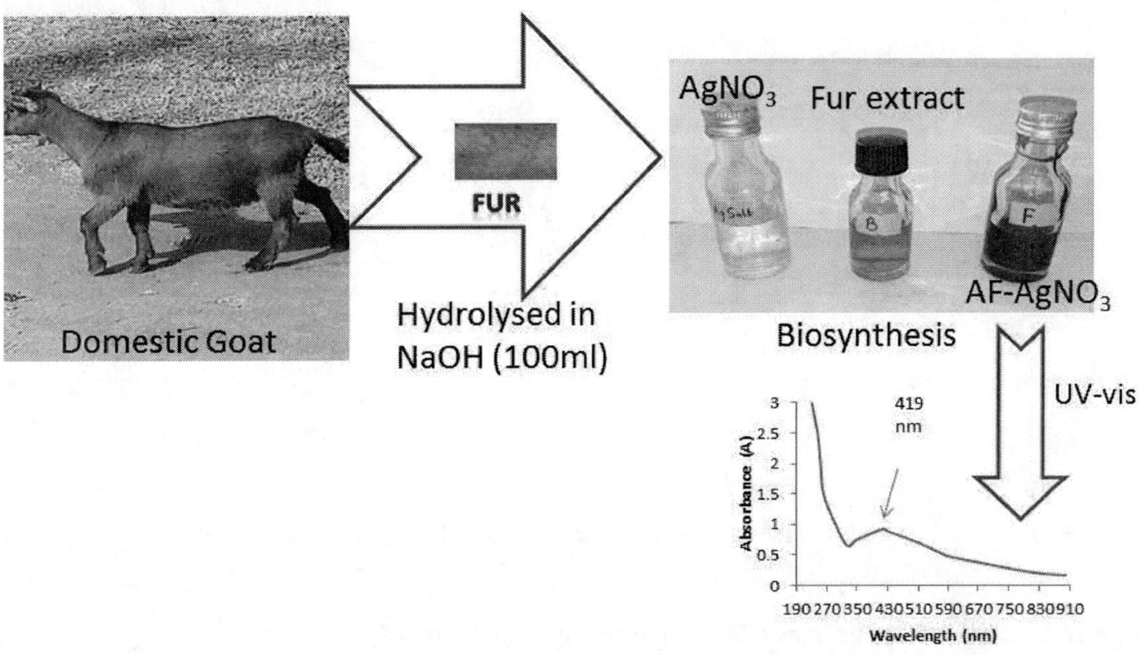

Figure 1. Green synthesis of AF-AgNPsfrom fur extract and its UV-visible spectrum

3.2. Characterization of biosynthesized AgNPs

The UV-vis spectrum of biosynthesized AF-AgNPs had broad absorbance peak at the wavelengths of 419 nm (Figure 1) which was within the range of 391-440 for AgNPs [18, 31]. The broad peak is an indication that the biosynthesized AF-AgNPs were polydispersed, observations similar this had been reported for various metal nanoparticles [33, 35].

The FTIR absorption spectrum showed distinct peaks at 3784, 3309, 1635cm^{-1} for the animal fur (Figure 2). The FTIR-spectra showed the transmittance of different chemical functional groups present in the fur extract treated with $AgNO_3$. The bands 3784 and 3309 cm^{-1} are typical of H bonded -OH and –OH inter H bond, respectively while 1635 cm^{-1} represents alkenyl C=C stretch. These indicate that proteins were the capping and stabilization biomolecules involved in the synthesis of AgNPs [36].

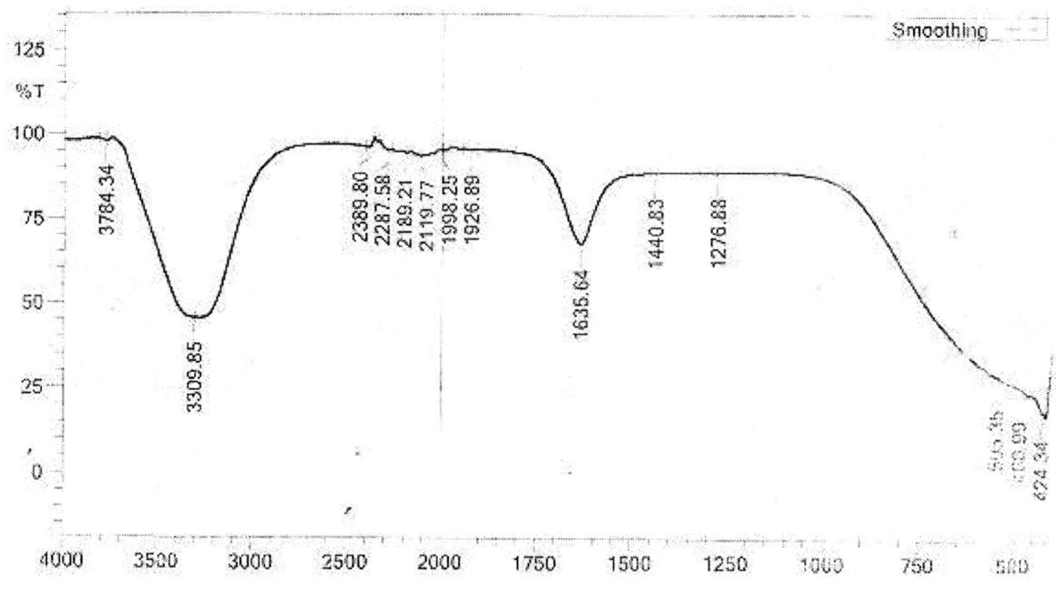

Figure 2. The FTIR spectrum of the animal fur-mediated AgNPs

The TEM micrograph revealed that the AF-AgNPs were spherical in shape with size ranging from approximately 11 to 32 nm (Figures 3A). These observations are similar to earlier reports described by Kannan *et al.* [37] where well dispersed particles were obtained. This indicates good stability of AF-AgNPs devoid of aggregation. The EDX pattern (Figure 3B) showed the predominance of silver in the AgNPs solution and is characterized with ring-like SAED pattern

(Figure 3C) typical of the face-centered cubic crystalline structure of silver [15, 38]. It can be deduced that animal fur extract that mediated the synthesis of AgNPs is a novel material for the biogenic and ecofriendly synthesis.

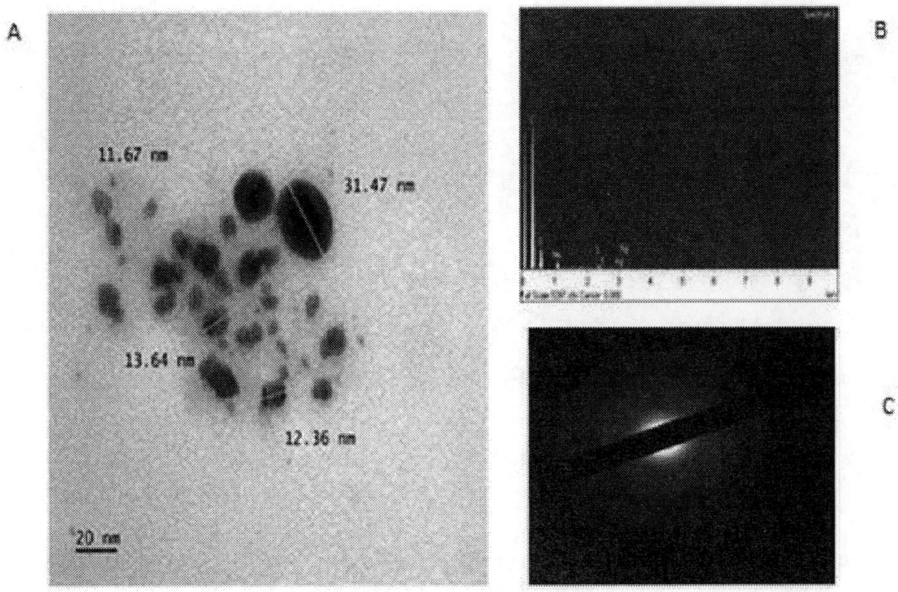

Figure 3. TEM micrograph (A), EDX spectrum (B) and SAED (C) of AF-AgNPs

3.3 Antioxidant activities

The DPPH-scavenging activities of the nanoparticles are presented in Figure 4. The synthesized AF-AgNPs showed inhibitions of 40-59 % while ascorbic acid used as the standard produced inhibitions of 41-60 % at same concentrations ranging from 1 to 40 μg/ml. The result of antioxidant activities of biosynthesized AgNPs showed significant free radical scavenging activities when compared with the standard ascorbic acid, which is similar to other reports [6, 7]. The hydrogen peroxide scavenging activities also displayed dose dependent reduction activities of 75-94 %, at concentrations of 1–80 μg/ml (Figure 4). The scavenging activities of the AgNPs observed in this work are similar to those that were previously reported [12, 39]. The free radical scavenging activity of the AgNPs has been credited to the functional groups of the biomolecules which adhered to the surface of the nanoparticles. Also, various researchers have proposed that antioxidants repress and inhibit damages caused via oxidative stress to components of the cell such as proteins, lipids and DNA which thus reduce the risk of chronic diseases related to aging [40].

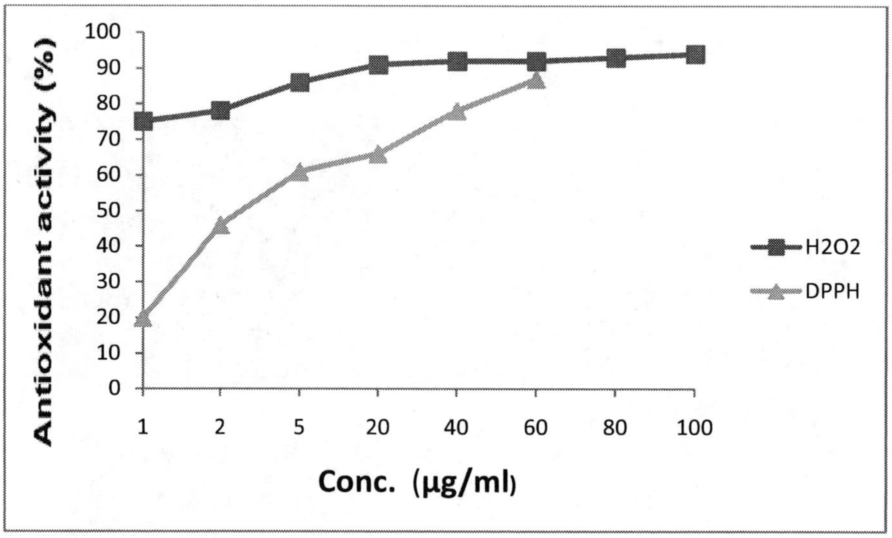

Figure 4. Antioxidant activities of AF-AgNPs

3.4 Anticoagulant activities of biosynthesized AgNPs

The AgNPs prevented formation of blood clot when used as an anticoagulation agent, which is comparable to result obtained with EDTA. The microscopic examination of both EDTA and AF-AgNPs treated blood samples showed typical disc-shaped red blood cells (Figure 5). Similar anticoagulation properties of biosynthesized silver nanoparticles had also been previously reported [7, 41-44]. The formation of blood clot arising from infection can damage tissues and cause organ failure [45] and this may be linked with various cardiovascular disorders, autoimmune reactions, allergic responses, injuries, and emergence of cancerous cells [46,47]. The uncertainties and complications associated with conventional anticoagulants can be appropriately resolved and avoided via the applications of nanotechnology. Nanoparticles can be contrived to exclusively work together with the blood coagulation system to circumvent various blood disorders [48].

IOP Conf. Series: Materials Science and Engineering **805** (2020) 012041 doi:10.1088/1757-899X/805/1/012041

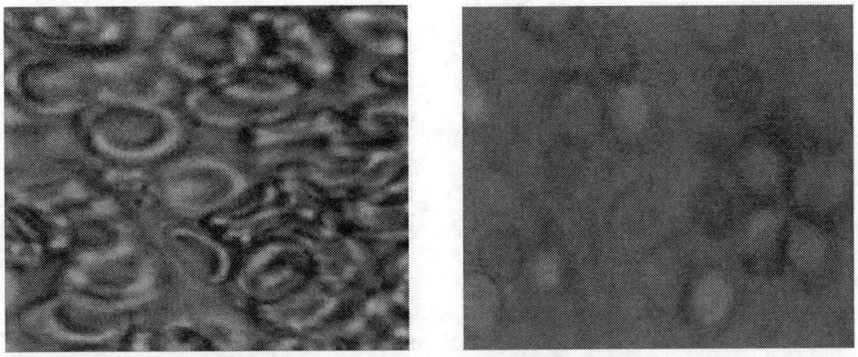

Figure 5. Microscopic view of anticoagulant activities of EDTA (left) and AF-AgNPs (right)

In order to avoid complications that may arise from the blood coagulation disorders, it is necessary to control the blood coagulation system to maintain a healthy state.This involves the prevention of aggregation of platelets, which can be achieved by AgNPs as previously demonstrated to inhibit the formation of thrombus [49]. The facts that AgNPs is non-toxic to platelets and its antimicrobial activities open a new way to treatment of thrombosis. Nanoparticles can as well impact varying degrees of influence on the blood coagulation system as a result of their size, charge, shape, and composition [48]. It can therefore, be inferred that the potency shown by AF-AgNPs obtained in this study can have useful applications in nanomedicine for the prevention of blood coagulation.

3.5 Thrombolytic activity of biosynthesized AgNPs

The AF-AgNPs dissolved the pre-formed blood clot after 90 min which revealed some amount of thrombolytic activity. The animal fur extract and the $AgNO_3$ solution used as the negative controls also yielded partial lyses of the blood clot. About 11, 13 and 25 % thrombolytic activities were observed for the precursor solution, animal fur extract and the biosynthesized AF-AgNPs (Figure 6). Also, the microscopic view of the lysed blood sample clearly showed red blood cells that were dispersed, an indication of the potency of the AF-AgNPs for application in nanomedicines. Similar observations had been reported of nanoparticles serving as anticoagulant and thrombolytic agents with potential for biomedical applications [7, 11, 12, 50].

IOP Conf. Series: Materials Science and Engineering **805** (2020) 012041 doi:10.1088/1757-899X/805/1/012041

3.6 Larvicidal activity of biosynthesized AgNPs

The AF-AgNPs displayed excellent larvicidal activities on mosquito larvae. Complete mortality (100 %) of the tested mosquito larvae was achieved at 12 h for 60-100 μg/ml, 24 h for 20-40 μg/ml, and 36 h for 10 μg/ml (Figure 7).The results obtained indicate potency of the AF-AgNPs on the larvae, which is also comparable with reports of larvicidal activities of biosynthesized AgNPs against mosquito larvae [51-54].

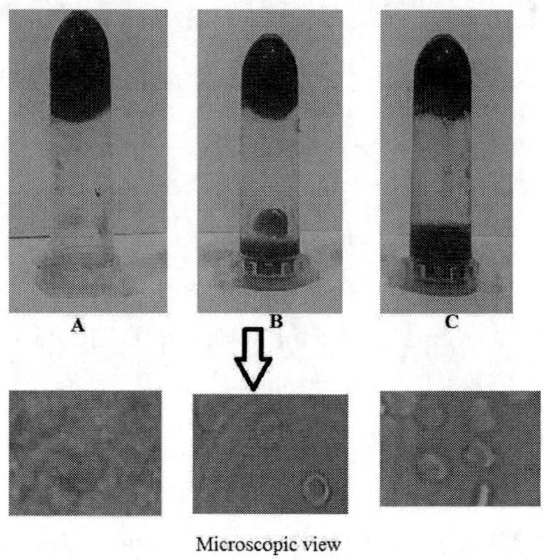

Figure 6. Thrombolytic activities (a, control; b, fur extract; c, AF-AgNPs)

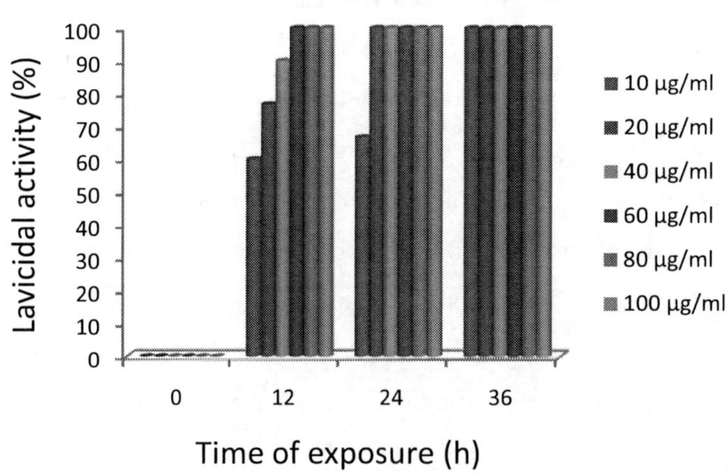

Figure 7. Larvicidal activity of AF-AgNPs on *Anopheles* larvae

IOP Conf. Series: Materials Science and Engineering **805** (2020) 012041 doi:10.1088/1757-899X/805/1/012041

3.7 Cytogenotoxic effects of AgNPs on *Allium cepa*

The cytotoxicity of the $AgNO_3$ solution, fur extract and AF-AgNPs at 24 h exposure revealed dose dependent reduction in dividing cells with mitotic index found less than half that of the control only at 100 µg/ml of the treatments (Table 1). Mitotic inhibition was also found to increase with increase in concentration across the concentrations. Numbers of dividing cell stages reflected dose dependent decrease with the lowest obtained at 100 µg/ml of the AF-AgNPs. Genotoxicity assessment revealed that all the three treatments showed potential to induce chromosomal aberrations with 100 µg/ml of $AgNO_3$ and fur extract inducing more aberrant cells than the AF-AgNPs. Trend in mitotic index and inhibition observed at 48 and 72 h were similar to that of 24 h, except that AF-AgNPs at 72 h showed more toxicity with only value obtained at 0.01 above half the value obtained for the control, suggesting the cytotoxicity of the AF-AgNPs (Tables 1-3). Observations similar to these were reported for *A. cepa* treated with different groups of pesticides [55-57]. However, unlike *Cola* pod mediated AgNPs [31], complete cell arrest was not observed in any of the equivalent concentrations used in the present study, indicating relatively low toxicity of the AF-AgNPs compared with those produced through plant sources. This probably suggests animal materials as better sources than plant materials for AgNPs synthesis due to reduced toxicity. The values of induced aberrations were higher in both $AgNO_3$ and fur extract at all concentrations compared to AF-AgNPs (Tables 2 and 3). Aberrant cells induced include stickiness, disturbed metaphase, c-mitosis, vagrant chromosomes and fragmentation (Figure 8). Exposure at all treatment concentrations revealed reduction in mitotic activities which could be due to the impairment of DNA synthesis [58, 59].

The growth inhibition revealed dose dependent decrease in mean root lengths with the value obtained for the control significantly higher compared to all concentrations in the three treatments (Table 4). The number of roots harvested also showed similar dose dependent decrease with highest percentage root inhibition observed at 100 µg/ml of each of the treatments. The observed results of growth inhibition corroborate that of cytogenotoxicity, which conform to the claim that before the occurrence of chromosome aberrations, some growth restrictions might have happened stemming from the cumulative response of all the damaging effects [60].

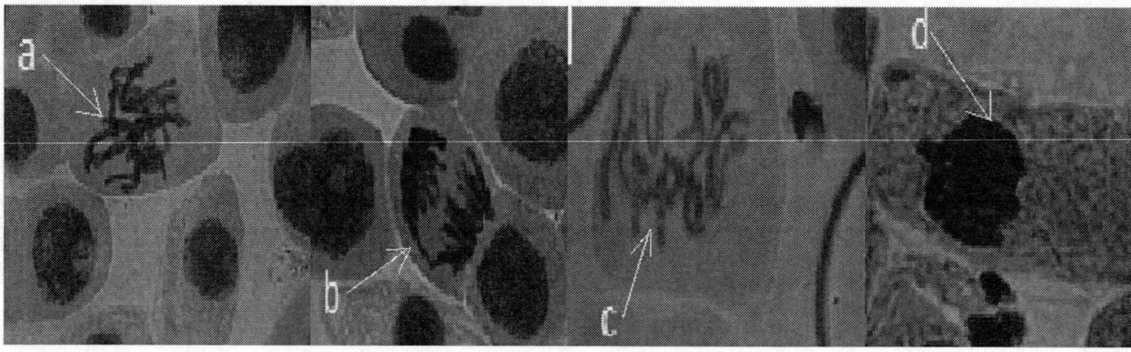

Figure 8. Various aberrations induced by the AF-AgNPs on *Allium cepa* (a, vagrant chromosome; b, chromosome bridge; c, c-mitosis; d, sticky chromosomes)

Table 1. Cytological effects of AF-AgNPs on *Allium cepa* after 24 h of treatment

Conc. (µg\ml)	No of dividing cells	Mitotic index	Mitotic inhibition	Pro	Meta	Ana	Telo	SK	DA	DM	VC	FG	%aberrant cells	%aberrant Per dividing cells	%aberrant per cells scored
$AgNO_3$ Solution															
Control	485	9.70	-	151	115	116	103	-	-	-	-	-	-	-	-
0.01	398	7.9	17.94	106	97	101	94	2	-	2	-	-	4	1.00	0.08
0.1	376	7.52	22.47	100	92	93	91	1	1	1	-	-	3	0.79	0.06
1	289	5.78	40.41	84	70	72	63	1	2	1	-	-	4	1.38	0.08
10	254	5.08	47.62	76	63	58	57	2	1	1	-	-	4	1.57	0.08
100	232	4.64*	52.16	67	57	53	55	2	-	1	-	2	5	2.15	0.10
Fur extract															
0.01	470	9.40	3.09	131	122	116	101	1	-	-	-	-	1	0.21	0.02
0.1	420	8.40	13.40	113	107	103	97	1	-	1	-	-	2	0.47	0.02
1	407	8.14	16.08	106	99	101	101	2	1	-	-	-	3	0.73	0.06
10	311	6.22	35.87	93	76	75	67	3	-	-	-	-	3	0.96	0.06
100	149	2.98*	69.27	48	37	35	29	5	-	-	-	-	5	3.35	0.10
AF-AgNPs															
0.01	468	9.36	3.50	146	112	104	106	1	-	-	-	-	1	0.21	0.02
0.1	415	8.30	14.43	113	96	105	101	1	-	-	-	-	1	0.24	0.02
1	403	8.06	16.90	121	106	91	85	-	-	1	-	-	1	0.24	0.02
10	305	6.10	37.11	105	79	55	66	-	1	-	-	-	1	0.33	0.02
100	147	2.94*	69.69	53	36	32	26	2	-	-	-	-	2	1.36	0.04

*Mitotic index less than half of the value of the control; SK, stickiness; DM, disturbed metaphase; DA, disturbed anaphase; VC, vagrant chromosomes; FG, fragmentation

IOP Conf. Series: Materials Science and Engineering **805** (2020) 012041 doi:10.1088/1757-899X/805/1/012041

Table 2. Cytological effects of AF-AgNPs on *Allium cepa* after 48 h of treatment

Conc. (µg\ml)	No of dividing cells	Mitotic index	Mitotic inhibition	Pro	Meta	Ana	Telo	SK	DM	CM	VC	FG	%aberrant cells	%aberrant Per dividing cells	%aberrant per cells scored
$AgNO_3$															
Control	509	10.18	-	163	122	108	116	-	-	-	-	-	-	-	-
0.01	474	9.48	6.87	148	110	109	107	1	1	-	-	-	2	0.42	0.04
0.1	449	8.98	11.78	138	108	105	98	3	-	1	-	1	5	1.11	0.10
1	363	7.26	28.68	97	89	91	86	1	1	2	-	-	4	1.10	0.08
10	276	5.52	45.77	78	68	66	64	2	2	2	-	1	7	2.54	0.14
100	229	4.58*	55.01	63	54	54	58	4	3	1	-	-	8	3.49	0.16
Fur Extract															
0.01	496	9.92	2.55	136	124	124	112	2	-	-	-	-	2	0.40	0.02
0.1	480	9.60	5.69	129	121	120	107	1	-	1	-	-	2	0.41	0.02
1	433	8.66	14.93	123	117	105	88	-	2	-	-	-	2	0.46	0.02
10	271	5.42	46.75	76	72	66	57	2	-	1	-	-	3	1.10	0.06
100	131	2.62*	74.26	39	34	30	28	3	-	-	-	-	3	2.29	0.06
AF-AgNPs															
0.01	492	9.84	3.33	157	115	117	103	1	-	-	-	-	1	0.20	0.02
0.1	474	9.48	6.87	147	111	113	103	1	-	-	-	-	1	0.21	0.02
1	430	8.60	16.67	131	105	104	90	-	2	-	-	-	2	0.46	0.04
10	268	5.36	47.34	83	69	60	56	-	2	-	-	-	2	0.75	0.04
100	130	2.60*	74.45	41	34	28	27	-	-	2	-	-	2	1.53	0.04

*Mitotic index less than half of the value of the control; SK, stickiness; DM, disturbed metaphase; VC, vagrant chromosomes; FG, fragmentation; CM, c-mitosis

IOP Conf. Series: Materials Science and Engineering **805** (2020) 012041 doi:10.1088/1757-899X/805/1/012041

Table 3. Cytological effects of AF-AgNPs on *Allium cepa* after 72 h of treatment

Conc. (µg\ml)	No of dividing cells	Mitotic index	Mitotic inhibition	Pro	Meta	Ana	Telo	SK	DM	CM	VC	FG	%aberrant cells	%aberrant Per dividing cells	%aberrant per cells scored
$AgNO_3$															
Control	490	9.80	-	155	115	114	106	-	-	-	-	-	-	-	-
0.01	392	7.84	20.00	106	99	93	94	2	3	1	-	-	6	1.53	0.12
0.1	350	7.00	28.57	99	90	83	78	2	4	2	-	1	9	2.57	0.18
1	239	4.78	51.22	67	56	54	62	5	-	2	-	-	7	2.93	0.14
10	221	4.42*	54.89	64	53	50	54	4	1	3	-	-	8	3.61	0.16
100	198	3.96*	59.59	61	49	46	42	2	4	3	-	-	9	4.55	0.18
Extract															
0.01	320	6.40	34.69	88	82	77	73	3	-	-	-	-	3	0.93	0.06
0.1	278	5.56	43.26	78	69	68	63	1	1	1	-	-	3	1.07	0.06
1	235	4.70	52.04	72	58	55	50	3	-	1	-	-	4	1.70	0.08
10	146	2.92*	70.20	44	36	35	31	4	-	1	-	-	5	3.42	0.10
100	104	2.04*	79.18	30	26	25	23	6	1	-	-	-	7	6.73	0.14
AgNPs															
0.01	226	4.52	53.87	74	59	51	42	-	1	-	-	-	1	0.44	0.02
0.1	205	4.10*	58.16	68	54	46	37	2	-	-	-	-	2	0.97	0.04
1	186	3.72*	62.04	64	49	41	32	-	2	-	-	-	2	1.07	0.04
10	141	2.82*	71.22	53	38	30	20	2	-	-	-	-	2	1.42	0.04
100	102	2.04*	79.18	42	28	20	12	2	-	1	1	-	4	3.92	0.08

*Mitotic index less than half of the value of the control; SK, stickiness; DM, disturbed metaphase; CM, c-mitosis; VC, vagrant chromosomes; FG, fragmentation

IOP Conf. Series: Materials Science and Engineering **805** (2020) 012041 doi:10.1088/1757-899X/805/1/012041

Table 4. Toxic effects of silver nitrate salt on *Allium cepa* root growth

Concentration (µg/ml)	**$AgNO_3$**				**Fur Extract only**				**AF-AgNPs**			
	Total (N)	Mean root length+SE	%Mean root length	%Root inhibition control	Total (N)	Mean root length+SE	%Mean root length	%Root inhibition control	Total (N)	Mean root length+SE	%Mean root length	%Root inhibition control
Control	255	$3.58^{a}±0.92$	100	-	255	$3.58^{a}±0.92$	100	-	255	$3.58^{a}±0.92$	100	-
0.01	239	$2.67^{b}±0.76$	74.58	25.42	157	$3.05^{b}±0.93$	85.19	14.80	180	$3.36^{b}±0.47$	93.85	6.14
0.1	242	$2.43^{c}±0.61$	67.87	32.12	150	$2.69^{c}±0.81$	75.13	24.86	153	$2.55^{c}±0.47$	71.22	28.77
1.0	169	$2.04^{d}±0.63$	56.98	43.02	130	$2.29^{d}±0.59$	63.96	36.03	165	$2.49^{c}±0.43$	69.55	30.44
10.0	213	$1.39^{e}±0.43$	38.83	61.17	123	$1.02^{e}±0.38$	28.49	71.50	112	$0.47^{d}±0.31$	13.12	86.87
100	124	$0.35^{f}±0.02$	9.78	90.22	131	$0.48^{f}±0.26$	13.40	86.59	111	$0.18^{e}±0.00$	5.02	94.97

Mean with different subscript letter a, b, c, d and e are significantly different from each other ($p< 0.05$)

IOP Conf. Series: Materials Science and Engineering **805** (2020) 012041 doi:10.1088/1757-899X/805/1/012041

4 Conclusion

The present study has clearly demonstrated the usefulness of animal fur extract as a cost-effective and eco-friendly bio-resource in the green synthesis of silver nanoparticles (AF-AgNPs). The biosynthesized particles were spherical with size ranging from 11-32 nm. The highly remarkable larvicidal activities of AF-AgNPs against larvae of the vector of *Plasmodium* parasites (60-100 % after 12 h of exposure), including their anticoagulant and thrombolytic activities, indicate their relevance in biomedical applications. Furthermore, the remarkable antioxidant activities (hydrogen peroxide and DPPH scavenging) suggest the biotechnological potential of the synthesized nanoparticles in the biomedical industry and treatment of industrial and domestic waste water. The success recorded in the use of animal fur, which in most instances are burnt or left to constitute nuisance in the environment, in the synthesis of AgNPs provides opportunity to turn waste to wealth. However, the synthesized AF-AgNPs exhibited some levels of cytogenotoxicity, which calls for caution in its application.

References

1. Aitken RJ, Chaudhry MQ, Boxall ABA and Hull M 2006 Manufacture and use of nanomaterials: current status in the UK and global trends. *Occup. Med.* **56** pp 300-306.
2. Benelli G 2016 Plant-mediated biosynthesis of nanoparticles as an emerging tool against mosquitoes of medical and veterinary importance: a review. *Parasitol Res.* **115** pp 23-34.
3. Boutonnet M, Lögdberg S, and Svensson EE 2008 Recent developments in the application of nanoparticles prepared from w/o microemulsions in heterogeneous catalysis. *Curr. Opin. Colloid Interf. Sci* **13** pp 270-286.
4. Koçak A and Karasu B 2018 General Evaluations of nanoparticles. *El-Cezerî J. Sci. Eng.* **5** (1) pp 191-236.
5. Oladipo IC, Lateef A, Azeez MA, Asafa TB, Yekeen TA, Akinboro A, Akinwale AS, Gueguim-Kana EB and Beukes LS 2017 Green synthesis and antimicrobial activities of silver nanoparticles using cell-free extracts of *Enterococcus* species. *Not. Sci. Biol.* **9** (2): 196-203.
6. Oladipo IC, Lateef A, Elegbede JA, Azeez MA, Asafa TB, Yekeen TA, Akinboro A, Gueguim-Kana EB, Beukes LS, Oluyide TO and Atanda OR 2017 *Enterococcus* species

for the one-pot biofabrication of gold nanoparticles: characterization and nanobiotechnological applications. *J. Photochem. Photobiol. B: Biol.* **173** pp 250-257.

7. Elegbede JA, Lateef A, Azeez MA, Asafa TB, Yekeen TA, Oladipo IC, Adebayo EA, Beukes LS and Gueguim-Kana EB 2018 Fungal xylanases-mediated synthesis of silver nanoparticles for catalytic and biomedical applications. *IET Nanobiotechnol.* **12** (6) pp 857-863.
8. Elegbede JA, Lateef A, Azeez MA, Asafa TB, Yekeen TA, Oladipo IC, Aina DA, Beukes LS and Gueguim-Kana EB 2020 Biofabrication of gold nanoparticles using xylanases through valorization of corncob by *Aspergillus niger* and *Trichoderma longibrachiatum*: antimicrobial, antioxidant, anticoagulant and thrombolytic activities. Waste *Biomass Valor.* **11** (3) pp 781-791.
9. Elegbede JA, Lateef A, Azeez MA, Asafa TB, Yekeen TA, Oladipo IC, Abbas SH, Beukes LS and Gueguim-Kana EB 2019 Silver-gold alloy nanoparticles biofabricated by fungal xylanases exhibited potent biomedical and catalytic activities. *Biotechnol. Progress.* **35** e2829. https://doi.org/10.1002/btpr.2829.
10. Rao KJ and Paria S 2015 *Aegle marmelos* leaf extract and plant surfactants mediated green synthesis of Au and Ag nanoparticles by optimizing process parameters using Taguchi method. *ACS Sustain. Chem. Eng.* **3** (3) pp 483-491.
11. Ojo SA, Lateef A, Azeez MA, Oladejo SM, Akinwale AS, Asafa TB, Yekeen TA, Akinboro A, Oladipo IC, Gueguim-Kana EB and Beukes LS 2016 Biomedical and catalytic applications of gold and silver-gold alloy nanoparticles biosynthesized using cell-free extract of *Bacillus safensis* LAU 13: antifungal, dye degradation, anti-coagulant and thrombolytic activities. *IEEE Trans. NanoBiosci.* **15** (5) pp 433-442.
12. Lateef A, Ojo SA, Elegbede JA, Azeez MA, Yekeen TA and Akinboro A 2017 Evaluation of some biosynthesized silver nanoparticles for biomedical applications: hydrogen peroxide scavenging, anticoagulant and thrombolytic activities. *J. Clust. Sci.* **28** (3) pp 1379-1392.
13. Ismail M, Gul S, Khan MI, Khan MA, Asiri AM Khan SB 2019 Green synthesis of zerovalent copper nanoparticles for efficient reduction of toxic azo dyes congo red and methyl orange. *Green Process Synth.* **8** pp 135-143.

14. Lateef A, Ojo SA, Azeez MA, Asafa TB, Yekeen TA, Akinboro A, Oladipo IC, Gueguim-Kana EB and Beukes LS 2016 Cobweb as novel biomaterial for the green and eco-friendly synthesis of silver nanoparticles. *Appl. Nanosci.* **6** (6) pp 863-874.
15. Lateef A, Akande MA, Ojo SA, Folarin BI, Gueguim-Kana EB and Beukes LS 2016 Paper wasp nest-mediated biosynthesis of silver nanoparticles for antimicrobial, catalytic, anticoagulant, and thrombolytic applications. *3Biotech.* **6** 140. http://dx.doi.org/10.1007/s13205-016-0459-x.
16. Khatami M, Iravani S, Varma RS, Mosazade F, Darroudi M and Borhani F 2019 Cockroach wings-promoted safe and greener synthesis of silver nanoparticles and their insecticidal activity. *Bioprocess Biosys. Eng.* **42** (12) pp 2007-2014.
17. Balasooriya ER, Jayasinghe CD, Jayawardena UA, Ruwanthika RW, Mendis de Silva R and Udagama PV 2017 Honey mediated green synthesis of nanoparticles: new era of safe nanotechnology. Journal of Nanomaterials Article ID 5919836. https://doi.org/10.1155/2017/5919836.
18. Lee KJ, Park SH, Govarthanan M, Hwang PH, Seo YS, Cho M, Lee WH, Lee JY, Kamala-Kannan S and Oh BT 2013 Synthesis of silver nanoparticles using cow milk and their antifungal activity against phytopathogens. *Mater Lett.* **105** pp 128-131.
19. Velmurugan P, Shim J, Kamala-Kannan S, Lee KJ, Oh BT, Balachandar V and Oh BT 2011 Crystallization of silver through reduction process using *Elaeis guineensis* biosolid extract. *Biotechnol. Progress* **27** (1) pp 273-279.
20. Tszydel M, Sztajnowski S, Michalak M, Wrzosek H, Kowalska S, Krucińska I and Lipp-Symonowicz B 2009 Structure and physical and chemical properties of fibres from the fifth larval instar of caddis-flies of the species *Hydropsyche angustipennis. Fibres Text. East. Eur.* **6** (77) pp 7-12.
21. Lateef A, Adelere IA, Gueguim-Kana EB, Asafa TB, and Beukes LS 2015 Green synthesis of silver nanoparticles using keratinase obtained from a strain of *Bacillus safensis* LAU 13. *Int. Nano Lett.* **5** (1) pp 29-35.
22. Morones JR, Elechiguerra JL, Camacho A, Holt K, Kouri JB, Ramírez JT and Yacaman MJ 2005 The bactericidal effect of silver nanoparticles. *Nanotechnol.* **16** (10) 2346. https://doi.org/10.1088/0957-4484/16/10/059.

23. Williams BW, Cuverlier, ME, Berset C 1995. Use of free radical method to evaluate antioxidant activity. *Food Sci. Technol.* LWT **28** pp 25-30.

24. Nabavi SM, Ebrahimzadeh MA, Nabavi SF, Hamidinia A and Bekhradnia AR 2008 Determination of antioxidant activity, phenol and flavonoids content of *Parrotia persica* Mey. *Pharmacol. Online.* **2** (9) pp 560-567.

25. Nabavi SM, Ebrahimzadeh MA, Nabavi SF, Fazelian M and Eslami B 2009 *In vitro* antioxidant and free radical scavenging activity of *Diospyros lotus* and *Pyrus boissieriana* growing in Iran. *Pharmacogn. Mag.* **5** (18) pp 122-126.

26. Sousa A, Ferreira IC, Barros L, Bento A and Pereira JA 2008 Effect of solvent and extraction temperatures on the antioxidant potential of traditional stoned table olives "alcaparras". *LWT-Food Sci. Technol.* **41** (4) pp 739-45.

27. Cicha I 2015 Thrombosis, novel nanomedical concepts of diagnosis and treatment. *World J. Cardiol.* **7** (8) pp 434-441.

28. Bansal SK, Singh KV and Kumar S 2009 Larvicidal activity of the extracts from different parts of the plant *Solanum xanthocarpum* against important mosquito vectors in the arid region. *J. Environ. Biol.* **30** (2) pp 221-226.

29. Bakare AA, Mosuro AA and Osibanjo O 2000 Effects of simulated leachates on chromosomes and mitosis of *Allium cepa* (L). *J. Environ. Biol.* **21** pp 263-271.

30. Yekeen TA, Ayandele AA, Akinboro A, Onipede AY and Alawode R 2013 Assessment of toxicity and microbial attributes of synthetic epoxy resin and acetaminophen effluents. *Pollut. Res.* **32** (4) pp 9-16.

31. Yekeen TA, Azeez MA, Akinboro A, Lateef A, Asafa TB, Oladipo IC, Oladokun SO and Ajibola AA 2017 Safety evaluation of green synthesized *Cola nitida* pod, seed and seed shell extract-mediated silver nanoparticles (AgNPs) using an *Allium cepa* assay. *J. Taibah Univ. Sci.* **11** (6) pp 895-909.

32. Yekeen TA, Azeez MA, Lateef A, Asafa TB, Oladipo IC, Badmus JA, Adejumo SA and Ajibola AA 2017 Cytogenotoxicity potentials of cocoa pod and bean-mediated green synthesized silver nanoparticles on *Allium cepa* cells. *Caryol.* **70** (4) pp 366-377.

33. Jeevan P, Ramya K and Rena AE 2012 Extracellular biosynthesis of silver nanoparticles by culture supernatant of *Pseudomonas aeruginosa*. *Indian J. Biotechnol.* **11** (1) pp 72-76.

34. Zakir S, El-Kady MF, and Abd-El-Haleem D 2011 Biosynthesis and structural characterization of silver nanoparticles from bacterial isolates. *Mater. Res. Bull.* **46** (2) pp 1571-1576.
35. Kalishwaralal K, Deepak V, Ramkumarpandian S, Nellaiah H and Sangiliyandi G 2008 Extracellular biosynthesis of silver nanoparticles by the culture supernatant of *Bacillus licheniformis*. *Mater. Lett.* **62** (29) pp 4411-413.
36. Shankar S, Jaiswal L, Aparna RSL and Prasad RGSV 2014 Synthesis, characterization, *in vitro* biocompatibility, and antimicrobial activity of gold, silver and gold silver alloy nanoparticles prepared from *Lansium domesticum* fruit peel extract. *Mater. Lett.* **137**: 75-78.
37. Kannan RR, Arumugam R, Ramya D, Manivannan K and Anantharaman P 2013 Green synthesis of silver nanoparticles using marine macroalga *Chaetomorpha linum*. *Appl. Nanosci.* **3** (1) pp 229-233.
38. Mehmood A, Murtaza G, Bhatti TM, Raffi M and Kausar R 2014 Antibacterial efficacy of silver nanoparticles synthesized by a green method using bark extract of *Melia azedarach* L. *J. Pharmaceut. Innov.* **9** (3) pp 238-245.
39. Shanmugam C, Sivasubramanian G, Parthasarathi B, Baskaran K, Balachander R and Parameswaran V 2016 Antimicrobial, free radical scavenging activities and catalytic oxidation of benzyl alcohol by nano-silver synthesized from the leaf extract of *Aristolochia indica* L.: a promenade towards sustainability. *Appl. Nanosci.* **6** (5) pp 711-723.
40. Elegbede JA and Lateef A 2019 Green synthesis of silver (Ag), gold (Au) and silver-gold (Ag-Au) alloy nanoparticles: A review on recent advances, trends and biomedical applications. In: Verma DK, Goyal MR and Suleria HAR (Eds.). *Nanotechnology and Nanomaterial Applications in Food, Health and Biomedical Sciences.* https://doi.org/10.1201/9780429425660-1. Apple Academic Press Inc. /CRC Press, Taylor and Francis Group, Oakville, Ontario, Canada. ISBN 978-1-77188-764-9. Pp. 3-89.
41. Jeyaraj M, Varadan S, Anthony KJ, Murugan M, Raja A and Gurunathan S 2013 Antimicrobial and anticoagulation activity of silver nanoparticles synthesized from the culture supernatant of *Pseudomonas aeruginosa*. *J. Ind. Eng. Chem.* **19** (4) pp 1299-1303.

42. Raja S, Ramesh V and Thivaharan V 2015 Antibacterial and anticoagulant activity of silver nanoparticles synthesised from a novel source–pods of *Peltophorum pterocarpum*. *J. Ind. Eng. Chem.* **29** pp 257-264.
43. Azeez MA, Lateef A, Asafa TB, Yekeen TA, Akinboro A, Oladipo IC, Gueguim-Kana EB and Beukes LS 2017 Biomedical applications of cocoa bean extract-mediated silver nanoparticles as antimicrobial, larvicidal and anticoagulant agents. *J. Clust. Sci.* **28** (1) pp 149-164.
44. Lateef A, Folarin BI, Oladejo SM, Akinola PO, Beukes LS and Gueguim-Kana EB 2018 Characterization, antimicrobial, antioxidant and anticoagulant activities of silver nanoparticles synthesized from *Petiveria alliacea* L. leaf extract. *Prep. Biochem. Biotechnol.* **48** (7) pp 646-652.
45. Levi M, Schultz M and van der Poll T 2010 Disseminated intravascular coagulation in infectious disease. *Sem. Thromb. Hemost.* **36** (4) pp 367-377.
46. Davalos D and Akassoglou K 2012 Fibrinogen as a key regulator of inflammation in disease. *Sem. Immunopathol.* **34** (1) pp 43-62.
47. Prandoni P, Falanga A and Piccioli A 2007 Cancer, thrombosis and heparin-induced thrombocytopenia. *Thromb. Res.* **120** pp S137-S140.
48. Ilinskaya AN and Dobrovolskaia MA 2013 Nanoparticles and the blood coagulation system. Part II: safety concerns. *Nanomed.* **8** (6) pp 969-981.
49. Shrivastava S, Bera T, Singh SK, Singh G, Ramachandrarao P and Dash D 2009 Characterization of antiplatelet properties of silver nanoparticles. *ACS Nano* **3** pp 1357-1364.
50. Harish BS, Uppuluri KB and Anbazhagan V 2015 Synthesis of fibrinolytic active nanoparticles using wheat bran xylan as a reducing and stabilizing agent. *Carbohydr. Polym.* **132** pp 104-110.
51. Gnanadesigan M, Anand M, Ravikumar S, Maruthupandy M, Vijayakumar V, Selvam S, Dhineshkumar M and Kumaraguru AK 2011 Biosynthesis of silver nanoparticles by using mangrove plant extract and their potential mosquito larvicidal property. *Asian Pacif. J. Trop. Med.* **4** (10) pp 799-803.
52. Priyadarshini KA, Murugan K, Panneerselvam C, Ponarulselvam S, Hwang JS and Nicoletti M 2012 Biolarvicidal and pupicidal potential of silver nanoparticles synthesized

using *Euphorbia hirta* against *Anopheles stephensi* Liston (Diptera: Culicidae). *Parasitol. Res.* **111** (3) pp 997-1006.

53. Lateef A, Ojo SA, Akinwale AS, Azeez L, Gueguim-Kana EB and Beukes LS 2015 Biogenic synthesis of silver nanoparticles using cell-free extract of *Bacillus safensis* LAU 13: antimicrobial, free radical scavenging and larvicidal activities. *Biologia* **70** (10) pp 1295-1306.
54. Aina DA, Owolo O, Lateef A, Aina FO, Abbas SH, Adeoye-Isijola M, Okon V, Asafa TB, Elegbede JA, Olukanni OD and Adediji I 2019 Biomedical applications of *Chasmanthera dependens* stem extract mediated silver nanoparticles as antimicrobial, antioxidant, anticoagulant, thrombolytic, and larvicidal agents. *Karbala Int. J. Modern Sci.* **5** (2) pp 71-80.
55. Asita AO and Matebesi LP 2010 Genotoxicity of hormoban and sevenother pesticides to onion root tip meristematic cells. *Afr. J. Biotechnol.* **9** (27) pp 4225-4232.
56. Shaikh S, Nazam N, Lone MI and Ahmad W 2012 Dichlorophen and Dichlorovos mediated genotoxic and cytotoxic assessment on root meristem cells of *Allium cepa. Sci. Diliman.* **24** (1) pp 13-22.
57. Yekeen TA and Adeboye MK 2013 Cytogenotoxic effects of cypermethrin, deltamethrin, lambdacyhalothrin and endosulfan pesticides on *Allium cepa* root cells. *Afr. J. Biotechnol.* **12** (41) pp 6000-6006.
58. Schneiderman MH, Dewey WC and Highfield DP 1971 Inhibition of DNA synthesis in synchronized Chinese hamster cells treated in G1 with cycloheximide. *Exp. Cell Res.* **67** (1) pp 147-155.
59. Sudhakar R, Gowda KNN and Venu G 2001. Mitotic abnormalities induced by silk dyeing industry effluents in the cells of *Allium cepa. Cytol.* **66** pp 235-239.
60. Fiskesjo G 1997 *Allium* test for screening chemical evaluation of cytological parameters. In Wang W, Gorsuch JW and Hughes JS (Eds), Plants for Environmental Studies, New York, NY: CRC Lewis Publishers, pp 307-333.

IOP Conf. Series: Materials Science and Engineering **805** (2020) 012042 doi:10.1088/1757-899X/805/1/012042

Facile synthesis of silver nanoparticles using leaf extract of *Hyptis suaveolens* (L.) Poit for environmental and biomedical applications

A Lateef[1, 2*], S M Oladejo[1], P O Akinola[1], D A Aina[3], L S Beukes[4], B I Folarin[1], and E B Gueguim-Kana[5]

[1]Department of Pure and Applied Biology, [2]Nanotechnology Research Group (*NANO*$^+$), Ladoke Akintola University of Technology, PMB 4000, Ogbomoso, Nigeria, [3]Department of Microbiology, Babcock University, Ilishan-Remo, Nigeria,
[4]Microscopy and Microanalysis Unit, [5]Department of Microbiology, School of Life Sciences, University of KwaZulu-Natal, Private Bag X01, Scottsville, Pietermaritzburg 3209, South Africa.
*Corresponding author: agbaje72@yahoo.com; alateef@lautech.edu.ng

Abstract

Aqueous leaf extract of *Hyptis suaveolens* was investigated to synthesize silver nanoparticles (Hs-AgNPs). Biomolecules such as phenolics, proteins, carbohydrates and esters facilitated bioformation of Hs-AgNPs with λmax at 420 nm. The nearly spherical Hs-AgNPs were polydispersed and sized 29.19-52.27 nm with yield of 85.8 %. Hs-AgNPs showed potent inhibition of 75.22-100 % against multidrug resistant bacteria, and fungal growth inhibition of 73.33-100 %. Hs-AgNPs scavenged DPPH and H_2O_2 by 77.75-83.19, and 54.21-70.11 % respectively. It also effectively inhibited coagulation of blood. The study established the nanobiotechnological importance of *H. suaveolens*, with bioformation of AgNPs that have potent activities for environmental and biomedical applications.
Keywords: Silver nanoparticles; antimicrobial; anticoagulant; antioxidant; *Hyptis suaveolens*

1 Introduction

There has been a continued expansion in the extent of synthesis and applications of nanomaterials (1-100 nm) in the field of nanoscience and nanotechnology relating to diverse areas of human endeavours. Its result has really bridged the gap that has existed between materials science and life sciences. The various efficacious uses of nanoparticles have

IOP Conf. Series: Materials Science and Engineering **805** (2020) 012042 doi:10.1088/1757-899X/805/1/012042

engineered a vital and conspicuous role in the search for the production of nanoparticles that are eco-friendly using different materials that are of biological origin. In the traditional mode of chemical synthesis; the application of noxious solvents, consumption of high energy and pressure have been found to be detrimental to the environment. However, the biological synthesis has resulted into the fabrication of nanoparticles with unique attributes, less consumption of energy and avoidance of hazardous procedures that have contributed to their expanding applications.

Due to the rich diversity of microorganisms as well as plants, the efficacy of bioresources in biofabrication of nanoparticles has not been fully realized. Relatively, limited plants from Nigeria have been reported for phytosynthesis of nanoparticles [1-13]. Other novel biological materials used for biosynthesis of various nanoparticles in different investigations comprise of cell-free bacterial extract, enzymes and metabolites of spider and paper wasp [14-23]. The sustained interest in the green synthesis of nanoparticles is consequent upon unfussiness of the process, reduced usage of chemicals and being eco-friendly [24-27]. Additionally, the existence of a lot of macromolecules or substances in biological resources that often act as excellent reducing, capping and stabilizing agents to mediate one-pot biosynthesis of nanoparticles furthermore contributed to the soaring popularity of biosynthesis as demonstrated by several researchers [28-33].

The genus *Hyptis* in the family Lamiaceae consists of about 400 different species. *Hyptis suaveolens* (Labiatae) is an average fragrant shrub that grows annually. It is commonly found both tropics and subtropics of the World, where it grows to reach height of about 2.5. The stem of the plant is quadrate and hairy, while its leaves are simple, entire, opposite, stalked, ovate or obovate and are about 3-5 cm long and 2-4 cm wide with serrulate margins and a long petiole [34]. It is a plant of medical importance with most of its parts exploited in local medicine for the treatment of variety of illnesses. Its leaves have found applications as stimulant, carminative as well as in the treatment of stomach-ache in Indian herbal medicine. In addition, the leaves are also used as anticancer and anti-fertility agents in females, and can play vital role in the management of anemia during pregnancy. The leaves also served as insect (mosquitoes) repellant [35], and used to treat colics, stomach ache, and fever.

The plant possesses significant pharmacological activities, including tumorigenic, mycotoxic, phytotoxic, antispasmodic, hepatoprotective, cytoprotective, anti-inflammatory, antioxidant activity, and antiseptic in wounds, burns as well as various skin diseases [36-40]. Moreover, suaveol, an isolated compound of this plant showed gastroprotection to fight nausea, indigestion, colds, flatulence, and infection of the gall bladder [41]. The extract of the roots is confirmed to be an appetizer with reports showing the presence of urosolic acid, a natural HIV integrase inhibitor [42, 43]. Chemically, the plant also contains saponins, carbohydrates, phenols, tannins, glycosides and steroids that account for its vast medicinal values [44]. However, only one report exists on its usage for the synthesis of nanoparticles and investigated for larvicidal activities [45]. This work therefore aims at advancing the nanobiotechnological utilization of the plant; to phytosynthesize silver nanoparticles for antimicrobial, radical scavenging and anticoagulant applications with consideration for applications in the environment and healthcare.

2 Materials and methods

2.1 Green synthesis of Hs-AgNPs and characterization

The extract of leaves *H. suaveolens* was prepared as earlier demonstrated [5] by heating 0.1 g of dried leaf sample in 10 ml of distilled water at 60 °C for 1 h. The centrifuged extract was utilized to synthesize Hs-AgNPs as stated previously [5] by reacting 40 ml of 1 mM $AgNO_3$ with 1 ml of the extract at ambient temperature (30 ± 2 °C). The transformation in colour was monitored, and the synthesis of Hs-AgNPs was assessed through measurement of its absorbance spectrum on a UV-visible spectrophotometer (Cecil, USA) at the range of 300-900 nm. Biomolecules that were involved in the phytosynthesis were resolved by FTIR spectroscopy (IR Affinity-1S spectrophotometer, Shimadzu, UK), while the size, morphology, selected area electron diffraction (SAED) and elemental fingerprints of Hs-AgNPs were revealed by transmission electron microscopic (JEM-1400, JEOL, USA) and energy dispersive X-ray (EDX) analyses.

2.2 Antibacterial actions of biosynthesized Hs-AgNPs

Antibacterial efficacy of Hs-AgNPs was investigated against strains of clinical bacterial isolates using the modified liquid culture method [46]. Multidrug resistant (MDR) clinical bacterial strains of *Klebsiella pneumoniae, Staphylococcus aureus,* and *Escherichia coli* that were originally sourced from LAUTECH Teaching Hospital, Ogbomoso served as test organisms to assess the potency of the nanoparticles. The isolates were inoculated into peptone water and

IOP Conf. Series: Materials Science and Engineering **805** (2020) 012042 doi:10.1088/1757-899X/805/1/012042

incubated at 37 °C for 24 h. Hs-AgNPs (1 ml each of graded concentrations from 5-15 μg/ml) were dispensed into test tubes that contained 9 ml of sterile peptone water. An aliquot of 0.1 ml (1×10^6 cfu/ml) served as inoculum and incubation was done at 37 °C for 24 h. The growth of the bacterial isolates was measured by determining the optical density ($OD_{600\ nm}$). The hindrance of bacterial growth was estimated accordingly;

Percentage growth inhibition = $\frac{OD\ control - OD\ test}{OD\ control} \times 100$ %, where OD is the optical density.

2.3 Antifungal activity of Hs-AgNPs

Antifungal activity of Hs-AgNPs was assayed using mycelial growth inhibition test [47]. Prepared Hs-AgNPs were included into potato dextrose agar (PDA), and thereafter inoculated with circular agar block of 7 mm obtained from 48 h-old cultures of *Aspergillus fumigatus, A. flavus* and *A. niger.* Hs-AgNPs was not included in the control plates. Incubation of the plates was done at 28 ± 2 °C and lasted till 72 h. At the end of 72 h, diameters of growth on plates were measure to calculate fungal growth inhibition as follows:

Percentage growth inhibition = $\frac{D\ control - D\ test}{D\ control} \times 100$ %; where D is the diameter of growth on plate.

2.4 Evaluation of antioxidant activities of Hs-AgNPs

2.4.1 Scavenging of DPPH

The ability of Hs-AgNPs to scavenge 2,2-diphenyl-1-picrylhydrazyl (DPPH, Sigma-Aldrich, Germany) was evaluated as stated by William *et al.* [48]. About 1 ml of different concentrations of Hs-AgNPs was separately reacted with 4.0 ml of methanolic solution of 0.1 mM DPPH in the dark box for 30 min at 30 ± 2 °C. Methanolic DPPH alone served as the control. At the end of the reaction, absorbance readings were taken at 517 nm on UV-vis spectrophotometer (Spectrumlab 752S, UK), and used to subsequently calculate the amount of DPPH scavenged:

Percentage DPPH scavenging activity = $\frac{A\ control - A\ sample}{A\ control} \times 100$; where A is the absorbance value.

2.4.2 Scavenging of hydrogen peroxide (H_2O_2)

To evaluate the capability of Hs-AgNPs at scavenging H_2O_2, the methods of Bhakya *et al.* [49] was employed, whereby 4 ml of graded concentrations of Hs-AgNPs was reacted with 0.6 ml of 40 mM H_2O_2 (pH 7.4), and held at 30 ± 2 °C for 20 min. The H_2O_2 solution and distilled water were used as the control and blank respectively. Following the reading of absorbance at 610 nm, the amount of H_2O_2 scavenged was calculated thus:

$$\% \ H_2O_2 \text{ scavenged} = \frac{A\ control - A\ sample}{A\ control} \times 100;$$ where A is the absorbance value.

2.5 Anticoagulant activity of Hs-AgNPs

The capability of Hs-AgNPs to inhibit coagulation of blood was determined by adding 1 ml of 100 µg/ml of Hs-AgNPs to 0.5 ml of blood unreservedly given by a hale and hearty donor. Three control samples that involved the use of EDTA, *H. suaveolens* leaf extract (HE), and solution of $AgNO_3$ were assayed for anticoagulation. The reaction mixes were incubated at 30 ± 2 °C for 1 h and subsequently observed both visually and microscopically for development of blood clot [19, 50].

3 Results and discussion

3.1 Phytosynthesis and characteristics of Hs-AgNPs

Leaf extract of *H. suaveolens* potentiated the bioformation of brown colloidal AgNPs within 2 min, when it was reacted with silver nitrate (Figure 1). The colour of biofabricated Hs-AgNPs stabilized within 10 minutes of formation. It is evident from the phytochemistry of the plant as earlier stated that it was rich in biomolecules that facilitated the reduction Ag^+ to Ag^0, thereby influencing the frequency of the surface plasmon resonance that led to the development of brown colour. Varying colours of AgNPs that include brown, dark brown, brownish-orange and light yellow exist [4, 15, 34, 51]. Biosynthesized Hs-AgPs had characteristic peak of absorbance at 420 nm (Figure 2), and this is within limit of various values of SPR ascribed to AgNPs by several authors [1, 2, 10, 15]. Elumalai et al. [45] reported SPR of 408 nm for the brown colloidal AgNPs phytosynthesized using the leaf extract of *H. suaveolens.*

FTIR spectrum (Figure 3), indicated the prominence of peaks at 3444.2, 2293.8, 1707.7, 1385.8, 1128.8 and 1109.3 cm^{-1}, that implied N-H found in amines or O-H stretch linked to alcohols/phenols, P-H phosphine stretch, C=O stretch of amides or esters; CH_3 absorption

IOP Conf. Series: Materials Science and Engineering **805** (2020) 012042 doi:10.1088/1757-899X/805/1/012042

bending of alkane, and C-O stretch of ethers, carboxylic acids, esters and alcohols. The incidence of C-H, C-O and C=O have been reported for the extract [45]. The presence of these peaks is a pointer to the richness of the plant extract in different phytochemicals. The TEM micrograph showed nearly spherical-shaped Hs-AgNPs that were polydispersed and varied in size from 29.19 to 52.27 nm (Figure 4a); which are in conformity with existing literature [4, 52]. However, Elumalai *et al.* [45] reported anisotropic shape of sphere, hexagon, triangle and polyhedral for AgNPs synthesized by leaf extract of *H. suaveolens.* The particles showed high level of dispersity, stability and devoid of aggregation on long storage. The predominance of Ag in EDX spectrum (Figure 4b) to tune of 85.8 % was indicated. Hs-AgNPs depicted the characteristic ring-like SAED pattern (Figure 4c) associated with face-centered cubic crystalline structure of Ag [5, 53].

Figure 1. The biosynthesis of Hs-AgNPs using the leaf extract of *H. suaveolens*

IOP Conf. Series: Materials Science and Engineering **805** (2020) 012042 doi:10.1088/1757-899X/805/1/012042

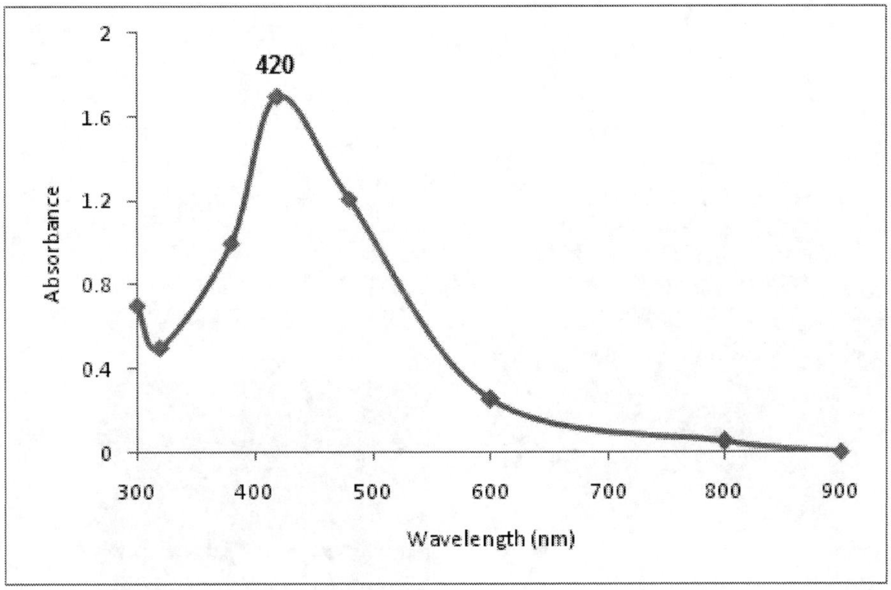

Figure 2. UV-visible absorption spectrum of Hs-AgNPs biosynthesized using the leaf extract of *H. suaveolens*

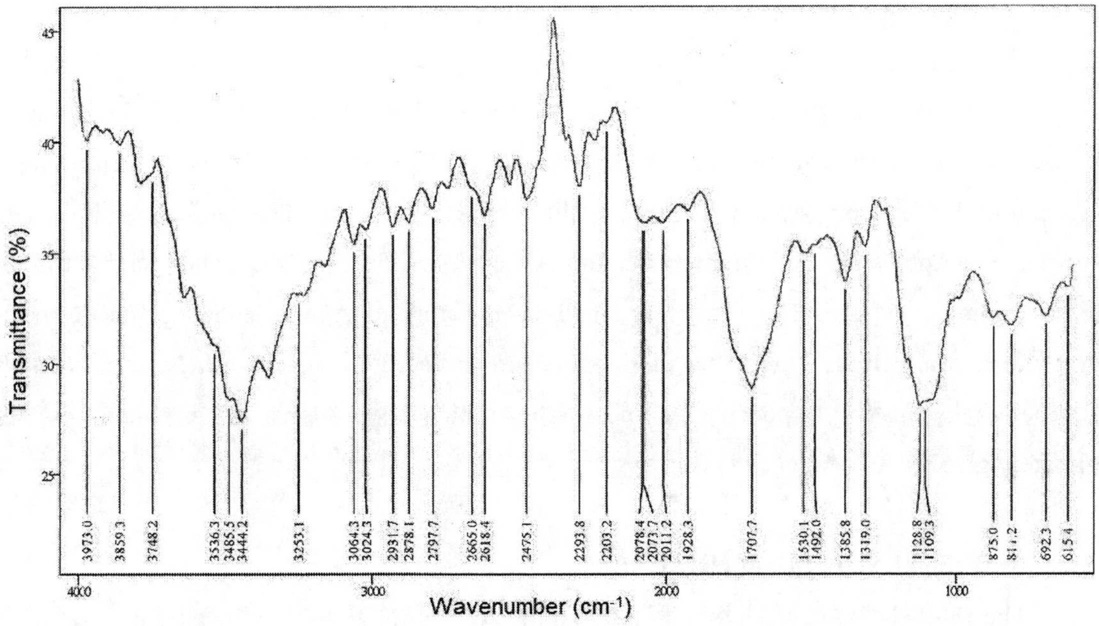

Figure 3. FTIR spectrum of Hs-AgNPs biosynthesized using the leaf extract of *H. suaveolens*

IOP Conf. Series: Materials Science and Engineering **805** (2020) 012042 doi:10.1088/1757-899X/805/1/012042

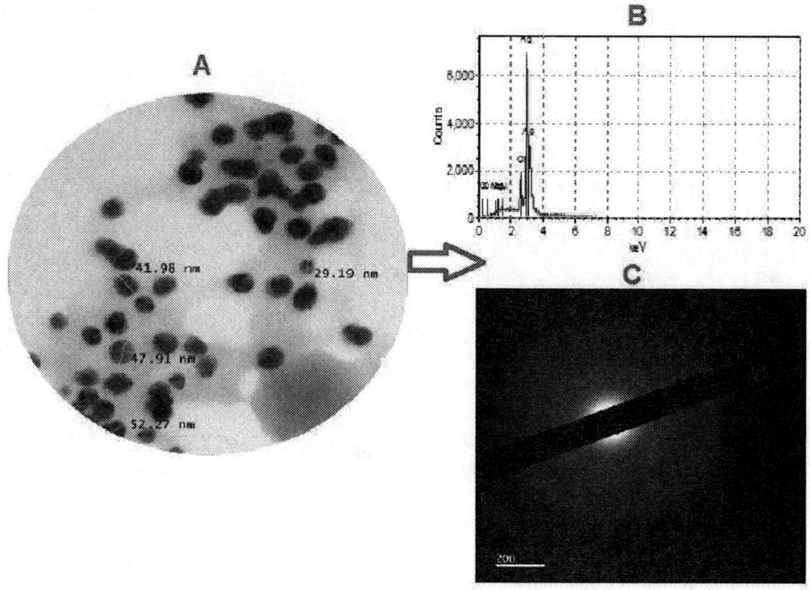

Figure 4. The TEM micrograph (A), EDX spectrum (B) and SAED pattern (C) of Hs-AgNPs biosynthesized by *H. suaveolens*

3.2 Hs-AgNPs action against MDR bacteria

The Hs-AgNPs effectively suppressed growth of MDR strains of *E. coli, S. aureus* and *K. pneumoniae* with percentage growth inhibition of 75.22-100 % (Figure 5a) at concentrations of 5-15 μg/ml, which conforms with some earlier reports [13, 46]. The antibacterial activities shown by Hs-AgNPs can be of tremendous importance in controlling MDR bacteria that thrive in the environment [54-56]. The extremely small-sized nature of AgNPs enhances its infiltration into the bacterial cell, and thereafter hinders metabolism activities through myriad of ways that may include by binding to phosphorus and sulphur molecules proteins, enzymes and DNA that ultimately give rise to death of the cell.

3.3 Antifungal Activities of Hs-AgNPs

The Hs-AgNPs greatly inhibited the fungi tested, although to varying extents (Figure 5b). While 100 % growth inhibition of *A. flavus* and *A. fumigatus* was recorded with Hs-AgNPs at concentration of 150 μg/ml, it was 78.57 % for *A. niger*. At 100 μg/ml of Hs-AgNPs, 100 %

inhibition of growth was also obtained for *A. fumigatus,* while it was 73.33 % for both *A. flavus* and *A. niger* at the same concentration. However, massive growths of fungi were observed in the control plates lacking Hs-AgNPs. These results revealed the potency of Hs-AgNPs against fungi of medical and agricultural importance, and thus can be used for their control in several biomedical and environmental applications. AgNPs have been variously reported to inhibit the growth of fungi [5, 47]. The ability of nanoparticles at inhibiting fungal growth is attributable to combination of assault on the fungal wall, and destruction of spores, that culminate into outflow of the constituents of cytoplasm that eventually cause the death of exposed fungi.

3.4 Antioxidant activities of Hs-AgNPs

3.4.1 Scavenging of DPPH

The synthesized Hs-AgNPs scavenged DPPH by 77.75-83.16 %, which was dose-independent at the tested concentrations of 5-40 μg/ml. In general, the free-radical scavenging potentials of AgNPs have been credited to the presence of reducing molecules that capped the nanoparticles thereby increasing the surface areas for antioxidant activity [49]. There are also reports of potentiation of improved antioxidant activities and phytomedicinal contents of plants treated with nanoparticles [57, 58]

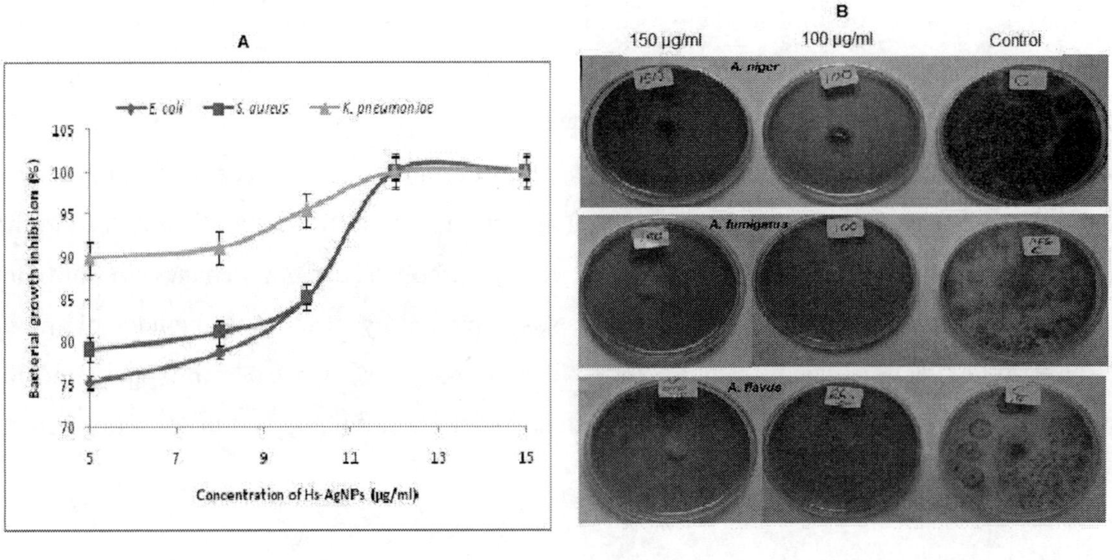

Figure 5. Antibacterial (A) and antifungal (B) activities of Hs-AgNPs biosynthesized using the leaf extract of *H. suaveolens*

IOP Conf. Series: Materials Science and Engineering **805** (2020) 012042 doi:10.1088/1757-899X/805/1/012042

3.4.2 Scavenging of H_2O_2

The phytosynthesized Hs-AgNPs scavenged H_2O_2 by 54.21-70.11 % with the spontaneous clearance of turbid solution of H_2O_2 which was prepared in phosphate buffer. Other researchers have reported H_2O_2 scavenging activities of 77.00-99.80 % for some biogenic AgNPs [49, 50, 59]. Human exposure to H_2O_2 is very soaring, because of its use as components of personal care products that include bleaching agents and disinfectants. It can also be encountered in industrial effluents, especially from paper manufacturing industry, where it is used to bleach paper and pulp. H_2O_2 is capable of initiating the production of strong reactive hydroxyl radicals ($OH^{\cdot}$) in living organisms [60] that may herald cellular injury and on-set of intestinal disorders, cancers, Alzheimer's disease, inflammation, aging amongst other debilitating conditions [61]. Environmental sources of H_2O_2 notably, drinking and wastewater could be treated with nano-based materials to effectively scavenge H_2O_2. Analytically, AgNPs-based sensors have been reported to detect as well as quantify H_2O_2 [62-64].

3.5 Hs-AgNPs as anticoagulant agent

The *in vitro* anticoagulation of human blood was ensured by Hs-AgNPs, simultaneously retaining the form of red blood cells as obtained in the fresh blood as well as the one collected in the EDTA bottle (Figure 6). Blood clot occurred in the control samples treated with extract of *H. suaveolens* and $AgNO_3$ solution due to their failure to prevent coagulation of blood. Some studies revealed that AgNPs possessed antiplatelet activity [7, 18, 65] and fibrinolytic activity [66], which did not permit blood coagulation. In addition, *in vitro* anticoagulation properties of AuNPs [20, 22], and Ag-AuNPs [7] have been recently reported. In the same way, AgNPs may inhibit fibrinogen to obstruct the formation of fibrin, thereby ensuring anticoagulation of blood. Additionally, anticoagulant activity of heparin was improved by 118.9 % through coupling with earthworm extract-mediated AuNPs [67]. The growing evidences imply that metallic nanoparticles could be potential sources of anticoagulant, thrombolytic and theranostic agents to manage blood coagulation illnesses [68-70].

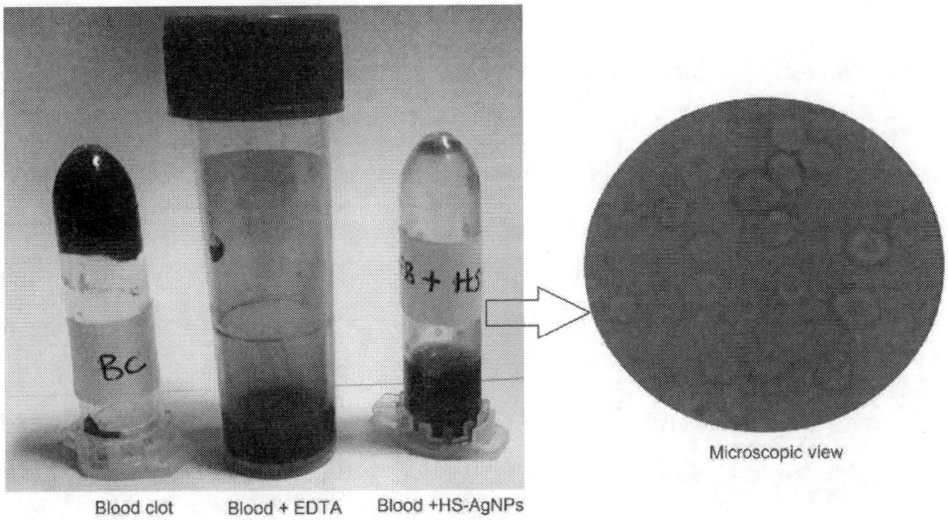

Figure 6. Anticoagulant activity of Hs-AgNPs biosynthesized using the leaf extract of *H. suaveolens*

4 Conclusion

This study has shown that leaf extract of *H. suaveolens* has the capacity to synthesize AgNPs in an eco-friendly manner. The synthesized Hs-AgNPs were spherical, sized 29.19-52.27 nm, and polydispersed in nature. Hs-AgNPs greatly inhibited MDR bacterial strains by 75.22-100 %, and equally displayed potent action against fungi (73.33-100 %), thus ascertaining the significance of Hs-AgNPs as antimicrobial agent. Correspondingly, Hs-AgNPs scavenged both DPPH and hydrogen peroxide with acceptable outcome, which can be deployed for scavenging free radicals in the environment. Hs-AgNPs also displayed potent blood anticoagulation property. Conclusively, the nanobiotechnological importance of *H. suaveolens,* most essentially, the cost-effective phytosynthesis of Hs-AgNPs has been established in this work. Until now, there is no report on antimicrobial, antioxidant and anticoagulant activities of AgNPs biofabricated by *H. suaveolens.*

Acknowledgment

The provision of some of the facilities used in this study by LAUTECH, Ogbomoso is duly acknowledged by AL.

IOP Conf. Series: Materials Science and Engineering **805** (2020) 012042 doi:10.1088/1757-899X/805/1/012042

References

[1] Emenike EE, Ojiefoh OC, Aleruchi C, Hassan LA, Christiana OM, Rebecca M, Dare EO and Temitope AE 2014 Evaluation of antibacterial activities of silver nanoparticles green-synthesized using pineapple leaf (*Ananas comosus*). *Micron* **57** pp 1-5.

[2] Dare EO, Oseghale CO, Labulo AH, Adesuji ET, Elemike EE, Onwuka JC and Bamgbose JT 2015 Green synthesis and growth kinetics of nanosilver under bio-diversified plant extracts influence. *J. Nanostruct. Chem.* **5** pp 85-94.

[3] Azeez L, Lateef A, Adebisi SA and Oyedeji AO 2018 Novel biosynthesized silver nanoparticles from cobweb as adsorbent for Rhodamine B: equilibrium isotherm, kinetic and thermodynamic studies. *Appl. Water. Sci.* **8**: 32. https://doi.org/10.1007/s13201-018-0676-z.

[4] Lateef A, Azeez MA, Asafa TB, Yekeen TA, Akinboro A, Oladipo IC, Ajetomobi FE, Gueguim-Kana EB and Beukes LS 2015 *Cola nitida*-mediated biogenic synthesis of silver nanoparticles using seed and seed shell extracts and evaluation of antibacterial activities. *Bionanosci.* **5** (4) pp 196-205.

[5] Lateef A, Azeez MA, Asafa TB, Yekeen TA, Akinboro A, Oladipo IC, Azeez L, Ajibade SE, Ojo SA, Gueguim-Kana EB, Beukes LS (2016) Biogenic synthesis of silver nanoparticles using a pod extract of *Cola nitida*: Antibacterial, antioxidant activities and application as a paint additive. *J. Taibah Univ. Sci.* **10** (4) pp 551-562.

[6] Lateef A, Azeez MA, Asafa TB, Yekeen TA, Akinboro A, Oladipo IC, Azeez L, Ojo SA, Gueguim-Kana EB and Beukes LS 2016 Cocoa pod husk extract-mediated biosynthesis of silver nanoparticles: its antimicrobial, antioxidant and larvicidal activities. *J. Nanostruct. Chem.* **6** (2) pp 159-169.

[7] Lateef A, Ojo SA, Folarin BI, Gueguim-Kana EB and Beukes LS (2016) Kolanut (*Cola nitida*) mediated synthesis of silver-gold alloy nanoparticles: antifungal, catalytic, larvicidal and thrombolytic applications. *J. Clust. Sci.* **27** (5) pp 1561-1577.

[8] Lateef A, Akande MA, Azeez MA, Ojo SA, Folarin BI, Gueguim-Kana EB and Beukes LS 2016 Phytosynthesis of silver nanoparticles (AgNPs) using miracle fruit plant (*Synsepalum dulcificum*) for antimicrobial, catalytic, anti-coagulant and thrombolytic applications. *Nanotechnol. Rev.* **5** (6) pp 507-520.

IOP Conf. Series: Materials Science and Engineering **805** (2020) 012042 doi:10.1088/1757-899X/805/1/012042

[9] Adelere IA and Lateef A 2016 A novel approach to the green synthesis of metallic nanoparticles: the use of agro-wastes, enzymes and pigments. *Nanotechnol. Rev.* **5** (6) pp 567-587.

[10] Adelere IA, Lateef A, Aboyeji DO, Abdulsalam R, Adabara NU and Bala JD 2017 Biosynthesis of silver nanoparticles using aqueous extract of *Buchholzia coriacea* (wonderful kola) seeds and their antimicrobial activities. *Annals Food Sci. Technol.* **18** pp 671-679.

[11] Azeez MA, Lateef A, Asafa TB, Yekeen TA, Akinboro A, Oladipo IC, Gueguim-Kana EB and Beukes LS 2017 Biomedical applications of cocoa bean extract-mediated silver nanoparticles as antimicrobial, larvicidal and anticoagulant agents. *J. Clust. Sci.* **28** (1) pp 149-164.

[12] Elemike EE, Onwudiwe DC, Fayemi OE, Ekenni, AC, Ebenso EE and Tiedt LR 2017 Biosynthesis, electrochemical, antimicrobial and antioxidant studies of silver nanoparticles mediated by *Talinum triangulare* aqueous leaf extract. *J. Clust. Sci.* **28** (1) pp 309-330.

[13] Lateef A, Folarin BI, Oladejo SM, Akinola PO, Beukes LS and Gueguim-Kana EB 2018 Characterization, antimicrobial, antioxidant and anticoagulant activities of silver nanoparticles synthesized from *Petiveria alliacea* L. leaf extract. *Prep. Biochem. Biotechnol.* **48** (7) pp 646-652.

[14] Lateef A and Adeeyo AO 2015 Green synthesis and antibacterial activities of silver nanoparticles using extracellular laccase of *Lentinus edodes. Not. Sci. Biol.* **7** (4) pp 405-411.

[15] Lateef A, Adelere IA, Gueguim-Kana EB, Asafa TB, and Beukes LS 2015 Green synthesis of silver nanoparticles using keratinase obtained from a strain of *Bacillus safensis* LAU 13. *Int. Nano Lett.* **5** (1) pp 29-35.

[16] Lateef A, Ojo SA, Azeez MA, Asafa TB, Yekeen TA, Akinboro A, Oladipo IC, Gueguim-Kana EB and Beukes LS 2016 Cobweb as novel biomaterial for the green and eco-friendly synthesis of silver nanoparticles. *Appl. Nanosci.* **6** (6) pp 863-874.

[17] Lateef A, Akande MA, Ojo SA, Folarin BI, Gueguim-Kana EB and Beukes LS 2016 Paper wasp nest-mediated biosynthesis of silver nanoparticles for antimicrobial, catalytic,

IOP Conf. Series: Materials Science and Engineering **805** (2020) 012042 doi:10.1088/1757-899X/805/1/012042

anticoagulant, and thrombolytic applications. *3Biotech* **6** 140. http://dx.doi.org/10.1007/s13205-016-0459-x.

[18] Lateef A, Ojo SA and Elegbede JA 2016 The emerging roles of arthropods and their metabolites in the green synthesis of metallic nanoparticles. *Nanotechnol. Rev.* 5 (6) pp 601-622.

[19] Lateef A, Ojo SA and Oladejo SM 2016 Anti-candida, anti-coagulant and thrombolytic activities of biosynthesized silver nanoparticles using cell-free extract of *Bacillus safensis* LAU 13. *Process Biochem.* **51** (10) pp 1406-1412.

[20] Ojo SA, Lateef A, Azeez MA, Oladejo SM, Akinwale AS, Asafa TB, Yekeen TA, Akinboro A, Oladipo IC, Gueguim-Kana EB and Beukes LS 2016 Biomedical and catalytic applications of gold and silver-gold alloy nanoparticles biosynthesized using cell-free extract of *Bacillus safensis* LAU 13: antifungal, dye degradation, anti-coagulant and thrombolytic activities. *IEEE Trans. NanoBiosci.* **15** (5) pp 433-442.

[21] Oladipo IC, Lateef A, Azeez MA, Asafa TB, Yekeen TA, Akinboro A, Akinwale AS, Gueguim-Kana EB and Beukes LS 2017 Green synthesis and antimicrobial activities of silver nanoparticles using cell-free extracts of *Enterococcus* species. *Not. Sci. Biol.* **9** (2): 196-203.

[22] Oladipo IC, Lateef A, Elegbede JA, Azeez MA, Asafa TB, Yekeen TA, Akinboro A, Gueguim-Kana EB, Beukes LS, Oluyide TO and Atanda OR 2017 *Enterococcus* species for the one-pot biofabrication of gold nanoparticles: characterization and nanobiotechnological applications. *J. Photochem. Photobiol. B: Biol.* **173** pp 250-257.

[23] Elegbede JA, Lateef A, Azeez MA, Asafa TB, Yekeen TA, Oladipo IC, Adebayo EA, Beukes LS and Gueguim-Kana EB 2018 Fungal xylanases-mediated synthesis of silver nanoparticles for catalytic and biomedical applications. *IET Nanobiotechnol.* **12** (6) pp 857-863.

[24] Salem WM, Haridy M, Sayed WF and Hassan NH 2014 Antibacterial activity of silver nanoparticles synthesized from latex and leaf extract of *Ficus sycomorus*. *Ind. Crops Prod.* **62** pp 228-234.

[25] Ahmed S, Ahmad M, Swami BL and Ikram S 2016 A review on plants extract mediated synthesis of silver nanoparticles for antimicrobial applications: a green expertise. *J. Adv. Res.* 7 (1) pp 17-28.

[26] Ahmadi O, Jafarizadeh-Malmiri H and Jodeiri N 2018 Eco-friendly microwave-enhanced green synthesis of silver nanoparticles using *Aloe vera* leaf extract and their physico-chemical and antibacterial studies. *Green Process. Synth.* 7 (3) pp 231-240.

[27] Ghanbari S, Vaghari H, Sayyar Z, Adibpour M and Jafarizadeh-Malmiri H 2018 Autoclave-assisted green synthesis of silver nanoparticles using *A. fumigatus* mycelia extract and the evaluation of their physico-chemical properties and antibacterial activity. *Green Process. Synth.* **7** (3) pp 217-224.

[28] Lateef A, Ojo SA, Akinwale AS, Azeez L, Gueguim-Kana EB and Beukes LS 2015 Biogenic synthesis of silver nanoparticles using cell-free extract of *Bacillus safensis* LAU 13: antimicrobial, free radical scavenging and larvicidal activities. *Biologia* **70** (10) pp 1295-1306.

[29] Augustine R, Kalarikkal N and Thomas S 2014 A facile and rapid method for the black pepper leaf mediated green synthesis of silver nanoparticles and the antimicrobial study. *Appl. Nanosci.* **4** (7) pp 809-818.

[30] Kumar B, Smita K, Cumbal L and Debut A 2014 Sacha inchi (*Plukenetia volubilis* L.) oil for one pot synthesis of silver nanocatalyst: An ecofriendly approach. *Ind. Crops Prod.* **58** pp 238-243.

[31] Nazeruddin GM, Prasad NR, Prasad SR, Shaikh YI, Waghmare SR and Adhyapak P (2014) *Coriandrum sativum* seed extract assisted *in situ* green synthesis of silver nanoparticle and its anti-microbial activity. *Ind. Crops Prod.* **60** pp 212-216.

[32] Singh J and Dhaliwal AS 2018 Novel green synthesis and characterization of the antioxidant activity of silver nanoparticles prepared from *Nepeta leucophylla* root extract. *Anal. Lett.* **52** (2) pp 213-230.

[33] Vijayan R, Joseph S and Mathew B 2019 Green synthesis of silver nanoparticles using *Nervalia zeylanica* leaf extract and evaluation of their antioxidant, catalytic, and antimicrobial potentials. *Particul. Sci. Technol.* **37** (7) pp. 809-819.

[34] Raizada P 2006 Ecological and vegetative characteristics of a potent invader, *Hyptis suaveolens* Poit. from India. *Lyonia* **11** (2) pp 115-120.

[35] Traore P 2009 Etude de l'activité antioxydante et de la toxicité de quatre plantes: *Calotropis procera, Centaurea perrottetii, Euphorbia sudanica, Hyptis suaveolens.* Université de Bamako, Doctorat en pharmacie, pp 37-38.

IOP Conf. Series: Materials Science and Engineering **805** (2020) 012042 doi:10.1088/1757-899X/805/1/012042

[36] Shirwaikar A, Shenoy R, Udupa AL, Udupa AL and Shetty S 2003 Wound healing property of ethanolic extract of leaves of *Hyptis suaveolens* with supportive role of antioxidant enzymes. *Indian J. Exp. Biol.* **41** pp 238-241.

[37] Gavani U and Paarakh PM 2008 Antioxidant activity of *Hyptis suaveolens* Poit. *Int. J. Pharmacol.* **4** (3) pp 227-229.

[38] Shenoy C, Patil MB and Kumar R 2009 Wound healing activity of *Hyptis suaveolens* (L.) Poit (Lamiaceae). *Int. J. Pharm. Technol. Res.* **1** (3) pp 737-744.

[39] Babalola OO, Ojo OE and Oloyede FA 2011 Hepatoprotective activity of aqueous extract of the leaves of *Hyptis suaveolens* (L.) Poit on acetaminophen induced hepatotoxicity in rabbits. *Res. J. Chem. Sci.* **1** pp 85-88.

[40] Ghaffari H, Ghassam BJ and Prakash HS 2012 Hepatoprotective and cytoprotective properties of *Hyptis suaveolens* against oxidative stress-induced damage by CCl_4 and H_2O_2. *Asian Pac. J. Trop. Med.* **5** (11) pp 868-874.

[41] Vera-Arzave C, Antonio LC, Arrieta J, Cruz-Hernández G, Velázquez-Méndez AM, Reyes-Ramírez A and Sánchez-Mendoza ME 2012 Gastroprotection of suaveolol, isolated from *Hyptis suaveolens*, against ethanol-induced gastric lesions in Wistar rats: role of prostaglandins, nitric oxide and sulfhydryls. *Mol.* **17** (8) pp 8917-8927.

[42] Mandal SM, Mondal KC, Dey S and Pati BR 2007 Antimicrobial activity of the leaf extracts of *Hyptis suaveolens* (L.) Poit. *Ind. J. Pharmaceut. Sci.* **69** (4) pp 568-569.

[43] Mozhiyarasi P and Anuradha R 2016 A study on phytochemical analysis and antimicrobial activity of *Hyptis suaveolens* (L.) Poit. *J. Chem. Pharmaceut. Res.* **8** (6) pp 438-442.

[44] Ghaffari H, Ghassam BJ, Nayaka SC, Kini KR and Prakash HS 2014 Antioxidant and neuroprotective activities of *Hyptis suaveolens* (L.) Poit. against oxidative stress-induced neurotoxicity. *Cellular Mol. Neurobiol.* **34** (3) pp 323-331.

[45] Elumalai D, Hemavathi M, Deepaa CV and Kaleena PK 2017 Evaluation of phytosynthesised silver nanoparticles from leaf extracts of *Leucas aspera* and *Hyptis suaveolens* and their larvicidal activity against malaria, dengue and filariasis vectors. *Parasite Epidemiol. Control* **2** (4) pp 15-26.

[46] Prasannaraj G and Venkatachalam P 2017 Enhanced antibacterial, anti-biofilm and antioxidant (ROS) activities of biomolecules engineered silver nanoparticles against

clinically isolated Gram positive and Gram negative microbial pathogens. *J. Clust. Sci.* **28** (1) pp 645-664.

[47] Elegbede JA, Lateef A, Azeez MA, Asafa TB, Yekeen TA, Oladipo IC, Abbas SH, Beukes LS and Gueguim-Kana EB 2019 Silver-gold alloy nanoparticles biofabricated by fungal xylanases exhibited potent biomedical and catalytic activities. *Biotechnol. Progress.* **35** e2829. https://doi.org/10.1002/btpr.2829.

[48] Williams BW, Curelier ME and Berset C 1995 Use of a free radical method to evaluate antioxidant activity. *LWT-Food Sci. Technol.* **28** (1) pp 25-30.

[49] Bhakya S, Muthukrishnan S, Sukumaran M and Muthukumar M 2015 Biogenic synthesis of silver nanoparticles and their antioxidant and antibacterial activity. *Appl. Nanosci.* **6** (5) pp 755-766.

[50] Lateef A, Ojo SA, Elegbede JA, Azeez MA, Yekeen TA and Akinboro A 2017 Evaluation of some biosynthesized silver nanoparticles for biomedical applications: hydrogen peroxide scavenging, anticoagulant and thrombolytic activities. *J. Clust. Sci.* **28** (3) pp 1379-1392.

[51] Aina DA, Owolo O, Lateef A, Aina FO, Abbas SH, Adeoye-Isijola M, Okon V, Asafa TB, Elegbede JA, Olukanni OD and Adediji I 2019 Biomedical applications of *Chasmanthera dependens* stem extract mediated silver nanoparticles as antimicrobial, antioxidant, anticoagulant, thrombolytic, and larvicidal agents. *Karbala Int. J. Modern Sci.* **5** (2) pp 71-80.

[52] Anandalakshmi K, Venugobal J and Ramasamy V 2015 Characterization of silver nanoparticles by green synthesis method using *Pedalium murex* leaf extract and their antibacterial activity. *Appl. Nanosci.* **6** (3) pp 399-408.

[53] Shankar S, Jaiswal L, Aparna RSL and Prasad RGSV 2014 Synthesis, characterization, *in vitro* biocompatibility, and antimicrobial activity of gold, silver and gold silver alloy nanoparticles prepared from *Lansium domesticum* fruit peel extract. *Mater. Lett.* **137**: 75-78.

[54] Lateef A, Oloke JK, Gueguim-Kana EB and Pacheco E 2006 The microbiological quality of ice used to cool drinks and foods in Ogbomoso metropolis, Southwest, Nigeria. *Internet J. Food Saf.* **8** pp 39-43.

[55] Lateef A, Davies TE, Adelekan A, Adelere IA, Adedeji AA and Fadahunsi AH 2010 Akara Ogbomoso: Microbiological examination and identification of hazards and critical control points. *Food Sci. Technol. Int.* **16** (5) pp 389-400.

[56] Zaki S, El-Kady MF and Abd-El-Haleem D 2011 Biosynthesis and structural characterization of silver nanoparticles from bacterial isolates. *Mater. Res. Bull.* **46** (10) pp 1571-1576.

[57] Azeez L, Adejumo AL, Lateef A, Adebisi SA, Adetoro RO, Adewuyi S, Tijani KO and Olaoye S 2019 Zero-valent silver nanoparticles attenuate Cd and Pb toxicities on *Moringa oleifera* via immobilization and induction of phytochemicals. *Plant Physiol. Biochem.* **139** pp 283-292.

[58] Azeez L, Lateef A and Adebisi SA 2017 Silver nanoparticles (AgNPs) biosynthesized using pod extract of *Cola nitida* enhances antioxidant activity and phytochemical composition of *Amaranthus caudatus* Linn. *Appl. Nanosci.* **7** (1-2) pp 59-66.

[59] Mata R, Nakkala JR and Sadras SR 2015 Catalytic and biological activities of green silver nanoparticles synthesized from *Plumeria alba* (frangipani) flower extract. *Mater. Sci. Eng. C* **51** pp 216-225.

[60] Gulcin I, Elmastas M and Aboul-Enein HY 2007 Determination of antioxidant and radical scavenging activity of Basil (*Ocimum basilicum* L. Family *Lamiaceae*) assayed by different methodologies. *Phytother. Res.* **21** (4) pp 354-361.

[61] Chang HF and Yang LL 2012 Radical-scavenging and rat liver mitochondria lipid peroxidative inhibitory effects of natural flavonoids from traditional medicinal herbs. *J. Med. Plants Res.* **6** (6) pp 997-1006.

[62] Nia PM, Meng WP and Alias Y 2015 Hydrogen peroxide sensor: uniformly decorated silver nanoparticles on polypyrrole for wide detection range. *Appl. Surf. Sci.* **357** pp 1565-1572.

[63] Tian Y, Liu Y, Wang W, Zhang X and Peng W 2015 Sulfur-doped graphene-supported Ag nanoparticles for nonenzymatic hydrogen peroxide detection. *J. Nanopart. Res.* **17** (4) pp 193. https://doi.org/10.1007/s11051-015-2976-7.

[64] Wang L, Ma S, Yang B, Cao W and Han X 2015 Morphology-controlled synthesis of Ag nanoparticle decorated poly (o-phenylenediamine) using microfluidics and its application for hydrogen peroxide detection. *Chem. Eng. J.* **268**: 102-108.

[65] Shrivastava S, Bera T, Singh SK, Singh G, Ramachandrarao P and Dash D 2009 Characterization of antiplatelet properties of silver nanoparticles. *Am. Chem. Soc. Nano* **3** (6) 1357-1364.

[66] Harish BS, Uppuluri KB and Anbazhagan V 2015 Synthesis of fibrinolytic active silver nanoparticle using wheat bran xylan as a reducing and stabilizing agent. *Carbohydr. Polym.* **132** pp 104-110.

[67] Kim HK, Choi MJ, Cha SH, Koo YK, Jun SH, Cho S and Park Y 2013 Earthworm extracts utilized in the green synthesis of gold nanoparticles capable of reinforcing the anticoagulant activities of heparin. *Nanoscale Res. Lett.* **8** 542. https://doi.org/10.1186/1556-276X-8-542.

[68] Lateef A, Ojo SA, Elegbede JA, Akinola PO and Akanni EO 2018 Nanomedical applications of nanoparticles for blood coagulation disorders. In: *Environmental Nanotechnology,* Volume 1. Eds: Dasgupta N, Ranjan S, Lichtfouse E. https://doi.org/10.1007/978-3-319-76090-2_8. Springer International Publishing AG, Cham, Switzerland. ISBN 978-3-319-76089-6. Pp. 243-277.

[69] Lateef A, Elegbede JA, Akinola PO and Ajayi VA 2019 Biomedical applications of green synthesized-metallic nanoparticles: a review. *Pan Afr. J. Life Sci.* **3** pp 157-182.

[70] Elegbede JA and Lateef A 2019 Green nanotechnology in Nigeria: The research landscape, challenges and prospects. *Annals Sci. Technol.* **4** (2) pp 6-38.

Green synthesized novel silver nanoparticles and their application as anticoagulant and thrombolytic agents: A perspective

M A Azeez[1,6*], F A Durodola[1], A Lateef[2,6], T A Yekeen[1,6], A O Adubi[1], I C.Oladipo[3,6], E A Adebayo[4,6] J A Badmus[5,6] and A O Abawulem[1]

[1]Laboratory of Ecotoxicology, Genetics and Nanobiotechnology, [2]Laboratory of Industrial Microbiology and Nanobiotechnology, Department of Pure and Applied Biology, [3]Department of Science Laboratory Technology, [4]Department of Pure and Applied Biology, [5]Department of Biochemistry, Ladoke Akintola University of Technology, PMB 4000, Ogbomoso, Nigeria

[6]Nanotechnology Research Group (*NANO*[+])

*Corresponding author: maazeez@lautech.edu.ng

Abstract

Effective management of haemostatic disorders in patients is pertinent in order to prevent complications attributable to incidences of thrombosis in atherosclerotic arteries to the heart, brain, limbs, stagnant blood in veins and heart chambers. The use of existing chemical agents are not without a number of challenges, varying from potential for unknown long-term adverse effect through lack of antidote for most agents, balance between cost and efficacy among others. However, advent of biogenic silver nanoparticles with anticoagulant and thrombolytic potentials has opened up a window of opportunity to address most of the shortcomings of the chemical synthesized anticoagulants. This review examined green synthesized novel silver nanoparticles, their characterization and possible application as anticoagulant and thrombolytic agents in the evolving area of nanomedicine.

Keywords: Silver nanoparticles, anticoagulant, thrombolysis, antiplatelet, green synthesis

1. Introduction

Nanotechnology is an emerging field of synthetic chemistry that is making tremendous impact in all aspects of human life. It is the technology of the future and has revolutionized all fields, ranging from medicines to agriculture, through environment to electronics by providing abilities that would never have been previously dreamt of [1]. Nanotechnology has endowed humans with the capacity to understand, control, design and manipulate matter with the dexterity

IOP Conf. Series: Materials Science and Engineering **805** (2020) 012043 doi:10.1088/1757-899X/805/1/012043

of a craftsman at the simplest level of atoms and molecules. It is a technology that enables working with atoms and molecules that are sized on a nanometer (nm) scale with dimensions less than 100nm-these are the nanoparticles. The particles are being synthesized using different physical techniques such as lithography, laser ablation, sputtering deposition, pulsed electrochemical etching, vapor deposition, laser pyrolysis, plasma or flame spraying synthesis and chemical processes (sol gel processing). Despite the increasing awareness in production and use of nanoparticles in many consumer based products, there is still growing interest in the biological and environmental safety of their production. These techniques suffer from a lot of setbacks including use of toxic solvents that are potentially hazardous to the environment and living organism [2,3], being labor intensive (consuming high energy) and very expensive [4,5]. The disadvantages of widely used physical and chemical processes have given rise to an alternative highly efficient, cost effective and environmentally safe method of nanoparticles production; green synthesis.

2. Green Synthesis of silver nanoparticles and its mechanisms

Green synthesis is the art of synthesizing nanoparticles using biological systems which may include whole organisms (microbes, plants and animals) or their metabolites; phytochemicals, protein and their derivatives; enzymes and templates like DNA or even membranes. Over the past decades, a lot of studies have demonstrated the synthesis of metal nanoparticles using different biological systems such as bacteria [6-8], fungi [9,10], plants and their metabolites [2-5,11-13], insects and their metabolites [14-17]; cobweb [15], honey [16] and earthworm [18] to mention a few. A detailed analysis of different methods of green synthesis, biological systems used and the special attributes of the synthesized particles have been compiled [14].

Green synthesis of silver nanoparticles (AgNPs) usually involves suitable source of biological system that have potential to elicit production of bioreductants, which manifest in form of enzymes or phytochemicals. Such bioreductants are known to be electron rich and therefore initiate the reduction of metal ions in solution to its nanoparticles (NPs). The reaction normally occur at ambient condition and the mix ratio of metal salt solution relative to the bioreductant in most cases is such that small quantity of bioreductant is added to large volume of

IOP Conf. Series: Materials Science and Engineering **805** (2020) 012043 doi:10.1088/1757-899X/805/1/012043

metal salt solution of low concentration (1×10^{-3} M) to initiate and effect subsequent formation of nanoparticles. For instance, in green synthesis of AgNPs, the ratio of bioreductant to metal salt solution ($AgNO_3$) by most researchers is usually around 1:40 [13] or 1:10 [19] with little variation. The process usually occur in two stages; (1) reduction stage in which the metal ions (Ag^+) are reduced to their nanoparticles (Ag^0) and (2) nucleation and stabilization stage in which each AgNP act as a nucleus surrounded by bioreductants to stabilize the AgNPs formed (Figure 1). By this later stage, the biogenic AgNPs synthesized can stay in storage without aggregation.

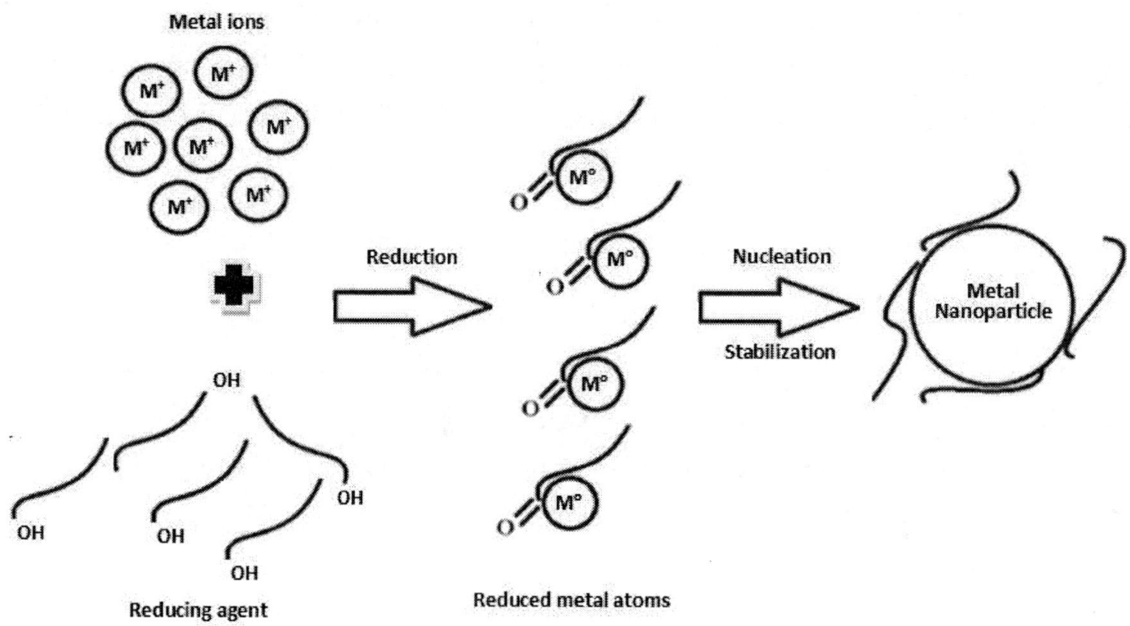

Figure 1. Mechanisms of metal ions bioreduction to nanoparticles [20]

When whole organisms (i.e. bacteria and fungi) are used in nanoparticles biosynthesis, the processes of metal ion bioreduction either occur outside the cell (extracellular accumulation of nanoparticles) or within the cell (cytoplasmic bioaccumulation of nanoparticles). In an extracellular accumulation of nanoparticles, membrane enzymes such as oxidoreductase or extracellular metabolites are produced which reduce metal ions to nanoparticles. The

IOP Conf. Series: Materials Science and Engineering **805** (2020) 012043 doi:10.1088/1757-899X/805/1/012043

nanoparticles produced are then obtained by simple purification. However, in cytoplasmic bioaccumulation of nanoparticles, enzymes or metabolites in the electron carrier system act as bioreductants that reduce metal ions to their nanoparticles (Figure 2). Such nanoparticles are then obtained by lyzing the cell either mechanically or otherwise to harvest the nanoparticles produced.

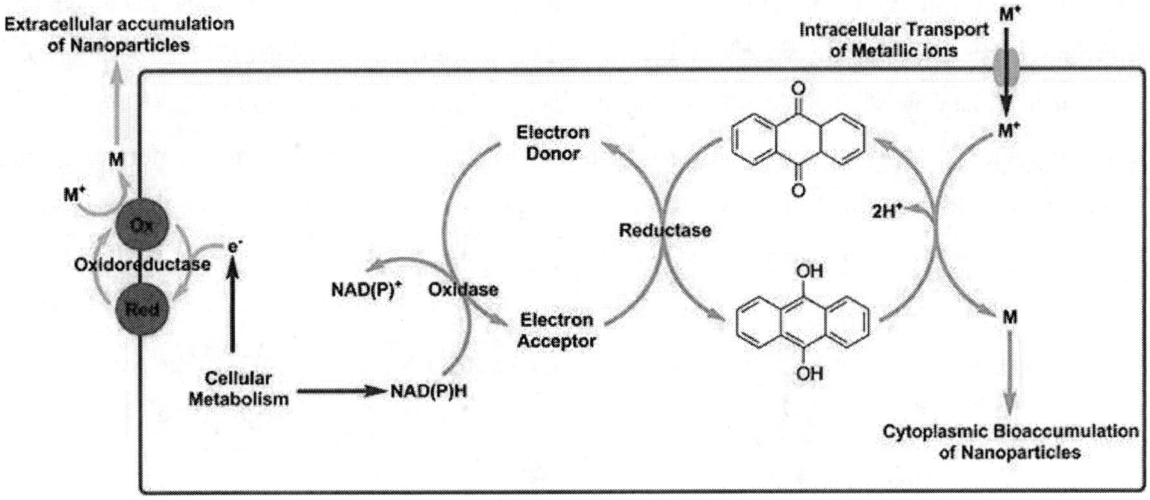

Figure 2. Illustration of whole organism biosynthesis of metal nanoparticles

Many animal metabolites such as derivatives of protein, starch, fat and oil are known to be rich sources of bioreductants. In addition, many plant extracts are known to contain phytochemicals such as alkaloids and phenolic compounds of different classes that also serve as rich sources of bioreductants. These biomolecules in their crude forms or when purified have potential to reduce metal ions to their nanoparticles. The roles of these biomolecules are both as reducing and stabilizing agents. These mechanisms of green synthesis of nanoparticles are favoured over both physical and chemical routes due to their cheap, less time and energy demanding, non-toxic and ecofriendly nature. Aside, green synthesis that makes use of plant materials is the most preferred because an elaborate system of cell culture and maintenance is not required.

IOP Conf. Series: Materials Science and Engineering **805** (2020) 012043 doi:10.1088/1757-899X/805/1/012043

In the green synthesis of metal nanoparticles, the shape, size, optical, catalytic and biological properties of the synthesized nanoparticles turn to depend on a number of factors which include reactant concentrations, mix ratio, pH, reaction kinetics, solution chemistry, and interaction time among others. All these can be varied and manipulated with a view to synthesis nanoparticles of various sizes and forms for various purposes. For instance, when nanoparticles with size and structural uniformity are synthesized, they are said to be monodispersed and isotropic respectively, otherwise, they are polydispersed and anisotropic. Biosynthesized nanoparticles can be engineered by varying the reaction conditions to produce well-define and uniform nanostructures as the situation demands (Figure 3). This also may offer an ample opportunity to scale up the bioprocess of metal nanoparticles production for wide applications in various fields of human endeavours.

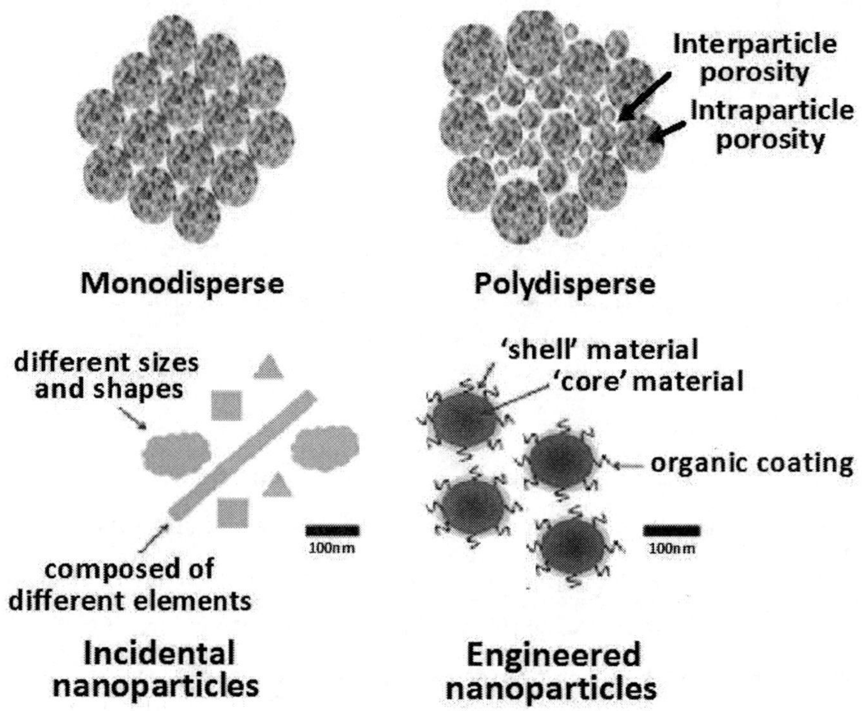

Figure 3. Different forms and size of nanoparticles

3. Characterization of biogenic metal nanoparticles

Most green synthesized nanoparticles especially AgNPs are usually accompanied by colour change when initiated. This colour change varies from light yellow to pinkish through ruby red, faint blue to purple and finally to dark brown when it finally stabilized, which is mostly reported in literature [15, 21, 22, 23, 24, 25, 26] depending on shape and size of the particles. The excitation of the reduced particles is greatly influenced by the nature of the bioreductant molecules, the consequences of which lead to manifestation of different shades of colour. The synthesis of green nanoparticles is monitored with UV-visible spectrophotometer to investigate the plasma resonance band for metal nanoparticles at λmax in the range of 200-800 nm (Figure 4). For AgNPs, the display of surface plasmon resonance (SPR) band is usually at 390-460 nm [27, 28]. This is followed with Fourier Transform-Infrared (FTIR) spectroscopy to identify the specific functional groups present in the synthesized metal nanoparticles in the range of 400-4000 cm^{-1} that might have played active roles in the reduction and capping of the nanoparticles. Scanning electron microscopy (SEM) is employed to study and unravel the surface topography of the particles usually expressed as function of shape and size (i.e. diameter) while energy dispersive spectroscopy is used to analyze the elemental constituents of the nanoparticles as facet of face-centered cubic lattice, while the transmission electron microscopy (TEM) analysis unravel the size and shape of the nanoparticles with selected area electron diffraction (SAED) to assess and discover the nature of the nanoparticles. X-Ray diffraction (XRD) analysis helps to confirm the crystalline nature of the nanoparticles.

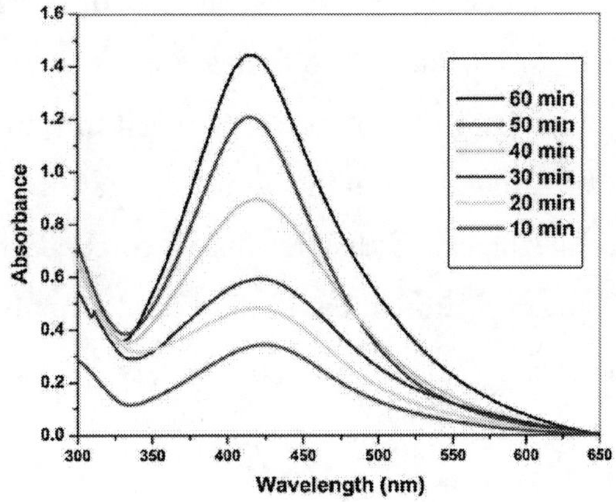

Figure 4. Typical UV-vis spectra for green synthesized AgNPs

IOP Conf. Series: Materials Science and Engineering **805** (2020) 012043 doi:10.1088/1757-899X/805/1/012043

4. Homeostatic disorder in human

Blood clotting is the nature's antidote to excessive bleeding. However, this portends a serious danger when such clot remains longer than necessary and obstructs blood flow in vessels supplying organs and tissues, resulting in clinical situations such as heart attacks and stroke. The formation of thrombus (a blood clot) in the circulatory system is part of the mechanism in human body that consolidates the repair of the injured or damaged blood vessel [29]. Consequent upon thrombus formation within blood vessels, the following clinical situations like embolism, ischemia, heart attack, stroke, and array of others may result [30].

Embolism is a phenomenon that precipitates blood clot formation in the circulatory system either within a vein or an artery, which may fully or partially impede free flow of blood to some parts of the body or body organs depending on its location. In such instance, a serious clinical situation may manifest with far reaching effects. For instance, occurrence of pulmonary embolism in which some arteries serving the lungs are blocked at the instance of embolus formation leads to inexplicable breathing difficulty, hemoptysis, and pain in and around the chest [31]. A blood clot that impedes free flow of blood or supply of oxygen to body tissues results in ischemia. A fully or partially restricted flow of blood to cardiac muscle otherwise called cardiac ischemia results in shortness of breath, syncope, angina, myocardial infarction, cardiac arrhythmia, or even death [32].

In addition, formation of blood clot in vessels innervating the brain tissues leads to an ischemic stroke as a consequence of restricted regular supply of blood to the brain [33]. Occurrence of ischemic stroke may be a direct consequences of restricted supply of blood to the brain (thrombolic stroke) due to blockage of blood vessel or alternatively, embolus produced elsewhere in the body from incidence of clot may be carried to some small arteries in the brain (embolic stroke). In some cases, blood clot initiated in the heart may find its way to the narrow arteries of the brain and create blockage that results in cerebral stroke. In whichever way it occurs, formation of blood clot in or around the blood vessels supplying the brain directly leads to starvation of brain tissue (i.e. of respiratory substrate), the consequences of which result in irreversible brain cell damage and death [34].

IOP Conf. Series: Materials Science and Engineering **805** (2020) 012043 doi:10.1088/1757-899X/805/1/012043

5. Anticoagulant and thrombolytic agents

Anticoagulants are agents that prevent and inhibit blood clot formation, whereas antiplatelet and thrombolytic agents are drugs used in reducing the incidence of blood clots or prevent existing clots from getting larger in the body. They are also called blood thinners. Drugs that interfere with blood coagulation (anticoagulants) are a main stay of cardiovascular therapy, of which atrial fibrillation, venous thromboembolism and artificial heart valves are the main indications for anticoagulation. An individual that is at risk of thrombotic condition is said to be prone to thromboembolism for each condition if not on anticoagulant therapy. Antithrombotic therapy is effectively used to manage atrial fibrillation closely connected to an increased risk of stroke and thromboembolic complications [35].

All over the world, venous thromboembolism (VTE) is the major cause of cardiovascular mortality, next in rank to myocardial infarction and stroke [36, 37]. This clinical situation affects patients in various settings and age groups, including the young children [38, 39]. The most prevalent manifestation of VTE is deep vein thrombosis (DVT), the expression of which may manifest in its most severe form known as acute pulmonary thromboembolism (PTE) [40]. Major treatment for both situations involves administration of full anticoagulation so as to minimize recurrence of VTE.

Over the years, it is a well-known and established fact that anticoagulation directly impacts or affects the mortality associated with VTE. The use of unfractionated or low-molecular-weight heparin and vitamin K antagonists, especially warfarin in the management of VTE was embraced and in use long before now. The traditional anticoagulants which were effective in the treatment of VTE have been found to be enshrouded with practical difficulties, necessitating the development and search for new classes of anticoagulants [40]. Two groups of new oral anticoagulant drugs such as factor Xa inhibitors (rivaroxaban, apixaban, and edoxaban) and direct thrombin inhibitors (dabigatran) have been reviewed for their efficacy and safety compared to the conventional or classical treatment using heparin and warfarin [40]. Despite their widespread use, they are still not without a number of shortcomings, prompting the need for development of an unprecedented number of new agents.

The classification, mechanism of action, route of administration and the attendant side effects or limitations of the various anticoagulants, antiplatelets and thrombolytics produced

IOP Conf. Series: Materials Science and Engineering **805** (2020) 012043 doi:10.1088/1757-899X/805/1/012043

synthetically were also reviewed [41, 42]. The side effects of the two drugs include bleeding (which may result to death if not managed properly), hemorrhage, osteoporosis, drug food reactions, drug induced hepatitis, alopecia, rashes and highly expensive cost. Traditional antithrombotic treatments like streptokinase also have limitations in their suitability for application. These may include short half-life, inactivation of the foreign agents by antibodies and possible incidence of excessive bleeding which may be life threatening. Nevertheless, application of liposomes that can resist degradation as carrier for streptokinase has greatly improved its thrombolytic activity [43]. Despite widespread use and efficiency, these anticoagulants still pose a threat to life and the problems besetting their administrations are still significant and are not easily discountenanced. Thus, there is need for development and design of new agents that are of biological origin, safe, ecofriendly with little or no side effects. The herald of nanotechnology has opened a window of opportunity for providing solution to these problems. This involves a process by which nontoxic and biocompatible nanoparticles are used as either stand-alone thrombolytic agents or carriers of active drugs [13].

6. Biogenic silver nanoparticles as anticoagulant and thrombolytic agents

Numerous and diverse metallic nanoparticles have been investigated, out of which AgNPs occupy prime position premised on their utility in several applications, which encompasses optical, electronic, catalytic, antimicrobial and electrochemical attributes. These have tremendously influenced and enhanced their relevance in vast areas of human endeavours such as food, healthcare, agriculture, biomedical, environmental, textile and catalytic applications [20]. Array of applications of AgNPs has remained endless as the improved compatibility with biological systems through avoidance of hazardous procedures and precursors are sought [13]. Of these new applications is an evolving and fast developing area of nanomedicine where biogenic AgNPs are employed and used to manage blood coagulation disorder.

It is clearly evident that nanoparticles can impart varying degree of influence on blood coagulation system, an attribute associated with their size, charge, shape and composition. They are also employed as carrier platform to redesign traditional drugs such as heparin with a potential to achieve reduced dosage, increased drug stability, shelf-life and overall cost benefit delivery of treatment [44]. Most green synthesized AgNPs generally display surface plasmon band at 390 - 460 [10, 45], they are mostly crystalline, fairly spherical and polydispersed with size range of 4 - 95 nm, while very few are uniformly spherical and monodispersed with size range of 5-85 nm [19, 46, 47] (Table 1). Nanoparticles of gold [48] and iron oxides [49] have demonstrated the potential to serve as carriers of active drugs to reduce the limitations of short half-life, prohibitive cost, incidences of bleeding and hemorrhagic stroke as consequences of preventing blood coagulation. In the same vein, AgNPs have manifested antiplatelet activity to prevent aggregation of platelets [50] thereby inhibiting blood coagulation.

Some nanoparticles synthesized from various sources of biological materials and employed as both reducing and stabilizing agents (Table 1) have exhibited potential to serve as anticoagulants. Kim *et al.* [18] reported an improvement of 118.9% in the anticoagulant activities of heparin coupled with AuNPs biosynthesized using earthworm extract. Green synthesized gold and silver-gold alloy nanoparticles using cell-free extracts of *Bacillus safensis* LAU13 have been shown to exhibit anticoagulant and thrombolytic properties [6], while Kolanut extract-mediated silver-gold alloy nanoparticles have demonstrated thrombolytic potential [12]. In addition, AgNPs synthesized by the use of cell-free extracts [51], Cobweb, Kolanut pod, seed and seed shell extracts [45] also exhibited anticoagulants and thrombolytic activities.

Furthermore, paper wasp nest extract-mediated biosynthesized AgNPs manifested anticoagulant and thrombolytic activities [52] in which the preformed blood clots were dissolved instantaneously (Figure 5) whereas the negative control silver nitrate treated and nest extract-treated blood clots did not lead to lyses of the blood clots. In another study, cocoa bean extract-mediated AgNPs (CBE-AgNPs) was found to impede the coagulation of human blood *in vitro* without any distortion to biological properties and essential attributes of red blood cells (Figure 6) [2]. Fungal xylanases-mediated silver nanoparticles (NEA-AgNPs and TEA-AgNPs) were reported to inhibit clotting of fresh blood from healthy human donor while they also initiate dissolution of clot within 5 min when added to preformed blood clot [10]. Dakshayani *et al.* [19] reported antiplatelet activity along with anticoagulant effect of *Selaginella bryopteris* plant leaf extract-mediated AgNPs and inferred the usefulness of the nanoparticles in thrombus disorder treatment, while the work of Raja *et al.* [46] further confirmed the utility of biogenic AgNPs as

IOP Conf. Series: Materials Science and Engineering **805** (2020) 012043 doi:10.1088/1757-899X/805/1/012043

Table 1. Some green synthesized nanoparticles with anticoagulant and thrombolytic properties

Biological resource	Particles synthesized	Organisms or part used	Size and special features	Reference
Bacteria; *Bacillus safensis*	Ag-Au	Cell-free extract	fairly spherical; 5-95 nm	[6]
Bacteria; *B. safensis*	Ag	Cell-free extract	Fairly spherical; 5-95 nm	[45]
Fungi; *Aspergillus niger* L3 and *Trichoderma longibrachiatum* L2	Ag, Au & Ag-Au	Crude xylanases extract	Spherical, cylindrical to oval; 15.21-77.49 nm for AgNPs, spherical and flower shaped; 4.88-123.99 nm for AuNPs, spherical, oval to irregular; 6.98-52-51 nm	[10, 57, 58]
Insect; *Polistes* sp	Ag	Paper wasp nest	crystalline polydisperse, 12.5- 95.55 nm, Anisotropic structure of sphere, triangle, hexagon, rod and rhombus	[14]
Insect: Spider	Ag	Cobweb;	Nearly spherical, polydispersed with size ranging 3-50 nm,	[15]
Earthworm; *Eisenia andrei*	Au	Cell-free extract	6.13 nm in size, spherical, cubic and block shaped	[18]
Plant; *Synsepalum dulcificum*	Ag	Leaf extract	5-22 nm well dispersed	[13]
S. dulcificum	Ag	Seed extract	4- 26 nm, spherical and crystalline in nature with little agglomeration	[13]
Cola nitida	Ag-Au	Leaf extract	polydispersed, near spherical Size range 17-90 nm	[12]
Cola nitida	Ag-Au	Seed extract	polydispersed, near spherical 17-90 nm	[12]
Theobroma cacao	Ag	Seed extract	8.96-54.22 nm, well dispersed and spherical	[2]
Cola nitida	Ag-Au	Seed shell	polydispersed, near spherical 17-90 nm	[6]
	Ag-Au	Pod extract	anisotropic mixture of near spherical, hexagonal, rod and triangular particles with size 12-91 nm	[6]
Plant: *Cola nitida*	Ag	Pod, Seed and Seed shell	Spherical, polydispersed with size 12-80 nm, 8-50 nm and 5-40 nm respectively	[12, 13]
Selaginella bryopteris	Ag	Leaf extract	Crystalline, fairly spherical and monodispersed, uniform distribution with size 5-10 nm	[19]
Peltophorum pterocarpum	Ag	Pods	Crystalline, uniformly spherical shape with size 70-85 nm	[46]
Petiveria alliacea	Ag	Leaf extract	Crystalline monodispersed with size range 16.70-33.74nm	[47]
Chasmanthera dependens	Ag	Stem extract	Cubic; 24.53-92.38 nm	[59]

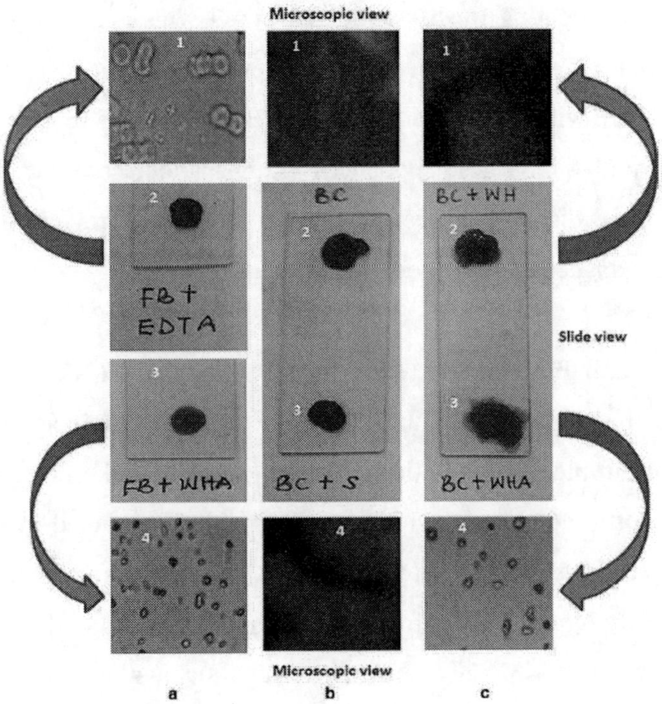

Figure 5. Anticoagulant (a), and thrombolytic activities (b, c) of biosynthesized AgNPs (*FB;* Fresh blood, *EDTA;* Anticoagulant, *WHA;* Wasp nest extract AgNPs, *BC;* Blood clot, *S;* $AgNO_3$ solution, *WH;* Wasp nest extract only) [48]

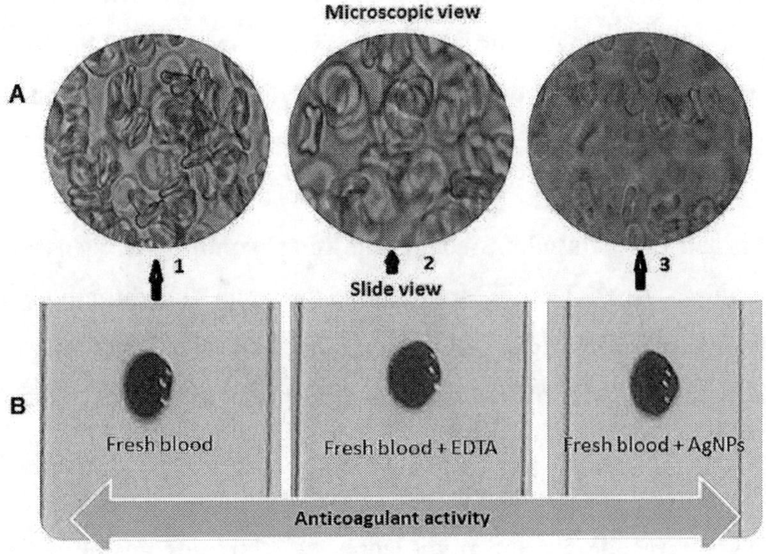

Figure 6. Anticoagulant activity of CBE-AgNPs [2]

IOP Conf. Series: Materials Science and Engineering **805** (2020) 012043 doi:10.1088/1757-899X/805/1/012043

anticoagulating agent by treating freshly collected blood with *Peltophorum pterocarpum* plant pods mediated AgNPs. Most of cited studies suggested nontoxic nature of green AgNPs to platelets and this coupled with established actions against microbes represent a paradigm shift in preventing blood coagulation. These evidences further underscore the potential of biogenic AgNPs to manage blood disorder. Lateef *et al.* [53] previously reviewed the contributions of nanoparticles for the management of blood coagulation disorders.

7.0 Proposed mechanism of action for thrombolysis by green AgNPs

AgNPs may be capable of inhibiting fibrinogen to prevent fibrin formation and consequently prohibiting blood coagulation. The whole events take place in a complex system involving series of clotting factors in which AgNPs might have acted in such a way to inhibit factors II, VII and X to prevent thrombin formation [2]. However, this is not yet fully understood but notwithstanding, the combined antiplatelet, fibrinolytic and possible fibrinogen-inhibiting activities undoubtedly play an active role in the anticoagulant activity of AgNPs. Two mechanisms were proposed: (1) that AgNPs probably act on plasminogen resulting in its activation to evoke the release of plasmin which then acts on the blood clot and breaks it or (2) that AgNPs may directly act on fibrin and cause it to break down as presented by Harish *et al.* [54]. In addition to mechanism 1, the nanoparticles may inhibit the activities of the inhibitors such that activation of plasminogen and plasmin may be prevented. The two mechanisms may also occur concurrently leading to pronounced thrombolytic activities as obtained in most studies (Figure 7).

8. Challenges of green silver based anticoagulant and thrombolytic agents

Biogenic AgNPs have clearly exhibited anticoagulant and thrombolytic activities with potential for biomedical applications. Of importance is the application of biologically synthesized AgNPs in this area of healthcare delivery where nontoxic and biocompatibility features are most desirable. However, some studies have highlighted the possibility of some green synthesized AgNPs not to be completely nontoxic and biocompatible as they have demonstrated cytotoxic, genotoxic and mito-depressive activities on the cells of *Allium cepa* roots [55, 56]. This safety concern and other number of hurdles have to be overcome before biogenic silver drug can be deplored for use in healthcare system. Most investigations on toxicity and compatibility were conducted *in vitro*, thus implying the need to explore *in vivo* investigations on potential biosynthesized AgNPs-based anticoagulants. In addition, the biochemical interaction with body cells and final fate of the metallic silver after the anticoagulation and thrombolytic events are yet to be fully understood. Moreover, the chemistry of action or pharmacokinetics of biogenic AgNPs should also be unraveled. Successful execution of these studies should be followed by design of potential AgNPs-based drugs for clinical trials. Nevertheless, the existing evidences here would be needed for effective and safe use of biogenic AgNPs as anticoagulant and thrombolytic agents in the immediate future.

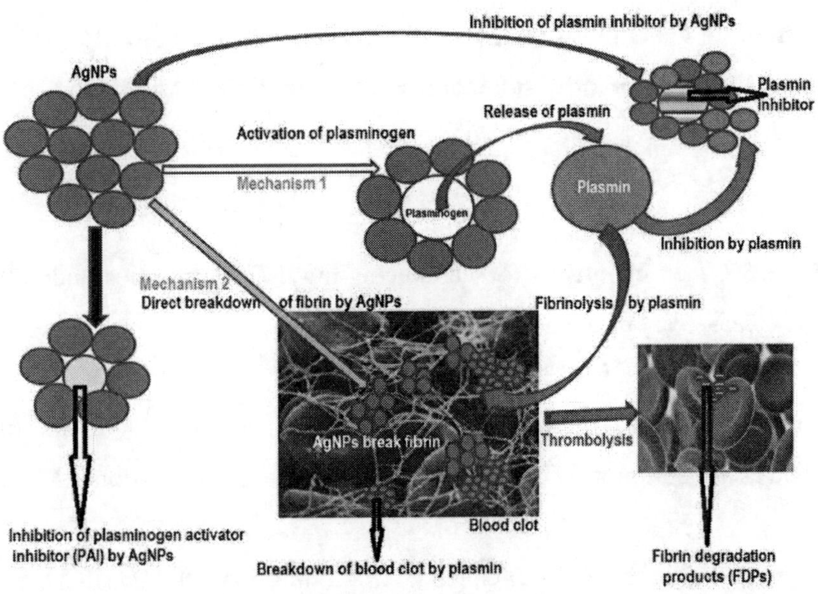

Figure 7. The proposed mechanisms for thrombolytic activities of biogenic AgNPs [13]

9. Conclusion

There are clear and unambiguous evidences that AgNPs have anticoagulant and thrombolytic potentials. Biosynthesized AgNPs have also demonstrated not only anticoagulant and thrombolytic potentials but also possibility of suitability in healthcare delivery because of implied nontoxic and biocompatible nature. Furthermore, probable modes of action of these nanoparticles have been proposed. However, the anticoagulant and thrombolytic potentials of biogenic nanoparticles are still underexploited due to limited reports in literature. There is therefore need for more intensive research effort on biologically synthesized metal nanoparticles with the view of exploring their application in biomedicine, most especially in the management and subsequent treatment of blood disorders. Also, the mechanism of action of these bio-nano agents coupled with their pharmacokinetics should be unraveled and well established. This becomes very important in a bid to ascertain their safe administration in healthcare delivery compare to the chemically synthesized anticoagulant and thrombolytic agents.

Acknowledgement

Authors are grateful to the authority of LAUTECH, Ogbomoso, Nigeria for provision of facility for the study and DBT-TWAS for Postdoctoral opportunity to MAA in India.

References

[1] Preeti NJ 2016 Green chemistry for nanotechnology: Opportunities and future challenges. *Res. Rev. Chem.* **5** (1) pp 2319-9849.

[2] Azeez MA, Lateef A, Asafa TB, Yekeen TA, Akinboro A, Oladipo IC, Gueguim-Kana EB and Beukes LS 2017 Biomedical applications of cocoa bean extract-mediated silver nanoparticles as antimicrobial, larvicidal and anticoagulant agents. *J. Clust. Sci.* **28** pp 149-164.

[3] Lateef A, Azeez MA, Asafa TB, Yekeen TA, Akinboro A, Oladipo IC, Azeez L, Ojo SA, Gueguim-Kana EB and Beukes LS 2016 Cocoa pod husk extract-mediated biosynthesis of silver nanoparticles: its antimicrobial, antioxidant and larvicidal activities. *J. Nanostruct. Chem.* **6** (2) pp 159-69.

[4] Shankar SS, Rai A, Ahmad A and Sastry M 2004 Rapid of Au, Ag and bimetallic Au core – Ag shell nanoparticles using Neem (*Azadirachta indica*) leaf broth. *J. Colloid Interf. Sci.* **275** (2) pp 496-502.

IOP Conf. Series: Materials Science and Engineering **805** (2020) 012043 doi:10.1088/1757-899X/805/1/012043

[5] Philip D 2011 *Mangifera indica* leaf-assisted biosynthesis of well-dispersed silver nanoparticles. *Spectrochim. Acta. A.* **78** (1) pp 327-331.

[6] Ojo SA, Lateef A, Azeez MA, Oladejo SM, Akinwale AS, Asafa TB, Yekeen TA, Akinboro A, Oladipo IC, Gueguim-Kana EB and Beukes LS 2016 biomedical and catalytic applications of gold and silver-gold alloy nanoparticles biosynthesized using cell-free extract of *Bacillus safensis* LAU 13: antifungal, dye degradation, anti-coagulant and thrombolytic activities. *IEEE Trans. Nanobiosci.* **15** (5) pp 433-42.

[7] Narayanan KB and Sakthivel N 2010 Biological synthesis of metal nanoparticles by microbes. *Adv. Colloid Interf. Sci.* **156** (1-2) pp 1-13.

[8] Lengke MF, Fleet ME and Southam G 2007 Biosynthesis of silver nanoparticles by filamentous cyanobacteria from nitrate complex. *Langmuir* **23** (5) pp 2694-2699.

[9] Rautaray D, Ahmad A and Sastry M 2003 Biosynthesis of $CaCO_3$ crystals of complex morphology using fungus and an actinomycete. *J. Am. Chem. Soc.* **125** (48) pp 14656-14657.

[10] Elegbede JA, Lateef A, Azeez MA, Asafa TB, Yekeen TA, Oladipo IC, Adebayo EA, Beukes LS and Gueguim-Kana EB 2018 Fungal xylanases-mediated synthesis of silver nanoparticles for catalytic and biomedical applications. *IET Nanobiotechnol.* **12** (6) pp 857-863.

[11] Lateef A, Azeez MA, Asafa TB, Yekeen TA, Akinboro A, Oladipo IC, Azeez L, Ojo SA, Gueguim-Kana EB and Beukes LS 2016 Cocoa pod husk extract-mediated biosynthesis of silver nanoparticles: Its antimicrobial, antioxidant and larvicidal activities. *J. Nanostruct. Chem.* **6** (2) pp 159-169.

[12] Lateef A, Ojo SA, Folarin BI, Gueguim-Kana EB and Beukes LS 2016 Kolanut (*Cola nitida*) mediated synthesis of silver-gold alloy nanoparticles: antifungal, catalytic, larvicidal and thrombolytic applications. *J. Clust. Sci.* **27** pp 1561-1577.

[13] Lateef A, Akande MA, Azeez MA, Ojo SA, Folarin BI, Gueguim-Kana EB and Beukes LS 2016 Phytosynthesis of silver nanoparticles (AgNPs) using miracle fruit plant (*Synsepalum dulcificum*) for antimicrobial, catalytic, anticoagulant, and thrombolytic applications. *Nanotechnol. Rev.* **5** (6) pp 507-520.

[14] Lateef A, Ojo SA and Elegbede JA 2016 The emerging roles of arthropods and their metabolites in the green synthesis of metallic nanoparticles. *Nanotechnol. Rev.* **5** (6) pp 601-622.

[15] Lateef A, Azeez MA, Asafa TB, Yekeen TA, Akinboro A, Oladipo IC, Gueguim-Kana EB and Beukes LS 2016 Cobweb as novel biomaterial for the green and eco-friendly synthesis of silver nanoparticles. *Appl. Nanosci.* **6** pp 863-874.

[16] Eranga RB, Chanika DJ, Uthpala AJ, Ranasinghe WD, Ruwanthika RM and Pretti VU 2017 Honey mediated green synthesis of nanoparticles: New era of safe Nanotechnology. *J. Nanomater.* http://doi.org/10.1155/2017/5919836.

[17] Ghosh S, Patil S, Ahire M, Kitture R, Gurav DD, Jabgunde AM, Kale S, Pardesi K, Shinde V, Bellare J, Dhavale DD and Chopade BA 2012 *Gnidia glauca* flower extract mediated synthesis of gold nanoparticles and evaluation of its chemocatalytic potential. *J. Nanobiotechnol.* **10** 17. https://doi.org/10.1186/1477-3155-10-17.

[18] Kim HK, Choi M, Cha S, Koo YK, Jun SH, Cho S and Park Y 2013 Earthworm extracts utilized in the green synthesis of gold nanoparticles capable of reinforcing the anticoagulant activities of heparin. *Nanoscale Res. Lett.* **8** 542. https://doi.org/10.1186/1556-276X-8-542.

[19] Dakshayani SS, Marulasiddeshwara MB, SharathKumar MN, Ramesh G, Kumar PR, Devaraja S and Hosamani R 2019 Antimicrobial, anticoagulant and antiplatelet activities of green synthesized silver nanoparticles using *Selaginella* (Sanjeevini) plant extract. *Int. J. Biol. Macromol.* https://doi.org/10.1016/j.ijbiomac.2019.01.222

[20] Keat CL, Aziz A, Eid AM and Elmarzugi NA 2015 Biosynthesis of nanoparticles and silver nanoparticles. *Bioresour. Bioprocess.* **2** 47. https://doi.org/10.1186/s40643-015-0076-2.

[21] Das VL, Thomas R, Varghese RT, Soniya EV, Mathew J and Radhakrishnan EK 2014 Extracellular synthesis of silver nanoparticles by the *Bacillus* strain CS 11 isolated from industrialized area. *3 Biotech.* **4** pp 121-126.

[22] Lateef A, Adelere IA, Gueguim-Kana EB, Asafa TB and Beukes LS 2015 Green synthesis of silver nanoparticles using keratinase obtained from a strain of *Bacillus safensis* LAU 13. *Int. Nano Lett.* **5** pp 29-35.

IOP Conf. Series: Materials Science and Engineering **805** (2020) 012043 doi:10.1088/1757-899X/805/1/012043

[23] Lateef A, Azeez MA, Asafa TB, Yekeen TA, Akinboro A, Oladipo IC, Ajetomobi FE, Gueguim-Kana EB and Beukes LS 2015 *Cola nitida* mediated biogenic synthesis of silver nanoparticles using seed and seed shell extracts and evaluation of antibacterial activities. *BioNanoSci.* **5** pp 196-205.

[24] Lateef A, Ojo SA, Akinwale AS, Azeez L, Gueguim-Kana EB and Beukes LS 2015 Biogenic synthesis of silver nanoparticles using cell-free extract of *Bacillus safensis* LAU 13: antimicrobial, free radical scavenging and larvicidal activities. *Biologia* **70** pp 1295-1306.

[25] Emeka EE, Ojiefoh OC, Aleruchi C, Hassan LA, Christiana OM, Rebecca M, Dare EO and Temitope AE 2014 Evaluation of antibacterial activities of silver nanoparticles green-synthesized using pineapple leaf (*Ananas comosus*). *Micron* **57** pp 1-5.

[26] Netala VR, Kotakadi VS, Domdi L, Gaddam SA, Bobbu P, Venkata SK, Ghosh SB and Tartte V 2016 Biogenic silver nanoparticles: efficient and effective antifungal agents. *Appl. Nanosci.* **6** pp 475-484.

[27] He Y, Wei F, Ma Z, Zhang H, Yang Q, Yao B et al. Green synthesis of silver nanoparticles using seed extract of *Alpinia katsumadai*, and their antioxidant, cytotoxic, and antibacterial activities. *RSC Adv.* **7** (63) pp 39842-39851.

[28] Das G, Patra JK, Debnath T, Ansari, A, Shin H-S 2019 Investigation of antioxidant, antibacterial, antidiabetic, and cytotoxicity potential of silver nanoparticles synthesized using the outer peel extract of *Ananas comosus* (L). *PLoS ONE* **14** (8) E0220950.

[29] Lillicrap D, Key N, Makris M and O'Shaughnessy D 2010 Practical Hemostasis and Thrombosis 3rd Edition. *Wiley-Blackwell Publishing Ltd.* 311pp.

[30] Shapiro SS 2003 Treating thrombosis in the 21st century. *The New Engl. J. Med.* **349** (18) pp 1762-1764.

[31] Kasper DL, Braunwald E, Fauci AS, Hauser SL, Longo DL and Jameson JL 2005 Harrison's Principles of Internal Medicine, 16th ed. *McGraw-Hill, Medical Publishing Division NY.* **2** 607pp.

[32] Maseri A, Severi S, Nes MD 1978 Variant angina: one aspect of a continuous spectrum of vasospastic myocardial ischemia; Pathogenetic mechanisms, estimated incidence and

clinical and coronary arteriographic findings in 138 patients. *Am. J. Cardiol.* **42** (6) pp 1019-1035.

[33] Shiber JR, Fontane E and Adewale A 2010 Stroke registry: hemorrhagic vsischemi strokes. *Am. J. Emerg. Med.* **28** (3) pp 331-333.

[34] Ohira T, Shahar TE, Chambless LE, Rosamond WD, Mosley TH and Folsom AR 2006 Risk factors for ischemic stroke subtypes: the atherosclerosis risk in communities study. *Stroke* **37** (10) pp 2493-2498.

[35] Thachil J, Gatt A and Martlew V 2008 Management of surgical patients receiving anticoagulation and antiplatelet agents. *Brit. J. Surg.* **95** pp 1437-1448.

[36] Cohen AT, Agnelli G, Anderson FA, Arcelus JI, Bergqvist D and Brecht JG 2007 Venous thromboembolism (VTE) in Europe. The number of VTE events and associated morbidity and mortality. *Thromb. Haemost.* **98** (4) pp 756-764.

[37] Heitt JA 2008 The epidemiology of venous thromboembolism in the community. *Arterioscl. Thromb. Vascul. Biol.* **28** (3) pp 370-372.

[38] Biss TT, Brandão LR, Kahr WH, Chan AK and Williams S 2008 Clinical features and outcome of pulmonary embolism in children. *Brit. J. Haematol.* **142** (5) pp 808-818.

[39] Andrade Ede O, Bindá FA, Silva AM, Costa TD and Fernandes MC 2009 Risk factors and prophylaxis for venous thromboembolism in hospitals in the city of Manaus. *J. Brasileiro de Pneumol.* 35 (2) pp 114-121.

[40] Santos Fernandes CJS, AlvesJúnior JL, Gavilanes F, Prada LF, Morinaga LK and Rogerio S 2016 New anticoagulants for the treatment of venous thromboembolism. *J. Brasileiro de Pneumol.* **42** (2) pp 146-154.

[41] Ramjan A, Hossain MS, Ariful Islam A, Arman SI, Raju GS, Dasgupta P, Noshin TF 2014 Aspect of thrombolytic therapy: A review. *Sci. World J.* Article ID 586510. https://doi.org/10.1155/2014/586510.

[42] Jhansi K and Vanita P 2014 A Review on Antiplatelet Drugs and Anticoagulants. *Adv. Pharmacoepidemiol. Drug Safety* **3** (3-R). https://doi.org10.4172/2167-1052.1000003-R.

[43] Vaidya B, Agrawal GP and Vyas SP 2011 Platelets directed liposomes for the delivery of streptokinase: development and characterization. *Eur. J. Pharmceut. Sci.* **44** pp 589-594.

[44] Ilinskaya AN and Dobrovolskaia MA 2013 Nanoparticles and the blood coagulation system Part I: benefits of nanotechnology. *Nanomed.* **8** pp 773-784.

[45] Lateef A, Ojo SA, Elegbede JA, Azeez MA, Yekeen TA and Akinboro A 2017 Evaluation of some biosynthesized silver nanoparticles for biomedical applications: hydrogen peroxide scavenging, anticoagulant and thrombolytic activities. *J. Clust. Sci.* **28** (3) pp 1379-1392.

[46] Raja S, Ramesh V and Thivaharan V 2015 Antibacterial and anticoagulant activity of silver nanoparticles synthesised from a novel source-pods of *Peltophorum pterocarpum*. *J. Ind. Eng. Chem.* **29** pp 257-264.

[47] Lateef A, Folarin BI, Oladejo SM, Akinola PO, Beukes LS and Gueguim-Kana EB 2018 Characterization, antimicrobial, antioxidant, and anticoagulant activities of silver nanoparticles synthesized from *Petiveria alliacea* L. leaf extract. *Prep. Biochem. Biotechnol.* **48** (7) pp 646-652.

[48] Shiang YC, Hsu CL, Huang CC and Chang HT 2011 Gold nanoparticles presenting hybridized self-assembled aptamers that exhibit enhanced inhibition of thrombin. *Angew. Chem.* **123** (33) pp 7802-7807.

[49] Hsu CL, Chang HT, Chen CT, Wei SC, Shiang YC and Huang CC 2011 Highly efficient control of thrombin activity by multivalent nanoparticles. *Chem.* **17** pp 10994-11000.

[50] Shrivastava S, Bera T, Singh SK, Singh G, Ramachandrarao P and Dash D 2009 Characterization of antiplatelet properties of silver nanoparticles. *ACS Nano.* **3** (6) pp 1357-1364.

[51] Lateef A, Ojo SA and Oladejo SM 2016 Anti-candida, anti-coagulant and thrombolytic activities of biosynthesized silver nanoparticles using cell-free extract of *Bacillus safensis* LAU 13. *Process Biochem.* **51** (10) pp 1406-1412.

[52] Lateef A, Akande MA, Ojo SA, Folarin BI, Gueguim-Kana EB and Beukes LS 2016 Paper wasp nest-mediated biosynthesis of silver nanoparticles for antimicrobial, catalytic, anticoagulant, and thrombolytic applications. *3Biotech.* **6** (2) pp 140. https://doi.org/10.1007/s13205-016-0459-x.

[53] Lateef A, Ojo SA, Elegbede JA, Akinola PO and Akanni EO 2018 Nanomedical applications of nanoparticles for blood coagulation disorders. In: *Environmental*

Nanotechnology, Volume 1. Eds: Dasgupta N, Ranjan S and Lichtfouse E. https://doi.org/10.1007/978-3-319-76090-2_8. Springer International Publishing AG, Cham, Switzerland. ISBN 978-3-319-76089-6. Pp. 243-277.

[54] Harish BS, Uppuluri KB and Anbazhagan V 2015 Synthesis of fibrinolytic active nanoparticles using wheat bran xylan as a reducing and stabilizing agent. *Carbohydr. Polymer* **132** pp 104-110.

[55] Yekeen TA, Azeez MA, Akinboro A, Lateef A, Asafa TB, Oladipo IC, Oladokun SO and Ajibola AA 2017 Safety evaluation of green synthesized *Cola nitida* pod, seed and seed shell extracts-mediated silver nanoparticles (AgNPs) using an *Allium cepa* assay. *J. Taibah Univ. Sci.* **11** (6) pp 895-909.

[56] Yekeen TA, Azeez MA, Lateef A, Asafa TB, Oladipo IC, Badmus JA, Adejumo SA and Ajibola AA 2017 Cytogenotoxicity potentials of cocoa pod and bean-mediated green synthesized silver nanoparticles on *Allium cepa* cells. *Caryol.* **70** (4) pp 366-377.

[57] Elegbede JA, Lateef A, Azeez MA, Asafa TB, Yekeen TA, Oladipo IC, Aina DA, Beukes LS and Gueguim-Kana EB 2020 Biofabrication of gold nanoparticles using xylanases through valorization of corncob by *Aspergillus niger* and *Trichoderma longibrachiatum*: antimicrobial, antioxidant, anticoagulant and thrombolytic activities. *Waste Biomass Valor.* **11** (3) pp 781-791.

[58] Elegbede JA, Lateef A, Azeez MA, Asafa TB, Yekeen TA, Oladipo IC, Abbas SH, Beukes LS and Gueguim-Kana EB 2019 Silver-gold alloy nanoparticles biofabricated by fungal xylanases exhibited potent biomedical and catalytic activities. *Biotechnol. Progr.* **35** e2829. https://doi.org/10.1002/btpr.2829.

[59] Aina DA, Owolo O, Lateef A, Aina FO, Abbas SH, Adeoye-Isijola M, Okon V, Asafa TB, Elegbede JA, Olukanni OD and Adediji I 2019 Biomedical applications of *Chasmanthera dependens* stem extract mediated silver nanoparticles as antimicrobial, antioxidant, anticoagulant, thrombolytic, and larvicidal agents. *Karbala Int. J. Modern Sci.* **5** (2) pp 71-80.

Nanotechnology in the built environment for sustainable development

JA Elegbede* and A Lateef

Laboratory of Industrial Microbiology and Nanobiotechnology, Department of Pure and Applied Biology, Ladoke Akintola University of Technology, Ogbomoso, Nigeria

*Email: josephadetunji18@gmail.com

Abstract

Nanotechnology as an emerging discipline of the 21st century has immense benefits for mankind due to the exceptional attributes of nanomaterials which can be used to render novel products and services. Amongst several fields of human endeavours, the built environment, largely composed of the disciplines of construction engineering, architecture, fine and applied arts, and urban and regional planning can make use of nanomaterials in creating green facilities with improved properties, aesthetics and innovation. Though there are reports of basic and applied investigations dealing with the production and evaluation of nanomaterials in the built environment, there is still a gap in harnessing the information together for the benefits of the practitioners in the built environment. Therefore, this treatise serves to x-ray the applications of nanotechnology in the built environment with the view of establishing nexus between the two areas. This compendium may stimulate the re-engineering of curricula in the built environment and stimulates new lines of research, product development and evaluation.

Keywords: Nanotechnology, built environment, green facilities, nanomaterials, composites, construction engineering, architecture, fine and applied arts

1 Introduction

The built environment, or the built world, is the human-made environment that grants the scenery for human activities which could range in scale from buildings to cities. It has been described as the human-made space in which humans reside, carry out work and refresh daily [1]. The built environment covers places and spaces formed or adapted by people to serve the

IOP Conf. Series: Materials Science and Engineering **805** (2020) 012044 doi:10.1088/1757-899X/805/1/012044

necessities for accommodation, organization as well as recreation. Architecture, building technology, urbanism and planning, civil engineering, fine and applied arts, landscaping and management of infrastructure and operations are areas covered by the sciences of the built environment. In public health, built environment is viewed as an act of building or renovation with the aim of improving the well-being of the community through the construction of improved, aesthetically and environmentally enhanced landscapes and structures for habitation [2].

Sustainability is of fundamental importance to all because it has to do with the continued existence of human species and more or less every living creature on the planet. Green buildings with respect to the environment reduce pollution effects, ensure conservation of natural resources and also prevent environmental degradation. In terms of the economy, they reduce the amount of funds spent by building operators on water and energy and also improve the productivity of the people using the facility. Also, socially, green buildings are beautiful and cause merely negligible strain on the local infrastructure [3]. Buildings which do not adapt with such challenges will become obsolete sooner than later or may require significant refurbishment or end up demolished, where neither option may produce a sustainable built environment [4]. It is estimated that by 2050, about 66% of the total world's population would dwell in urban areas [5] which will give rise to extensive challenges as regards to air pollution, waste management, congestion and human wellbeing [6]. Therefore, it becomes imperative to create smart solutions to surmount the consequences of urbanization. Exploitation of innovative and intelligent technologies is regarded as a major factor in decreasing greenhouse gas emissions and enhancing energy efficiency of cities. These technologies need to be smart, integrated, lean, resource-efficient and cost efficient, and they should have an all encompassing impact on environmental sustainability targets as well as wellbeing of the citizens and financial sustainability [7].

2 Nanotechnology and green synthesis of nanomaterials

Nanotechnology is a unique multidisciplinary idea of the 21st century that involves the study of matter at the nanoscale range. It has connected gaps in natural sciences, materials science, engineering, and medicine, with implausible relevance in diverse sectors [8]. Nanotechnology is concerned with the re-engineering or fabrication of materials and devices by controlling and

IOP Conf. Series: Materials Science and Engineering **805** (2020) 012044 doi:10.1088/1757-899X/805/1/012044

manipulating the matter at the atomic level. At the nanoscale, properties of materials are changed from that of the larger scales which represent the "nano-effects" that eventually determine the nature and characteristics that we are familiar with at the "macro-scale" and this is where the influence of nanotechnology comes in; if elements are manipulated at the nanoscale, this can affect their macro-properties and create significantly new materials and processes [9]. Various applications in nanoscience and nanotechnology have continue to increase the possibility of synthesis of nano-sized materials (1-100 nm) in various areas of human activities, thereby bridging the gap across the various frontiers of knowledge [10-15]. This field of technology has attracted and keeps attracting great interests owing to its extensive applications in diverse phases of life endeavors [16-18]. In broad-spectrum, nanotechnology deals basically with the production and stabilization of nanomaterials and nanoparticles of different types [19].

Nanomaterials are fabricated or engineered by different methods which include the physical, chemical, and biological methods. However, the physical and chemical techniques that have been extensively used long-ago result in environmental contaminations, as the processes cause a large volume of toxic by-products, and also use a great deal of energy. Therefore, there arose a claim for green nanotechnology that comprises clean, environmentally friendly, and safe systems of nanoparticles generation [20]. The biological technique, which is a major element of the green nanotechnology has expanded in the last decade to biosynthesize nanoparticles due to wealth of a wide-range of biological resources such as fungi [21], bacteria [22-26] plants [10, 27-35], as well as metabolites of arthropods [13, 15, 36-41] and enzymes [11, 22, 19, 42-48].

Moreover, green synthesis has become a credible alternative which has received unparalleled attention as a function of its reliability, cost effectiveness, less usage of chemicals and eco-friendliness [49], as well simplicity of the procedures and improved biocompatibility of nanoparticles that are biosynthesized [14, 15, 18]. These biosynthesized nanoparticles have found huge applications in various fields of life ranging from health care, biomedicine, drug–gene delivery, cosmetics, environment, food and feed, chemical industries, catalysis, single electron transistors, electronics, light emitters, space industries, energy science, nonlinear optical devices, mechanics, agriculture, remediation, and photo-electrochemical applications [14, 16-18, 50-55].

IOP Conf. Series: Materials Science and Engineering **805** (2020) 012044 doi:10.1088/1757-899X/805/1/012044

3 Nanotechnology in building and construction

The applications of nanotechnology in building and construction engineering have been on a steady rise (Table 1). The increasing interest in the relevance of nanomaterials in the building industry is largely due to their perceived positive characteristics which include moisture behavior, thermal properties, improved strength, energy efficiency, self-cleaning, improvement of air quality and antimicrobial effects [56]. At nano-level, gravity becomes unimportant as electrostatic forces take control, and quantum effects pull in. In addition, as particles tends to nano-size, the fraction of atoms on the surface increases compared to those inside, and this results in novel properties. These relatively new properties are what scientists are expansively exploring to take full advantage of the potentials vastly available [9].

Table 1. Some of the applications of nanotechnology in the built environment

Nanomaterials	**Characteristics**	**Applications/Potentials**	**References**
Nanosilica	Increased compressive strength, workability and setting time	Concrete production	[57]
Nanosilica/ silica fume	Compressive strength and long term hardening effect	Concrete production	[58]
Nanosilica	Increased compressive strength and modification of microstructure	Concrete production	[59]
TiO_2NPs	Photocatalytic properties/ depollution activity	Concrete production	[60]
TiO_2NPs	Improved water permeability, thermal and mechanical properties	Concrete production	[61]
Iron (III) oxide nanoparticles	Self detection compressive stress	Concrete production	[62]
Haematite	Reduced shrinkage and cracking effect	Concrete production	[63]
Iron (III) oxide nanoparticles	Improved water permeability resistance and reduced setting time	Concrete production	[64]
Iron (III) oxide nanoparticles	Increased compressive strength	Concrete production	[65]
Multi-walled Carbon nanotubes (MWCNTs)	Increased stiffness and strength		[66]
Silicone nanofilaments	Oil-repellent properties	Textile finishing (polyester fabric)	[67]
SiO_2 nanoparticles	Oleophobic property	Textile finishing (Cotton fibers)	[68]
Carbon nanotubes	Oleophobic property	Textile finishing (Cotton fibers)	[69]
ZnONPs	Antistatic property	Textile finishing	[70]
AgNPs	Antistatic property	Textile finishing (polyester fabric)	[71]
Carbon nanotubes	Enhanced toughness and strength,	Textile finishing	[72]

	and decrease weight		
ZnONPs	Absorb light in UV region	Textile finishing (polyester/ cotton fabrics)	[73]
SiO_2 nanoparticles and AgNPs	Antibacterial activities	Textile finishing (cotton fabrics)	[74]
AgNPs	Antibacterial activity	Textile finishing	[75]
Multi-walled Carbon nanotubes (MWCNTs)	Electrical conductive property	Paint additive	[76]
AgNPs	Antibacterial and antifungal property	Paint additive	[31]
AgNPs	Antibacterial and antifungal property	Paint additive	[28]
AgNPs	Antibacterial and antifungal property	Paint additive	[29]
AgNPs	Antibacterial and antifungal property	Paint additive	[37]
AgNPs	Antibacterial and antifungal property	Paint additive	[77]
Silica nanoparticles	Anti-ageing cycle	Anti-graffiti polyurethane coating	[78]
AgNPs	Antibacterial property and Photocatalytic degradation of safranin and methyl orange	Industrial waste water treatment	[79]
AgNPs	Degradation of malachite green	Industrial waste water treatment	[80]
AgNPs	Antibacterial property on blotting paper	Water treatment	[81]
CeO_2NPs, AgNPs, AuNPs, and TiO_2NPs	Antibacterial activities	Wastewater treatment	[82]
CuONPs	Reduction of COD and microbial populations	POME treatment	[83]
AuNPs and Ag-AuNPs	Catalytic degradation of malachite green dye	Wastewater treatment	[24]
Ag-AuNPs	Catalytic degradation of malachite green and methylene blue dye	Wastewater treatment	[30]
AgNPs	Catalytic degradation of malachite green	Wastewater treatment	[10]
AgNPs	Catalytic degradation of malachite green	Wastewater treatment	[13]
AgNPs	Catalytic degradation of malachite green and methylene blue dye	Wastewater treatment	[19]
Ag-AuNPs	Catalytic degradation of malachite green and methylene blue dye	Wastewater treatment	[43]
AgNPs	Adsorption of rhodamine B	Wastewater treatment	[50, 53]
AgNPs	Improved performance of DSSC	Solar cell efficiency	[84]
AgNPs	Improved performance of DSSC	Solar cell efficiency	[85]
TiO_2 nanoparticles	Improved solar conversion efficiency	Solar cell efficiency	[86]
AgNPs	Improved performance of DSSC	Solar cell efficiency	[87]
Nanocrystalline TiO_2	Improved DSSC conversion	Solar cell efficiency	[88]

IOP Conf. Series: Materials Science and Engineering **805** (2020) 012044 doi:10.1088/1757-899X/805/1/012044

	efficiency		
Nanofluids consortium of carbon nanotubes, graphite and silver nanoparticles	Improved efficiency	Solar thermal collector	[89]
TiO_2 nanowires coated with AuNPs or AgNPs.	Enhancement of visible light photocurrent	Renewable energy	[90]
Carbon nanoparticles	Higher sunlight absorption	Solar energy	[91]
Carbon nanospheres	Improved the photo-thermal properties	Solar energy	[92]
Al_2O_3 nanofluid	Improved thermal efficiency of solar collector	Solar water heating system	[93]
MgO nanofluid	Reduced heating loss time	Building heating system	[94]
Iron oxide nanoparticles and single-walled carbon nanotubes (SWCNTs)	Recycling of immobilized enzyme	Biofuel production	[95]
Ag/TiO_2 nanocomposite	High stability for more than one month	Hydrogen production	[96]

Nanotechnology provides a world of practically infinite possibilities in building technology; concrete can be much stronger, long-lasting and more easily set, steel can be tougher, while glass can produce some interesting properties including a self-cleaning effects. Thus, nanotechnology can enhance the performance of conventional construction materials, for instance concrete, glass, steel, paint, insulating materials, and coatings. Also, in recent times, nanotechnology is becoming popular to achieve sustainable built environment. The prospects of nanotechnology in terms of the improvement of building and construction materials are focused on the use of nanoparticles, nano-fibers, and carbon nanotubes to amplify the strength and resilience of cement composites, in addition to pollution reduction. Also, construction of inexpensive corrosion free steel, fabrication of thermal insulators (with performance 10 times of the existing commercial options), and production of thin films and coatings with self-cleansing capability and self-colour transformation to reduce energy consumption are some of the innovative interventions of nanotechnology in the built environment. The influence of nanotechnology on the built environment so far cannot be over emphasized, and this has led to the development of innovative structural materials with unique properties, stronger and lighter composites, sound absorber, fire insulator, low maintenance coating, nano-clay filled polymers, water repellants, self-disinfecting surfaces, UV-light protector, nano-sized sensors, air cleaners, ultra thin-strong-conductive wafers, solar cells amongst others.

IOP Conf. Series: Materials Science and Engineering **805** (2020) 012044 doi:10.1088/1757-899X/805/1/012044

It is well known that concrete is the foremost material in structural applications, where strength, firmness, and cost play a key role in the attributes of concrete. Concrete is a compound material at macroscale but its properties can equally be enhanced both at meso- and nanoscale [97], and nanotechnology has a high probability to add to the understanding of concrete's behaviour, to facilitate improvement of its mechanical properties and reduction of its ecological and construction materials' production costs [98]. According to Silvestre *et al.* [97], the most efficient nanoparticles for concrete production are nanosilica (NS) and silica fume (SF), titanium dioxide, Iron (III) oxide, chromium (III) oxide, nanoclay, calcium carbonate, alumina, carbon nanotubes (CNTs), and graphene oxide (GO). Nanosilica or silicon dioxide (SiO_2) nanoparticles and its colloidal form named as silica fume have attracted substantial interest from the scientific and technical communities. The advantages of using silica fume include high speedy compressive force, flexural strength, high tensile, modulus of elasticity, low permeability, enhanced durability amongst other. Behnood and Ziari [99] reported improvements of up to 25 % in compressive strength for heat and unheated samples with 10 wt % cement replacement. Rashad [57] presented a wide-ranging overview on the effects of nanosilica on the major properties of traditional cement materials and alkali-activated fly ash. It was presented that using nanosilica in the production of concrete has a high possibility to improve a variety of basic properties in concrete, such as strength, workability and setting time, fire and abrasion resistance or leaching, heat of hydration, and other behaviours under aggressive environments.

Shakhmenko *et al.* [58] tested the effect of inclusion of nanosilica and silica fume on the mechanical properties of cement. The results obtained showed that mixtures containing silica fume and silica fume compositions with nanosilica particles demonstrated higher values of compressive strength (over 3 times higher at early ages and around 15 % at 28 days) and better long term hardening effect when compared to pure cement paste. Jalal *et al.* [59] tested mechanical, durability, microstructural, and rheological properties of high-performance self-compacting concrete incorporating microSiO_2 (microsilica) and nanosilica which replaced a portion of Portland cement. It was concluded that the replacement by 2 % nanosilica in the binary mixes elevated the compressive strength for a binder content of 400 kg/m^3 for up to 73 % at 90 days. Also, the inclusion of nanosilica particles was found to control hydration behaviour and this led to modifications in the microstructure of the hardened paste. An excellent dispersion

IOP Conf. Series: Materials Science and Engineering **805** (2020) 012044 doi:10.1088/1757-899X/805/1/012044

of nanosilica particles in the cement mortar resulted into a denser microstructure and sped up the hydration course of the cement paste.

Moreover, it is renowned that the cement industry is accountable for the emission of high levels of pollutants, such as carbon dioxide and nitrogen oxides, which has led to a rising need for environmental policy to stimulate the development of new strategies to minimize the polluting agents [60]. Thus, combining TiO_2 nanoparticles with cement-based materials proffers a good solution, owing to its strong photocatalytic activity, which brings about environmental pollution remediation, high stability, self-cleaning, self-disinfection and comparatively low cost of production [100]. Titanium dioxide nanoparticles can attract photon energy resulting in the elevation of an electron from the valence band to the conduction band of titania when exposed to UV-irradiation, and this produce "holes" (h^+, electron vacancy) in the valence band. This electron–hole pairs may then re-combine in a short time to begin redox reactions. These reactions are dependent on the ambient conditions, producing several radicals, such as H_2O_2 or O_2 (from reduction), OH (from oxidation) when water vapour and oxygen are present around the activated titanium dioxide. Pollution is remediated when the radicals produced react with harmful substances that are absorbed on the TiO_2 surface, thus resulting in the degradation of these substances and release of harmless substances such as CO_2 and H_2O [101, 102]. The photocatalytic activity of TiO_2 is influenced by its crystallinity, crystal size, crystal structure and surface hydroxylation. Therefore, the application of TiO_2 nanoparticles may increase this activity compared to bulk TiO_2.

Cárdenas *et al.* [60] studied depollution activity of samples of cement pastes, added with TiO_2 nanoparticles by studying nitrogen oxide degradation. TiO_2 nanoparticles at different blend ratios were tested and measurements were made at two ageing times of 65 h and 28 days and the results showed that all cement pastes-containing TiO_2 nanoparticles had photocatalytic properties, irrespective of the ratio and the percentage of TiO_2 used. Also, TiO_2 nanoparticles can enhance the resistance to water permeability of concrete when it is added to the cement pastes with improvements upto a maximum replacement level of 2.0 wt % of TiO_2 nanoparticles [61]. A number of "depolluting" and "self-cleaning" concrete products with TiO_2 nanoparticles are already being produced and applied in some facades of buildings and in pavement materials for roads especially in Europe and Japan [103].

IOP Conf. Series: Materials Science and Engineering **805** (2020) 012044 doi:10.1088/1757-899X/805/1/012044

The opportunity of developing concrete that can sense or perceive its own strain and damage has opened new prospects in the subject of monitoring and controlling materials, and building performance during their life cycle. The merits of producing concrete having self-monitoring characteristics are immense; it provides absence of mechanical property degradation, and greater durability due to the embedment of sensors and fairly low cost [62]. Iron (III) oxide nanoparticles, also known as haematite, play a major role in this aspect. Li *et al.* [62] tested different samples of cement mortar with additions of iron (III) oxide nanoparticles. The results showed that cement mortar with the nanoparticles is able to detect its own compressive stress in the elastic and inelastic systems, and this is due to their capability to modify the volume electric resistance as the load applied changes.

There are also reports on radiation-shielding properties of concrete-containing haematite. Gencel *et al.* [63] investigated the physical and mechanical characteristics of concrete with haematite, with specific focus on workability and durability. It was found that haematite only had a minor effect on the compressive strength which was the same with that of plain concrete. However, haematite had high impact on other properties, such as shrinkage, and due to the reduction of stresses as a result of drying shrinkage, the cracking effect also diminished. Nazari and Riahi [64] examined the rate and percentage of water absorption, workability and setting time of binary blended concrete with fractional replacement of cement by iron (III) oxide nanoparticles at various concentrations. The results obtained showed that water permeability resistance increased as the concentration of the nanoparticles increased. However, the setting time and workability of fresh concrete decreased as concentration of the nanoparticles increased. Also, Nazari and Riahi [65] investigated the workability and compressive strength of concrete with addition of iron (III) oxide nanoparticles (average size of 15 nm) of various concentrations and results indicated an increase of 15 % in the ultimate concrete strength at maximum replacement level of 1.0 %.

Carbon nanotubes (CNTs) are almost certainly one of the most promising nanomaterials ever, owing to their superior mechanical properties compared to other forms of nanomaterials. CNTs have produced an enhancement of the mechanical properties of structural materials [104]. CNTs are tubular nanostructures having diameter of a few nanometers, a huge length/diameter aspect ratio and its atomic structure comprise of a single or several concentric hexagonal lattices of carbon atoms linked by sp2 bonds. An imperative area in the study of application of CNTs for

IOP Conf. Series: Materials Science and Engineering **805** (2020) 012044 doi:10.1088/1757-899X/805/1/012044

mechanical enhancement of materials is their dispersion in the reinforced matrix. The effective application of CNTs in nanocomposites is dependent on the ability to disperse CNTs within the matrix with no reduction their aspect ratio. Generally, two types of CNTs have been mostly reported, which are single-walled carbon nanotubes (SWCNT) and multi-walled carbon nanotubes (MWCNT), and the differences between them is mainly in their overall thickness (Figure 1). Theoretical calculations and experimental results reported on SWCNT showed that it has more desirable thermal, mechanical, photochemical, and electrical properties [105]. However, MWCNTs are less susceptible to buckling phenomena as a result of van der Waals forces in between different tube walls. Also, they are less expensive and can be effortlessly produced with its multiple walls, which undoubtedly justifies their more extensive applications [106] in strengthening other materials.

Studies on the reinforcing effect of MWCNTs in a cement matrix proposed that the cement paste matrix strengthened with CNTs can boost its flexural strength and Young's modulus by 25 and 50 % respectively. It was also specifically reported that little quantities of well-dispersed MWCNTs can significantly increase the stiffness and strength of the cementitious matrix [66]. CNTs can also enhance the conductivity of cementitious materials, besides the fact that cement-based materials have piezoresistive properties which indicates that excellent sensors can be generated for cement structures monitoring and the sensors can sense or detect microcracking and failure easily [107]. Additionally, as a result of bonding capacity of carbon atoms with other matter, materials modified with CNTs can be used as insulator, semi-conductor or conductor of electricity. Consequently, carbon nanotubes will have noteworthy influence on the building and architecture industry as such materials can variously act as source of light, generation of energy, switchable conduit, and conveyor system [108].

IOP Conf. Series: Materials Science and Engineering **805** (2020) 012044 doi:10.1088/1757-899X/805/1/012044

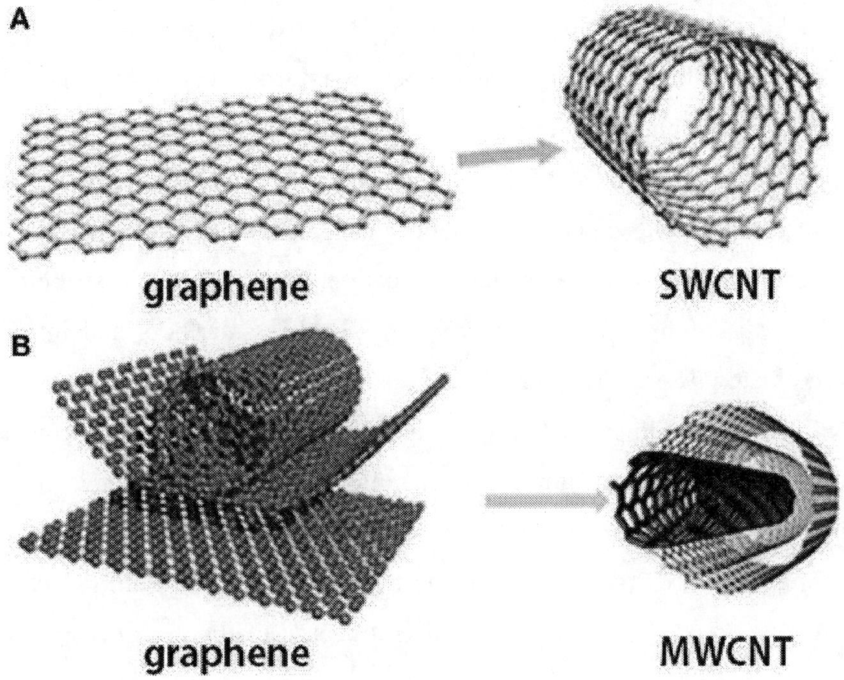

Figure 1. The transformation of graphene to SWCNT and MWCNT [109]

4 Nanoarchitecture

For the urban and regional planning, architecture and construction engineering profession, nanotechnology will influence the nature and properties of construction materials. As materials are manipulated at the nanoscale, so also they will behave in different ways that will impact their applications. Carbon nanotubes that are about one hundred times stronger than steel are a great example of how novel materials can be engineered through nanotechnology. Nanoarchitecture is the meeting point of nanotechnology and architecture, and it is predicted to totally revolutionize the architecture world in virtually every way; either the way architects think or how their ideas are inspired, the materials employed in building and finishing, or the way we express to the world and building users [108]. In essence, nanotechnology is bid to have profound effects on the ways humans live. Teaching of nanotechnology as a course of study would be required very soon in most architecture and engineering curricula because of the impact of nano-engineered materials in the construction industry. Therefore the need to widen an understanding of this broad-spectrum subject matter for architectures of the future cannot be underestimated. The possibility of manipulating matter at the scale of less than one billionth of a meter is a great

IOP Conf. Series: Materials Science and Engineering **805** (2020) 012044 doi:10.1088/1757-899X/805/1/012044

benefit presented by nanotechnology, and this presents the potential to transform the built environment in ways almost unimaginable today in return because as materials attain such transient features, new architectural design and construction will evolve.

Nanotechnology is already utilized in the production of everyday items from various clothing materials to sunscreen among others, and sooner than later, it will take over the production of building enclosure materials (such as panels, insulation, and coatings) to striking new levels of performance in terms of light, energy efficiency, security, smartness and intelligence. Many nano-engineered materials are already made accessible to builders and architects, and they have begun to transform our buildings, including what we can perform in them, and what they can do for us. Projecting into the nearest future, nanotechnology in research and development will likely have a massive impact on building designs over the next twenty to fifty years. Carbon nanotubes, for instance, could impact extraordinary strength and flexibility on buildings, which will lead to new forms, new functions, as well as new relationship between people, building and environment at large [108].

5 Applications of Nanotechnology in Fine and Applied Arts: Textiles, Coatings and Paints

5.1 Textiles

There has been a general increase in demands by customers for textiles with improved appearance, colour, texture, shape and functionality [110]. Functionalized textiles can be applied in various areas including medical monitoring of body function and metabolism, and rehabilitation [111, 112]. It is possible to integrate flexible electronics, optical devices and sensors into textiles [113]. Nanoengineered functional textile is new frontier in clothing technology [114]. The benefit of nanomaterials has to do with creating functions without changing the soothing properties of the substrate. The engineered materials are expected to be seamlessly incorporated into garments and be comfortable and flexible without allergic effect to the body. Also, the nano-engineered materials need to satisfy weight, appearance, colour and performance properties [110]. Textiles through nano-engineering can be made to have some specific functions including antibacterial properties, hydrophobicity, conductivity, anti-wrinkle properties, light guidance and scattering, and antistatic behavior.

IOP Conf. Series: Materials Science and Engineering **805** (2020) 012044 doi:10.1088/1757-899X/805/1/012044

Textiles with oil-repellent properties have also been produced. Artus *et al.* [67] reported that polyester fabric could be coated with silicone nanofilaments and then treated with plasma fluorination to convey super oleophobic properties onto the textiles, and the fabric samples thus produced had an oil repellency grade and repelled alkanes. It has been revealed that hydrophobic and oleophobic properties could be concurrently imparted to textiles. Hoefnagels *et al.* [68] reported that cotton fibers were impregnated with SiO_2 nanoparticles to create a dual-size surface roughness, this was followed by hydrophobization with poly dimethylsiloxane (PDMS) which resulted in a static water contact angle of 155° for a droplet. To stimulate oleophobic property, the SiO_2 nanoparticles on the fibers were treated with a perfluoroalkyl chain producing static contact angle of 140° and a roll-off angle of 24° for oil droplets. Liu *et al.* [69] reported that lotus leaf nanostructures also inspired biomimetic studies for potential application in textiles. Cotton fibers were coated with pristine and surface modified carbon nanotubes to imitate lotus leaves nanostructure, yielding fabrics with contact angle of more than 150°. In a similar study using lotus leaves, nanocoats with average size of 20 nm impacted hydrophobicity on textiles [115].

ZnO whiskers, TiO_2NPs, and antimony (Sb)-doped tin oxide (SnO_2) nanoparticles have been utilized to convey antistatic properties on synthetic fibers. These nanomaterials are electrically conductive and the static charges accumulated are dispersed on the textile. Silane nanosol have also been said to enhance antistatic properties, because it absorbs moisture in the air via hydroxyl groups [110]. Zhang and Yang [70] reported that ZnONPs have been utilized to induce antistatic properties on fabrics and Qiaozhen [71] also reported that the addition of AgNPs reduced static voltage of polyester fabric by 60.4%.

Nanocoatings that prevent wrinkle while conserving comfort are sought-after in textile products. Conventionally, fabrics are impregnated with resin to impart antiwrinkle properties to textiles. However, this method reduces the tensile strength of the fiber, dyeability, and abrasion resistance, while stimulating hydrophobicity. To pass on wrinkle resistance properties on fabrics, nanoparticles have been successfully used on cotton and silk. TiO_2NPs with carboxylic acid as a catalyst have been used to form cross-links in between cellulose molecules and the acidic groups. Therefore, carboxylic-acid treated fabrics with TiO_2NPs were reported to be softer compared to untreated fabrics [110]. Generally, TiO_2 through its inherent catalytic property can be utilized as a cocatalyst with sodium hypophosphite ($NaPO_2H_2$) to treat cotton with 1, 2, 3, 4-butane tetracarboxylic acid [116]. Carbon nanotube (CNT) reinforced polymer composite fibers have been developed to enhance toughness and strength, and decrease weight [72]. Nanoscale semiconductor oxides such as TiO_2 and ZnO can proficiently absorb and scatter UV radiation. Inorganic UV blockers are chemically stable and nontoxic, and can operate at high temperatures. ZnONPs synthesized were immobilized on dyed polyester/ cotton fabrics and the resulting fabric absorbed light in the UV region [73].

The demand for functional textiles with antimicrobial properties is expanding with increasing attention to health and hygiene of consumers. AgNPs, TiO_2NPs, and ZnONPs can be applied to induce fungicidal and antibacterial properties to textiles. AgNPs have large surface areas thus this leads to an increased contact with bacteria and fungi. The antiseptic system of AgNPs is based on interaction with proteins present in these microorganisms and adversely affecting their cellular function thus inhibiting cell growth. They also cut down microbial respiration, restraining the activities of the basal metabolism of the electron transfer system, and transportation of substrate into the cell membrane. When AgNPs make contact with moisture or bacteria, they stick on to the cell wall and membrane [110]. The antimicrobial efficacy of AgNPs additives depends on their concentration, surface area, as well as the rate of release of the Ag^+ ions [117]. In our laboratory, we have demonstrated the antibacterial properties of textile functionalized with biosynthesized AgNPs (Figure 2). TiO_2NPs can also be employed to induce textiles with antibacterial properties. The catalytic activity of TiO_2NPs is its inherent strength which is usually leveraged on to provide antibacterial properties in textiles. The photocatalytic activity can be enhanced by creating nanocomposites of TiO_2/SiO_2 or Au-doped TiO_2 nanocomposites in cotton fabrics with self-cleaning properties [118, 119].

IOP Conf. Series: Materials Science and Engineering **805** (2020) 012044 doi:10.1088/1757-899X/805/1/012044

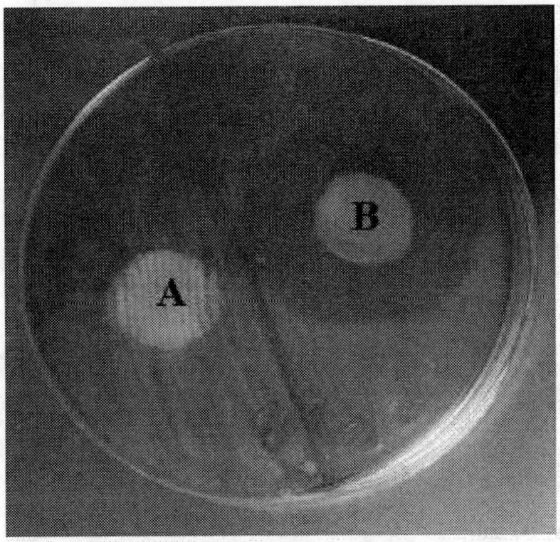

Figure 2. Antibacterial activity of textile functionalized with AgNPs against *Escherichia coli* (a, control; b, functionalized textile)

ZnONPs coated onto textiles was reported to show self-cleaning properties in the presence of Gram-negative *Escherichia coli* and aerobic Gram positive *Staphylococcus aureus*. Additionally, SiO_2 nanoparticles and AgNPs having a core-corona structure were electrostatically coated on cotton surfaces with high packing density to induce antibacterial properties to fabrics [120]. The nanoparticles were loaded with antibacterial moieties such as quaternary ammonia salts as well as metal coatings on the cotton fabrics [120]. SiO_2NPs can also be incorporated in polyimidoamide fibers through spinning and the incorporated of nanoparticles in fibers can create electrically conductive channels with enhanced antistatic and mechanical properties [110]. Yue *et al.* [75] reported the synthesis of silver nanoparticles-sericin (AgNPs-sericin) hybrid colloid which showed high bacterial reduction rates of 96.25 and 96.46 % for *S. aureus* and *E. coli* after 20 washing cycles. All these have shown that fabrics with special attributes can be created for applications in the built environment such as curtains and manufacture of upholstery and rugs.

5.2 Paints

Paints are widely used because their processes of application are relatively simple, inexpensive and robust compared with other coating techniques. At present, this area is focused on

IOP Conf. Series: Materials Science and Engineering **805** (2020) 012044 doi:10.1088/1757-899X/805/1/012044

finding innovative formulations with enhanced properties over microbial and environmental degradation, hydrophobicity, mechanical strength, adhesion and opacity, with consideration for the use of nanostructured materials as additives [121]. Yedra *et al.* [76] developed a new paint with electrical conductive characteristics from a polymer matrix in which are dispersed multi-walled carbon nanotubes (MWCNTs). Carbon nanotubes (CNT) have excellent and unique properties, such as electrical, optical, thermal and mechanical properties. The paints incorporated with dispersed MWCNTs displayed better resistance to degradation than the conventional pristine paint. Properties such as hardness and adhesion were not affected by the incorporation of nanofillers in the new paint. The potential application of nanoparticles as additives in paints have been widely reported and this indicates that the paint industry may benefit from the applications of these novel nanomaterials to improve the quality of the paint against biodegradation, microbial attack, and chemical deterioration [121, 122] which is an additional efforts towards achieving sustainable built environments.

In a study reported in our laboratory [31], impregnation of emulsion paints with AgNPs synthesized using cocoa bean extract effectively inhibited the growth of both bacterial and fungal isolates like; *E. coli*, *Klebsiella pneumoniae*, *Streptococcus pyogenes*, *Staphylococcus aureus*, *Pseudomonas aeruginosa*, *Aspergillus flavus*, *A. fumigatus* and *A. niger* . We alsoreported that incorporation of AgNPs synthesized using cocoa pod husk extract into emulsion paint led to effective inhibition of the growth of *E. coli*, *K. pneumoniae*, *S. pyogenes*, *S. aureus*, *P. aeruginosa*, *A. flavus*, *A. fumigatus* and *A. niger* compared to abundant growth on the control plates [29]. Lateef *et al.* [28] in a related research reported a total eradication of bacterial strains of *E. coli* and *P. aeruginosa* as well as fungal strains of *A. niger*, *A. fumigatus* and *A. flavus* in paint treated with AgNPs synthesized using pod extract of *Cola nitida*. AgNps synthesized using cobweb extract was shown to present total growth inhibition against *E. coli*, *P. aeruginosa*, *A. niger* and *A. flavus* when employed as paint additive [37] (Figure 3). Also, a complete suppression of the growth population of *S. aureus*, *P. aeruginosa*, *A. fumigatus* and *A. flavus* was reported when AgNPs synthesized using cell-free extract of *B. safensis* was incorporated as additive into emulsion paint [77].

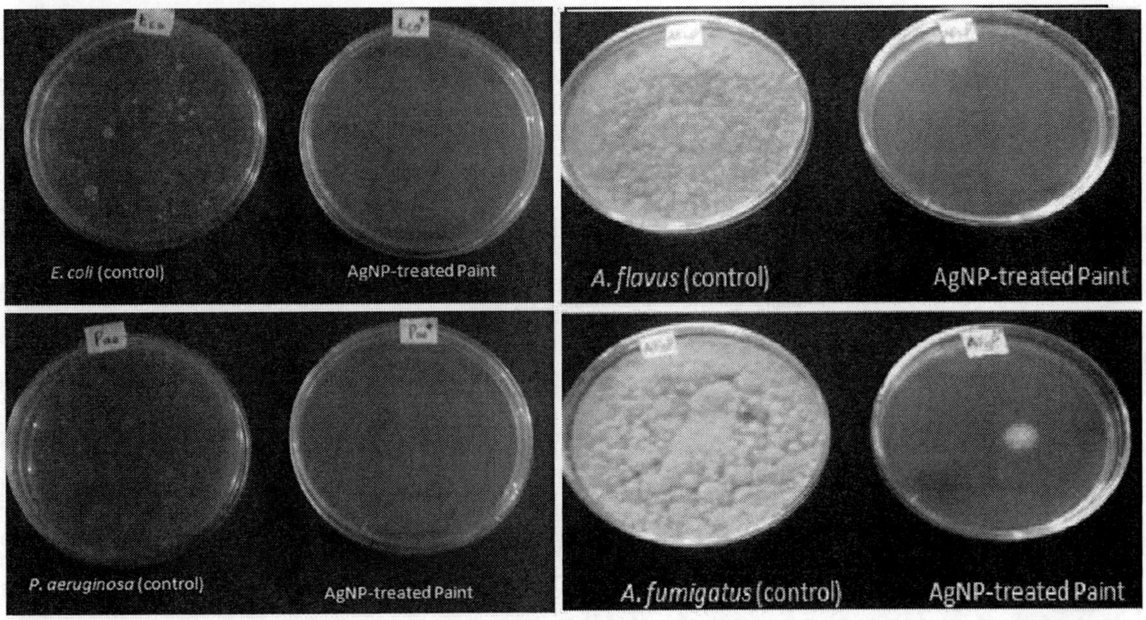

Figure 3. The antibacterial and antifungal activities of emulsion paint impregnated with AgNPs [37]

5.3 Nanocoatings

Nanocoatings are progressively used by the construction industries on building surfaces, such as walls, windows and doors, as they unlock new horizons for environment-friendly and sustainable buildings. Nanocoatings create a surface of the preferred protective or functional properties while providing a protective layer bound to the base material. The major mechanism of nanocoatings is their ability to produce self-healing properties through a process of self-assembly [123]. A self-cleaning glass mechanism based on a thin film titanium oxide (TiO_2) coating can be actualized [124] where the glass cleans itself in major two stages. The first stage is a photocatalytic process that breaks down the organic dirt on the glass using UV light and confers superhydrophilicity on the glass. At some point in the following superhydrophilic stage, rain water washes away the dirt, leaving more or less no streaks, because water spreads uniformly on superhydrophilic surfaces [125]. The films also have high-quality photo-induced antibacterial and antireflective properties. Both hydrophobic and hydrophilic coating systems are applicable for flat building surfaces and base materials such as tiles, woods, and stones.

IOP Conf. Series: Materials Science and Engineering **805** (2020) 012044 doi:10.1088/1757-899X/805/1/012044

Wang *et al.* [126] suggested that flame-retardant nanocoatings via the addition of nano-size magnesium-aluminium-layered double hydroxides (LDHs), titanium oxide (TiO_2) and silicon oxide (SiO_2). Quagliarini *et al.* [127, 128] and Munafo [129] reported that nanostructured titanium oxide (TiO_2) based coatings have promising properties on historical and stone surfaces respectively, while silica nanoparticles have been reportedly used on a permanent anti-graffiti polyurethane coating, which positively affected the anti-graffiti performance against ageing cycles [78].

6 Water pollution management

Human activities including industrialization and extreme use of pesticides for improving agricultural yield and production have negatively affected the ecosystem, causing pollutions in the available natural water reserves. Majority of the natural sources of drinking water such as surface water, lakes and reservoirs, groundwater, rivers and canals, including rainwater have been found to be polluted with ample variety of toxic materials as well as pathogenic microorganisms [130-135]. Remediation of contaminated and polluted water has been an area of interest with many techniques being applied for the improvement of quality of available natural water to the level that will be appropriate for human consumption. Conventional water disinfection procedures have some restrictions resulting in apprehensiveness about their applications at a large scale.

However, nanotechnology is a disruptive technology that is a capable of making an impact in the area of water purification as nanostructures offer large surface to volume ratios which is ideal for surface reactions. The prospect of creating photocatalytic membranes by impregnating semiconducting nanostructures on conventional membranes makes this idea even more striking [136]. Point-of-use water purification systems can be devised using antimicrobial nanomaterials such as silver (Ag) and zinc oxide (ZnO) [137]. Active functional membranes incorporated with photocatalytic or antimicrobial nanomaterials will be proficient in accomplishing multiple treatment targets in a singular course of action, and at the same time reducing fouling [137].

Nanostructures originating from metal oxide semiconductors such as zinc oxide (ZnO), titania (TiO_2), tungsten oxide (WO_3) and zinc stannate (Zn_2SnO_4) can be appealing means of water purification as it is capable of eliminating chemical and biological contaminants [138]. Higher

IOP Conf. Series: Materials Science and Engineering **805** (2020) 012044 doi:10.1088/1757-899X/805/1/012044

adsorption of the target molecules is possible because nanostructured photocatalysts will offer large surface to volume ratios. Concerted research over the past decade for implementation of nanotechnology in the purification of drinking water can be found in the literature [139-142]. Moreover, nanofiltration is a kind of membrane filtration used mostly with fresh groundwater and surface water having lesser amounts of dissolved solids. The major objectives of nanofiltration are elimination of DOB precursors such as natural and synthetic organic matter, and softening of water. Nanofiltration is a cross-flow filtration technology, that can be placed in between ultrafiltration and reverse osmosis, and the types of materials that can be filtered out depends upon the pore sizes of the filtration membranes. The pore size of the nanofiltration membrane can be as low as about 1 nm [142] which is adequate enough to screen out microorganisms.

Ramaswamy and Mani [79] reported that industrial waste water treated with AgNPs biosynthesized using *Allium fistulosum* displayed antibacterial efficiency against pathogenic bacteria *Proteus vulgaris*, *Shigella dysenteriae* and *Escherichia coli* isolated from industrial waste water. Also, the particles degraded safranin and methyl orange by 43.1 and 34.6% respectively. Allam *et al.* [80] described the biosynthesis of AgNPs by cell-free extracts from some bacterial species. The AgNPs was utilized as nanocatalyst for the elimination of malachite green dye (MG) from waste water with continuous reduction in dye absorbance at 617 nm until it disappeared over a period of 160 min. Dankovich and Gray [81] showed that filtration of contaminated water sample through paper impregnated with AgNPs could be an effective emergency water treatment technique in a study where AgNPs impregnated on blotting paper displayed antibacterial properties toward suspensions of *Enterococcus faecalis* and *Escherichia coli* in the effluent. García *et al.* [82] studied the antimicrobial activities of titanium dioxide, cerium dioxide, gold and silver nanoparticles in wastewater treatment. The activities of thermophilic, mesophilic, anaerobic, hetrotrophic and ammonia oxidizing bacteria in waste water were inhibited by the nanoparticles in this order; CeO_2NPs, AgNPs, AuNPs, and TiO_2NPs. Also, Kuppusamy *et al.* [83] reported the biosynthesis of *Commelina nudiflora* mediated copper nanoparticles as a novel bio-control agent and the CuONPs was employed for the treatment of palm oil mill effluent. It was shown that CuONPs sufficiently reduced COD and microbial populations in the treated effluent.

IOP Conf. Series: Materials Science and Engineering **805** (2020) 012044 doi:10.1088/1757-899X/805/1/012044

Ojo *et al.* [24] have reported that AuNPs and silver-gold alloy nanoparticles (Ag-AuNPs) biosynthesized using supernatant of *B. safensis* LAU 13 displayed excellent catalytic activities in the degradation of malachite green in the range of 86.1-92.7 % after 48 h at concentrations of 20-100 µg/ml. Lateef *et al.* [30] studied the catalytic degradation activities of leaf, seed, seed shell and pod extracts of *Cola nitida* mediated Ag-AuNPs against malachite green and methylene blue, with activities ranging from 63.2-97.1 %, and 2.1-73.2 % for malachite green and methylene blue respectively within 24 h. Lateef *et al.* [10] evaluated AgNPs mediated by leaf and seed extracts of *Synsepalum dulcificum* to degrade malachite green with activities of approximately 80 % within 24 h. Furthermore, AgNPs biosynthesized using nest extract of paper wasp (*Polistes* sp) degraded malachite green dye to the tune of 93.1 % [13]. Moreover, fungal xylanases have been used for synthesis of nanoparticles which have been reported to display dye degradation potentials. Fungal xylanases mediated AgNPs reportedly degraded malachite green and methylene blue with highest activities of 78.97 and 25.3 % respectively while both malachite green and methylene blue dye were degraded by fungal xylanases mediated Ag-AuNPs within 24 h at the range of 74.86-91.39 % and 12.1-47.1 % respectively within concentrations of 20-100 µg/ml [19, 43]. Similarly, nanoparticles have also been used for the adsorption of dyes in water [50, 53].

7. Nanotechnology and renewable energy sustainability

Renewable energies are that form of energy sources that can generate electricity and heat, as well as provide light, without polluting or contaminating the environment. Fossil fuels includes crude oil, natural gas and coal, they are not renewable, but are gone forever once burnt and more than 90 % of the world energy demand is supplied from these sources [143]. The sources of fossil fuel are generally believed to be exhaustible and they are also regarded as the major cause of emissions of greenhouse gas. Thus, there is a critical need to increase the portion of renewable energy sources in supplying the primary energy need of the world at large [144]. Moreover, consumption of petroleum was reportedly as 10^5 times faster than the capacity that nature can create and with this elevated consumption rate, the world's reserve for fossil fuel was predicted to have diminished by the year 2050 [145]. Also, the global demand for energy is projected to be more or less 30 and 46 TW by the year 2050 and 2100 respectively [143].

IOP Conf. Series: Materials Science and Engineering **805** (2020) 012044 doi:10.1088/1757-899X/805/1/012044

The expansion of renewable energy resources has attracted global attention in recent years. Energy sources which includes wind, solar, tidal wave, hydropower, geothermal and ocean thermal are regarded as the renewable energy resources. The sporadic fluctuations in the available energy from these sources make the advances of reliable and efficient storage systems crucial. These storage systems should have the capacity to collect and retain energy supplied during peak hours and supply the energy stored at periods when supply is low for necessary use [146]. According to Elegbede and Lateef [8], the roles of nanotechnology in improving the dependability of energy system covers every aspects of energy infrastructure, which includes production, storage, distribution, as well as the utilization. Creation of more resourceful products that can be utilized at the various levels of the energy architecture for more capable and optimal energy systems can be achieved via nanotechnology advances. In realizing an energy secured society, a more reliable energy system infrastructure will be required since traditional energy systems have fallen short in a lot of ways [147].

The prospect of applications of nanotechnology in provision of sustainable renewable energy cannot be underestimated. Devices that can store energy such as batteries and super-capacitors can be appreciably altered by the various applications of nanotechnology. Nanotechnology can be applied to re-engineer to ensure that certain relevant elements of lithium-ion batteries flexible, heat resistant and high-performance electrodes. Nanoporous materials like zeolites can be used to better improve the storage systems for thermal energy, which eventually could be employed as heat storages in for industrial and residential uses [143].

Solar energy is generally said to be one of most excellent supply of renewable energy but the major drawback in using most solar devices is that the apparent weakness in the absorption properties of the conventional fluids which ultimately leads to reduction in efficiency. However, with advances in nanotechnology, this can be easily solved. The amplified surface area to volume ratio of nanoparticles should practically increase the potentials of solar energy collection and efficiency by subjecting more conducting surfaces to sunlight. Nanotechnology will also boost solar cell efficiency when materials such as lead-selenide is used because they have capacity to cause more electrons to be released when hit by a photon of light. Dye-sensitized solar cell (DSSC) are solar cells that supports optical absorption and charge separation/injection

IOP Conf. Series: Materials Science and Engineering **805** (2020) 012044 doi:10.1088/1757-899X/805/1/012044

by connecting a dye sensitizer (light-absorbing material) to semiconductor (having a wide-band gap) of nanocrystalline morphology as the photoanode [143].

Improved performance of a DSSC using AgNPs-modified fluorine tin oxide (FTO) electrode via successive ion layer adsorption and reaction (SILAR) was reported by Eli *et al.* [84] and it was shown that the modification of fluorine tin oxide (FTO) electrode led to approximately 22 % improvement in photocurrent over that of bare FTO without AgNPs. Also, Eli *et al.* [86] presented the enhanced performance of DSSC modified with AgNPs through SILAR implementation. It was reported that AgNPs-modified DSSC produced 63 % improvement in efficiency, 48.4 % augmentation in short circuit current density and 8.5 % improvement in open circuit voltage which was predictably influenced by the absorption strength of AgNPs due to its surface plasmon resonance. Moreover, Eli *et al.* [85] noted the improved solar conversion efficiency (η) of 0.0067 % which was an influence of of TiO_2 nanoparticles in DSSC of *H. sabdariffa*. Similarly, Onimisi *et al.* [87] investigated the impacts of AgNPs on DSSC performance via SILAR. The results obtained indicated that 36 % improvement in efficiency was attained in the AgNPs modified DSSC. Bazargan [88] reported that dye sensitized solar cell (DSSC) was synthesized using natural pomegranate juice for sensitization of nanocrystalline TiO_2. Pulse current electron deposition method was used to prepare platinum and graphite coated electrodes while soot staining method for use as counter electrodes. The overall conversion efficiencies of fabricated (DSSC) were found to be 1.5 and 0.9 % for cell operated with platinum electrode and carbon coated counter electrodes respectively.

Moreover, Otanicar *et al.* [89] reported efficiency improvements of up to 5 % in solar thermal collectors by using nanofluids (from a variety of nanoparticles such as carbon nanotubes, graphite and silver) as an absorption mechanism. Liu *et al.* [90] reported the enhancement of visible light photocurrent by coating TiO_2 nanowires with gold or silver nanoparticles. The improvement was realized due to optical scattering from the plasmonic nanoparticles, which amplified the effective optical path of the thin film. The scattering and absorption characteristics of nanofluids composed of aqueous suspensions of single wall carbon nanohorns was in investigated by Mercatelli *et al.* [91] in order to use them as direct sunlight absorber fluids in solar devices. Considerably higher sunlight absorption was achieved induced by the carbon nanoparticles when compared to those of pure water. As a result, it was assumed that the carbon

IOP Conf. Series: Materials Science and Engineering **805** (2020) 012044 doi:10.1088/1757-899X/805/1/012044

nanohorns could be efficiently applied for overall efficiency improvement of sunlight exploiting devices. Also, Poinern *et al.* [92] studied the potentials of nanoparticles of functionalized carbon nanospheres (CNS) to improve the photo-thermal properties of the working fluid. The photo-thermal response of both films composed of (CNS) and nanofluids were examined under 1000 W/m^2 solar irradiation and the results obtained showed that the functionalized (CNS) nanofluids had the prospect to efficiently enhance the solar absorption capabilities of direct-absorption solar collectors.

Tiwari *et al.* [93], in a theoretically based study investigated the effect of applying Al_2O_3/ water nanofluid in a flat-plate solar collector as an absorbing medium and the results revealed that applying 1.5 % particle volume fraction of Al_2O_3 nanofluid improved the thermal efficiency of solar collector by about 31.64 % when compared with the conventional solar water heating system. Obaid *et al.* [94] investigated of the thermal energy storage efficiency by using nanofluid (adding magnesium oxide, MgO to distilled water). The system was intended to heat a building that has a reflector that concentrated solar energy to heat the fluid passing through it, then the fluid storage tank store the fluid and this was circulated by using a pump and solar radiator which transfer the heat into the air in the room. The results showed that addition small quantities of nanoparticles (MgO) (0.1 and 0.2 wt %) in the building heating system driven by solar energy reduced the heating loss time while the capability to heat gain was lower than pure water.

Zhang *et al.* [148] utilized carbon-based nanostructured catalyst with medium acid density for the production of biodiesel via catalytic distillation, and it was reported that the produced biodiesel was easy to amplify and promising, which was possible because some drawbacks of previous available methods were adequately avoided. Goh *et al.* [95] reported the incorporation of iron oxide nanoparticles into single-walled carbon nanotubes (SWCNTs) to create magnetic single-walled carbon nanotubes (mSWCNTs) which was used to recycle the immobilized enzyme, en route valuable use in biofuel production processes. This could lead to greater efficiency in carbon nanotube-enzyme bioreactors coupled with the intrinsic properties of the nanotubes, and reduction in capital costs in industrial enzyme systems can also be attained. The prospect of photoelectrochemical hydrogen production from water/methanol decomposition with Ag/TiO_2 nanocomposite thin films was demonstrated by Alenzi *et al.* [96]. The results obtained

IOP Conf. Series: Materials Science and Engineering **805** (2020) 012044 doi:10.1088/1757-899X/805/1/012044

revealed the nanocomposite films had high stability for hydrogen production for more than one month, also the films can be more suitably used than powders and they could be recycled easily.

8 Prospects of Nanotechnology in building and architecture

Nanotechnology has the possibility to drastically alter the built environment and how humans live generally. It is possibly the most transformative technology man has ever had which has generated more debate and research than space travel, computers, nuclear weapons, or any of the other technologies that have shaped our lives so far. Accompanied with it are enormous questions, concerns and consequences. Yet, its prospect to transform the built environment still remains largely unexplored. By transforming the critical features of matter, nanotechnology has capacity to change the way we build. Structures could be possibly constructed from the bottom-up because materials such as carbon nanotubes can self-assemble. In profoundly affecting the architectural industry, the designers' creative mind will be unleashed and more innovative ideas can spring up. Generally, all areas of architectural designs, building construction, as well as providing a safe environment at all levels, interior designs, and city design will all profit from advances in nanotechnology. Nanotechnology could give superior interactive functions to architectural designs as occupants may be able to select and communicate what transient states they would like to experience, this may include windows and walls with variable transparency, and mood/context sensitive clothing [108].

9 Conclusion

Nanotechnology gives unparalleled influence to architects and construction engineers in shaping and manipulating our world, and this could result in having buildings that shape us, as well as our relationships with one another and the environment at large. The actualization of sustainable built environment is achievable via the vast applications of nanotechnology in building constructions, as well as provision of enhanced finishes both in interior designs and city design itself. With advances in nanotechnology, architecture will have the aptitude to function at more optimal levels in revolutionizing the way inhabitants live. Therefore, ignorance about the vast potentials of nanotechnology in relations to the built environment can only predispose to being

IOP Conf. Series: Materials Science and Engineering **805** (2020) 012044 doi:10.1088/1757-899X/805/1/012044

suddenly left behind of technological advances as billions of dollars are poured into new research and development every year and innovative advances pour out.

References

[1] Roof K and Oleru N 2008 Public health: seattle and king county's push for the built environment. *J. Environ. Health* **71** (1) pp. 24-27.

[2] Aboelata 2004.The built environment and health - 11 Profiles of Neighborhood Transformation, Prevention Institute: Oakland, CA.

[3] Ragheb A, El-Shimy H and Ragheb G 2016 Green architecture: A concept of sustainability. *Procedia-Soc. Behav. Sci.* **216** pp 778-787.

[4] Manewa A, Siriwardena M, Ross A and Madanayake U 2016 Adaptable buildings for sustainable built environment. *Built Environ. Proj. Asset Manag*. **6** (2) Suppl 1 pp 139-158.

[5] United Nations Organization 2014 World urbanization prospects: The 2014 revision, highlights. Department of economic and social affairs. Population Division, United Nations. pp 32.

[6] OECD (Organisation for Economic Co-operation and Development), 2012. OECD Environmental Outlook to 2050: The consequences of inaction. Paris: OECD Publishing. https://doi.org/10.1787/9789264122246-en.

[7] Ahvenniemi H, Huovila A, Pinto-Seppä I and Airaksinen M 2017 What are the differences between sustainable and smart cities?. *Cities* **60** pp 234-245.

[8] Elegbede JA and Lateef A 2019 Green Nanotechnology in Nigeria: The research landscape, challenges and prospects. *Ann. Sci. Technol.* **4** (2) pp 6-38.

[9] Rawat P, Kumar A and Verma AK 2015 A review on nanotechnology in civil engineering. *Discov*. **39** (179) pp 152-158.

[10] Lateef A, Akande MA, Azeez MA, Ojo SA, Folarin BI, Gueguim-Kana EB and Beukes LS 2016 Phytosynthesis of silver nanoparticles (AgNPs) using miracle fruit plant (*Synsepalum dulcificum*) for antimicrobial, catalytic, anti-coagulant and thrombolytic applications. *Nanotechnol. Rev.* **5** (6) pp 507-520.

IOP Conf. Series: Materials Science and Engineering **805** (2020) 012044 doi:10.1088/1757-899X/805/1/012044

[11] Adelere IA and Lateef A 2016 A novel approach to the green synthesis of metallic nanoparticles: the use of agro-wastes, enzymes and pigments. *Nanotechnol. Rev.* **5** (6) pp 567-587.

[12] Adelere IA, Lateef A, Aboyeji DO, Abdulsalam R, Adabara NU and Bala JD 2017 Biosynthesis of silver nanoparticles using aqueous extract of *Buchholzia coriacea* (wonderful kola) seeds and their antimicrobial activities. *Annals Food Sci. Technol.* **18** pp 671-679.

[13] Lateef A, Akande MA, Ojo SA, Folarin BI, Gueguim-Kana EB and Beukes LS 2016 Paper wasp nest-mediated biosynthesis of silver nanoparticles for antimicrobial, catalytic, anticoagulant, and thrombolytic applications. *3Biotech* **6** 140. http://dx.doi.org/10.1007/s13205-016-0459-x.

[14] Lateef A, Elegbede JA, Akinola PO and Ajayi VA 2019 Biomedical applications of green synthesized-metallic nanoparticles: a review. *Pan Afr. J. Life Sci*

[15] Lateef A, Ojo SA and Elegbede JA 2016 The emerging roles of arthropods and their metabolites in the green synthesis of metallic nanoparticles. *Nanotechnol. Rev.* **5** (6) pp 601-622.

[16] Olajire AA, Abidemi JJ, Lateef A and Benson NU 2017 Adsorptive desulphurization of model oil by Ag nanoparticles-modified activated carbon prepared from brewer's spent grains. *J. Environ. Chem. Eng.* **5** (1) pp 147-159.

[17] Elegbede JA and Lateef A 2019 Green synthesis of silver (Ag), gold (Au) and silver-gold (Ag-Au) alloy nanoparticles: A review on recent advances, trends and biomedical applications. In: Verma DK, Goyal MR and Suleria HAR (Eds.). *Nanotechnology and Nanomaterial Applications in Food, Health and Biomedical Sciences.* https://doi.org/10.1201/9780429425660-1. Apple Academic Press Inc. /CRC Press, Taylor and Francis Group, Oakville, Ontario, Canada. ISBN 978-1-77188-764-9. Pp. 3-89.

[18] Lateef A, Ojo SA, Elegbede JA, Akinola PO and Akanni EO 2018 Nanomedical applications of nanoparticles for blood coagulation disorders. In: *Environmental Nanotechnology,* Volume 1. Eds: Dasgupta, N., Ranjan, S., and Lichtfouse, E.

https://doi.org/10.1007/978-3-319-76090-2_8. Springer International Publishing AG, Cham, Switzerland. ISBN 978-3-319-76089-6. Pp. 243-277.

[19] Elegbede JA, Lateef A, Azeez MA, Asafa TB, Yekeen TA, Oladipo IC, Adebayo EA, Beukes LS and Gueguim-Kana EB 2018 Fungal xylanases-mediated synthesis of silver nanoparticles for catalytic and biomedical applications. *IET Nanobiotechnol.* **12** (6) pp 857-863.

[20] Savithramma N, Rao ML, Rukmini K and Devi PS 2011 Antimicrobial activity of silver nanoparticles synthesized by using medicinal plants. *Int. J. Chem. Tech. Res.* **3** pp1394-1402.

[21] Verma VC, Kharwar RN and Gange AC 2010 Biosynthesis of antimicrobial silver nanoparticles by the endophytic fungus *Aspergillus clavatus*. *Nanomed.* **5** (1) pp 33-40.

[22] Lateef A, Adelere IA, Gueguim-Kana EB, Asafa TB, and Beukes LS 2015 Green synthesis of silver nanoparticles using keratinase obtained from a strain of *Bacillus safensis* LAU 13. *Int. Nano Lett.* **5** (1) pp 29-35.

[23] Lateef A, Ojo SA and Oladejo SM 2016 Anti-candida, anti-coagulant and thrombolytic activities of biosynthesized silver nanoparticles using cell-free extract of *Bacillus safensis* LAU 13. *Process Biochem.* **51** (10) pp 1406-1412.

[24] Ojo SA, Lateef A, Azeez MA, Oladejo SM, Akinwale AS, Asafa TB, Yekeen TA, Akinboro A, Oladipo IC, Gueguim-Kana EB and Beukes LS 2016 Biomedical and catalytic applications of gold and silver-gold alloy nanoparticles biosynthesized using cell-free extract of *Bacillus safensis* LAU 13: antifungal, dye degradation, anti-coagulant and thrombolytic activities. *IEEE Trans. NanoBiosci.* **15** (5) pp 433-442.

[25] Oladipo IC, Lateef A, Azeez MA, Asafa TB, Yekeen TA, Akinboro A, Akinwale AS, Gueguim-Kana EB and Beukes LS 2017 Green synthesis and antimicrobial activities of silver nanoparticles using cell-free extracts of *Enterococcus* species. *Not. Sci. Biol.* **9** (2): 196-203.

[26] Oladipo IC, Lateef A, Elegbede JA, Azeez MA, Asafa TB, Yekeen TA, Akinboro A, Gueguim-Kana EB, Beukes LS, Oluyide TO and Atanda OR 2017 *Enterococcus* species for the one-pot biofabrication of gold nanoparticles: characterization and nanobiotechnological applications. *J. Photochem. Photobiol. B: Biol.* **173** pp 250-257.

[27] Lateef A, Azeez MA, Asafa TB, Yekeen TA, Akinboro A, Oladipo IC, Ajetomobi FE, Gueguim-Kana EB and Beukes LS 2015 *Cola nitida*-mediated biogenic synthesis of silver nanoparticles using seed and seed shell extracts and evaluation of antibacterial activities. *Bionanosci.* **5** (4) pp 196-205.

[28] Lateef A, Azeez MA, Asafa TB, Yekeen TA, Akinboro A, Oladipo IC, Azeez L, Ajibade SE, Ojo SA, Gueguim-Kana EB, Beukes LS (2016) Biogenic synthesis of silver nanoparticles using a pod extract of *Cola nitida*: Antibacterial, antioxidant activities and application as a paint additive. *J. Taibah Univ. Sci.* **10** (4) pp 551-562.

[29] Lateef A, Azeez MA, Asafa TB, Yekeen TA, Akinboro A, Oladipo IC, Azeez L, Ojo SA, Gueguim-Kana EB and Beukes LS 2016 Cocoa pod husk extract-mediated biosynthesis of silver nanoparticles: its antimicrobial, antioxidant and larvicidal activities. *J. Nanostruct. Chem.* **6** (2) pp 159-169.

[30] Lateef A, Ojo SA, Folarin BI, Gueguim-Kana EB and Beukes LS (2016) Kolanut (*Cola nitida*) mediated synthesis of silver-gold alloy nanoparticles: antifungal, catalytic, larvicidal and thrombolytic applications. *J. Clust. Sci.* **27** (5) pp 1561-1577.

[31] Azeez MA, Lateef A, Asafa TB, Yekeen TA, Akinboro A, Oladipo IC, Gueguim-Kana EB and Beukes LS 2017 Biomedical applications of cocoa bean extract-mediated silver nanoparticles as antimicrobial, larvicidal and anticoagulant agents. *J. Clust. Sci.* **28** (1) pp 149-164.

[32] Lateef A, Folarin BI, Oladejo SM, Akinola PO, Beukes LS and Gueguim-Kana EB 2018 Characterization, antimicrobial, antioxidant and anticoagulant activities of silver nanoparticles synthesized from *Petiveria alliacea* L. leaf extract. *Prep. Biochem. Biotechnol.* **48** (7) pp 646-652.

[33] Adebayo EA, Ibikunle JB, Oke AM, Lateef A, Azeez MA, Oluwatoyin AO, Ajala VA, Olowoporoku TB, Okunlola OC, Ogundele OA, Badmus JA, Asafa TB, Beukes LS, Gueguim-Kana EB and Abbas SH 2019 Antimicrobial and antioxidant activity of silver, gold and silver-gold alloy nanoparticles phytosynthesized using extract of *Opuntia ficus-indica. Rev. Adv. Mater. Sci.* **58** (1) pp 313-326.

[34] Aina DA, Owolo O, Lateef A, Aina FO, Abbas SH, Adeoye-Isijola M, Okon V, Asafa TB, Elegbede JA, Olukanni OD and Adediji I 2019 Biomedical applications of *Chasmanthera*

dependens stem extract mediated silver nanoparticles as antimicrobial, antioxidant, anticoagulant, thrombolytic, and larvicidal agents. *Karbala Int. J. Modern Sci.* **5** (2) pp 71-80.

[35] Adebayo EA, Oke AM, Lateef A, Oyatokun AA, Abisoye OD, Adiji IP, Fagbenro DO, Amusan TV, Badmus JA, Asafa TB, Beukes LS, Gueguim-Kana EB and Abbas SH 2019 Biosynthesis of silver, gold and silver-gold alloy nanoparticles using *Persea americana* fruit peel aqueous extract for their biomedical properties. *Nanotechnol. Environ. Eng.* **4**: 13. https://doi.org/10.1007/s41204-019-0060-8.

[36] Singh N, Chaudhary A Abraham J 2014 Susceptibility testing of methicillin-resistant *Staphylococcus aureus* (MRSA) and biological role of silver nanoparticles of honey against MRSA. *J . Biol. Activ. Prod. Nat.* **4** pp 332-342.

[37] Lateef A, Ojo SA, Azeez MA, Asafa TB, Yekeen TA, Akinboro A, Oladipo IC, Gueguim-Kana EB and Beukes LS 2016 Cobweb as novel biomaterial for the green and eco-friendly synthesis of silver nanoparticles. *Appl. Nanosci.* **6** (6) pp 863-874.

[38] Philip D 2009 Honey mediated green synthesis of gold nanoparticles. *Spectrochim. Acta Part A: Mol. Biomol. Spectr.* **73** pp 650-653.

[39] Wu L, Cai X, Nelson K, Xing W, Xia J, Zhang R, Stacy AJ, Luderer M, Lanza GM, Wang LV and Shen, B 2013 A green synthesis of carbon nanoparticles from honey and their use in real-time photoacoustic imaging. *Nano Res.* **6** (5) pp 312-325.

[40] Xu S, Yong L and Wu P 2013 One-pot, green, rapid synthesis of flowerlike gold nanoparticles/reduced graphene oxide composite with regenerated silk fibroin as efficient oxygen reduction electrocatalysts. *ACS Appl. Mater. Interf.* **5** (3) pp 654-662.

[41] Obot IB, Umoren SA and Johnson AS 2013 Sunlight-mediated synthesis of silver nanoparticles using honey and its promising anticorrosion potentials for mild steel in acidic environments. *J. Mater. Environ. Sci.* **4** pp 1013-1018.

[42] Lateef A and Adeeyo AO 2015 Green synthesis and antibacterial activities of silver nanoparticles using extracellular laccase of *Lentinus edodes. Not. Sci. Biol.* **7** (4) pp 405-411.

[43] Elegbede JA, Lateef A, Azeez MA, Asafa TB, Yekeen TA, Oladipo IC, Abbas SH, Beukes LS and Gueguim-Kana EB 2019 Silver-gold alloy nanoparticles biofabricated by fungal

xylanases exhibited potent biomedical and catalytic activities. *Biotechnol. Progress.* **35** e2829. https://doi.org/10.1002/btpr.2829.

[44] Elegbede JA, Lateef A, Azeez MA, Asafa TB, Yekeen TA, Oladipo IC, Aina DA, Beukes LS and Gueguim-Kana EB 2020. Biofabrication of gold nanoparticles using xylanases through valorization of corncob by *Aspergillus niger* and *Trichoderma longibrachiatum*: antimicrobial, antioxidant, anticoagulant and thrombolytic activities. *Waste Biomass Valor.* **11** (3): 781-791.

[45] Durán M, Silveira CP and Durán N 2015 Catalytic role of traditional enzymes for biosynthesis of biogenic metallic nanoparticles: a mini-review. *IET Nanobiotechnol.* **9** pp 314-323.

[46] Durán N, Cuevas R, Cordi L, Rubilar O and Diez MC 2014 Biogenic silver nanoparticles associated with silver chloride nanoparticles (Ag@AgCl) produced by laccase from *Trametes versicolor*. *SpringerPlus* **3** 645. https://doi.org/10.1186/2193-1801-3-645.

[47] Khan S, Rizvi SMD, Avaish M, Arshad M, Bagga P and Khan MS 2015 A novel process for size controlled biosynthesis of gold nanoparticles using bromelain. *Mater. Lett.* **159** pp 373-376.

[48] Faramarzi MA and Forootanfar H 2011 Biosynthesis and characterization of gold nanoparticles produced by laccase from *Paraconiothyrium variabile*. *Colloids Surf. B: Biointerf.* **87** pp 23-27.

[49] Salem WM, Haridy M, Sayed WF and Hassan NH 2014 Antibacterial activity of silver nanoparticles synthesized from latex and leaf extract of *Ficus sycomorus*. *Ind. Crops Prod.* **62** pp 228-234.

[50] Azeez L, Lateef A, Adejumo AL, Adeleke JT, Adetoro RO and Mustapha Z 2019 Adsorption behaviour of Rhodamine B on hen feather and corn starch functionalized with green synthesized silver nanoparticles (AgNPs) mediated with cocoa pods extracts. *Chemistry Africa*. https://doi.org/10.1007/s42250-019-00113-7.

[51] Azeez L, Adejumo AL, Lateef A, Adebisi SA, Adetoro RO, Adewuyi S, Tijani KO and Olaoye S 2019 Zero-valent silver nanoparticles attenuate Cd and Pb toxicities on *Moringa oleifera* via immobilization and induction of phytochemicals. *Plant Physiol. Biochem.* **139** pp 283-292.

[52] Azeez L, Lateef A and Adebisi SA 2017 Silver nanoparticles (AgNPs) biosynthesized using pod extract of *Cola nitida* enhances antioxidant activity and phytochemical composition of *Amaranthus caudatus* Linn. *Appl. Nanosci.* **7** (1-2) pp 59-66.

[53] Azeez L, Lateef A, Adebisi SA and Oyedeji AO 2018 Novel biosynthesized silver nanoparticles from cobweb as adsorbent for Rhodamine B: equilibrium isotherm, kinetic and thermodynamic studies. *Appl. Water. Sci.* **8**: 32. https://doi.org/10.1007/s13201-018-0676-z.

[54] Azeez L, Lateef A, Wahab AA, Adejumo AL and Raji K 2019 Comparative effects of silver nanoparticles, sucrose and sodium chloride as osmotic solutions for tomato slices: antioxidant activity, microbial quality and modelling with polynomial regression model. *South Afr. J. Chem.* **72**: 21-31.

[55] Azeez L, Lateef A, Wahab AA, Rufai MA, Salau AK, Ajayi IO, Ajayi EM, Maryam AK and Adebisi B 2019 Phytomodulatory effects of silver nanoparticles on *Corchorus olitorius*: its antiphytopathogenic and hepatoprotective potentials. *Plant Physiol. Biochem.* 136: 109-117.

[56] Pacheco-Torgal F and Jalali S 2011 Nanotechnology: advantages and drawbacks in the field of construction and building materials. *Constr. Build. Mater.* **25** (2) pp 582-590.

[57] Rashad AM 2014 A comprehensive overview about the effect of nano-SiO_2 on some properties of traditional cementitious materials and alkali-activated fly ash. *Constr. Build. Mater.* **52** pp 437-464.

[58] Shakhmenko G, Juhnevica I and Korjakins A 2013 Influence of sol–gel nanosilica on hardening processes and physically-mechanical properties of cement paste. *Procedia Eng.* **57** pp 1013-1021.

[59] Jalal M, Mansouri E, Sharifipour M and Pouladkhan AR 2012 Mechanical, rheological, durability and microstructural properties of high performance self-compacting concrete containing SiO_2 micro and nanoparticles. *Mater. Design* **34** pp 389-400.

[60] Cárdenas C, Tobón JI, García C and Vila J 2012 Functionalized building materials: Photocatalytic abatement of NOx by cement pastes blended with TiO_2 nanoparticles. *Constr. Build. Mater.* **36** pp 820-825.

[61] Nazari A and Riahi S 2010 The effect of TiO_2 nanoparticles on water permeability and thermal and mechanical properties of high strength self-compacting concrete. *Mater. Sci. Eng.: A.* **528** (2) pp 756-763.

[62] Li H, Xiao H and Ou J 2004 A study on mechanical and pressure-sensitive properties of cement mortar with nanophase materials. *Cem. Concr. Res.* **34** pp 435-438.

[63] Gencel O, Brostow W, Ozel C and Filiz M 2010 Concretes containing hematite for use as shielding barriers. *Mater. Sci.* **16** pp 249-256.

[64] Nazari A and Riahi S 2010 Assessment of the effects of Fe_2O_3 nanoparticles on water permeability, workability and setting time of concrete. *J. Compos. Mater.* **45** pp 923-930.

[65] Nazari A and Riahi S 2010 Benefits of Fe_2O_3 nanoparticles in concrete mixing matrix. *J. Am. Sci.* **6** pp 102-106.

[66] Konsta-Gdoutos MS, Metaxa ZS and Shah SP 2010 Multi-scale mechanical and fracture characteristics and early-age strain capacity of high performance carbon nanotube/cement nanocomposites. *Cem. Concr. Compos.* **32** pp 110-115.

[67] Artus GR, Zimmermann J, Reifler FA, Brewer SA and Seeger SA 2012 Superoleophobic textile repellent towards impacting drops of alkanes. *Appl. Surf. Sci.* **258** pp 3835-3840.

[68] Hoefnagels H, Wu D, De With G and Ming W 2007 Biomimetic superhydrophobic and highly oleophobic cotton textiles. *Langmuir* **23** pp 13158-13163.

[69] Liu Y, Tang J, Wang R, Lu H, Li L, Kong Y, Qi K and Xin JH 2007 Artificial lotus leaf structures from assembling carbon nanotubes and their applications in hydrophobic textiles. *J. Mater. Chem.* **17** (11) pp 1071-1078.

[70] Zhang F and Yang J 2009 Preparation of nano-ZnO and its application to the textile on antistatic finishing. *Int. J. Chem.* **1** (1) 18-22.

[71] Qiaozhen Y 2007 Influence of nano-particles treatment on the antistatic property of polyester fabric. *J. Textile Res.* **12**:19

[72] Liu Y, Wang X, Qi K and Xin J 2008 Functionalization of cotton with carbon nanotubes. *J. Mater. Chem.* **18** pp 3454-3460.

[73] Kathirvelu S, D'souza L and Dhurai B 2009 UV protection finishing of textiles using ZnO nanoparticles. *Indian J. Fibre Textile Res.* **34** pp 267-273.

[74] Song J, Wang C and Hinestroza JP 2013 Electrostatic assembly of core-corona silica nanoparticles onto cotton fibers. *Cellulose* **20** pp 1727-1736.

[75] Yue X, Lin H, Yan T, Zhang D, Lin H and Chen Y 2014 Synthesis of silver nanoparticles with sericin and functional finishing to cotton fabrics. *Fibers Polym.* **15** (4) pp 716-722.

[76] Yedra A, Gutiérrez-Somavilla G, Manteca-Martínez C, González-Barriuso M and Soriano L 2016 Conductive paints development through nanotechnology. *Prog. Org. Coatings* **95** pp 85-90.

[77] Lateef A, Ojo SA, Akinwale AS, Azeez L, Gueguim-Kana EB, Beukes LS 2015 Biogenic synthesis of silver nanoparticles using cell-free extract of *Bacillus safensis* LAU 13: antimicrobial, free radical scavenging and larvicidal activities. *Biologia* **70** (10) pp 1295-1306.

[78] Rabea AM, Mohseni M, Mirabedini SM and Tabatabaei MH 2012 Surface analysis and anti-graffiti behavior of a weathered polyurethane-based coating embedded with hydrophobic nano silica. *Appl. Surf. Sci.* **258** (10) pp 4391-4396.

[79] Ramaswamy U and Mani V 2019 Synthesis of silver nanoparticle of aqueous extract of *Allium fistulosum* and its efficiency against bacterial contaminants from industrial waste water and its photocatalytic potential. In *Advances in Plant and Microbial Biotechnology* (pp. 121-127). Springer, Singapore.

[80] Allam NG, Ismail GA, El-Gemizy WM and Salem MA 2019 Biosynthesis of silver nanoparticles by cell-free extracts from some bacteria species for dye removal from wastewater. *Biotechnol. Lett.* **41** (3) pp 379-89.

[81] Dankovich TA and Gray DG 2019 Bactericidal paper impregnated with silver nanoparticles for point-of-use water treatment. *Environ. Sci. Technol.* **45** pp 1992-1998.

[82] García A, Delgado L, Torà JA, Casals E, González E, Puntes V, Font X, Carrera J and Sánchez A 2012 Effect of cerium dioxide, titanium dioxide, silver, and gold nanoparticles on the activity of microbial communities intended in wastewater treatment. *J. Hazard. Mater.* **199** pp 64-72.

[83] Kuppusamy P, Ilavenil S, Srigopalram S, Maniam GP, Yusoff MM, Govindan N and Choi KC 2016 Treating of palm oil mill effluent using *Commelina nudiflora* mediated copper nanoparticles as a novel bio-control agent. *J. Clean. Prod.* **141** pp.1023-1029.

[84] Eli D, Onimisi MY, Abdu SG, Gyuk PM, and Jonathan E 2016. Enhanced performance of a dye sensitized solar cell using silver nanoparticles modified photoanode. *J. Sci. Res. Reports* **10** pp 1-8.

[85] Eli D, Onimisi MY, Abdu SG, Jonathan E, and Yakubu SO 2016. Photoelectric characterization of a dye-sensitized solar cell based on natural pigment extracted from roselle (*Hibiscus sabdariffa*) flower and TiO_2 nanoparticles. *Curr. J. Appl. Sci. Technol.* **15** (2) pp 1-6.

[86] Eli D, Owolabi JA, Olowomofe GO, Onimisi MY and Francis A 2016. Enhancement in photovoltaic parameters of a dye sensitized solar cell by surface plasmon resonance of metallic silver nanoparticles. *Am. Chem. Sci. J.* **14** pp 1-8.

[87] Onimisi MY, Eli D, Abdu SG, Aboh HO and Jonathan E 2016 Size effects of silver nanoparticles on the photovoltaic performance of a dye sensitized solar cells. *Am. Chem. Sci. J.* **13** (3) pp 1-8.

[88] Bazargan MH 2009 Performance of nanostructured dye-sensitized solar cell utilizing natural sensitizer operated with platinum and carbon coated counter electrodes. *Digest J. Nanomater. Biostruct.* **4** (4) pp 723-727.

[89] Otanicar TP, Phelan PE, Prasher RS, Rosengarten G and Taylor RA 2010 Nanofluid-based direct absorption solar collector. *J. Renew. Sustain. Energ.* **2** (3) 033102. https://doi.org/10.1063/1.3429737.

[90] Liu CJ, Burghaus U, Besenbacher F and Wang ZL 2010 Preparation and characterization of nanomaterials for sustainable energy production. *ACS Nano.* **4** (10) pp 5517-5526.

[91] Mercatelli L, Sani E, Fontani D, Zaccanti G, Martelli F and Di Ninni P 2011 Scattering and absorption properties of carbon nanohorn-based nanofluids for solar energy applications. *J. Eur. Opt. Soc-Rapid* **6** 11025. https://doi.org/10.2971/jeos.2011.11025.

[92] Poinern GE, Brundavanam S, Shah M, Laava I and Fawcett D 2012 Photothermal response of CVD synthesized carbon (nano) spheres/aqueous nanofluids for potential application in direct solar absorption collectors: a preliminary investigation. *Nanotechnol. Sci. Appl.* **5** 49. https://doi.org/10.2147/NSA.S34166.

IOP Conf. Series: Materials Science and Engineering **805** (2020) 012044 doi:10.1088/1757-899X/805/1/012044

[93] Tiwari AK, Ghosh P and Sarkar J 2013 Solar water heating using nanofluids–a comprehensive overview and environmental impact analysis. *Int. J. Emerg. Technol. Adv. Eng.* **3** (3) pp 221-224.

[94] Obaid HN, Habeeb MA, Rashid F and Hashim A 2013 Thermal energy storage by nanofluids. *J. Energ. Technol. Policy* **3** pp 34-36.

[95] Goh WJ, Makam VS, Hu J, Kang L, Zheng M, Yoong SL, Udalagama CN and Pastorin G 2012 Iron oxide filled magnetic carbon nanotube–enzyme conjugates for recycling of amyloglucosidase: toward useful applications in biofuel production process. *Langmuir* **28** (49) pp 16864-16873.

[96] Alenzi N, Liao WS, Cremer PS, Sanchez-Torres V, Wood TK, Ehlig-Economides C and Cheng Z 2010 Photoelectrochemical hydrogen production from water/methanol decomposition using Ag/TiO2 nanocomposite thin films. *Int. J. Hydrogen Energ.* **35** (21) pp 11768-11775.

[97] Silvestre J, Silvestre N and de Brito J 2015 Review on concrete nanotechnology. *Eur. J. Environ. Civil Eng.* **20** (4) pp 455-485.

[98] Raki L, Beaudoin J, Alizadeh R, Makar J and Sato T 2010 Cement and concrete nanoscience and nanotechnology. *Mater.* **3** pp 918-942.

[99] Behnood A and Ziari H 2008 Effects of silica fume addition and water to cement ratio on the properties of high-strength concrete after exposure to high temperatures. *Cem. Concr. Compos.* **30** pp 106-112.

[100] Hashimoto K, Irie H and Fujishima A 2005 TiO_2 photocatalysis: A historical overview and Future Prospects. *Jpn. J. Appl. Phys.* **44** pp 8269-8285.

[101] Agrios AG and Pichat P 2005 State of the art and perspectives on materials and applications of photocatalysis over TiO_2. *J. Appl. Electrochem.* **35** pp 655-663.

[102] Chen J, Kou S and Poon C 2012 Hydration and properties of nano-TiO_2 blended cement composites. *Cem. Concr. Compos.* **34** pp 642-649.

[103] Sanchez F and Sobolev K 2010 Nanotechnology in concrete–A review. *Constr. Build.Mater.* **24** pp 2060-2071.

[104] Chaipanich A, Nochaiya T, Wongkeo W and Torkittikul P 2010 Compressive strength and microstructure of carbon nanotubes–fly ash cement composites. *Mater. Sci. Eng.: A* **527** pp 1063-1067.

[105] Lam CW, James JT, McCluskey R, Arepalli S and Hunter RL 2006 A review of carbon nanotube toxicity and assessment of potential occupational and environmental health risks. *Cri. Rev. Toxicol.***36** pp 189-217.

[106] Soliman EM, Kandil UF and Taha M 2012 The significance of carbon nanotubes on styrene butadiene rubber (SBR) and SBR modified mortar. *Mater. Struct.* **45** pp 803-816.

[107] Azhari F and Banthia N 2012 Cement-based sensors with carbon fibers and carbon nanotubes for piezoresistive sensing. *Cem. Concr. Compos.* **34** pp 866-873.

[108] Niroumand H, Zain MFM and Jamil M 2013 The role of nanotechnology in architecture and built environment. *Procedia-Soc. Behav. Sci.* **89** pp 10-15.

[109] Vidu R, Rahman M, Mahmoudi M, Enachescu M, Poteca TD and Opris I 2014 Nanostructures: a platform for brain repair and augmentation. *Front. Syst. Neurosci.* **8** 91. https://doi.org/10.3389/fnsys.2014.00091.

[110] Yetisen AK, Qu H, Manbachi A, Butt H, Dokmeci MR, Hinestroza JP, Skorobogatiy M, Khademhosseini A and Yun SH 2016 Nanotechnology in textiles. *ACS Nano* **10** (3) pp 3042-3068.

[111] Cheng MH, Chen LC, Hung YC and Yang CM 2008 A real-time maximum-likelihood heart-rate estimator for wearable textile sensors. In: *30th Annual International Conference of the IEEE Engineering in Medicine and Biology Society* pp. 254-257. https://ieeexplore.ieee.org/document/4649138/.

[112] Mattmann C, Amft O, Harms H, Troster G and Clemens F 2007 Recognizing upper body postures using textile strain sensors. In: *2007 11th IEEE International Symposium on Wearable Computers* pp 29-36. https://doi.org/10.1109/ISWC.2007.4373773.

[113] Cherenack K and van Pieterson L 2012 Smart textiles: challenges and opportunities. *J. Appl. Phys.* **112** (9) 091301. https://doi.org/10.1063/1.4742728.

[114] Hu L and Cui Y 2012 Energy and environmental nanotechnology in conductive paper and textiles. *Energ. Environ. Sci.* **5** pp 6423-6435.

[115] Ramaratnam K, Tsyalkovsky V, Klep V and Luzinov I 2007 Ultrahydrophobic textile surface via decorating fibers with monolayer of reactive nanoparticles and non-fluorinated polymer. *Chem. Commun.* **43** pp 4510-4512.

[116] Lam YL, Kan CW and Yuen CWM 2010 Effect of concentration of titanium dioxide acting as catalyst or co-catalyst on the wrinkle resistant finishing of cotton fabric. *Fibers Polym.***11** pp 551-558.

[117] Nowack B, Krug HF and Height M 2011 120 years of nanosilver history: implications for policy makers. *Environ. Sci. Technol.* **45** pp 1177-1183.

[118] Qi K, Chen X, Liu Y, Xin JH, Mak CL and Daoud WA 2007 Facile preparation of anatase/SiO_2 spherical nanocomposites and their application in self-cleaning textiles. *J. Mater. Chem.* **17** (33) pp 3504-3508.

[119] Uddin MJ, Cesano F, Scarano D, Bonino F, Agostini G, Spoto G, Bordiga S and Zecchina A 2008 Cotton textile fibres coated by Au/TiO_2 films: Synthesis, characterization and self cleaning properties. *J. Photochem. Photobiol.: A* **199** pp 64-72.

[120] Song J, Birbach NL and Hinestroza JP 2012 Deposition of silver nanoparticles on cellulosic fibers via stabilization of carboxymethyl groups. *Cellulose* **19** pp 411-424.

[121] Kaiser JP, Zuin S and Wick P 2013 Is nanotechnology revolutionizing the paint and lacquer industry? A critical opinion. *Sci. Total Environ.* **442** pp 282-289.

[122] Rajarathinam M, Dhanapal D, Morukattu G, Joseph S and Thangavelu KP 2014 Imparting potential antibacterial and antifungal activities to water based interior paint using nanoparticles of silver as an additive-an ecofriendly approach. *Adv. Sci. Eng. Med.* **6** (6) pp 676-682.

[123] Patel AS, Rathod HA and Sharma N 2013 An overview on application of nanotechnology in construction industry. *Int. J. Innov. Res. Sci. Eng. Technol.* **2** pp 6094-6098.

[124] Cannavale A, Fiorito F, Manca M, Tortorici G, Cingolani R and Gigli G 2010 Multifunctional bioinspired sol-gel coatings for architectural glasses. *Build. Env.* **45** (5) pp 1233-1243.

[125] Drelich J, Chibowski E, Meng DD and Terpilowski K 2011 Hydrophilic and superhydrophilic surfaces and materials. *Soft Matter* **7** (21) pp 9804-9828.

[126] Wang Z, Han E, Liu F and Ke W 2010 Fire and corrosion resistances of intumescent nano-coating containing nano-SiO_2 in salt spray condition. *J. Mater. Sci. Technol.* **26** (1) pp 75-81.

[127] Quagliarini E, Bondioli F, Goffredo GB, Cordoni C and Munafò P 2012 Self-cleaning and de-polluting stone surfaces: TiO_2 nanoparticles for limestone. *Constr. Build. Mater.* **37** pp 51-57.

[128] Quagliarini E, Bondioli F, Goffredo GB, Licciulli A and Munafò P 2012 Smart surfaces for architectural heritage: Preliminary results about the application of TiO_2-based coatings on travertine. *J. Cult. Herit.* **13** (2) pp 204-209.

[129] Munafò P, Quagliarini E, Goffredo GB, Bondioli F and Licciulli A 2014 Durability of nano-engineered TiO_2 self-cleaning treatments on limestone. *Constr. Build. Mater.* **65** pp 218-231.

[130] Baruah S and Dutta J 2009. Nanotechnology applications in pollution sensing and degradation in agriculture: a review. *Environ. Chem. Lett.* **7** (3)pp 191-204.

[131] Bakare AA, Lateef A, Amuda OS and Afolabi RO 2003 The aquatic toxicity and characterization of chemical and microbiological constituents of water samples from Oba River, Odo-Oba, Nigeria. *Asian J. Microbiol. Biotechnol. Environ. Sci.* **5** (1) pp 11-17.

[132] Lateef A, Oloke JK, Kana EG and Pacheco E 2006 The microbiological quality of ice used to cool drinks and foods in Ogbomoso Metropolis, Southwest, Nigeria. *Internet J. Food Saf.* **8** pp 39-43.

[133] Lateef A, Kolawole LL and Agosu SO 2012 Quality assessment of some groundwater samples in Ogbomoso metropolis, Southwest Nigeria. *J. Environ. Earth Sci.* **2** (6) pp 39-49.

[134] Lateef A 2004 The microbiology of a pharmaceutical effluent and its public health implications. *World J. Microbiol. Biotechnol.* **20** (2) pp 167-171.

[135] Lateef A, Yekeen TA and Ufuoma PE 2007 Bacteriology and genotoxicity of some pharmaceutical wastewaters in Nigeria. *Int. J. Environ. Health* **1** (4) pp 551-562.

[136] Baruah S and Dutta J 2011 Zinc stannate nanostructures: hydrothermal synthesis. *Sci. Technol. Adv. Mater.* **12** (1) 013004. https://doi.org/10.1088/1468-6996/12/1/013004.

[137] Li Q, Mahendra S, Lyon DY, Brunet L, Liga MV, Li D and Alvarez PJ 2008 Antimicrobial nanomaterials for water disinfection and microbial control: potential applications and implications. *Water Res.* **42** (18) pp 4591-4602.

[138] Baruah S, Jaisai M and Dutta J 2012 Development of a visible light active photocatalytic potable water purification unit using ZnO nanorods. *Catal. Sci.Technol.* **2** (5) pp 918-921.

[139] Mahmood MA, Baruah S and Dutta J 2011 Enhanced visible light photocatalysis by manganese doping or rapid crystallization with ZnO nanoparticles. *Mater. Chem.Phys.* **130** (1-2) pp 531-535.

[140] Makhal A, Sarkar S, Bora T, Baruah S, Dutta J, Raychaudhuri AK and Pal SK 2010 Role of resonance energy transfer in light harvesting of zinc oxide-based dye-sensitized solar cells. *J. Phys. Chem. C* **114** (23) pp 10390-10395.

[141] Baruah S, Mahmood MA, Myint MTZ, Bora T and Dutta J 2010 Enhanced visible light photocatalysis through fast crystallization of zinc oxide nanorods. *Beilstein J. Nanotechnol.* **1** (1) pp 14-20.

[142] Baruah S, Khan MN and Dutta J 2016 Perspectives and applications of nanotechnology in water treatment. *Environ. Chem. Lett.* **14** (1) pp 1-4.

[143] Hussein AK 2015 Applications of nanotechnology in renewable energies-A comprehensive overview and understanding. *Renew. Sustain. Energ. Rev.* **42** pp 460-476.

[144] Ahmadi MH, Ghazvini M, Nazari AM, Ahmadi MA, Pourfayaz F, Lorenzini G and Ming T 2019 Renewable energy harvesting with the application of nanotechnology: A review. *Int. J. Energ. Res.* **43** (4) pp 1387-1410.

[145] Satyanarayana KG, Mariano AB, and Vargas JVC 2011 A review on microalgae, a versatile source for sustainable energy and materials. *Int. J. Energ. Res.* **35** (4) pp 291-311.

[146] Imalerio TI, Amah AN and Onoja DA 2014 Capacitance and impedance evaluations in coconut shells and rice husks derived activated carbon electrodes. *J. Natural Sci. Res.* **4** (16) pp 104-108.

[147] Okeke I and Iloanusi O 2014 Application of nanotechnology for improving energy security. *Proceedings of the 1st African International Conference/Workshop on*

IOP Conf. Series: Materials Science and Engineering **805** (2020) 012044 doi:10.1088/1757-899X/805/1/012044

Applications of Nanotechnology to Energy, Health and Environment, UNN, March 23-29, 2014. pp 237-243.

[148] Zhang D, Wei D, Ding W and Zhang X 2014 Carbon-based nanostructured catalyst for biodiesel production by catalytic distillation. *Catal. Commun*. **43** pp 121-125.

AUTHOR INDEX